A Level
Biology
for OCR

Series Editor
Ann Fullick

Authors
Ann Fullick
Jo Locke
Paul Bircher

OXFORD
UNIVERSITY PRESS

OXFORD
UNIVERSITY PRESS

Great Clarendon Street, Oxford, OX2 6DP, United Kingdom

Oxford University Press is a department of the University of Oxford. It furthers the University's objective of excellence in research, scholarship, and education by publishing worldwide. Oxford is a registered trade mark of Oxford University Press in the UK and in certain other countries

British Library Cataloguing in Publication Data
Data available

978-0-19-835192-4

15

Paper used in the production of this book is a natural, recyclable product made from wood grown in sustainable forests. The manufacturing process conforms to the environmental regulations of the country of origin.

Printed in China by Shanghai Offset Printing Products Ltd

This resource is endorsed by OCR for use with specification H020 AS Level GCE Biology A and H420 A Level GCE Biology A. In order to gain endorsement this resource has undergone an independent quality check. OCR has not paid for the production of this resource, nor does OCR receive any royalties from its sale. For more information about the endorsement process please visit the OCR website www.ocr.org.uk

Acknowledgements

The authors would like to thank John Beazley for his reviewing, as well as Amy Johnson, Amie Hewish, Les Hopper, Clodagh Burke, Sharon Thorn for their tireless work and encouragement. In addition they would like to thank the teams at science and plants for schools (SAPS) and at the Wellcome Trust Sanger Institute for their valuable input to the project, and finally the help received from Dr Jeremy Pritchard and Jennifer Collins.

Ann Fullick would like to thank her partner Tony for his support and amazing photographs, and all of the boys William, Thomas, James, Edward, and Chris for their expert advice and for making her take time off.

Paul Bircher would like to thank his wife, Julie, and the rest of his family for their patience and support throughout the writing of this book. A special mention goes to his irrepressible grandchildren, Leo and Toby, who provided a welcome distraction. Their insanity has kept him sane.

Jo Locke would like to thank her husband Dave for all his support, encouragement, and endless cups of tea, as well as her girls Emily and Hermione who had to wait patiently for Mummy 'to just finish this paragraph'.

AS/A Level course structure

This book has been written to support students studying for OCR A Biology A. It covers all A level modules from the specification. The chapters within each module are shown in the contents list, which also shows you the page numbers for the main topics within each chapter. There is also an index at the back to help you find what you are looking for. If you are studying for OCR AS Biology A, you will only need to know the content in the blue box.

AS exam

A level exam

Year 1 content

1 Development of practical skills in biology
2 Foundations in biology
3 Exchange and transport
4 Biodiversity, evolution, and disease

Year 2 content

5 Communication, homeostasis, and energy
6 Genetics, evolution, and ecosystems

A Level exams will cover content from Year 1 and Year 2 and will be at a higher demand. You will also carry out practical activities throughout your course.

Contents

Learning outcomes

→ At the beginning of each topic, there is a list of learning outcomes.

→ These are matched to the specification and allow you to monitor your progress.

→ A specification reference is also included.
Specification reference: 2.1.3

Study Tips

Study tips contain prompts to help you with your understanding and revision.

Synoptic link

These highlight the key areas where topics relate to each other. As you go through your course, knowing how to link different areas of biology together becomes increasingly important. Many exam questions, particularly at A Level, will require you to bring together your knowledge from different areas.

This book contains many different features. Each feature is designed to support and develop the skills you will need for your examinations, as well as foster and stimulate your interest in biology.

Terms that you will need to be able to define and understand are highlighted by **bold text**.

Application features

These features contain important and interesting applications of biology in order to emphasise how scientists and engineers have used their scientific knowledge and understanding to develop new applications and technologies. There are also practical application features, with the icon 🧪, to support further development of your practical skills. There are also application features with the icon ⚙️ which help support development of your understanding of scientific issues and their impact in society.

1 All application features have a question to link to material covered with the concept from the specification.

Extension features

These features contain material that is beyond the specification. They are designed to stretch and provide you with a broader knowledge and understanding and lead the way into the types of thinking and areas you might study in further education. As such, neither the detail nor the depth of questioning will be required for the examinations. But this book is about more than getting through the examinations.

1 Extension features also contain questions that link the off-specification material back to your course.

Summary Questions

1 These are short questions at the end of each topic.

2 They test your understanding of the topic and allow you to apply the knowledge and skills you have acquired.

3 The questions are ramped in order of difficulty. The icon ⚙️ indicates where a question relates to scientific issues in society.

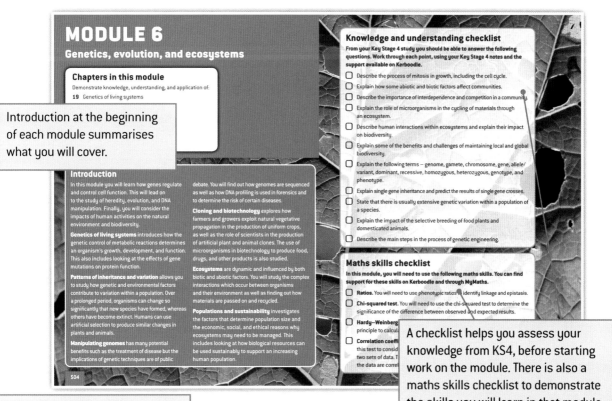

Introduction at the beginning of each module summarises what you will cover.

A checklist helps you assess your knowledge from KS4, before starting work on the module. There is also a maths skills checklist to demonstrate the skills you will learn in that module.

Visual summaries show how some of the key concepts of that module interlink with other modules, across the entire A Level course.

Application task brings together some of the key concepts of the module in a new context.

Extension task bring together some key concepts of the module and develop them further, leading you towards greater understanding and further study.

Practice questions at the end of each chapter and the end of each module, including questions that cover practical and math skills.

A dedicated Synoptic Concepts section at the end of the book, with help and advice on answering synoptic questions that cover multiple different topics. This section also contains further practice questions.

SYNOPTIC CONCEPTS

Analysing and answering a synoptic question

This book is supported by next generation Kerboodle, offering unrivalled digital support for independent study, differentiation, assessment, and the new practical endorsement.

If your school subscribes to Kerboodle, you will also find a wealth of additional resources to help you with your studies and with revision.

- Study guides
- Maths skills boosters and calculation worksheets
- On your marks activities to help you achieve your best
- Practicals and follow up activities to support the practical endorsement
- Interactive objective tests that give question-by-question feedback
- Animations and revision podcasts
- Self-assessment checklists

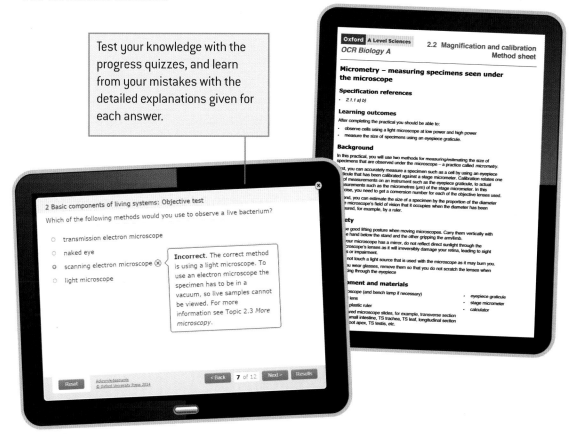

Test your knowledge with the progress quizzes, and learn from your mistakes with the detailed explanations given for each answer.

2 Basic components of living systems: Objective test

Which of the following methods would you use to observe a live bacterium?

○ transmission electron microscope

○ naked eye

◉ scanning electron microscope ⊗

○ light microscope

Incorrect. The correct method is using a light microscope. To use an electron microscope the specimen has to be in a vacuum, so live samples cannot be viewed. For more information see Topic 2.3 *More microscopy*.

Reset | Acknowledgements | © Oxford University Press 2014 | < Back | **7** of 12 | Next > | Results

Oxford A Level Sciences
OCR Biology A

2.2 Magnification and calibration
Method sheet

Micrometry – measuring specimens seen under the microscope

Specification references
- 2.1.1 a) b)

Learning outcomes
After completing the practical you should be able to:
- observe cells using a light microscope at low power and high power
- measure the size of specimens using an eyepiece graticule.

Background
In this practical, you will use two methods for measuring/estimating the size of specimens that are observed under the microscope – a practice called *micrometry*.

First, you can accurately measure a specimen such as a cell by using an eyepiece graticule that has been calibrated against a stage micrometer. Calibration relates one set of measurements on an instrument such as the eyepiece graticule, to actual measurements such as the micrometres (μm) of the stage micrometer. In this exercise, you need to get a conversion number for each of the objective lenses used.

Second, you can estimate the size of a specimen by the proportion of the diameter of the microscope's field of vision that it occupies when the diameter has been measured, for example, by a ruler.

Safety
Use good lifting posture when moving microscopes. Carry them vertically with one hand below the stand and the other gripping the arm/limb.

If your microscope has a mirror, do not reflect direct sunlight through the microscope's lenses as it will irreversibly damage your retina, leading to sight loss or impairment.

Do not touch a light source that is used with the microscope as it may burn you.

If you wear glasses, remove them so that you do not scratch the lenses when looking through the eyepiece

Equipment and materials
- microscope (and bench lamp if necessary)
- lens
- plastic ruler
- prepared microscope slides, for example, transverse section small intestine, TS trachea, TS leaf, longitudinal section root apex, TS testis, etc.
- eyepiece graticule
- stage micrometer
- calculator

For teachers, Kerboodle also has plenty of further assessment resources, answers to the questions in the book, and a digital markbook along with full teacher support for practicals and the worksheets, which include suggestions on how to support and stretch students. All of the resources are pulled together into teacher guides that suggest a route through each chapter.

MODULE 1
Development of practical skills in biology

Developing your practical skills is a fundamental part of a complete education in science. A good grounding of practical skills will help your understanding of Biology and help to prepare you for studying beyond A level.

You will carry out a number of practicals throughout the A level course. Practical skills knowledge will account for 15% of the marks in your written exams. There are no assessed practical components in the AS qualification, however, if you choose to study for the full A level qualification then these practicals will count towards the practical endorsement in your second year of study.

Practical coverage throughout this book

It is a good idea to keep a record of your practical work during your course, this will be very useful when you come to revise for your exams, and you can use it later as part of your practical endorsement. You can find more details of the practical endorsement from your teacher or from the specification.

In this book and its supporting materials, practical skills are covered in a number of ways – look out for the conical flask symbol. By studying **Application boxes** and **practice questions** in this student book you will have many opportunities to learn about scientific method and how to carry out practical activities. There are also more resources available on Kerboodle to help you practise your skills.

1.1 Practical skills assessed in written examinations

There are four key components of practical skill assessments. These are laid out here with a skills checklist to help you as you study.

1.1.1 Planning

- Experimental design
- Identification of variables
- Evaluation of experimental method

Skills checklist

- ☐ Forming a hypothesis
- ☐ Selecting suitable equipment
- ☐ Considering accuracy and precision
- ☐ Identifying dependent and independent variables
- ☐ Identifying variables which need to be controlled

1.1.2 Implementing

- Use of practical apparatus
- Appropriate units for measurement
- Presenting observations and data

Skills checklist

- ☐ Confidence using apparatus and techniques correctly
- ☐ Understanding S.I. units and prefixes
- ☐ Results table design
- ☐ Presenting data in the most suitable way:
  - Scatter graph
  - Line graph
  - Bar chart
  - Pie chart

1.1.3 Analysis

- Processing, analysing, and interpreting results
- Appropriate mathematical skills for data analysis
- Use of appropriate number of significant figures
- Plotting and interpreting graphs

Skills checklist

- ☐ Understanding results and using this to reach valid conclusions
- ☐ Using mathematical skills to process results
- ☐ Using significant figures correctly
- ☐ Plotting and interpreting graphs
  - labelling axes correctly
  - using appropriate scales
  - reading intercepts and gradients from graphs

1.1.4 Evaluation

- Evaluate results to draw conclusions
- Identify anomalies
- Explain limitations in method
- Precision and accuracy of measurements
- Uncertainties and errors
- Suggest improvements to help improve experimental design

Skills checklist

- ☐ Evaluate results to draw sound conclusions
- ☐ Understand and explain any limitations in the method, and make suggestions on how the method could be improved
- ☐ Understand accuracy of measurements and margins of error (including percentage error) and uncertainty in apparatus

3

Maths skills and How Science Works across A level Biology

Maths is a vital tool for scientists, and throughout this course you will become familiar with maths techniques and equations that support the development of your science knowledge. Each **module opener** in this book has an overview of the maths skills that relate to the theory in the chapter. There are also **questions using maths skills** throughout the book that will help you practice.

How Science Works are skills that will help you to apply your knowledge in a wider context, and the relevance of what you have learnt in the real world. This includes developing your critical and creative skills to help you solve problems in many different contexts. How Science Works is embedded throughout this book, particularly in **application boxes** and **practice questions**.

1.2 Practical skills assessed in practical endorsement

You are required to carry out 12 assessed practical activities over both years of your A level studies. The practical endorsement does not count towards your final A level grade but is reported alongside it as a pass or a fail. Universities and employers will look for this as evidence that you have a good level of practical competence. These practicals will be assessed by your teacher and will help to develop your skills and confidence.

The practicals you carry out should be recorded in a lab book or practical portfolio where the hypothesis, method, results, and conclusion are clearly displayed. The information here details the types of skills and equipment you should become familiar with.

1.2.1 Practical skills

Independent thinking
- Investigating and analysing the methods used in practicals in order to solve problems

Use and application of scientific methods and practices
- using practical equipment correctly and safely
- following written instructions, recording observations, taking measurements, and presenting data scientifically
- Using appropriate software and technology throughout.

Research and referencing
Using information available from a variety of different sources including websites, scientific journals, and textbooks to help provide context and background for the practical. It is important to use many sources of information as you can and to cite these correctly.

Instruments and equipment
Correct and appropriate use of a wide range of equipment, instruments, and techniques.

1.2.2 Use of apparatus and techniques

- Apparatus for quantitative measuring,
- Use of glassware apparatus
- Use of a light microscope at high and low power and the use of stage graticule
- Producing clear and well labelled scientific drawings
- Use of qualitative reagents
- Experience of carrying out electrophoresis or chromatography
- Ethical and safe use of organisms
- Microbial aseptic techniques
- Use of sampling techniques for fieldwork
- Use of IT to collect and process data.

A level PAG overview and Application features

The practical activity requirements (PAGs) for the OCR A Biology practical endorsement are listed in the table. You should take all opportunities to develop you practical skills and techniques in your first year of study to help build a greater understanding.

The table below shows the practical activity requirements, and which topics these are covered in.

Specification reference		Topic reference
PAG1	Microscopy	2.1, 2.2, 2.3, 13.9, 13.10, 14.2, 15.4
PAG2	Dissection	7.4, 8.2, 8.5, 9.1, 15.5, 22.1
PAG3	Sampling techniques	11.3, 23.5
PAG4	Enzyme controlled reactions	4.2, 18.6
PAG5	Colorimeter or potometer	3.4
PAG6	Chromatography or electrophoresis	3.6, 17.3, 21.1
PAG7	Microbial technique	22.6, 22.7
PAG8	Transport in and out of cells	5.2, 5.3, 5.5
PAG9	Qualitative testing	3.4, 15.7
PAG10	Investigation using a data logger or computer modelling	7.3, 17.4, 18.5
PAG11	Investigation into the measurement of plant or animal responses	9.3, 12, 13.8, 13.10, 14.6, 16.1, 16.4, 18.6
PAG12	Research skills	Throughout course

MODULE 2
Foundations in Biology

Introduction

Biology is the study of living organisms. Every living organism is made up of one or more cells, therefore understanding the structure and function of the cell is a fundamental concept in the study of biology. Since Robert Hooke coined the phrase 'cells' in 1665, careful observation using microscopes has revealed details of cell structure and ultrastructure and provided evidence to support hypotheses regarding the roles of cells and their organelles.

Basic components of living systems provides an introduction to cells and microscopy techniques. An understanding of cell biology is essential for most onward routes for biologists.

Biological molecules will begin to explore the biochemistry you need for your A Level course. An understanding of biochemistry provides a firm grounding for the study of key biological disciplines, such as medicine and disease research.

Enzymes are vital for many biological processes. In this chapter you will learn how they are structured and how they function.

Plasma membranes control the substances that move in and out of cells, and so a knowledge of how they work is essential for all areas of biology involving cellular processes.

Cell division explores the two processes by which cells divide – mitosis and meiosis. An understanding of these processes and how they can go wrong will help you understand health and disease, as well as explore new technologies in genetics and cloning.

Knowledge and understanding checklist

From your Key Stage 4 study you should be able to answer the following questions. Work through each point, using your Key Stage 4 notes and other resources. There is also support available on Kerboodle.

- [ ] Describe a cell as the basic structural unit of all organisms.
- [ ] Describe the main sub-cellular structures of eukaryotic and prokaryotic cells.
- [ ] Relate sub-cellular structures to their functions.
- [ ] Describe the cell cycle.
- [ ] Describe cell differentiation.
- [ ] Relate the adaptation of specialised cells to their function.
- [ ] Explain the mechanism of enzyme action.
- [ ] Recall the difference between intracellular enzymes and extracellular enzymes.
- [ ] Describe some anabolic and catabolic processes in living organisms including the importance of sugars, amino acids, fatty acids, and glycerol in the synthesis and breakdown of carbohydrates, lipids, and proteins.

Maths skills checklist

All biologists need to use maths in their studies and field of work. In this module, you will need to use the following maths skills.

- [ ] **Changing the subject of an equation.** You will need to be able to do this when working with microscopy.
- [ ] **Converting units.** You will need to be able to do this when working with microscopy.
- [ ] **Working with negative numbers.** You will need to be able to do this when calculating water potential when studying osmosis.
- [ ] **Working in standard form.** You will need to be able to do this throughout this chapter.

MyMaths.co.uk
Bringing Maths Alive

2 BASIC COMPONENTS OF LIVING SYSTEMS

2.1 Microscopy

Specification reference: 2.1.1

Learning outcomes

Demonstrate knowledge, understanding, and application of:

→ the use of **light microscopy**

→ the preparation of microscope slides for use in light microscopy

→ the use of staining in light microscopy

→ the representation of cell structure seen under light microscope using scientific annotated drawings.

Before the invention of microscopes, we knew nothing of bacteria, cells, sperm, pollen grains, chromosomes – the list is endless. Microscopes have given us the power to understand disease, see how a new life is formed, watch the dance of the chromosomes as cells divide, and manipulate the processes of life itself.

Seeing is believing

A microscope is an instrument which enables you to magnify an object hundreds, thousands and even hundreds of thousands of times. We can see many large organisms with the naked eye, but microscopes open up a whole world of unicellular organisms. By making visible the individual cells which make up multicellular organisms, microscopes allow us to discover how details of their structures relate to their functions.

The first types of microscopes to be developed were **light microscopes** in the 16th to 17th century. Since then they have continued to be developed and improved.

By the mid-19th century, scientists, for the first time, had access to microscopes with a high enough level of magnification to allow them to see individual cells. Cell theory was developed. It states that:

● both plant and animal tissue is composed of cells

● cells are the basic unit of all life

● cells only develop from existing cells.

Light microscopy continues to be important – it is easily available, relatively cheap and can be used out in the field, and it can be used to observe living organisms as well as dead, prepared specimens.

 History of the light microscope and the development of cell theory

Late in the Roman Empire the Romans began to develop and experiment with glass. They noted how objects looked bigger when viewed through pieces of glass that were thicker in the middle than at the edges.

There was little further development of glass lenses until around the 13th century and the invention of spectacles or eye glasses.

The credit for the invention of the light microscope is much disputed. Some accredit it to two Dutch spectacle makers who invented the telescope when experimenting with multiple glass lenses in a tube in the late 15th century. Others claim it was Galileo Galilei in 1609 who developed the first true or compound microscope (Table 1). Galileo's instrument was the first to be given the name 'microscope'.

Cell theory

The development of cell theory is a good example of how scientific theories change over time as new evidence is gained and as knowledge increases. Theories are proposed, accepted and can then be later disproved as new evidence comes to light. New evidence can arise in a number of ways, including as technology develops. This is the case with cell theory - as microscopes with higher magnification and resolution were developed, cells could be observed for the first time. Table 1 summarises some of the developments in cell theory.

▼ Table 1 *Cell theory timeline*

Timeline	Development of cell theory
1665	**Cell first observed** Robert Hooke, an English scientist, observed the structure of thinly sliced cork using an early light microscope. He described the compartments he saw as 'cells' – coining the term we still use today. As this was dead plant tissue he was observing only cell walls.
1674–1683	**First living cells observed** Anton van Leeuwenhoek, a Dutch biologist, developed a technique for creating powerful glass lenses and used his handcrafted microscopes to examine samples of pond water. He was the first person to observe bacteria and protoctista and described them as 'little animals' or 'animalcules' – today we call them microorganisms. He went on to observe red blood cells, sperm cells, and muscle fibres for the first time.
1832	**Evidence for the origin of new plant cells** Barthélemy Dumortier, a Belgian botanist, was the first to observe cell division in plants providing evidence against the theories of the time, that new cells arise from *within* old cells or that cells formed spontaneously from non-cellular material. However it was several more years until cell division as the origin of all new cells became the accepted theory.
1833	**Nucleus first observed** Robert Brown, an English botanist, was the first to describe the nucleus of a plant cell.
1837–1838	**The birth of a universal cell theory** Matthias Schleiden, a German botanist, proposed that *all* plant tissues are composed of cells. Jan Purkyně, a Czech scientist, was the first to use a microtome to make ultra-thin slices of tissue for microscopic examination. Based on his observations, he proposed that not only are animals composed of cells but also that the "basic cellular tissue is clearly analogous to that of plants". Not long after this, and independently, Theodor Schwann, a German physiologist, made a similar observation and declared that "all living things are composed of cells and cell products". He is the scientist credited with the 'birth' of cell theory.
1844 (1855)	**Evidence for the origin of new animal cells** Robert Remak, a Polish/German biologist, was the first to observe cell division in animal cells, disproving the existing theory that new cells originate from *within* old cells. He was not believed at the time however, and Rudolf Virchow, a German biologist, published these findings as his own a decade later in 1855.
1860	**Spontaneous generation disproved** Louis Pasteur disproved the theory of spontaneous generation of cells by demonstrating that bacteria would only grow in a sterile nutrient broth after it had been exposed to the air.

1 Outline the importance of microscopes in the study of living organisms.
2 Suggest, with reasons, why cell theory was not fully developed before the mid-19th century.

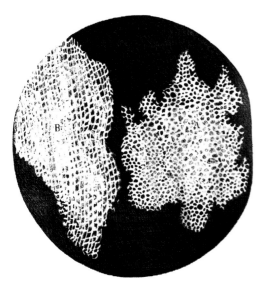

▲ Figure 1 *Drawing of cork from 1663 seen under an early microscope. Robert Hooke described the pores as cells, thus coining the term. He prepared the specimen by making thin slices with a razor blade, inventing the technique of sectioning*

▲ Figure 2 *Robert Hooke's drawing of his own compound microscope, which he used to see the 'cells' in a sample of cork*

How a light microscope works

A **compound light microscope** has two lenses – the objective lens, which is placed near to the specimen, and an eyepiece lens, through which the specimen is viewed. The objective lens produces a magnified image, which is magnified again by the eyepiece lens. This objective/eyepiece lens configuration allows for much higher magnification and reduced chromatic aberration than that in a simple light microscope.

Illumination is usually provided by a light underneath the sample. Opaque specimens can be illuminated from above with some microscopes.

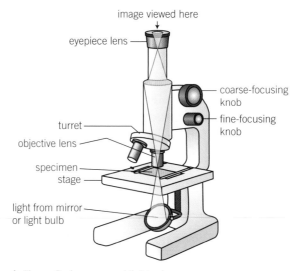

▲ Figure 3 *A compound light microscope*

Sample preparation

There are a number of different ways in which samples and specimens can be prepared for examination by light microscopy. The method chosen will depend on the nature of the specimen and the resolution that is desired.

- **Dry mount** – Solid specimens are viewed whole or cut into very thin slices with a sharp blade, this is called *sectioning*. The specimen is placed on the centre of the slide and a cover slip is placed over the sample. For example hair, pollen, dust and insect parts can be viewed whole in this way, and muscle tissue or plants can be sectioned and viewed in this way.

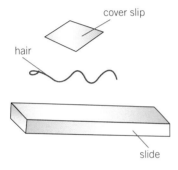

- **Wet mount** – Specimens are suspended in a liquid such as water or an immersion oil. A cover slip is placed on from an angle, as shown. For example, aquatic samples and other living organisms can be viewed this way.

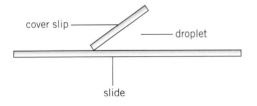

- **Squash slides** – A wet mount is first prepared, then a lens tissue is used to gently press down the cover slip. Depending on the material, potential damage to a cover slip can be avoided by squashing the sample between two microscope slides. Using squash slides is a good technique for soft samples. Care needs to be taken that the cover slip is not broken when being pressed. For example, root tip squashes are used to look at cell division.

- **Smear slides** – The edge of a slide is used to smear the sample, creating a thin, even coating on another slide. A cover slip is then placed over the sample. An example of a smear slide is a sample of blood. This is a good way to view the cells in the blood.

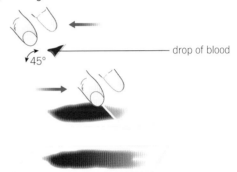

1 Suggest reasons for the following, with reference to slide preparation:
   a Specimens must be thin.
   b When preparing a wet mount the refractive index (ability to bend light) of the medium should be roughly the same as glass.
   c A cover slip must be placed onto a wet mount at an angle.

Using staining

In basic light microscopy the sample is illuminated from below with white light and observed from above (brightfield microscopy). The whole sample is illuminated at once (wide-field microscopy). The images tend to have low contrast as most cells do not absorb a lot of light. Resolution is limited by the wavelength of light and diffraction of light as it passes through the sample. Diffraction is the bending of light as it passes close to the edge of an object.

The cytosol (aqueous interior) of cells and other cell structures are often transparent. Stains increase contrast as different components within a cell take up stains to different degrees. The increase in contrast allows components to become visible so they can be identified (Figure 4).

To prepare a sample for staining it is first placed on a slide and allowed to air dry. This is then heat-fixed by passing through a flame. The specimen will adhere to the microscope slide and will then take up stains.

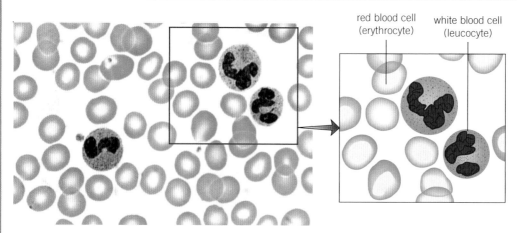

▲ **Figure 4** *Light micrograph and annotated diagram of a blood sample that has been stained using Wright's stain (a mixture of eosin red and methylene blue dyes). The nuclei of the white blood cells are stained purple*

Crystal violet or methylene blue are positively charged dyes, which are attracted to negatively charged materials in cytoplasm leading to staining of cell components.

Dyes such as nigrosin or Congo red are negatively charged and are repelled by the negatively charged cytosol. These dyes stay outside cells, leaving the cells unstained, which then stand out against the stained background. This is a negative stain technique.

Differential staining can distinguish between two types of organisms that would otherwise be hard to identify. It can also differentiate between different organelles of a single organism within a tissue sample.

● **Gram stain technique** is used to separate bacteria into two groups, Gram-positive bacteria and Gram-negative bacteria (Figure 5). Crystal violet is first applied to a bacterial specimen on a slide, then iodine, which fixes the dye. The slide is then washed with alcohol. The Gram-positive bacteria retain the crystal violet stain and will appear blue or purple under a microscope. Gram-negative bacteria have thinner cell walls and therefore lose the stain. They are then stained with safranin dye, which is called a **counterstain**. These bacteria will then appear red. Gram-positive bacteria are susceptible to the antibiotic penicillin, which inhibits the formation of cell walls. Gram-negative bacteria have much thinner cell walls that are not susceptible to penicillin.

● **Acid-fast technique** is used to differentiate species of *Mycobacterium* from other bacteria. A lipid solvent is used to carry carbolfuchsin dye into the cells being studied. The cells are then washed with a dilute acid-alcohol solution. Mycobacterium are not affected by the acid-alcohol and retain the carbolfuchsin stain,

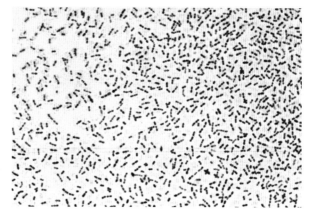

▲ **Figure 5** *Gram stain of Streptococcus pneumoniae, a Gram-positive bacteria which cause pneumonia (×200 magnification). Gram-positive bacteria retain a crystal violet dye during the Gram stain process and appear blue or violet under a microscope*

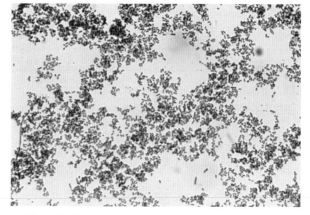

▲ **Figure 6** *Yersinia pestis, the Gram-negative bacteria which cause the bubonic plague infecting humans and animals (×1450 magnification). Gram-negative bacteria appear red or pink under a microscope*

which is bright red. Other bacteria lose the stain and are exposed to a methylene blue stain, which is blue.

You will often look at slides that have been pre-prepared. A number of stages are involved in the production of these slides:

- Fixing – chemicals like formaldehyde are used to preserve specimens in as near-natural a state as possible.
- Sectioning – specimens are dehydrated with alcohols and then placed in a mould with wax or resin to form a hard block. This can then be sliced thinly with a knife called a microtome.
- Staining – specimens are often treated with multiple stains to show different structures.
- Mounting – the specimens are then secured to a microscope slide and a cover slip placed on top.

Risk management

Many of the stains used in the preparation of slides are toxic or irritants. A risk assessment must carried out before any practical is started to identify any procedures involved that may result in harm.

CLEAPSS (Consortium of Local Education Authorities for the Provision of Science Services) is the organisation that provides support for the practical work carried out in schools. One of the main areas covered is health and safety, including risk assessment. Advice and support is provided to all types of educational establishments and their employees about all aspects of practical work such as the use of chemicals or living organisms, laboratory design, and even where to obtain the right equipment.

CLEAPSS provide student safety sheets that identify specific risks, advise on the measures to be taken to reduce these risks and the action to be taken in any emergency.

In fact, in schools many of the microscopy slides that are used are bought in ready-prepared and pre-stained, not only because of the harmful nature of the stains but also because of the long complex process needed to produce high quality sections.

1 Use your knowledge of the staining technique used to distinguish Gram-positive from Gram-negative bacteria and information from the paragraph above to answer the following question:
Suggest why Gram-negative infections are more difficult to treat than Gram-positive infections.

2 Crystal violet and potassium iodide are chemicals classed as irritants. Crystal violet is also toxic. Describe the precautions you should take when using these chemicals.

 ### Less is more

Scientific drawings are line drawings not pictures. They are used to highlight particular features and should not include unnecessary detail. The focus can be changed to help draw selected features.

The following is a list of rules for producing good scientific drawings:

- include a title
- state magnification
- use a sharp pencil for drawings and labels
- use white, unlined paper
- use as much of the paper as possible for the drawing
- draw smooth, continuous lines
- do not shade
- draw clearly defined structures
- ensure proportions are correct
- label lines should not cross and should not have arrow heads
- label lines should be parallel to the top of the page and drawn with a ruler

The light micrograph (Figure 7) shows a layer of onion cells as seen under a light microscope. Next to the micrograph is an example of a good scientific drawing of this image.

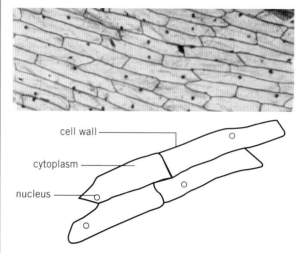

cell wall

cytoplasm

nucleus

× 18 magnification

▲ **Figure 7** *Top: Light micrograph of a layer of onion cuticle, showing the bands of large, rectangular cells. The dark spot in the centre of each cell is its nucleus. × 18 magnification. Bottom: A scientific drawing from the micrograph*

Below is an example of a poor scientific drawing:

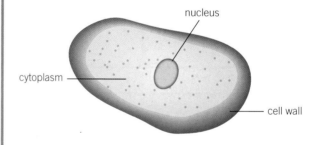

nucleus

cytoplasm

cell wall

1 Describe how this diagram is incorrect as a scientific drawing.

Summary questions

1 Outline the basic concepts of cell theory. *(3 marks)*

2 Explain why staining is used in microscopy. *(2 marks)*

3 Compound microscopes led to new discoveries essential for cell theory to be fully explained. Explain the benefit of having two lenses in a microscope. *(4 marks)*

4 a Calculate the low and high power magnifications of a microscope with the following lenses:
  i Eyepiece lens ×10
  ii Objective lenses ×10 and ×40. *(2 marks)*

b You are observing a specimen of squamous tissue under high power. The individual cells have an average diameter of 60 μm and the diameter of the field of view of the objective lens is 2 mm. Calculate the maximum number of whole cells that are visible in the field of view. *(3 marks)*

2.2 Magnification and calibration

Specification reference: 2.1.1

Magnification, resolution, and the magnification formula

Magnification is how many times larger the image is than the actual size of the object being viewed. Interchangeable objective lenses on a compound light microscope allow a user to adjust the magnification.

Simply magnifying an object does not increase the amount of detail that can be seen. The resolution also needs to be increased. The resolution of a microscope determines the amount of detail that can be seen – the higher the resolution the more details are visible.

Resolution is the ability to see individual objects as separate entities. Imagine a car coming towards you at night with its headlights on. When it is a long way off you will only see one light but as the car gets closer you eventually see that there are, in fact, two headlights – they have been resolved.

Resolution is limited by the diffraction of light as it passes through samples (and lenses). Diffraction is the tendency of light waves to spread as they pass close to physical structures such as those present in the specimens being studied. The structures present in the specimens are very close to each other and the light reflected from individual structures can overlap due to diffraction. This means the structures are no longer seen as separate entities and detail is lost. In optical microscopy structures that are closer than half the wavelength of light cannot be seen separately (resolved).

Resolution can be increased by using beams of electrons which have a wavelength thousands of times shorter than light (Topic 2.3, More microscopy). Electron beams are still diffracted but the shorter wavelength means that individual beams can be much closer before they overlap. This means objects which are much smaller and closer together can be seen separately without diffraction blurring the image.

Calculation for magnification

The magnification of an object can be calculated using the magnification formula:

$$\text{magnification} = \frac{\text{size of image}}{\text{actual size of object}}$$

In practice, the size of the image refers to the length of the image as measured, for example with a ruler. You may need to change the units of measurement to that of the actual size of the object. The magnification formula, like all mathematical equations, can be rearranged to find any of the unknowns, where the remaining values are known. To help you, you can imagine this three-part formula in a standard formula triangle (Figure 2).

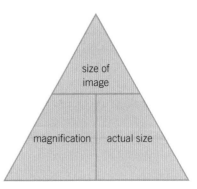

▲ Figure 2 *Formula triangle for magnification calculations*

Study tip

When solving magnification problems, carry out your calculations with all measurements having the same units.

Measure the image size in mm and convert it to the smallest unit present in the problem, usually micrometres.

1000 nanometres (nm) = 1 micrometre (µm)

1000 micrometres (µm) = 1 millimetre (mm)

1000 millimetres (mm) = 1 metre (m)

Magnification itself does not have units of measurement.

Students should also practise rearranging the magnification formula.

So, if the actual size of the object isn't known but the magnification is known, the actual size can be calculated by rearranging the formula to give:

$$\text{actual size of object} = \frac{\text{size of image}}{\text{magnification}}$$

Or, the size of the image can be calculated by rearranging the formula as below:

$$\text{size of image} = \text{magnification} \times \text{actual size of object}$$

 ### Worked example: Magnification calculation

Calculate the magnification of the image of the nuclear pore shown here.

To calculate the magnification you first convert all figures to the smallest unit, in this case nm.

24 millimetres is equal to $(24 \times 1000 \times 1000)$ nanometres or 24 000 000 nanometres.

$$\text{Magnification} = \frac{\text{size of image}}{\text{actual size of object}}$$

$$= \frac{24\,000\,000\,\text{nm}}{120\,\text{nm}}$$

= 200 000 or 2 hundred thousand times.

120 nm

 ### Using a graticule to calibrate a light microscope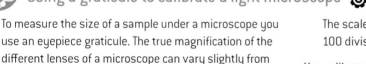

To measure the size of a sample under a microscope you use an eyepiece graticule. The true magnification of the different lenses of a microscope can vary slightly from the magnification stated so every microscope, and every lens, has to be calibrated individually using an eyepiece graticule and a slide micrometer.

- An *eyepiece graticule* is a glass disc marked with a fine scale of 1 to 100. The scale has no units and remains unchanged whichever objective lens is in place. The relative size of the divisions, however, increases with each increase in magnification. You need to know what the divisions represent at the different magnifications so you can measure specimens. The scale on the graticule at each magnification is calibrated using a stage micrometer.

- A *stage micrometer* is a microscope slide with a very accurate scale in micrometres (µm) engraved on it.

The scale marked on the micrometer slide is usually 100 divisions = 1 mm, so 1 division = 10 µm.

You calibrate the eyepiece graticule scale for each objective lens separately. Once all three lenses are calibrated, if you measure the same cell using the three different lenses you should get the same actual measurement each time.

For example:

Calibrating a ×4 objective lens:

1 Put the stage micrometer in place and the eyepiece graticule in the eyepiece.

2 Get the scale on the micrometer slide in clear focus.

3 Align the micrometer scale with the scale in the eyepiece. Take a reading from the two scales – see next page:

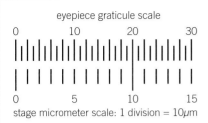

eyepiece graticule scale

stage micrometer scale: 1 division = 10μm

20 divisions on the eyepiece graticule = 10 divisions on the stage micrometer

Use these readings to calculate the calibration factor for the ×4 objective lens.

100 micrometer divisions = 1 mm

So each small division is 1/100 mm = 0.01 mm or 10.0 μm

20 graticule divisions = 10 micrometer divisions

10 micrometer divisions = 10 × 10 = 100 μm

$$1 \text{ graticule division} = \frac{\text{number of micrometres}}{\text{number of graticule divisions}}$$

20 graticule divisions = 100 μm so 1 graticule division = 100/20 = 5.0 μm

The magnification factor is 5.0

To use this magnification factor remove the stage micrometer and place a prepared slide on the stage. Measure the size of an object in graticule units. To find the actual size multiply the number of graticule units measured by the magnification factor to give you the length in μm

graticule divisions × magnification factor = measurement (μm)

e.g., the diameter of a cell seen using the ×4 objective lens measures 10 graticule divisions. Each graticule division = 5.0 μm so the cell diameter = 10 × 5.0 = 50.0 μm.

Calibration example

A student was asked to calibrate the ×10 lens of a light microscope and then to determine the diameter of a pollen grain from a sample slide provided.

1 The student placed a scale on the stage of the microscope and focused on it with the ×10 lens. 100 small divisions of the scale are 1 mm long, so 1 division is 10 μm.

2 The student then aligned the scale in the eyepiece with the scale on the microscope stage and calibrated it.

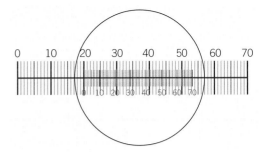

3 The student replaced the micrometer slide with pollen sample slides and used the calibrated scale in the eyepiece to measure the diameter of the pollen grains.

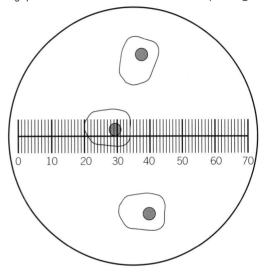

Results

The student decided to use the ×10 objective lens.

Result	1	2	3
Diameter of pollen grain/divisions	11	16	

1 State the correct names of the scale used on the stage of the microscope and the scale in the eyepiece.

2 Calculate the calibration factor of the ×10 objective lens.

3 a Fill in the missing reading in the table for the pollen grain shown in the artwork above.

   b Using the table calculate the diameter of the different pollen grains using the calibration factor you have calculated in question 2.

   c Calculate the mean diameter of the pollen grains.

   d Suggest why it is important to calibrate the lens that is going to be used to view the pollen grains.

Worked example: Using calibrated scales to measure specimens

Using the ×4 objective lens

Step 1: The diagram shows the graticule scale and the eyepiece micrometer readings used to calibrate the ×4 lens described in the previous application box, along with the measurements seen using the ×40 lens. On this micrometer 100 divisions = 1 mm so each graticule unit = 10 µm.

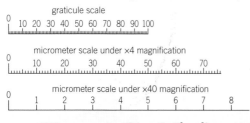

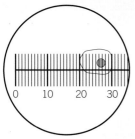

Step 2: The diameter of this pollen grain seen using the ×4 objective lens measures 10 graticule divisions. The calibration calculations on the previous page tell you that the magnification factor for the ×4 lens is 5.0

the diameter of the pollen grain = graticule units x magnification factor = 10 × 5.0 = 50 µm

Using the ×40 objective lens:

Step 1: 20 divisions on the eyepiece graticule = 1 division on the stage micrometer
Each division of the micrometer is 1/100 mm = 10 µm

Step 2: Using our observations
20 graticule divisions = 1.0 micrometer division
1.0 micrometre division = 1 × 10 µm = 10 µm
20 graticule divisions = 10 µm
1 graticule division = 10/20 = 0.5 µm
The magnification factor is 0.5

Step 3: The stage micrometer is removed and the same prepared slide placed on the stage. In this example you measure the same pollen grain: Calibrate the eyepiece graticule using the ×40 objective lens – see previous diagram.

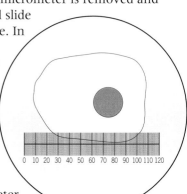

Step 4: The diameter of the pollen grain seen using the ×40 objective lens measures 100 graticule divisions

the pollen grain diameter = graticule units × magnification factor = 100 × 0.5 = 50 µm

The diameter of the pollen grain is the same measured using lenses of two very different magnifications. This is what you would expect but it is reassuring to see the theory confirmed and gives confidence in any measurements you take, regardless of which objective lens is in place.

Summary questions

1 Suggest why you should put all measurements into the same units before carrying out calculations.
(2 marks)

2 Calculate how many nanometres are present in 3846 centimetres.
Give your answer in standard form.

3 Explain the difference between contrast and resolution. *(2 marks)*

4 Calculate the magnification of the micrograph showing human cheek cells (Figure 3).
The average diameter of a cheek cell is 60 µm. *(2 marks)*

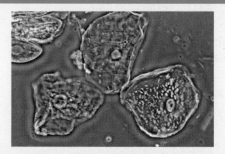

▲ Figure 3 *Light micrograph of squamous epithelium cells from the inside of a human cheek*

5 Explain how diffraction limits resolution. *(5 marks)*

6 Explain why eyepiece graticules do not have units. *(2 marks)*

Light microscopy started the science of cell biology, but it has limitations. In the middle of the 20th century a new invention, the electron microscope, revolutionised the study of cells and enabled biologists to see deep inside structures that were invisible under a light microscope.

Electron microscopy

In light microscopy, increased magnification can be achieved easily using the appropriate lenses, but if the image is blurred no more detail will be seen. Resolution is the limiting factor.

In **electron microscopy**, a beam of electrons with a wavelength of less than 1 nm is used to illuminate the specimen. More detail of cell **ultrastructure** can be seen because electrons have a much smaller wavelength than light waves. They can produce images with magnifications of up to ×500 000 and still have clear resolution.

Electron microscopes have changed the way we understand cells but there are some disadvantages to this technique. They are very expensive pieces of equipment and can only be used inside a carefully controlled environment in a dedicated space. Specimens can also be damaged by the electron beam and because the preparation process is very complex, there is a problem with **artefacts** (structures that are produced due to the preparation process). However, as techniques improve a lot of theses artefacts can be eliminated.

There are two types of electron microscope:

- In a **transmission electron microscope (TEM)** a beam of electrons is transmitted through a specimen and focused to produce an image. This is similar to light microscopy. This has the best resolution with a resolving power of 0.5 nm (Figure 1).

- In a **scanning electron microscope (SEM)** a beam of electrons is sent across the surface of a specimen and the reflected electrons are collected. The resolving power is from 3–10 nm, so the resolution is not as good as with transmission electron microscopy but stunning three-dimensional images of surfaces are produced, giving us valuable information about the appearance of different organisms (see Figure 2).

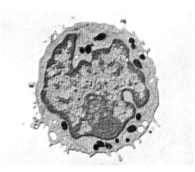

▲ Figure 1 *Coloured transmission electron micrograph of a lymphocyte (white blood cell). Magnification ×1600*

▲ Figure 2 *Coloured scanning electron micrograph of a lymphocyte (white blood cell). Magnification ×2000*

 Sample preparation for electron microscopes

The inside of an electron microscope is a vacuum to ensure the electron beams travel in straight lines. Because of this, samples need to be processed in a specific way.

Specimen preparation involves fixation using chemicals or freezing, staining with heavy metals and dehydration with solvents. Samples for a TEM will then be set in resin and may be stained again. Samples for a SEM may be fractured to expose the inside and will then need to be coated with heavy metals.

1 Suggest reasons for the following steps in the preparation of samples for electron microscopy:
- fixation
- dehydration
- embedding in resin
- staining with heavy metals.

Table 1 summarises the differences between light and electron microscopy.

▼ Table 1 *A comparison of light and electron microscopy*

Light microscope	Electron microscope
inexpensive to buy and operate	expensive to buy and operate
small and portable	large and needs to be installed
simple sample preparation	complex sample preparation
sample preparation does not usually lead to distortion	sample preparation often distorts material
vacuum is not required	vacuum is required
natural colour of sample is seen (or stains are used)	black and white images produced (but can be coloured digitally)
up to ×2000 magnification	over ×500 000 magnification
resolving power is 200 nm	resolving power of transmission electron microscope is 0.5 nm and a scanning electron microscope is 3–10 nm
specimens can be living or dead	specimens are dead

Scientific drawings from electron micrographs

Electron microscopes produce images with much greater resolution than light microscopes and therefore much more detail can be seen.

Producing good scientific drawings from electron micrographs takes practice. The same rules used in producing drawings from light micrographs must still be observed (Topic 2.1, Microscopy).

a

starch granules — — chloroplasts

cell wall — — nucleus

b

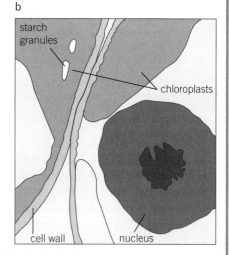

starch granules

chloroplasts

cell wall   nucleus

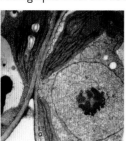

◀ Figure 3 *Transmission electron micrograph of a section through two leaf cells at their junction. Their cell walls run from top centre to lower left. A nucleus is seen at lower right. Starch granules (pale ovals) can be seen in the chloroplast (dark grey, upper left and right). ×18 700 magnification*

1   Study the two drawings above and state, with reasons, which of them represents the best scientific drawing.

Creation of artefacts

An artefact is a visible structural detail caused by processing the specimen and not a feature of the specimen. Artefacts appear in both light and electron microscopy. The bubbles that get trapped under the cover slip as you prepare a slide for light microscopy are artefacts. When preparing specimens for electron microscopy, changes in the ultrastructure of cells are inevitable during the processing that the samples must undergo. They are seen as the loss of continuity in membranes, distortion of organelles and empty spaces in the cytoplasm of cells

Experience enables scientists to distinguish between an artefact and a true structure.

 ## Identifying artefacts

Identifying artefacts in microscopy preparation can cause much discussion and controversy. 'Mesosome' was the name given to invaginations (inward foldings) of cell membranes that were observed using electron microscopes after bacterial specimens had been chemically fixed. They were thought to be a normal structure, or organelle, found within prokaryotes. The large surface area of the folded membrane was considered to be an important site for the process of oxidative phosphorylation. However, when specimens were fixed by the more recently developed, non-chemical technique called cryofixation, the mesosomes were no longer visible.

It is now widely thought that the majority of mesosomes observed are actually artefacts produced by the chemicals used in the fixation process in electron microscopy preparation, which damage bacterial cell membranes. However, there are still a number of scientists who believe

that some species of bacteria do have mesosomes as part of their normal structure, but this is not the general consensus.

This is a good example of how the scientific community accepted an idea based on the evidence available at the time and as techniques improved and more evidence became available, the collective knowledge and understanding developed and changed. New evidence can either provide further support for a theory or disprove an earlier theory. Scientific knowledge is constantly developing.

1 Structures that look similar to mesosomes have recently been observed in bacteria after treatment with certain types of antibiotics.
Suggest, with reasons, whether this information is evidence to support the current theory that mesosomes are artefacts or the theory that they are, in fact, organelles.

Laser scanning confocal microscopy

Light microscopy has also continued to develop. Some of the latest technology produces images that are very different from electron micrographs but are just as useful.

Conventional optical microscopes use visible light to illuminate specimens and a lens to produce a magnified image. In fluorescent microscopes a higher light intensity is used to illuminate a specimen that has been treated with a fluorescent chemical (a fluorescent 'dye'). Fluorescence is the absorption and re-radiation of light. Light of a longer wavelength and lower energy is emitted and used to produce a magnified image.

A **laser scanning confocal microscope** moves a single spot of focused light across a specimen (point illumination). This causes fluorescence from the components labelled with a 'dye'. The emitted light from the specimen is filtered through a pinhole aperture. Only light radiated from very close to the focal plane (the distance that gives the sharpest focus) is detected (Figure 4).

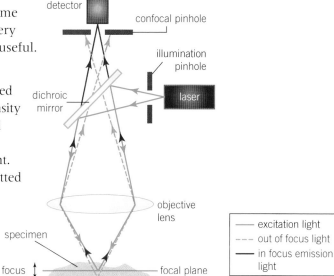

▲ Figure 4 *The light rays from the laser and the fluorescing sample follow the same path and have the same focal plane*

Light emitted from other parts of the specimen would reduce the resolution and cause blurring. This unwanted radiation does not pass through the pinhole and is not detected. A laser is used instead of light to get higher intensities, which improves the illumination.

As very thin sections of specimen are examined and light from elsewhere is removed, very high resolution images can be obtained.

The spot illuminating the specimen is moved across the specimen and a two dimensional image is produced. A three dimensional image can be produced by creating images at different focal planes.

Laser scanning confocal microscopy is non-invasive and is currently used in the diagnosis of diseases of the eye and is also being developed for use in endoscopic procedures. The fact that it can be used to see the distribution of molecules within cells means it is also used in the development of new drugs.

The future uses for advanced optical microscopy include virtual biopsies, particularly in cases of suspected skin cancer.

The beamsplitter is a dichroic mirror, which only reflects one wavelength (from the laser) but allows other wavelengths (produced by the sample) to pass through.

The positions of the two pinholes means the light waves from the laser (illuminating the sample) follow the same path as the light waves radiated when the sample fluoresces. This means they will both have the same focal plane, hence the term *confocal*.

Synoptic link

You will learn more about antibodies in Topic 12.6, The specific immune system.

Synoptic link

You will learn about genetic engineering in Topic 21.4, Genetic engineering.

Fluorescent tags

By using antibodies with fluorescent 'tags', specific features can be targeted and therefore studied by confocal microscopy with much more precision than when using staining and light microscopy.

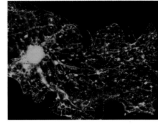

Green fluorescent protein (GFP) is produced by the jellyfish *Aequorea victoria*. The protein emits bright green light when illuminated by ultraviolet light. GFP molecules have been engineered to fluoresce different colours, meaning different components of a specimen can be studied at the same time. The gene for this protein has been isolated and can be attached, by genetic engineering, to genes coding for proteins under investigation. The fluorescence indicates that a protein is being made and is used to see where it goes within the cell or organism. Bacterial, fungal, plant, and human cells have all been modified to express this gene and fluoresce. The use of these fluorescing proteins provides a non-invasive technique to study the production and distribution of proteins in cells and organisms.

a   Define the term resolution with reference to microscopy.
b   Suggest whether fluorescent microscopy has a higher resolution than normal light microscopy. Explain your answer.

Atomic force microscopy

The atomic force microscope (AFM) gathers information about a specimen by 'feeling' its surface with a mechanical probe. These are scanning microscopes that generate three-dimensional images of surfaces.

An AFM consists of a sharp tip (probe) on a cantilever (a lever supported at one end) that is used to scan the surface of a specimen. When this is brought very close to a surface, forces between the tip and the specimen cause deflections of the cantilever. These deflections are measured using a laser beam reflected from the top of the cantilever into a detector.

Fixation and staining are not required and specimens can be viewed in almost normal cell conditions without the damage caused during the preparation of specimens for electron microscopy. Living systems can even be examined.

The resolution of AFM is very high, in the order of 0.1 nm. Information can be gained at the atomic level, even about the bonds within molecules.

The pharmaceutical industry in particular uses AFM to identify potential drug targets on cellular proteins and DNA. These microscopes can lead to a better understanding of how drugs interact with their target molecule or cell.

AFM is also being employed to identify new drugs. Finding and identifying new chemical compounds from the natural world, which may have medical applications, takes a long time, and is expensive. The molecular structures need to be understood before their potential use in medicine is known. Atomic force microscopes can speed up this process, saving money and, potentially, lives. The case study below is a good example of the importance of AFM.

Case Study: Deep sea molecules

In 2010, scientists working on a species of bacterium from a mud sample taken from the Mariana Trench – the deepest place on the planet located nearly 11 000 metres beneath the Pacific Ocean, found that the bacteria produced an unknown chemical compound.

The chemical composition (the number and type of atoms present) was easily determined. However,

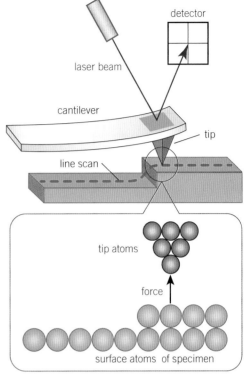

▲ Figure 5 *Top: The principle of atomic force microscopy. Bottom: A nuclear pore as shown by atomic force microscopy*

the molecular structure, the way in which the atoms were joined together, was not so easy to work out and would have taken months using conventional techniques.

Using atomic force microscopy the scientists were able to image the molecules at very high, atomic level resolution within one week, giving them the molecular structure they needed.

This was the first time this method had been used in this way. This new approach could lead to much faster identification of unknown compounds and ultimately speed up the process of the development of new medicines.

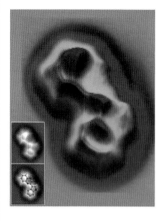

◀ **Figure 6** *Atomic force microscopy unveiled the previously unknown structure of cephalandole A, a chemical compound that could lead to the development of new drugs*

1 Explain why atomic force microscopy has a greater resolution than traditional light microscopy.
2 AFM is capable of producing magnifications equal or better than electron microscopes (this is demonstrated in Figure 7). Explain why.
3 Discuss why, despite the comparable magnification, atomic force microscopy could not have resulted in the same advances in the study of cell function as electron microscopy.

▲ **Figure 7** *Atomic force microscopy image of* Staphylococcus aureus *bacteria, commonly known for causing MRSA infections*

 ## Super resolved fluorescence microscopy

Electron microscopes cannot be used to examine living cells and it was always believed that the maximum resolution for light microscopes was 0.2 μm, about half the wavelength of light. This limits the detail that can be seen in living cells. In 2014 Eric Betzig, Stefan W. Hell, and William E. Moerner were awarded the Nobel Prize in Chemistry for achieving resolutions greater than 0.2 μm using light microscopy.

Two principles were involved, both forms of super resolution fluorescent microscopy (SRFM). One involved building up a very high resolution image by combining many very small images. The other involved superimposing many images with normal resolution to create one very high resolution image.

Stefan Hell developed stimulated emission depletion (STED) which involves the use of two lasers which are slightly offset. The first laser scans a specimen causing fluorescence, followed by the second laser which negates the fluorescence from all but a molecular sized area. A picture is built up with a resolution much greater than

that produced normally in light microscopy. In this way, individual strands of DNA become visible.

Eric Betzig and William E. Moerner independently developed the second principle which relies on the ability to control the fluorescence of individual molecules. Specimens are scanned multiple times but each time different molecules are allowed to fluoresce. The images are then superimposed and the resolution of the combined image is at the molecular level, much greater than 0.2 μm.

It is now possible to follow individual molecules during cellular processes. Proteins involved in Parkinson's and Alzheimer's diseases can be observed interacting and fertilised eggs dividing into embryos can be studied at a molecular level.

1 a Explain why electron microscopes cannot be used to examine living cells.
  b Describe how the ability to control the fluorescence of individual molecules helped uncover cell processes.

Summary questions

1 Explain why you would see more detail with an electron microscope than with a light microscope. *(2 marks)*

2 a Define the term artefact with reference to microscopy. *(2 marks)*
   b Explain why artefacts are more likely to be produced when preparing samples for electron microscopy than for light microscopy. *(3 marks)*

3 Study the two images below.

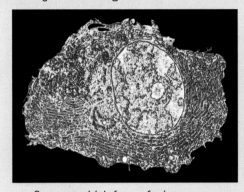 

   a Suggest which form of microscopy was used to produce each image. *(1 mark)*
   b Explain the reasons for your choices. *(3 marks)*
   c Outline the advantages and disadvantages of each technique. *(6 marks)*

4 Confocal microscopy is used in medicine to study the cornea of the eye and the progression of skin cancer.
   a Explain the meaning of the term fluorescence. *(2 marks)*
   b State why lasers are used to provide illumination. *(1 mark)*
   c Explain the purpose of the pinhole aperture in confocal microscopy. *(3 marks)*
   d One limitation of confocal microscopy is that it can not be used for deep tissue imaging. Suggest why. *(1 mark)*

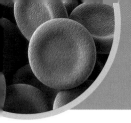

2.4 Eukaryotic cell structure

Specification reference: 2.1.1

Learning outcomes

Demonstrate knowledge, understanding, and application of:

→ the ultrastructure and function of eukaryotic cellular components

→ the importance of the cytoskeleton. The interrelationship between the organelles involved in the production and secretion of proteins.

Microscopes not only make cells visible – they also enable us to look deep inside individual cells. Using different types of microscopes you can discover how cells are organised and investigate the ways in which the structures you can see relate to their function. Microscopy allows you to see what goes on in a healthy cell, and to observe some of the changes which take place if the cell is attacked or diseased.

Relative sizes of molecules, organelles and cells

The diagram below demonstrates how the development of microscopes has allowed biologists to discover increasing amounts of detail of cell ultrastructure. The increased knowledge of structure has led to a better understanding of cell function.

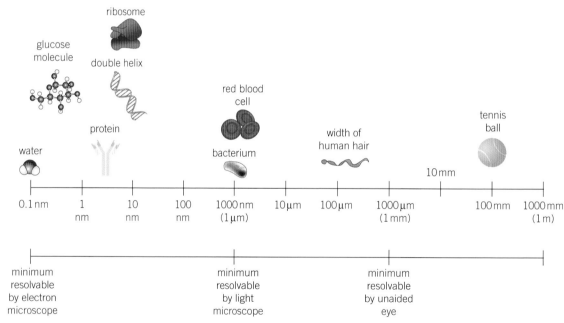

▲ Figure 1 *This diagram illustrates the relative sizes of the different components of living organisms*

Cells

The basic unit of all living things is the cell – but not all cells are the same. There are two fundamental types of cell – **prokaryotic** and **eukaryotic**. Prokaryotes are single-celled organisms with a simple structure of just a single undivided internal area called the **cytoplasm** (composed of cytosol, which is made up of water, salts and organic molecules). Eukaryotic cells make up multicellular organisms like animals, plants, and fungi. Eukaryotic cells have a much more complicated internal structure, containing a membrane-bound nucleus (nucleoplasm) and cytoplasm, which contains many membrane-bound cellular components. You will learn more about the differences between prokaryotic and eukaryotic cells in Topic 2.6.

Synoptic link

You will learn about the role of enzymes in cellular metabolism and how they are affected by cellular conditions in Chapter 4, Enzymes.

In this topic you will learn about the ultrastructure of eukaryotic cells. The ultrastructure of a cell is those features that can be seen using an electron microscope.

Compartments for life

Chemical reactions are the fundamental processes of life and in cells they require both enzymes and specific reaction conditions. **Metabolism** involves both the synthesis (building up) and the breaking down of molecules. Different sets of reactions take place in different regions of the ultrastructure of the cell.

The reactions take place in the cytoplasm. The cell cytoplasm is separated from the external environment by a cell-surface membrane. In eukaryotic cells the cytoplasm is divided into many different membrane-bound compartments, known as **organelles**. These provide distinct environments and therefore conditions for the different cellular reactions.

Membranes are selectively permeable and control the movement of substances into and out of the cell and organelles. Membranes are effective barriers in controlling which substances enter and exit cells but they are fragile.

There are a number of organelles that are common to all eukaryotic cells. Each type has a distinct structure and function. They are clearly seen in animal cells, the focus of this topic. The ultrastucture specific to plant cells is discussed in the next topic.

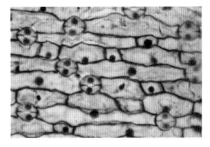

▲ Figure 2 *The detail you can see inside a cell depends on the type of microscope used to produce the image. This photomicrograph is of onion cells as seen under a light microscope. Only the nuclei can be seen inside the cells (x200 magnification)*

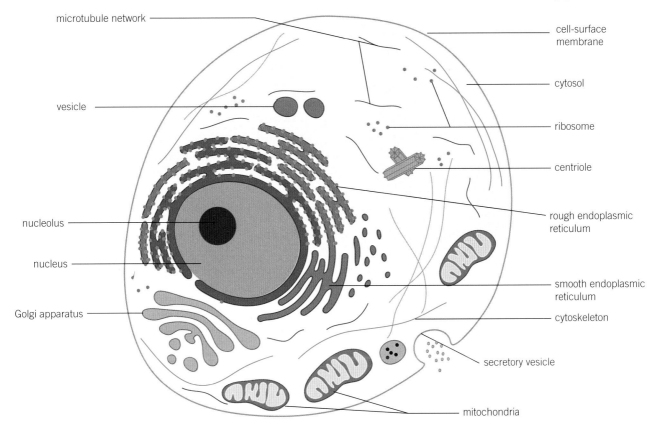

▲ Figure 3 *A drawing of a eukaryotic animal cell showing the many other components that are not visible with a light microscope*

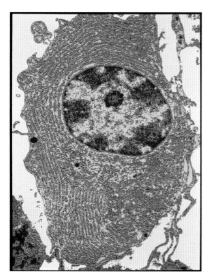

▲ Figure 4 *Coloured transmission electron micrograph of a human cell showing the nucleus (large oval) and endoplasmic reticulum. Magnification: ×10 000*

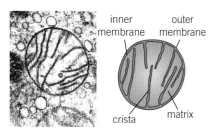

▲ Figure 5 *Electron micrograph and drawing of a mitochondrion x30 000 magnification*

Nucleus

The **nucleus** (plural nuclei) contains coded genetic information in the form of DNA molecules. DNA directs the synthesis of all proteins required by the cell (although this protein synthesis occurs outside of the nucleus at ribosomes). In this way the DNA controls the metabolic activities of the cell, as many of these proteins are the enzymes necessary for metabolism to take place. Not surprisingly, the nucleus is often the biggest single organelle in the cell (Figure 4).

DNA is contained within a double membrane called a *nuclear envelope* to protect it from damage in the cytoplasm. The nuclear envelope contains *nuclear pores* (Topic 2.2, Magnification and calibration) that allow molecules to move into and out of the nucleus. DNA itself is too large to leave the nucleus to the site of protein synthesis in the cell cytoplasm. Instead it is transcribed into smaller RNA molecules, which are exported via the nuclear pores.

DNA associates with proteins called **histones** to form a complex called **chromatin**. Chromatin coils and condenses to form structures known as **chromosomes**. These only become visible when cells are preparing to divide.

Nucleolus

The nucleolus is an area within the nucleus and is responsible for producing ribosomes. It is composed of proteins and RNA. RNA is used to produce ribosomal RNA (rRNA) which is then combined with proteins to form the ribosomes necessary for protein synthesis.

Mitochondria

Mitochondria (singular mitochondrion) are essential organelles in almost all eukaryotic cells. They are the site of the final stages of cellular respiration, where the energy stored in the bonds of complex, organic molecules is made available for the cell to use by the production of the molecule ATP. The number of mitochondria in a cell is generally a reflection of the amount of energy it uses, so very active cells usually have a lot of mitochondria.

Mitochondria have a double membrane. The inner membrane is highly folded to form structures called **cristae** and the fluid interior is called the **matrix**. The membrane forming the cristae contains the enzymes used in aerobic respiration. Interestingly, mitochondria also contain a small amount of DNA, called **mitochondrial (mt)DNA**. Mitochondria can produce their own enzymes and reproduce themselves.

Vesicles and lysosomes

Vesicles are membranous sacs that have storage and transport roles. They consist simply of a single membrane with fluid inside. Vesicles are used to transport materials inside the cell.

Lysosomes are specialised forms of vesicles that contain hydrolytic enzymes. They are responsible for breaking down waste material in cells, including old organelles. They play an important role in the immune system as they are responsible for breaking down pathogens ingested by phagocytic cells. They also play an important role in programmed cell death or apoptosis.

Synoptic link

You will learn about apoptosis in Topic 19.3, Body plans.

The cytoskeleton

The **cytoskeleton** is present throughout the cytoplasm of all eukaryotic cells. It is a network of fibres necessary for the shape and stability of a cell. Organelles are held in place by the cytoskeleton and it controls cell movement and the movement of organelles within cells.

The cytoskeleton has three components:

- Microfilaments – contractile fibres formed from the protein **actin**. These are responsible for cell movement and also cell contraction during cytokinesis, the process in which the cytoplasm of a single eukaryotic cell is divided to form two daughter cells.

- Microtubules – globular tubulin proteins polymerise to form tubes that are used to form a scaffold-like structure that determines the shape of a cell. They also act as tracks for the movement of organelles, including vesicles, around the cell. Spindle fibres, which have a role in the physical segregation of chromosomes in cell division, are composed of microtubules.

- Intermediate fibres – these fibres give mechanical strength to cells and help maintain their integrity.

Synoptic link

You will learn about the role of spindle fibres in cell division in Chapter 6, Cell division.

Cell movement

The movement of cells like phagocytes depends on the activity of the actin filaments in the cytoskeleton. The filament lengths change with the addition and removal of monomer subunits. The rate at which these subunits are added is different at each end of a filament. The subunits are not symmetrical and can only be added if they are in the correct orientation.

The subunits have to change shape before they are added to one end (the minus end) of the filament but not the other end (the plus end). This means that the subunits are added at a faster rate at the plus end. The filaments therefore increase in length at a faster rate in one particular direction.

Whether subunits are added or removed, at either end, is determined by the concentration of subunits in the cytoplasm. Due to the different rates of addition at either end, at certain concentrations subunits will be added at one end and removed at the other. This called treadmilling.

The increasing length of the filaments at one edge of a cell, the leading edge, leads to cells such as phagocytes moving in a particular direction.

1 Suggest, giving your reasons, which components of the cytoskeleton undergo treadmilling and which components do not.

Centrioles

Centrioles are a component of the cytoskeleton present in most eukaryotic cells with the exception of flowering plants and most fungi. They are composed of microtubules. Two associated centrioles form the *centrosome*, which is involved in the assembly and organisation of the spindle fibres during cell division.

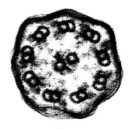

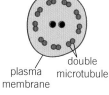

plasma membrane / double microtubule

▲ Figure 6 *TEM of a cross-section through a single cilium from a protozoan (× 70 000 magnification)*

Synoptic link

You will learn about the role of cilia in Topic 7.2, The mammalian gaseous exchange system.

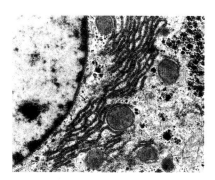

▲ Figure 7 *Coloured TEM showing the rough endoplasmic reticulum (folds, centre). The cell nucleus is partially seen to the left. The round structures are vesicles that are being used to transport proteins from the rough endoplasmic reticulum to elsewhere in the cell. Magnification × 20 000*

Synoptic link

You will learn about proteins in Topic 3.6, Structure of proteins and Topic 3.7, Types of proteins. You will learn about enzymes in Chapter 4, Enzymes.

In organisms with flagella and cilia, centrioles are thought to play a role in the positioning of these structures.

Flagella and cilia

Both flagella (whip-like) and cilia (hair-like) are extensions that protrude from some cell types. Flagella are longer than cilia but cilia are usually present in much greater numbers.

Flagella are used primarily to enable cells motility. In some cells they are used as a sensory organelle detecting chemical changes in the cell's environment.

Cilia can be mobile or stationary. Stationary cilia are present on the surface of many cells and have an important function in sensory organs such as the nose. Mobile cilia beat in a rhythmic manner, creating a current, and cause fluids or objects adjacent to the cell to move. For example, they are present in the trachea to move mucus away from the lungs (helping to keep the air passages clean), and in fallopian tubes to move egg cells from the ovary to the uterus.

Each cilium contains two central microtubules (black circles) surrounded by nine pairs of microtubules arranged like a "wheel". This is known as the 9+2 arrangement (Figure 6). Pairs of parallel microtubules slide over each other causing the cilia to move in a beating motion.

Organelles of protein synthesis

A key function of a cell is to synthesise proteins (including enzymes) for internal use and for **secretion** (transport out of the cell). A significant proportion of the internal structure of a cell is required for this process. The ribosomes, the endoplasmic reticulum, and the Golgi apparatus are all closely linked and coordinate the production of proteins and their preparation for different roles within the cell. The cytoskeleton plays a key role in coordinating protein synthesis.

Endoplasmic reticulum

The **endoplasmic reticulum (ER)** is a network of membranes enclosing flattened sacs called cisternae. It is connected to the outer membrane of the nucleus. There are two types:

- **Smooth endoplasmic reticulum** is responsible for lipid and carbohydrate synthesis, and storage.
- **Rough endoplasmic reticulum** has ribosomes bound to the surface and is responsible for the synthesis and transport of proteins.

Secretory cells, which release hormones or enzymes, have more rough endoplasmic reticulum than cells that do not release proteins.

Ribosomes

Ribosomes can be free-floating in the cytoplasm or attached to endoplasmic reticulum, forming rough endoplasmic reticulum. They are not surrounded by a membrane. They are constructed of RNA molecules made in the nucleolus of the cell. Ribosomes are the site of protein synthesis.

Mitochondria and chloroplasts also contain ribosomes, as do prokaryotic cells.

Golgi apparatus

The **Golgi apparatus** is similar in structure to the smooth endoplasmic reticulum. It is a compact structure formed of cisternae and does not contain ribosomes. It has a role in modifying proteins and 'packaging' them into vesicles. These may be secretory vesicles, if the proteins are destined to leave the cell, or lysosomes, which stay in the cell.

Protein production

Proteins are synthesised on the ribosomes bound to the endoplasmic reticulum (1). They then pass into its cisternae and are packaged into transport vesicles (2). Vesicles containing the newly synthesised proteins move towards the Golgi apparatus via the transport function of the cytoskeleton (3). The vesicles fuse with the cis face of the Golgi apparatus and the proteins enter. The proteins are structurally modified before leaving the Golgi apparatus in vesicles from its trans face (4).

Secretory vesicles carry proteins that are to be released from the cell. The vesicles move towards and fuse with the cell-surface membrane, releasing their contents by exocytosis. Some vesicles form lysosomes – these contain enzymes for use in the cell (5).

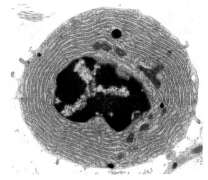

▲ Figure 8 *Transmission electron micrograph of a plasma cell with a large central nucleus surrounded by large amounts of rough endoplasmic reticulum. Plasma cells, which are found in the blood and lymph, produce and secrete antibodies (which are made of protein) during an immune response. × 6000 magnification*

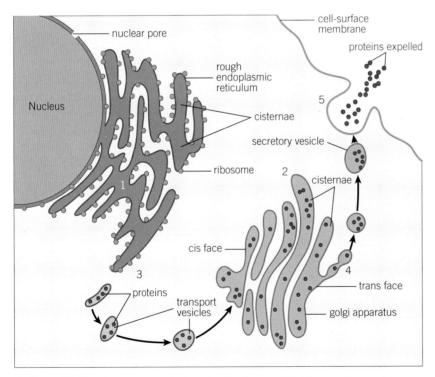

▲ Figure 10 *The ribosomes, endoplasmic reticulum, and Golgi apparatus work together to synthesise, modify and then transport proteins, including enzymes and hormones, out of the cell*

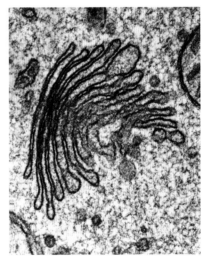

▲ Figure 9 *Transmission electron micrograph of the Golgi apparatus. Golgi are membrane-bound organelles that modify and package proteins for onward transport. × 8000 magnification*

Synoptic link

You will learn more about the details of protein synthesis at the ribosome in Topic 3.10, Protein synthesis.

Summary questions

1 What is a lysosome and why is the membrane that surrounds it so important? *(3 marks)*

2 Explain why cells need to be compartmentalised, and describe three examples of compartmentalisation within an animal cell. *(4 marks)*

3 Compare the structure and function of the rough and smooth endoplasmic reticulum. *(3 marks)*

4 Describe the structure and function of the cytoskeleton. *(5 marks)*

5 Given the following information about a eukaryotic cell

7×10^7 base pairs of DNA per chromosome 0.34×10^{-9} m per base pair diploid number is 46

a Calculate the length of DNA in a single cell. Give your answer in metres.

b Suggest how this DNA is packed into a cell only 50 μm in diameter.

6 Discuss how the structure of microfilaments and microtubules means these components of the cytoskeleton are involved in the movement of cells but the intermediate fibres are not. *(6 marks)*

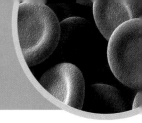

Plant cells have all of the cellular components you have just seen in animal cells. However, there are some structures that are only seen in plant cells, that carry out photosynthesis.

Cellulose cell wall

Plant cells, unlike animal cells, are rigid structures. They have a cell wall surrounding the cell-surface membrane.

Plant cell walls are made of cellulose, a complex carbohydrate. They are freely permeable so substances can pass into and out of the cell through the cellulose wall. The cell walls of a plant cell give it shape. The contents of the cell press against the cell wall making it rigid. This supports both the individual cell and the plant as a whole. The cell wall also acts as a defence mechanism, protecting the contents of the cell against invading pathogens. All plant cells have cellulose cell walls.

Synoptic link
You will learn about carbohydrates, including cellulose, in Topic 3.3, Carbohydrates.

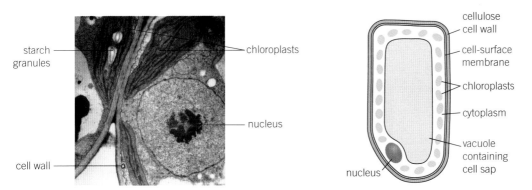

▲ Figure 1 *Left: A transmission electron micrograph of the junction of two leaf cells × 18 700 magnification. Right: A representation of a plant cell as seen under a light microscope*

Plant cell organelles

Plant cells, unlike animal cells, are rigid structures. Structures which are unique to plant cells include:

Vacuoles

Vacuoles are membrane lined sacs in the cytoplasm containing cell sap. Many plant cells have large permanent vacuoles which are very important in the maintenance of turgor, so that the contents of the cell push against the cell wall and maintain a rigid framework for the cell. The membrane of a vacuole in a plant cell is called the **tonoplast**. It is selectively permeable, which means that some small molecules can pass through it but others cannot. If vacuoles appear in animal cells, they are small and transient (not permanent).

Synoptic link

You will learn more about the role of chloroplasts in photosynthesis in Chapter 17, Energy for biological processes.

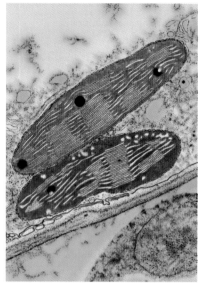

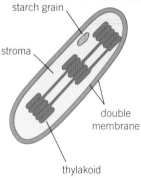

▲ Figure 2 *Transmission electron micrograph (top) and drawing of chloroplasts seen in the leaf of a pea plant (bottom). The chloroplasts are seen cut lengthways so the grana are visible. Starch produced during photosynthesis is seen as dark circles (starch grains) within each chloroplast. ×13 000 magnification*

Chloroplasts

Chloroplasts are the organelles responsible for photosynthesis in plant cells. They are found in the cells in the green parts of plants such as the leaves and the stems but not in the roots. They have a double membrane structure, similar to mitochondria. The fluid enclosed in the chloroplast is called the **stroma**. They also have an internal network of membranes, which form flattened sacs called thylakoids. Several thylakoids stacked together are called a **granum** (plural grana). The grana are joined by membranes called lamellae. The grana contain the chlorophyll pigments, where light-dependent reactions occur during photosynthesis. Starch produced by photosynthesis is present as starch grains. Like mitochondria, chloroplasts also contain DNA and ribosomes. Chloroplasts are therefore able to make their own proteins.

The internal membranes provide the large surface area needed for the enzymes, proteins and pigment molecules necessary in the process of photosynthesis.

Summary questions

1 Using your knowledge of cell ultrastructure, identify the structures visible in the micrograph below. State, with reasons, whether the cell is a plant or animal cell *(4 marks)*

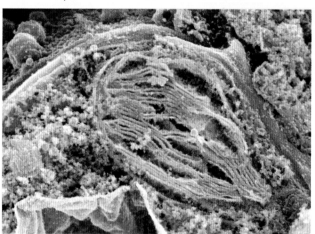

2 a Many different organisms have cell walls including fungi and bacteria. What is unique about plant cell walls? *(1 mark)*
  b Give three functions of plant cell walls *(3 marks)*

3 Describe the similarities and differences between a human cell and a plant root cell *(3 marks)*

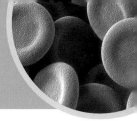

2.6 Prokaryotic and eukaryotic cells
Specification reference: 2.1.1

Animals, plants, and fungi are all complex multicellular organisms. The cells making up these organisms are eukaryotic. There is a lot of evidence that suggests that eukaryotic cells evolved from less complex prokaryotic cells. These prokaryotic cells, present in great numbers, live in an incredibly diverse range of habitats. These unicellular organisms can be classed into two evolutionary domains – Archaea and Bacteria, which evolved from an ancient common ancestor.

Prokaryotic cells

Prokaryotic cells may have been among the earliest forms of life on Earth. They first appeared around 3.5 billion years ago when the surface of the Earth was a very hostile environment. Scientists believe that these early cells were adapted to living in extremes of salinity, pH and temperature.

These organisms are known as extremophiles and they still exist today. They can be found in hydrothermal vents and salt lakes – similar environments to those believed to have made up the early Earth. They are usually of the domain Archaea and more recently they have been found in more hospitable environments such as soil and the human digestive system.

Prokaryotic organisms are always unicellular with a relatively simple structure. Their DNA is not contained within a nucleus, they have few organelles and the organelles they do have are not membrane-bound.

DNA

The structure of the DNA contained within prokaryotes is fundamentally the same as in eukaryotes but it is packaged differently. Prokaryotes generally only have one molecule of DNA, a chromosome, which is supercoiled to make it more compact. The genes on the chromosome are often grouped into operons, meaning a number of genes are switched on or off at the same time.

Ribosomes

The ribosomes in prokaryotic cells are smaller than those in eukaryotic cells. Their relative size is determined by the rate at which they settle, or form a sediment, in solution. The larger eukaryotic ribosomes are designated 80S and the smaller prokaryotic ribosomes, 70S. They are both necessary for protein synthesis, although the larger 80S ribosomes are involved in the formation of more complex proteins.

Cell wall

Prokaryotic cells have a cell wall made from peptidoglycan, also known as murein. It is a complex polymer formed from amino acids and sugars.

Learning outcomes

Demonstrate knowledge, understanding, and application of:

→ the structure and ultrastructure of prokaryotic cells and eukaryotic cells.

Synoptic link

You will learn more about the classification of prokaryotes into the domains Archaea and Bacteria in Topic 10.2, The five kingdoms.

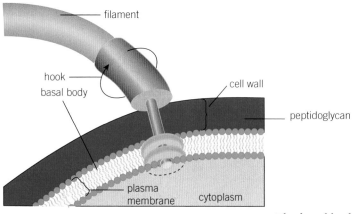

▲ Figure 1 *A prokaryotic flagellum*

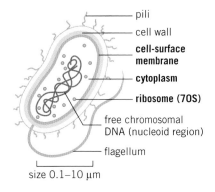

size 0.1–10 µm

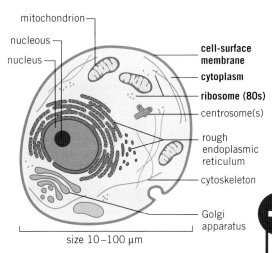

size 10–100 µm

▲ Figure 2 *Top: Features of prokaryotic cell (bacterium). Bottom: features of a typical eukaryotic cell (animal). Common features are highlighted in* **bold**

Flagella

The flagella of prokaryotes is thinner than the equivalent structure of eukaryotes and does not have the 9 + 2 arrangement. The energy to rotate the filament that forms the flagellum is supplied from the process of chemiosmosis, not from ATP as in eukaryotic cells.

The flagellum is attached to the cell membrane of a bacterium by a basal body and rotated by a molecular motor.

The basal body attaches the filament comprising the flagellum to the cell-surface membrane of a bacterium. A molecular motor causes the hook to rotate giving the filament a whip like movement, which propels the cell.

A comparison with eukaryotic cells

The first eukaryotic cells appeared about 1.5 billion years ago. As you have learned eukaryotic cells are much more complex than prokaryotic cells. Their DNA is present within a nucleus and exists as multiple **chromosomes**, which are supercoiled, and each one wraps around a number of proteins called **histones**, forming a complex for efficient packaging. This complex is called **chromatin** and chromatin coils and condenses to form chromosomes. Eukaryotic genes are generally switched on and off individually.

As you learnt earlier, eukaryotic cells have membrane-bound organelles including mitochondria and chloroplasts (Topic 2.4, Eukaryotic cell structure and Topic 2.4, The ultrastructure of plant cells).

Organisms from the plant, animal, fungi, and protoctista kingdoms are all composed of eukaryotic cells. Many are multicellular.

➕ Endosymbiosis

The theory of endosymbiosis is that mitochondria and chloroplasts, and possibly other eukaryotic organelles, were formerly free-living bacteria, that is, prokaryotes. The theory is that these prokaryotes were taken inside another cell as an endosymbiont – an organism that lives within the body or cells of another organism. This eventually led to the evolution of eukaryotic cells.

1 Discuss, using information from this topic, any evidence that supports the endosymbiotic theory.

The similarities and differences between prokaryotic and eukaryotic cells

The similarities and differences between prokaryotic and eukaryotic cells are summarised in Table 1.

▼ Table 1 *Prokaryotic and eukaryotic cells compared*

Feature	Prokaryotic	Eukaryotic
nucleus	not present	present
DNA	circular	linear
DNA organisation	proteins fold and condense DNA	associated with proteins called histones
extra chromosomal DNA	circular DNA called plasmids	only present in certain organelles such as chloroplasts and mitochondria
organelles	non membrane-bound	both membrane-bound and non membrane-bound
cell wall	peptidoglycan	chitin in fungi, cellulose in plants, not present in animals
ribosomes	smaller, 70S	larger, 80S
cytoskeleton	present	present, more complex
reproduction	binary fission	asexual or sexual
cell type	unicellular	unicellular and multicellular
cell-surface membrane	present	present

➕ Prokaryotic cell study

Here is a transmission electron micrograph image of a slice through a rod-shaped Gram-negative *Escherichia coli* bacterium. The cell wall can be seen as a double line around the cell. The darker area inside is the nucleoid which contains the DNA. Look closely at the photo and answer the questions below

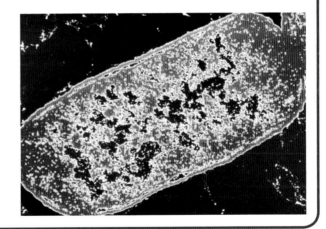

1 a Use the micrograph shown here to produce a scientific drawing of the bacterium.
   b Describe the differences you would see if you were observing a eukaryotic cell with the same microscope, at the same magnification.

Summary questions

1 List three structural differences between prokaryotic cells and eukaryotic cells.   (*3 marks*)

2 Suggest why the lack of membrane-bound organelles does not stop prokaryotic cells making proteins.   (*4 marks*)

3 Some antibiotics kill bacteria by disrupting the formation of peptidoglycan molecules. Explain why these antibiotics kill bacteria and why they do not have any effect on eukaryotic cells.   (*4 marks*)

Practice questions

1   The cytoskeleton is present throughout the cytoplasm of all eukaryotic cells.

Which of the following statements is/are correct with respect to the structure of the cytoskeleton?

Statement 1: Intermediate fibres - these fibres give mechanical strength to cells and help maintain their integrity.

Statement 2: Microtubules - contractile fibres formed from the protein actin. Responsible for cell movement

Statement 3: Microfilaments - formed from the cylindrical-shaped protein tubulin. They form a scaffold-like structure determining the shape of a cell.

A   1, 2 and 3 are correct

B   Only 1 and 2 are correct

C   Only 2 and 3 are correct

D   Only 1 is correct   (*1 mark*)

2   Serous cells are present in the salivary glands of animals. They are responsible for the production of the enzyme amylase which begins the breakdown of starch.

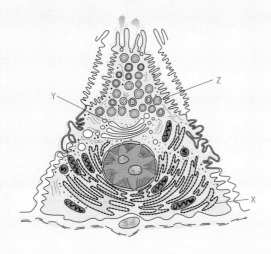

a   (i)   identify the structures labelled in the diagram.

     x   (*1 mark*)

     y   (*1 mark*)

     z   (*1 mark*)

                                   (*3 marks*)

     (ii)   State whether the cell is eukaryotic or prokaryotic giving the reason for your decision.   (*2 marks*)

b   (i)   State which group of enzymes contains amylase.   (*1 mark*)

     (ii)   Outline the stages and organelles involved in the production and release of amylase.   (*5 marks*)

c   Explain the process of exocytosis.

                                   (*3 marks*)

d   Discuss the different roles of vesicles, vacuoles and lysosomes.   (*4 marks*)

3   The photo below shows a transmission electron micrograph of plankton. These single-celled marine micro-organisms are thought to be the most abundant photosynthetic organisms on Earth.

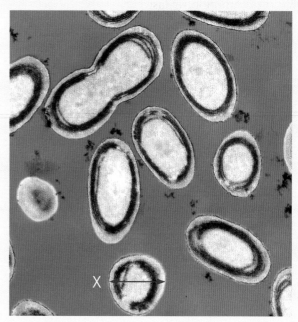

a   Calculate the magnification of the Plankton labelled X. The actual diameter of the plankton is 2.6 μm.   (*2 marks*)

The amount of detail that can be seen with a microscope depends on both the magnification and resolution possible with the microscope being used. Any increase in magnification beyond the limit of resolution results in 'empty magnification'.

   b Define the following terms

     (i) *resolution*   *(2 marks)*

     (ii) *magnification*   *(1 mark)*

     (iii) Suggest what is meant by the term 'empty magnification'   *(1 mark)*

   c Outline how a compound light microscope magnifies an image of a specimen.

                                    (4 marks)

   d Describe three different ways of preparing microscope slides for light microscopy.

                                    (6 marks)

4 a Explain the meaning of the term artefact with reference to microscopy.   *(2 marks)*

   b Discuss the advantages and disadvantages of using an electron microscope to study the ultrastructure of cells.   *(4 marks)*

   c Outline how laser scanning confocal microscopes produce an image.   *(4 marks)*

5 a Complete and complete the table below.

    The first row has been done for you.

                          (5 marks)

Feature	Prokaryotic	Eukaryotic
DNA	*circular*	*linear*
Extra chromosomal DNA		only present in certain organelles such as chloroplasts and mitochondria
Organelles	non membrane bound	
Cell wall	peptidoglycan	
Ribosomes		large, 80 s
Cell surface membrane		present

   b Define the term 'cell ultrastructure'.

                                  (2 marks)

6 Human genomes contain many more genes than bacterial genomes, and they are much longer.

   Discuss the way in which this affects the packing of DNA in eukaryotes and prokaryotes.   *(6 marks)*

3 BIOLOGICAL MOLECULES
3.1 Biological elements
Specification reference: 2.1.2

Learning outcomes

Demonstrate knowledge, understanding, and application of:

→ the chemical elements that make up biological molecules

→ the key inorganic ions that are involved in biological processes

→ monomers and polymers as biological molecules.

The building blocks of life

In section 2 you looked at cells and cellular components, however, these too are composed from smaller components called molecules. Molecules are built from even smaller components called atoms. In fact, atoms are built from yet smaller components including protons, neutrons, and electrons, which you will have learned about in your GCSE science or chemistry studies.

Knowledge of biochemistry is essential, as it underpins a proper understanding of the metabolic processes, structures, and limitations of biology, for example the complex series of reactions and molecules involved in cellular respiration. You do not need to be a chemist to understand basic biochemistry, as you will see in coming topics, but you do need to understand some essential chemical concepts and rules, which will be explored in this chapter.

Elements

Different types of atoms are called **elements**. Elements are distinguished by the number of protons in their atomic nuclei. There are over a hundred known elements in the universe but only a small percentage of these are present in the living world.

If you have ever built a model using interlocking bricks you will know how useful the bricks that make the most connections are, at the start of a new build.

In the same way that complex models can be built from a small range of simple bricks, all living things are made primarily from four key elements – carbon (C), hydrogen (H), oxygen (O) and nitrogen (N). In addition, phosphorus (P) and sulfur (S) also have important roles in the biochemistry of cells. These six elements are by far the most abundant elements present in biological molecules.

Other elements, including sodium (Na), potassium (K), calcium (Ca), and iron (Fe), also have important roles in biochemistry. You will learn about some of the roles of these elements later in this chapter.

Bonding

Atoms connect with each other by forming bonds. Atoms can bond to other atoms of the same element, or atoms of different elements, provided this follows the 'bonding rules' (described on the next page). When two or more atoms bond together the complex is called a molecule.

A covalent bond occurs when two atoms share a pair of electrons. The electrons used to form bonds are unpaired and present in the outer orbitals of the atoms.

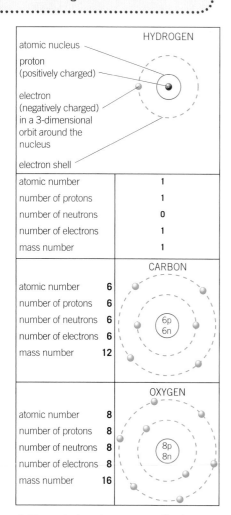

HYDROGEN	
atomic number	1
number of protons	1
number of neutrons	0
number of electrons	1
mass number	1

atomic nucleus
proton (positively charged)
electron (negatively charged) in a 3-dimensional orbit around the nucleus
electron shell

CARBON	
atomic number	6
number of protons	6
number of neutrons	6
number of electrons	6
mass number	12

6p
6n

OXYGEN	
atomic number	8
number of protons	8
number of neutrons	8
number of electrons	8
mass number	16

8p
8n

▲ Figure 1 *Atomic structure of hydrogen, carbon and oxygen which are common biological elements*

Bonding follows some simple rules, determined by the number of unpaired electrons present in the outer orbitals of different elements:

- Carbon atoms can form four bonds with other atoms.
- Nitrogen atoms can form three bonds with other atoms.
- Oxygen atoms can form two bonds with other atoms.
- Hydrogen atoms can only form one bond with another atom.

▼ Table 1 *Hydrogen atoms can only form one bond with other atoms. Carbon atoms can form four bonds, nitrogen three and oxygen two. The dots and crosses represent electrons and which atom they belong to*

Molecule	Electron diagram	Displayed formula	'Ball and stick' model
Hydrogen (H_2)		H —— H	
Water (H_2O)			
Carbon dioxide (CO_2)		O = C = O	
Methane (CH_4)			
Ammonia (NH_3)			

The number of bonds formed by these elements can be no more or less than stated. There are, however, exceptions to this rule, which you will learn about in later sections. Life on this planet is often referred to as being 'carbon-based' because carbon, which can form four bonds, forms the backbone of most biological molecules.

Ions

An atom or molecule in which the total number of electrons is not equal to the total number of protons is called an ion. If an atom or molecule loses one or more electrons it has a net positive charge and is known as a cation. If an atom or molecule gains electrons, it has a net negative charge and is known as an anion.

In ionic bonds, one atom in the pair donates an electron and the other receives it. This forms positive and negative ions that are held together by the attraction of the opposite charges.

Ions in solution are called electrolytes. The following tables list *some* of the important roles of ions in living organisms.

▼ Table 2 *Roles of cations*

Cations	Necessary for
calcium ions (Ca^{2+})	nerve impulse transmission muscle contraction
sodium ions (Na^+)	nerve impulse transmission kidney function
potassium ions (K^+)	nerve impulse transmission stomatal opening
hydrogen ions (H^+)	catalysis of reactions pH determination
ammonium ions (NH_4^+)	production of nitrate ions by bacteria

▼ Table 3 *Roles of anions*

Anions	Necessary for
nitrate ions (NO_3^-)	nitrogen supply to plants for amino acid and protein formation
hydrogen carbonate ions (HCO_3^-)	maintenance of blood pH
chloride ions (Cl^-)	balance positive charge of sodium and potassium ions in cells
phosphate ions (PO_4^{3-})	cell membrane formation nucleic acid and ATP formation bone formation
hydroxide ions (OH^-)	catalysis of reactions pH determination

Biological molecules

Below is a summary of the elements present in some of the key biological molecules. You will learn more about each of these classes of molecule in the coming topics of this chapter.

- **Carbohydrates** – carbon, hydrogen, and oxygen, usually in the ratio $C_x(H_2O)_x$.
- **Lipids** – carbon, hydrogen, and oxygen.
- **Proteins** – carbon, hydrogen, oxygen, nitrogen, and sulfur.
- **Nucleic acids** – carbon, hydrogen, oxygen, nitrogen, and phosphorus.

Polymers

Biological molecules are often **polymers**. Polymers are long-chain molecules made up by the linking of multiple individual molecules (called **monomers**) in a repeating pattern. In carbohydrates the monomers are sugars (saccharides) and in proteins the monomers are amino acids.

Summary questions

1 Explain how atoms join together to form molecules.   (*2 marks*)

2 Explain the difference between a cation and an anion.   (*4 marks*)

3 Explain how the bonds between the atoms in both water *and* carbon dioxide molecules fulfil the 'bonding rules'.   (*4 marks*)

4 The image below, obtained in 1953, helped confirm the recently proposed structure of DNA. The equipment that was used to obtain the image is also shown, X-ray diffraction photograph of DNA (deoxyribonucleic acid). This image was obtained in 1953 and results from a beam of X-rays being scattered onto a photographic plate by the DNA. Various features about the structure of the DNA can be determined from the pattern of spots and bands. The cross of bands indicates the helical nature of DNA.

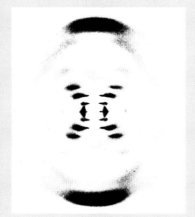

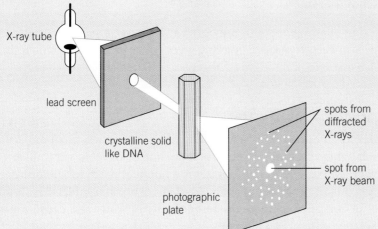

X-ray tube

lead screen

crystalline solid like DNA

photographic plate

spots from diffracted X-rays

spot from X-ray beam

a Suggest why the x-ray diffraction technique used to produce this image was not considered a form of microscopy but the use of electrons to produce images is called electron microscopy.   (*3 marks*)

b Explain why cells are visible with light microscopes but electron microscopes are needed to see ribosomes.   (*3 marks*)

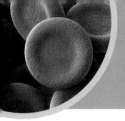

3.2 Water

Learning outcomes

Demonstrate knowledge, understanding, and application of:

→ how hydrogen bonding occurs between water molecules

→ how the properties of water relate to its roles in living organisms.

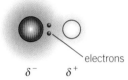

▲ **Figure 1** *Covalent bond between oxygen and hydrogen. The unequal sharing of the electrons leads to oxygen being more negative compared with hydrogen*

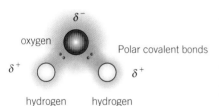

▲ **Figure 2** *The polar covalent bonds make water a polar molecule*

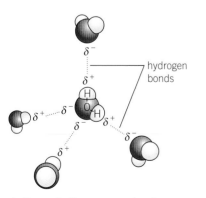

▲ **Figure 3** *Five water molecules interacting via hydrogen bonds*

The bonds of life

Atoms join together to form molecules by making bonds with each other. As you learnt in the previous topic, in ionic bonds, atoms give or receive electrons. They form negative or positive ions that are held together by the attraction of the opposite charges. Covalent bonds occur when atoms *share* electrons. However, the negative electrons are not always shared equally by the atoms of different elements. In many covalent bonds, the electrons will spend more time closer to one of the atoms than to the other. The atom with the greater share of negative electrons will be slightly negative (δ^-) compared with the other atom in the bond, which will therefore be slightly positive (δ^+) (Figure 1).

Molecules in which this happens are said to be **polar** – they have regions of negativity and regions of positivity.

Oxygen and hydrogen are examples of elements that do not share electrons equally in a covalent bond. Oxygen always has a much greater share of the electrons in an O—H bond. Many organic molecules contain oxygen and hydrogen bonded together in what are called hydroxyl (OH) groups and so they are slightly polar. Water (H_2O) is an example of such a molecule, in fact, water contains two of these hydroxyl groups (Figure 2).

Polar molecules, including water, interact with each other as the positive and negative regions of the molecule attract each other and form bonds, called hydrogen bonds. Hydrogen bonds are relatively weak interactions, which break and reform between the constantly moving water molecules.

Although hydrogen bonds are only weak interactions, they occur in high numbers. Hydrogen bonding gives water its unique characteristics, which are essential for life on this planet. These characteristics are explored further below.

Characteristics of water

Water has an unusually high boiling point. Water is a small molecule, much lighter than the gases carbon dioxide or oxygen, yet unlike oxygen and carbon dioxide, water is a liquid at room temperature. This is due to the hydrogen bonding between water molecules. It takes a lot of energy to increase the temperature of water and cause water to become gaseous (evaporate).

When water freezes it turns to ice. Most substances are more dense in their solid state than in their liquid state, but when water turns to ice it becomes less dense. This is because of the hydrogen bonds formed. As water is cooled below 4 °C the hydrogen bonds fix the positions of the polar molecules slightly further apart than the average distance in the liquid state. This produces a giant, rigid but open structure, with every oxygen atom at the centre of a tetrahedral arrangement of hydrogen atoms, resulting in a solid that is less dense than liquid water. For this reason, ice floats.

Water therefore has *cohesive* properties. It moves as one mass because the molecules are attracted to each other (cohesion). It is in this way that plants are able to draw water up their roots and how you are able to drink water through a straw. Water also has *adhesive* properties – this is where water molecules are attracted to other materials. For example, when you wash your hands your hands become wet, the water doesn't run straight off.

Water molecules are more strongly cohesive to each other than they are to air, this results in water having a 'skin' of surface tension. (Figure 4)

Water for life

The characteristics and properties of water are critical in sustaining life. In this way, water is unique. Some of the ways in which water is vital for life are summarised below.

Because it is a polar molecule, water acts as a *solvent* in which many of the solutes in an organism can be dissolved. The cytosol of prokaryotes (bacterial) and eukaryotes is mainly water. Many solutes are also polar molecules, amino acids, proteins (Topic 3.6, Structure of proteins) and nucleic acids (Topic 3.8, Nucleic acids). Water acts as a medium for chemical reactions and also helps transport dissolved compounds into and out of cells.

Water makes a very efficient *transport medium* within living things. Cohesion between water molecules means that when water is transported though the body, molecules will stick together. Adhesion occurs between water molecules and other polar molecules and surfaces. The effects of adhesion and cohesion result in water exhibiting **capillary action**. This is the process by which water can rise up a narrow tube against the force of gravity.

Water acts as a coolant, helping to buffer temperature changes during chemical reactions in prokaryotic and eukaroytic cells because of the large amounts of energy required to overcome hydrogen bonding. Maintaining constant temperatures in cellular environments is important as enzymes are often only active in a narrow temperature range.

Many organisms, such as fish, live in water and cannot survive out of it. Water is stable – it does not change temperature or become a gas easily, therefore providing a constant environment. Because ice floats, it forms on the surface of ponds and lakes, rather than from the bottom up. This forms an insulating layer above the water below. Aquatic organisms would not be able to survive freezing temperatures if their entire habitat froze solid. Some organisms also inhabit the surface of water. Surface tension is strong enough to support small insects such as pond skaters.

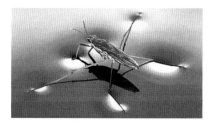

▲ **Figure 4** *This pond skater (Gerris lacustris) inhabits the surface of water, supported by surface tension*

Study tip

Do not confuse polarity with 'charged' or 'ionic'. In polar bonds electrons are shared, albeit unequally, but if atoms actually lose an electron to another atom they both become charged and are called ions. In ionic bonding, the atom that gains the electron becomes a negative ion and the atom that loses the electron becomes a positive ion.

Synoptic link

You will learn more about how temperature affects enzyme activity in Topic 4.2, Factors affecting enzyme activity.

Summary questions

1 Explain how hydrogen bonds form. (*3 marks*)

2 Explain why water is a polar molecule. (*2 marks*)

3 Suggest, with reasons, which properties of water make it such an important component of blood. (*5 marks*)

4 Water forms the basis of the stroma in chloroplasts and the matrix in mitochondria.

 Describe which properties of water make it such an important component of these particular organelles. (*5 marks*)

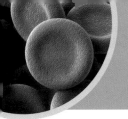

3.3 Carbohydrates

Specification reference: 2.1.2

Learning outcomes

Demonstrate knowledge, understanding, and application of:

→ the ring structure and properties of glucose

→ the structure of ribose

→ the synthesis and breakdown of a disaccharide and polysaccharide by the formation and breakage of glycosidic bonds

→ the structures and properties of glucose, starch, glycogen and cellulose molecules.

Carbohydrates are molecules that only contain the elements carbon, hydrogen, and oxygen. Carbohydrate literally means 'hydrated carbon' (carbon and water). The elements in carbohydrates usually appear in the ratio $C_x(H_2O)_y$. This is known as the general formula of carbohydrates.

Carbohydrates are also known as saccharides or sugars. A single sugar unit is known as a **monosaccharide**, examples include glucose, fructose, and ribose. When two monosaccharides link together they form a disaccharide, for example lactose and sucrose. When two or more (usually many more) monosaccharides are linked they form a polymer called a **polysaccharide**. Glycogen, cellulose, and starch are examples of polysaccharides.

Glucose

The basic building blocks, or monomers, of some biologically important large carbohydrates are **glucose** molecules, which have the chemical formula $C_6H_{12}O_6$. Glucose is a monosaccharide composed of six carbons and therefore is a **hexose monosaccharide** (hexose sugar) (Figure 1).

In molecular structure diagrams, the carbons are numbered clockwise, beginning with the carbon to the right (clockwise) of the oxygen atom within the ring.

There are two structural variations of the glucose molecule, alpha (α) and beta (β) glucose, in which the OH (hydroxyl) group on carbon 1 is in opposite positions, as shown in Figure 1.

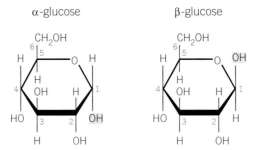

▲ Figure 1 *The sugars alpha and beta glucose, which are examples of monosaccharides (single sugar units). Note the different position of the OH group on carbon 1*

Glucose molecules are polar and soluble in water. This is due to the hydrogen bonds that form between the hydroxyl groups and water molecules. This solubility in water is important, because it means glucose is dissolved in the cytosol of the cell.

Condensation reactions

When two alpha glucose molecules are side by side, two hydroxyl groups interact (react). When this happens bonds are broken and new bonds reformed in different places producing new molecules.

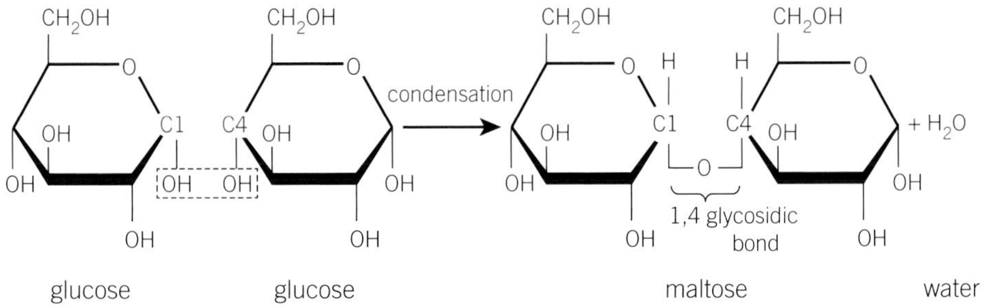

▲ Figure 2 *As the two OH are so close they react, forming a covalent bond called a glycosidic bond between the two glucose molecules*

As you can see in Figure 2, two hydrogen atoms and an oxygen atom are removed from the glucose monomers and join to form a water molecule. A bond forms between carbons 1 and 4 on the glucose molecules and the molecules are now joined.

A covalent bond called a **glycosidic bond** is formed between two glucose molecules. The reaction is called a **condensation reaction** because a water molecule is formed as one of the products of the reaction. Because in this reaction carbon 1 of one glucose molecule is joined to carbon 4 of the other glucose molecule, the bond is known as a 1,4 glycosidic bond. In this reaction the new molecule is called **maltose**. This is an example of a **disaccharide** (a molecule made up of two monosaccharides).

Other sugars
Fructose and galactose are also hexose monosaccharides. Fructose naturally occurs in fruit, often in combination with glucose forming the disaccharide **sucrose**, commonly known as cane sugar or just sugar.

Galactose and glucose form the disaccharide **lactose**. Lactose is commonly found in milk and milk products.

Fructose is sweeter than glucose and glucose is sweeter than galactose.

Pentose monosaccharides are sugars that contain five carbon atoms. Two pentose sugars are important components of biological molecules – **ribose** is the sugar present in RNA nucleotides and deoxyribose is the sugar present in DNA nucleotides.

Synoptic link
You will find out more about the structure of nucleotides in Topic 3.8, Nucleic acids.

Starch and glycogen
Many alpha glucose molecules can be joined by glycosidic bonds to form two slightly different polysaccharides known collectively as **starch**. Glucose made by photosynthesis in plant cells is stored as starch. It is a chemical energy store.

One of the polysaccharides in starch is called amylose. Amylose is formed by alpha glucose molecules joined together only by 1–4 glycosidic bonds. The angle of the bond means that this long chain of glucose twists to form a helix which is further stabilised by hydrogen bonding within the molecule. This makes the polysaccharide more compact, and much less soluble, than the glucose molecules used to make it.

▲ Figure 3 *The characteristic helix shape of amylose*

Another type of starch is formed when glycosidic bonds form in condensation reactions between carbon 1 and carbon 6 on two glucose molecules.

The other starch polysaccharide is called amylopectin. Amylopectin is also made by 1-4 glycosidic bonds between alpha glucose molecules, but (unlike amylose) in amylopectin there are also some glycosidic bonds formed by condensation reactions between carbon 1 and carbon 6 on two glucose molecules. this means that amylopectin has a branched structure, with the 1-6 branching points occurring approximately once in every 25 glucose subunits.

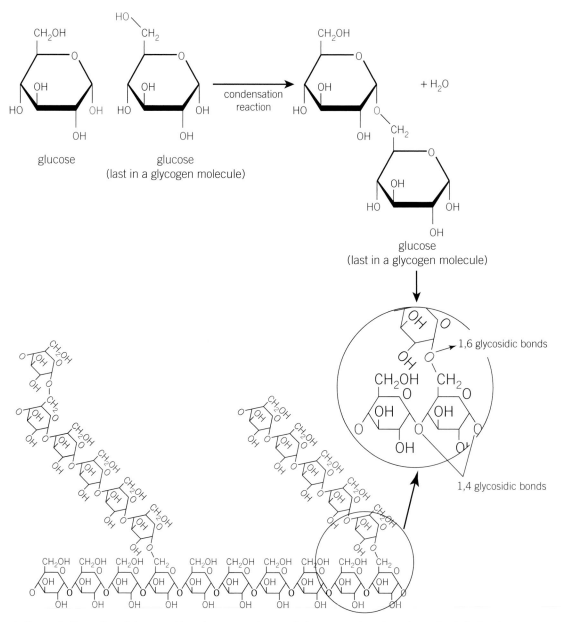

▲ Figure 4 Formation of glycogen. Note that a new glucose chain starts to form from the main chain forming a branch

The functionally equivalent energy storage molecule to starch in animals and fungi is called **glycogen** (Figure 4). Glycogen forms more branches than amylopectin, which means it is more compact and less space is needed for it to be stored. This is important as animals are mobile, unlike plants. The coiling or branching of these polysaccharides makes them very compact, which is ideal for storage. The branching also means there are many free ends where glucose molecules can be added or removed. This speeds up the processes of storing or releasing glucose molecules required by the cell.

So, the key properties of amylopectin and glycogen are that they are insoluble, branched, and compact. These properties mean they are ideally suited to the storage roles that they carry out.

Hydrolysis reactions

Glucose is stored as starch by plants or glycogen by animals and fungi, until it is needed for respiration – the process in which biochemical energy in these stored nutrients is converted into a useable energy source for the cell.

To release glucose for respiration, starch or glycogen undergo **hydrolysis reactions**, requiring the addition of water molecules. The reactions are catalysed by enzymes. These are the reverse of the condensation reactions that form the glycosidic bonds.

Cellulose

Beta glucose molecules are unable to join together in the same way that alpha glucose molecules can. As you can see in Figure 5, the hydroxyl groups on carbon 1 and carbon 4 of the two glucose molecules are too far from each other to react.

The only way that beta glucoses molecules can join together and form a polymer is if alternate beta glucose molecules are turned upside down as in Figure 6.

> **Synoptic link**
>
> You will learn about cellular respiration in more detail in Chapter 18, Respiration.

> **Synoptic link**
>
> You will learn more about ATP in Topic 3.11, ATP.

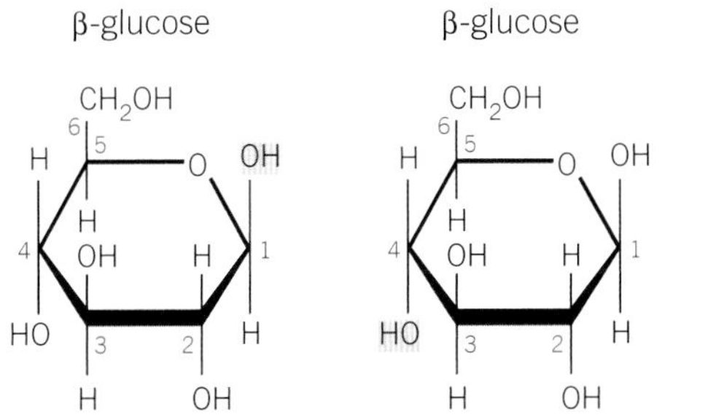

▲ Figure 5 *Note how far apart the OH groups are on these two β-glucose molecules*

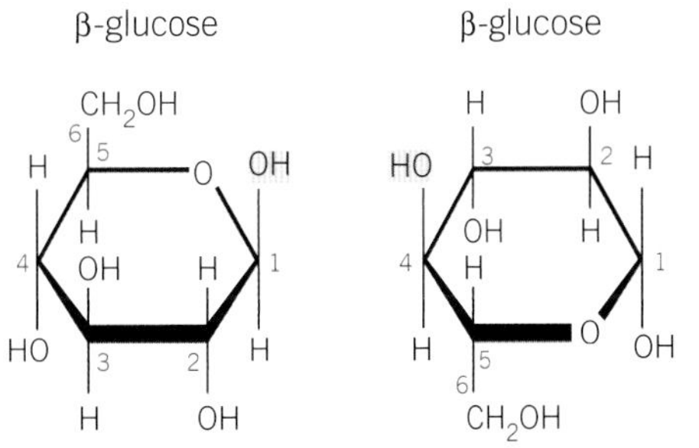

▲ Figure 6 *The OH groups of the two β-glucoses are now close enough to react and a 1,4 glycosidic bond is formed*

When a polysaccharide is formed from glucose in this way it is unable to coil or form branches. A straight chain molecule is formed called **cellulose** (Figure 7).

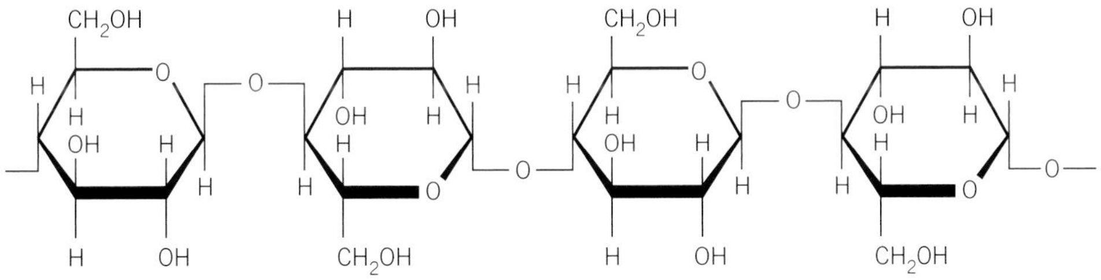

▲ Figure 7 *The cellulose molecule is straight and unbranched*

Cellulose molecules make hydrogen bonds with each other forming microfibrils. These microfibrils join together forming macrofibrils, which combine to produce fibres (Figure 8). These fibres are strong and insoluble and are used to make cell walls. Cellulose is an important part of our diet, it is very hard to break down into its monomers and forms the 'fibre' or 'roughage' necessary for a healthy digestive system.

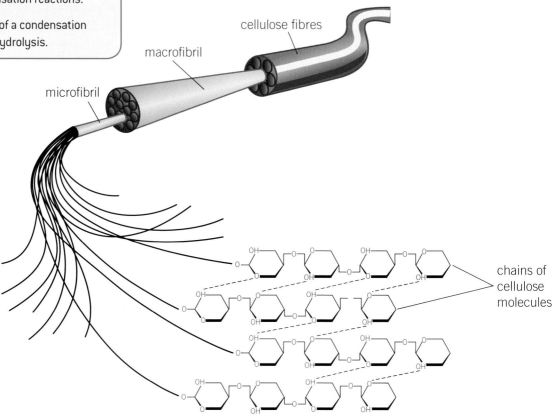

▲ Figure 8 *Formation of cellulose fibres*

Summary questions

1 Describe the difference between alpha and beta glucose.   (*2 marks*)

2 Describe the formation of a glycosidic bond.   (*4 marks*)

3 Explain how the structure of cellulose is related to its function.   (*4 marks*)

4 Explain why beta glucose, when polymerised, leads to the production of cellulose instead of starch.   (*6 marks*)

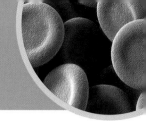

3.4 Testing for carbohydrates

Specification reference: 2.1.2

Chemical tests

Benedict's test for reducing sugars

In chemistry reduction is a reaction involving the gain of electrons. All monosaccharides and some disaccharides (for example maltose and lactose) are **reducing sugars**. This means that they can donate electrons, or reduce another molecule or chemical.

In the chemical test for a reducing sugar, this chemical is **Benedict's reagent**, an alkaline solution of copper(II)sulfate.

The test is carried out as follows:

1   Place the sample to be tested in a boiling tube. If it is not in liquid form, grind it up or blend it in water.

2   Add an equal volume of Benedict's reagent.

3   Heat the mixture gently in a boiling water bath for five minutes.

Reducing sugars will react with the copper ions in Benedict's reagent. This results in the addition of electrons to the blue Cu^{2+} ions, reducing them to brick red Cu^+ ions. When a reducing sugar is mixed with Benedict's reagent and warmed, a brick-red precipitate is formed indicating a positive result.

The more reducing sugar present, the more precipitate formed and the less blue Cu^{2+} ions are left in solution, so the actual colour seen will be a mixture of brick-red (precipitate) and blue (unchanged copper ions) and will depend on the concentration of the reducing sugar present (Figure 1). This makes the test qualitative.

Using Benedict's test for non-reducing sugars

Non-reducing sugars do not react with Benedict's solution and the solution will remain blue after warming, indicating a negative result. Sucrose is the most common non-reducing sugar.

If sucrose is first boiled with dilute hydrochloric acid it will then give a positive result when warmed with Benedict's solution. This is because the sucrose has been hydrolysed by the acid to glucose and fructose, both reducing sugars.

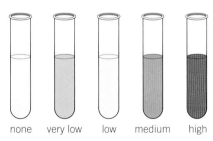

none   very low   low   medium   high

▲ Figure 1 *Colour changes in a Benedict's test according to the concentration of reducing sugar present*

Iodine test for starch

The **iodine test** is used to detect the presence of starch. To carry out the test, a few drops of iodine dissolved in potassium iodide solution are mixed with a sample. If the solution changes colour from yellow/brown to purple/black starch is present in the sample.

▲ Figure 2 *A positive test result for starch using iodine*

If the iodine solution remains yellow/brown it is a negative result and starch is not present.

Reagent strips

Manufactured reagent test strips can be used to test for the presence of reducing sugars, most commonly glucose. The advantage is that, with the use of a colour-coded chart, the concentration of the sugar can be determined.

Quantitative methods to determine concentration:

Colorimetry

In a Benedict's test, the colour produced is dependent on the concentration of reducing sugar present in the sample.

A colorimeter is a piece of equipment used to quantitatively measure the absorbance, or transmission, of light by a coloured solution. The more concentrated a solution is the more light it will absorb and the less light it will transmit. This can be used to calculate the concentration of reducing sugar present.

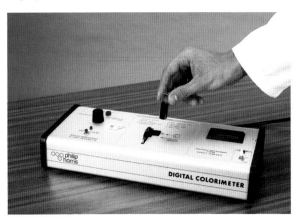

▲ Figure 3 *A colorimeter measures the colour of a liquid. It passes filtered light through the sample and the results can be transmitted to a graph plotter or computer*

A student was asked to determine the concentration of a solution of glucose.

The procedure was carried out as follows:

1 A filter was placed in the colorimeter.
2 The colorimeter was calibrated using distilled water.
3 Benedict's test was performed on a range of known concentrations of glucose.

4 The resulting solutions were filtered to remove the precipitate.
5 The % transmission of each of the solutions of glucose was measured using the colorimeter.
6 Using this information a calibration curve was plotted. Steps 3–6 were repeated using the solution with the unknown concentration of glucose.

▼ Table 1 *Shows the results of the experiment*

Transmission / %	Concentration of glucose / mM
68	1.0
56	2.0
47	3.0
40	4.0
27	5.0
17	6.0
7	7.0
44	Unknown solution

1 Describe how you would calculate % absorbance from a % transmission reading.
2 Explain why it is important to use the correct filter (step 1).
3 Describe how you calibrate a colorimeter (step 2).
4 Describe what you have after the solutions have been filtered (step 4).
5 Plot a graph of the results from Table 1 and draw a calibration curve.
6 Estimate the concentration of glucose in the unknown solution.

Biosensors

Biosensors use biological components to determine the presence and concentration of molecules such as glucose.

The basic components of a biosensor are shown in Figure 4.

The analyte is the compound under investigation.

- Molecular recognition – a protein (enzyme or antibody) or single strand of DNA (ssDNA) is immobilised to a surface, for example a glucose test strip. This will interact with, or bind to, the specific molecule under investigation.

- Transduction – this interaction will cause a change in a transducer. A transducer detects changes, for example in pH, and produces a response such as the release of an immobilised dye on a test strip or an electric current in a glucose-testing machine.

- Display – this then produces a visible, qualitative or quantitative signal such as a particular colour on a test strip or reading on a test machine.

1 Canaries used to be used in the coal mining industry to detect the presence of harmful gases such as carbon monoxide. Miners would take the canaries, in cages, into the mines where they were working, and if the birds started to show signs of distress this signalled the presence of harmful gas. Discuss whether a canary in a cage is a biosensor and suggest a disadvantage of the use of this method to detect toxic gases.

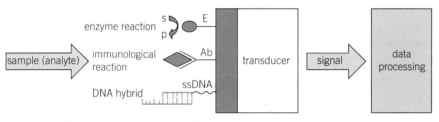

▲ Figure 4 The main components in a biosensor

Summary questions

1 Describe the feature of enzymes essential to their role as components in a biosensor. (3 marks)

2 Why does Benedict's reagent turn red when warmed with a reducing sugar? (2 marks)

3 Explain why an iodine test is used in experiments to show that plants require light for photosynthesis. (3 marks)

4 Suggest how reagent strips might be useful in the management of the medical condition diabetes, where a person's blood sugar level can become too high. (4 marks)

3.5 Lipids

Specification reference: 2.1.2

Learning outcomes

Demonstrate knowledge, understanding, and application of:

→ the structure of a triglyceride and a phospholipid

→ the synthesis and breakdown of triglycerides

→ properties of triglyceride, phospholipid, and cholesterol molecules

→ how to carry out and interpret the results of an emulsion test for lipids.

Lipids, commonly known as fats and oils, are molecules containing the elements carbon, hydrogen, and oxygen. Generally, fats are lipids that are solid at room temperature and oils are lipids that are liquid at room temperature.

Lipids are non-polar molecules as the electrons in the outer orbitals that form the bonds are more evenly distributed than in polar molecules. This means there are no positive or negative areas within the molecules and for this reason lipids are not soluble in water. Oil and water do not mix.

Lipids are large complex molecules known as **macromolecules**, which are built from repeating units, or monomers, like polysaccharides. In this topic we will be looking at the lipids, triglycerides, phospholipids, and sterols.

Triglycerides

A **triglyceride** is made by combining one **glycerol** molecule with three **fatty acids**. Glycerol is a member of a group of molecules called alcohols. Fatty acids belong to a group of molecules called carboxylic acids – they consist of a carboxyl group (–COOH) with a hydrocarbon chain attached.

As you can see in Figure 1, both of these molecules contain hydroxyl (OH) groups. The hydroxyl groups interact, leading to the formation of three water molecules and bonds between the fatty acids and the glycerol molecule. These are called ester bonds and this reaction is called esterification. Esterification is another example of a condensation reaction, which you learnt about in Topic 3.3, Carbohydrates.

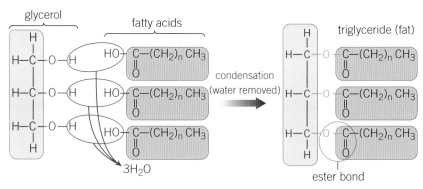

▲ Figure 1 *Synthesis of a triglyceride from glycerol and three fatty acids by the formation of three ester bonds, producing three water molecules*

When triglycerides are broken down, three water molecules need to be supplied to reverse the reaction that formed the triglyceride. This is another example of a hydrolysis reaction (Topic 3.3, Carbohydrates).

Saturated and unsaturated

Fatty acid chains that have *no* double bonds present between the carbon atoms are called saturated, because all the carbon atoms form the maximum number of bonds with hydrogen atoms (i.e., they are saturated with hydrogen atoms).

A fatty acid *with* double bonds between some of the carbon atoms is called unsaturated. If there is just one double bond it is called monounsaturated. If there are two or more double bonds it is called polyunsaturated. The presence of double bonds causes the molecule to kink or bend (Figure 2) and they therefore cannot pack so closely together. This makes them liquid at room temperature rather than solid, and they are therefore described as oils rather than fats.

Plants contain unsaturated triglycerides, which normally occur as oils, and tend to be more healthy in the human diet than saturated triglycerides, or (solid) fats. There has been some evidence that in excess, saturated fats can lead to coronary heart disease, however the evidence remains inconclusive. An excess of any type of fat can lead to obesity, which also puts a strain on the heart.

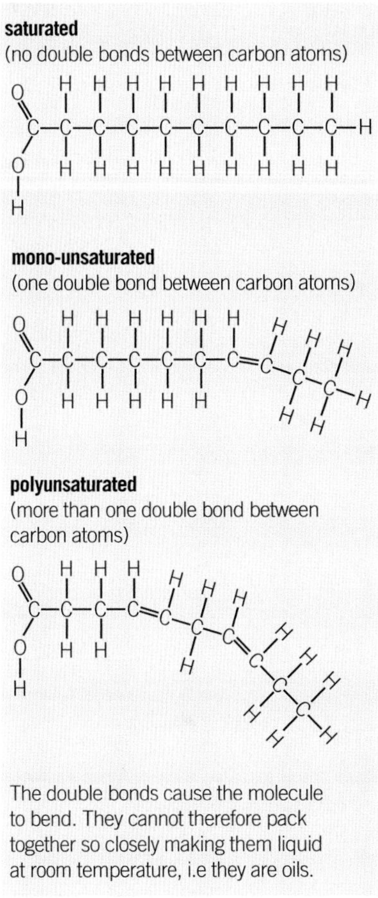

Phospholipids

Phospholipids are modified triglycerides and contain the element phosphorus along with carbon, hydrogen, and oxygen. Inorganic phosphate ions (PO_4^{3-}) are found in the cytoplasm of every cell. The phosphate ions have extra electrons and so are negatively charged, making them soluble in water.

One of the fatty acid chains in a triglyceride molecule is replaced with a phosphate group to make a phospholipid.

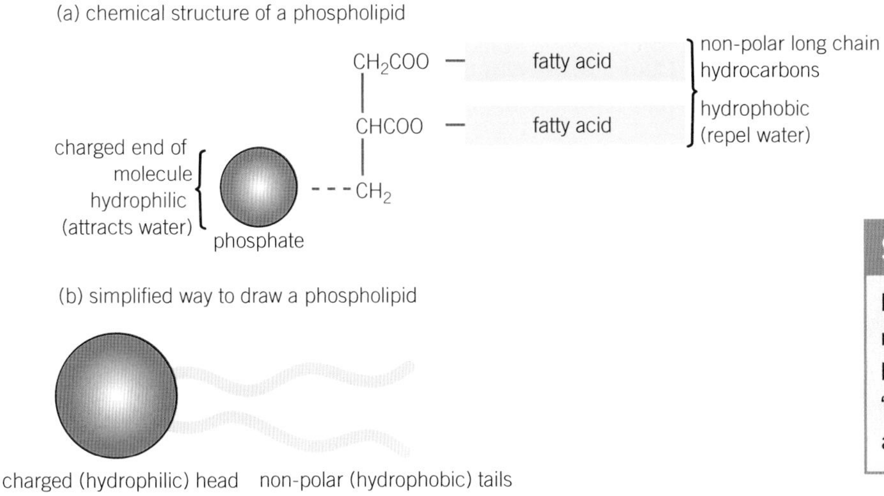

(a) chemical structure of a phospholipid

CH₂COO — fatty acid } non-polar long chain hydrocarbons

CHCOO — fatty acid } hydrophobic (repel water)

charged end of molecule hydrophilic (attracts water) { ---CH₂

phosphate

(b) simplified way to draw a phospholipid

charged (hydrophilic) head   non-polar (hydrophobic) tails

▲ Figure 3 *Structure of a phospholipid*

Phospholipids are unusual because, due to their length, they have a non-polar end or tail (the fatty acid chains) and a charged end or head (the phosphate group). The non-polar tails are repelled by water (but mix readily with fat). They are **hydrophobic**. The charged heads (often incorrectly called polar ends) will interact with, and are attracted to, water. They are **hydrophilic**.

saturated
(no double bonds between carbon atoms)

mono-unsaturated
(one double bond between carbon atoms)

polyunsaturated
(more than one double bond between carbon atoms)

The double bonds cause the molecule to bend. They cannot therefore pack together so closely making them liquid at room temperature, i.e they are oils.

▲ Figure 2 *Saturated and unsaturated fatty acids*

Synoptic link

You will learn about the role of surfactants in the lungs in Topic 6.2, The mammalian gaseous exchange system.

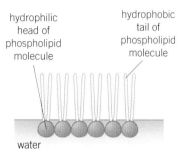

hydrophilic head of phospholipid molecule

hydrophobic tail of phospholipid molecule

water

▲ Figure 4 *A phospholipid monolayer in water*

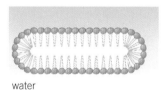

water

▲ Figure 5 *Phospholipid bilayer structure in water*

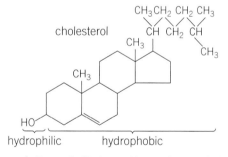

cholesterol

CH₃CH₂ CH₂ CH₃

HO

hydrophilic   hydrophobic

▲ Figure 6 *Cholesterol has a characteristic hydroxyl group and four carbon rings*

Synoptic link

You will learn about the roles of lipids in cell membranes in topic 5.1, The structure and function of membranes.

As a result of their dual hydrophobic/hydrophilic structure, phospholipids behave in an interesting way when they interact with water.

They will form a layer on the surface of the water with the phosphate heads in the water and the fatty acid tails sticking out of the water (Figure 4). Because of this they are called surface active agents or **surfactants** for short.

They can also form structures based on a two-layered sheet formation (a bilayer) with all of their hydrophobic tails pointing toward the centre of the sheet, protected from the water by the hydrophilic heads (Figure 5).

It is as a result of this bilayer arrangement that phospholipids play a key role in forming cell membranes. They are able to separate an aqueous environment in which cells usually exist from the aqueous cytosol within cells. It is thought that this is how the first cells were formed and, later on, membrane-bound organelles within cells.

Sterols

Sterols, also known as steroid alcohols, are another type of lipid found in cells. They are not fats or oils and have little in common with them structurally. They are complex alcohol molecules, based on a four carbon ring structure with a hydroxyl (OH) group at one end. Like phospholipids, however, they have dual hydrophilic/hydrophobic characteristics. The hydroxyl group is polar and therefore hydrophilic and the rest of the molecule is hydrophobic.

Cholesterol is a sterol. The body manufactures cholesterol primarily in the liver and intestines. It has an important role in the formation of cell membranes, becoming positioned between the phospholipids with the hydroxyl group at the periphery of the membrane. This adds stability to cell membranes and regulates their fluidity by keeping membranes fluid at low temperatures and stopping them becoming too fluid at high temperatures.

Vitamin D, steroid hormones, and bile are all manufactured using cholesterol.

Roles of lipids

Due to their non-polar nature, lipids have many biological roles. These include:

● membrane formation and the creation of hydrophobic barriers
● hormone production
● electrical insulation necessary for impulse transmission
● waterproofing, for example in birds' feathers and on plant leaves.

Lipids, triglycerides in particular, also have an important role in long-term energy storage. They are stored under the skin and around vital organs, where they also provide:

● thermal insulation to reduce heat loss for example, in penguins
● cushioning to protect vital organs such as the heart and kidneys
● buoyancy for aquatic animals like whales.

Identification of lipids

Lipids can be identified in the laboratory by a simple test known as the **emulsion test**. First, the sample is mixed with ethanol. The resulting solution is mixed with water and shaken. If a white emulsion forms as a layer on top of the solution this indicates the presence of a lipid. If the solution remains clear the test is negative.

Changing health advice

It can be confusing because health advice constantly changes. The way that new advice is issued in the media from new findings is partly responsible. The validity of the research has not usually been evaluated, the science is often not easy to explain, and as the majority of the general public do not have scientific background they are not aware of the fluid nature of scientific understanding. Scientific knowledge is also constantly changing as technology develops and so our understanding of biological processes increases.

It is often difficult to isolate the effect of just one nutrient and, in fact, it is now generally believed that nutrients do not work in isolation but as part of the combined effect of a whole range of nutrients. This is called food synergy. For example whole grains are believed to have a greater beneficial effect than any of their individual components and it is the combined effect of fish, fruit and vegetables that help prevent certain types of heart disease.

The data used in reports is often flawed, particularly where diet is concerned, as the subjects involved in studies often do not provide accurate information. People tend to underestimate what they eat, they forget what they have eaten and don't often know the exact ingredients of meals, particularly if they are eating out. People are also different due to their genetic make up and therefore respond differently to different nutrients.

The studies that catch the headlines often involve small numbers of subjects and these inherent differences distort the findings. The resulting headlines can be eye-catching, but not very accurate, and are often contradicted by the next study.

 Fats in our diet

The presence of a double bond in a fatty acid leads to a kink in the chain causing the lipid to be more liquid in nature and, as you will discover in later sections, a more healthy component of the diet than saturated fats.

Plants contain unsaturated triglycerides, which normally occur as oils.

Animals (but generally not fish) contain saturated triglycerides, or (solid) fats. As mentioned, the evidence that saturated fats cause heart diseases is inconclusive.

Previously it was thought that saturated fats did cause heart disease, but more recent evidence has contradicted this.

Margarine versus butter

Butter is an animal fat made from cows' milk and is therefore high in saturated lipids. Various alternatives to butter have been developed over the last 200 years with the focus initially being to find cheaper or longer-lasting substitutes. More recently the aim has been to produce a more 'healthy' substitute for butter.

The main problem faced initially by food scientists was that the vegetable oils used to produce the 'substitute butters' are more liquid than the animal fats in milk.

This was overcome by using hydrogen to saturate, or removing the double bonds from, the unsaturated fatty acids in the vegetable oils. Solid hydrogenated fat was produced and the oil was said to be have been hardened.

The fat was then coloured and sometimes mixed with butter to improve the taste. Different degrees of hardening, colouring and mixing with butter gave rise to the many different margarines on the market.

An unwanted byproduct of the hardening process was the production of trans fats. These are unsaturated lipids in which the kinks that the double bonds naturally form in the fatty acid chains have been reversed. Trans fats, which actually increase the shelf life of baked products, have more recently been linked with the development of coronary heart disease and are now usually removed from foods.

With more focus on producing healthy alternatives to butter, and improvements in the manufacturing process, many spreads now contain less, if any, hydrogenated fats. Mono- and polyunsaturated plant oils are used instead and these have been shown to reduce high cholesterol levels, which are a factor in the development of coronary heart disease.

Reduced fat spreads

Lipids release the same quantity of energy gram for gram when respired whether saturated or unsaturated, so butter and margarine have always had the same calorific value. More recently the focus has been to reduce the overall fat content in such spreads.

1   Explain how hardening vegetable oils produces solid fats.
2   Explain why it is considered more healthy to have a low overall fat content as well as a low saturated fat content in a spread.

Summary questions

1   Using your knowledge of the structure of fatty acids, describe why oils are liquid and fats are solid at room temperature. *(4 marks)*

2   Describe the formation and the hydrolysis of an ester bond. *(4 marks)*

3   Some bacteria are extremophiles meaning they live in extreme environments that are very acidic or have very high temperatures. The phospholipids present in other bacteria or eukaryotic cells would be broken down in such extreme conditions.

   Extremophiles have membranes composed of modified phospholipids.

   a   Identify which of the phospholipids in the diagram is present in the cell membrane of extremophiles. *(1 mark)*
   b   Outline the similarities and differences between the two types of phospholipid. *(3 marks)*
   c   Suggest why the phospholipids in the membranes of extremophiles can withstand extremes of temperature and pH. *(2 marks)*

4   Read the following statements.

   Lipids are not soluble in water.
   Lipids and ethanol are soluble in water.
   Water is more soluble than lipids in ethanol.
   Use the information to explain how the emulsion test for lipids works. *(4 marks)*

Peptides are polymers made up of **amino acid** molecules (the monomers). **Proteins** consist of one or more polypeptides arranged as complex macromolecules and they have specific biological functions. All proteins contain the elements carbon, hydrogen, oxygen, and nitrogen.

Amino acids

All amino acids have the same basic structure (Figure 1). Different **R-groups** (variable groups) result in different amino acids. Twenty different amino acids are commonly found in cells. Five of these are said to be non-essential as our bodies are able to make them from other amino acids. Nine are essential and can only be obtained from what we eat. A further six are said to be conditionally essential as they are only needed by infants and growing children.

Synthesis of peptides

Amino acids join when the amine and carboxylic acid groups connected to the central carbon atoms react. The R-groups are not involved at this point. The hydroxyl in the carboxylic acid group of one amino acid reacts with a hydrogen in the amine group of another amino acid. A **peptide bond** is formed between the amino acids and water is produced (this is another example of a condensation reaction, which you learnt about in Topic 3.3, Carbohydrates). The resulting compound is a dipeptide.

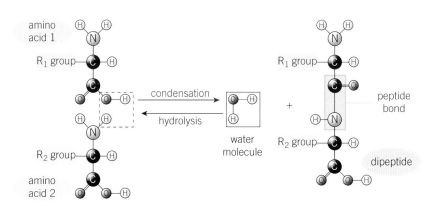

▲ Figure 2 *Condensation reaction to form a peptide bond*

When many amino acids are joined together by peptide bonds a **polypeptide** is formed. This reaction is catalysed by the enzyme peptidyl transferase present in ribosomes, the sites of protein synthesis.

The different R-groups of the amino acids making up a protein are able to interact with each other (R-group interactions) forming different types of bond. These bonds lead to the long chains of amino acids (polypeptides) folding into complex structures (proteins). The presence of different sequences of amino acids leads to different structures with

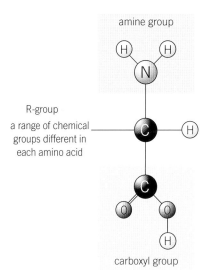

▲ Figure 1 *The general structure of an amino acid*

different shapes being produced. The very specific shapes of proteins are vital for the many functions proteins have within living organisms.

If you look at the way protein structures are built up in stages, it is easier to understand what is happening.

 ## Separating amino acids using thin layer chromatography

Thin layer chromatography (TLC) is a technique used to separate the individual components of a mixture. The technique can be used to separate and identify a mixture of amino acids in solution. There are two phases, the stationary phase and the mobile phase which involves an organic solvent. The mobile phase picks up the amino acids and moves through the stationary phase and the amino acids are separated.

In the stationary phase a thin layer of silica gel (or another adhesive substance) is applied to a rigid surface, for example a sheet of glass or metal. Amino acids are then added to one end of the gel. This end is then submerged in organic solvent. The organic solvent then moves through the silica gel, this is known as the mobile phase.

The rate at which the different amino acids in the organic solvent move through the silica gel depends on the interactions (hydrogen bonds) they have with the silica in the stationary phase, and their solubility in the mobile phase. This results in different amino acids moving different distances in the same time period resulting in them separating out from each other.

 Remember, when working with chemicals to take care, wear safety glasses and report any spillages/breakages to the teacher.

A student carried out the following procedure to separate and identify a mixture of amino acids in solution.

1 Wearing gloves, the student drew a pencil line on the chromatography plate about 2 cm from the bottom edge. The plate was only handled by the edges.
2 Four equally spaced points were marked at along the pencil line.
3 The amino acid solution was spotted onto the first pencil mark using a capillary tube. The spot was allowed to dry and then spotted again. The spot was labelled using a pencil.
4 The three remaining marks were spotted with solutions of three known amino acids.
5 The plate was then placed into a jar containing the solvent. The solvent was no more than 1 cm deep. The jar was then closed.

6 The plate was left in the solvent until it had reached about 2 cm from the top. The plate was then removed and a pencil line drawn along the solvent front. The plate was then allowed to dry.
7 The plate was then sprayed, in a fume cupboard, with ninhydrin spray. Amino acids react with ninhydrin and a purple/brown colour is produced. The centre of each spot present was then marked with a pencil.

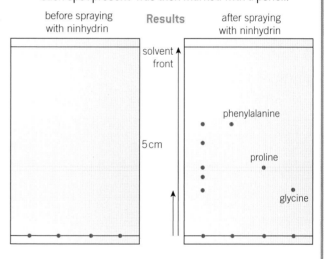

Here you can see the TLC plate showing the separated amino acids appearing purple after spraying with ninhydrin.

1 a Suggest why gloves were worn by the student and the plate was only handled by the edges.
  b A mixture of solvents (such as hexane, water, acetic acid, and butanol) is usually used as the mobile phase when separating an unknown mixture of amino acids. Suggest why.
  c Explain why the solvent was no more than 1 cm deep.
  d Suggest why the jar was sealed.
2 a Using the information provided identify as many of the amino acids present in the solution as you can. The distance an amino acid travels is determined by the interactions it has with both the mobile phase and the stationary phase. Different amino acids will therefore move different distances in a set time. As long as the conditions are kept

the same, the same amino will always travel the same distance in the same time.

The retention value (Rf) for each amino acid is the distance travelled by the pigment divided by the distance travelled by the solvent, it can be calculated using the formula:

$$R_f = \frac{\text{distance travelled by component}}{\text{distance travelled by solvent}}$$

This will be constant for each amino acid tested under identical conditions.

b Calculate the Rf value for the two unidentified spots and, using the standard Rf values below, identify the amino acid.

Alanine 0.31   Cysteine 0.40
Aspartic acid 0.24   Methionine 0.49
Phenylalanine 0.59   Glutamine 0.13

Levels of protein structure

Primary structure – this is the *sequence* in which the amino acids are joined. It is directed by information carried within DNA (discussed further in Topic 3.8, Nucleic acids and Topic 3.9, DNA replication and the genetic code). The particular amino acids in the sequence will influence how the polypeptide folds to give the protein's final shape. This in turn determines its function. The only bonds involved in the primary structure of a protein are peptide bonds.

Secondary structure – the oxygen, hydrogen, and nitrogen atoms of the basic, repeating structure of the amino acids (the variable groups are not involved at this stage) interact. Hydrogen bonds may form within the amino acid chain, pulling it into a coil shape called an alpha helix (Figure 3a).

Polypeptide chains can also lie parallel to one another joined by hydrogen bonds, forming sheet-like structures. The pattern formed by the individual amino acids causes the structure to appear pleated, hence the name beta pleated sheet (Figure 3b).

Secondary structure is the result of hydrogen bonds and forms at regions along long protein molecules depending on the amino sequences.

Tertiary structure – this is the folding of a protein into its final shape. It often includes sections of secondary structure. The coiling or folding of sections of proteins into their secondary structures brings R-groups of different amino acids closer together so they are close enough to interact and further folding of these sections will occur. The following interactions occur between the R-groups:

- hydrophobic and hydrophilic interactions – weak interactions between polar and non-polar R-groups
- hydrogen bonds – these are weakest of the bonds formed
- ionic bonds – these are stronger than hydrogen bonds and form between oppositely charged R-groups
- disulfide bonds (also known as disulfide bridges) – these are covalent and the strongest of the bonds but only form between R-groups that contain sulfur atoms.

(a) alpha helix

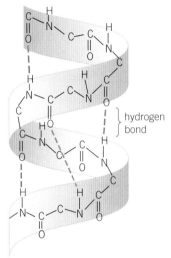

hydrogen bond

(b) beta pleated sheet

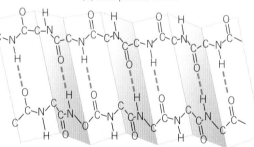

▲ Figure 3 *Depending on the amino acid composition, polypeptides initially form either (a) alpha helices or beta (b) pleated sheets– types of secondary structure*

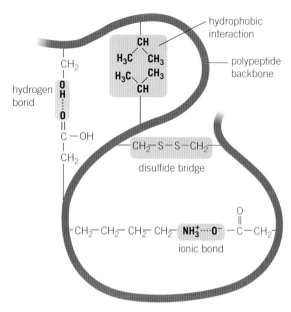

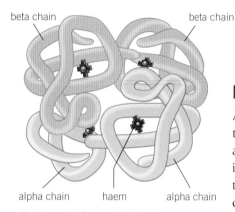

▲ **Figure 4** *The tertiary structures of proteins are very complex shapes involving multiple types of bonds and interactions between R-groups*

beta chain

beta chain

alpha chain   haem   alpha chain

▲ **Figure 5** *Haemoglobin has a quaternary structure made up of four individual proteins (two alpha and two beta) as well as haem groups containing iron*

Synoptic link

You will learn more about enzymes in Chapter 4, Enzymes.

Synoptic link

You will learn more about haemoglobin in Topic 8.4, Transport of oxygen and carbon dioxide in the blood.

This produces a variety of complex-shaped proteins, with specialised characteristics and functions (Figure 4).

Quaternary structure – this results from the association of two or more individual proteins called subunits. The interactions between the subunits are the same as in the tertiary structure except that they are between different protein molecules rather than within one molecule.

The protein subunits can be identical or different. Enzymes often consist of two identical subunits whereas insulin (a hormone) has two different subunits. Haemoglobin, a protein required for oxygen transport in the blood, has four subunits, made up of two sets of two identical subunits (Figure 5).

Hydrophilic and hydrophobic interactions

Proteins are assembled in the aqueous environment of the cytoplasm. So the way in which a protein folds will also depend on whether the R-groups are hydrophilic or hydrophobic. Hydrophilic groups are on the outside of the protein while hydrophobic groups are on the inside of the molecule shielded from the water in the cytoplasm.

Breakdown of peptides

As you have learned, peptides are created by amino acids linking together in condensation reactions to form peptide bonds. Proteases are enzymes that catalyse the reverse reaction – turning peptides back into their constituent amino acids. A water molecule is used to break the peptide bond in a hydrolysis reaction, reforming the amine and carboxylic acid groups.

 Identification of proteins

Biuret test

Peptide bonds form violet coloured complexes with copper ions in alkaline solutions. This can be used as the basis of a test for proteins.

Remember, when working with chemicals:

 Take care, wear safety glasses and report any spillages/breakages to the teacher.

A student carried out the following procedure to test a sample for the presence of protein.

1   $3\,cm^3$ of a liquid sample was mixed with an equal volume of 10% sodium hydroxide solution.

2   1% copper sulfate solution was then added a few drops at a time until the sample solution turned blue.

3   The solution was mixed and left to stand for five minutes.

This test is known as the biuret test. A mixture of an alkali and copper sulfate solution is called biuret reagent and can be used instead of adding the solutions individually.

1   State the colour you would expect to see on addition of the copper sulfate solution if protein is present in the sample.
2   State the colour you would expect to see if the sample contained amino acids instead of proteins.
    Explain the reason for this colour.
3   Suggest why this test is not used quantitatively.

Summary questions

1   Draw the structure of an amino acid.   (*3 marks*)

2   Describe the formation of a peptide bond.   (*3 marks*)

3 a   Draw a box identifying the peptide bond in the diagram below.
                                                                          (*1 mark*)

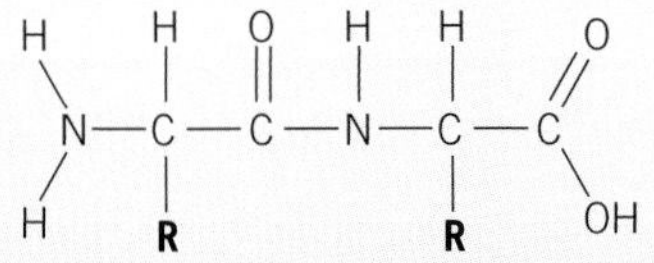

  b   Describe how hydrogen bonds form within the secondary
      structure of proteins.   (*2 marks*)
  c   Alpha keratin, a protein found in sheep's wool, is primarily
      composed of alpha helices. Explain why alpha keratin
      has a more regular structure than the quaternary protein
      haemoglobin.   (*3 marks*)

4   Compare and contrast the role of R-group interactions in the
    formation of the tertiary and quaternary structures of proteins. (*6 marks*)

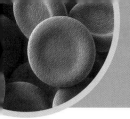

3.7 Types of proteins

Learning outcomes

Demonstrate knowledge, understanding, and application of:

→ the structure and function of globular proteins including a conjugated protein

→ the properties and functions of fibrous proteins.

▲ Figure 1 *Insulin. The complex shape of a globular protein formed from the folding of the primary structure into secondary structures (helices), which are further folded into the tertiary structure*

Synoptic link

You will learn about hormones in Chapter 14, Hormonal communication.

▲ Figure 2 *Haemoglobin. Four subunits are each wrapped around a haem group (red), protecting it from being oxidised and destroyed by the oxygen it is intended to transport. The iron within each haem group reversibly bonds to oxygen in the blood. Four haemoglobin monomers (green, beige, purple, blue) usually bind together to form one large haemoglobin molecule*

In the previous topic you saw how the complex tertiary and quaternary structures of proteins are built up. These structures determine the role the protein will play in the body. The two main groups are globular proteins and fibrous proteins.

Globular proteins

Globular proteins are compact, water soluble, and usually roughly spherical in shape. They form when proteins fold into their tertiary structures in such a way that the hydrophobic R-groups on the amino acids are kept away from the aqueous environment. The hydrophilic R-groups are on the outside of the protein. This means the proteins are soluble in water.

This solubility is important for the many different functions of globular proteins. They are essential for regulating many of the processes necessary to life. As you will see in later sections, these include processes such as chemical reactions, immunity, muscle contraction, and many more.

Insulin

Insulin is a globular protein. It is a hormone involved in the regulation of blood glucose concentration. Hormones are transported in the bloodstream so need to be soluble. Hormones also have to fit into specific receptors on cell-surface membranes to have their effect and therefore need to have precise shapes.

Conjugated proteins

Conjugated proteins are globular proteins that contain a non-protein component called a **prosthetic group**. Proteins without prosthetic groups are called simple proteins.

There are different types of prosthetic groups. Lipids or carbohydrates can combine with proteins forming lipoproteins or glycoproteins. Metal ions and molecules derived from vitamins also form prosthetic groups.

Haem groups are examples of prosthetic groups. They contain an iron II ion (Fe^{2+}). Catalase and haemoglobin both contain haem groups.

Haemoglobin

Haemoglobin is the red, oxygen-carrying pigment found in red blood cells. It is a quaternary protein made from four polypeptides, two alpha and two beta subunits (Figure 5, Topic 3.6, Structure of proteins). Each subunit contains a prosthetic haem group. The iron II ions present in the haem groups are each able to combine reversibly with an oxygen molecule. This is what enables haemoglobin to transport oxygen around the body. It can pick oxygen up in the lungs and transport it to the cells that need it, where it is released.

Catalase

Catalase is an enzyme. Enzymes catalyse reactions, meaning they increase reaction rates, and each enzyme is *specific* to a particular reaction or type of reaction.

Catalase is a quaternary protein containing four haem prosthetic groups. The presence of the iron II ions in the prosthetic groups allow catalase to interact with hydrogen peroxide and speed up its breakdown. Hydrogen peroxide is a common byproduct of metabolism but is damaging to cells and cell components if allowed to accumulate. Catalase makes sure this doesn't happen.

Fibrous proteins

Fibrous proteins are formed from long, insoluble molecules. This is due to the presence of a high proportion of amino acids with hydrophobic R-groups in their primary structures. They contain a limited range of amino acids, usually with small R-groups. The amino acid sequence in the primary structure is usually quite repetitive. This leads to very organised structures reflected in the roles fibrous proteins often have. Keratin, elastin, and collagen are examples of fibrous proteins.

Fibrous proteins tend to make strong, long molecules which are *not* folded into complex three-dimensional shapes like globular proteins.

Keratin

Keratin is a group of fibrous proteins present in hair, skin, and nails. It has a large proportion of the sulfur-containing amino acid, cysteine. This results in many strong disulfide bonds (disulfide bridges) forming strong, inflexible, and insoluble materials. The degree of disulfide bonds determines the flexibility – hair contains fewer bonds making it more flexible than nails, which contain more bonds. The unpleasant smell produced when hair or skin is burnt is due to the presence of relatively large quantities of sulfur in these proteins.

Elastin

Elastin is a fibrous protein found in elastic fibres (along with small protein fibres). Elastic fibres are present in the walls of blood vessels and in the alveoli of the lungs – they give these structures the flexibility to expand when needed, but also to return to their normal size. Elastin is a quaternary protein made from many stretchy molecules called tropoelastin (see the Extension, The structure of fibrous proteins, for further detail of structure).

Collagen

Collagen is another fibrous protein. It is a connective tissue found in skin, tendons, ligaments and the nervous system. There are a number of different forms but all are made up of three polypeptides wound together in a long and strong rope-like structure. Like rope, collagen has flexibility (see the Extension, The structure of fibrous proteins, for further detail of structure).

Synoptic link

You will learn more about haemoglobin and its role in the transport of oxygen in Topic 8.4, Transport of oxygen and carbon dioxide in the blood.

Synoptic link

You will learn about enzymes in Chapter 4, Enzymes.

Study tip

You are not required to learn the detailed structure of fibrous proteins, but an overview is useful in understanding their properties and functions. Details of the structure of fibrous proteins are given in the Extension.

▲ Figure 3 *A coloured SEM image of eyelash hairs growing from the surface of human skin. The shafts of hair are made up of the fibrous protein, keratin. × 50 magnification*

Synoptic link

You will learn more about the role of elastin in blood vessels in Topic 8.2, Blood vessels.

The structure of fibrous proteins

Elastin

Elastin is made by linking many soluble tropoelastin protein molecules to make a very large, insoluble, and stable, cross-linked structure (Figure 4).

Tropoelastin molecules are able to stretch and recoil without breaking, acting like small springs. They contain alternate hydrophobic and lysine-rich areas.

Elastin is formed when multiple tropoelastin molecules aggregate via interactions between the hydrophobic areas. The structure is stabilised by cross-linking covalent bonds involving the amino acid lysine, but the polypeptide structure still has flexibility.

Elastin confers strength and elasticity to the skin and other tissues and organs in the body.

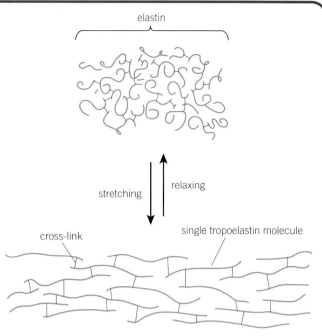

▲ Figure 4 *The structure of elastin allows it to be stretched without breaking*

Collagen

Collagen molecules have three polypeptide chains wound around each other in a triple helix structure to form a tough, rope-like protein (Figure 5d and e).

Every third amino acid in the polypeptide chains is glycine, which is a small amino acid. Its small size allows the three protein molecules to form a closely packed triple helix. Many hydrogen bonds form between the polypeptide chains forming long quaternary proteins with staggered ends (Figure 5d). These allow the proteins to join end to end, forming long fibrils called tropocollagen (Figure 5c). The tropocollagen fibrils cross-link to produce strong fibres.

Collagen also contains high proportions of the amino acids proline and hydroxyproline. The R-groups in these amino acids repel each other and this adds to the stability of collagen.

In some tissues, multiple fibres of collagen aggregate into larger bundles (Figure 5a). This is the structure found in ligaments and tendons. In skin, collagen fibres form a mesh that is resistant to tearing.

▶ Figure 5 *(a) A ligament is composed of a bundle of (b) collagen microfibrils. Each collagen microfibril is composed of (c) multiple triple-helix proteins wound together with staggered ends so that the individual polypeptides overlap, forming a very strong structure*

(a)

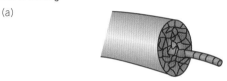

ligament tissue

(b)

multiple triple-helix proteins wound to form a microfibril

(c)

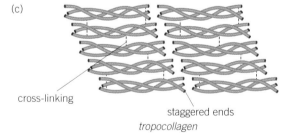

cross-linking

staggered ends
tropocollagen

(d)

three polypeptide chains form a triple helix – collagen

(e)

single polypeptide chain

1 Suggest what property the arrangement of collagen fibres into large bundles gives to tendons.

2 As we age the collagen in our skin starts to break down. This leads to the loss of skin structure and the formation of wrinkles. Many beauty products are available that contain collagen in the form of creams and capsules. Using your knowledge of the structure of collagen, suggest why these products are unlikely to have any beneficial effect in reducing or preventing wrinkles.

3 Which of the following sequences is correct in terms of increasing bond strength?
   a ionic bonds, disulfide bonds, hydrogen bonds
   b hydrogen bonds, ionic bonds, disulfide bonds
   c disulfide bonds, ionic bonds, hydrogen bonds

Summary questions

1 Explain the difference between a simple protein and a conjugated protein. *(3 marks)*

2 Describe the differences in properties and functions of insulin, a hormone, and keratin present in hair and nails. *(4 marks)*

3 Describe why globular proteins are soluble in water but fibrous proteins are not. *(3 marks)*

4 Myoglobin is an oxygen carrying molecule found primarily in muscle tissue. It is formed from a single polypeptide chain which is folded to form eight alpha helices. This chain is further folded around a central prosthetic group which binds reversibly with oxygen. The hydrophobic R groups of the amino acids are positioned towards the centre of the molecule.

   Discuss the similarities and differences in the structures of haemoglobin and myoglobin. *(6 marks)*

Study tip

Make sure you are able to compare globular and fibrous proteins using named examples.

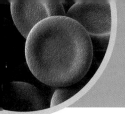

3.8 Nucleic acids

Specification reference: 2.1.2, 2.1.3

Nucleic acids are large molecules that were discovered in cell nuclei – hence their name. There are two types of nucleic acid – DNA and RNA, and both have roles in the storage and transfer of genetic information and the synthesis of polypeptides (proteins). They are the basis for heredity.

Nucleotides and nucleic acids

Nucleic acids contain the elements carbon, hydrogen, oxygen, nitrogen and phosphorus. They are large polymers formed from many **nucleotides** (the monomers) linked together in a chain.

An individual nucleotide is made up of three components, as shown in Figure 1:

● a pentose monosaccharide (sugar), containing five carbon atoms

● a phosphate group, $-PO_4^{2-}$, an inorganic molecule that is acidic and negatively charged

● a nitrogenous base – a complex organic molecule containing one or two carbon rings in its structure as well as nitrogen.

Nucleotides are linked together by condensation reactions to form a polymer called a polynucleotide. The phosphate group at the fifth carbon of the pentose sugar (5′) of one nucleotide forms a covalent bond with the hydroxyl (OH) group at the third carbon (3′) of the pentose sugar of an adjacent nucleotide. These bonds are called **phosphodiester bonds**. This forms a long, strong sugar-phosphate 'backbone' with a base attached to each sugar (Figure 2). The phosphodiester bonds are broken by **hydrolysis**, the reverse of condensation, releasing the individual nucleotides.

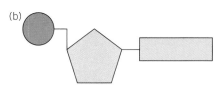

▲ Figure 1 (a) A nucleotide and its three component parts: a sugar, a phosphate and a base (the numbers in blue denote the standard numbering of the five carbons in the sugar). (b) A simple representation of a nucleotide

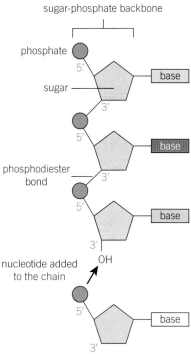

▲ Figure 2 Nucleotides (monomers) are linked together to form the polymer, nucleic acid (polynucleotide). Note the sugar-phosphate backbone with all the bases projecting from the opposite side

Deoxyribonucleic acid (DNA)

As the name suggests, the sugar in **deoxyribonucleic acid (DNA)** is deoxyribose – a sugar with one fewer oxygen atoms than ribose, as shown in Figure 3.

The nucleotides in DNA each have one of four different bases. This means there are four different DNA nucleotides, see Figure 4. The four bases can be divided into two groups:

- **Pyrimidines** – the smaller bases, which contain single carbon ring structures – thymine (T) and cytosine (C)
- **Purines** – the larger bases, which contain double carbon ring structures – adenine (A) and guanine (G).

deoxyribose

ribose

◀ **Figure 3** *The sugar in DNA is deoxyribose (top) and the sugar in RNA is ribose (bottom). Deoxyribose has one less oxygen atom than ribose*

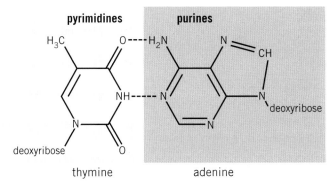

thymine   adenine

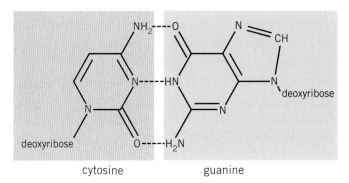

cytosine   guanine

▲ **Figure 4** *The chemical structure of the four different bases in DNA showing hydrogen bonding between complementary pairs*

Study tip

You do *not* need to remember the chemical structure of the bases, but you should remember that:

The comparative sizes of pyrimidines and purines is due to the presence of either a single ring or a double ring. The complementary pair thymine and adenine form *two* hydrogen bonds and the complementary pair cytosine and guanine form *three* hydrogen bonds *and* purines pair with pyrimidines.

The double helix

The DNA molecule varies in length from a few nucleotides to millions of nucleotides. It is made up of two strands of polynucleotides coiled into a helix, known as the DNA double helix, see Figure 6.

The two strands of the double helix are held together by hydrogen bonds between the bases, much like the rungs of a ladder. Each strand has a phosphate group (5') at one end and a hydroxyl group (3') at the other end. The two parallel strands are arranged so that they run in opposite directions (Figure 5) – they are said to be **antiparallel**.

The pairing between the bases allows DNA to be copied and transcribed – key properties required of the molecule of heredity.

Study tip

When referring to DNA, the molecule is the double helix composed of two antiparallel strands. Each strand is one chain of nucleotides (a polynucleotide).

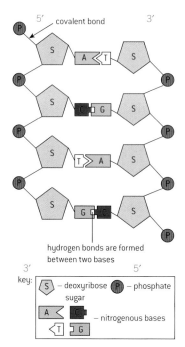

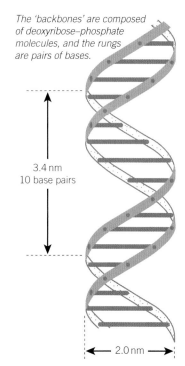

The 'backbones' are composed of deoxyribose–phosphate molecules, and the rungs are pairs of bases.

3.4 nm
10 base pairs

2.0 nm

▲ Figure 5 *A simplified diagram of base pairing in DNA*

▲ Figure 6 *The double helix structure of DNA*

Base pairing rules

The bases bind in a very specific way (Figure 7). Adenine and thymine are both able to form *two* hydrogen bonds and always join with each other. Cytosine and guanine form *three* hydrogen bonds and so also only bind to each other. This is known as **complementary base pairing**.

These rules mean that a small pyrimidine base always binds to a larger purine base. This arrangement maintains a constant distance between the DNA 'backbones', resulting in parallel polynucleotide chains.

Complementary base pairing means that DNA always has equal amounts of adenine and thymine *and* equal amounts of cytosine and guanine. This was known long before the detailed structure of DNA was determined by Watson and Crick in 1953.

It is the *sequence* of bases along a DNA strand that carries the genetic information of an organism in the form of a code. In the next topic we will examine how the sequence of bases 'codes' for the sequences of amino acids that are needed to make different proteins.

Ribonucleic acid (RNA)

Ribonucleic acid (RNA) plays an essential role in the transfer of genetic information from DNA to the proteins that make up the enzymes and tissues of the body. DNA stores all of the genetic information needed by an organism, which is passed on from generation to generation. However, the DNA of each eukaryotic chromosome is a very long molecule, comprising many hundreds of genes, and is unable to leave the nucleus in order to supply the information directly to the sites of protein synthesis.

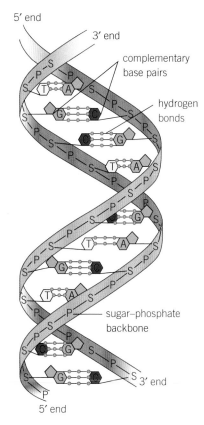

▲ Figure 7 *The DNA double helix – showing complementary base pairing and antiparallel strands*

To get around this problem, the relatively short section of the long DNA molecule corresponding to a single gene is transcribed into a similarly short messenger RNA (mRNA) molecule. Each individual mRNA is therefore much shorter than the whole chromosome of DNA. It is a polymer composed of many nucleotide monomers.

RNA nucleotides are different to DNA nucleotides as the pentose sugar is ribose rather than deoxyribose (Figure 3) and the thymine base is replaced with the base uracil (U) (see Figure 9). Like thymine, uracil is a pyrimidine that forms two hydrogen bonds with adenine. Therefore the base pairing rules still apply when RNA nucleotides bind to DNA to make copies of particular sections of DNA.

The RNA nucleotides form polymers in the same way as DNA nucleotides – by the formation of phosphodiester bonds in condensation reactions. The RNA polymers formed are small enough to leave the nucleus and travel to the ribosomes, where they are central in the process of protein synthesis.

After protein synthesis the RNA molecules are degraded in the cytoplasm. The phosphodiester bonds are hydrolysed and the RNA nucleotides are released and reused.

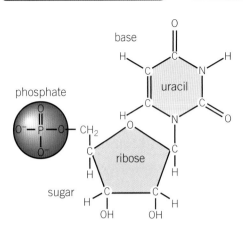

▲ Figure 8 *The thymine base is replaced with uracil in RNA nucleotides. The sugar is ribose not deoxyribose*

Synoptic link

You will learn more about transcription and translation in protein synthesis in Topic 3.9, DNA replication and the genetic code and Topic 3.10, Protein synthesis.

DNA extraction

DNA can be extracted from plant material using the following procedure:

- Grind sample in a mortar and pestle – this breaks down the cell walls.
- Mix sample with detergent – this breaks down the cell membrane, releasing the cell contents into solution.
- Add salt – this breaks the hydrogen bonds between the DNA and water molecules.
- Add protease enzyme – this will break down the proteins associated with the DNA in the nuclei.
- Add a layer of alcohol (ethanol) on top of the sample – alcohol causes the DNA to precipitate out of solution.

▲ Figure 9 *DNA as a white precipitate*

- The DNA will be seen as white strands forming between the layer of sample and layer of alcohol (Figure 9). The DNA can be picked up by 'spooling' it onto a glass rod.

1 The temperature should be kept low throughout this DNA extraction procedure. Suggest why.
2 Explain why detergent breaks down cell membranes.

Summary questions

1 Describe the differences between DNA and RNA nucleotides. *(2 marks)*

2 Explain the base pairing rule. *(3 marks)*

3 Explain how the structure of DNA is ideally suited to its role. *(4 marks)*

4 A sample of DNA was tested and 17% of the total bases present were found to be adenine.

Calculate the percentages of each of the other three bases present in this sample. Show all of your working. *(3 marks)*

3.9 DNA replication and the genetic code

Specification reference: 2.1.3

Learning outcomes

Demonstrate knowledge, understanding, and application of:

→ semi-conservative DNA replication

→ the nature of the genetic code.

Synoptic link

You will learn more about cell division in Topic 6.2, Mitosis.

Cells divide to produce more cells needed for growth or repair of tissues. The two daughter cells produced as a result of cell division are genetically identical to the parent cell and to each other. In other words, they contain DNA with a base sequence identical to the original parent cell.

When a cell prepares to divide, the two strands of DNA double helix separate and each strand serves as a template for the creation of a new double-stranded DNA molecule. The complementary base pairing rules, which you learnt about in the previous topic, ensure that the two new strands are identical to the original. This process is called **DNA replication**.

Semi-conservative replication

For DNA to replicate, the double helix structure has to unwind and then separate into two strands, so the hydrogen bonds holding the complementary bases together must be broken (Figure 1c). Free DNA nucleotides will then pair with their complementary bases, which have been exposed as the strands separate. Hydrogen bonds are formed between them. Finally, the new nucleotides join to their adjacent nucleotides with phosphodiester bonds (Figure 1d).

In this way, two new molecules of DNA are produced. Each one consists of one old strand of DNA and one new strand. This is known as **semi-conservative** (meaning half the same) replication.

Roles of enzymes in replication

DNA replication is controlled by enzymes, a class of proteins that act as catalysts for biochemical reactions. Enzymes are only able to carry out their function by recognising and attaching to specific molecules or particular parts of the molecules.

Synoptic link

You will learn more about enzymes in Chapter 4, Enzymes.

Before replication can occur, the unwinding and separating of the two strands of the DNA double helix is carried out by the enzyme **DNA helicase**. It travels along the DNA backbone, catalysing reactions that break the hydrogen bonds between complementary base pairs as it reaches them. This can be thought of as the strand 'unzipping'.

Free nucleotides pair with the newly exposed bases on the template strands during the 'unzipping' process. A second enzyme, **DNA polymerase** catalyses the formation of phosphodiester bonds between these nucleotides.

a A representative portion of DNA, which is about to undergo replication.

b An enzyme, DNA helicase, causes the two strands of the DNA to separate.

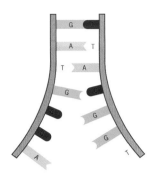

c DNA helicase completes the separation of the strand. Meanwhile, free nucleotides that have been activated are attracted to their complementary bases.

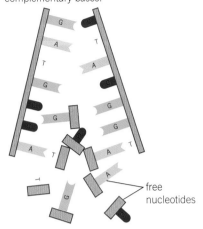

free nucleotides

d Once the activated nucleotides are lined up, they are joined together by DNA polymerase (bottom three nucleotides). The remaining unpaired bases continue to attract their complementary nucleotides.

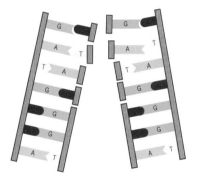

e Finally, all the nucleotides are joined to form a complete polynucleotide chain using DNA polymerase. In this way, two identical molecules of DNA are formed. Each new molecule of DNA is composed of one original strand and one newly formed molecule – semi-conservative replication.

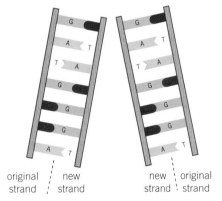

original strand ⦙ new strand   new strand ⦙ original strand

▲ Figure 1 *The semi-conservative replication of DNA*

Continuous and discontinuous replication

DNA polymerase always moves along the template strand in the same direction. It can only bind to the 3' (OH) end, so travels in the direction of 3' to 5'. As DNA only unwinds and unzips in one direction, DNA polymerase has to replicate each of the template strands in opposite directions. The strand that is unzipped from the 3' end can be continuously replicated as the strands unzip. This strand is called the *leading strand* and is said to undergo *continuous replication*.

The other strand is unzipped from the 5' end, so DNA polymerase has to wait until a section of the strand has unzipped and then work back along the strand.

This results in DNA being produced in sections (called *Okazaki fragments*), which then have to be joined. This strand is called the *lagging strand* and is said to undergo *discontinuous replication*.

1 Describe the difference between continuous and discontinuous replication. *(4 marks)*

2 Using your knowledge of enzymes, explain why DNA polymerase does not catalyse the joining of Okazaki fragments into a single strand but a different enzyme (DNA ligase) is used. *(3 marks)*

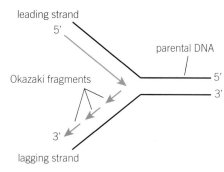

▲ **Figure 2** *As DNA polymerase only travels in the 3' to 5' direction, only the leading strand of DNA can be replicated continuously as the DNA unwinds but the lagging strand has to be replicated in the opposite direction in short sections called Okazaki fragments*

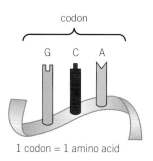

▲ **Figure 3** *A codon is a triplet of bases that code for an amino acid*

Replication errors

Sequences of bases are not always matched exactly, and an incorrect sequence may occur in the newly-copied strand. These errors occur randomly and spontaneously and lead to a change in the sequence of bases, known as a **mutation**.

Genetic code

DNA is contained within the cells of *all* organisms and scientists determined that this molecule was the means by which genetic information was passed from one generation to the next. But how does this happen? Scientists understood that DNA must carry the 'instructions' or 'blueprint' needed to synthesise the many different proteins needed by these organisms. Proteins are the foundation for the different physical and biochemical characteristics of living things. They are made up of a sequence of amino acids, folded into complex structures. Therefore DNA must *code* for a sequence of amino acids. This is called the **genetic code**.

A triplet code

The instructions that DNA carries are contained in the sequence of bases along the chain of nucleotides that make up the two strands of DNA. The code in the base sequences is a simple **triplet code**. It is a sequence of three bases, called a **codon**. Each codon codes for an amino acid.

A section of DNA that contains the complete sequence of bases (codons) to code for an entire protein is called a **gene**.

The genetic code is universal – all organisms use this same code, although the sequences of bases coding for each individual protein will be different.

Degenerate code

As you have learnt, there are four different bases, which means there are 64 different base triplets or codons possible (4^3 or $4 \times 4 \times 4$). This includes one codon that acts as the start codon when it comes at the beginning of a gene, signalling the start of a sequence that codes for a protein. If it is in the middle of a gene, it codes for the amino acid methionine. There are also three 'stop' codons that do not code for any amino acids and signal the end of the sequence.

Having a single codon to signal the start of a sequence ensures that the triplets of bases (codons) are read 'in frame'. In other words the DNA base sequence is 'read' from base 1, rather than base 2 or 3. So the genetic code is non-overlapping.

As there are only 20 different amino acids that regularly occur in biological proteins, there are a lot more codons than amino acids. Therefore, many amino acids can be coded for by more than one codon, see Figure 4. Due to this, the code is known as degenerate.

Study tip

Remember, the genetic code is a triplet code. Three bases = one codon. One codon codes for one amino acid.

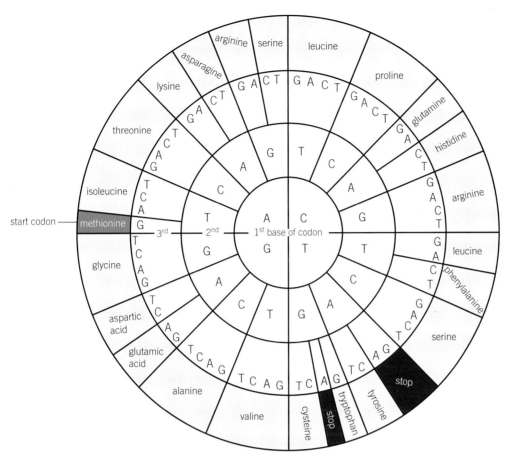

▲ Figure 4 *The DNA code wheel shows how the genetic code is degenerate – different combinations of bases can code for the same amino acid*

Summary questions

1 Explain why DNA replication is described as semi-conservative.

(*3 marks*)

2 Explain what is meant by the triplet code. (*2 marks*)

3 Enzymes are cellular proteins that catalyse reactions – they have active sites in which specific substrates fit precisely. Suggest how a genetic mutation may result in an enzyme becoming non-functional. (*4 marks*)

4 The sequences of bases in specific sections of DNA and the sequences of amino acids in specific proteins are both used to compare how closely related different species are. The fewer differences in the sequences the more closely related the species. Explain why there are likely to be more differences, overall, between base sequences of DNA than between amino acid sequences of proteins. (*3 marks*)

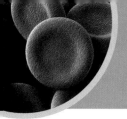

3.10 Protein synthesis

Specification reference: 2.1.3

As you learnt in the previous topic, the amino acid sequence of a protein is coded for by triplets (codons) in the sequence of bases along a strand of a DNA molecule. A section of DNA that contains the complete sequence of codons to code for an entire protein is called a gene. In this topic you will explore how the information in a gene is 'transcribed' into RNA molecules and then 'translated' into a specific amino acid sequence.

Transcription

In a eukaryotic cell, DNA is contained within a double membrane called the nuclear envelope that encloses the nucleus. This protects the DNA from being damaged in the cytoplasm. Protein synthesis occurs in the cytoplasm at ribosomes, but a chromosomal DNA molecule is too large to leave the nucleus to supply the coding information needed to determine the protein's amino acid sequence.

To get around this problem, the base sequences of genes have to be copied and transported to the site of protein synthesis, a ribosome. This process is called **transcription** and produces shorter molecules of RNA.

Although transcription results in a different polynucleotide, it has many similarities with DNA replication. The section of DNA that contains the gene unwinds and unzips under the control of a DNA helicase, beginning at a start codon. This involves the breaking of hydrogen bonds between the bases.

Only one of the two strands of DNA contains the code for the protein to be synthesised. This is the **sense strand** and it runs from 5' to 3'. The other strand (3' to 5') is a complementary copy of the sense strand and does not code for a protein. This is the **antisense strand** and it acts as the **template strand** during transcription, so that the complementary RNA strand formed carries the same base sequence as the sense strand.

Free RNA nucleotides will base pair with complementary bases exposed on the antisense strand when the DNA unzips. As you learnt in Topic 3.8 Nucleic acids, the thymine base in RNA nucleotides is replaced with the base uracil (U). So RNA uracil binds to adenine on the DNA template strand.

Phosphodiester bonds are formed between the RNA nucleotides by the enzyme **RNA polymerase**. Transcription stops at the end of the gene and the completed short strand of RNA is called **messenger (m)RNA**. It has the same base sequence as the sequence of bases making up the gene on the DNA, except that it has uracil in place of thymine.

The mRNA then detaches from the DNA template and leaves the nucleus through a nuclear pore. The DNA double helix reforms.

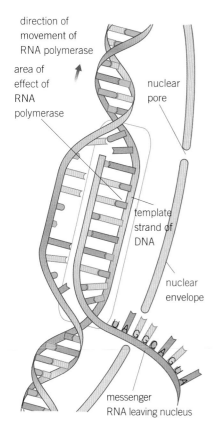

direction of movement of RNA polymerase

area of effect of RNA polymerase

nuclear pore

template strand of DNA

nuclear envelope

messenger RNA leaving nucleus

▲ Figure 1 *Summary of transcription*

This mRNA molecule then travels to a ribosome in the cell cytoplasm for the next step in protein synthesis.

Translation

In eukaryotic cells, ribosomes are made up of two subunits, one large and one small. These subunits are composed of almost equal amounts of protein and a form of RNA known as **ribosomal (r)RNA**. rRNA is important in maintaining the structural stability of the protein synthesis sequence and plays a biochemical role in catalysing the reaction.

After leaving the nucleus, the mRNA binds to a specific site on the small subunit of a ribosome. The ribosome holds mRNA in position while it is decoded, or translated, into a sequence of amino acids. This process is called **translation**.

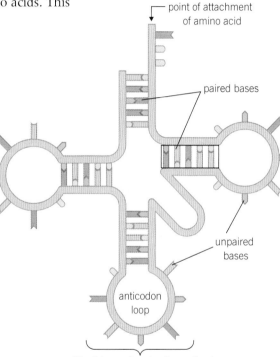

The 3 bases forming the anticodon

▲ Figure 3 *The structure of tRNA*

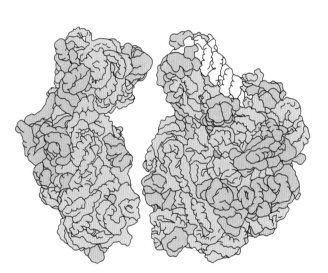

▲ Figure 2 *Large and small subunits of the ribosome with proteins shown in purple, ribosomal RNA in pink and yellow and the site that catalyses the formation of peptide bonds in green*

Transfer (t)RNA is another form of RNA, which is necessary for the translation of the mRNA. It is composed of a strand of RNA folded in such a way that three bases, called the anticodon, are at one end of the molecule (Figure 3). This anticodon will bind to a complementary codon on mRNA following the normal base pairing rules. The tRNA molecules carry an amino acid corresponding to that codon (Figure 4).

When the tRNA anticodons bind to complementary codons along the mRNA, the amino acids are brought together in the correct sequence to form the primary structure of the protein coded for by the mRNA.

This cannot happen all at once. Instead amino acids are added one at a time and the polypeptide chain (protein) grows as this happens. Ribosomes act as the binding site for mRNA and tRNA and catalyse the assembly of the protein. The sequence of events in translation is summarised on the next page and in Figure 5.

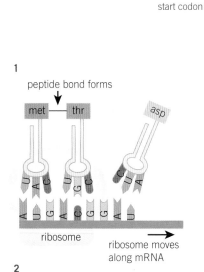

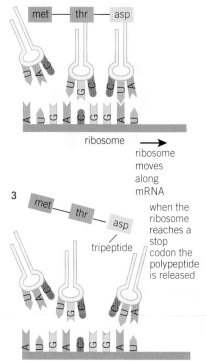

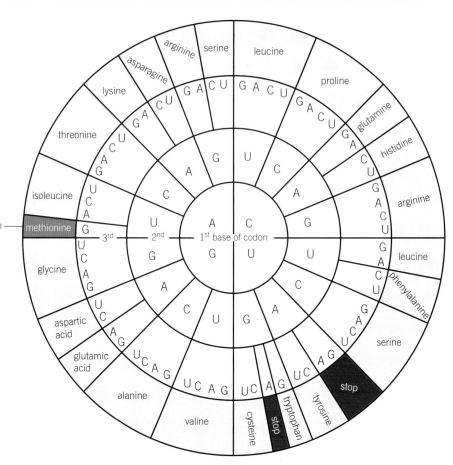

▲ Figure 4 *The codon-amino acid wheel, as seen in Topic 3.9, DNA replication and the genetic code. This time RNA codons are shown*

1 The mRNA binds to the small subunit of the ribosome at its start codon (AUG).

2 A tRNA with the complementary anticodon (UAC) binds to the mRNA start codon. This tRNA carries the amino acid methionine.

3 Another tRNA with the anticodon UGC and carrying the corresponding amino acid, threonine, then binds to the next codon on the mRNA (ACG). A maximum of two tRNAs can be bound at the same time.

4 The first amino acid, methionine, is transferred to the amino acid (threonine) on the second tRNA by the formation of a peptide bond. This is catalysed by the enzyme *peptidyl transferase*, which is an rRNA component of the ribosome.

5 The ribosome then moves along the mRNA, releasing the first tRNA. The second tRNA becomes the first.

Stages 3–5 are repeated, with another amino acid added to the chain each time. The process keeps repeating until the ribosome reaches the end of the mRNA at a stop codon and the polypeptide is released.

As the amino acids are joined together forming the primary structure of the protein, they fold into secondary and tertiary structures. This folding and the bonds that are formed are determined by the sequence

▲ Figure 5 *A summary of translation*

of amino acids in the primary structure. The protein may undergo further modifications at the Golgi apparatus (Topic 2.4, Eukaryotic cell structure) before it is fully functional and and ready to carry out the specific role for which it has been synthesised.

Many ribosomes can follow on the mRNA behind the first, so that multiple identical polypeptides can be synthesised simultaneously (Figure 6).

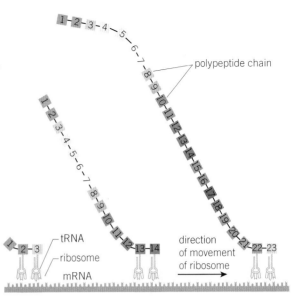

▶ **Figure 6** *The formation of multiple identical proteins simultaneously by translation of one mRNA by several ribosomes*

Summary questions

1 Copy and complete the diagram below to show the missing bases forming the tRNA anticodons. *(4 marks)*

DNA	TAC	CGG	AGT	GCA
mRNA	AUG	GCC	UCA	CGU
tRNA	A met	C ala	ser	arg

2 Describe the roles of mRNA, tRNA and rRNA in protein synthesis *(5 marks)*

3 a An enzyme forms part of the structure of ribosome. Suggest the role of this enzyme. *(2 marks)*

 b rRNA also forms part of the structure. Suggest why RNA needs to be present in a ribosome. *(2 marks)*

 c Ribosomes are either free floating within the cytoplasm or bound to endoplasmic reticulum. Suggest a reason for the different ribosomal sites. *(2 marks)*

4 Post-transcriptional modification of mRNA is carried out before it can leave the nucleus. This involves 'capping' each end to protect the mRNA from degradation in the cytoplasm and the removal of introns. Introns are non-coding sections of DNA and have no role in the formation of proteins.

 a Explain why unnecessary base sequences must be removed before protein synthesis begins. *(4 marks)*

 b Suggest an advantage of being able to edit mRNA *(2 marks)*

 c Suggest a reason for the presence of introns within genes. *(4 marks)*

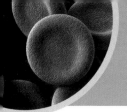

Universal energy currency

Muscle contraction, cell division, the transmission of nerve impulses, and even memory formation are just some of the many biological processes that require energy. Energy comes in many forms, such as heat, light, and the energy in chemical bonds. Energy has to be supplied in the right form and quantity to the processes that require it.

Cells require energy for three main types of activity:

● synthesis – for example of large molecules such as proteins

● transport – for example pumping molecules or ions across cell membranes by active transport

● movement – for example protein fibres in muscle cells that cause muscle contraction.

Inside cells, molecules of **adenosine triphosphate (ATP)** are able to supply this energy in such a way that it can be used.

An ATP molecule is composed of a nitrogenous base, a pentose sugar and three phosphate groups, as shown in Figure 1 – it is a nucleotide. You will notice that the structure of ATP is very similar to that of the nucleotides involved in the structure of DNA and RNA (Topic 3.8, Figure 1). However, in ATP the base is always adenine and there are three phosphate groups instead of one. The sugar in ATP is ribose, as in RNA nucleotides.

$$O^- - \overset{\overset{\displaystyle O}{\|}}{\underset{\underset{\displaystyle O^-}{|}}{P}} - O^-$$

Phosphate group

Ribose

Adenine

▲ Figure 2 The structure of a phosphate group, the pentose sugar ribose, and the nitrogenous base adenine

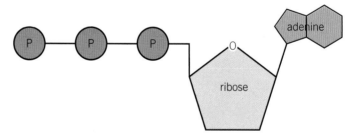

▲ Figure 1 The structure of an ATP molecule

ATP is used for energy transfer in all cells of all living things. Hence it is known as the universal energy currency.

How ATP releases energy

Energy is needed to break bonds and is released when bonds are formed. A small amount of energy is needed to break the relatively weak bond holding the last phosphate group in ATP. However, a large amount of energy is then released when the liberated phosphate undergoes other reactions involving bond formation. Overall a lot more energy is released than used, approximately $30.6 \, \text{kJ mol}^{-1}$.

As water is involved in the removal of the phosphate group this is another example of a **hydrolysis reaction**.

$$\text{ATP} \quad + \quad H_2O \quad \rightarrow \quad \text{ADP} \quad + \quad P_i \quad + \quad \text{energy}$$

| adenosine triphosphate | water | adenosine diphosphate | inorganic phosphate |

The hydrolysis of ATP does not happen in isolation but in association with energy-requiring reactions. The reactions are said to be 'coupled' as they happen simultaneously.

ATP is hydrolysed into **adenosine diphosphate (ADP)** and a phosphate ion, releasing energy.

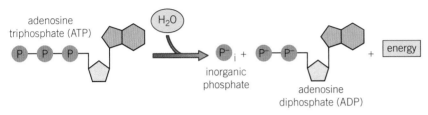

▲ Figure 3 *The hydrolysis of ATP to ADP releases energy*

The instability of the phosphate bonds in ATP, however, means that it is not a good *long-term* energy store. Fats and carbohydrates are much better for this. The energy released in the breakdown of these molecules (a process called cellular respiration) is used to create ATP. This occurs by reattaching a phosphate group to an ADP molecule. The process is called *phosphorylation*. As water is removed in this process, the reaction is another example of a **condensation reaction**.

Due to the instability of ATP, cells do not store large amounts of it. However, ATP is rapidly reformed by the phosphorylation of ADP (Figure 4). This interconversion of ATP and ADP is happening constantly in all living cells, meaning cells do not need a large store of ATP. ATP is therefore a good *immediate* energy store.

Properties of ATP

The structure and properties of ATP mean that it is ideally suited to carry out its function in energy transfer. A summary of these properties is given below.

- Small – moves easily into, out of and within cells.
- Water soluble – energy-requiring processes happen in aqueous environments.
- Contains bonds between phosphates with intermediate energy: large enough to be useful for cellular reactions but not so large that energy is wasted as heat.
- Releases energy in small quantities – quantities are suitable to most cellular needs, so that energy is not wasted as heat.
- Easily regenerated – can be recharged with energy.

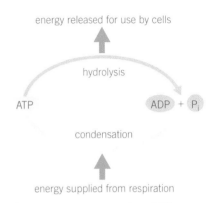

▲ Figure 4 *Interconversion of ATP and ADP*

Study tip

Make sure you know the characteristics that make ATP so useful as the universal energy currency.

Summary questions

1 Describe the structure of ATP.
(3 marks)

2 Describe why ATP is called the universal energy currency. *(3 marks)*

3 People and other animals store excess energy in the form of fat. Explain why fat is stored, not ATP? *(4 marks)*

4 a Outline how energy is transferred with reference to bond formation and cleavage. *(2 marks)*

 b Discuss the validity of the statement: ATP is the universal energy currency. *(4 marks)*

Practice questions

Water has a very simple molecular structure but is a vital component of living organisms. Water has many roles such as metabolite, solvent and reaction medium. It is also the environment in which many organisms live.

1 **a** Define the word 'molecule'.   (*2 marks*)

b Explain what is meant by the term 'polar' with reference to water.   (*2 marks*)

c Water has an essential role as a transport medium in both plants and animals.

Explain why water makes an ideal transport medium using **two** examples.
(*4 marks*)

2 A colorimeter can be used to estimate the concentration of a solution of reducing sugar. This is done by carrying out Benedict's test on the solution, allowing the precipitate to settle and measuring the absorbance of the liquid portion or supernatant.

Describe how you could estimate the concentration of a solution of glucose without using a colorimeter.   (*5 marks*)

3 Copy and fill in the table below.

The first row has been done for you.
(*3 marks*)

Feature	Cellulose	Collagen
Fibrous	✓	✓
Monomers joined by condensation reactions		
Monomers identical		
Branching		

4 **a** Describe the difference between hydrophobic and hydrophilic using phospholipid molecules as an example.
(*4 marks*)

b Describe how you would identify the presence of a lipid in an unknown sample.
(*4 marks*)

5 Explain the meaning of prosthetic group with reference to haemoglobin.   (*4 marks*)

6 **a** Identify A, B, C and D on the diagram below.   (*4 marks*)

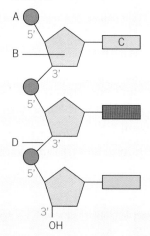

b Explain why nucleic acids are described as polymers.   (*2 marks*)

c Outline the differences between DNA and RNA.   (*2 marks*)

d Chargaff's rule states that the ratios of adenine to thymine and guanine to cytosine in DNA are equal.

(i) Explain, using your knowledge of the structure of DNA why the bases in DNA always obey this rule.
(*4 marks*)

(ii) Suggest what has happened during replication if a section of DNA does not obey this rule.   (*3 marks*)

7 The graph below shows how the ratio of ribosome synthesis to total protein synthesis changes as the rate of cell division in a bacterial colony increases.

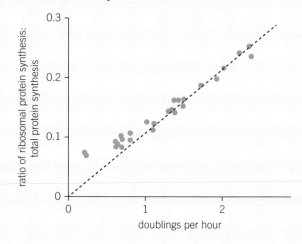

a Describe the relationship shown by the graph. *(4 marks)*

b Explain why ribosomes are necessary in protein synthesis. *(3 marks)*

c Assuming a constant rate of translation explain the trend shown by the graph. *(4 marks)*

8 The diagram below summarises the flow of information between DNA, RNA and protein.

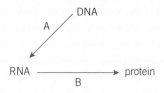

a Suggest which processes are happening at A and B on the diagram. *(2 marks)*

b Explain why there is no arrow between DNA and protein. *(3 marks)*

Viruses are often incorporated into DNA within the nucleus of infected cells.

c Suggest why an arrow could also be drawn going from RNA to DNA. *(2 marks)*

9 Compare the structure and function of ATP and a DNA nucleotide. *(5 marks)*

10 The diagram represents a water molecule

Water molecules are polar. As a result, they attract each other.

Draw a second water molecule on the diagram

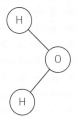

Your drawing should show:

- The bond(s) between the two molecules
- The name of the bond

  The charges on each atom *(3 marks)*

  OCR F212 2010

11 a Ponds provide a very stable environment for aquatic organisms.

Three properties that contribute to this stability are as follows:

- The density of water decreases as the temperature falls below 4 °C so ice floats on the top of the bond

- It acts as a solvent for ions such as nitrates (NO_3)

- A large quantity of energy is required to raise the temperature of water by 1 °C.

Explain how these three properties help organisms survive in the pond.

In you answer you should make clear the links between the behaviour of the water molecules and the survival of the organisms. *(8 marks)*

b Water is important in many biological questions.

Complete table 1 by writing an appropriate term next to each description.

description	term
The type of reaction that occurs when water is added to break a bond in a molecule	
The phosphate group of a phospholipid that readily attracts water molecules	

OCR F212 2010

4 ENZYMES

4.1 Enzyme action

Specification reference: 2.1.4

Synoptic link

You learned about globular proteins in Topic 3.7, Types of proteins.

Study tip

Enzymes are often named for the reaction they are involved in, or the substrate that they act upon. Many enzyme names end in 'ase', but not all. Some enzyme names indicate both substrate and function. For example:

- Phosphorylases catalyse the addition of a phosphate groups to molecules.

- Lactate dehydrogenase catalyses the transfer of a hydrogen ion to and from lactate.

- Pepsins catalyse the breakdown of proteins (peptides) in the stomach.

- ATPase uses ATP as its substrate.

Why are enzymes important?

Most of the processes necessary to life involve chemical reactions, and these reactions need to happen very fast. In the laboratory or in industry this would demand very high temperatures and pressures. These extreme conditions are not possible in living cells – they would damage the cell components. Instead, the reactions are catalysed by **enzymes**.

Enzymes are biological catalysts. They are globular proteins that interact with **substrate** molecules causing them to react at much faster rates without the need for harsh environmental conditions. Without enzymes many of the processes necessary to life would not be possible.

The role of enzymes in reactions

Living organisms need to be built and maintained. This involves the synthesis of large polymer-based components. For example, cellulose forms the walls of plants cells and long protein molecules form the contractile filaments of muscles in animals. The different cell components are synthesised and assembled into cells, which then form tissues, organs, and eventually the whole organism. The chemical reactions required for growth are **anabolic** (building up) reactions and they are all catalysed by enzymes.

Energy is constantly required for the majority of living processes, including growth. Energy is released from large organic molecules, like glucose, in metabolic pathways consisting of many **catabolic** (breaking down) reactions. Catabolic reactions are also catalysed by enzymes.

These large organic molecules are obtained from the digestion of food, made up of even larger organic molecules, like starch. Digestion is also catalysed by a range of enzymes.

Reactions rarely happen in isolation but as part of multi-step pathways. Metabolism is the sum of all of the different reactions and reaction pathways happening in a cell or an organism, and it can only happen as a result of the control and order imposed by enzymes.

Just like reactions in a laboratory, the speed at which different cellular reactions proceed varies considerably and is usually dependent on environmental conditions. The temperature, pressure, and pH may all have an effect on the rate of a chemical reaction. Enzymes can only increase the rates of reaction up to a certain point called the V_{max} (maximum initial velocity or rate of the enzyme-catalysed reaction).

Mechanism of enzyme action

Molecules in a solution move and collide randomly. For a reaction to happen, molecules need to collide in the right orientation. When high temperatures and pressures are applied the speed of the molecules will increase, therefore so will the number of successful collisions and the overall rate of reaction.

Many different enzymes are produced by living organisms, as each enzyme catalyses one biochemical reaction, of which there are thousands in any given cell. This is termed the *specificity* of the enzyme.

Energy needs to be supplied for most reactions to start. This is called the **activation energy**. Sometimes, the amount of energy needed is so large it prevents the reaction from happening under normal conditions. Enzymes help the molecules collide successfully, and therefore reduce the activation energy required. There are two hypotheses for how enzymes do this.

Lock and key hypothesis

An area within the tertiary structure of the enzyme has a shape that is complementary to the shape of a specific substrate molecule. This area is called the **active site**.

In the same way that only the right key will fit into a lock, only a specific substrate will 'fit' the active site of an enzyme. This is the lock and key hypothesis.

When the substrate is bound to the active site an **enzyme-substrate complex** is formed. The substrate or substrates then react and the product or products are formed in an **enzyme-product complex**. The product or products are then released, leaving the enzyme unchanged and able to take part in subsequent reactions.

The substrate is held in such a way by the enzyme that the right atom-groups are close enough to react. The R-groups within the active site of the enzyme will also interact with the substrate, forming temporary bonds. These put strain on the bonds within the substrate, which also helps the reaction along.

Induced-fit hypothesis

More recently, evidence from research into enzyme action suggests the active site of the enzyme actually changes shape slightly as the substrate enters. This is called the **induced-fit hypothesis** and is a modified version of the lock and key hypothesis. The initial interaction between the enzyme and substrate is relatively weak, but these weak interactions rapidly induce changes in the enzyme's tertiary structure that strengthen binding, putting strain on the substrate molecule. This can weaken a particular bond or bonds in the substrate, therefore lowering the activation energy for the reaction.

Intracellular enzymes

Enzymes have an essential role in both the structure and the function of cells and whole organisms. The synthesis of polymers from monomers, for example making polysaccharides from glucose, requires enzymes. Enzymes that act within cells are called intracellular enzymes.

Hydrogen peroxide is a toxic product of many metabolic pathways. The enzyme *catalase* ensures hydrogen peroxide is broken down to oxygen

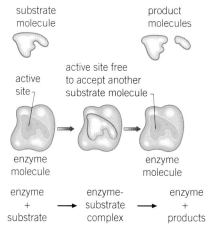

▲ Figure 1 *The substrate fits into the active site of the enzyme forming the enzyme-substrate complex where the reaction takes place and the products are released*

Study tip

The substrate should be described as having a complementary shape to its active site, not the same shape.

and water quickly, therefore preventing its accumulation. It is found in both plant and animal tissues.

Extracellular enzymes

All of the reactions happening within cells need substrates (raw materials) to make products needed by the organism. These raw materials need to be constantly supplied to cells to keep up with the demand. Nutrients (components necessary for survival and growth) present in the diet or environment of the organism supply these materials.

Nutrients are often in the form of polymers such as proteins and polysaccharides. These large molecules cannot enter cells directly through the cell-surface membrane. They need to be broken down into smaller components first.

Enzymes are released from cells to break down these large nutrient molecules into smaller molecules in the process of digestion. These enzymes are called extracellular enzymes. They work outside the cell that made them. In some organisms, for example fungi, they work outside the body.

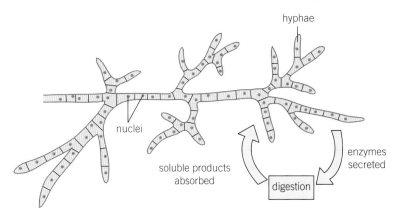

▲ **Figure 2** *Cells in fungal hyphae synthesise enzymes that are secreted outside the cell. The enzymes breakdown organic matter, so it can be absorbed and used for growth*

Both single-celled and multicellular organisms rely on extracellular enzymes to make use of polymers for nutrition.

Single-celled organisms, such as bacteria and yeast, release enzymes into their immediate environment. The enzymes break down larger molecules, such as proteins, and the smaller molecules produced, such as amino acids and glucose, are then absorbed by the cells.

Many multicellular organisms eat food to gain nutrients. Although the nutrients are taken into the digestive system the large molecules still have to be digested so smaller molecules can be absorbed into the bloodstream. From there they are transported around the body to be used as substrates in cellular reactions. Examples of extracellular enzymes involved in digestion in humans are amylase and trypsin.

Digestion of starch

The digestion of starch begins in the mouth and continues in the small intestine. Starch is digested in two steps, involving two different enzymes. Different enzymes are needed because each enzyme only catalyses one specific reaction.

1   Starch polymers are partially broken down into maltose, which is a disaccharide. The enzyme involved in this stage is called *amylase*. Amylase is produced by the salivary glands and the pancreas. It is released in saliva into the mouth, and in pancreatic juice into the small intestine.

2 Maltose is then broken down into glucose, which is a monosaccharide. The enzyme involved in this stage is called *maltase*. Maltase is present in the small intestine.

Glucose is small enough to be absorbed by the cells lining the digestive system and subsequently absorbed into the bloodstream.

Digestion of proteins

Trypsin is a **protease**, a type of enzyme that catalyses the digestion of proteins into smaller peptides, which can then be broken down further into amino acids by other proteases. Trypsin is produced in the pancreas and released with the pancreatic juice into the small intestine, where it acts on proteins. The amino acids that are produced by the action of proteases are absorbed by the cells lining the digestive system and then absorbed into the bloodstream.

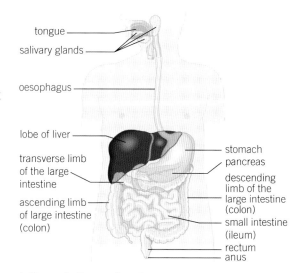

▲ Figure 3 *Human digestive system*

Summary questions

1 a State the type of biological molecule used to form enzymes. (*1 mark*)
   b Name the monomers that form this biological molecule. (*1 mark*)
   c Describe how the structure(s) of this biological molecule determines enzyme activity. (*4 marks*)

2 Explain how catabolism and anabolism are related to metabolism. (*3 marks*)

3 There are two theories explaining enzyme substrate interaction. The lock and key model and induced fit model of enzyme action.
   a Explain what is meant by the term model in the sentence above. (*2 marks*)
   b Explain how the following terms are relevant to each of the models. *complementary, flexibility, R group interactions, bond strain* (*4 marks*)

4 The transition state model adds more detail to the induced fit model, in the same way that the induced fit model was more detailed than the original lock and key model.

[Graph: energy (y-axis) vs time (x-axis), showing "reaction without enzyme" (black curve), "activation energy without enzyme", "activation energy with enzyme", "reaction catalysed by enzyme"]

All reactions go through a transition state as the different chemical components interact with each other. The formation of this transition sate determines the activation energy of the reaction.

   a Explain the meaning of the term activation energy. (*2 marks*)
   b Explain the energy changes occurring resulting in the transition state model (the red line). (*4 marks*)
   c Discuss why the models of enzyme action have changed over time. (*4 marks*)

Synoptic link

You learnt about the structure of proteins in Topic 3.6, Structure of proteins.

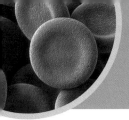

4.2 Factors affecting enzyme activity

Specification reference: 2.1.4

Learning outcomes

Demonstrate knowledge, understanding, and application of:

→ the effects of pH, temperature, enzyme concentration and substrate concentration on enzyme activity

→ practical investigations into factors effecting enzyme activity.

For enzymes to catalyse a reaction, they must come into contact with the substrate, and the enzyme must be the right shape (complementary) for the substrate. Enzymes are complex proteins and their structure can be affected by factors such as temperature and pH. These can cause changes in the shape of their active site. Enzymes are more likely to come into contact with the substrate if temperature and substrate concentration are increased.

Factors affecting enzyme action can be investigated by measuring the rate of the reactions they catalyse.

Temperature

Increasing the temperature of a reaction environment increases the kinetic energy of the particles. As temperature increases, the particles move faster and collide more frequently. In an enzyme-controlled reaction an increase in temperature will result in more frequent successful collisions between substrate and enzyme. This leads to an increase in the rate of reaction.

The **temperature coefficient**, Q_{10}, of a reaction (or process) is a measure of how much the rate of a reaction increases with a 10 °C rise in temperature. For enzyme-controlled reactions this is usually taken as two, which means that the rate of reaction doubles with a 10 °C temperature increase.

Denaturation from temperature

As enzymes are proteins their structure is affected by temperature. At higher temperatures the bonds holding the protein together vibrate more. As the temperature increases the vibrations increase until the bonds strain and then break. The breaking of these bonds results in a change in the precise tertiary structure of the protein. The enzyme has changed shape and is said to have been **denatured**.

When an enzyme is denatured the active site changes shape and is no longer complementary to the substrate. The substrate can no longer fit into the active sites and the enzyme will no longer function as a catalyst.

Optimum temperature

The optimum temperature is the temperature at which the enzyme has the highest rate of activity. The optimum temperature of enzymes can vary significantly. Many enzymes in the human body have optimum temperatures of around 40 °C, meanwhile thermophilic bacteria (found in hot springs) have enzymes with optimum temperatures of 70 °C, and psychrophilic organisms (that live in areas that are cold such as the antarctic and arctic regions) have enzymes with optimum temperatures below 5 °C.

Once the enzymes have denatured above the optimum temperature, the decrease in rate of reaction is rapid. There only needs to be a slight

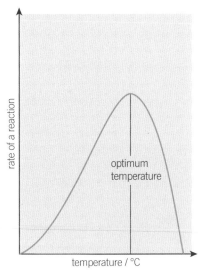

▲ Figure 1 *The rate of reaction increases up to the optimum temperature and then decreases rapidly*

change in shape of an active site for it to no longer be complementary to its substrate. This happens to all of the enzyme molecules at about the same temperature so the loss of activity is relatively abrupt. At this point in an enzyme-controlled reaction the *temperature coefficient*, Q_{10} does not apply any more as the enzymes have denatured.

The decrease in the rate of reaction below the optimum temperature is less rapid. This is because the enzymes have not denatured, they are just less active.

Temperature extremes

The majority of living organisms have evolved to cope with living within a certain temperature range. Some organisms can also cope with extremes.

Examples of extremely cold environments are deep oceans, high altitudes, and polar regions. The enzymes controlling the metabolic activities of organisms living in these environments need to be adapted to the cold. Enzymes adapted to the cold tend to have more flexible structures, particularly at the active site, making them less stable than enzymes that work at higher temperatures. Smaller temperature changes will denature them.

Thermophiles are organisms adapted to living in very hot environments such as hot springs and deep sea hydrothermal vents. The enzymes present in these organisms are more stable than other enzymes due to the increased number of bonds, particularly hydrogen bonds and sulfur bridges, in their tertiary structures. The shapes of these enzymes, and their active sites, are more resistant to change as the temperature rises.

Enzymes in action

Siamese cats provide visual evidence of the effect of temperature on enzyme activity. Tyrosinase is an enzyme responsible for catalysing the production of melanin, the pigment responsible for dark coloured fur. Due to a mutation, Siamese cats produce a form of the enzyme tyrosinase that is denatured and therefore inactive at normal body temperature meaning that their fur is primarily white or cream coloured. The extremities of these cats – the tails, ears, and limbs – are at a slightly lower temperature, too low to denature mutant tyrosinase. This leads to the distinctive point coloration of these cats as melanin is produced in these areas.

1   Suggest why Siamese kittens are born completely white.

pH

Proteins, and so enzymes, are also affected by changes in pH. Hydrogen bonds and ionic bonds between amino acid R-groups hold proteins in their precise three-dimensional shape. These bonds result from

Study tip

Remember that enzymes are denatured not 'killed'.

A denatured protein is still a protein – it just has a different shape and can no longer function as it normally would.

interactions between the polar and charged R-groups present on the amino acids forming the primary structure. A change in pH refers to a change in hydrogen ion concentration. More hydrogen ions are present in low pH (acid) environments and fewer hydrogen ions are present in high pH (alkaline) environments.

The active site will only be in the right shape at a certain hydrogen ion concentration. This is the **optimum pH** for any particular enzyme. When the pH changes from the optimum – becoming more acidic or alkaline – the structure of the enzyme, and therefore the active site, is altered. However, if the pH returns to the optimum then the protein will resume its normal shape and catalyse the reaction again. This is called renaturation.

When the pH changes more significantly (beyond a certain pH) the structure of the enzyme is irreversibly altered and the active site will no longer be complementary to the substrate. The enzyme is now said to be denatured and substrates can no longer bind to the active sites. This will reduce the rate of reaction.

Hydrogen ions interact with polar and charged R-groups. Changing the concentration of hydrogen ions therefore changes the degree of this interaction. The interaction of R-groups with hydrogen ions also affects the interaction of R-groups with each other.

The more hydrogen ions present (low pH), the less the R-groups are able to interact with each other. This leads to bonds breaking and the shape of the enzyme changing. The reverse is true when fewer hydrogen ions (high pH) are present. This means the shape of an enzyme will change as the pH changes and therefore it will only function within a narrow pH range. Table 1 shows the pH conditions under which the various enzymes of the human digestive system function. These are the optimum pHs for these enzymes.

> **Study tip**
>
> Different enzymes in different organisms also have a whole range of optimum pHs from very acid to very alkaline.

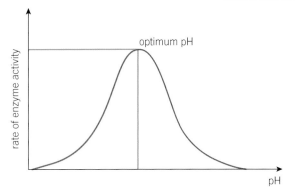

▲ Figure 2 *Effect of pH on the rate of an enzyme-controlled reaction. The rate of reaction decreases as the pH moves away from the optimum, becoming higher or lower. Note how this graph is symmetrical, unlike the graph showing rate of reaction against temperature. It shows reversible reduction in enzyme activity as the pH moves away from the optimum*

▼ Table 1 *The action of different enzymes at various points in the digestive system*

	Site of action	pH	Enzymes	Function
Saliva	mouth/throat	neutral (pH 7–8)	amylase	starch → maltose
Gastric juice	stomach	acidic (pH 1–2)	pepsin	proteins → polypeptides
Pancreatic juice	small intestine/ duodenum	slightly alkaline (pH 8)	trypsin	proteins → polypeptides
			lipase	triglycerides → glycerol + fatty acids
			amylase	starch → maltose
			maltase	maltose → glucose

Substrate and enzyme concentration

When the concentration of substrate is increased the number of substrate molecules, atoms, or ions in a particular area or volume increases. The increased number of substrate particles leads to a higher collision rate with the active sites of enzymes and the formation of more enzyme–substrate complexes. The rate of reaction therefore increases.

This is also true when the concentration of the enzyme increases. This will increase the number of available active sites in a particular area or volume, leading to the formation of enzyme-substrate complexes at a faster rate.

The rate of reaction increases up to its maximum (V_{max}). At this point all of the active sites are occupied by substrate particles and no more enzyme-substrate complexes can be formed until products are released from active sites. The only way to increase the rate of reaction would be to add more enzyme or increase the temperature.

If the concentration of the enzyme is increased more active sites are available so the reaction rate can rise towards a higher V_{max}. The concentration of substrate becomes the limiting factor again and increasing this will once again allow the reaction rate to rise until the new V_{max} is reached.

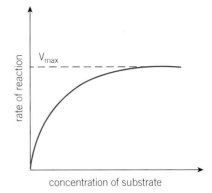

▲ **Figure 3** *Effect of substrate concentration on the rate of an enzyme-controlled reaction. The rate of reaction increases as the substrate concentration increases up to a maximum rate, V_{max}*

 ## Investigations into the effects of different factors on enzyme activity

Investigating the effects of different factors on enzyme activity provides an insight into how enzymes work.

Catalase is an enzyme present in plant tissue and animal tissue, making it a good choice for use in investigations because it is readily available. Catalase catalyses the breakdown of hydrogen peroxide into water and oxygen. The volume of oxygen gas collected in a set length of time can be used as a measure of the rate of reaction.

In the second experiment the liver was boiled for five minutes before being placed in the hydrogen peroxide solution.

The graph below shows the results from the experiment.

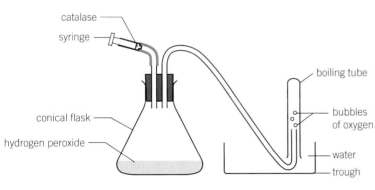

▲ **Figure 4** *Apparatus used to collect oxygen released when catalase reacts with hydrogen peroxide*

A student conducted a series of experiments to determine the effect of temperature on enzymes.

In the first experiment, using apparatus similar to that shown in the diagram above, liver tissue was put into hydrogen peroxide solution and the volume of oxygen released every five seconds was measured.

1 Explain why the student used liver tissue.
2 Show the chemical reaction that occurs resulting in the release of oxygen.
3 Explain the shape of the graph.
4 Describe and explain what has happened in the second experiment.

The student then investigated the effect of substrate concentration on enzyme activity.

Suggest a hypothesis for this experiment.

The results are shown below.

Percentage of Hydrogen peroxide	Volume of gas collected (cm^3) after … (s)											
	5	10	15	20	25	30	35	40	45	50	55	60
Test 1												
100	43	62	68	74	76	77	77	77	77	77	77	77
30	12	21	24	26	27	27	27	27	27	27	27	27
10	10	14	16	18	18	18	18	18	18	18	18	18
Test 2												
100	42	62	70	65	76	77	77	77	77	77	77	77
30	13	21	25	27	28	28	28	28	28	28	28	28
10	8	15	17	19	19	19	19	19	19	19	19	19
Test 3												
100	43	63	68	74	75	76	77	77	77	77	77	77
30	12	20	24	26	27	27	27	27	27	27	27	27
10	8	14	17	18	18	18	18	18	18	18	18	18
Mean of three tests												
100	42.7	62.3	68.7	71.0	75.7	76.7	77.0	77.0	77.0	77.0	77.0	77.0
30	12.3	20.7	24.3	26.3	27.3	27.3	27.3	27.3	27.3	27.3	27.3	27.3
10	8.7	14.3	16.7	18.3	18.3	18.3	18.3	18.3	18.3	18.3	18.3	18.3

5 Describe how the student would have carried out the experiment at different temperatures.

6 a Name the independent variable and the dependent variable in this experiment.

   b List the controlled variables.

7 Explain the importance of keeping these variables (controlled) constant.

8 a Explain the term anomaly.

   b Identify any anomalies in the results above.

9 Explain why the student did the experiment three times and then calculated the means of the results.

10 Plot a graph of the results.

11 Explain the shape of the graph you have produced.

12 Write an evaluation for this experiment.

 Serial dilutions

A serial dilution is a repeated, stepwise dilution of a stock solution of known concentration. A serial dilution is usually done by factors of ten, to produce a range of concentrations. Serial dilutions are useful even if the concentration of the initial solution is unknown as they give us relative concentrations. Serial dilutions are used in many different ways – for example to investigate the effect of changing the concentration of an enzyme or a substrate in a reaction, and in determining the numbers of microorganisms in a culture.

Figure 5 shows how a serial dilution might be set up. Adding 1 ml of stock solution to 9 ml of distilled water gives 10 ml of dilute solution in which there is 1 ml stock/10 ml hence a 1/10 or 10 fold dilution. This step is repeated a number of times to give a serial dilution.

Catalase is an enzyme that catalyses the breakdown of hydrogen peroxide. To investigate the effect of different concentrations of catalase on the rate of breakdown of hydrogen peroxide, catalase-rich tissues such as liver or potato can be ground down to make a solution. The solution will contain catalase released from cells. Serial dilution of this solution will produce a range of solutions with different relative concentrations of catalase. The effect of the different concentrations on the rate of reaction can be investigated by adding equal volumes of a given concentration of hydrogen peroxide to each solution.

1   You are provided with a stock solution of enzyme X with a concentration of 20 mmol/dm^{-3}. Explain how you would prepare a range of five solutions with different concentrations of enzyme X using the stock solution. Show your working and state the concentrations produced.

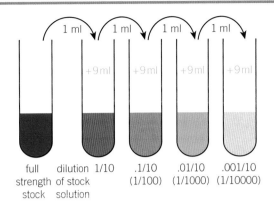

| full strength stock | dilution of stock solution | 1/10 | .1/10 (1/100) | .01/10 (1/1000) | .001/10 (1/10000) |

▲ **Figure 5** *This figure shows how a serial dilution might be set up. Adding 1 ml of stock solution to 9 ml of distilled water gives 10 ml of dilute solution in which there is 1 ml stock/10 ml hence a 1/10 or 10 fold dilution. This step is repeated a number of times to give a serial dilution.*

Summary questions

1   Explain the term 'denatured' with reference to enzymes.   *(3 marks)*

2   The graph below shows the optimum pHs of three enzymes.

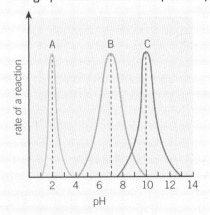

Pepsin is an enzyme that, along with stomach acid, digests proteins in the stomach.

State which of the curves, A, B or C, is likely to represent the activity of the enzyme pepsin over a range of pHs. Explain the reasons for your choice.   *(2 marks)*

3   Bacteria that colonise hydrothermal vents, where temperatures are very high, have enzymes with very high optimum temperatures.

Suggest why these bacteria are unlikely to cause infections in humans.   *(3 marks)*

4   Enzymes with very low optimum temperatures tend to have quite flexible structures. Using your knowledge of collision theory, explain why this flexibility is necessary.   *(6 marks)*

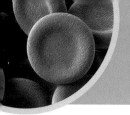

4.3 Enzyme inhibitors

Specification reference: 2.1.4

As you have seen, it is important that cellular conditions such as pH and temperature are kept within narrow limits so that enzyme activity is not delayed. This ensures that reactions can happen at a rate fast enough to sustain living processes, for example, respiration.

It is also important that reactions do not happen too fast as this could lead to the build-up of excess products. Living processes rarely involve just one reaction but are complex, multi-step reaction pathways. These pathways need to be closely regulated to meet the needs of living organisms without wasting resources.

Control of metabolic activity within cells

The different steps in reaction pathways are controlled by different enzymes. Controlling the activity of enzymes at crucial points in these pathways regulates the rate and quantity of product formation.

Enzymes can be activated with cofactors, (Topic 4.4 Cofactors, coenzymes, and prosthetic groups), or inactivated with **inhibitors**.

Inhibitors are molecules that prevent enzymes from carrying out their normal function of catalysis (or slow them down). There are two types of enzyme inhibition – competitive and non-competitive.

Competitive inhibition

Competitive inhibition works in the following way:

- A molecule or part of a molecule that has a similar shape to the substrate of an enzyme can fit into the active site of the enzyme.
- This blocks the substrate from entering the active site, preventing the enzyme from catalysing the reaction.
- The enzyme cannot carry out its function and is said to be inhibited.
- The non-substrate molecule that binds to the active site is a type of inhibitor. Substrate and inhibitor molecules present in a solution will compete with each other to bind to the active sites of the enzymes catalysing the reaction. This will reduce the number of substrate molecules binding to active sites in a given time and slows down the rate of reaction. For this reason such inhibitors are called **competitive inhibitors** and the degree of inhibition will depend on the relative concentrations of substrate, inhibitor, and enzyme.

competitive inhibitor interferes with active site of enzyme so substrate cannot bind

substrate

enzyme

▲ **Figure 1** *Competitive inhibition. The shape of the inhibitor is similar in shape to the part of the substrate that binds to the active site. This means the inhibitor can temporarily bind to the active site and block the substrate from binding*

Most competitive inhibitors only bind temporarily to the active site of the enzyme, so their effect is reversible. However there are some exceptions such as aspirin.

Effect on rates of reaction

A competitive inhibitor reduces the rate of reaction for a given concentration of substrate, but it does not change the V_{max} of the enzyme it inhibits. If the substrate concentration is increased enough there will be so much more substrate than inhibitor that the original V_{max} can still be reached.

Examples of competitive inhibition

Statins are competitive inhibitors of an enzyme used in the synthesis of cholesterol. Statins are regularly prescribed to help people reduce blood cholesterol concentration. High blood cholesterol levels can result in heart disease.

Aspirin irreversibly inhibits the active site of COX enzymes, preventing the synthesis of prostaglandins and thromboxane, the chemicals responsible for producing pain and fever.

Non-competitive inhibition

Non-competitive inhibition works in the following way:

- The inhibitor binds to the enzyme at a location other than the active site. This alternative site is called an allosteric site.

- The binding of the inhibitor causes the tertiary structure of the enzyme to change, meaning the active site changes shape.

- This results in the active site no longer having a complementary shape to the substrate so it is unable to bind to the enzyme.

- The enzyme cannot carry out its function and it is said to be inhibited.

As the inhibitor does not compete with the substrate for the active site it is called a **non-competitive inhibitor**.

Effect on rates of reaction

Increasing the concentration of enzyme or substrate will not overcome the effect of a non-competitive inhibitor. Increasing the concentration of inhibitor, however, will decrease the rate of reaction further as more active sites become unavailable.

Examples of irreversible non-competitive inhibitors

As mentioned, the binding of the inhibitor may be reversible or non-reversible. Irreversible inhibitors cannot be removed from the part of the enzyme they are attached to. They are often very toxic, but not always. Organophosphates used as insecticides and herbicides irreversibly inhibit the enzyme acetyl cholinesterase, an enzyme necessary for nerve impulse transmission. This can lead to muscle cramps, paralysis, and even death if accidentally ingested.

Proton pump inhibitors (PPIs) are used to treat long-term indigestion. They irreversibly block an enzyme system responsible for secreting hydrogen ions into the stomach. This makes PPIs very effective in reducing the production of excess acid which, if left untreated, can lead to formation of stomach ulcers.

End-product inhibition

End-product inhibition is the term used for enzyme inhibition that occurs when the product of a reaction acts as an inhibitor to the enzyme that produces it. This serves as a *negative-feedback*

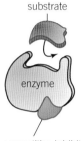

▲ **Figure 2** *Non-competitive inhibition. The inhibitor binds to an allosteric site on the enzyme, changing the shape of the active site. This means the substrate can no longer bind the active site*

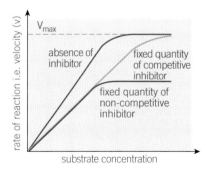

▲ **Figure 3** *Comparison of competitive and non-competitive inhibition of an enzyme-controlled reaction. With a competitive inhibitor, as the substrate concentration is increased the effect of the inhibitor is almost overcome. With a non-competitive inhibitor, as the substrate concentration is increased the effect of the inhibitor is not overcome*

Synoptic link

You learnt about ATP in Topic 3.11, ATP and you will learn more about cellular respiration in Chapter 18, Respiration.

control mechanism for the reaction. Excess products are not made and resources are not wasted. It is an example of non-competitive reversible inhibition.

Respiration is a metabolic pathway resulting in the production of ATP. Glucose is broken down in a number of steps. The first step involves the addition of two phosphate groups to the glucose molecule. The addition of the second phosphate group, which results in the initial breakdown of the glucose molecule, is catalysed by the enzyme phosphofructokinase (PFK). This enzyme is competitively inhibited by ATP. ATP therefore regulates its own production.

- When the levels of ATP are high, more ATP binds to the allosteric site on PFK, preventing the addition of the second phosphate group to glucose. Glucose is not broken down and ATP is not produced at the same rate.

- As ATP is used up, less binds to PFK and the enzyme is able to catalyse the addition of a second phosphate group to glucose. Respiration resumes, leading to the production of more ATP.

Summary questions

1   Explain why a non-competitive inhibitor does not need to have a similar shape to a substrate molecule.   *(3 marks)*

2   Explain why increasing the concentration of substrate will never produce the V_{max} of a reaction after the addition of a non-competitive inhibitor.   *(2 marks)*

3   End-product inhibition is likely to be competitive rather than non-competitive. Suggest reasons for this, and give an example of end-product inhibition.   *(4 marks)*

4   Ethylene glycol present in antifreeze is poisonous when ingested. Ethylene glycol is oxidised using the same enzymes used to oxidise ethanol (alcohol). The products made during the breakdown of ethylene glycol, rather than ethylene glycol itself, are responsible for the toxic effects. Ethylene glycol is able to leave the body unchanged in urine.

Suggest why ethanol is often used in emergency departments as an antidote to antifreeze poisoning.   *(6 marks)*

4.4 Cofactors, coenzymes, and prosthetic groups

Specification reference: 2.1.4

The difference between cofactors and coenzymes

Some enzymes need a non-protein 'helper' component in order to carry out their function as biological catalysts. They may transfer atoms or groups from one reaction to another in a multi-step pathway or they may actually form part of the active site of an enzyme. These components are called **cofactors**, or if the cofactor is an organic molecule it is called a coenzyme.

Inorganic cofactors are obtained via the diet as minerals, including iron, calcium, chloride, and zinc ions. For example, the enzyme amylase, which catalyses the breakdown of starch (Topic 4.1, Enzyme action), contains a chloride ion that is necessary for the formation of a correctly shaped active site.

Many coenzymes are derived from vitamins, a class of organic molecule found in the diet. For example, vitamin B3 is used to synthesise NAD (nicotinamide adenine dinucleotide), a coenzyme responsible for the transfer of hydrogen atoms between molecules involved in respiration. NADP, which plays a similar role in photosynthesis, is also derived from vitamin B3.

Another example is vitamin B5, which is used to make coenzyme A. Coenzyme A is essential in the breakdown of fatty acids and carbohydrates in respiration.

Prosthetic groups

You have previously met prosthetic groups when you learnt about the globular protein, haemoglobin, in which the prosthetic group is an iron (Fe) ion. Prosthetic groups are cofactors – they are required by certain enzymes to carry out their catalytic function. While some cofactors are loosely or temporarily bound to the enzyme protein in order to activate them, prosthetic groups are tightly bound and form a permanent feature of the protein. For example zinc ions (Zn^{2+}) form an important part of the structure of carbonic anhydrase, an enzyme necessary for the metabolism of carbon dioxide.

Precursor activation

Many enzymes are produced in an inactive form, known as inactive precursor enzymes, particularly enzymes that can cause damage within the cells producing them or to tissues where they are released, or enzymes whose action needs to be controlled and only activated under certain conditions.

Precursor enzymes often need to undergo a change in shape (tertiary structure), particularly to the active site, to be activated. This can be achieved by the addition of a cofactor. Before the cofactor is added,

Learning outcomes

Demonstrate knowledge, understanding, and application of:

→ the need for coenzymes, cofactors, and prosthetic groups in some enzyme-controlled reactions.

→ the role of inactive precursors

▲ Figure 1 *A molecular model of the enzyme alpha amylase, which catalyses the breakdown of starch*

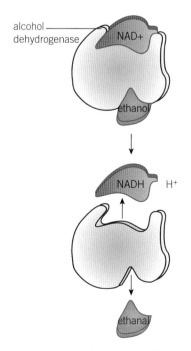

alcohol dehydrogenase — NAD+

ethanol

NADH   H+

ethanal

ethanal, H+, and NADH are released from the active site

▲ Figure 2 *Alcohol dehydrogenase needs NAD+ to accept hydrogen produced when ethanal is formed from ethanol*

Synoptic link

You will learn about NAD and coenzyme A in cellular respiration in more depth in Chapter 18, Respiration.

Synoptic link

You learnt about prosthetic groups in proteins in Topic 3.6, Structure of proteins.

the precursor protein is called an apoenzyme. When the cofactor is added and the enzyme is activated, it is called a holoenzyme.

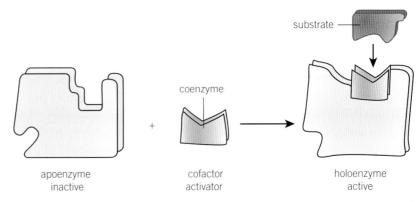

▲ Figure 3 *Coenzymes, or cofactors, are often needed to produce the specific shape of the active site*

Sometimes the change in tertiary structure is brought about by the action of another enzyme, such as a protease, which cleaves certain bonds in the molecule. In some cases a change in conditions, such as pH or temperature, results in a change in tertiary structure and activates a precursor enzyme. These types of precursor enzymes are called zymogens or proenzymes.

When inactive pepsinogen is released into the stomach to digest proteins, the acid pH brings about the transformation into the active enzyme pepsin. This adaptation protects the body tissues against the digestive action of pepsin.

Synoptic link

You will learn more about blood clotting in Topic 12.5, Non-specific animal defences against pathogens.

Enzyme activation and the blood-clotting mechanism

Blood clotting, or coagulation, is an important biological response to tissue damage. The blood-clotting process only begins when platelets aggregate at the site of tissue damage. The aggregated platelets release clotting factors, including factor X.

Factor X is an important component in the blood-clotting mechanism. It is an enzyme that is dependent on the cofactor vitamin K for activation. Activated factor X catalyses the conversion of prothrombin into the enzyme thrombin by cleaving certain bonds in the molecule, thus altering its tertiary structure. Thrombin is a protease and catalyses the conversion of soluble fibrinogen to insoluble fibrin fibres. Fibrin molecules, together with platelets, form a blood clot.

This series of successive enzyme activations in blood clotting is called the coagulation cascade.

1   Explain the importance of enzyme activation in controlling blood clotting.

Summary questions

1   Describe two ways in which cofactors are necessary for the catalytic role of some enzymes.   (*2 marks*)

2   Explain, using an appropriate example of each, how prosthetic groups are different from coenzymes.   (*4 marks*)

3   Using blood clotting as your example, explain the different ways in which enzymes can be activated.   (*4 marks*)

Practice questions

1 Compounds containing some metal ions such as lead, mercury, copper or silver are poisonous. This is because ions of these metals are non-competitive inhibitors of several enzymes.

Other metal ions are required for enzymes to function. Magnesium ions, for example, are required by DNA polymerase.

Which of the following statements correctly describes the nature and function of magnesium ions?

Statement 1: minerals and coenzymes

Statement 2: vitamins and cofactors

Statement 3: minerals and cofactors

Statement 4: vitamins and coenzymes

(*1 mark*)

2 a Describe what is happening at the points labelled on the graph below. You should use the terms active site, substrate and enzyme substrate complex in your answer. (*5 marks*)

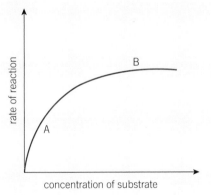

Amylase catalyses the hydrolysis of starch to maltose in the mouth. Amylase stops working as the food passes into the stomach. Maltase then catalyses the hydrolysis of maltose into glucose in the small intestine.

b (i) State which group of biological molecules amylase and maltase belong to. (*1 mark*)

(ii) Describe the meaning of the term hydrolysis. (*2 marks*)

(iii) Explain why maltase, not amylase, hydrolyses maltose completing the digestion of starch. (*3 marks*)

(iv) Explain why amylase stops working in the stomach. (*4 marks*)

The acid stomach contents are neutralised by alkaline pancreatic juice in the small intestine.

c Suggest why this is necessary to ensure digestion of maltose. (*3 marks*)

d Explain what is meant by the term Vmax with respect to enzymes. (*3 marks*)

3 a State the **type** of monomers that make up proteins. (*1 mark*)

b Malonate inhibits cellular respiration. Hans Krebs added malonate to minced pigeon muscle tissue when trying to work out the different steps in Krebs cycle, a series of reactions that occur in respiration.

$$COO^-$$
$$|$$
$$CH_2 \quad \xrightarrow[\text{dehydrogenase}]{\text{succinate}} \quad {}^-OOC-C-H$$
$$| \qquad\qquad\qquad\qquad\qquad \|$$
$$CH_2 \qquad\qquad\qquad\qquad\quad H-C-COO^-$$
$$|$$
$$COO^-$$
succinate $\qquad\qquad\qquad\qquad\qquad$ fumarate

$$COO^-$$
$$|$$
$$CH_2 \quad \xrightarrow[\text{dehydrogenase}]{\text{succinate}} \quad \text{no reaction}$$
$$|$$
$$COO^-$$
malonate

In step 6 of Krebs cycle, succinate dehydrogenase catalyzes the oxidation of succinate to fumarate.

Explain why malonate inhibits this reaction. (*4 marks*)

c Identify which curve on the graph below demonstrates competitive inhibition giving the reasons for your choice.

(*1 mark*)

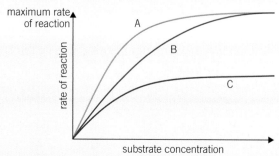

d (i) Explain why end product inhibitors have to be both reversible and competitive (*3 marks*)

(ii) Explain why end product enzyme inhibition is essential for the control of cellular activity. (*3 marks*)

4 Amylase is an enzyme that hydrolyses amylose to maltose. Maltose, like glucose, is a reducing sugar.

A student investigated the action of amylase on amylose. She mixed amylase with amylose and placed the mixture in a water bath.

Describe how she could measure the change in concentration of maltose (reducing sugar) as the reaction proceeds.

In you answer, you should ensure that the steps in the procedure are sequenced correctly (*7 marks*)

OCR F212 2011

5 Figure 4 shows the results that the student obtained from a practical procedure in which the rate of formation of maltose was measured in the presence and absence of chloride ions.

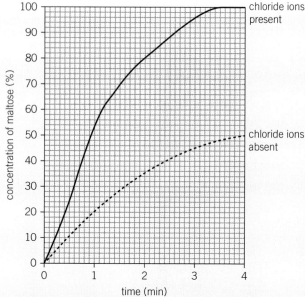

a Describe the effect of chloride ions on the rate of reaction. (*2 marks*)

OCR F212 2011

b State **three** variables that need to be controlled in this practical procedure in order to produce valid results.

(*3 marks*)

OCR F212 2011

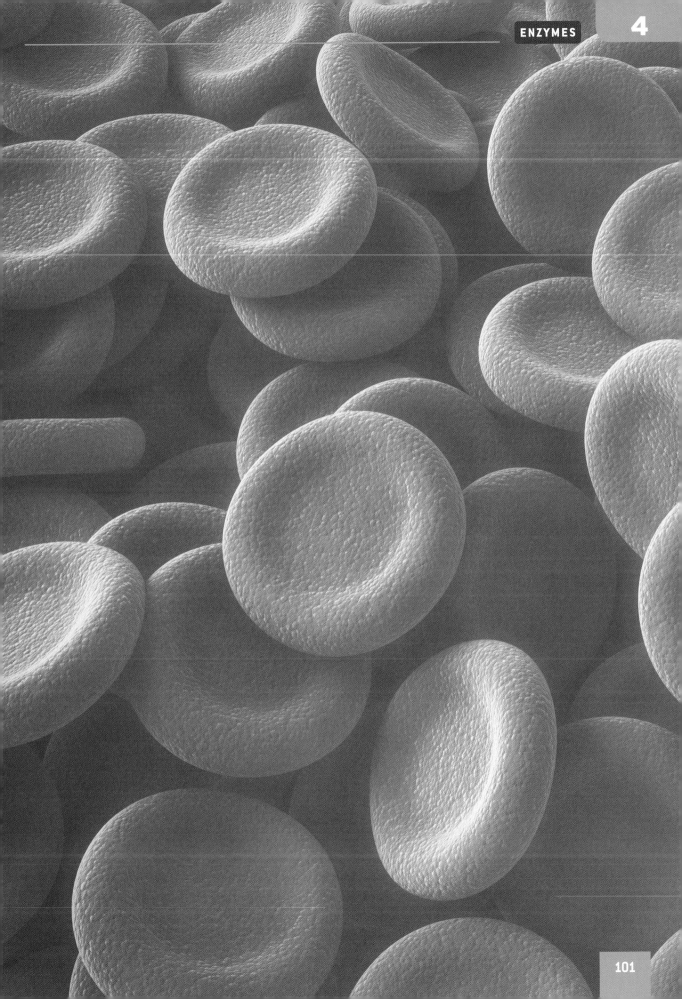

PLASMA MEMBRANES
5.1 The structure and function of membranes
Specification reference: 2.1.5

You learnt about the structure of cells in Chapter 2, Basic components of living systems. Membranes are the structures that separate the contents of cells from their environment. They also separate the different areas within cells (organelles) from each other and the cytosol. Some organelles are divided further by internal membranes.

The formation of separate membrane-bound areas in a cell is called **compartmentalisation**. Compartmentalisation is vital to a cell as metabolism includes many different and often incompatible reactions. Containing reactions in separate parts of the cell allows the specific conditions required for cellular reactions, such as chemical gradients, to be maintained, and protects vital cell components.

Membrane structure

All the membranes in a cell have the same basic structure. The cell surface membrane which separates the cell from its external environment is known as the **plasma membrane**.

Membranes are formed from a **phospholipid bilayer**. The hydrophilic phosphate heads of the phospholipids form both the inner and outer surface of a membrane, sandwiching the fatty acid tails of the phospholipids to form a hydrophobic core inside the membrane.

Cells normally exist in aqueous environments. The inside of cells and organelles are also usually aqueous environments. Phospholipid bilayers are perfectly suited as membranes because the outer surfaces of the hydrophilic phosphate heads can interact with water.

Synoptic link

You will learn about the importance of chemical gradients in biological processes, including gas exchange in mammals and fish in Topic 7.2, Mammalian gaseous exchange system and Topic 7.4 Ventilation and gas exchange in other organisms). You will learn about the transport of oxygen in the blood in Topic 8.1, Transport systems in multicellular animals and about translocation in plants in Topic 9.4, Translocation.

Synoptic link

In Chapter 13, Neuronal communication you will learn about the role of chemical gradients in neuronal communication and in Chapter 18, Respiration, the role of gradients in respiration.

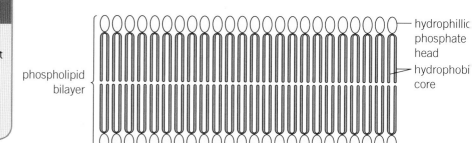

▲ Figure 1 *Phospholipids arranged in a double layer forming a hydrophobic core*

Synoptic link

You learnt about phospholipids and how they interact in aqueous environments to form bilayers in Topic 3.5, Lipids.

Cell membrane theory

Membranes were seen for the first time following the invention of electron microscopy, which allowed images to be taken with higher magnification and resolution. Images taken in the 1950s showed the membrane as two black parallel lines – supporting an earlier theory that membranes were composed of a lipid bilayer.

In 1972 American scientists Singer and Nicolson proposed a model, building upon an earlier lipid-bilayer model, in which proteins occupy various positions in the membrane. The model is known as the **fluid-mosaic model** because the phospholipids are free to move within the layer relative to each other (they are fluid), giving the membrane flexibility, and because the proteins embedded in the bilayer vary in shape, size, and position (in the same way as the tiles of a mosaic). This model forms the basis of our understanding of membranes today.

A closer look at cell membrane components

Plasma membranes contain various proteins and lipids – the type and number of which are particular to each cell type.

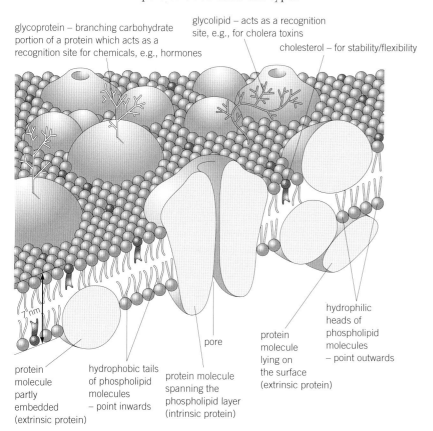

glycoprotein – branching carbohydrate portion of a protein which acts as a recognition site for chemicals, e.g., hormones

glycolipid – acts as a recognition site, e.g., for cholera toxins

cholesterol – for stability/flexibility

hydrophilic heads of phospholipid molecules – point outwards

protein molecule lying on the surface (extrinsic protein)

pore

protein molecule spanning the phospholipid layer (intrinsic protein)

hydrophobic tails of phospholipid molecules – point inwards

protein molecule partly embedded (extrinsic protein)

▲ Figure 2 *The fluid-mosaic model of a plasma membrane showing the many components, including cholesterol, glycolipids and glycoproteins, and membrane proteins*

The components of plasma membranes play an important role in the functions of the membrane and the cell or organelle they are part of.

Membrane proteins

Membrane proteins have important roles in the various functions of membranes. There are two types of proteins in the cell-surface membrane – intrinsic and extrinsic proteins.

Study tip

When discussing transport proteins be specific – use the term 'channel protein' (passive) or 'carrier protein' (active or passive).

Intrinsic proteins

Intrinsic proteins, or integral proteins, are transmembrane proteins that are embedded through both layers of a membrane. They have amino acids with hydrophobic R-groups on their external surfaces, which interact with the hydrophobic core of the membrane, keeping them in place.

Channel and carrier proteins are intrinsic proteins. They are both involved in transport across the membrane.

- **Channel proteins** provide a hydrophilic channel that allows the passive movement (Topic 5.3, Diffusion) of polar molecules and ions down a concentration gradient through membranes. They are held in position by interactions between the hydrophobic core of the membrane and the hydrophobic R-groups on the outside of the proteins.

- **Carrier proteins** have an important role in both passive transport (down a concentration gradient) and active transport (against a concentration gradient) into cells (Topic 5.4, Active transport). This often involves the shape of the protein changing.

Glycoproteins

Glycoproteins are intrinsic proteins. They are embedded in the cell-surface membrane with attached carbohydrate (sugar) chains of varying lengths and shapes. Glycoproteins play a role in cell adhesion (when cells join together to form tight junctions in certain tissues) and as **receptors** for chemical signals.

When the chemical binds to the receptor, it elicits a response from the cell. This may cause a direct response or set off a cascade of events inside the cell. This process is known as cell communication or **cell signalling**. Examples include:

- receptors for neurotransmitters such as acetylcholine at nerve cell synapses. The binding of the neurotransmitters triggers or prevents an impulse in the next neurone

- receptors for peptide hormones, including insulin and glucagon, which affect the uptake and storage of glucose by cells.

Some drugs act by binding to cell receptors. For example, β blockers are used to reduce the response of the heart to stress.

Synoptic link

You will learn about cell communication in the nervous and endocrine (hormone) systems in Chapter 13, Neuronal communication and Chapter 14, Hormonal communication.

Glycolipids

Glycolipids are similar to glycoproteins. They are lipids with attached carbohydrate (sugar) chains. These molecules are called cell markers or antigens and can be recognised by the cells of the immune system as self (of the organism) or non-self (of cells belonging to another organism).

Synoptic link

You will learn about antigens and cell recognition in an immune response in Chapter 12, Communicable diseases.

Extrinsic proteins

Extrinsic proteins or peripheral proteins are present in one side of the bilayer. They normally have hydrophilic R-groups on their outer surfaces and interact with the polar heads of the phospholipids or with intrinsic proteins. They can be present in either layer and some move between layers.

Cholesterol

Cholesterol is a lipid with a hydrophilic end and a hydrophobic end, like a phospholipid. It regulates the fluidity of membranes.

Cholesterol molecules are positioned between phospholipids in a membrane bilayer, with the hydrophilic end interacting with the heads and the hydrophobic end interacting with the tails, pulling them together. In this way cholesterol adds stability to membranes without making them too rigid. The cholesterol molecules prevent the membranes becoming too solid by stopping the phospholipid molecules from grouping too closely and crystallising.

Sites of chemical reactions

Like enzymes, proteins in the membranes forming organelles, or present within organelles, have to be in particular positions for chemical reactions to take place. For example, the electron carriers and the enzyme ATP synthase have to be in the correct positions within the cristae (inner membrane of mitochondrion) for the production of ATP in respiration. The enzymes of photosynthesis are found on the membrane stacks within the chloroplasts.

Summary questions

1 Define the term 'compartmentalisation'. (2 marks)

2 Describe the difference between intrinsic and extrinsic proteins. State two examples of each. (4 marks)

3 Alcohol, caffeine, and nicotine are all lipid-soluble molecules – they have an almost instant and widespread effect on the body. Explain why. (2 marks)

4 Membranes, particularly those present within mitochondria, are often highly folded. Suggest what advantages this folding provides. (6 marks)

5.2 Factors affecting membrane structure

Specification reference: 2.1.5

Membranes control the passage of different substances into and out of cells (and organelles). If membranes lose their structure, they lose control of this and cell processes will be disrupted. A number of factors affect membrane structure including temperature and the presence of solvents.

Temperature

Phospholipids in a cell membrane are constantly moving. When temperature is increased the phospholipids will have more kinetic energy and will move more. This makes a membrane more fluid and it begins to lose its structure. If temperature continues to increase the cell will eventually break down completely.

This loss of structure increases the permeability of the membrane, making it easier for particles to cross it.

Carrier and channel proteins in the membrane will be denatured at higher temperatures. These proteins are involved in transport across the membrane so as they denature, membrane permeability will be affected.

Solvents

Water, a polar solvent, is essential in the formation of the phospholipid bilayer. The non-polar tails of the phospholipids are orientated away from the water, forming a bilayer with a hydrophobic core. The charged phosphate heads interact with water, helping to keep the bilayer intact.

Many organic solvents are less polar than water for example alcohols, or they are non-polar like benzene. Organic solvents will dissolve membranes, disrupting cells. This is why alcohols are used in antiseptic wipes. The alcohols dissolve the membranes of bacteria in a wound, killing them and reducing the risk of infection.

Pure or very strong alcohol solutions are toxic as they destroy cells in the body. Less concentrated solutions of alcohols, such as alcoholic drinks, will not dissolve membranes but still cause damage. The non-polar alcohol molecules can enter the cell membrane and the presence of these molecules between the phospholipids disrupts the membrane.

When the membrane is disrupted it becomes more fluid and more permeable. Some cells need intact cell membranes for specific functions, for example, the transmission of nerve impulses by neurones (nerve cells). When neuronal membranes are disrupted, nerve impulses are no longer transmitted as normal.

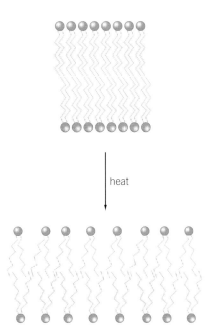

heat

▲ Figure 1 *The increase in kinetic energy of the phospholipids disrupts the structure of the membrane, creating gaps and making it more permeable*

This also happens to neurones in the brain, explaining the changes seen in peoples' behaviour after consuming alcoholic drinks.

 Investigating membrane permeability

Beetroot cells contain betalain, a red pigment that gives them their distinctive colour, because of this they are useful for investigating the effects of temperature and organic solvents on membrane permeability. When beetroot cells membranes are disrupted the red pigment is released and the surrounding solution is coloured. The amount of pigment released into a solution is related to the disruption of the cell membranes.

To investigate the effect of temperature on the permeability of cell membranes a student carried out the following procedure. Five small pieces of beetroot of equal size were cut using a cork borer. The beetroot pieces were thoroughly washed in running water, they were then placed in 100 ml of distilled water in a water bath. The temperature of the water bath was increased in 10 °C intervals. Samples of the water containing the beetroot were taken five minutes after each temperature was reached. The absorbance of each sample was measured using a colorimeter with a blue filter. The experiment was done three times, each time with fresh beetroot pieces and a mean calculated for each temperature. Their results are shown in the graph below.

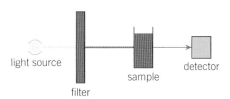

▲ Figure 3 *Light first passes through a filter and then the sample. The intensity of light hitting the detector is recorded*

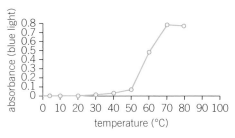

▲ Figure 4

1 Suggest why:
   a The beetroot pieces were washed in running water.
   b Samples of the water containing the beetroot were taken five minutes after each temperature was reached.
   c The experiment was repeated three times.
   d The absorbance of the samples was measure using a colorimeter with a blue filter
2 The absorbance of the solution can be calculated from the amount of light transmitted. Suggest how and explain why the absorbance would change as the amount of pigment increases.
3 Look at the graph and suggest at what point the membrane what disrupted.
4 Suggest how you would carry out an investigation to see the effects of organic solvents on membrane permeability.

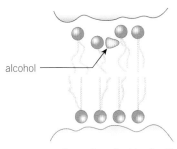

▲ Figure 2 *The presence of alcohol molecules between the phospholipids disrupts the structure of the membrane*

Synoptic link

You will learn about nerve impulse transmission in Topic 13.4, Nervous transmission.

Summary questions

1 Explain why solvents like water do not disrupt cell membranes. (*2 marks*)

2 Describe how the absorbance of light could be measured quantitatively as the concentration of released pigment increased with increasing temperature. Describe the graph you would plot to show these results. (*4 marks*)

3 Suggest how an excessive consumption of alcohol could lead to liver cell death and ultimately be fatal. (*6 marks*)

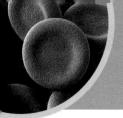

5.3 Diffusion

Specification reference: 2.1.5

Study tip

Diffusion only occurs between different concentrations of the *same* substance.

The exchange of substances between cells and their environment or between membrane-bound compartments within cells and the cell cytosol is defined as either active (requiring metabolic energy) or passive. All movement requires energy. Passive movement, however, utilises energy from the natural motion of particles, rather than energy from an another source. This topic will focus on **passive transport** methods.

Diffusion

Diffusion is the net, or overall, movement of particles (atoms, molecules or ions) from a region of higher concentration to a region of lower concentration. It is a passive process and it will continue until there is a concentration equilibrium between the two areas. Equilibrium means a balance or no difference in concentrations.

Diffusion happens because the particles in a gas or liquid have kinetic energy (they are moving). This movement is random and an unequal distribution of particles will eventually become an equal distribution. Equilibrium doesn't mean the particles stop moving, just that the movements are equal in both directions.

Particles move at high speeds and are constantly colliding, which slows down their overall movement. This means that over short distances diffusion is fast, but as diffusion distance increases the rate of diffusion slows down because more collisions have taken place.

For this reason cells are generally microscopic – the movement of particles within cells depends on diffusion and a large cell would lead to slow rates of diffusion. Reactions would not get the substrates they need quickly enough or ATP would be supplied too slowly to energy-requiring processes.

Factors affecting rate of diffusion

- temperature – the higher the temperature the higher the rate of diffusion. This is because the particles have more kinetic energy and move at higher speeds.

- concentration difference – the greater the difference in concentration between two regions the faster the rate of diffusion. because the overall movement from the higher concentration to lower concentration will be larger.

A concentration difference is said to be a concentration gradient, which goes from high to low concentration. Diffusion takes place *down* a concentration gradient. It takes a lot more energy to move substances *up* a concentration gradient.

So far diffusion in the absence of a barrier or membrane has been considered. This is **simple diffusion**.

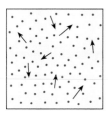

▲ Figure 1 *The random movement of particles means the initial unequal distribution eventually evens out*

Rate of diffusion and surface area

The rate of diffusion can be calculated in two ways – by distance travelled/time and volume filled/time. Distance travelled/time is not affected by changes in surface area, whilst volume/time varies depending on the surface area.

A student used different sized agar blocks to investigate how the rate of diffusion was affected by surface area.

The agar used to make the blocks contained the indicator phenolphthalein with turns pink in the presence of an alkali.

The agar blocks were immersed in a solution of sodium hydroxide for ten minutes. The blocks were removed and distance the sodium hydroxide had diffused was measured with a ruler.

Results

Cube size (cm)	Surface area (cm²)	Volume (cm³)	Surface area / volume	Diffusion distance (cm)	Rate of diffusion using distance (cm / min)	Rate of diffusion using volume (cm³ / min)	Rate of diffusion using volume per 64 cm³ agar
$4 \times 4 \times 4$				0.3			
$2 \times 2 \times 2$				0.3			
$1 \times 1 \times 1$				0.3			

1 Copy and complete the table.
2 Explain what has happened in order for the diffusion distances to be measured.
3 Describe and explain the results.

Diffusion across membranes

Diffusion across membranes involves particles passing through the phospholipid bilayer. It can only happen if the membrane is permeable to the particles – non-polar molecules such as oxygen (O_2) diffuse through freely down a concentration gradient.

The hydrophobic interior of the membrane repels substances with a positive or negative charge (ions), so they cannot easily pass through. Polar molecules, such as water (H_2O) with partial positive and negative charges can diffuse through membranes, but only at a very slow rate. Small polar molecules pass through more easily than larger ones. Membranes are therefore described as **partially permeable**.

The rate at which molecules or ions diffuse across membranes is affected by:

- surface area – the larger the area of an exchange surface, the higher the rate of diffusion
- thickness of membrane – the thinner the exchange surface, the higher the rate of diffusion.

Synoptic link

You will learn about surface area to volume ratio in Topic 7.1, Specialised exchange surfaces.

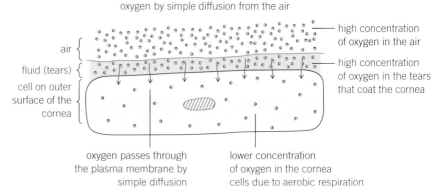

the cornea has no blood supply so its cells obtain oxygen by simple diffusion from the air

high concentration of oxygen in the air

high concentration of oxygen in the tears that coat the cornea

air

fluid (tears)

cell on outer surface of the cornea

oxygen passes through the plasma membrane by simple diffusion

lower concentration of oxygen in the cornea cells due to aerobic respiration

▲ Figure 2 *Passive diffusion of oxygen into a cell of the cornea*

Facilitated diffusion

As you have learnt, the phospholipid bilayers of membranes are barriers to polar molecules and ions. However, membranes contain channel proteins through which polar molecules and ions can pass. Diffusion across a membrane through protein channels is called **facilitated diffusion**.

Membranes with protein channels are **selectively permeable** as most protein channels are specific to one molecule or ion.

Facilitated diffusion can also involve carrier proteins (Topic 5.4, Active transport), which change shape when a specific molecule binds. In facilitated diffusion the movement of the molecules is down a concentration gradient and does not require external energy.

The rate of facilitated diffusion is dependent on the temperature, concentration gradient, membrane surface area and thickness, but is also affected by the number of channel proteins present. The more protein channels, the higher the rates of diffusion overall.

 Investigations into the factors affecting diffusion rates in model cells

As you have already seen, cell membranes are highly complex structures involved in the active and passive transport of ions and molecules. The hydrophobic hydrocarbon core of the membrane is a barrier to ions and large polar molecules, but it allows the passage of non-polar molecules. Cells are too small and cell membranes too thin to use in practical investigations so dialysis tubing is used as a substitute membrane. This model enables us to investigate the effects of temperature and concentration on the rate of diffusion across membranes.

Dialysis tubing is partially permeable, with pores a similar size to those on a partially permeable membrane. This means that small molecules like water can pass through it, but larger molecules like starch cannot fit through the pores. The tubing is therefore a barrier to large molecules.

A model cell can be simulated by tying one end of a section of tubing, filling with a solution and then tying the other end. The 'cell' is then placed into another solution. The solutions could contain different sizes, or concentrations, of solute molecules.

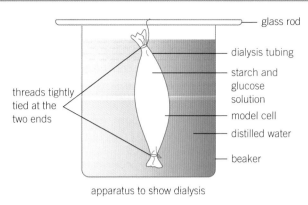

glass rod

dialysis tubing

starch and
glucose
solution

threads tightly
tied at the
two ends

model cell

distilled water

beaker

apparatus to show dialysis

▲ Figure 3 *The apparatus above is used to demonstrate
that glucose molecules are small enough to diffuse out of the
'cell' but the starch molecules are too large. After a set time
the water is tested for the presence of starch and glucose*

The changes in concentration of solute molecules, both
inside and outside the model cells, can be measured over
time. Rates of diffusion across the tubing can then be
calculated.

Glucose is a small molecule which can cross the tubing.
Benedict's solution is used to test for the presence of
glucose, and can also be used to estimate concentration.

Starch molecules are large and will not cross the tubing.
Iodine is used to test for the presence of starch.

Water is a small molecule which will pass through the tubing
while other solutes such as sucrose will not. Model cells can
be placed in solutions with different solute concentrations.
The rates of osmosis can be calculated using changes in
volume or mass of the model cells over time.

Rates of diffusion at different temperatures can also be
calculated using a water bath to change the temperature
of the model cell. Other variables such as concentration
must be then be kept constant.

1   Explain why Benedict's test is both quantitative
    and qualitative.
2   Explain what is meant by the term model cell.
3   a   Describe the differences between dialysis
        tubing and cell membranes with reference to
        transport across membranes.
    b   Explain why some ions can pass through dialysis
        tubing by diffusion but can only pass through cell
        membranes by facilitated diffusion.

Summary questions

1   Explain why the rate of diffusion increases as temperature
    increases.   (*2 marks*)

2   State two changes to the structure of a cell-surface membrane
    that would increase the rate at which polar molecules diffuse
    into a cell.   (*2 marks*)

3   Movement requires energy and yet the movement of molecules
    in diffusion is described as passive (not requiring energy).
    Explain this statement and state the source of the energy
    involved in diffusion.   (*2 marks*)

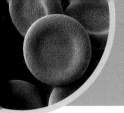

5.4 Active transport

Specification reference: 2.1.5

Diffusion, by its nature, will ultimately result in concentration gradients being reduced until particles (atoms, molecules or ions) in the different regions reach equilibrium. However, many biological processes depend on the presence of a concentration gradient, for example, the transmission of nerve impulses. To maintain this concentration gradient, particles must be moved up it at a rate faster than the rate of diffusion. This is an energy-requiring process called **active transport.**

Active transport

Active transport is the movement of molecules or ions into or out of a cell from a region of lower concentration to a region of higher concentration. The process requires energy and carrier proteins. Energy is needed as the particles are being moved up a concentration gradient, in the opposite direction to diffusion. Metabolic energy is supplied by ATP.

Carrier proteins span the membranes and act as 'pumps'. The general process of active transport is described below – in this example transport is from outside to inside a cell (Figure 1).

1   The molecule or ion to be transported binds to receptors in the channel of the carrier protein on the outside of the cell.

2   On the inside of the cell ATP binds to the carrier protein and is hydrolysed into ADP and phosphate.

3   Binding of the phosphate molecule to the carrier protein causes the protein to change shape – opening up to the inside of the cell.

4   The molecule or ion is released to the inside of the cell.

5   The phosphate molecule is released from the carrier protein and recombines with ADP to form ATP.

6   The carrier protein returns to its original shape.

The process is selective – specific substances are transported by specific carrier proteins.

Bulk transport

Bulk transport is another form of active transport. Large molecules such as enzymes, hormones, and whole cells like bacteria are too large to move through channel or carrier proteins, so they are moved into and out of cell by bulk transport.

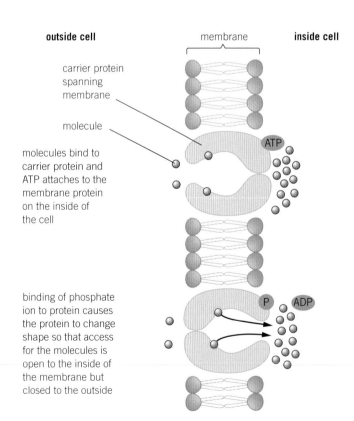

outside cell   membrane   inside cell

carrier protein spanning membrane

molecule

molecules bind to carrier protein and ATP attaches to the membrane protein on the inside of the cell

ATP

binding of phosphate ion to protein causes the protein to change shape so that access for the molecules is open to the inside of the membrane but closed to the outside

P   ADP

▲ Figure 1 *Active transport. The shape of the carrier protein changes to move a particle from one side of the membrane to the other*

- **Endocytosis** is the bulk transport of material *into* cells. There are two types of endocytosis, **phagocytosis** for solids and **pinocytosis** for liquids – the process is the same for both. The cell-surface membrane first invaginates (bends inwards) when it comes into contact with the material to be transported. The membrane enfolds the material until eventually the membrane fuses, forming a vesicle. The vesicle pinches off and moves into the cytoplasm to transfer the material for further processing within the cell. For example, vesicles containing bacteria are moved towards lysosomes, where the bacteria are digested by enzymes.

- **Exocytosis** is the reverse of endocytosis. Vesicles, usually formed by the Golgi apparatus, move towards and fuse with the cell surface membrane. The contents of the vesicle are then released *outside* of the cell.

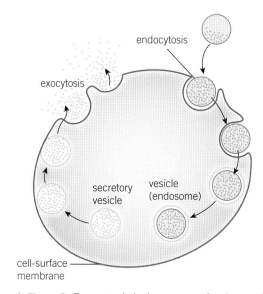

▲ Figure 2 *Exocystosis is the reverse of endocytosis*

Energy in the form of ATP is required for movement of vesicles along the cytoskeleton, changing the shape of cells to engulf materials, and the fusion of cell membranes as vesicles form or as they meet the cell-surface membrane.

Study tip

It is important to understand the difference between facilitated diffusion and active transport. Both use carrier proteins but facilitated diffusion occurs *down* a concentration gradient and therefore does not require energy.

Synoptic link

You learnt about vesicles in Topic 2.4, Eukaryotic cell structure.

Synoptic link

You will learn about phagocytosis in Topic 12.5, Non-specific defences against pathogens.

Summary questions

1 Explain why facilitated diffusion is not a form of active transport.
(*2 marks*)

2 Cells that carry out active transport usually have more mitochondria than cells that do not. Explain why. (*2 marks*)

3 Plant roots take up mineral ions from the soil. The concentration of mineral ions in the soil water is very low. Suggest why active transport is very important in root hair cells. (*4 marks*)

5.5 Osmosis

Specification reference: 2.1.5

Learning outcomes

Demonstrate knowledge, understanding, and application of:

→ the movement of water across membranes by osmosis

→ the effects that solutions of different water potential can have on plant and animal cells

→ practical investigations into the effects of solutions of different water potential on plant and animal cells.

Study tip

Remember that all water potential values are negative. Pure water has a water potential of zero.

Osmosis is a particular type of diffusion – specifically the diffusion of water across a partially permeable membrane. As with all types of diffusion it is a passive process and energy is not required.

Water potential

A solute is a substance dissolved in a solvent (for example water) forming a solution.

The amount of solute in a certain volume of aqueous solution is the concentration. **Water potential** is the pressure exerted by water molecules as they collide with a membrane or container. It is measured in units of pressure pascals (Pa) or kilopascals (kPa). The symbol for water potential is the Greek letter psi Ψ.

Pure water is defined as having a water potential of 0 kPa (at standard temperature and atmospheric pressure – 25 °C and 100 kPa). This is the highest possible value for water potential, as the presence of a solute in water lowers the water potential below zero. All solutions have negative water potentials – the more concentrated the solution the more negative the water potential.

When solutions of different concentrations, and therefore different water potentials, are separated by a partially permeable membrane, the water molecules can move between the solutions but the solutes usually cannot. There will be a net movement of water from the solution with the higher water potential (less concentrated) to the solution with the lower water potential (more concentrated). This will continue until the water potential is equal on both sides of the membrane (equilibrium).

Effects of osmosis on plant and animal cells

The diffusion of water into a solution leads to an increase in volume of this solution. If the solution is in a closed system, such as a cell, this results in an increase in pressure. This pressure is called **hydrostatic pressure** and has the same units as water potential, kPa. At the cellular level this pressure is relatively large and potentially damaging.

Animal cells

If an animal cell is placed in a solution with a higher water potential than that of the cytoplasm, water will move into the cell by osmosis, increasing the hydrostatic pressure inside the cell. All cells have thin cell-surface membranes (around 7 nm) and no cell walls. The cell-surface membrane cannot stretch much and cannot withstand the increased pressure. It will break and the cell will burst, an event called **cytolysis**.

water molecules

partially permeable membrane

net movement of water

solute molecules

dilute solution
low concentration of solute
high water potential (ψ)

concentrated solution
high concentration of solute
low water potential (ψ)

▲ Figure 1 *Due to the greater number of water molecules on the left-hand side of the partially permeable membrane, diffusion occurs until the number of water molecules is equal on both sides of the membrane. This movement is called osmosis*

If an animal cell is placed in a solution that has a lower water potential than the cytoplasm it will lose water to the solution by osmosis down the water potential gradient. This will cause a reduction in the volume of the cell and the cell-surface membrane to 'pucker', referred to as crenation (Figure 2).

To prevent either cytolysis or crenation, multicellular animals usually have control mechanisms to make sure their cells are continuously surrounded by aqueous solutions with an equal water potential (isotonic). In blood the aqueous solution is blood plasma.

▼ Table 1 *Osmosis in a red blood cell*

Water potential (Ψ) of external solution compared to cell solution	Higher (less negative)	Equal	Lower (more negative)
Net movement of water	Enters cell	Water constantly enters and leaves, but at equal rates	Leaves cell
State of cell	Swells and bursts	No change	Shrinks
	contents, including haemoglobin, are released — remains of cell surface membrane	normal red blood cell	haemoglobin is more concentrated, giving cell a darker appearance — cell shrunken and shrivelled

Plant cells

Like animal cells, plant cells contain a variety of solutes, mainly dissolved in a large vacuole. However, unlike animals, plants are unable to control the water potential of the fluid around them, for example, the roots are usually surrounded by almost pure water.

Plants cells have strong cellulose walls surrounding the cell-surface membrane. When water enters by osmosis the increased hydrostatic pressure pushes the membrane against the rigid cell walls. This pressure against the cell wall is called **turgor**. As the turgor pressure increases it resists the entry of further water and the cell is said to be turgid.

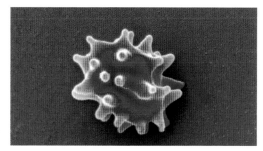

▲ Figure 2 *Scanning electron micrograph of a red blood cell that has been placed in a solution of lower water potential than the cytoplasm and become crenated by osmosis (× 5000 magnification)*

When plant cells are placed in a solution with a lower water potential than their own, water is lost from the cells by osmosis. This leads to a reduction in the volume of the cytoplasm, which eventually pulls the cell-surface membrane away from the cell wall – the cell is said to be plasmolysed.

▼ Table 2 *Osmosis in a plant cell*

Water potential (Ψ) of external solution compared to cell solution	Higher (less negative)	Equal	Lower (more negative)
Net movement of water	Enters cell	Water constantly enters and leaves, but at equal rates	Leaves cell
Condition of protoplast	Swells and becomes turgid	No change	Plasmolysis, contents shrink
	protoplast pushed against cell wall nucleus cellulose cell wall protoplast	protoplast beginning to pull away from the cell wall	protoplast completely pulled away from the cell wall space filled with external solution of lower water potential

Osmosis investigations

The effect of solutions with different water potentials can be observed in both plant and animal cells.

Plant cells

Pieces of potato or onion can be placed into sugar or salt solutions with different concentrations, and therefore different water potentials. Water will move into or out of cells depending on the water potential of the solution relative to the water potential of the plant tissue. As the plant tissue gains or loses water it will increase or decrease in mass and size, and vice versa.

A student used potato cores and their knowledge of osmosis to investigate the water potential of potato cells.

The following results were obtained.

Sugar concentration mol dm^{-3}	Original mass (g)	Final mass (g)	Difference in mass (g)	% mass change	Mean % mass change
0.0	3.0	4.0			
	3.0	4.1			
	3.3	4.2			
0.1	3.0	3.5			
	3.2	3.6			
	2.9	3.3			
0.3	3.0	3.0			
	2.9	3.0			
	3.2	3.2			
0.5	3.2	2.8			
	3.0	2.6			
	3.1	2.7			
0.7	3.1	2.2			
	3.3	2.4			
	3.0	2.0			

1   Copy and complete the table to show final masses, percentage mass changes and mean percentage changes.
2   Plot a graph of the results.
3   Describe and explain the shape of the graph.
4   Write a short evaluation using the information given and suggest any improvements to the investigation.

Animal cells

Eggs can be used to demonstrate osmosis in animal cells. A chicken's egg is not exactly a single cell, but with the shell removed a single membrane-bound structure remains and it will behave in the same way as a cell when placed in solutions of varying water potentials.

To investigate osmosis, eggs without their shells are placed in different concentrations of sugar syrup. Over time, osmosis takes place and there will have been a net movement of water into or out of the eggs, depending on the concentration of the syrup they were in. (Note that if the egg is hard boiled for easier handling that this will damage the membrane.)

Summary questions

1 Copy the diagram below and use arrows to show the net movement of water. *(2 marks)*

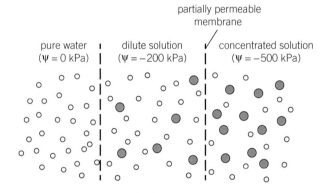

partially permeable membrane

pure water ($\Psi = 0$ kPa)   dilute solution ($\Psi = -200$ kPa)   concentrated solution ($\Psi = -500$ kPa)

2 Explain why it is not possible to have a positive water potential. *(2 marks)*

3 At which point on the graph in Figure 3 is the water potential of the solution equal to the water potential of the cells? *(1 mark)*

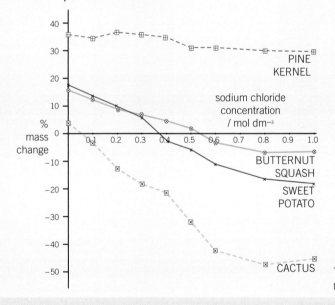

sodium chloride concentration / mol dm^{-3}

PINE KERNEL

BUTTERNUT SQUASH

SWEET POTATO

CACTUS

% mass change

◀ Figure 3 *Mass changes in plant tissues bathed in salt solutions*

4 Explain why it is important to keep the concentrations of electrolytes (solutes) in body tissues at the correct level to ensure proper hydration. *(3 marks)*

5 Look at Figure 3 again. State which plant tissue in these data has the highest solute concentration (lowest water potential). Suggest a reason for this. *(4 marks)*

Practice questions

1 Solution A has a more negative water potential than solution B.

Which of the following statements is/are correct?

Statement 1: Solution B has a higher water potential than solution A

Statement 2: There would be a net movement of water from solution A to solution B

Statement 3: There would be a net movement of water from solution A to distilled water

A 1, 2 and 3 are correct

B Only 1 and 2 are correct

C Only 2 and 3 are correct

D Only 1 is correct   (*1 mark*)

2 Membrane proteins are essential in order for cells to interact with their environment and other cells. About a third of the genes in a human genome code for membrane proteins.

  a (i) Explain what is meant by the term simple protein.   (*2 marks*)

    (ii) List the different levels of protein structure.   (*3 marks*)

  b Outline the roles of proteins in a cell surface membrane.   (*5 marks*)

3 A plasma membrane has a very complex structure which is affected by changes in the cells environment. Changes in pH or temperature affect the permeability of plasma membranes.

  a Describe how you would use a colorimeter to investigate how cell membrane permeability changes with temperature.   (*6 marks*)

  b Describe what you would do to ensure your results were:

    (i) *valid*   (*2 marks*)

    (ii) *reliable*   (*2 marks*)

  c Suggest why alcohol is used in antiseptic wipes.   (*3 marks*)

4 The graph below shows the permeability of a plasma membrane to different organic solvents.

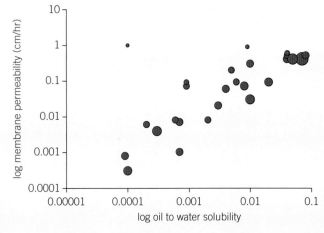

  a Describe the relationship between lipid solubility and membrane permeability of organic molecules.   (*2 marks*)

  b The size of the circles is proportional to the size of the organic molecule.

    It is known that with two organic molecules of the same lipid solubility, the one with greater molecular weight, or size, will cross the membrane more slowly.

    State whether the data in the graph above agrees with this fact. Suggest a reason for your answer.   (*3 marks*)

5 a A student stated the following answer about membranes

    The cell membrane is a thin semi-permeable membrane that surrounds the cytoplasm of a cell

    (i) Suggest why the student should have used the term 'selectively permeable' rather than 'semi-permeable' to describe a cell membrane.   (*2 marks*)

    (ii) Identify one other error in this statement   (*1 mark*)

  b Explain how water molecules cross selectively permeable membranes by **simple** diffusion.   (*2 marks*)

6 A classic experiment investigated the effect of temperature on the rate of sugar transport in a potted plant.

Aphid mouthparts were used to take samples of sugar solution from the transport tissue in the stem. The sugary solution dripped from the mouthparts. The number of drips per minute was counted.

The procedure was repeated at different temperatures.

Table 1 shows the results obtained

Temperature (°C)	Number of drips per minute
5	3
10	6
20	14
30	26
40	19
50	0

Suggest brief explanations for these results.

(*3 marks*)

OCR F211 2012

7 The bilayer is the fundamental structure of all cell membranes. The bilayer is composed of two lipid layers which provide an effective barrier to aqueous environments. This allows for compartmentalisation and the formation of cells and organelles.

Discuss the roles of the different lipid components of plasma membranes. (*6 marks*)

6 CELL DIVISION
6.1 The cell cycle

Specification reference: 2.1.6

The **cell cycle** is a highly ordered sequence of events that takes place in a cell, resulting in division of the cell, and the formation of two genetically identical daughter cells.

Phases of the cell cycle

In eukaryotic cells the cell cycle has two main phases – interphase and mitotic (division) phase.

Interphase

Cells do not divide continuously – long periods of growth and normal working separate divisions. These periods are called **interphase** and a cell spends the majority of its time in this phase.

Interphase is sometimes referred to as the resting phase as cells are not actively dividing. However, this is not an accurate description – interphase is actually a very active phase of the cell cycle, when the cell is carrying out all of its major functions such as producing enzymes or hormones, while also actively preparing for cell division.

During interphase:

- DNA is replicated and checked for errors in the nucleus
- protein synthesis occurs in the cytoplasm
- mitochondria grow and divide, increasing in number in the cytoplasm
- chloroplasts grow and divide in plant and algal cell cytoplasm, increasing in number
- the normal metabolic processes of cells occur (some, including cell respiration, also occur throughout cell division).

The three stages of interphase, as shown in Figure 1 are:

- G_1 – the first growth phase: proteins from which organelles are synthesised are produced and organelles replicate. The cell increases in size.
- S – synthesis phase: DNA is replicated in the nucleus.
- G_2 – the second growth phase: the cell continues to increase in size, energy stores are increased and the duplicated DNA is checked for errors.

Mitotic phase

The mitotic phase is the period of cell division. Cell division involves two stages:

▲ Figure 1 *The cell cycle showing the stages of interphase*

- **Mitosis** – the nucleus divides.
- **Cytokinesis** – the cytoplasm divides and two cells are produced.

The processes that take place during mitosis and cytokinesis (division of the cell into two separate cells) are discussed in more detail in Topic 6.2, Mitosis.

G_0

G_0 is the name given to the phase when the cell leaves the cycle, either temporarily or permanently. There are a number of reasons for this including:

- Differentiation – A cell that becomes specialised to carry out a particular function (differentiated) is no longer able to divide. It will carry out this function indefinitely and not enter the cell cycle again (you will learn more about cell specialisation and differentiation in Topic 6.4, The organisation and specialisation of cells).

- The DNA of a cell may be damaged, in which case it is no longer viable. A damaged cell can no longer divide and enters a period of permanent cell arrest (G_0). The majority of normal cells only divide a limited number of times and eventually become senescent.

- As you age, the number of these cells in your body increases. Growing numbers of senescent cells have been linked with many age related diseases, such as cancer and arthritis.

A few types of cells that enter G_0 can be stimulated to go back into the cell cycle and start dividing again, for example lymphocytes (white blood cells) in an immune response.

Control of the cell cycle

It is vital to ensure a cell only divides when it has grown to the right size, the replicated DNA is error-free (or is repaired) and the chromosomes are in their correct positions during mitosis. This is to ensure the fidelity of cell division – that two identical daughter cells are created from the parent cell.

Checkpoints are the control mechanisms of the cell cycle. They monitor and verify whether the processes at each phase of the cell cycle have been accurately completed before the cell is allowed to progress into the next phase.

Checkpoints occur at various stages of the cell cycle:

- G_1 checkpoint – This checkpoint is at the end of the G_1 phase, before entry into S phase. If the cell satisfies the requirements of this checkpoint (Figure 2) it is triggered to begin DNA replication. If not, it enters a resting state (G_0).

- G_2 checkpoint – This checkpoint is at the end of G_2 phase, before the start of

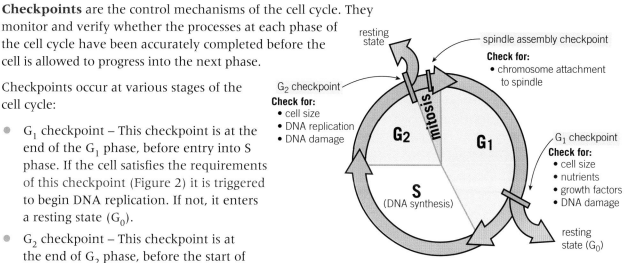

▲ Figure 2 *Checkpoints controlling the cell cycle*

the mitotic phase. In order for this checkpoint to be passed, the cell has to check a number of factors (Figure 2), including whether the DNA has been replicated without error. If this checkpoint is passed, the cell initiates the molecular processes that signal the beginning of mitosis (Topic 6.2, Mitosis).

• Spindle assembly checkpoint (also called **metaphase** checkpoint): This checkpoint is at the point in mitosis where all the chromosomes should be attached to spindles and have aligned (metaphase – Topic 6.2, Mitosis). Mitosis cannot proceed until this checkpoint is passed.

Cell-cycle regulation and cancer

The passing of a cell-cycle checkpoint is brought about by kinases. These are a class of enzyme that catalyse the addition of a phosphate group to a protein (phosphorylation). Phosphorylation changes the tertiary structure of checkpoint proteins, activating them at certain points in the cell cycle.

Kinases involved in cell-cycle regulation are activated by binding to a variety of checkpoint proteins called cyclins. Binding of the correct cyclin to the appropriate kinase forms a cyclin-dependent kinase (CDK) complex. These complexes are activated by enzymes.

CDK complexes catalyse the activation of key cell-cycle proteins by phosphorylation. This ensures a cell progresses through the different phases of its cycle at the appropriate times. Different enzymes break down cyclins when they are not needed, signalling a cell to move into the next phase of the cycle.

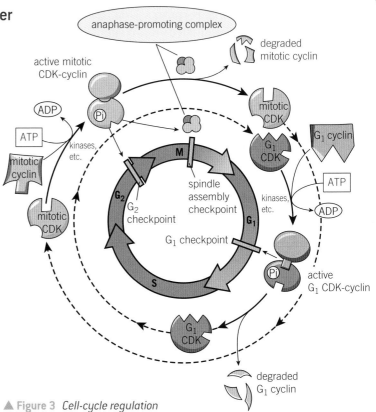

▲ Figure 3 Cell-cycle regulation

Cancer is a group of many different diseases caused by the uncontrolled division of cells. An abnormal mass of cells is called a tumour. Tumours can be benign, meaning that they stop growing and do not travel to other locations in the body. If a tumour continues to grow unchecked and uncontrolled, it is termed malignant. A malignant tumour is the basis of cancer.

Tumours are often the result of damage or spontaneous mutation of the genes that encode the proteins that are involved in regulating the cell cycle, including the checkpoint proteins.

For example, if overexpression of a cyclin gene results from mutation, the abnormally large quantity of cyclins produced would disrupt the regulation of the cell cycle, resulting in uncontrolled cell division, tumour formation, and possibly leading to cancer.

Cyclin-dependent kinases can be used as a possible target for chemical inhibitors in the treatment of cancer. If the activity of CDKs can be reduced it may reduce or stop cell division and therefore cancer formation.

1   Compare the roles of cyclins and enzyme inhibitors. Describe any similarities and/or differences between their roles.

Summary questions

1 Mitosis and cytokinesis are processes involved in the production of new cells. Explain the difference between mitosis and cytokinesis. *(2 marks)*

2 Explain, with reference to the structure and function of proteins, the importance of G_2 checkpoint. *(3 marks)*

3 Telomeres are repetitive sequences of DNA at the ends of chromosomes. They protect the genes at the end of chromosomes and stop the ends of chromosomes fusing. DNA is not replicated all the way to the end so every time DNA replication occurs the telomeres shorten. This limits the number of times a cell can replicate, it is called the Hayflick limit.

  a Suggest a disadvantage of indefinite cell division. *(3 marks)*

  b Telomerases are enzymes that result in the elongation of telomeres. They are not usually present in differentiated cells. Describe what the presence of telomerases could cause. *(2 marks)*

4 A typical human cell contains 3×10^9 base pairs of DNA divided into 46 chromosomes.

DNA replication in eukaryotic cells takes place at the rate of about 50 base pairs per minute.

  a Suggest why the length of DNA is usually given by the number of base pairs rather than number of nucleotides. *(1 mark)*

  b Calculate the time it would take to replicate a section of DNA of this length, assuming replication started at one end and didn't stop until reaching the other end. *(3 marks)*

  c The DNA of a eukaryotic cell is usually replicated in eight hours. Explain how this is possible. *(2 marks)*

  d Suggest why it takes a much shorter time to replicate the genome of a prokaryotic cell. *(1 mark)*

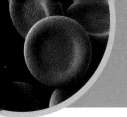

6.2 Mitosis

Specification reference: 2.1.6

The importance of mitosis

Mitosis is the term usually used to describe the entire process of cell division in eukaryotic cells. It actually refers to nuclear division (division of the nucleus), an essential stage in cell division. Mitosis ensures that both daughter cells produced when a parent cell divides are genetically identical (except in the rare events where mutations occur). Each new cell will have an exact copy of the DNA present in the parent cell and the same number of chromosomes.

Mitosis is necessary when all the daughter cells have to be identical. This is the case during growth, replacement and repair of tissues in multicellular organisms such as animals, plants, and fungi. Mitosis is also necessary for **asexual reproduction**, which is the production of genetically identical offspring from one parent in multicellular organisms including plants, fungi, and some animals, and also in eukaryotic single-celled organisms such as *Ameoba* species. Prokaryotic organisms, including bacteria, do not have a nucleus and they reproduce asexually by a different process known as binary fission.

Chromosomes

Before mitosis can occur, all of the DNA in the nucleus is replicated during interphase (Topic 6.1, The cell cycle). Each DNA molecule (chromosome) is converted into two identical DNA molecules, called **chromatids**.

The two chromatids are joined together at a region called the **centromere**. It is necessary to keep the chromatids together during mitosis so that they can be precisely manoeuvred and segregated equally, one each into the two new daughter cells (Figures 1 and 2).

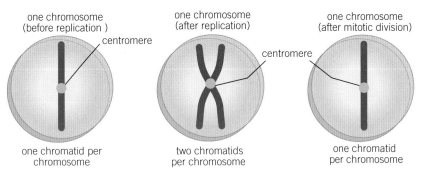

one chromosome (before replication)　　centromere

one chromatid per chromosome

one chromosome (after replication)　　centromere

two chromatids per chromosome

one chromosome (after mitotic division)

one chromatid per chromosome

▲ Figure 1 *Each chromosome replicates during cell division to form an identical copy, or chromatid. The chromatids of a pair are joined at the centromere*

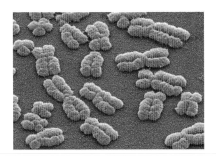

▲ Figure 2 *Scanning electron micrograph of replicated human chromosomes consisting of chromatid pairs joined at the centromere (× 3000 magnification)*

The stages of mitosis

There are four stages of mitosis – **prophase**, **metaphase**, **anaphase**, and **telophase**. We describe them separately but in fact they flow seamlessly from one to another. Each of these phases can be viewed and indentified using a light microscope. Dividing cells can be easily obtained from growing root tips of plants. The root tips can be treated with a chemical to allow the cells to be separated – then they can be squashed to form a single layer of cells on a microscope slide. Stains that bind DNA are used to make the chromosomes clearly visible.

The description of the four stages with example micrographs and labelled diagrams will help you to identify each phase in your own cell sections:

Prophase

1 During prophase, chromatin fibres (complex made up of various proteins, RNA and DNA) begin to coil and condense to form chromosomes that will take up stain to become visible under the light microscope. The nucleolus, a distinct area of the nucleus responsible for RNA synthesis, disappears. The nuclear membrane begins to break down.

2 Protein microtubules form spindle-shaped structures linking the poles of the cell. The fibres forming the spindle are necessary to move the chromosomes into the correct positions before division.

3 In animal cells and some plant cells, two centrioles migrate to opposite poles of the cell. The centrioles are cylindrical bundles of proteins that help in the formation of the spindle.

4 The spindle fibres attach to specific areas on the centromeres and start to move the chromosomes to the centre of the cell.

5 By the end of prophase the nuclear envelope has disappeared.

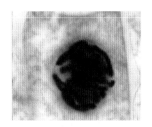

▲ **Figure 3** *Light micrograph of an onion (*Allium *spp.) root tip cell during* **prophase***. The chromosomes are beginning to condense into defined units, which will allow organised and equal segregation into daughter cells*

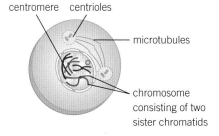

▲ **Figure 4** *Early prophase*

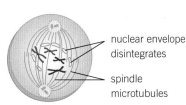

▲ **Figure 5** *Late prophase*

Metaphase

During metaphase the chromosomes are moved by the spindle fibres to form a plane in the centre of the cell, called the **metaphase plate**, and then held in position.

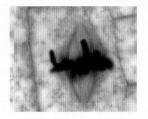

▲ **Figure 6** *Light micrograph of an onion root tip cell during* **metaphase***. The chromosomes are organized in a plane across the centre of the cell*

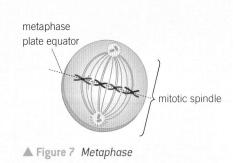

▲ **Figure 7** *Metaphase*

Anaphase

The centromeres holding together the pairs of chromatids in each chromosome divide during anaphase. The chromatids are separated – pulled to opposite poles of the cell by the shortening spindle fibres.

The characteristic 'V' shape of the chromatids moving towards the poles is a result of them being dragged by their centromeres through the liquid cytosol.

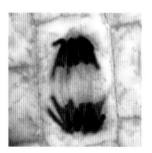

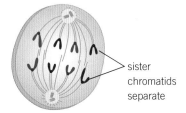

▲ Figure 9 *Anaphase*

▲ Figure 8 *Light micrograph of an onion root tip cell during anaphase. The sister chromatids are moved to separate poles by shortening spindle fibres*

Telophase

In telophase the chromatids have reached the poles and are now called chromosomes. The two new sets of chromosomes assemble at each pole and the nuclear envelope reforms around them. The chromosomes start to uncoil and the nucleolus is formed. Cell division – or cytokinesis, begins.

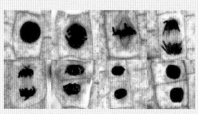

▲ Figure 10 *Light micrograph of onion root tip cells during* **telophase**. *Sister chromatids have been moved to separate poles by spindle fibres and cell division begins*

▲ Figure 11 *Telophase*

Cytokinesis

Cytokinesis, the actual division of the cell into two separate cells, begins during telophase.

Animal cells

In animal cells a cleavage furrow forms around the middle of the cell. The cell-surface membrane is pulled inwards by the cytoskeleton until it is close enough to fuse around the middle, forming two cells (Figure 13).

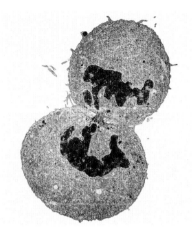

▲ Figure 12 *Cytokinesis in a human embryonic kidney cell. Coloured transmission electron micrograph × 800 magnification*

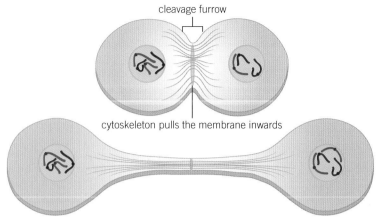

cleavage furrow

cytoskeleton pulls the membrane inwards

▲ Figure 13 *Cytokinesis in an animal cell*

Plant cells

Plant cells have cell walls so it is not possible for a cleavage furrow to be formed. Vesicles from the Golgi apparatus begin to assemble in the same place as where the metaphase plate was formed. The vesicles fuse with each other and the cell surface membrane, dividing the cell into two (Figure 15).

New sections of cell wall then form along the new sections of membrane (if the dividing cell wall were formed before the daughter cells separated they would immediately undergo osmotic lysis from the surrounding water).

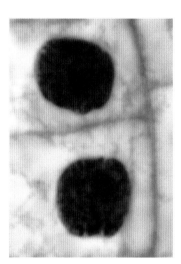

▲ Figure 14 *Light micrograph of cytokinesis in an onion root tip cell*

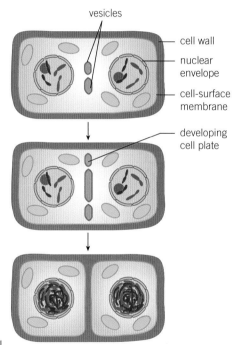

vesicles

cell wall
nuclear envelope
cell-surface membrane

developing cell plate

▲ Figure 15 *Cytokinesis in plant cells*

Summary questions

1 Explain why we normally see chromosomes as a double structure containing two chromatids. *(2 marks)*

2 Explain why it is essential that DNA replication results in two exact copies of the genetic material. *(2 marks)*

3 Describe the differences between cytokinesis in animal and plant cells and give reasons for these differences. *(4 marks)*

4 How many chromatids would be present in a human cell at prophase and at G_1 during interphase? *(2 marks)*

5 Explain why plant root tips are a good source of cells to examine for mitosis. *(4 marks)*

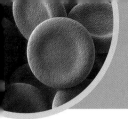
Normal cells have two chromosomes of each type (termed **diploid**) – one inherited from each parent. During mitosis the nucleus divides once following DNA replication. This results in two genetically identical diploid daughter cells.

In sexual reproduction two sex cells (**gametes**), one from each parent, fuse to produce a fertilised egg. The fertilised egg (**zygote**) is the origin of all the cells that the organism develops. Gametes must therefore only contain half of the standard (diploid) number of chromosomes in a cell or the chromosome number of an organism would double with every round of reproduction.

Gametes are formed by another form of cell division known as **meiosis**. Unlike in mitosis, the nucleus divides twice to produce four daughter cells – the gametes. Each gamete contains half of the chromosome number of the parent cell – it is **haploid**. Meiosis is therefore known as **reduction division**.

Homologous chromosomes

As you will remember, each characteristic of an organism is coded for by two copies of each gene, one from each parent. Each nucleus of the organism's cells contains two full sets of genes, a pair of genes for each characteristic. Therefore each nucleus contains matching sets of chromosomes, called **homologous chromosomes**, and is termed diploid. Each chromosome in a homologous pair has the same genes at the same loci.

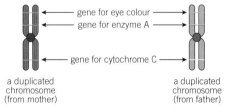

▲ **Figure 1** *A homologous pair of chromosomes. The same genes and all the alleles of that gene will appear at the same position (locus) on the chromosome*

Alleles

Genes for a particular characteristic may vary, leading to differences in the characteristic, for example blue eyes and brown eyes. The genes are still the same type as they both code for eye colour but the colour is different, meaning they are different versions of the same gene. Different versions of the same gene are called **alleles** (also known as gene variants). The different alleles of a gene will all have the same locus (position on a particular chromosome).

As homologous chromosomes have the same genes in the same positions, they will be the same length and size when they are visible in prophase. The centromeres will also be in the same positions.

The stages of meiosis

As discussed at the start of this topic, meiosis involves two divisions:

● **Meiosis I** – the first division is the reduction division when the pairs of homologous chromosomes are separated into two cells. Each intermediate cell will only contain one full set of genes instead of two, so the cells are haploid.

- **Meiosis II** – the second division is similar to mitosis, and the pairs of chromatids present in each daughter cell are separated, forming two more cells. Four haploid daughter cells are produced in total.

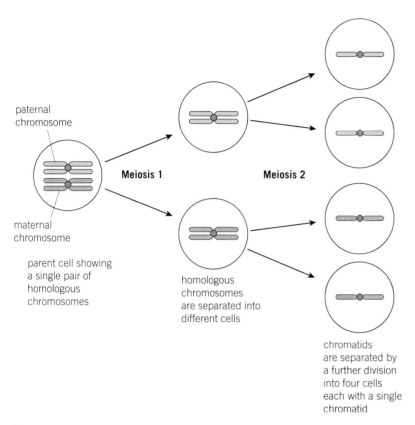

paternal chromosome

maternal chromosome

parent cell showing a single pair of homologous chromosomes

Meiosis 1

homologous chromosomes are separated into different cells

Meiosis 2

chromatids are separated by a further division into four cells each with a single chromatid

▲ Figure 2   *Summary of meiosis simplified to a single homologous pair*

Meiosis I

Prophase 1

During prophase 1, chromosomes condense, the nuclear envelope disintegrates, the nucleolus disappears and spindle formation begins, as in prophase of mitosis.

The difference in **prophase 1** is that the homologous chromosomes pair up, forming **bivalents**. Chromosomes are large molecules of DNA and moving them through the liquid cytoplasm as they are brought together results in the chromatids entangling. This is called **crossing over** (Figure 3).

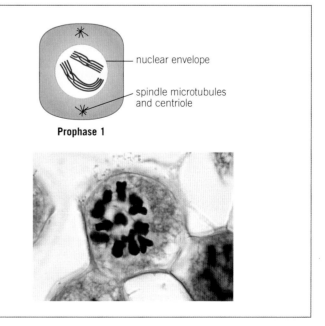

nuclear envelope

spindle microtubules and centriole

Prophase 1

Metaphase 1

Metaphase 1 is the same as metaphase in mitosis except that the homologous pairs of chromosomes assemble along the metaphase plate instead of the individual chromosomes.

The orientation of each homologous pair on the metaphase plate is random and independent of any other homologous pair. The maternal or paternal chromosomes can end up facing either pole. This is called **independent assortment**, and can result in many different combinations of alleles facing the poles (Figure 5). Independent assortment of chromosomes in metaphase 1 results in genetic variation.

bivalents aligned on the equator

Metaphase 1

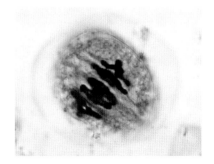

Anaphase 1

Anaphase 1 is different from anaphase of mitosis as the homologous chromosomes are pulled to the opposite poles and the chromatids stay joined to each other.

Sections of DNA on 'sister' chromatids, which became entangled during crossing over, now break off and rejoin – sometimes resulting in an exchange of DNA. The points at which the chromatids break and rejoin are called **chiasmata**.

When exchange occurs this forms **recombinant** chromatids, with genes being exchanged between chromatids. The genes being exchanged may be different alleles of the same gene, meaning the combination of alleles on the recombinant chromatids will be different from the allele combination on either the original chromatids (Figure 4). **Genetic variation** arises from this new combinations of alleles – the sister chromatids are no longer identical.

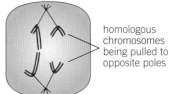

homologous chromosomes being pulled to opposite poles

Anaphase 1

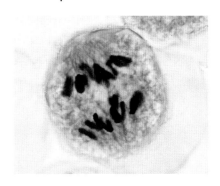

Telophase 1

Telophase 1 is essentially the same as telophase in mitosis. The chromosomes assemble at each pole and the nuclear membrane reforms. Chromosomes uncoil.

The cell undergoes cytokinesis and divides into two cells. The reduction of chromosome number from diploid to haploid is complete.

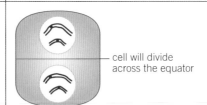

cell will divide across the equator

Telophase 1

Meiosis II

Prophase 2 In **prophase 2** the chromosomes, which still consist of two chromatids, condense and become visible again. The nuclear envelope breaks down and spindle formation begins.	 Prophase II	
Metaphase 2 Metaphase 2 differs from metaphase 1, as the individual chromosomes assemble on the metaphase plate, as in metaphase in mitosis. Due to crossing over, the chromatids are no longer identical so there is **independent assortment** again and more **genetic variation** produced in metaphase II.	 Metaphase II	
Anaphase 2 Unlike anaphase 1, **anaphase 2** results in the chromatids of the individual chromosomes being pulled to opposite poles after division of the centromeres – the same as in anaphase of mitosis.	 Anaphase II	
Telophase 2 The chromatids assemble at the poles at telophase 2 as in telophase of mitosis. The chromosomes uncoil and form chromatin again. The nuclear envelope reforms and the nucleolus becomes visible. Cytokinesis results in division of the cells forming four daughter cells in total. The cells will be *haploid* due to the *reduction division*. They will also be genetically different from each other, and from the parent cell, due to the processes of *crossing over* and *independent assortment*.	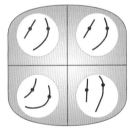 Telophase II	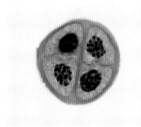

chromatids of homologous chromosomes twist around one another, crossing over many times

simplified representation of a single cross over

point of breakage (chiasmata)

result of a single cross over showing equivalent portions of the chromatid having been exchanged

(recombinant chromosomes)

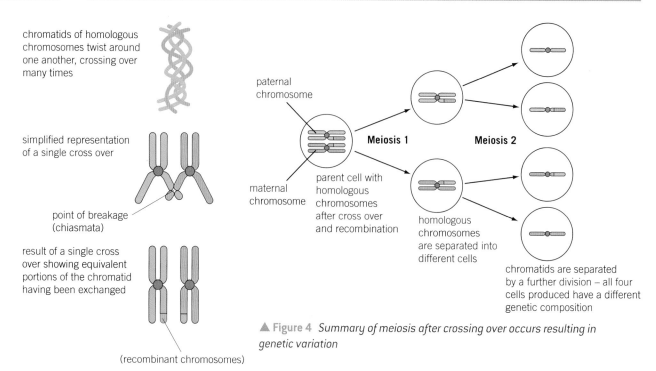

paternal chromosome

Meiosis 1   **Meiosis 2**

maternal chromosome

parent cell with homologous chromosomes after cross over and recombination

homologous chromosomes are separated into different cells

chromatids are separated by a further division – all four cells produced have a different genetic composition

▲ Figure 4 *Summary of meiosis after crossing over occurs resulting in genetic variation*

▲ Figure 3 *Crossing over*

During meiosis I, chromosomes can line up in different ways before the homologous chromosomes separate.

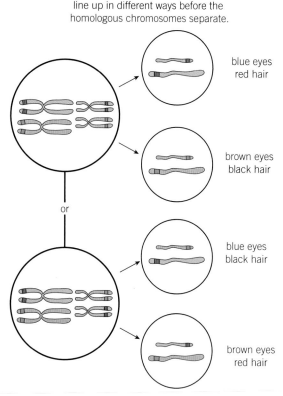

blue eyes red hair

brown eyes black hair

or

blue eyes black hair

brown eyes red hair

▲ Figure 5 *The pole that maternal or paternal homologous chromosomes face is due to random independent assortment and can result in many different combinations of alleles on either side of the metaphase plate*

Summary questions

1 **a** State which division in meiosis is a reduction division. *(1 mark)*
   b Explain why a reduction division is necessary in the production of gametes. *(2 marks)*

2 Explain the meaning of the term homologous chromosomes. *(2 marks)*

3 Outline how you could observe meiosis in a plant cell. *(4 marks)*

4 Explain how crossing over and independent assortment lead to genetic variation. *(4 marks)*

5 **a** Copy the diagram below and label the alleles on the recombinant chromosomes. *(3 marks)*

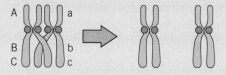

   b Suggest the importance of the creation of different allele combinations in populations. *(4 marks)*

6.4 The organisation and specialisation of cells

Specification reference: 2.1.6

As you have learnt, the basic unit of life is a cell. But many organisms are multicellular – they are made up of not one but hundreds, thousands or millions of cells. Although these cells within a single organism have common features such as membranes, organelles, and nuclei, they are not all identical. Different cells within an organism are **specialised** for different roles and organised into efficient biological structures, each with a particular function.

Levels of organisation in multicellular organisms

The organisation of a multicellular organism can be summarised as:

specialised cells → tissues → organs → organ systems → whole organism

Specialised cells

The cells within a multicellular organism are **differentiated**, meaning they are specialised to carry out very specific functions. You will explore cell differentiation further in the next topic.

Some examples of specialised cells are given in the following tables.

▼ Table 1 *Specialised animal cells. Dimensions given are according to cells in a human*

Erythrocytes or red blood cells have a flattened biconcave shape, which increases their surface area to volume ratio. This is essential to their role of transporting oxygen around the body. In mammals these cells do not have nuclei or many other organelles, which increases the space available for haemoglobin, the molecule that carries oxygen. They are also flexible so that they are able to squeeze through narrow capillaries.	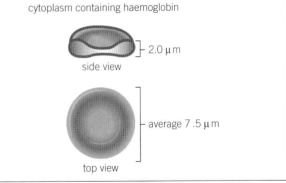 cytoplasm containing haemoglobin / side view / 2.0 μm / top view / average 7.5 μm
Neutrophils (a type of white blood cell) play an essential role in the immune system. They have a characteristic multi-lobed nucleus, which makes it easier for them to squeeze through small gaps to get to the site of infections. The granular cytoplasm contains many lysosomes that contain enzymes used to attack pathogens.	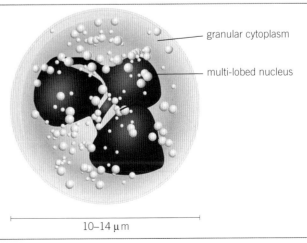 granular cytoplasm / multi-lobed nucleus / 10–14 μm

Sperm cells are male gametes. Their function is to deliver genetic information to the female gamete, the ovum (or egg). Sperm have a tail or flagellum, so they are capable of movement and contain many mitochondria to supply the energy needed to swim. The acrosome on the head of the sperm contains digestive enzymes, which are released to digest the protective layers around the ovum and allow the sperm to penetrate, leading to fertilisation.

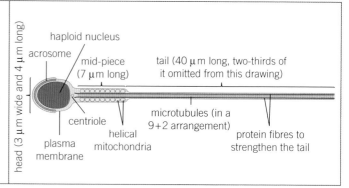

▼ Table 2 *Specialised plant cells*

Palisade cells present in the mesophyll contain chloroplasts to absorb large amounts of light for photosynthesis. The cells are rectangular box shapes, which can be closely packed to form a continuous layer. They have thin cell walls, increasing the rate of diffusion of carbon dioxide. They have a large vacuole to maintain turgor pressure. Chloroplasts can move within the cytoplasm in order to absorb more light.

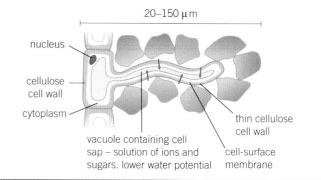

Root hair cells, present at the surfaces of roots near the growing tips, have long extensions called root hairs, which increase the surface area of the cell. This maximises the uptake of water and minerals from the soil.

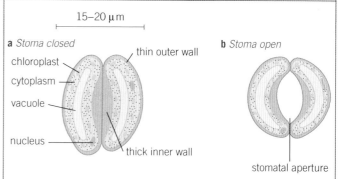

Pairs of *guard cells* on the surfaces of leaves form small openings called stomata. These are necessary for carbon dioxide to enter plants for photosynthesis. When guard cells lose water and become less swollen as a result of osmotic forces, they change shape and the stoma closes to prevent further water loss from the plant. The cell wall of a guard cell is thicker on one side so the cell does not change shape symmetrically as its volume changes.

Tissues

A **tissue** is made up of a collection of differentiated cells that have a specialised function or functions. As a result, each tissue is adapted for a particular function within the organism.

There are four main categories of tissues in animals:

- nervous tissue, adapted to support the transmission of electrical impulses
- epithelial tissue, adapted to cover body surfaces, internal and external
- muscle tissue, adapted to contract
- connective tissue, adapted either to hold other tissues together or as a transport medium.

Some examples of specialised tissues in animals are given in Table 3.

Synoptic link

You will learn about water transport in plants in Topic 9.2, Water transport in multicellular plants.

▼ Table 3 *Specialised animal tissues*

Squamous epithelium, made up of specialised squamous epithelial cells, is sometimes known as pavement epithelium due to its flat appearance. It is very thin due to the squat or flat cells that make it up and also because it is only one cell thick. It is present when rapid diffusion across a surface is essential. It forms the lining of the lungs and allows rapid diffusion of oxygen into the blood.	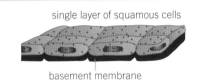 single layer of squamous cells basement membrane
Ciliated epithelium is made up of ciliated epithelial cells. The cells have 'hair-like' structures called cilia on one surface that move in a rhythmic manner. Ciliated epithelium lines the trachea, for example, causing mucus to be swept away from the lungs. Goblet cells are also present, releasing mucus to trap any unwanted particles present in the air. This prevents the particles, which may be bacteria, from reaching the alveoli once inside the lungs.	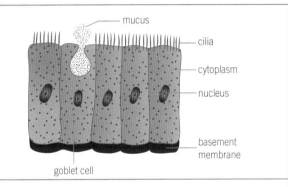 mucus — cilia — cytoplasm — nucleus — basement membrane — goblet cell
Cartilage is a connective tissue found in the outer ear, nose and at the ends of (and between) bones. It contains fibres of the proteins elastin and collagen. Cartilage is a firm, flexible connective tissue composed of chondrocyte cells embedded in an extracellular matrix. Cartilage, among other things, prevents the ends of bones from rubbing together and causing damage. Many fish have whole skeletons made of cartilage, not bone.	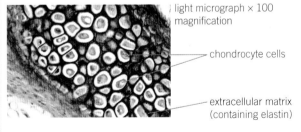 light micrograph × 100 magnification — chondrocyte cells — extracellular matrix (containing elastin)
Muscle is a tissue that needs to be able to shorten in length (contract) in order to move bones, which in turn move the different parts of the body. There are different types of muscle fibres. Skeletal muscle fibres (muscles which are attached to bone) contain myofibrils (dark pink bands on the micrograph) which contain contractile proteins. The skeletal muscle micrograph shown here has several individual muscle fibres (pink) separated by connective tissue (thin white strips).	light micrograph × 100 magnification (longitudinal section) — muscle fibres — connective tissue

Synoptic link

You will learn more about the structure and function of the xylem and phloem in Topic 9.1, Transport systems in dicotylyedonous plants.

There are a number of different tissues in plants, including:

- epidermis tissue, adapted to cover plant surfaces
- vascular tissue, adapted for transport of water and nutrients.

Some examples of specialised tissues in plants are given in the Table 4.

▼ Table 4 *Specialised plant tissues*

The *epidermis* is a single layer of closely packed cells covering the surfaces of plants. It is usually covered by a waxy, waterproof cuticle to reduce the loss of water. Stomata, formed by a pair of guard cells that can open and close (Table 2) are present in the epidermis. They allow carbon dioxide in and out, and water vapour and oxygen in and out.	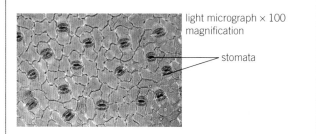 light micrograph × 100 magnification — stomata
Xylem tissue is a type of vascular tissue responsible for transport of water and minerals throughout plants. The tissue is composed of vessel elements, which are elongated dead cells. The walls of these cells are strengthened with a waterproof material called lignin (pink rings in the micrograph), which provides structural support for plants.	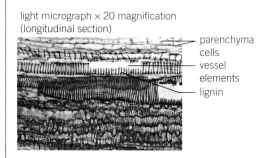 light micrograph × 20 magnification (longitudinal section) — parenchyma cells — vessel elements — lignin
Phloem tissue is another type of vascular tissue in plants, responsible for the transport of organic nutrients, particularly sucrose, from leaves and stems where it is made by photosynthesis to all parts of the plant where it is needed. It is composed of columns of sieve tube cells separated by perforated walls called sieve plates.	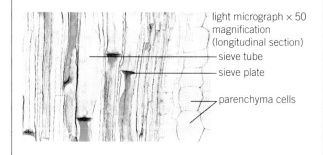 light micrograph × 50 magnification (longitudinal section) — sieve tube — sieve plate — parenchyma cells

Synoptic link

You will learn about the structure of the heart in Topic 8.5 The heart.

Organs

An **organ** is a collection of tissues that are adapted to perform a particular function in an organism. For example, the mammalian heart is an organ that is adapted for pumping blood around the body. It is made up of muscle tissue and connective tissue. The leaf is a plant organ that is adapted for photosynthesis. it contains epidermis tissues and vascular tissue (Figure 1).

Organ systems

Large multicellular organisms have coordinated **organ systems**. Each organ system is composed of a number of organs working together to carry out a major function in the body. Animal examples include:

- the digestive system, which takes in food, breaks down the large insoluble molecules into small soluble ones, absorbs the nutrients into the blood, retains water needed by the body and removes any undigested material from the body

- the cardiovascular system, which moves blood around the body to provide an effective transport system for the substances it carries

- the gaseous exchange system, which brings air into the body so oxygen can be extracted for respiration, and carbon dioxide can be expelled.

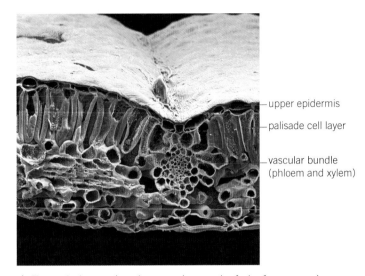

— upper epidermis

— palisade cell layer

— vascular bundle (phloem and xylem)

▲ **Figure 1** *A scanning electron micrograph of a leaf cross-section showing the arrangement of different tissues in this organ. × 150 magnification*

Summary questions

1 State two examples of epithelial tissue and describe how each is adapted for its function. *(6 marks)*

2 Describe two specialised cells, one that is not usually part of a tissue and one that is usually found as part of a tissue. In each case explain how they are adapted to their functions. *(4 marks)*

3 The cardiac muscle that makes up most of the heart is a tissue, but the heart itself is an organ. Explain the difference. *(4 marks)*

4 Using the digestive system as an example, explain the relationship between organs in an organ system. *(6 marks)*

6.5 Stem cells

Specification reference: 2.1.6

Learning outcomes

Demonstrate knowledge, understanding, and application of:

→ the features and differentiation of stem cells

→ the production of erythrocytes and neutrophils

→ the production of xylem vessels and phloem sieve tubes

→ the potential uses of stem cells in research and medicine.

As you explored in the previous topic, the different cells in a multicellular organism are specialised for different functions. The process of a cell becoming specialised is called **differentiation**. Despite being differentiated in structure and function, all body cells within an organism have the same DNA (except those like erythrocytes and sieve tube elements which don't have a nucleus). Differentiation involves the expression of some genes but not others in the cell's genome.

Stem cells

All cells in plants and animals begin as **undifferentiated** cells and originate from mitosis or meiosis. They are not adapted to any particular function (they are unspecialised) and they have the potential to differentiate to become any one of the range of specialised cell types in the organism. These undifferentiated cells are called **stem cells**.

Stem cells are able to undergo cell division again and again, and are the source of new cells necessary for growth, development, and tissue repair. Once stem cells have become specialised they lose the ability to divide, entering the G_0 phase of the cell cycle.

The activity of stem cells has to be strictly controlled. If they do not divide fast enough then tissues are not efficiently replaced, leading to ageing. However, if there is uncontrolled division then they form masses of cells called tumours, which can lead to the development of cancer.

Stem cell potency

A stem cell's ability to differentiate into different cell types is called **potency**. The greater the number of cell types it can differentiate into, the greater its potency. Stem cells differ depending on the type of cell they can turn into:

● **Totipotent** – these stem cells can differentiate into any type of cell. A fertilised egg, or zygote and the 8 or 16 cells from its first few mitotic divisions are totipotent cells, which are destined eventually to produce a whole organism. They can also differentiate into extra-embryonic tissues like the amnion and umbilicus.

● **Pluripotent** – these stem cells can form all tissue types but not whole organisms. They are present in early embryos and are the origin of the different types of tissue within an organism.

● **Multipotent** – these stem cells can only form a range of cells within a certain type of tissue. Haematopoetic stem cells in bone marrow are multipotent because this gives rise to the various types of blood cell.

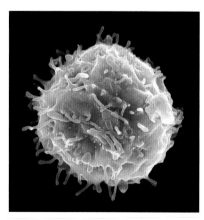

▲ Figure 1 *The stem cell in this coloured scanning electron micrograph is destined to become a blood cell. Some differentiated red blood cells can also be seen, × 100 magnification*

Differentiation

Multicellular organisms like animals and plants have evolved from unicellular (single-celled) organisms because groups of cells with different functions working together as one unit can make use of resources more efficiently than single cells operating on their own.

In multicellular organisms cells have to specialise to take on different roles in tissues and organs. They may be required to form barriers such as skin or be motile such as sperm cells. Cells have adapted to different roles in an organism and so have many shapes (and sizes) and often contain different organelles.

Erythrocytes (red blood cells) and neutrophils (white blood cells) are both present in blood (Topic 6.4, Table 1). They look very different because they have different functions. When cells differentiate they become adapted to their specific role. What form this adaptation takes is dependent on the function of the tissue, organ and organ system to which the cell belongs.

All blood cells are derived from stem cells in the **bone marrow**.

Replacement of red and white blood cells

Mammalian erythrocytes are essential for the transport of oxygen around the body. They are adapted to maximise their oxygen-carrying capacity by having only a few organelles so there is more room for haemoglobin.

Due to the lack of nucleus and organelles they only have a short lifespan of around 120 days. They therefore need to be replaced constantly. The stem cell colonies in the bone marrow produce approximately three billion erythrocytes per kilogram of body mass per day to keep up with the demand.

Neutrophils have an essential role in the immune system. They live for only about 6 hours and the colonies of stem cells in bone marrow produce in the region of 1.6 billion per kg per hour. This figure will increase during infection.

Sources of animal stem cells

- Embryonic stem cells – these cells are present at a very early stage of embryo development and are totipotent. After about seven days a mass of cells, called a blastocyst, has formed and the cells are now in a pluripotent state. They remain in this state in the fetus until birth.

- Tissue (adult) stem cells – these cells are present throughout life from birth. They are found in specific areas such as bone marrow. They are multipotent, although there is growing evidence that they can be artificially triggered to become pluripotent. Stem cells can also be harvested from the umbilical cords of newborn babies. The advantages of this source are the plentiful supply of umbilical cords and that invasive surgery is not needed. These stem cells can be stored in case they are ever

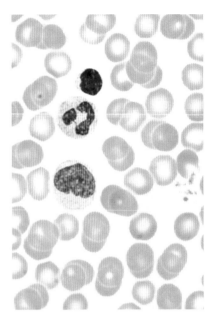

▲ Figure 2 Light micrograph of human blood – red blood cells (erythrocytes) are pink and the purple cells are white blood cells (from left to right a monocyte, a neutrophil and a lymphocyte), × 500 magnification

▲ Figure 3 Scanning electron micrograph of human embryonic stem cell, × 500 magnification

needed by the individual in the future, and tissues cultured from such stem cells would not be rejected in a transplant to the umbilicus' owner.

Sources of plant stem cells

Stem cells are present in **meristematic tissue** (**meristems**) in plants. This tissue is found wherever growth is occurring in plants, for example at the tips of roots and shoots (termed apical meristems).

Meristematic tissue is also located sandwiched between the phloem and xylem tissues and this is called the vascular cambium. Cells originating from this region differentiate into the different cells present in xylem and phloem tissues (Topic 6.4, The organisation and specialisation of cells). In this way the vascular tissue grows as the plant grows. The pluripotent nature of stem cells in the meristems continues throughout the life of the plant.

Uses of stem cells

Stem cells transplanted into specific areas have the potential to treat certain diseases, such as:

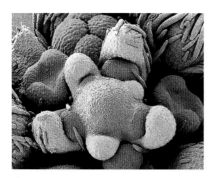

▲ Figure 4 *Scanning electron micrograph of a shoot apical meristem (purple) on a flowering plant. Floral buds (red) are appearing between developing young leaves (green). The coloration seen here is computer generated, × 300 magnification*

- heart disease – muscle tissue in the heart is damaged as a result of a heart attack, normally irreparably – this has been tried experimentally with some success already

- type 1 diabetes – with insulin-dependent diabetes the body's own immune system destroys the insulin-producing cells in the pancreas; patients have to inject insulin for life – this has been tried experimentally with some success already

- Parkinson's disease – the symptoms (shaking and rigidity) are caused by the death of dopamine-producing cells in the brain; drugs currently only delay the progress of the disease

- Alzheimer's disease – brain cells are destroyed as a result of the build up of abnormal proteins; drugs currently only alleviate the symptoms

- macular degeneration – this condition is responsible for causing blindness in the elderly and diabetics; scientists are currently researching the use of stem cells in its treatment and early results are very encouraging

- birth defects – scientists have already successfully reversed previously untreatable birth defects in model organisms such as mice.

- spinal injuries – scientists have restored some movement to the hind limbs of rats with damaged spinal cords using stem cell implants.

Stem cells are already used in such diverse areas as:

- the treatment of burns – stem cells grown on biodegradable meshes can produce new skin for burn patients, this is quicker than the normal process of taking a graft from another part of the body

- drug trials – potential new drugs can be tested on cultures of stem cells before being tested on animals and humans

- developmental biology – with their ability to divide indefinitely and differentiate into almost any cell within an organism, stem cells have become an important area of study in developmental biology. This is the study of the changes that occur as multicellular organisms grow and develop from a single cell, such as a fertilised egg – and why things sometimes go wrong.

Ethics

Stem cells have been used in medicine for many years in the form of bone marrow transplants. More recently, the use of embryonic stem cells in therapies and research has lead to controversy and debates regarding the ethics of such use.

The embryos used originally were donated from those left over after fertility treatment. More recently the law in the UK has changed so that embryos can be specifically created in the laboratory as a source of stem cells.

The removal of stem cells from embryos normally results in the destruction of the embryos, although techniques are being developed that will allow stem cells to be removed without damage to embryos.

There are not only religious objections to the use of embryos in this way but moral objections too – many people believe that life begins at conception and the destruction of embryos is, therefore, murder. There is a lack of consensus as to when the embryo itself has rights, and also who owns the genetic material that is being used for research.

This controversy is holding back progress that could lead to the successful treatment of many incurable diseases. The use of umbilical cord stem cells overcomes these issues to a large extent, but these cells are merely multipotent, not pluripotent like embryonic stem cells, thus restricting their usefulness. Adult tissue stem cells can also be used but they do not divide as well as umbilical stem cells and are more likely to have acquired mutations. Developments are being made towards artificially transforming tissue stem cells into pluripotent cells. Induced pluripotent stem cells (iPSCs) are adult stem cells that have been genetically modified to act like embryonic stem cells and so are pluripotent.

The use of plant stem cells does not raise the same ethical issues as animal cells.

> **Synoptic link**
>
> You will learn more about ethical issues surrounding research in Topic 21.5, Gene technology and ethics.

> **Synoptic link**
>
> You will learn about T cells and B cells in immunity in Topic 12.6, The specific immune system.

 Gene therapy using stem cells

Children born with the rare genetic condition Severe Combined Immunodeficiency (SCID) are extremely vulnerable to all infections and without treatment are unlikely to live for more than a year. They produce no T cells, and without T cells the B cells do not function either (T cells and B cells are types of white blood cell).

Normally SCID is treated with a bone marrow transplant, which depends on finding a matching donor. The transplanted stem cells divide and differentiate into the different types of white blood cells needed for a healthy immune system.

More recently experimental gene therapy has been used to treat SCID. The aim is that stem cells from the patient's own bone marrow are removed and genetically modified so that they function normally to produce the white blood cells needed. These are then put back into the patient and the condition should be corrected.

This treatment was initially successful in a small number of children, but in some of the children another gene was damaged in the process and they went on to develop leukaemia. However, gene therapy is still seen as having the most potential for treating SCID in the future.

1 Explain how this condition can be cured using a bone marrow transplant.
2 Explain the risks of a bone marrow transplant from a donor, and how the gene therapy described removes this risk.
3 Suggest why some patients receiving this gene therapy developed cancer.

 ## Plant stem cells and medicines

Plant stem cells have a huge potential role to play in medicine. Many drugs used in medicines are derived from plants. Plant stem cells can be cultured, leading to an unlimited, and cheap, supply of plant-based drugs.

Paclitaxel is a common drug used in the treatment of breast and lung cancer. It cannot be chemically synthesised and must be obtained from the bark of yew trees (*Taxus brevifolia*). The trees have to be mature, which means the supply is limited and the extraction process difficult and expensive. An alternative way of producing the drug was developed using a related plant but it is still a difficult and expensive process. Recently stem cells from the yew tree have been used to produce paclitaxel cheaply and in sustainable quantities.

1 Suggest an environmental benefit of the use of plant stem cells in medicine production.

Summary questions

1 Describe the difference between pluripotent and multipotent with regard to stem cells, and state a source for both cell types in animals. (4 marks)

2 State where you would find meristematic tissue in a plant and explain the importance of the position of meristematic tissue to a plant. (5 marks)

3 Evaluate the advantages and disadvantages of using embryonic stem cells in medical research. (4 marks)

4 Alzheimer's is a progressive disease resulting in the loss of neurones in the brain. An area known as the cerebral cortex is primarily affected. There is a reduction in the quantity of the neurotransmitter acetylcholine released, which is necessary for memory and learning. This leads to the symptoms of dementia such as memory loss, mood swings, and confusion which get progressively worse over time. At the present time there is no cure and the few drugs available only temporarily relieve the symptoms.

Parkinson's disease is due to neurones in a part of the brain called the substantia nigra dying. This leads to a reduction in the amount of the neurotransmitter dopamine released. The lack of dopamine leads to loss of the fine control of movement causing shaking, slowness, and rigidity. There are drugs available which treat the symptoms effectively but as the disease progresses the increasing doses needed result in more and more side effects.

a Explain why the use of stem cells is called regenerative medicine. (2 marks)
b Evaluate the use of stem cells in the potential treatment of the two diseases described above. (5 marks)

Practice questions

1   The cell cycle is a highly ordered sequence of events that takes place in a cell, resulting in division of the cell.

Which of the following statements is/are correct with respect to the cell cycle?

Statement 1: S – DNA is replicated in the nucleus.

Statement 2: G_1 – the cell continues to increase in size, energy stores are increased and the DNA is checked for errors.

Statement 3: G_2 – proteins from which organelles are synthesised are produced and organelles replicate.

A   1, 2 and 3 are correct

B   Only 1 and 2 are correct

C   Only 2 and 3 are correct

D   Only 1 is correct   (*1 mark*)

2   Fertilised eggs are transported along a structure called the fallopian tube to the uterus. The fallopian tubes are lined with ciliated epithelium.

a   Define the term 'differentiation'. (*2 marks*)

b   Describe how ciliated epithelial cells are adapted to their function   (*3 marks*)

c   Explain the difference between ciliated cells and ciliated epithelium   (*4 marks*)

d   Suggest how the ciliated epithelium is involved in the transport of fertilised eggs.
                                        (*2 marks*)

e   List the letters in the correct order to show correctly increasing levels of organisation in a living organism.

   A   is a collection of tissues that are adapted to perform a particular function in an organism

   B   is made up of a collection of differentiated cells that have a specialised function

   C   is made up of not one but hundreds, thousands or millions of cells

   D   is the smallest structural and functional unit of **B**

   E   is composed of more than one **A** working together to carry out a major function in the organism   (*1 mark*)

3   a   The figure shows some drawings of a cell during different stages of mitosis.

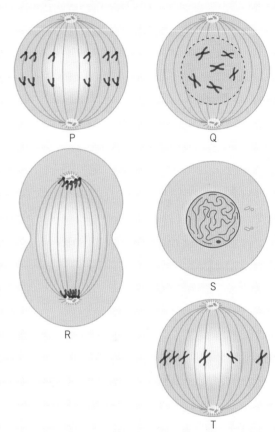

P   Q

R   S

T

Place stages P, Q, R, S and T in the correct sequence

The first stage has been identified for you.

S...................................... (*4 marks*)

b Mitosis is part of the cell cycle. The figure shows a diagram of the cell cycle

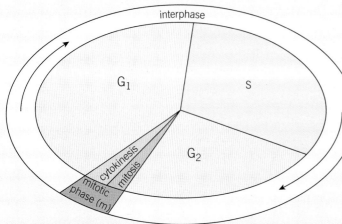

(i) Name one process that occurs during stages G_1 and G_2 *(1 mark)*

(ii) During stage S, the genetic information is copied and checked.

Suggest what might happen if the genetic information is not checked. *(2 marks)*

c During **meiosis** a cell undergoes two divisions.

Suggest how cells produced by meiosis may differ from those produced by mitosis. *(2 marks)*

OCR F211 2009

4 a A fertilised egg cell undergoes mitosis producing two cells in 36 hours. The cells continue to undergo mitosis completing each cycle in eight hours.

Calculate how many cells there will be present three days after fertilisation. *(2 marks)*

b (i) Explain why the nuclei of most human cells contain 46 pairs of chromosomes. *(2 marks)*

(ii) Name the type of cell which contain half the number of chromosomes present in a diploid cell. *(1 mark)*

c Copy and complete the table below.

(4 marks)

	Meiosis	Mitosis
Homologous chromosomes	pair up	
Daughter cells n/2n		
Number of cell divisions		
Crossing over		✗

The figure below shows drawings of the six chromosomes inside an animal cell viewed during later prophase of mitosis.

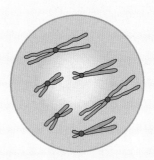

d (i) Identify **one pair** of homologous chromosomes in the diagram by drawing around each chromosome in the pair **on the diagram**. *(1 mark)*

The nucleus of a sperm cell is produced by **meiosis.**

(ii) Draw a diagram in the space below to represent the chromosomes that are present in the nucleus of a sperm cell from **the same animal**. *(2 marks)*

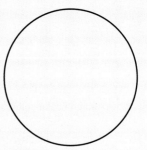

OCR F211 2010

e Explain why prokaryotic cells do not undergo meiosis. *(3 marks)*

5 The diagram below shows the relative times that cells can spend in each stage of the cell cycle.

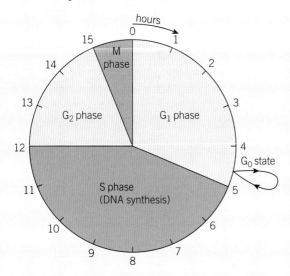

a Calculate the percentage of the overall time that the cell spends in interphase.

(*2 marks*)

b Outline the importance of strict regulation of the cell cycle. (*3 marks*)

6 Stem cells are being investigated as a possible cure for certain types of diabetes. In type 1 diabetes the immune system recognises the beta cells in the pancreas, which produce insulin, as foreign and destroys them. Stem cells, which can be obtained from different sources, can be used to replace the beta cells that have been destroyed.

Some of these stem cells are particular useful as they do not trigger a response by the immune system.

a State what is meant by the term stem cell.

(*2 marks*)

b List three sources of stem cells. (*3 marks*)

c Explain the meaning of stem cell potency.

(*5 marks*)

d Describe three examples of the use of stem cells to reverse disease. (*6 marks*)

Module 2 summary

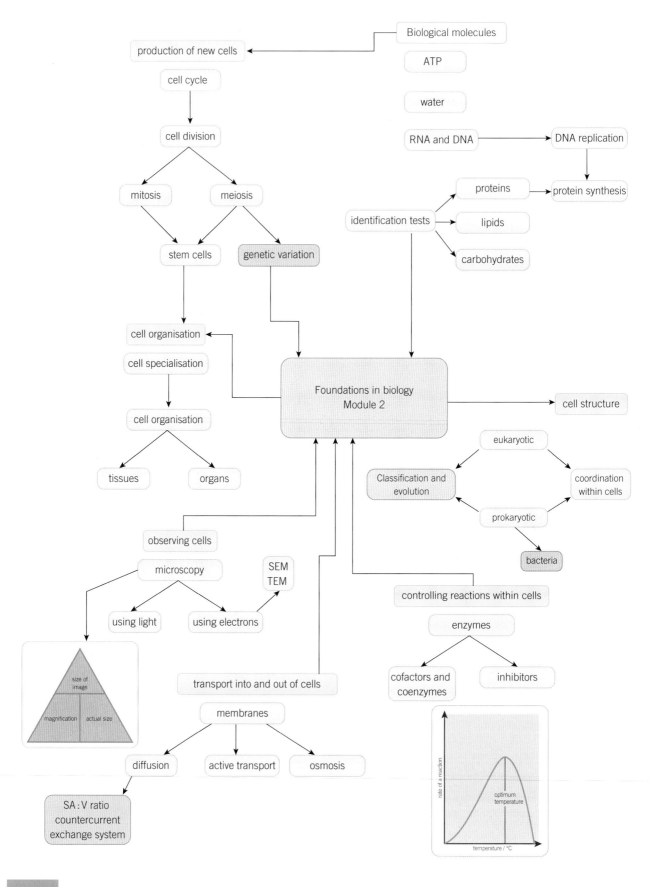

Biological molecules

production of new cells

cell cycle

cell division

mitosis

meiosis

stem cells

genetic variation

ATP

water

RNA and DNA → DNA replication

protein synthesis

identification tests → proteins → protein synthesis

lipids

carbohydrates

cell organisation

cell specialisation

cell organisation

tissues

organs

Foundations in biology
Module 2

cell structure

eukaryotic

Classification and evolution

coordination within cells

prokaryotic

bacteria

observing cells

microscopy

SEM TEM

using light

using electrons

controlling reactions within cells

enzymes

cofactors and coenzymes

inhibitors

size of image / magnification / actual size

transport into and out of cells

membranes

diffusion

active transport

osmosis

SA : V ratio
countercurrent
exchange system

rate of a reaction / optimum temperature / temperature / °C

Application

A study published in the New England Journal of Medicine in 2013 showed that a Mediterranean diet, high in olive oil, nuts, fruit and vegetables and low in red meat and pastries gave a substantial reduction in the risk of death from heart attacks, strokes and cardiovascular disease. 7447 people took part in the study. All were between 55 and 80 and judged to be at relatively high risk of cardiovascular disease. The mean participation time was 4.8 years. One group had a Mediterranean diet with extra olive oil, one group had a Mediterranean diet with added nuts and the control group had a low fat diet.

One set of results is shown below:

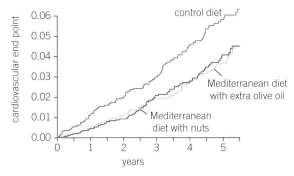

▲ **Figure 1** *Graph to show the effect of diet on the cardiovascular end point, which is the occurrence of myocardial infarctions (heart attacks), stroke and deaths from cardiovascular disease.*

1 In this study one of the major components controlled in the diets of the people taking part was their lipid consumption.
  a What are lipids?
  b Where would you expect to find lipids in the ultrastructure of a cell?
  c What has been the role of the electron microscope in helping us understand the role of lipids in the ultrastructure of the cell?

2 In the study, people on the Mediterranean diet had to eat lots of olive oil or nuts. These both contain monounsaturated fatty acids. What are these, and how do they differ from other fatty acids?

3 a Summarise the data from Figure 1 to describe the effect of the three different diets on the risk of cardiovascular disease.
  b Suggest two additional pieces of evidence you might want to see before accepting the findings of this study.
  c In 2014 a meta-analysis study comparing the results from 72 studies showed no statistically relevant link between saturated fats and heart disease, and no protective effect of cetain types of polyunsaturated fats. Write a news report on this large study, explaining the potential impact of the findings and the importance of the study to the general public.

Extension

1 If a patient has a suspected myocardial infarction there are a number of biochemical markers which occur in the blood which can help doctors make a diagnosis of what is happening inside the body. Research each of these key markers and write a brief report, describing the biochemistry of the molecule and explaining why they are an indicator that a patient has had a heart attack:
  a Cardiac troponins I and T
  b Creatine kinase
  c myoglobin
  d natriuretic peptides (peptides)

2 Obesity is closely linked with many health issues, including cardiovascular health. Prepare a poster presentation looking at the biochemistry and cell biology of obesity – show the biochemistry of lipids and carbohydrates, summarise how they are linked to obesity and investigate fat cells, including their microscopic appearance and adaptations to their functions in the body.

1 a Explain the meaning of the terms *cation*, *anion* and *electrolyte*. (*3 marks*)

b Copy and complete the table (*5 marks*)

Ion	Function
calcium ions (Ca^{2+})	
phosphate ions (PO_4^{3-})	
sodium ions (Na^+)	
potassium ions (K^+)	
ammonium ions, (NH_4^+)	

2 The diagram summarises the process of DNA replication.

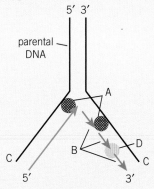

a State the names of A, B, C, and D. (*4 marks*)

The diagram outlines three different models of DNA replication.

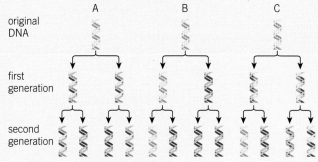

b **(i)** Outline what is meant by the term 'model'. (*2 marks*)

(ii) Identify which of the models in the diagram shows semi-conservative replication. (*1 mark*)

(iii) Discuss why the semi-conservative model is the most efficient way for DNA to be replicated. (*4 marks*)

3 The activity of an enzyme, laccase, was investigated using different concentrations of copper solution. The results are shown in the graph.

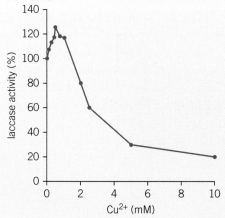

a **(i)** Describe the change in enzyme activity shown in the graph (*3 marks*)

(ii) Suggest reasons for the changes that you have described. (*4 marks*)

The activity of the same enzyme was investigated at different temperatures. The results are shown in the graph.

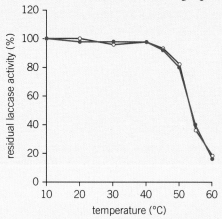

b Explain the the reason for the trend shown in the graph. (*4 marks*)

4 Consider the scientific drawings shown here.

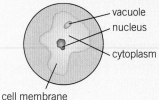

cell membrane

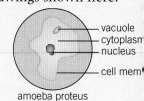

amoeba proteus

a State, giving your reasons, which is the better scientific drawing. *(4 marks)*

The photomicrograph shown was taken during the process of translation.

The magnification of the image is ×362 500.

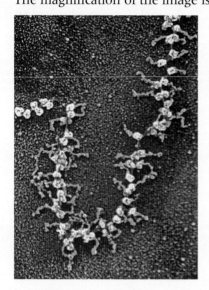

b (i) Calculate, using the photo, the mean size of a ribosome. *(3 marks)*

(ii) Suggest the name of the microscope used to capture this image. *(2 marks)*

(iii) Outline the main steps involved in translation. *(4 marks)*

5 The diagram shows four cells that have been placed in different solutions.

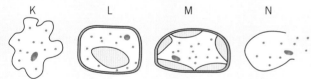

a In the table write the letter **K**, **L**, **M** or **N** next to the description that best matches the diagram. One has been done for you.

description	letter
an animal cell that has been placed in distilled water	
an animal cell that has been placed in a concentrated sugar solution	
a plant cell that has been placed in distilled water	
a plant cell that has been placed in concentrated sugar solution	**M**

(3 marks)

b Explain, using the term **water potential**, what has happened to cell **M**. *(3 marks)*

c Small non-polar substances enter cells in different ways to large or polar substances. Outline the ways in which the substances below, can enter a cell through the plasma (cell surface) membrane.

● Small, non-polar substances

● Large substances

● Polar substances *(5 marks)*

OCR F211 2009
From the AS paper 1 practice questions

6 A student carried out an investigation into the effect of the enzyme inhibitor, lead nitrate, on the activity of the enzyme amylase.

The results that were obtained are shown in the table.

percentage concentration of lead nitrate solution	transmission of light / arbitrary units			mean transmission of light / arbitrary units
	first run	second run	third run	
0	84	87	82	84.3
0.2	55	53	52	53.3
0.4	36	37	36	36.3
0.6	27	24	27	26.0
0.8	22	20	21	21.0
1.0	20	21	18	

a (i) Outline a procedure that the student could have followed to obtain the results. *(4 marks)*

(ii) State the missing value from the table *(1 mark)*

b (i) The third run for 1.0% solution of lead nitrate initially produced a reading of 70 arbitrary units.

The student discarded this result and repeated this run. Explain why. *(2 marks)*

(ii) Plot a graph using the results in the table. *(3 marks)*

c (i) Describe and explain the shape of the graph. (*4 marks*)

(ii) State an appropriate conclusion for these results. (*1 mark*)

The student stated in their conclusion:

'*the enzyme amylase was inhibited at all concentrations of lead nitrate*'

d Discuss, whether or not, the student was correct in making this statement. (*4 marks*)

e Evaluate the validity of this investigation. (*4 marks*)

f Suggest how this investigation could be improved, (*1 mark*)

From the AS paper 2 practice questions

7 a Copy and complete the following paragraph about cells by using the most appropriate term(s).

Cell that are not specialised but still have the ability to divide are called …………… cells. Such cells can be found in the …………….. of the long bones of mammals. These cells can …………… into other types of cell, such as erythrocytes that carry oxygen in the blood. In plants, ………………….. tissues also contains cells that are not specialised. (*4 marks*)

b Sponges are simple eukaryotic multicellular organisms that live underwater on the surface of rocks. Sponges have a cellular level of organisation. This means they have no tissues.

Each cell type is specialised to perform a particular function. One type of cell found in a sponge is a collar cell. Collar cells are held in positions on the inner surface of the body of the sponge. Figure 3 is a diagram showing a vertical section through the body of a sponge and an enlarged drawing of a collar cell.

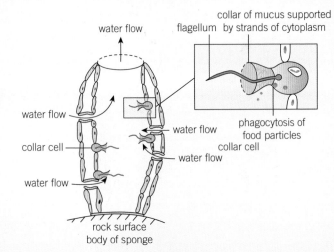

(i) Suggest **one** function of the flagellum in the collar cell. (*1 mark*)

(ii) Suggest **one** possible role for the collar of mucus in the cell (*1 mark*)

c In more advanced organisms, cells are organised into tissues consisting of one or more types of specialised cells.

Describe how cells are organised into tissues, using **xylem** and **phloem** as examples (*4 marks*)

OCR F211 2012

From the AS paper 1 practice questions

8 A student investigating how different concentrations of sucrose solution affect the size of animal cells obtained the results shown in the table.

	Concentration of sucrose solution / mol dm⁻³				
	0.05	0.10	0.20	0.40	0.80
diameter of cell 1 / μm	8.4	7.2	6.6	5.7	2.3
diameter of cell 2 / μm	7.8	7.3	6.8	5.7	2.5
diameter of cell 3 / μm	8.1	7.4	7.0	5.5	2.4
mean diameter / μm	8.1	7.3	6.8		2.4
mean change in diameter / μm	+1.1	+0.3	−0.2		−4.6

a The original mean diameter of the cells was 7.0 µm. Copy and complete the table. *(2 marks)*

b Plot a graph using the results obtained in the table. *(4 marks)*

c Describe and explain the trend shown by the graph. *(4 marks)*

d Suggest the type of cells the student used in the investigation. *(1 mark)*

From the AS paper 1 practice questions

9 A student investigating rate diffusion carried out the following procedure.

1 Prepared a petri-dish containing a layer of agar.

2 Cut a 1 cm well in the centre of the agar.

3 Placed 10 drops of a coloured solution in the well.

4 Measured the distance travelled by the colored solution from the edge of the well every 15 minutes.

The results obtained by the student are shown in the table.

time / min	distance diffused from well by coloured solution / mm
0	0
15	14
30	22
45	26
60	28
75	29

a Plot a graph using the results obtained in the table. *(4 marks)*

b Calculate, using the graph, the rate of diffusion of the solution between 10 minutes and 20 minutes. *(4 marks)*

c Describe and explain the trend shown by the graph. *(4 marks)*

The smallest units on the ruler used to measure the distances diffused by the coloured solution were 1 mm.

d **(i)** State the uncertainty of the measurements obtained by the student. *(1 mark)*

(ii) Calculate the percentage error for the measurement taken at 45 minutes. *(4 marks)*

(iii) Suggest how the student could have improved the precision of the measurements. *(1 mark)*

(iv) Suggest how the student could have improved the reliability of the investigation. *(2 marks)*

e Evaluate the validity of the investigation. *(4 marks)*

From the AS paper 1 practice questions

10 A student carried out an investigation into the effect of pH on the activity of the enzyme amylase.

a Suggest a hypothesis that the student made before starting the investigation. *(2 marks)*

b **(i)** State the independent and dependent variables in this investigation. *(2 marks)*

(ii) State two variables that should be controlled in this investigation. *(2 marks)*

The student obtained the results shown in the table.

pH		5	6	7	8	9
Time taken for starch to be broken down/min	First run	11	7	3	4	10
	Second run	10	6	4	5	9
	Third run	8	7	3	6	10

c Draw a table to present the raw data correctly. Calculate and include the mean values in the table that you produce. *(4 marks)*

d Describe how the student increased the reliability of the investigation. *(2 marks)*

e Discuss whether the results obtained by the student support your hypothesis. *(4 marks)*

From the AS paper 2 practice questions

MODULE 3
Exchange and transport

Introduction

All living organisms need to move materials into and out of their cells. Commonly this includes getting nutrients and oxygen in and carbon dioxide and other waste products out. Simple diffusion works well for single celled organisms but larger, multi-cellular organisms need to transport materials from the outside word into their bodies before they can pass into the individual cells. Exchange surfaces and transport systems are vital for these exchanges to take place in plants and animals alike.

Exchange surfaces explores the need for specialised exchange surfaces, and what makes an effective one, before moving on to specific examples. You will learn about the mammalian gaseous exchange system, how air is moved in and out of the system and the interrelationships between the volume of the lungs and the rate of breathing. You will compare the gas exchange surfaces with those of insects and fish, adapted for very different bodies and lifestyles.

Transport in animals explains why, as animals become larger and more active, transport systems become essential to supply nutrients and oxygen to and waste products from individual cells. The key roles of the blood, the blood vessels and the heart in this transport are fully explored including the electrical control of the heart beat and how this can be recorded using an ECG.

Transport in plants describes the key transport systems in plants. You will learn how both the supply of nutrients from the soil and the movement of the products of photosynthesis around the plant depend on the flow of water through the vascular system made up of the xylem and the phloem.

Knowledge and understanding checklist

From your Key Stage 4 study you should be able to answer the following questions. Work through each point, using your Key Stage 4 notes and other resources. There is also support available on Kerboodle.

☐ Explain how substances are transported into and out of cells though diffusion, osmosis, and active transport.

☐ Explain the need for exchange surfaces and a transport system in terms of surface area: volume ratio.

☐ Describe some of the substances transported into and out of a range of organisms in terms of the requirements of those organisms to include oxygen, carbon dioxide, water, dissolved food molecules, mineral ions, and urea.

☐ Describe the human circulatory system, including the relationship with the gaseous exchange system, and explain how the structure of the heart and the blood vessels are adapted to their functions.

☐ Explain how red blood cells, white blood cells, platelets and plasma are adapted to their functions in the blood.

☐ Explain how the structure of the xylem and the phloem are adapted to their functions in the plant.

☐ Explain how water and mineral ions are taken up by plants, relating the structure of the root hair cells to their function.

☐ Describe the processes of transpiration and translocation, including the structure and function of the stomata.

☐ Explain the effect of a variety of environmental factors on the rate of water uptake by a plant, to include light intensity, air movement, and temperature.

Maths skills checklist

In this module, you will need to use the following maths skills.

☐ **Recognise and make use of appropriate units in calculations.** You will need to be able to do this in all your calculations of volume, breathing rates, etc.

☐ **Recognise and use expressions in decimal and standard form.** You will need this to analyse and interpret primary and secondary data relating to lung volumes and breathing rates.

☐ **Use ratios, fractions, and percentages.** You will need this to calculate surface area:volume ratios.

☐ **Estimate results.** You will need this to understand surface area: volume ratios.

MyMaths.co.uk
Bringing Maths Alive

EXCHANGE SURFACES AND BREATHING
7.1 Specialised exchange surfaces
Specification reference: 3.1.1

▲ **Figure 1** *Getting oxygen into the muscle cells of a dolphin isn't as easy as getting it into an* <u>Amoeba</u>

Synoptic link

You have already learnt about diffusion in Topic 5.3, Diffusion.

Imagine a microscopic single-celled organism such as *Amoeba* drifting in the ocean currents. Now think of a dolphin, swimming at high speeds, hunting fish or playing with other dolphins. They both need glucose and oxygen for cellular respiration and produce waste carbon dioxide, which must be removed. However, the quantities involved and the distances the substances need to travel, are very different.

The need for specialised exchange surfaces

In microscopic organisms such as *Amoeba* all of the oxygen needed by the organism, and the waste carbon dioxide produced, can be exchanged with the external environment by diffusion through the cell surface. The distances the substances have to travel are very small.

There are two main reasons why diffusion alone is enough to supply the needs of single-celled organisms:

- The metabolic activity of a single-celled organism is usually low, so the oxygen demands and carbon dioxide production of the cell are relatively low.
- The surface area to volume (SA:V) ratio of the organism is large (see below).

As organisms get larger they can be made up of millions or even billions of cells arranged in tissues, organs, and organ systems. Their metabolic activity is usually much higher than most single-celled organisms. Think of the dolphin in Figure 1. The amount of energy used in moving through the water means the oxygen demands of the muscle cells deep in the body will be high and they will produce lots of carbon dioxide. The distance between the cells where the oxygen is needed and the supply of oxygen is too far for effective diffusion to take place. What's more, the bigger the organism, the smaller the SA:V ratio. So gases can't be exchanged fast enough or in large enough amounts for the organism to survive.

Surface area : volume ratio – modelling an organism

The SA:V ratio is important in many areas of biology. A sphere is a useful shape for modelling cells or organisms. A series of simple calculations shows clearly how the SA:V ratio changes as the organism gets bigger, and why size matters so much.

Worked example: SA : V ratios

Compare the SA : V ratios of organisms with a radius of 1, 3 and 10 arbitrary units (au) and explain how this affects their ability to exchange materials with the environment.

The surface area of a sphere is calculated using the formula $4\pi r^2$

The volume of a sphere is calculated using the equation $\frac{4}{3}\pi r^3$

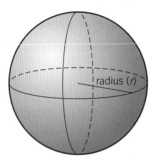

▲ **Figure 2** *The radius of a sphere*

To do these calculations you use 3.14 as the value of π.

Example 1: Radius = 1 au

Surface area is $4\pi r^2$ so: $4 \times 3.14 \times 1 \times 1 = 12.6$ (1dp)

Volume is $\frac{4}{3}\pi r^3$ so: $\frac{(4 \times 3.14 \times 1 \times 1 \times 1)}{3} = 4.2$ (1dp)

SA : V ratio is $\frac{12.6}{4.2} = 3 : 1$

Example 2: Radius = 3 au

Surface area is $4\pi r^2$ so: $4 \times 3.14 \times 3 \times 3 = 113.0$ (1dp)

Volume is $\frac{4}{3}\pi r^3$ so: $\frac{(4 \times 3.14 \times 3 \times 3 \times 3)}{3} = 113.0$ (1dp)

SA : V ratio is $\frac{113.0}{113.0} = 1 : 1$

Example 3: Radius = 10 au

Surface area is $4\pi r^2$ so: $4 \times 3.14 \times 10 \times 10 = 1256.0$ (1dp)

Volume is $\frac{4}{3}\pi r^3$ so: $\frac{(4 \times 3.14 \times 10 \times 10 \times 10)}{3} = 4186.7$ (1dp)

SA : V ratio is $\frac{1256}{4186} = 1 : 3$ (to the nearest whole number)

The bigger the organism, the smaller the surface area to volume ratio becomes, and the distances that substances need to travel from the outside to reach the cells at the centre of the body get longer. This makes it harder and ultimately impossible to absorb enough oxygen through the available surface area to meet the needs of the body.

Synoptic link

You will apply the same principles of SA : V ratio when you consider how nutrients are supplied to the cells of multicellular organisms in Chapter 8, Transport in animals and Chapter 9, Transport in plants. You will also consider SA : V ratio when looking at adaptations to reduce or increase heat loss in Topics 15.2 and 15.3, Thermoregulation in ectotherms and endotherms, respectively.

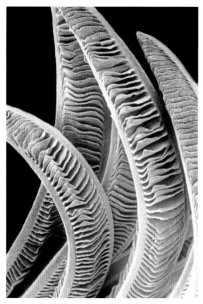

▲ Figure 3 *This scanning electron micrograph shows the large surface area of the gills of a fish, × 200 magnification*

Synoptic link

You will learn more about specialised exchange surfaces in many other topics. More detail on the increased surface area of root hair cells in plants and the villi in the small intestine can be found in Topic 9.2, Water transport in multicellular animals and Topic 8.3, Blood, tissue fluid, and lymph respectively.

Specialised exchange surfaces

Large, multicellular organisms have evolved specialised systems for the exchange of the substances they need and the substances they must remove.

All effective **exchange surfaces** have certain features in common. You will be looking at many of them in detail in this chapter. Here is a summary of the characteristic features of effective exchange surfaces, along with some examples:

● **Increased surface area** – provides the area needed for exchange and overcomes the limitations of the SA : V ratio of larger organisms. Examples include root hair cells in plants and the villi in the small intestine of mammals.

● **Thin layers** – these mean the distances that substances have to diffuse are short, making the process fast and efficient. Examples include the alveoli in the lungs (see next topic) and the villi of the small intestine.

● **Good blood supply** – the steeper the concentration gradient, the faster diffusion takes place. Having a good blood supply ensures substances are constantly delivered to and removed from the exchange surface. This maintains a steep concentration gradient for diffusion. For example the alveoli of the lungs, the gills of a fish and the villi of the small intestine.

● **Ventilation to maintain diffusion gradient** – for gases, a ventilation system also helps maintain concentration gradients and makes the process more efficient, for example the alveoli and the gills of a fish where ventilation means a flow of water carrying dissolved gases (see Topic 7.4, Ventilation and gas exchange in other organisms).

Summary questions

1 Explain why single-celled organisms do not need specialised exchange surfaces. *(4 marks)*

2 Describe the main features of any efficient exchange surface and explain how the structures relate to their functions. *(6 marks)*

3 One roughly spherical organism has a radius of 2 au. Another has a radius of 6 au. Compare the SA : V ratios of the organisms and use these to explain why the larger organisms need specialised exchange surfaces. *(6 marks)*

Animals that live on the land face a continual conflict between the need for gaseous exchange and the need for water. Gaseous exchange surfaces are moist, so oxygen dissolves in the water before diffusing into the body tissues. As a result the conditions needed to take in oxygen successfully are also ideal for the evaporation of water. Mammals have evolved complex systems that allow them to exchange gases efficiently but minimise the amount of water lost from the body. You are going to look at the human **gaseous exchange system** as an example of the specialised systems common to all mammals.

The human gaseous exchange system

Mammals are relatively big – they have a small SA : V ratio and a very large volume of cells. They also have a high metabolic rate because they are active and maintain their body temperature independent of the environment. As a result they need lots of oxygen for cellular respiration and they produce carbon dioxide, which needs to be removed. This exchange of gases takes place in the lungs.

> ### Learning outcomes
> Demonstrate knowledge, understanding, and application of:
> → the structures and functions of the components of mammalian gaseous exchange system
> → the mechanism of ventilation in mammals.

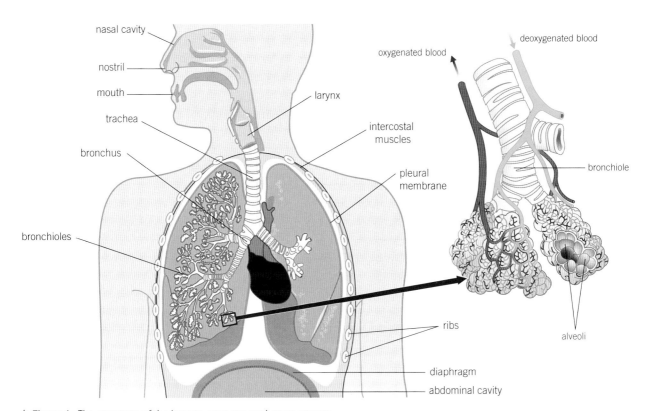

▲ Figure 1 *The structure of the human gaseous exchange system*

Key structures

Nasal cavity

The nasal cavity has a number of important features:

- a large surface area with a good blood supply, which warms the air to body temperature
- a hairy lining, which secretes mucus to trap dust and bacteria, protecting delicate lung tissue from irritation and infection
- moist surfaces, which increase the humidity of the incoming air, reducing evaporation from the exchange surfaces.

After passing through the nasal cavity, the air entering the lungs is a similar temperature and humidity to the air already there.

Trachea

The **trachea** is the main airway carrying clean, warm, moist air from the nose down into the chest. It is a wide tube supported by incomplete rings of strong, flexible **cartilage**, which stop the trachea from collapsing. The rings are incomplete so that food can move easily down the oesophagus behind the trachea.

The trachea and its branches are lined with a **ciliated epithelium**, with **goblet cells** between and below the epithelial cells (Figure 2). Goblet cells secrete mucus onto the lining of the trachea, to trap dust and microorganisms that have escaped the nose lining. The cilia beat and move the mucus, along with any trapped dirt and microorganisms, away from the lungs. Most of it goes into the throat and is swallowed and digested. One of the effects of cigarette smoke is that it stops these cilia beating.

Synoptic link

You learnt about cilia in Topic 2.4, Eukaryotic cell structure.

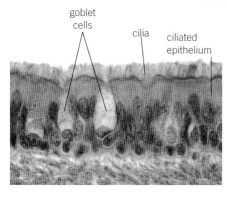
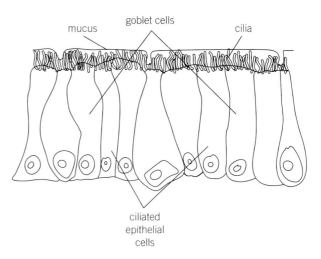

▲ **Figure 2** Left: *A light micrograph and.* Right: *a diagrammatic representation of a section through the lining of a trachea. The lining consists of mucus-secreting goblet cells and ciliated epithelial cells, × 300 magnification*

Bronchus

In the chest cavity the trachea divides to form the left bronchus (plural bronchi), leading to the left lung, and the right bronchus leading to the right lung. They are similar in structure to the trachea, with the same supporting rings of cartilage, but they are smaller.

Bronchioles

In the lungs the bronchi divide to form many small bronchioles. The smaller bronchioles (diameter 1 mm or less) have no cartilage rings. The walls of the bronchioles contain smooth muscle. When the smooth muscle contracts, the bronchioles constrict (close up). When it relaxes, the bronchioles dilate (open up). This changes the amount of air reaching the lungs. Bronchioles are lined with a thin layer of flattened epithelium, making some gaseous exchange possible.

Alveoli

The alveoli (singular alveolus) are tiny air sacs, which are the main gas exchange surfaces of the body. Alveoli are unique to mammalian lungs. Each alveolus has a diameter of around 200–300 μm and consists of a layer of thin, flattened epithelial cells, along with some collagen and elastic fibres (composed of elastin). These elastic tissues allow the alveoli to stretch as air is drawn in. When they return to their resting size, they help squeeze the air out. This is known as the **elastic recoil** of the lungs.

> **Synoptic link**
>
> You learnt about specialised epithelial cells in Topic 6.4, The organisation and specialisation of cells.

> **Synoptic link**
>
> You learnt about collagen and elastin in Topic 3.6, Structure of proteins.

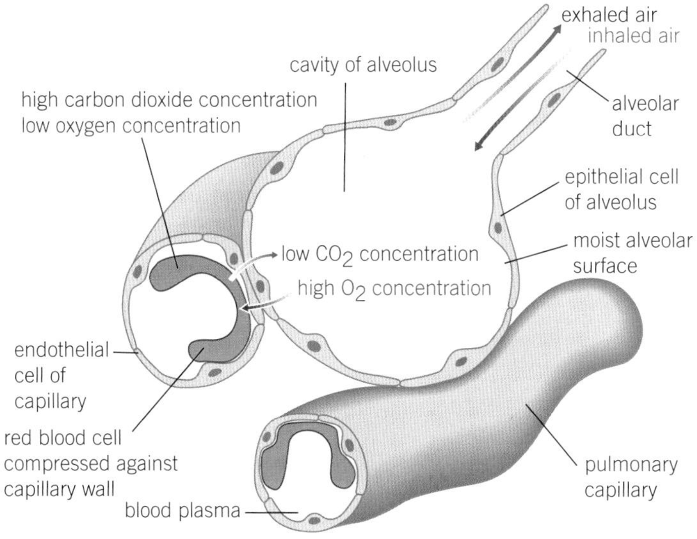

exhaled air
inhaled air
cavity of alveolus
high carbon dioxide concentration
low oxygen concentration
alveolar duct
epithelial cell of alveolus
low CO_2 concentration
high O_2 concentration
moist alveolar surface
endothelial cell of capillary
red blood cell compressed against capillary wall
blood plasma
pulmonary capillary

▲ **Figure 3** *Gaseous exchange in an alveolus*

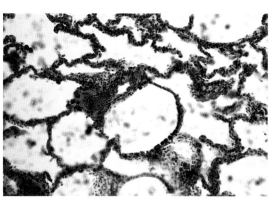

▲ **Figure 4** *The single cell walls of the alveoli and the capillaries (containing red blood cells) that make up the structure of the lung can be seen clearly in this light micrograph, × 60 magnification*

The main adaptations of the alveoli for effective gaseous exchange include:

- Large surface area – there are 300–500 million alveoli per adult lung. The alveolar surface area for gaseous exchange in the two lungs combined is around 50–75 m^2. The average floor area of a 4-bedroom

> **Synoptic link**
>
> You learnt about cell membranes and how they are freely permeable to gases such as oxygen and carbon dioxide in Topic 5.3, Diffusion.

house in the UK is only 67 m². If the lungs were simple, balloon-like structures, the surface area would not be big enough for the amount of oxygen needed to diffuse into the body. This demonstrates again the importance of the SA:V ratio (Topic 7.1, Specialised exchange surfaces).

- Thin layers – both the alveoli and the capillaries that surround them have walls that are only a single epithelial cell thick, so the diffusion distances between the air in the alveolus and the blood in the capillaries are very short.

- Good blood supply – the millions of alveoli in each lung are supplied by a network of around 280 million capillaries. The constant flow of blood through these capillaries brings carbon dioxide and carries off oxygen, maintaining a steep concentration gradient for both carbon dioxide and oxygen between the air in the alveoli and the blood in the capillaries.

- Good ventilation – breathing moves air in and out of the alveoli, helping maintain steep diffusion gradients for oxygen and carbon dioxide between the blood and the air in the lungs.

The inner surface of the alveoli is covered in a thin layer of a solution of water, salts and **lung surfactant**. It is this surfactant that makes it possible for the alveoli to remain inflated. Oxygen dissolves in the water before diffusing into the blood, but water can also evaporate into the air in the alveoli. Several of the adaptations of the human gas exchange system are to reduce this loss of water.

Ventilating the lungs

Air is moved in and out of the lungs as a result of pressure changes in the thorax (chest cavity) brought about by the breathing movements. This movement of air is called ventilation.

The rib cage provides a semi-rigid case within which pressure can be lowered with respect to the air outside it. The diaphragm is a broad, domed sheet of muscle, which forms the floor of the thorax. The external intercostal muscles and the internal intercostal muscles are found between the ribs. The thorax is lined by the pleural membranes, which surround the lungs. The space between them, the pleural cavity, is usually filled with a thin layer of lubricating fluid so the membranes slide easily over each other as you breathe.

Inspiration

Inspiration (taking air in or inhalation) is an energy-using process.

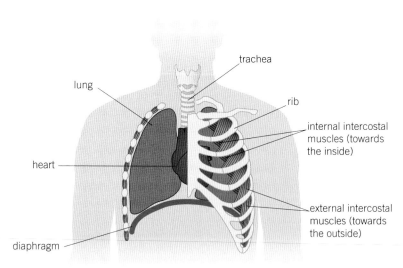

trachea

lung

rib

internal intercostal muscles (towards the inside)

heart

external intercostal muscles (towards the outside)

diaphragm

▲ **Figure 5** *The arrangement of the rib cage, diaphragm, and intercostal muscles*

The dome-shaped diaphragm contracts, flattening, and lowering. The external intercostal muscles contract, moving the ribs upwards and outwards. The volume of the thorax increases so the pressure in the thorax is reduced. It is now lower than the pressure of the atmospheric air, so air is drawn through the nasal passages, trachea, bronchi, and bronchioles into the lungs. This equalises the pressures inside and outside the chest.

Expiration

Normal expiration (breathing out or exhalation) is a passive process.

The muscles of the diaphragm relax so it moves up into its resting domed shape. The external intercostal muscles relax so the ribs move down and inwards under gravity. The elastic fibres in the alveoli of the lungs return to their normal length. The effect of all these changes is to decrease the volume of the thorax. Now the pressure inside the thorax is greater than the pressure of the atmospheric air, so air moves out of the lungs until the pressure inside and out is equal again.

You can exhale forcibly using energy. The internal intercostal muscles contract, pulling the ribs down hard and fast, and the abdominal muscles contract forcing the diaphragm up to increase the pressure in the lungs rapidly.

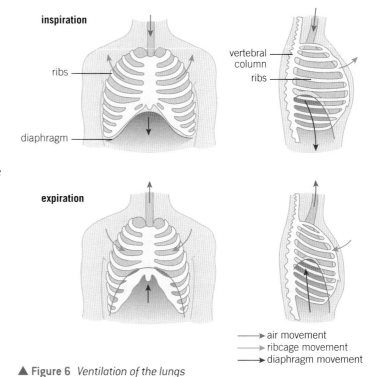

▲ **Figure 6** *Ventilation of the lungs*

 Attacking asthma

5.4 million people in the UK are currently being treated for asthma. They have airways that are sensitive to everyday triggers including house dust mites, cigarette smoke, pollen, and stress.

During an asthma attack, the cells lining the bronchioles release histamines, chemicals that make the epithelial cells become inflamed and swollen. Histamines stimulate the goblet cells to make excess mucus, and the smooth muscle in the bronchiole walls to contract. As a result, the airways narrow and fill with mucus, making it difficult to breathe.

Asthma medicines have been developed to reduce the symptoms and even prevent attacks. The drugs are delivered straight into the breathing system using an inhaler.

There are two main ways of treating asthma:

Relievers give immediate relief from the symptoms. They are chemicals similar to the hormone adrenaline. They attach to active sites on the surface membranes of smooth muscle cells in the bronchioles, making them relax and dilating the airways.

Preventers are often steroids, which are taken every day to reduce the sensitivity of the lining of the airways.

1 Explain how reliever medicines would overcome the symptoms of an asthma attack.
2 Explain how steroids could reduce the likelihood of an asthma attack.

The first breath

The first breath a newborn baby takes needs a force 15–20 times greater than any normal inhalation to inflate the lungs. The lungs are enormously stretched as the air flows in, and the elastic tissue never returns to its original length. This intake of breath is only possible because of special chemicals called lung surfactants containing phospholipids and both hydrophilic and hydrophobic proteins. The surfactant stops the alveoli collapsing and sticking together as the baby exhales. Without it, the second breath would be as difficult as the first, and continued breathing impossible.

Babies born at full term have alveoli coated in lung surfactant all ready for breathing. However, the cells of the alveoli do not produce enough surfactant for the lungs to work properly until around the 30th week of pregnancy. This is one reason why premature babies can struggle to breathe and may die. In recent years artificial lung surfactants have been produced. A tiny amount sprayed into the lungs of a premature baby coats the alveoli just like the natural surfactant, making breathing easier, helping to prevent lung damage and enabling many more babies to survive.

1 Discuss why the first breath taken is so much harder than any subsequent breaths and outline how artificial lung surfactants have improved survival rates for premature babies.

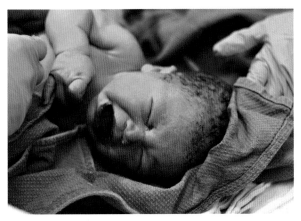

▲ **Figure 7** *The first breath a newborn takes is the hardest – and without lung surfactant it would be impossible*

Synoptic link

You learnt about phospholipids and proteins in Chapter 2, Basic components of living systems.

Summary questions

1 Explain how the structures below are adapted to make gaseous exchange as effective as possible.
   a nose (*3 marks*)
   b trachea (*3 marks*)
   c bronchioles (*3 marks*)

2 a Explain how the alveoli are adapted for gaseous exchange. (*4 marks*)
   b In some diseases, the structure of the alveoli breaks down to give much bigger air sacs. Explain (with example calculations) how this reduces their effectiveness for gaseous exchange. (*5 marks*)

3 Smokers get more infections of the breathing system than non-smokers. Suggest why. (*6 marks*)

The amount of gaseous exchange that needs to take place in your lungs will vary a lot depending on your size and level of activity. The gaseous exchange system has to be able to respond to the differing demands of your body.

Measuring the capacity of the lungs

The volume of air that is drawn in and out of the lungs can be measured in a variety of different ways:

- A peak flow meter (Figure 1) is a simple device that measures the rate at which air can be expelled from the lungs. People who have asthma often use these to monitor how well their lungs are working.

- Vitalographs are more sophisticated versions of the peak flow meter. The patient being tested breathes out as quickly as they can through a mouthpiece, and the instrument produces a graph of the amount of air they breathe out and how quickly it is breathed out. This volume of air is called the forced expiratory volume in 1 second.

- A spirometer is commonly used to measure different aspects of the lung volume, or to investigate breathing patterns. There are many different forms of the spirometer but they all use the principle shown in Figure 2.

Components of the lung volume

There are several different aspects of the lung volume that can be measured.

- **Tidal volume** is the volume of air that moves into and out of the lungs with each resting breath. It is around 500 cm^3 in most adults at rest, which uses about 15% of the vital capacity of the lungs.

- **Vital capacity** is the volume of air that can be exhaled when the deepest possible intake of breath is followed by the strongest possible exhalation.

- **Inspiratory reserve volume** is the maximum volume of air you can breathe in over and above a normal inhalation.

- **Expiratory reserve volume** is the extra amount of air you can force out of your lungs over and above the normal tidal volume of air you breathe out.

- **Residual volume** is the volume of air that is left in your lungs when you have exhaled as hard as possible. This cannot be measured directly.

Learning outcomes

Demonstrate knowledge, understanding, and application of:

→ the relationship between vital capacity, tidal volume, breathing rate and oxygen uptake.

▲ **Figure 1** *Peak flow meters give a useful quick measure of how much air can be moved out of (and therefore into) the lungs*

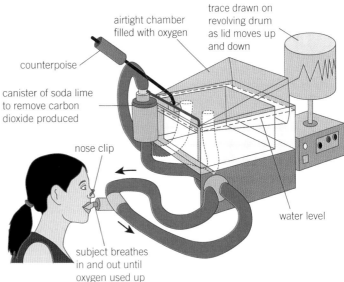

▲ **Figure 2** *Spirometers can be used to measured the volumes of gas breathed in and out under different conditions*

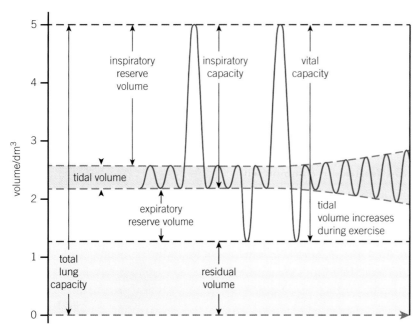

▲ **Figure 3** *These volumes are averages taken from many individuals of different sexes, sizes, and levels of fitness*

- **Total lung capacity** is the sum of the vital capacity and the residual volume.

Recordings from a spirometer show the different volumes of air moved in and out of the lungs.

Breathing rhythms

The pattern and volume of breathing changes as the demands of the body change. The **breathing rate** is the number of breaths taken per minute. The ventilation rate is the total volume of air inhaled in one minute.

ventilation rate = tidal volume × breathing rate (per minute)

When the oxygen demands of the body increase, for example during exercise, the tidal volume of air moved in and out of the lungs with each breath can increase from 15% to as much as 50% of the vital capacity. The breathing rate can also increase. In this way the ventilation of the lungs and so the oxygen uptake during gaseous exchange can be increased to meet the demands of the tissues.

Worked example: Breathing calculations

- The normal tidal volume of a male is 500 cm³. His ventilation rate is 6 dm³ per minute. What is his resting breathing rate?

ventilation rate = tidal volume × breathing rate (per minute)

$6 \, dm^3 = 6000 \, cm^3$

$6000 = 500 \times$ breathing rate

Breathing rate = $\frac{6000}{500}$

= 12 breaths per minute

- During physical exertion, his breathing rate goes up to 20 breaths per minute and the ventilation rate to 15 dm³. What is the new tidal volume?

$15 \, dm^3 = 15000 \, cm^3$

$15000 =$ tidal volume × 20

tidal volume = $\frac{15000}{20}$

= 750 cm³

Summary questions

1 Describe how you could investigate breathing rates in a school laboratory. *(3 marks)*

2 Describe the relationships between tidal volume, breathing rate, and oxygen uptake. *(4 marks)*

3 A dog is under stress during a visit to the vet and pants but the ventilation rate remains steady. Suggest a possible explanation for these observations. *(2 marks)*

4 The normal breathing rate of a healthy 50 year old woman is 18 breaths per minute and her tidal volume is 500 cm³.
   a During strenuous exercise her ventilation rate goes up to 45 000 cm³ per minute and she is breathing 30 times a minute. What is her tidal volume during exercise and what increase is this over her normal tidal volume? *(6 marks)*
   b She develops a chest infection and her breathing rate increases to 25 breaths per minute, but her tidal volume falls to 300. By what percentage does her ventilation rate fall compared wirh her normal resting rate as a result of the infection? *(6 marks)*

Internal gas-exchange systems such as the lungs in mammals are not the only way for multicellular organisms to get the oxygen needed by the cells. The gaseous exchange systems of insects are effective but very different, yet they have many key features in common with mammalian systems.

Gaseous exchange systems in insects

Many insects are very active during parts of their life cycles. They are mainly land-dwelling animals with relatively high oxygen requirements. However, they have a tough **exoskeleton** through which little or no gaseous exchange can take place. They do not usually have blood pigments that can carry oxygen. They need a different way of exchanging gases. The gaseous exchange system of insects has evolved to deliver the oxygen directly to the cells and to remove the carbon dioxide in the same way (Figure 1).

How does gas exchange take place in insects?

Along the thorax and abdomen of most insects are small openings known as **spiracles**. Air enters and leaves the system through the spiracles, but water is also lost. Just like mammals, insects need to maximise the efficiency of gaseous exchange, but minimise the loss of water. In many insects the spiracles can be opened or closed by sphincters. The spiracle sphincters are kept closed as much as possible to minimise water loss.

When an insect is inactive and oxygen demands are very low, the spiracles will all be closed most of the time. When the oxygen demand is raised or the carbon dioxide levels build up, more of the spiracles open.

Leading away from the spiracles are the **tracheae**. These are the largest tubes of the insect respiratory system, up to 1 mm in diameter, and they carry air into the body. They run both into and along the

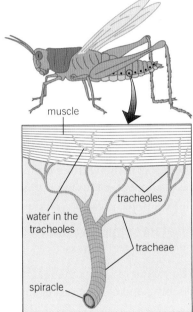

▲ Figure 1 *The gaseous exchange system in an insect*

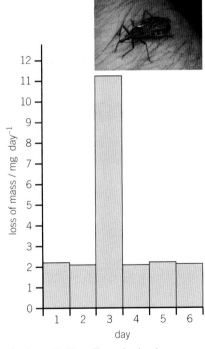

▲ Figure 2 *The effect of spiracle opening on water loss in the blood-sucking bug* Rhodnius *(top). The insect was kept in dry air for 6 days. It was not fed, to keep it relatively inactive. On day 3 carbon dioxide levels were raised, which resulted in the spiracles opening. The air was returned to normal on day 4*

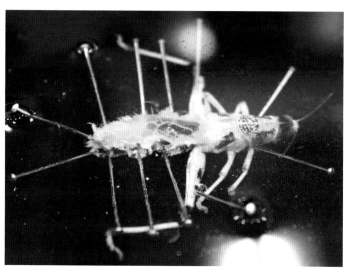

▲ **Figure 3** *Careful dissection of an insect such as this immature desert locust (Schistocerca gregaria) can show the network of tracheae which spread out from the spiracles. Here the locust dissection has been submerged in water to help show the tracheae which are filled with air and appear a pearly white. Very fine tracheae can be seen floating up into the water*

body of the insect. The tubes are lined by spirals of chitin, which keep them open if they are bent or pressed. Chitin is the material that makes up the cuticle. It is relatively impermeable to gases and so little gaseous exchange takes place in the trachea.

The tracheae branch to form narrower tubes until they divide into the tracheoles, minute tubes of diameter 0.6–0.8 μm. Each tracheole is a single, greatly elongated cell with no chitin lining so they are freely permeable to gases. Because of their very small size they spread throughout the tissues of the insect, running between individual cells. This is where most of the gaseous exchange takes place between the air and the respiring cells.

In most insects, for most of the time, air moves along the tracheae and tracheoles by diffusion alone, reaching all the tissues. The vast numbers of tiny tracheoles give a very large surface area for gaseous exchange. Oxygen dissolves in moisture on the walls of the tracheoles and diffuses into the surrounding cells. Towards the end of the tracheoles there is **tracheal fluid**, which limits the penetration of air for diffusion. However, when oxygen demands build up – when the insect is flying, for example – a lactic acid build up in the tissues results in water moving out of the tracheoles by osmosis. This exposes more surface area for gaseous exchange.

All of the oxygen needed by the cells of an insect is supplied to them by the tracheal system.

The extent of gas exchange in most insects is controlled by the opening and closing of the spiracles.

Some insects, for example larger beetles, locusts and grasshoppers, bees, wasps and flies, have very high energy demands. To supply the extra oxygen they need, these insects have alternative methods of increasing the level of gaseous exchange. These include:

- mechanical ventilation of the tracheal system – air is actively pumped into the system by muscular pumping movements of the thorax and/or the abdomen. These movements change the volume of the body and this changes the pressure in the tracheae and tracheoles. Air is drawn into the tracheae and tracheoles, or forced out, as the pressure changes

- collapsible enlarged tracheae or air sacs, which act as air reservoirs – these are used to increase the amount of air moved through the gas exchange system. They are usually inflated and deflated by the ventilating movements of the thorax and abdomen.

Discontinuous gas exchange cycles in insects

Discontinuous gas exchange cycles (DGC) have been found to be relatively common in many species of insects. In DCG spiracles have three states – closed, open, and fluttering.

- When the spiracles are closed no gases move in or out of the insect. Oxygen moves into the cells by diffusion from the tracheae and carbon dioxide diffuses into the body fluids of the insect where it is held in a process called buffering.

- When the spiracles flutter, they open and close rapidly. This moves fresh air into the tracheae to renew the supply of oxygen, while minimising water loss.

- When carbon dioxide levels build up really high in the body fluids of the insect, the spiracles open widely and carbon dioxide diffuses out rapidly. There may also be pumping movements of the thorax and abdomen when the spiracles are open to maximise gaseous exchange.

Originally scientists thought discontinuous gas exchange was an adaptation for water conservation in insects. Now the evidence suggests this is not the case and there are a number of conflicting theories about the adaptive advantages of discontinuous gas exchange for insects, which include helping gaseous exchange in insects that spend at least part of their lives in enclosed spaces such as burrows, or reducing the entry of fungal spores, which can parasitise an insect. This is still an area of very active research and argument.

1 Suggest one way in which each of the given theories on the value of DGC as an adaptation might be investigated.

Respiratory systems in bony fish

Animals that get their oxygen from water do not need to try and prevent water loss from their gaseous exchange surfaces as land animals do, but there are other difficulties to overcome.

Water is 1000 times denser than air. It is 100 times more viscous (thick) and has a much lower oxygen content. To cope with the viscosity of water and the slow rate of oxygen diffusion, fish have evolved very specialised respiratory systems that are different from those of land-dwelling animals. It would use up far too much energy to move dense, viscous water in and out of lung-like respiratory organs. Moving water in one direction only is much simpler and more economical in energy terms.

Gills

Bony fish such as trout and cod are relatively big, active animals that live almost exclusively in water. Because they are very active, their cells have a high oxygen demand. Their SA : V ratio means that diffusion would not be enough to supply their inner cells with the oxygen they need, and their scaly outer covering does not allow gaseous exchange. However, bony fish have evolved a ventilatory system adapted to take oxygen from the water and get rid of carbon dioxide into the water. They maintain a flow of water in one direction over the **gills**, which are their organs of gaseous exchange. Gills have the large surface area, good blood supply, and thin layers needed for successful gaseous exchange. In bony fish they are contained in a gill cavity and covered by a protective **operculum** (a bony flap), which is also active in maintaining a flow of water over the gills (Figure 4).

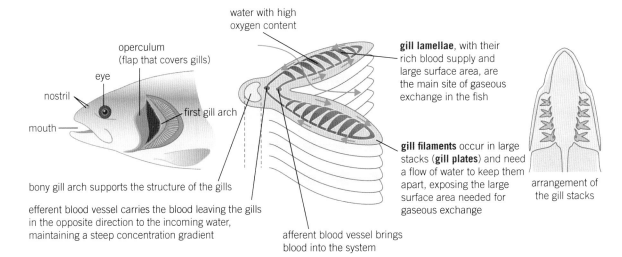

water with high oxygen content

operculum (flap that covers gills)

eye

nostril

mouth

first gill arch

gill lamellae, with their rich blood supply and large surface area, are the main site of gaseous exchange in the fish

gill filaments occur in large stacks (**gill plates**) and need a flow of water to keep them apart, exposing the large surface area needed for gaseous exchange

arrangement of the gill stacks

bony gill arch supports the structure of the gills

efferent blood vessel carries the blood leaving the gills in the opposite direction to the incoming water, maintaining a steep concentration gradient

afferent blood vessel brings blood into the system

▲ **Figure 4** *The gills are the site of gaseous exchange in bony fish and have a number of adaptations for their role in supplying oxygen and removing carbon dioxide*

The gills make up the gaseous exchange surface of the fish. They have many features in common with both mammalian and insect gaseous exchange surfaces. However, they also have particular challenges. To allow efficient gas exchange at all times, fish need to maintain a continuous flow of water over the gills, even when they are not moving. They also need to carry out gaseous exchange as effectively as possible in water, a medium where diffusion is slower than in air.

Water flow over the gills

When fish are swimming they can keep a current of water flowing over their gills simply by opening their mouth and operculum. However, when the fish stops moving, the flow of water also stops. The more primitive cartilaginous fish such as the sharks and rays often rely on continual movement to ventilate the gills. This is known as ram ventilation – they just ram the water past the gills. However, most bony fish do not rely on movement-generated water flow over the gills. They have evolved a sophisticated system involving the operculum, which allows them to move water over their gills all the time (Figure 5).

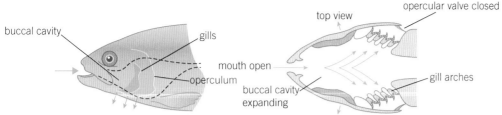

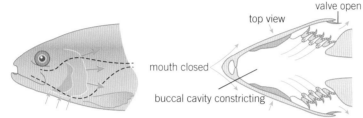

The mouth is opened and the floor of the buccal cavity (mouth) is lowered. This increases the volume of the buccal cavity. As a result the pressure in the cavity drops and water moves into the buccal cavity. At the same time the opercular valve is shut and the opercular cavity containing the gills expands. This lowers the pressure in the opercular cavity containing the gills. The floor of the buccal cavity starts to move up, increasing the pressure there so water moves from the buccal cavity over the gills.

The mouth closes, the operculum opens and the sides of the opercular cavity move inwards. All of these actions increase the pressure in the opercular cavity and force water over the gills and out of the operculum. The floor of the buccal cavity is steadily moved up, maintaining a flow of water over the gills.

▲ Figure 5 *Although this diagram shows the ventilation of the gills in stages, this is actually a continuous process, which ensures that water is constantly flowing over the gills*

Effective gaseous exchange in water

Gills have a large surface area for diffusion, a rich blood supply to maintain steep concentration gradients for diffusion, and thin layers so that diffusing substances have only short distances to travel. Gills have two extra adaptations that help to ensure the most effective possible gaseous exchange occurs in the water:

- The tips of adjacent gill filaments overlap. This increases the resistance to the flow of water over the gill surfaces and slows down the movement of the water. As a result there is more time for gaseous exchange to take place.

- The water moving over the gills and the blood in the gill filaments flow in different directions. A steep concentration gradient is needed for fast, efficient diffusion to take place. Because the blood and water flow in opposite directions, a countercurrent exchange system is set up. This adaptation ensures that steeper concentration gradients are maintained than if blood and water flowed in the same direction (known as a parallel system). As a result, more gaseous exchange can take place. The bony fish, with their countercurrent systems, remove about 80% of the oxygen from the water flowing over them. The cartilaginous fish have parallel systems and can only extract about 50% of the oxygen from the water flowing over them (Figure 6).

▲ Figure 6 *Careful dissection of a trout (genera* Onchorhynchus*). The operculum (top) helps to protect the gills. Removal of the operculum shows thin feathery gills which have a large surface area (middle). Here you can see that the water flows in through the mouth over the gills and out (bottom)*

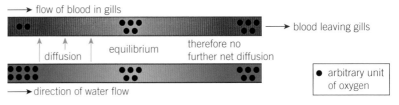

parallel system: blood in the gills and water flowing over the gills travel in the same direction, which gives an initial steep oxygen concentration gradient between blood and water. Diffusion takes place until the oxygen concentration of the blood and water are in equilibrium, then no net movement of oxygen into the blood occurs.

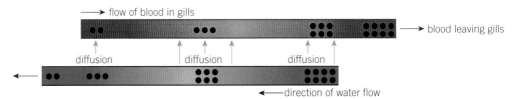

countercurrent system: blood and water flow in opposite directions so an oxygen concentration gradient between the water and the blood is maintained all along the gill. Oxygen continues to diffuse down the concentration gradient so a much higher level of oxygen saturation of the blood is achieved.

▲ **Figure 7** *The advantages of a countercurrent exchange system are clear to see. The system also enables bony fish to remove more carbon dioxide from the blood than a parallel system*

 ## Dissecting, examining, and drawing gaseous exchange systems

Dissecting an animal gives you a unique insight into the complexity of a multicellular living organism. The process of evolution over millions of years has resulted in internal systems of great efficiency and elegance.

When you dissect an animal it can be confusing. You will be looking at one particular body system but they do not exist in isolation. You cannot see the gas exchange system, for example, without getting through the body wall.

You will need specialist equipment including boards and pins with which to display your dissection. The tools needed for successful dissections include sharp scissors and scalpels along with tweezers and mounted needles to lift and tease out tissues.

When you carry out a dissection the aim is to be as precise and clean in your work as possible. You should observe and display the relevant features of an organism to the best of your ability, and then record what you have seen in a clear and well-labelled diagram. It may be useful to take a photograph of your dissection and so preserve what it actually looked like alongside your labelled diagram. In Figure 3 and Figure 6 you can see examples of dissections of the gas exchange system of an insect and the gas exchange system of a bony fish.

1 Drawings from dissections are always done in pencil – suggest why.
2 Make careful labelled drawings of the dissections shown in Figure 3 and Figure 6.

 ## The histology of exchange surfaces

Some of the key features of exchange surfaces – the large surface area, the short diffusion distances, and the proximity of the blood supply often cannot be seen when you look at the whole organ system.

However using prepared slides with the light microscope can give you an insight into the detailed adaptations of these surfaces for their role in gaseous exchange.

Example 1 – The network of tiny air sacs with walls that are a single cell thick that make up the surface area of the alveoli can clearly be seen in light micrographs of lung tissue , × 40 magnification.

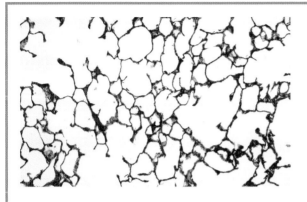

Example 2 – The delicate rings of chitin which support the trachea in insects, holding them open so that air can diffuse through them to the tracheoles and the tissues themselves are revealed in this light micrograph , × 37 magnification.

Example 3 – The structure of the gills in a bony fish show clearly in a light micrograph, showing the delicate structures and large surface area as well as the major blood vessels , × 278 magnification.

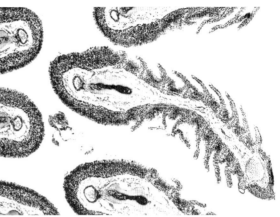

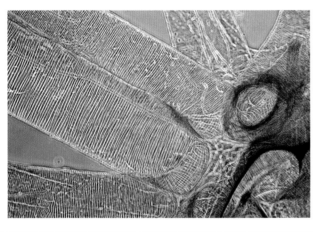

1   Make carefully labelled drawings of each of these micrographs, using the skills you learned in Chapter 2.
2   What are the advantages of using the light microscope to look at these tissues?

Summary questions

1   Suggest why a fish will die when it is left out of water for too long.   (2 marks)

2   Compile a table to compare the gaseous exchange systems of a human, an insect and a bony fish.   (5 marks)

3   Explain how insects that have particularly high energy requirements can increase the amount of gaseous exchange taking place in their bodies.   (6 marks)

4   Explain how the structure of the gas exchange system of a bony fish maximises the amount of oxygen that can be taken from the water.   (6 marks)

Study tip

Make sure you are clear about the key features of a successful gaseous exchange surface and can recognise those common features in the gaseous exchange systems of a variety of organisms.

Practice questions

1 There are several different aspects of the lung volume that can be measured.

 Which of the following statements is/are correct with respect to lung volumes?

 Statement 1: tidal volume is the volume of air which moves into and out of the lungs with each resting breath

 Statement 2: vital capacity is the maximum amount of air you can breathe in over and above a normal inhalation

 Statement 3: inspiratory reserve volume is the volume of air which can be breathed out by the strongest possible exhalation followed by the deepest possible intake of breath

 A 1, 2 and 3 are correct

 B Only 1 and 2 are correct

 C Only 2 and 3 are correct

 D Only 1 is correct *(1 mark)*

2 **a** State what a peak flow meter is used to measure *(2 marks)*

 b The chart below shows the range of normal peak flow for both men and women of different ages.

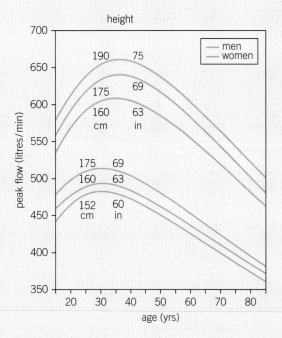

 Using the graph, describe how lung function varies with both age and gender.
  (4 marks)

 c Draw another line on the graph for a man who had untreated asthma from birth. Explain the reasons for the position of your line and state why it would be unlikely to see someone produce this line. *(3 marks)*

 d Explain why inhalation is an active process but normal exhalation is a passive process. *(3 marks)*

3 Fick's law states that:

$$\text{Rate of diffusion} = \frac{\text{Area of diffusion surface} \times \text{Difference in concentration}}{\text{Thickness of surface over which diffusion takes place}}$$

 a Explain how the human respiratory system is designed to obey this law. *(3 marks)*

 b The gaseous exchange surfaces in fish and mammals are designed to perform the same function: the uptake of oxygen and removal of carbon dioxide.

 Discuss how they are each adapted to carry out this function in different environments; fish in water and mammals in air. *(3 marks)*

4 Insects have an exoskeleton.

 a State the main component of this exoskeleton. *(1 mark)*

 b The component named in a is a long-chain polymer of a N-acetylglucosamine, a derivative of glucose.

 Explain what is meant by the term polymer, with reference to this component. *(2 marks)*

 c Although relatively small, insects have a high oxygen demand. The exoskeleton is impermeable to oxygen.

 Describe how insects are adapted to ensure their tissues receive oxygen at a sufficient rate. *(4 marks)*

 d Describe one similarity and one difference between the trachea of a mammal and the trachea of an insect. *(2 marks)*

5 The diagram below shows the lung volumes at different phases of the respiratory cycle.

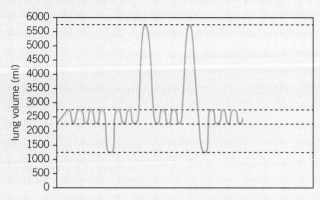

a Calculate, using the diagram, the inspiratory reserve volume and tidal volume of the lungs. (*2 marks*)

b Copy the diagram and indicate the residual volume and expiratory reserve volume. (*2 marks*)

c The diagram below compares the lung volumes of a woman before and during pregnancy.

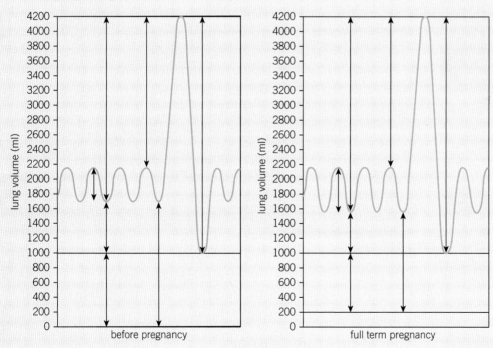

Describe the changes in lung volumes that occur during pregnancy and explain the reasons for these changes. (*6 marks*)

8 TRANSPORT IN ANIMALS
8.1 Transport systems in multicellular animals

Specification reference: 3.1.2

Learning outcomes

Demonstrate knowledge, understanding, and application of:

→ the need for transport systems in multicellular animals

→ the different types of circulatory systems found in multicellular animals.

▲ **Figure 1** *When a multicellular animal takes in food, the nutrients have to be transported to all the cells in the body*

Synoptic link

You learnt about surface area to volume ratio in Topic 7.1, Specialised exchange surfaces.

Synoptic link

It will be useful now to recap your understanding of the various mechanisms for the movement of molecules across membranes in Topics 4.3–4.5 Diffusion, Active transport, and Osmosis.

In the previous chapter you looked at why multicellular organisms need exchange surfaces. They also need **transport systems** to supply oxygen and nutrients to the sites where they are needed and to remove waste products from the individual cells.

The need for specialised transport systems in animals

In single-celled organisms, processes such as diffusion, osmosis, active transport, endocytosis and exocytosis can supply everything the cell needs to import or export. These processes are also important in multicellular organisms, transporting substances within and between individual cells. However, as organisms get bigger, the distances between the cells and the outside of the body get greater. Diffusion would transport substances into and out of the inner core of the body, but it would be so slow that the organism would not survive. Specialised transport systems are needed because:

● the metabolic demands of most multicellular animals are high (they need lots of oxygen and food, they produce lots of waste products) so diffusion over the long distances is not enough to supply the quantities needed

● the surface area to volume (SA : V) ratio gets smaller as multicellular organisms get bigger so not only do the diffusion distances get bigger but the amount of surface area available to absorb or remove substances becomes relatively smaller

● molecules such as hormones or enzymes may be made in one place but needed in another

● food will be digested in one organ system, but needs to be transported to every cell for use in respiration and other aspects of cell metabolism

● waste products of metabolism need to be removed from the cells and transported to excretory organs.

Types of circulatory systems

Most large, multicellular animals have specialised **circulatory systems** (transport systems) which carry gases such as oxygen and carbon dioxide, nutrients, waste products and hormones around the body. Most circulatory systems have features in common:

● They have a liquid transport medium that circulates around the system (blood).

- They have vessels that carry the transport medium.
- They have a pumping mechanism to move the fluid around the system.

When substances are transported in a mass of fluid with a mechanism for moving the fluid around the body it is known as a **mass transport system**. Large, multicellular animals usually have either an open circulatory system or a closed circulatory system.

Open circulatory systems

In an **open circulatory system** there are very few vessels to contain the transport medium. It is pumped straight from the heart into the body cavity of the animal. This open body cavity is called the haemocoel. In the haemocoel the transport medium is under low pressure. It comes into direct contact with the tissues and the cells. This is where exchange takes place between the transport medium and the cells. The transport medium returns to the heart through an open-ended vessel (Figure 2).

These open-ended circulatory systems are found mainly in invertebrate animals, including most insects and some molluscs. Remember that in insects, gas exchange takes place in the tracheal system. Insect blood is called **haemolymph**. It doesn't carry oxygen or carbon dioxide. It transports food and nitrogenous waste products and the cells involved in defence against disease. The body cavity is split by a membrane and the heart extends along the length of the thorax and the abdomen of the insect. The haemolymph circulates but steep diffusion gradients cannot be maintained for efficient diffusion. The amount of haemolymph flowing to a particular tissue cannot be varied to meet changing demands.

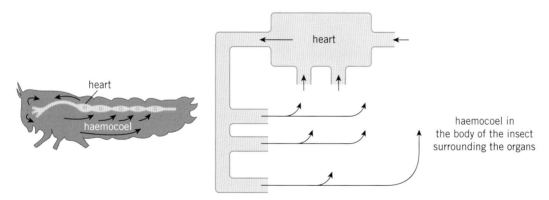

▲ Figure 2 *The open circulatory system of a locust supplies the cells with food and removes the nitrogenous waste products*

Closed circulatory systems

In a **closed circulatory system**, the blood is enclosed in blood vessels and does not come directly into contact with the cells of the body. The heart pumps the blood around the body under pressure and relatively quickly, and the blood returns directly to the heart. Substances leave and enter the blood by diffusion through the walls of the blood vessels.

Synoptic link

You learnt about mechanisms of ventilation and gaseous exchange in insects in Topic 7.4, Ventilation and gas exchange in other organisms.

The amount of blood flowing to a particular tissue can be adjusted by widening or narrowing blood vessels. Most closed circulatory systems contain a blood pigment that carries the respiratory gases.

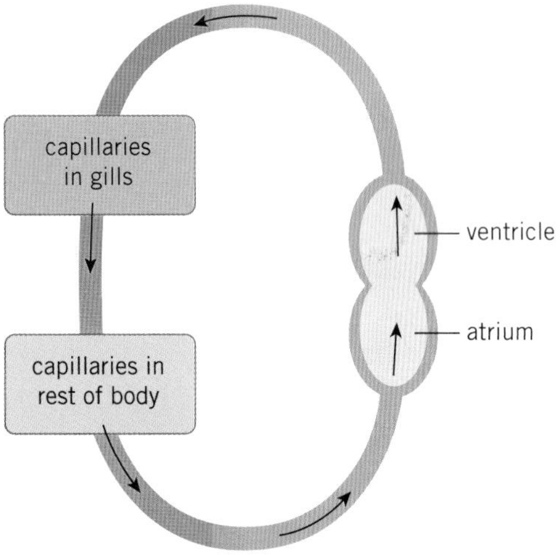

Closed circulatory systems are found in many different animal phyla, including echinoderms (animals such as sea urchins and starfish), cephalopod molluscs including the octopods and squid, annelid worms including the common earthworm, and all of the vertebrate groups, including the mammals.

Single closed circulatory systems

Single closed circulatory systems are found in a number of groups including fish and annelid worms. In **single circulatory systems** (Figure 3) the blood flows through the heart and is pumped out to travel all around the body before returning to the heart. In other words, the blood travels only once through the heart for each complete circulation of the body.

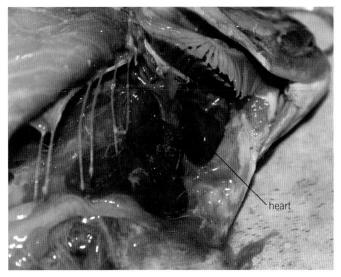

▲ **Figure 3** *The single closed circulatory system of a fish*

In a single closed circulation, the blood passes through two sets of capillaries (microscopic blood vessels) before it returns to the heart. In the first, it exchanges oxygen and carbon dioxide. In the second set of capillaries, in the different organ systems, substances are exchanged between the blood and the cells. As a result of passing through these two sets of very narrow vessels, the blood pressure in the system drops considerably so the blood returns to the heart quite slowly. This limits the efficiency of the exchange processes so the activity levels of animals with single closed circulations tends to be relatively low.

Fish are something of an exception. They have a relatively efficient single circulatory system, which means they can be very active. They have a countercurrent gaseous exchange mechanism in their gills that allows them to take a lot of oxygen from the water. Their body weight is supported by the water in which they live and they do not maintain their own body temperature. This greatly reduces the metabolic demands on their bodies and, combined with their efficient gaseous exchange, explains how fish can be so active with a single closed circulatory system.

▲ **Figure 4** *The single closed circulatory system of a fish. Dissection of this trout shows the close proximity of the fish heart to the gills. Fish hearts have one atrium and one ventricle*

Double closed circulatory systems

Birds and most mammals are very active land animals that maintain their own body temperature. This way of life is made possible in part by their double closed circulatory system (Figure 5). This is the most efficient system for transporting substances around the body. It involves two separate circulations:

Synoptic link

You learnt about the mechanisms of ventilation and gaseous exchange in fish in Topic 7.4, Ventilation and gas exchange in other organisms.

- Blood is pumped from the heart to the lungs to pick up oxygen and unload carbon dioxide, and then returns to the heart.
- Blood flows through the heart and is pumped out to travel all around the body before returning to the heart again.

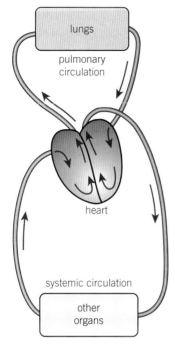

▲ **Figure 5** *The double closed circulatory system of a mammal*

So in a **double circulatory system**, the blood travels twice through the heart for each circuit of the body. Each circuit – to the lungs and to the body – only passes through one capillary network, which means a relatively high pressure and fast flow of blood can be maintained.

Summary questions

1 Describe the function of a circulatory system.   (*3 marks*)

2 Explain why circulatory systems are found in multicellular organisms but not unicellular organisms.   (*2 marks*)

3 Compare open and closed circulatory systems.   (*6 marks*)

4 Land predators such as foxes have a double closed circulatory system. Aquatic predators such as pike are effective with a single closed circulatory system. Explain why these two types of predator have different circulatory systems.   (*6 marks*)

Study tip

Make sure you are clear about the differences between open and closed circulatory systems, and between single and double closed circulatory systems.

Be aware of the advantages and disadvantages of the different systems.

8.2 Blood vessels

Specification reference: 3.1.2

Circulation in humans is typical of a mammalian circulatory system. It is estimated that if all the blood vessels of an average adult human were laid end to end they would stretch to 100 000 miles – that is the equivalent of about four times around the circumference of the Earth.

There are several different types of blood vessels in the body and their structural composition is closely related to their function. Some examples of different components utilised in some blood vessels are:

- Elastic fibres – these are composed of elastin and can stretch and recoil, providing vessel walls with flexibility.

- Smooth muscle – contracts or relaxes, which changes the size of the lumen (the channel within the blood vessel).

- Collagen – provides structural support to maintain the shape and volume of the vessel.

The sections below detail which components are found in each of the blood vessel types.

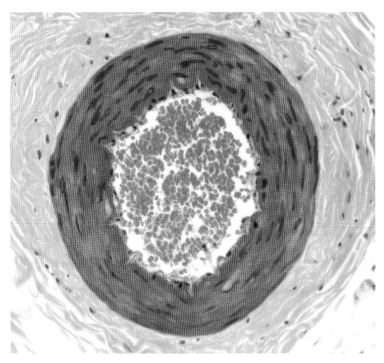

Arteries and arterioles

The arteries carry blood away from the heart to the tissues of the body. They carry **oxygenated blood** *except* in the pulmonary artery, which carries deoxygenated blood from the heart to the lungs, and (during pregnancy) the umbilical artery, which carries deoxygenated blood from the fetus to the placenta. The blood in the arteries is under higher pressure than the blood in the veins.

Artery walls contain elastic fibres, smooth muscle and collagen (Figure 1). The elastic fibres enable them to withstand the force of the blood pumped out of the heart and stretch (within limits maintained by collagen) to take the larger blood volume. In between the contractions of the heart, the elastic fibres recoil and return to their original length. This helps to even out the surges of blood pumped from the heart to give a continuous flow. However, you can still feel a pulse (surge of blood) when the heart contracts, which the elastic fibres cannot completely eliminate. The lining of an artery (endothelium) is smooth so the blood flows easily over it.

Arterioles link the arteries and the capillaries. They have more smooth muscle and less elastin in their walls than arteries, as they have little pulse surge, but can constrict or dilate to control the flow of blood into individual organs. When the smooth muscle in the arteriole contracts it constricts the vessel and prevents blood flowing into a

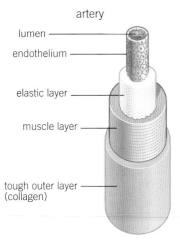

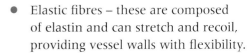

lumen
endothelium

elastic layer

muscle layer

tough outer layer
(collagen)

▲ **Figure 1** *The structure of the artery wall is closely related to its functions, and changes as the arteries get smaller, × 250 magnification* capillary bed. This is *vasoconstriction*. When the smooth muscle in the wall of an arteriole relaxes, blood flows through into the capillary bed. This is *vasodilation*.

Synoptic link

You learnt about the structure of elastin and collagen in Topic 3.6 The structure of proteins.

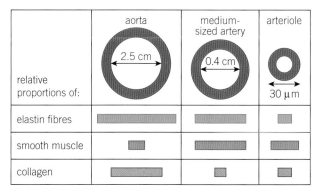

relative proportions of:	aorta	medium-sized artery	arteriole
	2.5 cm	0.4 cm	30 μm
elastin fibres	▬	▬	▪
smooth muscle	▪	▬	▬
collagen	▬	▪	▪

▲ **Figure 2** *The differing proportions of components of the artery wall depending on how far the artery is from the heart and thus its role in the body*

➕ Collagen, elastin, and aortic aneurysms

An aneurysm is a bulge or weakness in a blood vessel. The most common places for aneurysms are in the aorta and in the arteries of the brain. Most people do not know they have an aneurysm until it bursts. This is very serious and can be fatal. High blood pressure is one factor that increases the risk of an aneurysm. However, scientists have also discovered changes in the proportion of collagen to elastin in the aorta wall. The ratio of collagen : elastin in a normal aorta is 1.85 : 1. In a small aneurysm it increases to around 3.75 : 1 and in large aortic aneurysms it is 7.91 : 1. Research is continuing to see if this apparent link is real – and, if so, whether it can be used to predict who is at risk so they can have regular aortic screening.

1 What hypothesis for the formation of aneurysms can you develop from these data?
2 Suggest two possible ways in which patients might be treated to reduce the possibility of aneurysms developing or enlarging.

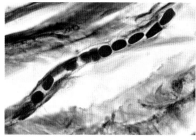

red blood cell

7-8 μm

capillary 10μm

lumen

capillary wall

▲ **Figure 3** *The structure of capillaries means that they are small enough to form the immense networks needed to exchange substances between the blood and the tissues*

Capillaries

The **capillaries** are microscopic blood vessels that link the arterioles with the venules. They form an extensive network through all the tissues of the body. The lumen of a capillary is so small that red blood cells (which have a diameter of only 7.5–8 μm) have to travel through in single file (Figure 3). Substances are exchanged through the capillary walls between the tissue cells and the blood. The gaps between the endothelial cells that make up the capillary walls in most areas of the body are relatively large. This is where many substances pass out of the capillaries into the fluid surrounding the cells. The exception is the capillaries in the central nervous system, which have very tight junctions between the cells.

In most organs of the body the blood entering the capillaries from the arterioles is oxygenated. By the time it leaves the capillaries for the venules it has less oxygen and more carbon dioxide (it is deoxygenated). Again, the lungs and the placenta are the exceptions, with deoxygenated blood entering the capillaries and oxygenated blood leaving in the venules.

Ways in which capillaries are adapted for their role:

- They provide a very large surface area for the diffusion of substances into and out of the blood.

- The total cross-sectional area of the capillaries is always greater than the arteriole supplying them so the rate of blood flow falls. The relatively slow movement of blood through capillaries gives more time for the exchange of materials by diffusion between the blood and the cells.

- The walls are a single endothelial cell thick, giving a very thin layer for diffusion.

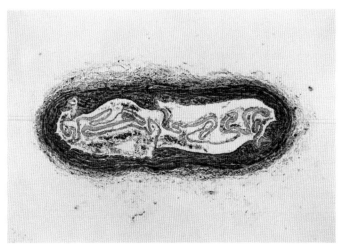

Veins and venules

The veins carry blood away from the cells of the body towards the heart and, with two exceptions, they carry deoxygenated blood. The pulmonary vein carries oxygenated blood from the lungs to the heart, and (during pregnancy) the umbilical vein carries oxygenated blood from the placenta to the fetus.

Deoxygenated blood flows from the capillaries into very small veins celled venules and then into larger veins. Finally it reaches the two main vessels carrying deoxygenated blood back to the heart – the inferior vena cava from the lower parts of the body and the superior vena cava from the head and upper body.

Veins do not have a pulse – the surges from the heart pumping are lost as the blood passes through the narrow capillaries. However, they do hold a large reservoir of blood – up to 60% of your blood volume is in your veins at any one time.

The blood pressure in the veins is very low compared with the pressure in the arteries. Medium-sized veins (the majority of the venous system) have valves to prevent the backflow of blood (see next page).

The walls contain lots of collagen and relatively little elastic fibre, and the vessels have a wide lumen and a smooth, thin lining (known as the endothelium) so the blood flows easily (Figure 4).

Venules link the capillaries with the veins. They have very thin walls with just a little smooth muscle. Several venules join to form a vein.

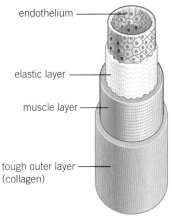

endothelium

elastic layer

muscle layer

tough outer layer (collagen)

▲ Figure 4 *Veins do not have to withstand high pressures like arteries do but they need a big capacity and this is reflected in their structure*

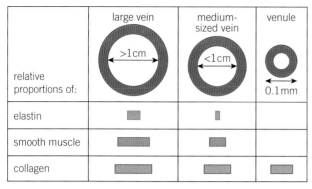

relative proportions of:	large vein	medium-sized vein	venule
elastin	▬	▮	
smooth muscle	▭	▭	
collagen	▭	▭	▭

▲ **Figure 5** *The differing proportions of components of the vein and venule walls depending on how far the vein is from the heart and thus its role in the body*

Deoxygenated blood in the veins must be returned to the heart to be pumped to the lungs and oxygenated again. However, the blood is under low pressure and needs to move against gravity. There are three main adaptations that enable the body to overcome this problem:

- The majority of the veins have one-way valves at intervals. These are flaps or infoldings of the inner lining of the vein. When blood flows in the direction of the heart, the valves open so the blood can pass through. If the blood starts to flow backwards, the valves close to prevent this from happening.

- Many of the bigger veins run between the big, active muscles in the body, for example in the arms and legs. When the muscles contract they squeeze the veins, forcing the blood towards the heart. The valves prevent backflow when the muscles relax.

- The breathing movements of the chest act as a pump. The pressure changes and the squeezing actions move blood in the veins of the chest and abdomen towards the heart.

In combination, these adaptations assist in the return of deoxygenated blood to the heart.

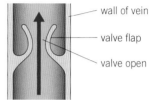

blood flowing towards the heart passes easily through the valves

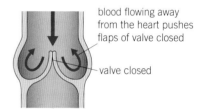

blood flowing away from the heart pushes valves closed and so blood is prevented from flowing any further in this direction

▲ **Figure 6** *The valves in a vein prevent the backflow of blood*

Summary questions

1 Veins have valves but arteries do not – explain why. *(3 marks)*

2 The structure of an arteriole is different from the structure of an artery. Describe the differences in structure and explain how they are related to the functions of the vessels. *(2 marks)*

3 a Explain how the different structures of large veins, medium veins and venules are related to their functions in the body. *(6 marks)*
   b Explain why the venules and veins in the lungs are so unusual. *(2 marks)*

Learning outcomes

Demonstrate knowledge, understanding, and application of:

→ the differences in the composition of blood, tissue fluid and lymph

→ the formation of tissue fluid from plasma.

Blood is the main transport medium of the human circulatory system, but it is only part of the story. Tissue fluid is the other important player in the exchange of substances between the blood and the cells. A third liquid, lymph, is also part of the complex system that makes up the circulation of the body.

Blood

Blood consists of a yellow liquid – **plasma** – which carries a wide variety of other components including dissolved glucose and amino acids, mineral ions, hormones, and the large plasma proteins including albumin (important for maintaining the osmotic potential of the blood), fibrinogen (important in blood clotting) and globulins (involved in transport and the immune system). Plasma also transports red blood cells (which carry oxygen to the cells and also give the blood its red appearance) and the many different types of white blood cells. It also carries platelets. Platelets are fragments of large cells called megakaryocytes found in the red bone marrow, and they are involved in the clotting mechanism of the blood. Plasma makes up 55% of the blood by volume – and much of that volume is water. Only the plasma and the red blood cells are involved in the transport functions of the blood. The other components have different functions.

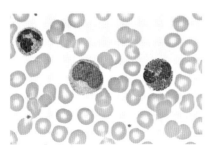

▲ Figure 1 *Human blood under the light microscope*

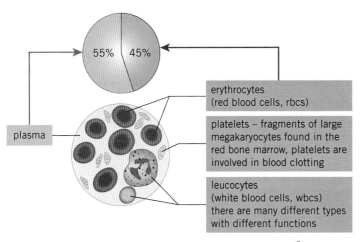

▲ Figure 2 *The main cellular components of the blood. 1 mm³ of healthy blood contains around 5 million erythrocytes, 7000 leucocytes and 250 000 platelets, which are cell fragments*

Labels:
- erythrocytes (red blood cells, rbcs)
- platelets – fragments of large megakaryocytes found in the red bone marrow, platelets are involved in blood clotting
- leucocytes (white blood cells, wbcs) there are many different types with different functions
- plasma

Functions of the blood

The composition of the blood is closely related to its functions in the body, many of which involve transport. They include transport of:

- oxygen to, and carbon dioxide from, the respiring cells
- digested food from the small intestine
- nitrogenous waste products from the cells to the excretory organs
- chemical messages (hormones)
- food molecules from storage compounds to the cells that need them
- platelets to damaged areas
- cells and antibodies involved in the immune response.

The blood also contributes to maintenance of a steady body temperature and acts as a buffer, minimising pH changes.

Synoptic link

You will learn more about clotting of the blood, phagocytosis, and the immune system in Topic 12.5, The specific immune system.

Tissue fluid

The substances dissolved in plasma can pass through the fenestrations in the capillary walls, with the exception of the large plasma proteins. The plasma proteins, particularly albumin, have an osmotic effect. They give the blood in the capillaries a relatively high solute potential (and so a relatively low water potential) compared with the surrounding fluid. As a result, water has a tendency to move into the blood in the capillaries from the surrounding fluid by osmosis. The tendency of water to move into the blood by osmosis is termed **oncotic pressure** and it is about −3.3 kPa.

However, as blood flows through the arterioles into the capillaries, it is still under pressure from the surge of blood that occurs every time the heart contracts. This is **hydrostatic pressure**. At the arterial end of the capillary, the hydrostatic pressure forcing fluid out of the capillaries is relatively high at about 4.6 kPa (Figure 3). It is higher than the oncotic pressure attracting water in by **osmosis**, so fluid is squeezed out of the capillaries. This fluid fills the spaces between the cells and is called **tissue fluid**. Tissue fluid has the same composition as the plasma, without the red blood cells and the plasma proteins. Diffusion takes place between the blood and the cells through the tissue fluid.

As the blood moves through the capillaries towards the venous system, the balance of forces changes. The hydrostatic pressure falls to around 2.3 kPa in the vessels as fluid has moved out and the pulse is completely lost. The oncotic pressure is still −3.3 kPa, so it is now stronger than the hydrostatic pressure, so water moves back into the capillaries by osmosis as it approaches the venous end of the capillaries. By the time the blood returns to the veins, 90% of the tissue fluid is back in the blood vessels.

Synoptic link

You will learn about temperature control in endotherms and ectotherms and about excretion in Chapter 15, Homeostasis. You will learn about hormones in Topic 14.1, Hormonal communication.

Synoptic link

You learnt about hydrostatic pressure and osmosis in Topic 5.5, Osmosis.

Study tip

Filtration pressure = hydrostatic pressure − oncotic pressure

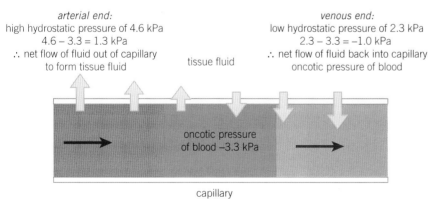

arterial end:
high hydrostatic pressure of 4.6 kPa
4.6 − 3.3 = 1.3 kPa
∴ net flow of fluid out of capillary to form tissue fluid

tissue fluid

venous end:
low hydrostatic pressure of 2.3 kPa
2.3 − 3.3 = −1.0 kPa
∴ net flow of fluid back into capillary
oncotic pressure of blood

oncotic pressure of blood −3.3 kPa

capillary

▲ **Figure 3** *Diagram showing differences in hydrostatic pressure at arterial and venous end and how this results in movement into or out of the capillary*

Lymph

Some of the tissue fluid does not return to the capillaries. 10% of the liquid that leaves the blood vessels drains into a system of blind-ended tubes called lymph capillaries, where it is known as **lymph**. Lymph is similar in composition to plasma and tissue fluid but has less oxygen and fewer nutrients. It also contains fatty acids, which have been absorbed into the lymph from the villi of the small intestine. The lymph capillaries join up to form larger vessels. The fluid is transported through them by the squeezing of the body muscles. One-way valves like those in veins prevent the backflow of lymph. Eventually the lymph returns to the blood, flowing into the right and left subclavian veins (under the clavicle, or collar bone).

Along the lymph vessels are the lymph nodes. **Lymphocytes** build up in the lymph node when necessary and produce antibodies, which are then passed into the blood. Lymph nodes also intercept bacteria and other debris from the lymph, which are ingested by phagocytes found in the nodes. The lymphatic system plays a major role in the defence mechanisms of the body.

Enlarged lymph nodes are a sign that the body is fighting off an invading pathogen. This is why doctors often examine the neck, armpits, stomach or groin of their patients – these are the sites of some of the major lymph nodes (which people often refer to as 'lymph glands').

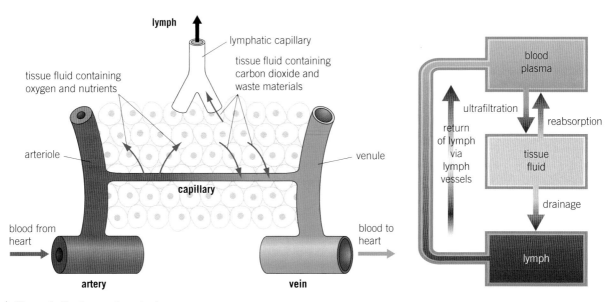

▲ **Figure 4** *The human lymphatic system*

Summary questions

1 Describe the main functions of the blood. (*4 marks*)

2 a What are platelets? (*3 marks*)
  b What is the role of platelets in the body? (*3 marks*)
  c What percentage of the cells/cell fragments in the blood is made up of platelets? (*3 marks*)

3 Summarise the similarities and differences between plasma, tissue fluid, and lymph. (*5 marks*)

4 Explain how hydrostatic and oncotic pressure affect the movement of fluids into and out of capillaries. (*6 marks*)

8.4 Transport of oxygen and carbon dioxide in the blood

Specification reference: 3.1.2

The most specialised transport role of the blood is the transport of oxygen from the lungs to the cells of the body by the erythrocytes (red blood cells). The erythrocytes are also involved in the removal of carbon dioxide from the cells and its transport to the lungs for gaseous exchange.

Transporting oxygen

The erythrocytes are very specialised, with a number of adaptations to their main function of transporting oxygen.

Erythrocytes have a biconcave shape. This shape has a larger surface area than a simple disc structure or a sphere, increasing the surface area available for diffusion of gases. It also helps them to pass through narrow capillaries. In adults, erythrocytes are formed continuously in the red bone marrow. By the time mature erythrocytes enter the circulation they have lost their nuclei, which maximises the amount of haemoglobin that fits into the cells. It also limits their life, so they only last for about 120 days in the bloodstream.

Erythrocytes contain **haemoglobin**, the red pigment that carries oxygen and also gives them their colour. Haemoglobin is a very large globular conjugated protein made up of four peptide chains, each with an iron-containing haem prosthetic group. There are about 300 million haemoglobin molecules in each red blood cell and each haemoglobin molecule can bind to four oxygen molecules.

The oxygen binds quite loosely to the haemoglobin forming **oxyhaemoglobin**. The reaction is reversible.

$$Hb + 4O_2 \rightleftharpoons Hb(O_2)_4$$
$$\text{haemoglobin} + \text{oxygen} \rightleftharpoons \text{oxyhaemoglobin}$$

Carrying oxygen

When the erythrocytes enter the capillaries in the lungs, the oxygen levels in the cells are relatively low. This makes a steep concentration gradient between the inside of the erythrocytes and the air in the alveoli. Oxygen moves into the erythrocytes and binds with the haemoglobin. The arrangement of the haemoglobin molecule means that as soon as one oxygen molecule binds to a haem group, the molecule changes shape, making it easier for the next oxygen molecules to bind. This is known as positive cooperativity. Because the oxygen is bound to the haemoglobin, the free oxygen concentration in the erythrocyte stays low, so a steep diffusion gradient is maintained until all of the haemoglobin is saturated with oxygen.

When the blood reaches the body tissues, the situation is reversed. The concentration of oxygen in the cytoplasm of the body cells

Learning outcomes

Demonstrate knowledge, understanding, and application of:

→ the role of haemoglobin in transporting oxygen and carbon dioxide

→ the oxygen dissociation curve for fetal and adult haemoglobin.

Synoptic link

You learnt about haemoglobin as a globular conjugated protein in Topic 3.7, Types of proteins.

is lower than in the erythrocytes. As a result, oxygen moves out of the erythrocytes down a concentration gradient. Once the first oxygen molecule is released by the haemoglobin, the molecule again changes shape and it becomes easier to remove the remaining oxygen molecules.

An **oxygen dissociation curve** (Figure 1) is an important tool for understanding how the blood carries and releases oxygen. The percentage saturation haemoglobin in the blood is plotted against the partial pressure of oxygen (pO_2). Oxygen dissociation curves show the affinity of haemoglobin for oxygen. A very small change in the partial pressure of oxygen in the surroundings makes a significant difference to the saturation of the haemoglobin with oxygen, because once the first molecule becomes attached, the change in the shape of the haemoglobin molecule means other oxygen molecules are added rapidly. The curve levels out at the highest partial pressures of oxygen because all the haem groups are bound to oxygen and so the haemoglobin is saturated and cannot take up any more.

This means that at the high partial pressure of oxygen in the lungs the haemoglobin in the red blood cells is rapidly loaded with oxygen. Equally, a relatively small drop in oxygen levels in the respiring tissues means oxygen is released rapidly from the haemoglobin to diffuse into the cells. This effect is enhanced by the relatively low pH in the tissues compared with the lungs.

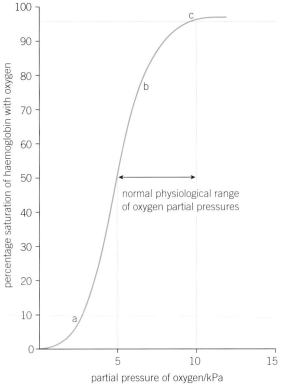

a at low pO_2, few haem groups are bound to oxygen, so haemoglobin does not carry much oxygen

b at higher pO_2, more haem groups are bound to oxygen, making it easier for more oxygen to be picked up

c the haemoglobin becomes saturated at very high pO_2, as all the haem groups become bound

▲ Figure 1 *The oxygen dissociation curve for human haemoglobin*

When you are not very active, only about 25% of the oxygen carried in your erythrocytes is released into the body cells. The rest acts as a reservoir for when the demands of the body increase suddenly.

The effect of carbon dioxide

As the partial pressure of carbon dioxide rises (in other words, at higher partial pressures of CO_2), haemoglobin gives up oxygen more easily (Figure 2). This change is known as the **Bohr effect**. The Bohr effect is important in the body because as a result:

- in active tissues with a high partial pressure of carbon dioxide, haemoglobin gives up its oxygen more readily

- in the lungs where the proportion of carbon dioxide in the air is relatively low, oxygen binds to the haemoglobin molecules easily.

Fetal haemoglobin

When a fetus is developing in the uterus it is completely dependent on its mother to supply it with oxygen. Oxygenated blood from the mother runs close to the deoxygenated fetal blood in the placenta. If the blood of the fetus had the same affinity for oxygen as the blood of the mother, then little or no oxygen would be transferred to the blood of the fetus. However, fetal haemoglobin has a higher affinity for oxygen than adult haemoglobin at each point along the dissociation curve (Figure 3). So it removes oxygen from the maternal blood as they move past each other.

Transporting carbon dioxide

Carbon dioxide is transported from the tissues to the lungs in three different ways:

- About 5% is carried dissolved in the plasma.

- 10–20% is combined with the amino groups in the polypeptide chains of haemoglobin to form a compound called **carbaminohaemoglobin**.

- 75–85% is converted into hydrogen carbonate ions (HCO_3^-) in the cytoplasm of the red blood cells.

Most of the carbon dioxide that diffuses into the blood from the cells is transported to the lungs in the form of hydrogen carbonate ions.

Carbon dioxide reacts slowly with water to form carbonic acid ($H_2CO_3^-$). The carbonic acid then dissociates to form hydrogen ions and hydrogen carbonate ions.

$$CO_2 + H_2O \rightleftharpoons H_2CO_3 \rightleftharpoons H^+ + HCO_3^-$$

In the blood plasma this reaction happens slowly. However, in the cytoplasm of the red blood cells there are high levels of the enzyme **carbonic anhydrase**. This enzyme catalyses the reversible reaction between carbon dioxide and water to form carbonic acid. The carbonic acid then dissociates to form hydrogen carbonate ions and hydrogen ions.

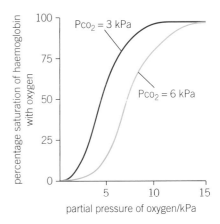

▲ **Figure 2** *The Bohr shift – as the proportion of carbon dioxide increases, the oxygen dissociation curve for haemoglobin moves to the right*

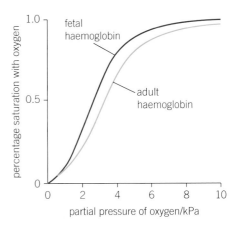

▲ **Figure 3** *The oxygen dissociation curves for adult and fetal haemoglobin show how the fetus can gain oxygen from the mother*

The negatively charged hydrogen carbonate ions move out of the erythrocytes into the plasma by diffusion down a concentration gradient and negatively charged chloride ions move into the erythrocytes, which maintains the electrical balance of the cell. This is known as the **chloride shift**.

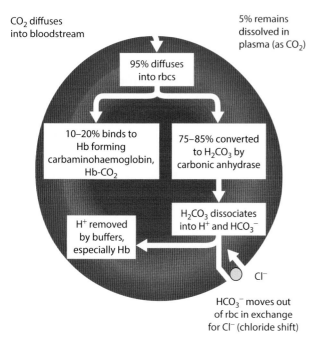

▲ **Figure 4** *The transport of carbon dioxide in the blood is a complex process*

By removing the carbon dioxide and converting it to hydrogen carbonate ions, the erythrocytes maintain a steep concentration gradient for carbon dioxide to diffuse from the respiring tissues into the erythrocytes.

When the blood reaches the lung tissue where there is a relatively low concentration of carbon dioxide, carbonic anhydrase catalyses the reverse reaction, breaking down carbonic acid into carbon dioxide and water. Hydrogen carbonate ions diffuse back into the erythrocytes and react with hydrogen ions to form more carbonic acid. When this is broken down by carbonic anhydrase it releases free carbon dioxide, which diffuses out of the blood into the lungs. Chloride ions diffuse out of the red blood cells back into the plasma down an electrochemical gradient.

Haemoglobin in the erythrocytes also plays a role in this process. It acts as a buffer and prevents changes in the pH by accepting free hydrogen ions in a reversible reaction to form **haemoglobinic acid**.

Summary questions

1 Explain how the structure of erythrocytes is adapted to their function in the body. *(3 marks)*

2 a Draw the oxygen dissociation curve for normal adult haemoglobin and fetal haemoglobin. *(2 marks)*

 b What is the difference in the oxygen saturation of adult and fetal haemoglobin at a partial pressure of oxygen of 6 kPa? Explain the importance of this difference in the survival of the fetus. *(3 marks)*

3 Myoglobin is an oxygen-binding molecule found in the muscles.

 a Sketch a graph to showing myoglobin oxygen affinity and haemoglobin oxygen affinity. *(2 marks)*

 b Explain the differences between the two curves. *(2 marks)*

4 Use a flow chart to summarise how the carbon dioxide produced in the cells is carried to the lungs in the red blood cells. *(6 marks)*

8.5 The heart

Specification reference: 3.1.2

The **heart** is the organ that moves the blood around the body. In some animal groups it is no more than a simple muscular tube. In mammals the heart is a complex, four-chambered muscular 'bag' found in the chest, enclosed by the ribs and sternum.

The human heart

The heart consists of two pumps, joined and working together. Deoxygenated blood from the body flows into the right side of the heart, which pumps it to the lungs. Oxygenated blood from the lungs returns to the left side of the heart, which pumps it to the body. The blood from the two sides of the heart does not mix.

The heart is made of cardiac muscle, which contracts and relaxes in a regular rhythm. It does not get fatigued and need to rest like skeletal muscle. The coronary arteries supply the cardiac muscle with the oxygenated blood it needs to keep contracting and relaxing all the time. The heart is surrounded by inelastic pericardial membranes, which help prevent the heart from over-distending with blood.

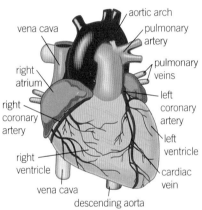

▲ **Figure 1** *Diagram showing the external structure of the heart*

Labels on Figure 1: vena cava, aortic arch, pulmonary artery, right atrium, pulmonary veins, right coronary artery, left coronary artery, left ventricle, right ventricle, cardiac vein, vena cava, descending aorta

Labels on Figure 2: carotid arteries, aorta, superior vena cava, left pulmonary artery, right pulmonary artery, pulmonary veins, left atrium, semilunar valves, cardiac muscle, bicuspid valves (left atrioventricular valve), right atrium, septum, tricuspid valve (right atrioventricular valve), left ventricle, tendinous cords, right ventricle, very thick muscular wall of left ventricle, inferior vena cava

▲ **Figure 2** *The structure of the human heart is closely related to its function. In an average lifetime it will beat around 3×10^9 times, and each ventricle will pump around 200 million litres of blood*

Dissecting a heart

The heart of a sheep or a pig is similar in shape and size to a human heart and is often used in dissection. By careful examination of a heart you can identify many of the important structures in the mammalian heart – although the real thing is much more complicated than the standard diagram in Figure 3.

The external view of the heart enables you to see and trace the coronary arteries which supply the heart muscle with the blood it needs to beat. It is the narrowing or blockage of these blood vessels that cause the symptoms of coronary heart disease and even heart attacks.

However hearts obtained from the butcher are often not intact. The major blood vessels will have been cut right back and often the atria have been removed – people don't want to eat all the tubes. So when you examine, dissect and draw a heart, you have to be aware of which, if any, parts are missing.

1 How does the dissected heart in Figure 4 differ from the diagram of the structures of the heart in Figure 2? Compare the two and explain the differences.

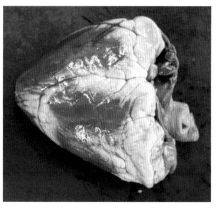

▲ **Figure 3** *Anterior view of a sheep's heart*

▲ **Figure 4** *A dissected sheep's heart, as you can see a dissected mammalian heart shows its more complex intricacies*

The structure and function of the heart

Deoxygenated blood enters the *right atrium* of the heart from the upper body and head in the superior vena cava, and from the lower body in the inferior vena cava, at relatively low pressure. The atria have thin muscular walls. As the blood flows in, slight pressure builds up until the atrio-ventricular valve (the tricuspid valve) opens to let blood pass into the right ventricle. When both the atrium and ventricle are filled with blood the atrium contracts, forcing all the blood into the **right ventricle** and stretching the ventricle walls. As the right ventricle starts to contract, the tricuspid valve closes, preventing any backflow of blood to the atrium. The tendinous cords make sure the valves are not turned inside out by the pressures exerted when the ventricle contracts. The right ventricle contracts fully and pumps deoxygenated blood through the semilunar valves into the pulmonary artery, which transports it to the capillary beds of the lungs. The semilunar valves prevent the backflow of blood into the heart.

At the same time oxygenated blood from the lungs enters the *left atrium* from the pulmonary vein. As pressure in the atrium builds the bicuspid valve opens between the left atrium and the

left ventricle so the ventricle also fills with oxygenated blood. When both the atrium and ventricle are full the atrium contracts, forcing all the oxygenated blood into the left ventricle. The left ventricle then contracts and pumps oxygenated blood through semilunar valves into the **aorta** and around the body. As the ventricle contracts the tricuspid valve closes, preventing any backflow of blood.

The muscular wall of the left side of the heart is much thicker than that of the right. The lungs are relatively close to the heart, and the lungs are also much smaller than the rest of the body so the right side of the heart has to pump the blood a relatively short distance and only has to overcome the resistance of the pulmonary circulation. The left side has to produce sufficient force to overcome the resistance of the aorta and the arterial systems of the whole body and move the blood under pressure to all the extremities of the body.

The septum is the inner dividing wall of the heart which prevents the mixing of deoxygenated and oxygenated blood.

The right and left side of the heart fill and empty together.

 A hole in the heart

The development of the septum is not completed until after birth. In the fetus the blood is oxygenated in the placenta, not in the lungs. As a result, all the blood in the heart is very similar and so mixes freely. In the days after birth the gap in the septum closes to ensure that the deoxygenated and oxygenated bloods are kept completely separate. Any gap remaining in the septum after the first few weeks of life is referred to as a 'hole in the heart' and it can often be heard with a stethoscope as a heart murmur. Many people have a small hole in their septum without knowing about it. However, if the hole is large it can lead to severe health problems unless it is diagnosed and repaired by surgery.

1 Explain why a small hole in the septum of the heart might not have any adverse effects but a large hole will cause health problems.

The cardiac cycle and the heartbeat

The **cardiac cycle** describes the events in a single heartbeat, which lasts about 0.8 seconds in a human adult.

- In **diastole** the heart relaxes. The atria and then the ventricles fill with blood. The volume and pressure of the blood in the heart build as the heart fills, but the pressure in the arteries is at a minimum.

- In **systole** the atria contract (atrial systole), closely followed by the ventricles (ventricular systole). The pressure inside the heart increases dramatically and blood is forced out of the right side of the heart to the lungs and from the left side to the main body circulation. The volume and pressure of the blood in the heart are low at the end of systole, and the blood pressure in the arteries is at a maximum.

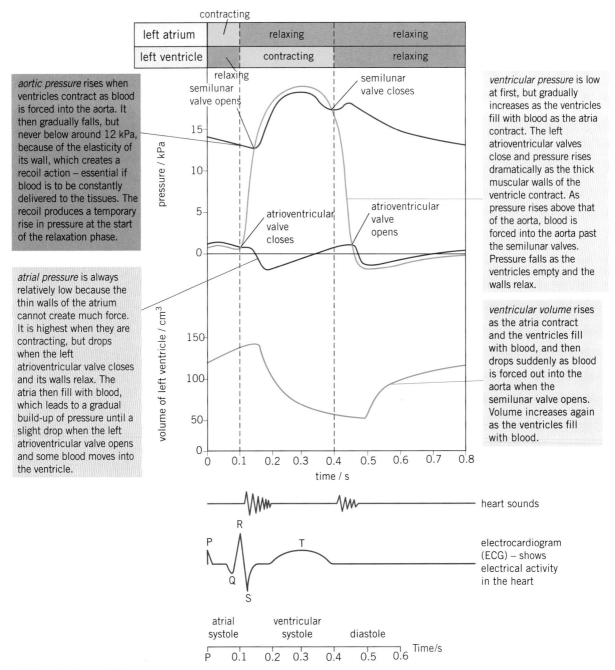

aortic pressure rises when ventricles contract as blood is forced into the aorta. It then gradually falls, but never below around 12 kPa, because of the elasticity of its wall, which creates a recoil action – essential if blood is to be constantly delivered to the tissues. The recoil produces a temporary rise in pressure at the start of the relaxation phase.

atrial pressure is always relatively low because the thin walls of the atrium cannot create much force. It is highest when they are contracting, but drops when the left atrioventricular valve closes and its walls relax. The atria then fill with blood, which leads to a gradual build-up of pressure until a slight drop when the left atrioventricular valve opens and some blood moves into the ventricle.

ventricular pressure is low at first, but gradually increases as the ventricles fill with blood as the atria contract. The left atrioventricular valves close and pressure rises dramatically as the thick muscular walls of the ventricle contract. As pressure rises above that of the aorta, blood is forced into the aorta past the semilunar valves. Pressure falls as the ventricles empty and the walls relax.

ventricular volume rises as the atria contract and the ventricles fill with blood, and then drops suddenly as blood is forced out into the aorta when the semilunar valve opens. Volume increases again as the ventricles fill with blood.

▲ Figure 5 Some of the pressure changes in the heart during the cardiac cycle, as well as the heart sounds and an ECG trace

Heart sounds

The sounds of the heartbeat, which can be heard through a stethoscope, are made by blood pressure closing the heart valves. The two sounds of a heartbeat are described as 'lub-dub'. The first sound comes as the blood is forced against the atrio-ventricular valves as the ventricles contract, and the second sound comes as a backflow of blood closes the semilunar valves in the aorta and pulmonary artery as the ventricles relax.

The basic rhythm of the heart

Cardiac muscle is **myogenic** – it has its own intrinsic rhythm at around 60 beats per minute (bpm). This prevents the body wasting resources maintaining the basic heart rate. The average resting heart rate of an adult is higher, at around 70 bpm. This is because other factors including exercise, excitement, and stress also affect our heart rate.

The basic rhythm of the heart is maintained by a wave of electrical excitation, rather like a nerve impulse (Figure 6).

- A wave of electrical excitation begins in the pacemaker area called the **sino-atrial node (SAN)**, causing the atria to contract and so initiating the heartbeat. A layer of non-conducting tissue prevents the excitation passing directly to the ventricles.

- The electrical activity from the SAN is picked up by the **atrio-ventricular node (AVN)**. The AVN imposes a slight delay before stimulating the **bundle of His**, a bundle of conducting tissue made up of fibres **(Purkyne fibres)**, which penetrate through the septum between the ventricles.

- The bundle of His splits into two branches and conducts the wave of excitation to the apex (bottom) of the heart.

- At the apex the Purkyne fibres spread out through the walls of the ventricles on both sides. The spread of excitation triggers the contraction of the ventricles, starting at the apex. Contraction starting at the apex allows more efficient emptying of the ventricles.

The way in which the wave of excitation spreads through the heart from the SAN, with AVN delay, makes sure that the atria have stopped contracting before the ventricles start.

Study tip

Don't just describe the sino-atrial node (SAN) as the pacemaker of the heart. It initiates the heartbeat by producing a wave of electrical excitation which causes both of the atria to contract followed by the ventricles as a result of the AVN delay.

Study tip

Be clear about the role of the SAN – explain the two effects of the wave of excitation generated there.

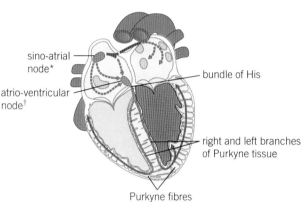

*The sino-atrial node (SAN) or natural pacemaker sets up a wave of electrical excitation.

†The atrio-ventricular node (AVN) is excited by the SAN. From here the excitation passes into the Purkyne tissue.

▲ **Figure 6** *The SAN initiates the heartbeat*

Electrocardiograms

You can measure the spread of electrical excitation through the heart as a way of recording what happens as it contracts. This recording of the electrical activity of the heart is called an **electrocardiogram (ECG)**. An ECG doesn't directly measure the electrical activity of your heart. It measures tiny electrical differences in your skin, which result from the electrical activity of the heart.

To pick up these tiny changes, electrodes are stuck painlessly to clean skin to get the good contacts needed for reliable results. The signal from each of the electrodes is fed into the machine, which produces an ECG. A normal ECG is shown in Figure 5 and Figure 7(a). ECGs are used to help diagnose heart problems. For example, if someone is having a heart attack, recognisable changes take place in the electrical activity of their heart, which can be used to diagnose the problem and treat it correctly and fast.

Study tip

Learn to recognise normal and abnormal ECG patterns.

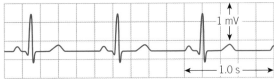

(a) Normal ECG – beats evenly spaced, rate 60–100/min

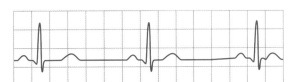

(b) Bradycardia – slow heart rate – beats evenly spaced, rate <60/min

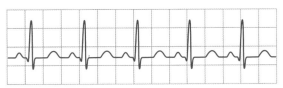

(c) Tachycardia – fast heart rate – beats evenly spaced, rate >100/min

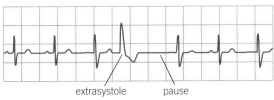

extrasystole   pause

(d) Ectopic heart beat – altered rhythm, extra beat followed by longer than normal gap before the next beat

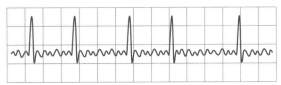

(e) Atrial fibrillation – abnormal irregular rhythm from atria, ventricles lose regular rhythm

▲ **Figure 7** *Normal and abnormal ECGs*

Heart rhythm abnormalities that commonly show up on ECGs include:

- **Tachycardia** – when the heartbeat is very rapid, over 100 bpm. This is often normal, for instance when you exercise, if you have a fever, if you are frightened or angry. If it is abnormal it may be caused by problems in the electrical control of the heart and may need to be treated by medication or by surgery.

- **Bradycardia** – when the heart rate slows down to below 60 bpm. Many people have bradycardia because they are fit – training makes the heart beat more slowly and efficiently. Severe bradycardia can be serious and may need an artificial pacemaker to keep the heart beating steadily.

- **Ectopic heartbeat** – extra heartbeats that are out of the normal rhythm. Most people have at least one a day. They are usually normal but they can be linked to serious conditions when they are very frequent.

- **Atrial fibrillation** – this is an example of an **arrhythmia**, which means an abnormal rhythm of the heart. Rapid electrical impulses are generated in the atria. They contract very fast (fibrillate) up to 400 times a minute. However, they don't contract properly and only some of the impulses are passed on to the ventricles, which contract much less often. As a result the heart does not pump blood very effectively.

Blood pressure

The blood travels through the arterial system at pressures that vary as the ventricles contract. The blood pressure is also affected by the diameter of the blood vessels themselves. Narrowing the arteries is one way in which the body affects and controls local blood flow, but permanent changes can cause severe health problems.

Most people will have their blood pressure taken at some point in their lives. Blood pressure is expressed as two figures, the first higher than the second. But what is being measured? Traditionally blood pressure is measured using a manual sphygmomanometer. A cuff, which is connected to a mercury manometer (a way of measuring pressure using the height of a column of mercury), is placed around the upper arm. The cuff is then inflated until the blood supply to the lower arm is completely cut off.

A stethoscope is positioned over the blood vessels at the elbow. Air is slowly let out of the cuff. The pressure at which the blood sounds first reappear as a slight tapping sound is recorded. The first blood to get through the cuff is that under the highest pressure – in other words, when the left ventricle of the heart is contracting strongly. The height of the mercury at this point gives the *systolic blood pressure* in mmHg (the height of the mercury column). The blood sounds return to normal at the point when even the lowest pressure during diastole is sufficient to get through the cuff. This gives the *diastolic blood pressure*. A reading of 120/80 mmHg is regarded as being normal. More recently a simpler, digital sphygmomanometer is often used – but the same principles apply. The stethoscope is simply built into the cuff applied around the arm.

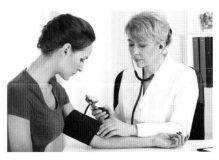

▲ **Figure 8** *Measuring blood pressure with a aneroid sphygmomanometer*

1   Blood pressure is used as an indicator of the health of both the heart and the blood vessels. A weakened heart may produce a low blood pressure, whereas damaged blood vessels that are closing up or becoming less elastic will give a raised blood pressure. Explain how these symptoms might be produced.

Summary questions

1   Explain why healthy coronary arteries are important for maintaining a regular heart rhythm. *(2 marks)*

2   a   What causes the heart sounds? *(2 marks)*
    b   Explain the relationship between the heart sounds and the events of the cardiac cycle. *(6 marks)*

3   Explain the following responses:
    a   Bradycardia is common in diving mammals such as whales and seals. *(5 marks)*
    b   Many people experience tachycardia when they travel to high altitudes. *(5 marks)*

4   Look carefully at Figure 7:
    a   Work out the heart rate of the individuals with i) a normal heart rhythm ii) bradycardia and iii) tachycardia. *(4 marks)*
    b   What is the percentage decrease or increase in the heart rate over the normal rate in these patients? *(3 marks)*

Practice questions

1 In mammals, the lungs are adapted to enable efficient gaseous exchange.

The table below lists some of the adaptations of the lungs.

Complete the table explaining how each adaptation improves the efficiency of gaseous exchange.

Adaptation	How this adaptation improves efficiency of gaseous exchange
squamous epithelium	
large number of alveoli	
good blood supply	
good ventilation	

(4 marks)

OCR F211/01 2013

2 Animals and fish live in completely different environments. However their cells respire in the same way and depend on a constant supply of oxygen.

a Explain how both gills and lungs are designed to maximise the rate of diffusion in gaseous exchange. *(4 marks)*

b Using the simple representations of the two systems below, discuss the different way that gills and lungs are designed to maximise the rate of gaseous exchange.

(3 marks)

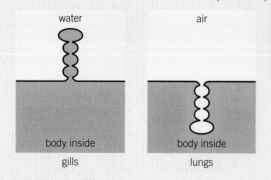

3 The diagram below summarises the formation of tissue fluid.

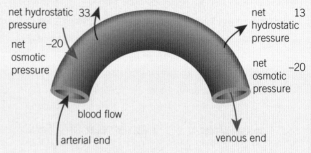

a (i) Calculate the net filtration pressures (NFP) at either end of the capillary in the diagram above:

arterial end

venous end *(2 marks)*

Interstitial fluid is constantly being formed at the arterial ends of capillary networks. Approximately 90% of this fluid is reabsorbed at the venous end.

(ii) Describe what happens to the other 10% and explain what could happen if this process did not occur.

(4 marks)

b Transport systems are essential in multicellular organisms to overcome the limitations of diffusion.

Using a mammalian transport system as an example, explain why diffusion is still an essential process. *(3 marks)*

c The diagram shows the circulatory system of a mollusc.

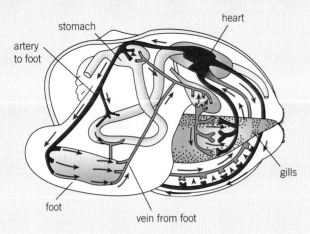

State, with reasons, which type of circulatory system is present in a mollusc. (*4 marks*)

4 Varicose veins occur when the valves in the veins fail to work correctly and blood starts to pool in the veins. This occurs most commonly in the legs and feet.

   a State the function of the valves in the veins. (*1 mark*)

   b Suggest why you do not see 'varicose arteries'. (*2 marks*)

   c (i) Explain why haemoglobin is known as a conjugate protein. (*2 marks*)

      (ii) Explain why an oxygen dissociation curve is sigmoidal (s shaped). (*4 marks*)

   d Name the molecule formed when carbon dioxide binds to haemoglobin. (*1 mark*)

   e Myoglobin is a molecule with one haem group that binds to oxygen and only releases it at low oxygen levels. It effectively acts as store of oxygen in muscles.

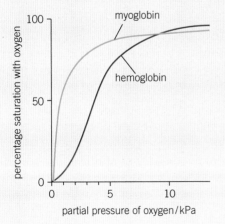

   (i) Explain why the position of the dissociation curve for myoglobin is

different than that of haemoglobin. (*2 marks*)

      (ii) Suggest why the curve for myoglobin is not sigmoidal. (*2 marks*)

      (iii) Draw a curve on the graph for fetal haemoglobin. (*2 marks*)

5 The majority of carbon dioxide released by respiring cells diffuses into red blood cells. Carbon dioxide can bind to haemoglobin forming a compound called carbaminohaemoglobin. About 10% of carbon dioxide is transported to the lungs by this method.

The remaining carbon dioxide undergoes an enzyme controlled reaction once inside a red blood cell.

   a State what type of molecule forms an enzyme. (*2 marks*)

   b Describe what an enzyme does. (*2 marks*)

   c Outline the fate of carbon dioxide after it has entered a red blood but does not bind to haemoglobin. (*3 marks*)

TRANSPORT IN PLANTS
9.1 Transport systems in dicotyledonous plants
Specification reference: 3.1.3

▲ **Figure 1** *This saguaro cactus (Carnegiea gigantea) in the Sonoran Desert is as tall as many trees*

Plants have transport systems to move substances between leaves, stems, and roots. These transport systems work at tremendous pressures. The pressure in the phloem, one of the main transport tissues of a plant, is around 2000 kPa. For comparison, the systolic blood in your main arteries is at a pressure of around 16 kPa, while the steam turbines in a power station work at around 4000 kPa. The higher pressures in plants, however, are confined to much smaller spaces than in arteries and turbines.

The need for plant transport systems

There are three main reasons why multicellular plants need transport systems:

● Metabolic demands – The cells of the green parts of the plant make their own glucose and oxygen by photosynthesis – but many internal and underground parts of the plant do not photosynthesise. They need oxygen and glucose transported to them and the waste products of cell metabolism removed. Hormones made in one part of a plant need transporting to the areas where they have an effect. Mineral ions absorbed by the roots need to be transported to all cells to make the proteins required for enzymes and the structure of the cell.

● Size – Some plants are very small but because plants continue to grow throughout their lives, many perennial plants (plants that live a long time and reproduce year after year) are large and some of them are enormous. The tallest trees in the world include the coastal redwood (*Sequoia sempervirens*) and giant redwood (*Sequoiadendron giganteum*) in the USA (up to around 115 m tall) and the mountain ash (*Eucalyptus regnans*) in Australia (up to around 114 m tall). This means plants need very effective transport systems to move substances both up and down from the tip of the roots to the topmost leaves and stems.

● Surface area : volume ratio (SA : V) – Surface area : volume ratios are not simple in plants. Leaves are adapted to have a relatively large SA : V ratio for the exchange of gases with the air. However, the size and complexity of multicellular plants means that when the stems, trunks, and roots are taken into account they still have a relatively small SA : V ratio. This means they cannot rely on diffusion alone to supply their cells with everything they need.

Transport systems in dicotyledonous plants

Dicotyledonous plants (dicots) make seeds that contain two cotyledons, organs that act as food stores for the developing embryo plant and form the first leaves when the seed germinates. There are herbaceous dicots, with soft tissues and a relatively short life cycle (leaves and stems that die down at the end of the growing season to the soil level), and woody (arborescent) dicots, which have hard, lignified tissues and a long life cycle (in some cases hundreds of years). You will be looking at herbaceous dicots in this section.

Dicotyledonous plants have a series of transport vessels running through the stem, roots, and leaves. This is known as the **vascular system**. In herbaceous dicots this is made up of two main types of transport vessels, the xylem and the phloem, described later in this topic. These transport tissues are arranged together in **vascular bundles** in the leaves, stems, and roots of herbaceous dicots. The pattern of the vascular tissue is easily recognised and is shown in transections (TS) in Table 1.

Synoptic link

It will help to look back at your work on surface area : volume ratios in Topic 7.1, Specialised exchange surfaces and Topic 8.1, Transport systems in multicellular animals.

▼ Table 1

TS stem of young herbaceous plant	
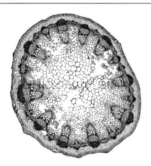	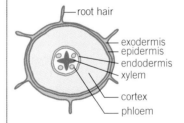 epidermis cortex phloem xylem vascular bundle parenchyma (packing and supporting tissue) In the stem, the vascular bundles are around the edge to give strength and support.
TS root of young herbaceous plant	
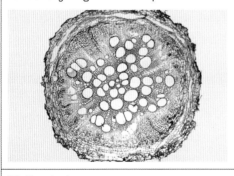	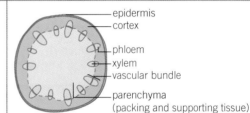 root hair exodermis epidermis endodermis xylem cortex phloem In the roots, the vascular bundles are in the middle to help the plant withstand the tugging strains that result as the stems and leaves are blown in the wind.
TS dicot leaf	
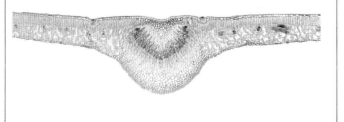	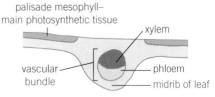 palisade mesophyll–main photosynthetic tissue xylem vascular bundle phloem midrib of leaf In the leaves, the midrib of a dicot leaf is the main vein carrying the vascular tissue through the organ. It also helps to support the structure of the leaf. Many small, branching veins spread through the leaf functioning both in transport and support.

Observing xylem vessels in living plant stems

Xylem vessels can be seen clearly stained in transverse and longitudinal sections of plant stems and roots on prepared slides. However, it is also possible to observe these vessels in living tissue.

If plant material, for example celery stalks with plenty of leaves, flowers such as a gerberas, or the roots of germinating seeds, are put in water containing a strongly coloured dye for at least 24 hours, you can remove the plant from the dye, rinse it and look for the xylem vessels which should have been stained by the dye.

- In one specimen, make clean transverse cuts across the stem with a sharp blade on a white tile. Take great care with the blade.
- Observe and draw the position of the xylem vessels which should show up as coloured spots.
- In another specimen make a careful longitudinal cut through a region where you expect there to be xylem vessels.
- Observe and draw the xylem vessels which may show up as coloured lines.

1   How does the information from your dissection compare with the micrographs of transverse and longitudinal sections of a dicot stem in Table 1 on the previous page?
2   What are the limitations of this method of observing the transport tissues in plants?

The structure and functions of the xylem

The **xylem** is a largely non-living tissue that has two main functions in a plant – the transport of water and mineral ions, and support. The flow of materials in the xylem is up from the roots to the shoots and leaves. Xylem is made up of several types of cells, most of which are dead when they are functioning in the plant. The xylem vessels are the main structures. They are long, hollow structures made by several columns of cells fusing together end to end.

There are two other tissues associated with xylem in herbaceous dicots. Thick-walled xylem parenchyma packs around the xylem vessels, storing food, and containing tannin deposits. Tannin is a bitter, astringent-tasting chemical that protects plant tissues from attack by herbivores. Xylem

Xylem vessel showing lignin spirals

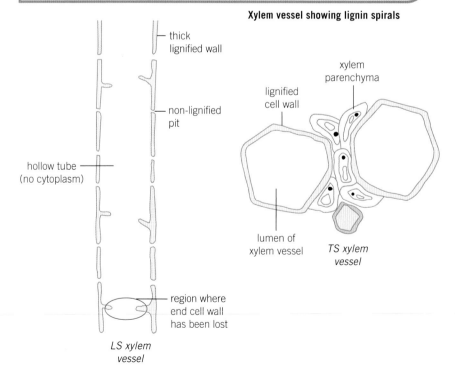

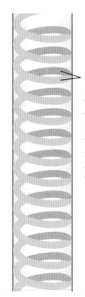

spirals of lignin running around the lumen of the xylem, lignin helps reinforce the xylem vessels so that they do not collapse under the transpiration pull

▲ **Figure 2** *Xylem tissue is both strong and vital for the transport of water and minerals around plants, lignified xylem vessels not shown*

fibres are long cells with lignified secondary walls that provide extra mechanical strength but do not transport water. Lignin can be laid down in the walls of the xylem vessels in several different ways. It can form rings, spirals or relatively solid tubes with lots of small unlignified areas called bordered pits. This is where water leaves the xylem and moves into other cells of the plant.

The structure and functions of the phloem

Phloem is a living tissue that transports food in the form of organic solutes around the plant from the leaves where they are made by photosynthesis. The phloem supplies the cells with the sugars and amino acids needed for cellular respiration and for the synthesis of all other useful molecules. The flow of materials in the phloem can go both up and down the plant.

The main transporting vessels of the phloem are the **sieve tube elements**. Like xylem, the phloem sieve tubes are made up of many cells joined end to end to form a long, hollow structure. Unlike xylem tissue, the phloem tubes are not lignified. In the areas between the cells, the walls become perforated to form **sieve plates**, which look like sieves and let the phloem contents flow through. As the large pores appear in these cell walls, the tonoplast (vacuole membrane), the nucleus and some of the other organelles break down. The phloem becomes a tube filled with phloem sap and the mature phloem cells have no nucleus.

Closely linked to the sieve tube elements are **companion cells**, which form with them. These cells are linked to the sieve tube elements by many plasmodesmata – microscopic channels through the cellulose cell walls linking the cytoplasm of adjacent cells. They maintain their nucleus and all their organelles. The companion cells are very active cells and it is thought that they function as a 'life support system' for the sieve tube cells, which have lost most of their normal cell functions.

Phloem tissue also contains supporting tissues including fibres and sclereids, cells with extremely thick cell walls.

Synoptic link

You learnt about the cells of the xylem and phloem in Topic 6.4, The organisation and specialisation of cells.

▲ Figure 3 *Plants such as these desert cholla grow big but do not form wood like a tree, so they are very dependent on the complex lignified xylem structures to act as a supporting skeleton*

Study tip

Be clear about the structural differences between xylem and phloem. Neither xylem vessels nor mature sieve tubes have nuclei, but xylem vessels are dead and phloem vessels are living tissue.

Summary questions

1 Explain why multicellular plants need transport systems. *(4 marks)*

2 State three differences between transport systems in multicellular plants and multicellular animals. *(3 marks)*

3 State the positioning of the transport tissue in herbaceous dicot stems, roots and leaves, and explain how this positioning is related to their functions. *(6 marks)*

4 Compare and contrast the structure and function of the main cell types in xylem and phloem tissues. *(6 marks)*

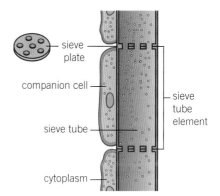

▲ Figure 4 *Phloem tissue transports carbohydrates and other solutes around a plant*

9.2 Water transport in multicellular plants

Specification reference: 3.1.3

Learning outcomes

Demonstrate knowledge, understanding, and application of:

→ the transport of water into the plant, through the plant and to the air surrounding the leaves

→ the mechanisms of water movement in plants.

Synoptic link

You learnt about osmosis, water potential and turgor pressure in Topic 5.5, Osmosis.

Synoptic link

You learnt about meristems in plant roots and shoots (the tissues of a plant where growth can take place) in Topic 6.5, Stem cells. You will learn more about plant growth in Topic 16.1, Plant hormones and growth in plants.

Unlike some other multicellular organisms, plants do not have a muscular heart beating to move fluids through the vessels of the xylem and phloem. In this topic you are going to discover exactly how they manage to move the substances they need through bodies that may be many metres tall and weigh many tonnes.

Water transport in plants

Water is key in both the structure and in the metabolism of plants:

- Turgor pressure (or hydrostatic pressure) as a result of osmosis in plant cells provides a hydrostatic skeleton to support the stems and leaves. So, for example, the turgor pressure in leaf cells is around 1.5 MPa – that is 11 251 mm Hg (unit of pressure still commonly used in medicine) (human systolic blood pressure is around 120 mm Hg).

- Turgor also drives cell expansion – it is the force that enables plant roots to force their way through tarmac and concrete.

- The loss of water by evaporation helps to keep plants cool.

- Mineral ions and the products of photosynthesis are transported in aqueous solutions.

- Water is a raw material for photosynthesis.

To understand the role of water in plants we need to look at how water moves into a plant from the soil and how it moves around the plant.

Movement of water into the root

Root hair cells are the exchange surface in plants where water is taken into the body of the plant from the soil. A root hair is a long, thin extension from a root hair cell, a specialised epidermal cell found near the growing root tip.

Root hairs are well adapted as exchange surfaces:

- Their microscopic size means they can penetrate easily between soil particles.

- Each microscopic hair has a large SA : V ratio and there are thousands on each growing root tip.

- Each hair has a thin surface layer (just the cell wall and cell-surface membrane) through which diffusion and osmosis can take place quickly.

- The concentration of solutes in the cytoplasm of root hair cells maintains a water potential gradient between the soil water and the cell.

Soil water has a very low concentration of dissolved minerals so it has a very high water potential. The cytoplasm and vacuolar sap of the root hair cell (and the other root cells) contain many different solvents including sugars, mineral ions, and amino acids so the water potential in the cell is lower. As a result water moves into the root hair cells by osmosis.

Movement of water across the root

Once the water has moved into the root hair cell it continues to move across the root to the xylem in one of two different pathways:

The symplast pathway

Water moves through the **symplast** – the continuous cytoplasm of the living plant cells that is connected through the plasmodesmata – by osmosis. The root hair cell has a higher water potential than the next cell along. This is the result of water diffusing in from the soil, which has made the cytoplasm more dilute. So water moves from the root hair cell into the next door cell by osmosis. This process continues from cell to cell across the root until the xylem is reached.

As water leaves the root hair cell by osmosis, the water potential of the cytoplasm falls again, and this maintains a steep water potential gradient to ensure that as much water as possible continues to move into the cell from the soil.

The apoplast pathway

This is the movement of water through the **apoplast** – the cell walls and the intercellular spaces. Water fills the spaces between the loose, open network of fibres in the cellulose cell wall. As water molecules move into the xylem, more water molecules are pulled through the apoplast behind them due to the cohesive forces between the water molecules. The pull from water moving into the xylem and up the plant along with the cohesive forces between the water molecules creates a tension that means there is a continuous flow of water through the open structure of the cellulose wall, which offers little or no resistance.

Movement of water into the xylem

Water moves across the root in the apoplast and symplast pathways until it reaches the endodermis – the layer of cells surrounding the vascular tissue (xylem and phloem) of the roots (Figure 4). The endodermis is particularly noticeable in the roots because of the effect of the Casparian strip. The Casparian strip is a band of waxy material called suberin that runs around each of the endodermal cells forming a waterproof layer (Figure 3). At this point, water in the apoplast pathway can go no further and it is forced into the cytoplasm of the cell, joining the water in the symplast pathway. This diversion to the cytoplasm is significant as to get to get there, water must pass through the selectively permeable cell surface membranes, this excludes any potentially-toxic solutes in the soil water from reaching living tissues, as the membranes would have no carrier proteins to admit them. Once forced into the cytoplasm the water joins the symplast pathway.

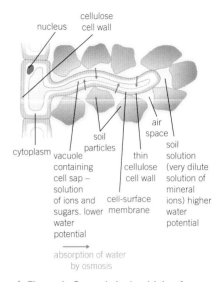

▲ **Figure 1** *Osmosis is the driving force behind the absorption of water by a root hair cell*

Synoptic link

You learned about water potential in Topic 5.5, Osmosis.

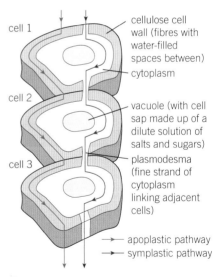

▲ **Figure 2** *The apoplast and symplast pathways along which water moves across the root*

Synoptic link

You learnt about the properties of water in Topic 3.2, Water.

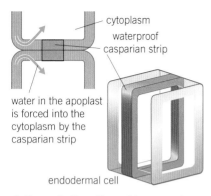

▲ Figure 3 *The effect of the casparian strip on the movement of water across the endodermis*

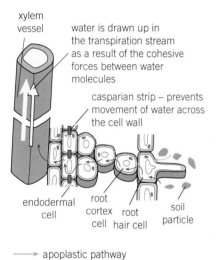

▲ Figure 4 *The root cells provide a very efficient system for moving water from the soil into the xylem*

▲ Figure 5 *Guttation – evidence for root pressure independent of transpiration*

The solute concentration in the cytoplasm of the endodermal cells is relatively dilute compared to the cells in the xylem. In addition, it appears that the endodermal cells move mineral ions into the xylem by active transport. As a result the water potential of the xylem cells is much lower than the water potential of the endodermal cells. This increases the rate of water moving into the xylem by osmosis down a water potential gradient from the endodermis through the symplast pathway.

Once inside the vascular bundle, water returns to the apoplast pathway to enter the xylem itself and move up the plant. The active pumping of minerals into the xylem to produce movement of water by osmosis results in **root pressure** and it is independent of any effects of transpiration. Root pressure gives water a push up the xylem, but under most circumstances it is not the major factor in the movement of water up from the roots to the leaves.

Evidence for the role of active transport in root pressure

It has taken some time and many different investigations to determine the role of active transport in moving water from the root endodermis into the xylem. There are several strands of evidence supporting the current model:

- Some poisons, such as cyanide, affect the mitochondria and prevent the production of ATP. If cyanide is applied to root cells so there is no energy supply, the root pressure disappears.
- Root pressure increases with a rise in temperature and falls with a fall in temperature, suggesting chemical reactions are involved.
- If levels of oxygen or respiratory substrates fall, root pressure falls.
- Xylem sap may exude from the cut end of stems at certain times. In the natural world, xylem sap is forced out of special pores at the ends of leaves in some conditions – for example overnight, when transpiration is low. This is known as *guttation*.

Summary questions

1   Explain how a root hair cell is adapted for its role in the uptake of water from the soil. **(4 marks)**

2   Explain the differences between the apoplast and symplast pathways for water movement across the root of a plant into the xylem. **(6 marks)**

3   a   Suggest why the effect of temperature on root pressure is not sufficient to prove that active transport is involved in the development of root pressure. **(2 marks)**

   b   Explain why the effects of cyanide, oxygen levels, and respiratory substrates on root pressure are taken as evidence that active transport is involved in the development of root pressure. **(5 marks)**

Photosynthesis, the process by which green plants make their own food, takes place mainly in the leaves. Carbon dioxide (CO_2) and water (H_2O) are both needed so, for successful photosynthesis to take place in a leaf, water must be transported there from the roots and carbon dioxide must be taken into the cells of the leaf from the air.

Carbon dioxide diffuses into the leaf cells down a concentration gradient from the air spaces within the leaf. In a process of gaseous exchange, oxygen (O_2) also moves out of the leaf cells into the air spaces by diffusion down a concentration gradient (oxygen is a waste product of photosynthesis). At the same time water evaporates from the surfaces of the leaf cells into the air spaces.

The process of transpiration

Leaves have a very large surface area for capturing sunlight and carrying out photosynthesis. Their surfaces are covered with a waxy cuticle that makes them waterproof. This is an important adaptation that prevents the leaf cells losing water rapidly and constantly by evaporation from their surfaces. However, it is also important that gases can move into and out of the air spaces of the leaf so that photosynthesis is possible.

Carbon dioxide moves from the air into the leaf and oxygen moves out of the leaf by diffusion down concentration gradients through microscopic pores in the leaf (usually on the underside of the leaf) called **stomata** (singular stoma). The stomata can be opened and closed by **guard cells**, which surround the stomatal opening (further detail is given later in this topic).

When the stomata are open to allow an exchange of carbon dioxide and oxygen between the air inside the leaf and the external air, water vapour also moves out by diffusion and is lost. This loss of water vapour from the leaves and stems of plants is called **transpiration**. Transpiration is an inevitable consequence of gaseous exchange.

Stomata open and close to control the amount of water lost by a plant, but during the day a plant needs to take in carbon dioxide for photosynthesis and at night when no oxygen is being produced by photosynthesis it needs to take in oxygen for cellular respiration, so at least some stomata need to be open all the time.

It has been estimated that an acre of corn loses around 11 500–15 000 litres of water through transpiration every day, whilst a single large tree can lose more than 700 litres a day.

The transpiration stream

As you learn in Topic 9.2, water enters the roots of the plant by osmosis and is transported up in the xylem until it reaches the leaves. Here, it moves by osmosis across membranes and by diffusion in the apoplast pathway from the xylem through the cells of

Synoptic link

You will learn more about photosynthesis in Topic 17.3, Photosynthesis.

the leaf where it evaporates from the freely permeable cellulose cell walls of the mesophyll cells in the leaves into the air spaces. The water vapour then moves into the external air through the stomata along a diffusion gradient. This is the **transpiration stream**.

The transpiration stream moves water up from the roots of a plant to the highest leaves – a height which can be up to 100 m or more in the tallest trees. As you learn in Topic 9.1, xylem vessels are non-living, hollow tubes so the process must be passive. How does it work?

- Water molecules evaporate from the surface of mesophyll cells into the air spaces in the leaf and move out of the stomata into the surrounding air by diffusion down a concentration gradient.

- The loss of water by evaporation from a mesophyll cell lowers the water potential of the cell, so water moves into the cell from an adjacent cell by osmosis, along both apoplast and symplast pathways.

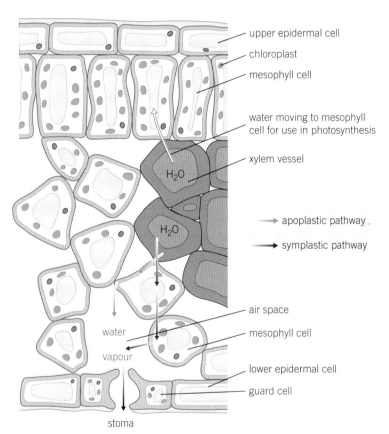

▲ **Figure 1** *Movement of water across a leaf*

- This is repeated across the leaf to the xylem. Water moves out of the xylem by osmosis into the cells of the leaf.

- Water molecules form hydrogen bonds with the carbohydrates in the walls of the narrow xylem vessels – this is known as adhesion. Water molecules also form hydrogen bonds with each other and so tend to stick together – this is known as cohesion. The combined effects of adhesion and cohesion result in water exhibiting capillary action. This is the process by which water can rise up a narrow tube against the force of gravity. Water is drawn up the xylem in a continuous stream to replace the water lost by evaporation. This is the *transpiration pull*.

- The transpiration pull results in a tension in the xylem, which in turn helps to move water across the roots from the soil.

This model of water moving from the soil in a continuous stream up the xylem and across the leaf is known as the **cohesion-tension theory**.

Synoptic link

You learn about the adhesive and cohesive properties of water and capillary action in Topic 3.2, Water.

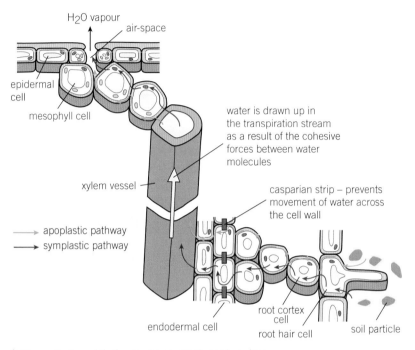

▲ **Figure 2** *A model of water transport through a plant*

Evidence for the cohesion-tension theory

Several pieces of evidence support the cohesion-tension theory for the movement of water up the xylem of a plant. These include the following:

- Changes in the diameter of trees. When transpiration is at its height during the day, the tension in the xylem vessels is at its highest too. As a result the tree shrinks in diameter. At night, when transpiration is at its lowest, the tension in the xylem vessels is at its lowest and the diameter of the tree increases. This can be tested by measuring the circumference of a suitably sized tree at different times of the day.

- When a xylem vessel is broken – for example when you cut flower stems to put them in water – in most circumstances air is drawn in to the xylem rather than water leaking out.

- If a xylem vessel is broken and air is pulled in as described in the previous bullet, the plant can no longer move water up the stem as the continuous stream of water molecules held together by cohesive forces has been broken.

In summary, transpiration delivers water, and the mineral ions dissolved in that water, to the cells where they are needed. The evaporation of water from the leaf cell surfaces also helps to cool the

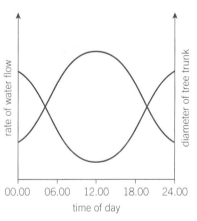

▲ **Figure 3** *The relationship between water flow in the xylem and the diameter of a tree trunk*

Study tip

Read questions about water movement in plants carefully. A question about how water in the xylem in the root reaches the cells of the leaves of a plant does not require you to give any information about the uptake of water from the soil by the root hairs or about the movement of water across the root.

leaf down and prevent heat damage (refer back to Topic 3.2, Water). However, transpiration, specifically the water loss, is also a problem for a plant because the amount of water available is often limited. In high intensity sunlight when the plant is photosynthesising rapidly, there will be a high rate of gaseous exchange, the stomata will all be open and the plant may lose so much water through transpiration that the supply cannot meet the demand.

Measuring transpiration

It is very difficult to make direct measurements of transpiration because of the practical difficulties with condensing and collecting all of the water that evaporates from the surfaces of the leaves and stems of a plant without also collecting water that evaporates from the soil surface. It is also very difficult to separate water vapour from transpiration and water vapour produced as a waste product of respiration.

However, it is relatively easy to measure the uptake of water by a plant. As around 99% of the water taken up by a plant is then lost by transpiration, water uptake gives us a good working model of transpiration losses. By measuring factors that affect the uptake of water by a plant, you are effectively also measuring the factors that affect the rate of transpiration.

The rate of water uptake can be measured in a variety of ways. The most common is using a **potometer**. The apparatus has to be set up with great care, and all joints must be sealed with waterproof jelly to make sure that

any water loss measured is as a result of transpiration from the stem and leaves (Figure 4).

Rate of water uptake = distance moved by air bubble/ time taken for air bubble to move that distance. The units are cm s^{-1}.

A plant was placed in a potometer in bright light. The time taken for the bubble to move 5 cm was recorded under different conditions:

Conditions	Time in seconds	Rate of bubble movement cm s^{-1}
Normal lab conditions	35.0	0.14
Bright light directed at leaves	18.0	0.28
Bottom surfaces of leaves covered with vaseline	175.0	0.028

Data like these can be used to help demonstrate the effect of different factors on transpiration

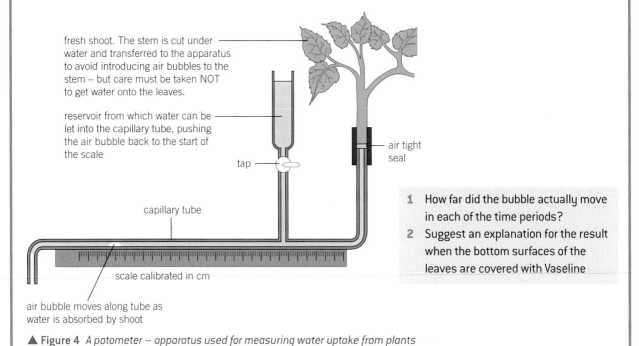

fresh shoot. The stem is cut under water and transferred to the apparatus to avoid introducing air bubbles to the stem – but care must be taken NOT to get water onto the leaves.

reservoir from which water can be let into the capillary tube, pushing the air bubble back to the start of the scale

tap

air tight seal

capillary tube

scale calibrated in cm

air bubble moves along tube as water is absorbed by shoot

1 How far did the bubble actually move in each of the time periods?
2 Suggest an explanation for the result when the bottom surfaces of the leaves are covered with Vaseline

▲ **Figure 4** *A potometer – apparatus used for measuring water uptake from plants*

Stomata – controlling the rate of transpiration

As you learn at the start of this topic, the main way in which the rate of transpiration is controlled by the plant is by the opening and closing of the stomatal pores. This is a turgor-driven process. When turgor is low the asymmetric configuration of the guard cell walls closes the pore. When the environmental conditions are favourable guard cells pump in solutes by active transport, increasing their turgor. Cellulose hoops prevent the cells from swelling in width, so they extend lengthways. Because the inner wall of the guard cell is less flexible than the outer wall, the cells become bean-shaped and open the pore. When water becomes scarce, hormonal signals from the roots can trigger turgor loss from the guard cells, which close the stomatal pore and so conserve water.

Factors affecting transpiration

Any factor affecting the rate of water loss from the leaves of a plant will affect the rate of transpiration. Factors that affect water loss from the leaf must either act on the opening/closing of the stomata, the rate of evaporation from the surfaces of the leaf cells or the diffusion gradient between the air spaces in the leaves and the air surrounding the leaf. The effects of some of these factors are described below:

- Light is required for photosynthesis and in the light the stomata open for the gas exchange needed. In the dark, most of the stomata will close. Increasing light intensity gives increasing numbers of open stomata, increasing the rate of water vapour diffusing out and therefore increasing the evaporation from the surfaces of the leaf. So, increasing light intensity increases the rate of transpiration (Figure 6).

- Relative humidity is a measure of the amount of water vapour in the air (humidity) compared to the total concentration of water the air can hold. A very high relative humidity will lower the rate of transpiration because of the reduced water vapour potential gradient between the inside of the leaf and the outside air. Very dry air has the opposite effect and increases the rate of transpiration.

- Temperature affects the rate of transpiration in two ways:
  - An increase in temperature increases the kinetic energy of the water molecules and therefore increases the rate of evaporation from the spongy mesophyll cells into the air spaces of the leaf.
  - An increase in temperature increases the concentration of water vapour that the external air can hold before it becomes saturated (so decreases its relative humidity and its water potential).

Both factors increase the diffusion gradient between the air inside and outside the leaf, thus increasing the rate of transpiration.

- Air movement – Each leaf has a layer of still air around it trapped by the shape of the leaf and features such as hairs on the surface of

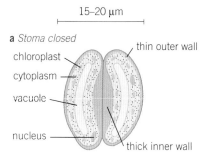

15–20 µm

a *Stoma closed*

chloroplast · cytoplasm · vacuole · nucleus · thin outer wall · thick inner wall

b *Stoma open*

stomatal aperture

▲ **Figure 5** *The opening and closing of the stomata is vital to the control of transpiration in a plant*

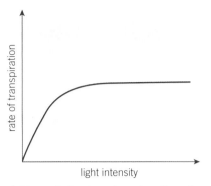

▲ **Figure 6** *Graph to show the effect of light intensity on transpiration*

the leaf decrease air movement close to the leaf. The water vapour that diffuses out of the leaf accumulates here and so the water vapour potential around the stomata increases, in turn reducing the diffusion gradient. Anything that increases the diffusion gradient will increase the rate of transpiration. So air movement or wind will increase the rate of transpiration, and conversely a long period of still air will reduce transpiration (Figure 7).

● Soil-water availability – The amount of water available in the soil can affect transpiration rate. If it is very dry the plant will be under water stress and the rate of transpiration will be reduced.

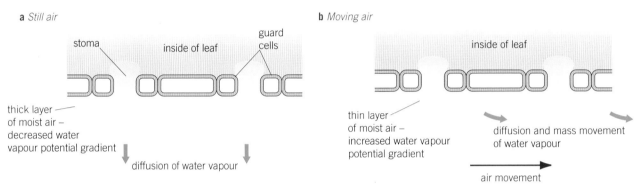

▲ **Figure 7** *The effect of air movements on the rate of transpiration*

Summary questions

1   Explain the difference between transpiration and the transpiration stream.   *(2 marks)*

2   Compare root pressure and the transpiration pull.   *(3 marks)*

3   Describe how you might use a potometer to investigate the effect of air movements on transpiration rates.   *(6 marks)*

4   In an investigation into water uptake by a shoot using a bubble potometer, a group recorded how far the air bubble moved in a set time period of 40 seconds, in different conditions.
     Their results are as follows:
     a   In the dark: 0.4 cm   *(5 marks)*
     b   In normal daylight: 5.6 cm   *(5 marks)*
     c   With bright lights shining on the plant shoot: 8.0 cm   *(5 marks)*
     d   With bright light shining and a fan blowing on the shoot: 12 cm   *(6 marks)*

     Calculate the rate of water movement under each set of conditions and explain each result.

5   a   Describe the cohesion-tension theory of transpiration.   *(6 marks)*
     b   Use the theory to explain:
          i   the changes in diameter measured in the trunk of a tree between midday and midnight   *(4 marks)*
          ii the fact that cut flowers placed in water may droop and die very quickly.   *(4 marks)*

9.4 Translocation

Specification reference: 3.1.3

The leaves of a plant produce large amounts of glucose, which is needed for respiration by all the cells of the plant. The glucose is converted to sucrose for transport. When it reaches the cells where it is needed it is converted back to glucose for respiration, or to starch for storage, or used to produce the amino acids and other compounds needed within the cell.

From source to sink

Plants transport organic compounds in the phloem from **sources** to **sinks** (the tissues that need them) in a process called **translocation**. In many plants translocation is an active process that requires energy to take place and substances can be transported up or down the plant. The products of photosynthesis that are transported are known as **assimilates**. Although glucose is made in the process of photosynthesis, the main assimilate transported around the plant is **sucrose**. The sucrose content of most cell sap is only around 0.5%, but it can be 20–30% of the phloem sap content.

The main **sources** of assimilates in a plant are:

- green leaves and green stems
- storage organs such as tubers and tap roots that are unloading their stores at the beginning of a growth period
- food stores in seeds when they germinate.

Some of these need to transport resources downwards, and some need to move materials up the plant.

The main **sinks** in a plant include:

- roots that are growing and/or actively absorbing mineral ions
- meristems that are actively dividing
- any parts of the plant that are laying down food stores, such as developing seeds, fruits or storage organs.

▼ Table 1 *Carbohydrates in cyclamen. The data in this table suggest that sucrose is transported in the phloem*

Plant part	Mean carbohydrate content ($\mu g\,g^{-1}$ fresh mass ± standard error of mean)	
	sucrose	starch
leaf blade	1312 ± 212	62 ± 25
vascular bundle in the leaf stalk, consisting of xylem and phloem	5757 ± 1190	< 18
tissue surrounding the vascular bundle in the leaf stalk	417 ± 96	< 18
buds, roots, and tubers (underground storage organs)	2260 ± 926	152 ± 242

Learning outcomes

Demonstrate knowledge, understanding, and application of:

→ the mechanism of translocation including the transport of assimilates between sources and sinks.

Synoptic link

You learnt about glucose, sucrose and starch in Topic 3.3, Carbohydrates.

The process of translocation

Translocation is a vital and effective process – a large tree can transport around 250 kg of sucrose down its trunk in a year, and substances move at speeds of around 0.15–7 metres per hour. The details of how substances are moved in the phloem of plants are still the subject of active investigation but the main steps are described here:

Phloem loading

In many plants the soluble products of photosynthesis are moved into the phloem from the sources by an active process. Sucrose is the main carbohydrate transported – it is not used in metabolism as readily as glucose and is therefore less likely to be metabolised during the transport process.

There are two main ways in which plants load assimilates into the **phloem** (phloem loading) for transport. One is largely passive, the other is active. Active phloem loading by the apoplast route is the most widely studied.

> ### The symplast route
>
> In some species of plants the sucrose from the source moves through the cytoplasm of the mesophyll cells and on into the sieve tubes by diffusion through the plasmodesmata (known as the **symplast** route). Although phloem loading and translocation are often referred to as active processes, this route is largely *passive*. The sucrose ends up in the sieve elements and water follows by osmosis. This creates a pressure of water that moves the sucrose through the phloem by mass flow.

The apoplast route

In many plant species sucrose from the source travels through the cell walls and inter-cell spaces to the companion cells and sieve elements (known as the **apoplast** route) by diffusion down a concentration gradient, maintained by the removal of sucrose into the phloem vessels. In the companion cells sucrose is moved into the cytoplasm across the cell membrane in an *active process*. Hydrogen ions (H^+) are actively pumped out of the companion cell into the surrounding tissue using ATP. The hydrogen ions return to the companion cell down a concentration gradient via a co-transport protein. Sucrose is the molecule that is co-transported. This increases the sucrose concentration in the companion cells and in the sieve elements through the many plasmodesmata between the two linked cells.

Companion cells have many infoldings in their cell membranes to give an increased surface area for the active transport of sucrose into the cell cytoplasm. They also have many mitochondria to supply the ATP needed for the transport pumps.

As a result of the build up of sucrose in the companion cell and sieve tube element, water also moves in by osmosis. This leads to a build up of turgor pressure due to the rigid cell walls. The water carrying the assimilates moves into the tubes of the sieve elements, reducing the pressure in the companion cells, and moves up or down the plant by mass flow to areas of lower pressure (the sinks).

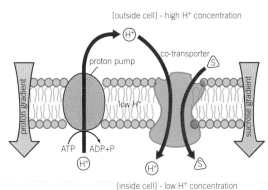

▲ **Figure 1** *The active movement of sucrose (S) into a companion cell or sieve tube across the cell membrane*

Solute accumulation in source phloem leads to an increase in turgor pressure that forces sap to regions of lower pressure in the sinks. The pressure generated in the phloem is around 2 MPa (15 000 mm Hg) – considerably higher than the 0.016 MPa (120 mm Hg) of pressure in a human artery. These pressure differences in plants can transport solutes and water rapidly over many metres. Solutes are translocated either up or down the plant, depending on the positions of the source.

Phloem unloading

The sucrose is unloaded from the phloem at any point into the cells that need it. The main mechanism of phloem unloading seems to be by diffusion of the sucrose from the phloem into the surrounding cells. The sucrose rapidly moves on into other cells by diffusion or is converted into another substance (for example glucose for respiration, starch for storage) so that a concentration gradient of sucrose is maintained between the contents of the phloem and the surrounding cells.

The loss of the solutes from the phloem leads to a rise in the water potential of the phloem. Water moves out into the surrounding cells by osmosis. Some of the water that carried the solute to the sink is drawn into the transpiration stream in the xylem.

Looking at the evidence

There is still a lot of research to be done to determine all the details of translocation. But there is a body of evidence that supports the main principles:

● Advances in microscopy allow us to see the adaptations of the companion cells for active transport.

● If the mitochondria of the companion cells are poisoned, translocation stops.

● The flow of sugars in the phloem is about 10 000 times faster than it would be by diffusion alone, suggesting an active process is driving the mass flow.

● Aphids can be used to demonstrate the translocation of organic solutes in the phloem. Using evidence from aphid studies, it has been shown that there is a positive pressure in the phloem that forces the sap out through the stylet. The pressure and therefore the flow rate in the phloem is lower closer to the sink than it is near the source. The concentration of sucrose in the phloem sap is also higher near to the source than near the sink.

However, some questions remain. Not all solutes in the phloem move at the same rate. On the other hand, sucrose always seems to move at the same rate regardless of the concentration at the sink. And no one is yet completely sure about the role of the sieve plates in the process.

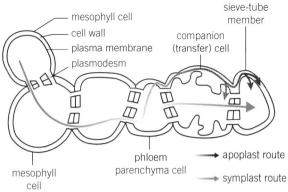

▲ **Figure 2** *The apoplast and symplast routes by which sucrose is loaded into the phloem*

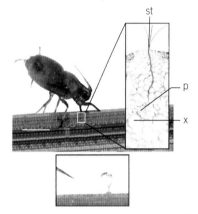

▲ **Figure 3** *Aphids (order Hemiptera) are a simple but very effective research tool for plant biologists. Top: Aphids penetrate the plant tissue with their mouth parts or stylet (st) to reach the phloem (p). Bottom: If the aphid is anaesthetised and removed from the stylet, phloem continues to flow out of the stylet due to pressure from the phloem contents. The rate of flow and composition of sap can be analysed*

Summary questions

1 Explain the difference between a source and a sink. (*4 marks*)

2 Describe how aphids can be used to demonstrate translocation, including the effect light intensity on the process. (*5 marks*)

3 a Explain the role of active transport in translocation in the phloem. (*5 marks*)
   b Describe the evidence that supports this model. (*5 marks*)

9.5 Plant adaptations to water availability

Specification reference: 3.1.3

Learning outcomes

Demonstrate knowledge, understanding, and application of:

→ adaptations of plants to the availability of water in their environment, including xerophytes and hydrophytes.

Synoptic link

You will learn more about the adaptations of marram grass in Topic 10.7, Adaptations.

Synoptic link

You will learn about the classification and naming of species in Topic 10.1, Classification.

Synoptic link

You will learn about habitat biodiversity in Topic 11.1, Biodiversity and about ecosystems in Chapter 23, Ecosystems.

Land plants exist in a state of constant compromise between getting the carbon dioxide they need for photosynthesis and losing the water they need for turgor pressure and transport. They must have a large SA : V ratio for gaseous exchange and the capture of light for photosynthesis, but this greatly increases their risk of water loss by transpiration.

Xerophytes

Most plants have adaptations to conserve water. These include a waxy cuticle to reduce transpiration from the leaf surfaces, stomata found mainly on the underside of the leaf that can be closed to prevent the loss of water vapour, and roots that grow down to the water in the soil. However, in habitats where water is often in very short supply, this is not enough. In hot conditions, particularly hot, dry, and breezy conditions – water will evaporate from the leaf surfaces very rapidly. Plants in dry habitats have evolved a wide range of adaptations that enable them to live and reproduce in places where water availability is very low indeed. They are known as **xerophytes**.

Conifers (class Pinopsida) are xerophytes. So is marram grass (*Ammophila* spp.), a plant found widely on sand dunes and coastal areas, in dry and salty conditions.

Many plants that survive in very cold and icy conditions are also xerophytes – the water in the ground is not freely available to them because it is frozen.

Perhaps the best known xerophytes are the cacti (members of the plant family Cactaceae).

▲ Figure 1 *Successful desert plants like these have many adaptations to a xerophytic way of life because the gap between rains may be months or even years so water is rarely available*

Ways of conserving water

Xerophytes use a range of strategies for conserving water:

- A thick waxy cuticle – in most plants up to 10% of the water loss by transpiration is actually through the cuticle. Some plants have a particularly thick waxy cuticle to help minimise water loss. This adaptation is common in evergreen plants and helps them survive both hot dry summers and cold winters when water can be hard to absorb from the frozen ground. Holly (*Ilex* spp.) is an example commonly seen in the UK.

- Sunken stomata – many xerophytes have their stomata located in pits, which reduce air movement, producing a microclimate of still, humid (moist) air that reduces the water vapour potential gradient and so reduces transpiration. These are seen clearly in xerophyes such as marram grass, cacti, and conifers.

- Reduced numbers of stomata – Many xerophytes have reduced numbers of stomata, which reduce their water loss by transpiration but also reduce their gas exchange capabilities.

- Reduced leaves – by reducing the leaf area, water loss can be greatly reduced. The leaves of conifers are reduced to thin needles. These narrow leaves, which are almost circular in cross-section, have a greatly reduced SA : V ratio, minimising the amount of water lost in transpiration.

- Hairy leaves – some xerophytes have very hairy leaves that, like the spines of some cacti, create a microclimate of still, humid air, reducing the water vapour potential gradient and minimising the loss of water by transpiration from the surface of the leaf. Some plants – such as marram grass – even have microhairs in the sunken stomatal pits (Figure 4).

- Curled leaves – another adaptation that greatly reduces water loss by transpiration, especially in combination with other adaptations, is the growth of curled or rolled leaves. This confines all of the stomata within a microenvironment of still, humid air to reduce diffusion of water vapour from the stomata. Marram grass is a good example of a plant with this strategy.

- Succulents – succulent plants store water in specialised parenchyma tissue in their stems and roots. They get their name because, unlike other plants, they often have a swollen or fleshy appearance. Water is stored when it is in plentiful supply and then used in times of drought. *Salicornia* spp. (edible samphire), which grows on UK salt marshes, and desert cacti are examples of succulents, as are aloes, which include *Aloe vera*, a plant often used in cosmetics.

- Leaf loss – some plants prevent water loss through their leaves by simply losing their leaves when water is not available. Palo verde (*Parkinsonia* spp.) is a desert tree that loses all of its leaves in the long dry seasons. The trunk and branches turn green and photosynthesise with minimal water loss to keep it alive. Its name is derived from the Spanish words meaning 'green pole'.

Synoptic link

You learned about the structure and function of stomata in Topic 6.4, The organisation and specialisation of cells and Topic 9.3, Transpiration.

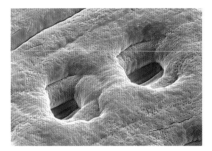

▲ **Figure 2** *The stomata of this sitka spruce (*Picea stichensis*) are in sunken pits and the thick waxy cuticle shows up as a pale green layer – two clear adaptations to minimise water loss*

▲ **Figure 3** *Reducing the leaves to spines in this Barrel cactus prevents transpiration from the leaves but also helps create a microclimate to prevent water loss from the stem*

▲ **Figure 4** *This Borage flower growing in the desert needs these very hairy leaves to prevent water loss and survive*

▲ **Figure 5** *The folds in the stem of this saguaro cactus (*Carnegiea gigantea*) allow it to expand in the rare times when water is freely available. A fully hydrated saguaro can weigh over a tonne.*

▲ **Figure 6** *A leafless palo verde (*Parkinsonia sp.*) not only photosynthesises but also flowers without any leaves*

- Root adaptations – many xerophytes have root adaptations that help them to get as much water as possible from the soil. Long tap roots growing deep into the ground can penetrate several metres, so they can access water that is a long way below the surface. A mass of widespread, shallow roots with a large surface area able to absorb any available water before a rain shower evaporates is another adaptation. Many cacti show both of these adaptations, including the giant saguaro (*Carnegiea gigantea*), which can get enough water to grow to around 12–18 metres tall and live for around 200 years. The root system of marram grass consists of long vertical roots that penetrate metres into the sand. They also have a mat of horizontal rhizomes (modified stems) from which many more roots develop to form an extensive network that helps to change their environment and enable the sand to hold more water.

- Avoiding the problems – some plants are adapted to cope with the problems of low water availability by avoiding the situation entirely. Plants may lose their leaves and become dormant, or die completely, leaving seeds behind to germinate and grow rapidly when rain falls again. Others survive as storage organs such as bulbs (onions, daffodils), corms (crocuses) or tubers (potatoes, dahlias). A few plants can withstand complete dehydration and recover – they appear dead but when it rains the cells recover, the plant becomes turgid and green again and begins to photosynthesise. The ability to survive in this way is linked to the disaccharide trehalose, which appears to enable to the cells to survive unharmed.

 Investigating stomatal numbers

You can compare the numbers of stomata on the leaves and stems of plants by taking an impression of the epidermis of the leaf or the stem and looking at it under a microscope.

You can use clear nail varnish, Germolene New Skin or DIY water based varnish.

You need to observe the numbers of stomata, and whether they are open or closed over the same area each time so take care to use the same magnification with your microscope. You may choose to use a graticule.

This technique can be used to compare stomatal numbers in different areas of a plant and in different types of plants. It can also be used to investigate the opening and closing of the stomata under different conditions.

1   How would you predict the number and position of the stomata might vary between xerophytes and normal plants?
2   Suggest why it can be difficult to investigate the stomata of cacti, even though they are some of the most effective xerophytes.

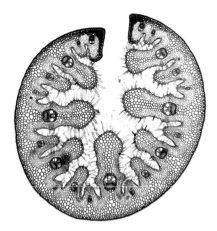

▲ Figure 7 *Marram grass shows many adaptations to reduced water availability including both vertical and horizontal roots as well as the stomatal pits, hairs, and curled leaves visible in the light micrograph (× 10 magnification)*

Hydrophytes

Not all plants have to conserve water. In fact the **hydrophytes** – plants that actually live in water (submerged, on the surface or at the edges of bodies of water) – need special adaptations to cope with growing in water or in permanently saturated soil.

Examples of hydrophytes include water lilies (plants of the family Nymphaeaceae) and water cress (*Nasturtium officinale*), which grow at the surface, duckweeds (genus *Lemna*), which are submerged or free-floating plants, and marginals such bulrushes (*Typha latifolia*) and yellow iris (*Iris pseudacorus*), which grow at the edge of the water.

It is important in surface water plants that the leaves float so they are near the surface of the water to get the light needed for photosynthesis.

Water-logging is a major problem for all hydrophytes. The air spaces of the plant need to be full of air, not water, for the plant to survive.

▲ Figure 8 *Hydrophytes like this water lily (family Nymphaeaceae) have no need to conserve water, but they have other problems to overcome*

Adaptations of hydrophytes:

- Very thin or no waxy cuticle – hydrophytes do not need to conserve water as there is always plenty available so water loss by transpiration is not an issue.

- Many always-open stomata on the upper surfaces – maximising the number of stomata maximises gaseous exchange. Unlike other plants there is no risk to the plant of loss of turgor as there is always an abundance of water available, so the stomata are usually open all the time for gaseous exchange and the guard cells are inactive. In plants with floating leaves such as water lilies the stomata need to be on the upper surface of the leaf so they are in contact with the air.

- Reduced structure to the plant – the water supports the leaves and flowers so there is no need for strong supporting structures.

- Wide, flat leaves – some hydrophytes, including the water lilies, have wide, flat leaves that spread across the surface of the water to capture as much light as possible.

- Small roots – water can diffuse directly into stem and leaf tissue so there is less need for uptake by roots.

- Large surface areas of stems and roots under water – this maximises the area for photosynthesis and for oxygen to diffuse into submerged plants.

- Air sacs – some hydrophytes have air sacs to enable the leaves and/or flowers to float to the surface of the water.

- Aerenchyma – specialised parenchyma (packing) tissue forms in the leaves, stems and roots of hydrophytes. It has many large air spaces, which seem to be formed at least in part by apoptosis (programmed cell death) in normal parenchyma. It has several different functions within the plants, including:

  — making the leaves and stems more buoyant

  — forming a low-resistance internal pathway for the movement of substances such as oxygen to tissues below the water. This helps the plant to cope with anoxic (extreme low oxygen conditions) conditions in the mud, by transporting oxygen to the tissues.

Aerenchyma is found in crop species that grow in water, such as rice (*Oryza sativa* and *Oryza glaberrima*). Studies suggest that aerenchyma may provide a low resistance pathway by which methane produced by the rice plants can be vented into the atmosphere. This is part of a major problem. Atmospheric methane, which contributes of the greenhouse effect and the resulting climate change, has doubled over the past two centuries and flooded rice paddies represent a major source.

In situations where there is plenty of water – for example in mangrove swamps, the roots can become waterlogged. It is air rather than water that is in short supply. Special aerial roots called pneumatophores grow upwards into the air. They have many lenticels, which allow the entry of air into the woody tissue.

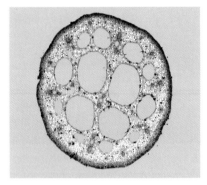

▲ **Figure 9** *Light micrograph showing aerenchyma tissue in the stem of a water lily leaf (Nymphaea alba) – compare this to the TS stem of a land plant you looked at in Topic 9.1 Transport systems in dicotyledonous plants.*

Summary questions

1. State three structural adaptations seen in the leaves of xerophytes and explain how these are related to their functions. *(3 marks)*

2. State three structural adaptations of hydrophytes and explain how they are related to their functions. *(3 marks)*

3. Compare the challenges of a plant of living in dry conditions to a plant living in water or waterlogged conditions. *(4 marks)*

4. Based on your knowledge of plant adaptations to water availability, suggest two characteristics that might be targeted by scientists in their search for suitable genes to use in future crops to help them withstand drought and two characteristics that might help them withstand flooding. Explain the advantages they confer on the plants and why you have chosen those characteristics. *(6 marks)*

Practice questions

1 The term vascular is derived from the Latin word vas meaning vessel.

   a State what is meant by the term vascular bundle. *(2 marks)*

   b Draw what you would expect to see in a dicotyledonous plant stem. *(2 marks)*

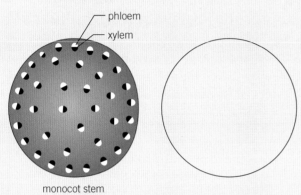

monocot stem

   c Describe and explain what happens when a ring of bark is removed from a tree. *(4 marks)*

2 The movement of water and solutes through a plant is known as the transpiration stream. It is the way essential minerals are transported from roots to leaves.

   a (i) Define transpiration. *(2 marks)*

      (ii) Explain why transpiration is an inevitable consequence of photosynthesis. *(4 marks)*

   b (i) Name the piece of apparatus used to measure transpiration. *(1 mark)*

      (ii) Explain why the measurements obtained from this apparatus are only an estimate of the rate of transpiration. *(2 marks)*

   c Table 1 below shows the readings obtained using the apparatus named in bi for a plant exposed to different environments.

The distance moved by a column of water in a capillary tube was measured every three minutes for 30 minutes as the plant was exposed to each environment.

Water loss ml/m²												
Time/minutes		0	3	6	9	12	15	18	21	24	27	30
treatment	fan											
	mist	0.00	4.17	4.17	4.17	4.17	2.08	0.00	2.08	2.08	0.00	2.08

      (i) Given the following information, copy and complete the first table and fill in the missing figures. *(2 marks)*

Water loss ml												
Time/minutes fan	0	3	6	9	12	15	18	21	24	27	30
	0.00	0.01	0.01	0.01	0.01	0.01	0.02	0.01	0.01	0.01	0.00

Mass of leaves = 1.1 g

Leaf Surface Area = 0.0044 m²

      (ii) Suggest how the mass of leaves could have been used to calculate the surface area of the leaves. *(3 marks)*

      (iii) State the independent variable, the dependent variable and two variables that would have been controlled in the experiment. *(4 marks)*

      (iv) Plot a graph to display the results. *(4 marks)*

      (v) Identify any anomalous results giving the reason for your choice. *(2 marks)*

      (vi) Explain the trends shown on the graph. *(4 marks)*

   d Describe how water is moving at A, B and C in the diagram below.

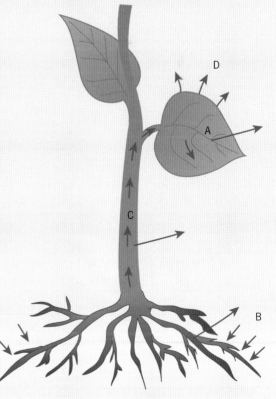

(6 marks)

3 In an experiment to measure the rate of diffusion, a student placed cubes of agar jelly containing an indicator into dilute hydrochloric acid. The indicator changes from pink to colourless in acidic conditions.

The student used cubes of different sizes and recorded the time taken for the pink colour of each cube to disappear completely.

The students results are recorded in the table

Length of side of cube (mm)	Surface area of cube (mm^2)	Volume of cube (mm^3)	Surface area to volume ratio	Time taken for pink colour to disappear (s)	Rate of diffusion (mm s^{-1})
2	24	8	3.0:0	50	0.020
5	150	125	1.2:1	120	0.021
10	600	1000		300	0.017
20	2400	8000	0.3:1	700	0.014
30	5400	27000	0.2:1	1200	0.013

a Define diffusion. *(1 mark)*

b Calculate the surface area to volume ratio of the cube with 10 mm sides.

Show your working. *(2 marks)*

c (i) Using the data in the table describe the relationship between the rate of diffusion and the surface area to volume ratio. *(2 marks)*

(ii) Explain the significance of the relationship between rate of diffusion and the surface area to volume ratio for large plants. *(2 marks)*

d Another student used the same raw data obtained in the experiment but calculated a different rate of diffusion for each cube. This student's results are shown in the table.

Length of side of cube (mm)	Time taken for pink colour to disappear (s)	Rate of diffusion (mm s^{-1})
2	50	0.040
5	120	0.042
10	300	0.033
20	700	0.029
30	1200	0.025

In this student's table, the calculation of the rate of diffusion is incorrect.

(i) suggest the method used to calculate the rate of diffusion in the table *(1 mark)*

(ii) state why the method in **d**(i) is not correct *(1 mark)*

OCR F211/01 2013

4 Xerophytes are plants adapted to living in environments where little water is available. These environments can be as varied as deserts where there is a complete lack of water or the antarctic where all the water is present as ice or snow.

a Explain what is meant by the term xerophyte. *(1 mark)*

b Describe how a cactus plant is adapted to the environment in which it lives. *(3 marks)*

5 a Explain the difference between active transport and facilitated transport. *(3 marks)*

b Describe the difference in the direction of movement in the phloem and xylem. *(2 marks)*

c The following statements are evidence that support the explanation of translocation in phloem.

1 *there is an exudation of solution from the phloem when the stem is cut or punctured by the mouthparts of an aphid*

2 *concentration gradients of organic solutes are proved to be present between the sink and the source.*

3 *when viruses or growth chemicals are applied to a well-illuminated leaf, they are translocated downwards to the roots. This does not happen if the leaves are shaded.*

Explain how each statement supports this explanation. *(6 marks)*

Module 3 summary

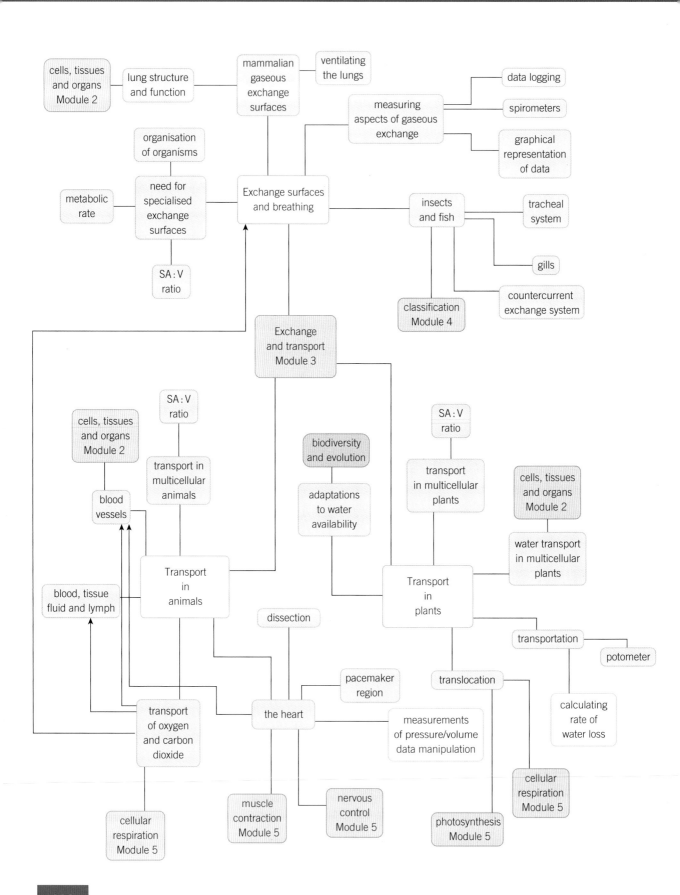

cells, tissues and organs Module 2 — lung structure and function — mammalian gaseous exchange surfaces — ventilating the lungs

measuring aspects of gaseous exchange — data logging — spirometers — graphical representation of data

organisation of organisms

metabolic rate — need for specialised exchange surfaces — Exchange surfaces and breathing

SA : V ratio

insects and fish — tracheal system — gills — countercurrent exchange system

classification Module 4

Exchange and transport Module 3

cells, tissues and organs Module 2

SA : V ratio

transport in multicellular animals

biodiversity and evolution — adaptations to water availability

SA : V ratio

transport in multicellular plants

cells, tissues and organs Module 2 — water transport in multicellular plants

blood vessels

blood, tissue fluid and lymph

Transport in animals

dissection

Transport in plants

transportation — potometer

pacemaker region

translocation

calculating rate of water loss

transport of oxygen and carbon dioxide

the heart

measurements of pressure/volume data manipulation

cellular respiration Module 5

muscle contraction Module 5

nervous control Module 5

photosynthesis Module 5

cellular respiration Module 5

Application

Aphids feed by tapping into the phloem of plants. They push their stylets through the tissues of the plant stem or leaf until they reach the phloem. The stylet has a food canal to transport the sugary sap from the phloem into the insect. It also has a saliva canal to carry saliva down the stylet into the plant.

How does the plant respond to this invasion? There is now evidence that plants have a number of protein–based systems which act in seconds to block off damaged phloem vessels and prevent the loss of sap.

▲ **Figure 1** *Why doesn't a lawn 'bleed' to death every time it is cut?*

- the plastids may burst and release their contents which then coagulate and block the sieve elements

- certain proteins associated with the endoplasmic reticulum of the sieve elements will coagulate if the cells are damaged, blocking the vessel

- some plants, for example, broad beans, contain special proteins called *forisomes* in the phloem. As soon as a sieve element is damaged, calcium ions flow into the cells and the forisomes expand and block the vessel.

Researcher have hypothesised that the saliva pumped down the stylet of an aphid into the phloem of a plant helps prevent this coagulation from taking place.

1   When a lawn is cut, the ends of the leaves of the grass are chopped off, exposing the cut surfaces of the phloem. Suggest why the grass plants do not die as a result of losing all of the assimilates from photosynthesis through the cut transport vessels.

2   Why is it so important for aphids to prevent the protein coagulation which occurs in damaged phloem vessels if they are to successfully transport of food into their guts?

3   Find out more – produce a powerpoint presentation EITHER on phloem tissue and the mechanisms which protect the plant when the phloem is damaged OR looking at aphids and the ways in which they overcome the defence mechanisms of the plant phloem. Make sure you use lots of resources to help compile your presentation.

Extension

In mammals such as human beings, the blood is transported around the body in the blood vessels. If the blood vessels are damaged, there is a clotting mechanism which involves both proteins and calcium ions in a cascade which rapidly forms a clot to block the damaged blood vessel and prevent blood loss.

Investigate the clotting mechanism of the blood. Produce large, illustrated flow diagrams to compare the way in which the plant transport system prevents the loss of sugar-rich sap when the phloem is damaged with the clotting mechanism seen in the human circulatory system when a blood vessel is damaged. Write a commentary to highlight the similarities and differences between the two systems.

1 Groundkeepers have an essential role during the winter ensuring football pitches don't get too muddy and are kept in a condition suitable to play on. It is important that they use the most suitable species of grass.

Some species of grass grow more in the winter and therefore transpire at a greater rate.

a (i) Describe what is meant by the term transpiration. (*2 marks*)

(ii) Explain why transpiration rates are affected by the rate of growth of grass. (*4 marks*)

An investigation was carried to determine the difference in transpiration rates of a winter active grass and a grass that is relatively dormant in the winter.

The results are shown in the graph.

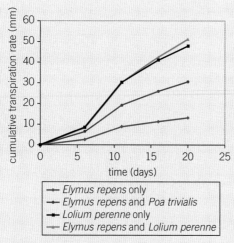

Key:
- *Elymus repens* only
- *Elymus repens* and *Poa trivialis*
- *Lolium perenne* only
- *Elymus repens* and *Lolium perenne*

b (i) Describe the results of the investigation shown in the graph. (*4 marks*)

(ii) Suggest, giving your reasons, which grass, or combination of grasses, the groundkeeper would pick based on the results in the graph. (*3 marks*)

2 The diagram shows the diagram of a transverse section of the heart viewed from the top.

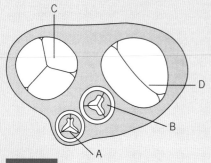

a (i) State the name of the valves labelled A, B, C and D in the diagram (*4 marks*)

(ii) Outline the functions of the four valves. (*5 marks*)

The diagram shows a normal and diseased aortic heart valve in the open and closed positions.

healthy aortic valve – closed

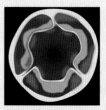
healthy aortic valve – open

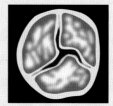
diseased aortic valve – closed

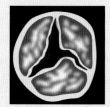
diseased aortic valve – open

b (i) Describe the differences between the normal and diseased valves. (*4 marks*)

(ii) Explain how the presence of the diseased valves would affect the flow of blood through the heart. (*5 marks*)

3 Stomatal conductance is a measure of the rate of diffusion of carbon dioxide through the stomata of a leaf.

Five-month-old seedlings grown in planting bags were not watered for 21 days.

The stomatal conductance (top graph) and leaf water potential (bottom graph) were measured during this period.

The results are shown in the graphs.

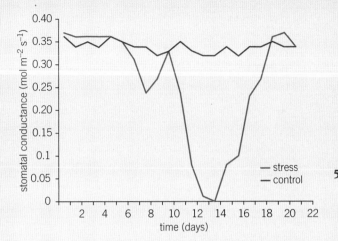

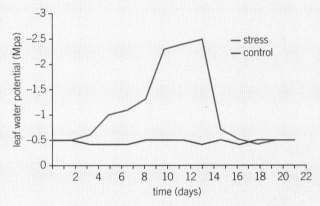

a Describe the changes in rate of diffusion of carbon dioxide and water potential over the 22 day period. *(4 marks)*

b Explain the link between the change in rate of diffusion and the change in water potential. *(4 marks)*

c Outline why controls were used in the investigations. *(2 marks)*

4 The diagram shows a section through the root of a typical plant.

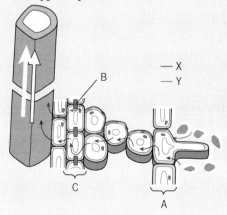

a **(i)** State the names of the pathways, taken by water, labelled X and Y. *(2 marks)*

(ii) State the names of the structures labelled A, B and C. *(3 marks)*

(iii) Describe what happens at B. *(3 marks)*

b Outline how the structure of a root hair cell is adapted to its function. *(4 marks)*

5 Various measurements of lung function are used to help diagnose lung disease and to monitor its treatment.

a State what is meant by the following terms:

(i) vital capacity *(1 mark)*

(ii) forced expiratory volume 1 (FEV1). *(1 mark)*

One measure of lung function is:

$$\text{Percentage lung function} = \frac{\text{FEV 1}}{100} \times 100$$

This is particularly useful in identifying possible obstructive disorders of the airways and lungs, such as asthma or chronic obstructive pulmonary disease (COPD).

● Asthma is a condition that responds to, and can be controlled by, the use of bronchodilators.
These are drugs that dilate the airways and improve airflow.

● COPD lasts for a long period of time and is caused by progressive and permanent damage to the lung tissue.

When the value calculated for the percentage lung function is less than or equal to 70%, this indicates an obstructive disorder. A 'normal' value is approximately 80%.

The table shows data relating to three patients, **C**, **D** and **E**, before and after treatment with a bronchodilator drug.

patient	age (years)	before treatment			after treatment		
		vital capacity (dm³)	FEV1 (dm³)	percentage lung function	vital capacity (dm³)	FEV1 (dm³)	percentage lung function
C	18	5.5	3.8	69	5.6	4.5	
D	45	5.3	3.6	68	5.5	4.0	73
E	78	3.8	2.2	58	3.8	2.2	58

b **(i)** Calculate the percentage lung function for patient **C** after treatment with the bronchodilator drug.

Show your working and give your answer in percentage **to the nearest whole number**. *(2 marks)*

(ii) Using the information in the table and your answer to **(c)(i)**, copy and complete the table, indicating with a tick (✓) the diagnosis for each patient.

Patient	Diagnosis	
	Asthma	COPD
C		
D		
E		

(3 marks)

OCR F221 2009

From the AS paper 1 practice questions

6 **a** Define the following terms:

cardiac output *(2 marks)*

stroke volume *(2 marks)*

heart rate *(1 mark)*

b Copy these terms and arrange them in the correct order using the formula:

.... × .... = ... *(1 mark)*

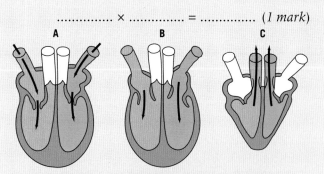

c Describe, using the diagrams in Figure 1, what is happening at each stage in the cardiac cycle shown. *(6 marks)*

From the AS paper 1 practice questions

7 Xylem vessels have two walls, a primary wall made of cellulose and a secondary wall composed of lignin. Lignin, a strong inflexible polymer, is not usually deposited uniformly but laid down in rings or spirals.

a Describe the role of lignin in xylem vessels. *(4 marks)*

b Suggest the benefit to a plant of way that lignin is deposited in the walls of xylem vessels. *(2 marks)*

From the AS paper 1 practice questions

8 The graph shows oxygen dissociation curves for both myoglobin and haemoglobin.

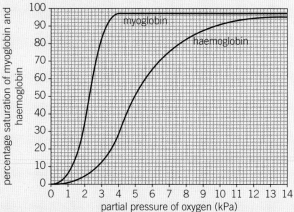

Calculate the decrease in percentage saturation of both myoglobin and haemoglobin between 4 kPa and 2 kPa partial pressure of oxygen. *(1 mark)*

OCR F224 2010

From the AS paper 2 practice questions

9 **a** Explain, using the term **surface area to volume ratio**, why large, active organisms need a specialised surface for gaseous exchange. *(2 marks)*

b The table describes some of the key features of the mammalian gas exchange system.

Copy and complete the table by explaining how each feature improves the efficiency of gaseous exchange. The first two have been completed for you.

Feature of gas exchange system	How feature improves efficiency of gaseous exchange system
Many alveoli	This increases the surface across which oxygen and carbon dioxide can diffuse
The epithelium of the alveoli is very thin	
There are capillaries running over the surface of the alveoli	
The lungs are surrounded by the diaphragm and intercostal muscles	

(3 marks)

c Outline how the diaphragm **and** intercostal muscles cause **inspiration**.

(4 marks)

d The graph shows the trace from a spirometer recorded from a 16-year old student.

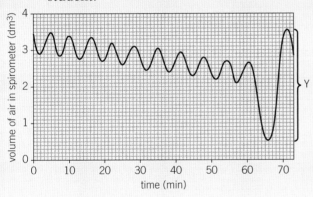

(i) Roughly copy the graph and label on the trace using the letter X, a point that indicates when the student was inhaling. *(1 mark)*

(ii) At the end of the trace the student measured their vital capacity. This is indicated by the letter Y.
State the vital capacity of the student.
(1 mark)

OCR F211 2009
From the AS paper 2 practice questions

10 The diagram shows a normal ECG trace.

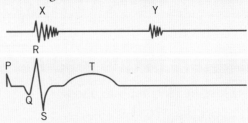

a (i) State which valves are closing at points X and Y. *(2 marks)*

(ii) Describe the function of the atrio-ventricular valves in the heart.
(2 marks)

(iii) Explain why the right ventricular wall of the heart is less muscular than the left ventricular wall. *(5 marks)*

The graphs show an ECG trace during a heart attack (a) and during fibrillation (b).

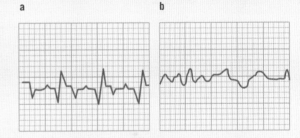

b Describe the differences between the ECG traces shown during a heart attack and fibrillation. *(3 marks)*

c 'A cardiac arrest occurs when there is a problem with the electrical activity in the heart, whereas a heart attack happens when there is a problem with the plumbing'.
Discuss what you understand by this statement. *(6 marks)*

From the AS paper 2 practice questions

MODULE 4

Biodiversity, evolution, and disease

Chapters in this module

Introduction

In this module you will learn about the vast biodiversity of organisms – how they are classified and the ways in which biodiversity can be measured. Classification is an attempt to impose a hierarchy on the complex and dynamic variety of life on Earth. The way in which organisms are classified has changed many times as our understanding of biological molecules and genetics increases.

The module also serves as an introduction to ecology, emphasising practical techniques used to study biodiversity and an appreciation of the need to maintain biodiversity.

Finally you will gain an understanding of the variety of organisms that are pathogenic and the way in which plants and animals have evolved defences to deal with disease. The impact of the evolution of pathogens on the treatment of disease is also considered.

Classification and evolution introduces you to the current system of classification used by scientists. It also explains historically how organisms were classified, and why the system has changed as our knowledge of the biology of organisms develops. It also covers how organisms are adapted to their environment and how, as a result of naturally occurring variation organisms have evolved, and continue to evolve.

Biodiversity is an important indicator in the study of habitats. You will learn how to sample habitats to measure and monitor biodiversity. You will also study the importance of maintaining biodiversity for ecological, economic and aesthetic reasons. To ensure biodiversity is maintained you will learn about how conservation action must be taken at local, national and global levels.

Communicable diseases explores how organisms are surrounded by pathogens and have evolved defences against them. You will discover how plants defend themselves and the role of the mammalian immune system. You will also learn how medical intervention can be used to support these natural defences such as the role of vaccinations and antibiotics.

Knowledge and understanding checklist

From your Key Stage 4 study you should be able to answer the following questions. Work through each point, using your Key Stage 4 notes and other resources. There is also support available on Kerboodle.

- [ ] Describe how to carry out a field investigation into the distribution and abundance of organisms in an ecosystem and explain how to determine their numbers in a given area.
- [ ] Describe both positive and negative human interactions within ecosystems and explain their impact on biodiversity.
- [ ] Explain some of the benefits and challenges of maintaining local and global biodiversity.
- [ ] Explain how evolution occurs through natural selection of variants that give rise to phenotypes best suited to their environment and may result in the formation of new species.
- [ ] Describe the evidence for evolution, including fossils and antibiotic resistance in bacteria.
- [ ] Describe the impact of developments in biology on classification systems.
- [ ] Explain how communicable diseases are spread in animals and plants.
- [ ] Describe the non-specific defence systems and the role of the immune system in the human bodies defence against disease.
- [ ] Explain the use of vaccines and medicines in the prevention and treatment of disease.

Maths skills checklist

In this module, you will need to use the following maths skills.

- [ ] **Standard deviation.** You will need to calculate the standard deviation to measure the spread in a set of data.
- [ ] **Student's t test.** You will need to use this test to compare the means of data values of two populations.
- [ ] **Correlation coefficient.** You will need to use this test to consider the relationship between two sets of data. This will determine if and how the data is correlated.
- [ ] **Simpson's Index of Diversity.** You will use this formula to measure biodiversity in a habitat. The higher the value of Simpson's Index of Diversity, the more diverse the habitat.
- [ ] **Proportion of polymorphic gene loci.** You will use this formula to measure genetic biodiversity.

10 CLASSIFICATION AND EVOLUTION

10.1 Classification

Specification reference: 4.2.2

No one knows how many different types of organism currently exist on the Earth. Through studying evolutionary relationships and performing mathematical calculations, in 2011 a team of scientists from the UK, USA, and Canada arrived at a widely accepted estimate of 8.7 million. However, the vast majority of these organisms have not been identified, and cataloguing them all could take more than 1000 years. The team also warned that many species would become extinct before they were studied.

Classification systems

Classification is the name given to the process by which living organisms are sorted into groups. The organisms within each group share similar features.

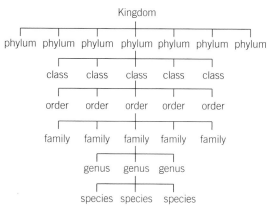

▲ **Figure 1** *This diagram shows how the seven taxonomic groups are arranged into a hierarchy. An organism can only belong to one group at each level of the hierarchy*

A number of different classification systems exist. Until recently the most widely used system contained seven groups ordered in a hierarchy – these are referred to as **taxonomic groups**. The seven groups are: **kingdom**, phylum (plural phyla), class, order, family, genus (plural genera) and **species** (Figure 1). Kingdoms are the biggest and broadest taxonomic group, with species being the smallest and most specific classification. Similar or related groups at one hierarchical level are combined into more inclusive groups at the next higher level.

Hierarchical classification systems are often referred to as Linnaean classification, after the 18th century Swedish botanist, Carl Linnaeus who was the first to propose such a system.

Based on recent studies of genetic material many scientists now add a further level of classification into the hierarchy. It is known as a *domain* (you will find out more about the three domain system of classification in Topic 10.2, The five kingdoms). This level of classification is placed at the top of the hierarchy. As new scientific discoveries are made (for example, through genome sequencing), the current system of classification may change again.

The development of classification systems

The system of classification of living organisms provides a good example of how our scientific knowledge and understanding has developed over time, as new information is gathered or discovered. Advances in scientific techniques have provided more detailed information on the genetic and biological make up of organisms, which has led to several revisions in the way organisms are classified. Our current system of classification may well change in the future, as further discoveries are made.

Some of the key steps in the development of the current system of classification are shown here:

1 Explain why the classification system we use today is different from the one originally proposed by Aristotle around the year 350 BC.

Ernst Haeckel

| Konrad Gesner, a Swiss botanist, produces a four volume encyclopaedia of the then-known animal world, consisting of over 4500 pages | Ernst Haeckel, a German biologist and naturalist, proposes the third kingdom of protoctista, in addition to Linnaeus' animal and plant kingdoms | Following genetic analysis, Carl Woese introduces the six kingdom classification model, sub-dividing the prokaryotes into archaebacteria and eubacteria |

1551   1866   1977

350 BC   1758   1969

| Aristotle compiles his book 'A history of animals', the first comprehensive study of animals | Carl Linnaeus publishes the tenth edition of his book Systema Naturae, now considered the starting point of binomial nomenclature | A five kingdom classification system is proposed by Robert Whittaker, consisting of prokaryotae, protoctista, fungi, plantae, and animalia |

Why do scientists classify organisms?

- To identify species – by using a clearly defined system of classification, the species an organism belongs to can be easily identified.

- To predict characteristics – if several members in a group have a specific characteristic, it is likely that another species in the group will have the same characteristic.

- To find evolutionary links – species in the same group probably share characteristics because they have evolved from a common ancestor.

By using a single classification system, scientists worldwide can share their research. Links between different organisms can be seen, even if they live on different continents. Remember, though, that classification systems have been created to order observed organisms. This form of hierarchical organisation is not defined by 'nature'.

How are organisms classified?

The classification system begins by separating organisms into the three domains – Archaea, Bacteria, and Eukarya (discussed further in Topic 10.2, The five kingdoms). These are the broadest groups. As you move down the hierarchy there are more groups at each level, but fewer organisms in each group. The organisms in each group become more similar and share more of the same characteristics.

The system ends with organisms being classified as individual species. These are the smallest units of classification – each group contains only one type of organism. A species is defined as a group of organisms that are able to reproduce to produce fertile offspring. For example, donkeys can reproduce with other donkeys, the offspring of which can subsequently breed. Likewise, horses can breed with other horses to produce fertile offspring. However, when a horse is bred with a donkey, the offspring produced (a mule or a hinny) is infertile. Therefore, donkeys and horses are classified as belonging to different species. Mules or hinnies are not a species.

Mules and hinnies are infertile because their cells contain an odd number of chromosomes (63). This means that meiosis and gamete production cannot take place correctly as all chromosomes must pair up. This chromosome number is created because horses have 64 chromosomes (32 pairs) whereas donkeys have 62 chromosomes (31 pairs).

▲ **Figure 2** *Mules (left) are produced by crossing a horse (Equus caballus, middle) and a donkey (Equus asinus, right). Because a mule is infertile, it is not classified as a species.*

To show how the system works, the classification of three organisms is given in Table 1.

▼ **Table 1** *Classification of three organisms, you do not need to learn these examples*

Level of hierarchy	Brewer's yeast	English oak tree	European badger
Domain	Eukaryote	Eukaryote	Eukaryote
Kingdom	Fungi	Plantae	Animalia
Phylum	Ascomycota	Angiosperms	Chordata
Class	Saccharomycetes	Eudicots	Mammalia
Order	Saccharomycetales	Fagales	Carnivora
Family	Saccharomycetaceae	Fagaceae	Mustelidae
Genus	*Saccharomyces*	*Quercus*	*Meles*
Species	*cerevisiae*	*robur*	*meles*

Classification of humans

You belong to a species named *Homo sapiens*. This is the scientific name for humans. Humans are classified as shown in Table 2.

Naming organisms

Before classification systems were widely used, many organisms were given names according to certain physical characteristics, behaviour or habitat. Examples are 'blackbirds' for their colour, 'song thrushes' for their song and 'fieldfares' for their habitat. These are called their 'common names'.

This was not a very useful system for scientists working internationally, as organisms may have more than one common name, and different names in different languages. Another problem is that common names do not provide information about relationships between organisms. For example, the blackbird, song thrush, and fieldfare all belong to the genus *Turdus*, meaning that they have all evolved from a common ancestor, but you wouldn't know this from their common names, nor necessarily from their observable characteristics.

▼ **Table 2** *Classification of humans*

Level of hierarchy	Human
Domain	Eukarya
Kingdom	Animalia
Phylum	Chordata
Class	Mammalia
Order	Primates
Family	Hominidae
Genus (plural – genera)	*Homo*
Species	*sapiens*

▲ **Figure 3** *The blackbird* (Turdus merula), *song thrush* (Turdus philomelos) *and fieldfare* (Turdus pilaris) *all belong to the genus* Turdus

To ensure scientists the world over are discussing the same organism we now use a system developed in the 18th century, also by Carl Linnaeus, a Swedish botanist. This system is known as **binomial nomenclature**.

All species are given a scientific name consisting of two parts:

- The first word indicates the organism's genus. It is called the generic name; you can think of this as being equivalent to your surname or family name, as it is shared by close relatives.

- The second word indicates the organism's species. It is called the specific name.

- Unlike people, no two species have the same generic and specific name. Two different species could have the same specific name, however their genus would be different. An example of this is *Anolis cuvieri* (a lizard) and *Oplurus cuvieri* (a bird). The only link between them is that they are both named after the famous French naturalist and zoologist Georges Cuvier (1769–1832). Many of these scientific names derive from Latin.

When naming an organism using its scientific name the word should be presented in italics. As it is difficult to handwrite in italics, the

Study tip

The abbreviation for species 'sp' is used after genus, when not identifying the species fully. For example, you may only know the willow tree in your garden to be *Salix* sp.

The plural 'spp.' is used to refer to multiple species within a genus.

▲ **Figure 4** *This is a jungle cat (*<u>Felis</u> <u>chaus</u>*). It is closely related to the domestic cat (*<u>Felis</u> <u>catus</u>*) and both belong to the genus Felis*

▲ **Figure 5** *The domestic cat is also related to lions, tigers and leopards, but not as closely as the jungle cat. All these cats belong to the family Felidae*

standard procedure in handwritten documents is to underline the name. The name should be written in lowercase, with the exception of the first letter of the genus name, which should be uppercase.

Some examples of scientific names are included in Table 3. Split the name into two parts and you can easily work out which genus and species the organism belongs to.

▼ **Table 3** *Examples of scientific names*

Common name	Scientific name	Genus	Species
dog	*Canis familiaris*	*Canis*	*familiaris*
lion	*Panthera leo*	*Panthera*	*leo*
daisy	*Bellis perennis*	*Bellis*	*perennis*
Christmas tree (Norway spruce)	*Picea abies*	*Picea*	*abies*
E.coli	*Escherichia coli*	*Escherichia*	*coli*

Summary questions

1   State two reasons why classification is important.   (*2 marks*)

2   Ligers are the offspring of male lions (*Panthera leo*) and female tigers (*Panthera tigris*). Suggest two reasons why ligers are not classified as a species, but their parents are.   (*2 marks*)

3   *Erithacus rubecula* is the scientific name for the robin. Copy and complete the table to show its full classification   (*3 marks*)

Kingdom	Animalia
a	Chordata
Class	Aves
Order	Passeriformes
Family	Muscicapidae
Genus	b
Species	c

4   The loganberry (*Rubus loganobaccus*) is the fertile offspring of the blackberry (*Rubus ursinus*) and the raspberry (*Rubus idaeus*). Explain why the loganberry is difficult to classify into a taxonomic group.   (*3 marks*)

Study tip

As an organism's scientific name may be long or difficult to pronounce, the genus name is often shortened to the first letter after the first mention in full. For example, baker's yeast, *Saccharomyces cerevisiae* is often shortened to *S. cerevisiae* and the bacterium *Escherichia coli* to *E. coli*.

Originally living organisms were classified into just two kingdoms – animals and plants. Aristotle (384–322 BC) classified animal species in his text *History of Animals*, while his pupil Theophrastus (371–287 BC) wrote a parallel work, *History of Plants*. The animal kingdom included every living thing that moved, ate and grew to a certain size then stopped growing. The plant kingdom included every living thing that did not move or eat and that continued to grow throughout life.

As more was discovered about organisms and more species were discovered, it became increasingly difficult to divide living organisms into just two kingdoms. For example, the introduction of the microscope in the 16th to 17th century enabled scientists to study the cells of an organism and showed that bacteria have a very different cell structure to that of other organisms. From the 1960s, scientists classified organisms into five kingdoms. This classification system was introduced by Robert Whittaker, an American plant ecologist, based on the principles developed by Carl Linnaeus.

What are the five kingdoms?

Living organisms can be classified into five kingdoms:

- **Prokaryotae** (bacteria) — the **prokaryotes**
- **Protoctista** (the unicellular eukaryotes)
- **Fungi** (e.g., yeasts, moulds, and mushrooms) — the **eukaryotes**
- Plantae (the plants)
- Animalia (the animals)

Organisms were originally classified into these kingdoms based on similarities in their observable features, as described below.

Prokaryotae

General features:

Examples include the bacteria *Escherichia coli*, *Staphylococcus aureus*, and *Bacillus anthracis*.

- unicellular
- no nucleus or other membrane-bound organelles – a ring of 'naked' DNA – small ribosomes
- no visible feeding mechanism – nutrients are absorbed through the cell wall or produced internally by photosynthesis.

Protoctista

General features:

Examples include species belonging to the genera *Paramecium* and *Amoeba*.

- (mainly) unicellular
- a nucleus and other membrane-bound organelles
- some have chloroplasts

▲ **Figure 1** *This is* Streptococcus thermophilus, *a type of bacteria used in yoghurt production. All bacteria belong to the Prokaryotae kingdom. Scanning electron micrograph, × 8000 magnification.*

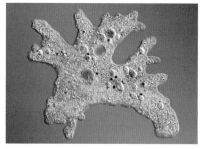

▲ **Figure 2** *This is an amoeba (*Amoeba proteus*), it is member of the Protoctista kingdom. Light micrograph × 100 magnification*

- some are sessile, but others move by cilia, flagella, or by amoeboid mechanisms
- nutrients are acquired by photosynthesis (**autotrophic** feeders), ingestion of other organisms (**heterotrophic** feeders), or both – some are parasitic.

Fungi

General features:

Examples include mushrooms, moulds, and yeast.

- unicellular or multicellular
- a nucleus and other membrane-bound organelles and a cell wall mainly composed of chitin
- no chloroplasts or chlorophyll
- no mechanisms for locomotion
- most have a body or mycelium made of threads or hyphae
- nutrients are acquired by absorption – mainly from decaying material – they are **saprophytic** feeders – some are parasitic
- most store their food as glycogen.

Plantae

With over 250 000 species, the plant kingdom is the second largest of the kingdoms.

Examples include flowering plants such as roses, trees such as oak, and grasses.

▲ **Figure 3** *Two different types of fungi. Top: Scanning electron micrograph of microscopic yeast,* Saccharomyces cerevisiae *× 2000 magnification. Bottom: Large mushrooms*

General features:

- multicellular
- a nucleus and other membrane-bound organelles including chloroplasts, and a cell wall mainly composed of cellulose
- all contain chlorophyll
- most do not move, although gametes of some plants move using cilia or flagella
- nutrients are acquired by photosynthesis – they are **autotrophic** feeders – organisms that make their own food
- store food as starch.

Animalia

The animal kingdom is the largest kingdom with over 1 million known species.

General features:

Examples include mammals such as cats, reptiles such as lizards, birds, insects, molluscs, worms, sponges, and anemones.

- multicellular
- a nucleus and other membrane-bound organelles (no cell walls)
- no chloroplasts
- move with the aid of cilia, flagella, or contractile proteins, sometimes in the form of muscular organs

Study tip

You will need to be able to explain how classification systems have changed over time as an example of advances in science.

- nutrients are acquired by ingestion – they are heterotrophic feeders
- food stored as glycogen.

Recent changes to classification systems

As scientists learn more about organisms, classification systems change. Originally classification systems were based on observable features. Through the study of genetics and other biological molecules, scientists are now able to study the evolutionary relationships between organisms. These links can then be used to classify organisms.

When organisms evolve, their internal and external features change, as does their DNA. This is because their DNA determines the proteins that are made, which in turn determines the organism's characteristics. In order for their characteristics to have changed, their DNA must also have changed. By comparing the similarities in the DNA and proteins of different species, scientists can discover the evolutionary relationships between them. You will learn more about DNA sequencing and its use in studying evolutionary relationships in Topic 10.4, Evidence for evolution.

An example of a protein that has changed in structure is haemoglobin. Haemoglobin has four polypeptide chains, each made up of a fixed number of amino acids. The haemoglobin of humans differs from chimpanzees in only one amino acid, from gorillas in three amino acids and from gibbons in eight amino acids. As the structure of haemoglobin is remarkably similar, it indicates a common ancestry between the various primate groups.

> **Synoptic link**
>
> You learnt about haemoglobin in Topic 3.7, Types of protein ans Topic 8.4, Transport of oxygen and carbon dioxide in the blood.

Are there now six kingdoms?

The current classification system used by scientists is known as the 'Three Domain System', and was proposed by Carl Woese, an American microbiologist in 1977, reusing the word 'Kingdom'. In 1990 it was renamed 'Domain'. Domains are a further level of classification at the top of the hierarchy.

Woese's system groups organisms using differences in the sequences of nucleotides in the cells' ribosomal RNA (rRNA), as well as the cells' membrane lipid structure and their sensitivity to antibiotics. Observation of these differences was made possible through advances in scientific techniques.

> **Synoptic link**
>
> You learnt about ribosomal RNA and nucleotides in Topic 3.8 Nucleic acids.

Under the Three Domain System, organisms are classified into three domains and six kingdoms. The three domains are Archaea, Bacteria, and Eukarya. The organisms in the different domains contain a unique form of rRNA and different ribosomes:

- Eukarya – have 80s ribosomes
  - RNA polymerase (responsible for most mRNA transcription) contains 12 proteins.

- Archaea – have 70s ribosomes
  - ▪ RNA polymerase of different organisms contains between eight and 10 proteins and is very similar to eukaryotic ribosome.
- Bacteria – have 70s ribosomes
  - ▪ RNA polymerase contains five proteins.

The organisation of this system is shown in Figure 4.

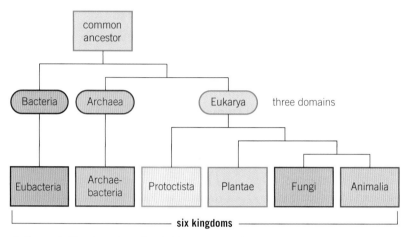

▲ **Figure 4** *The three domain, six kingdom classification system*

In Woese's system the Prokaryotae kingdom becomes divided into two kingdoms – Archaebacteria and Eubacteria. The six kingdoms are therefore: Archaebacteria, Eubacteria, Protoctista, Fungi, Plantae and Animalia.

Although both Archaebacteria and Eubacteria are single-celled prokaryotes, Eubacteria are classified in their own kingdom because their chemical makeup is different from Archaebacteria. For example, they contain peptidoglycan (a polymer of sugars and amino acids) in their cell wall whereas Archaebacteria do not.

Archaebacteria

Archaebacteria, also known as ancient bacteria, can live in extreme environments. These include hot thermal vents, anaerobic conditions, and highly acidic environments. For example, methanogens live in anaerobic environments such as sewage treatment plants and make methane.

Eubacteria

Eubacteria, also known as true bacteria, are found in all environments and are the ones you will be most familiar with. Most bacteria are of the Eubacteria kingdom.

Some scientists still use the traditional five kingdom system, but since Archaebacteria have been found to be different chemically from Eubacteria, most scientists now use the three domain, six kingdom system.

▲ **Figure 5** *The hot springs of Yellow stone National Park, USA, were among the first places Archaebacteria were discovered. These types of bacteria are called thermophiles. They can survive in extreme heat*

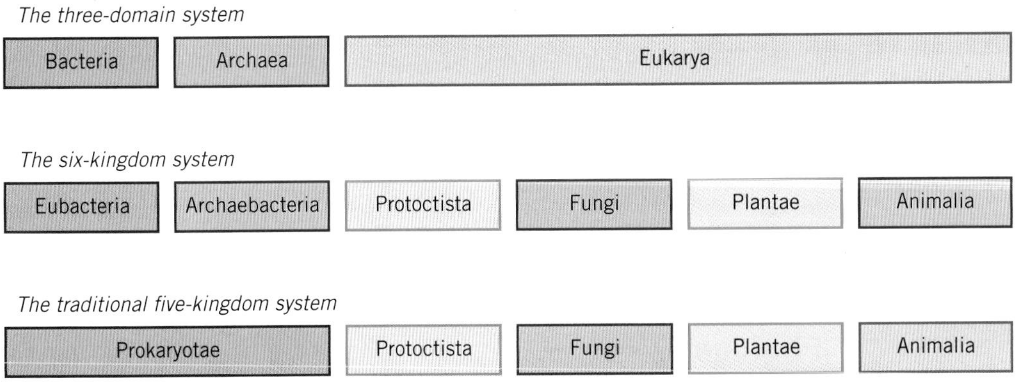

The three-domain system

Bacteria	Archaea	Eukarya

The six-kingdom system

Eubacteria	Archaebacteria	Protoctista	Fungi	Plantae	Animalia

The traditional five-kingdom system

Prokaryotae	Protoctista	Fungi	Plantae	Animalia

▲ **Figure 6** *This diagram shows how the three commonly used classification systems are related*

Summary questions

1 State two differences between fungi and plants. *(2 marks)*

2 Using the information in this topic, classify the following organisms into the correct kingdom using your knowledge of the Five kingdom system of classification:

   a *Escherichia coli* – an organism that lives in the human intestine. It is unicellular and has no nucleus.

   b *Saccharomyces cerevisiae* – an organism used in the manufacture of beer. It is unicellular and has a nucleus and a cell wall made of chitin.

   c *Euglena* – an organism that lives in fresh water. It is unicellular and contains chloroplasts. *(3 marks)*

3 Explain why prokaryotes are now classified as two separate domains. *(3 marks)*

4 Describe how and why classification systems have changed over time. *(6 marks)*

In the last topic you studied how the current classification system is based on both shared physical characteristics between organisms and on evolutionary relationships. To discover the links between organisms and common ancestors, scientists study the organisms' DNA, proteins, and the fossil record.

Phylogeny

Phylogeny is the name given to the *evolutionary* relationships between organisms. The study of the evolutionary history of groups of organisms is known as phylogenetics. It reveals which group a particular organism is related to, and how closely related these organisms are. You will learn more about the evidence that scientists use to study evolutionary relationships in the next topic.

Classification can occur without any knowledge of phylogeny, as occurred in the past. However, it is the objective of many scientists to develop a classification system that also correctly takes into account the phylogeny of an organism.

Phylogenetic trees

A phylogenetic tree (or evolutionary tree) is a diagram used to represent the evolutionary relationships between organisms. They are branched diagrams, which show that different species have evolved from a common ancestor.

The diagram is similar in structure to that of a branching tree – the earliest species is found at the base of the tree and the most recent species are found at the tips of the branches.

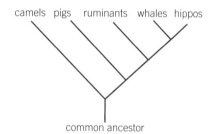

The closer the branches, the closer the evolutionary relationship. Hippos and whales are more closely related than hippos and ruminants.

▲ **Figure 1** *A phylogenetic tree showing the evolutionary relationships between certain mammals*

Phylogenetic trees are produced by looking at similarities and differences in species' physical characteristics and genetic makeup. Much of the evidence has been gained from fossils.

How do you interpret phylogenetic trees?

The tips of the phylogenetic tree represent groups of descendent organisms (often species). The nodes on the tree (the points where the new lines branch off) represent the common ancestors of those descendants. Two descendants that split from the same node are called sister groups. The closer the branches of the tree are, the closer the evolutionary relationship.

Study Figure 2. Begin by looking at the base of the tree. The organism at this point is the common ancestor of all the organisms on the tree. The letters A–F represent six different species that have evolved from this ancestor.

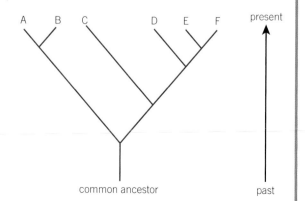

▲ **Figure 2** *A phylogenetic tree structure*

Then look at the top of the tree. You will see that species A and B are sister groups as these share a common ancestor. Species E and F are also sister groups that share their own common ancestor, which itself shared a common ancestor with species D further back in time. Further back in time again, C shared a common ancestor with D, E and F.

1 Another species exists (species X) which shares a common ancestor with species D. Add a line onto Figure 1 to show the correct evolutionary relationships of species X.
2 Species Y is now extinct but shared a common ancestor with species A and Species B. Add another line onto Figure 1 line to show where this species should be placed on the phylogenetic tree.

Advantages of phylogenetic classification

Phylogeny can be done without reference to Linnaean classification. Classification uses knowledge of phylogeny in order to confirm the classification groups are correct or causes them to be changed. For example, a dolphin has many of the same characteristics as a fish, so in theory a dolphin could be classified as a fish. However, knowledge of the phylogeny of dolphins confirms its classification as a mammal.

Other advantages:

- Phylogeny produces a continuous tree whereas classification requires discrete taxonomical groups. Scientists are not forced to put organisms into a specific group that they do not quite fit.

- The hierarchal nature of Linnaean classification can be misleading as it implies different groups within the same rank are equivalent. For example, the cats (Felidae) and the orchids (Orchidaceae) are both families. However, the two groups are not comparable – one has a longer history than the other (cats have existed for around 30 million years, but orchids have been in existence for over 100 million years). The two families also have different levels of diversity (with approximately 35 cat species and 20 000 orchid species) and different degrees of biological differentiation (many orchids of different genera are able to hybridise, but cats cannot).

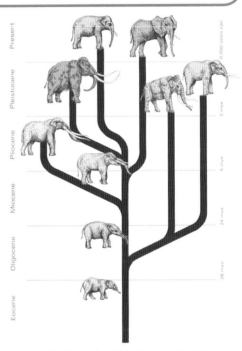

▲ **Figure 3** *All species of elephant evolved from a palaeomastodon, which lived around 40 million years ago. Over time their appearance changed as the elephant adapted to the current environment. Today, only two species of elephant survive – the Indian and African elephant. The timeline stops for the other species of elephant as they are now extinct*

Summary questions

1 State the main difference between early classification systems and systems based on phylogeny. (*1 mark*)

2 Describe the advantages of phylogenetic classification over the Linnaean system. (*3 marks*)

3 Use the information in the phylogenetic tree to answer the following questions:
a State which group of organisms is most closely related to lizards. (*1 mark*)
b Explain why dinosaurs are listed lower in the diagram than the other organisms. (*2 marks*)

c Explain how the diagram shows that birds are more closely related to crocodiles than turtles. (*2 marks*)

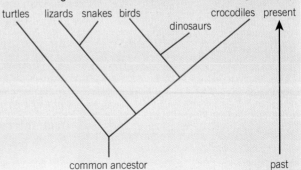

Charles Darwin.

▲ **Figure 1** *Charles Darwin, author of 'On the Origin of Species'*

1. Geospiza magnirostris   2. Geospiza fortis,
3. Geospiza parvula.   4. Certhidea olivasea.

▲ **Figure 2** *Darwin 'noticed that the shape of the finches' beaks were adapted to the food available on the island*

You are probably aware that Charles Darwin is credited with formulating the theory of **evolution**. However, it was studying the work of a number of other scientists as well as his own observations that led him to develop his theory.

Evolution is the theory that describes the way in which organisms evolve, or change, over many many years as a result of natural selection. Darwin realised that organisms best suited to their environment are more likely to survive and reproduce, passing on their characteristics to their offspring. Gradually, a species changes over time to have a more advantageous phenotype for the environment in which it lives. We now know that the advantageous characteristics are passed on from one generation to the next by genes in DNA molecules.

Developing the theory of evolution

When Charles Darwin was born in 1809, most people in Europe believed, in a literal sense, in the Christian Bible. They believed God directly created all life on Earth, including human beings. The Bible doesn't state how far in the past this occurred – in Darwin's day the common belief was that this creation had occurred only a few thousand years before.

In 1831 aboard the HMS Beagle, Darwin read '*Principles of Geology*'. This book was written by his friend Charles Lyell, a Scottish geologist. He suggested that fossils were actually evidence of animals that had lived millions of years ago. We now have scientific evidence that supports this.

In it Lyell also popularised the principle of uniformitarianism (the concept itself was originally proposed by another Scottish geologist, James Hutton). This is the idea that in the past, the Earth was shaped by forces that you can still see in action today, such as sedimentation in rivers, wind erosion, and deposition of ash and lava from volcanic eruptions. In emphasising these natural processes, he challenged the claims of earlier geologists who had tried to explain geological formations as a result of biblical events such as floods. This concept prompted Darwin to think of evolution as a slow process, one in which small changes gradually accumulate over very long periods of time.

Darwin carried out some of his most famous observations on finches in the Galapagos Islands. He noticed that different islands had different finches. The birds were similar in many ways and thus must be closely related, but their beaks and claws were different shapes and sizes.

Through these observations Darwin realised that the design of the finches' beaks was linked to the foods available on each island. He concluded that a bird born with a beak more suited to the food available would survive longer than a bird whose beak was less suited. Therefore, it would have more offspring, passing on its characteristic beak. Over time the finch population on that island would all share this characteristic.

Throughout his trip Darwin sent specimens of organisms back to the UK for other scientists to preserve and classify. This enabled scientists not only to see specimens first hand but also enabled them to spot characteristics and links between organisms that Darwin had not. For example, Darwin did not notice that the tortoises (which the Galapagos islands are named after) present on different islands were different subspecies. Before this was pointed out to him he had simply stacked their shells randomly in the hold.

Upon his return to England, Darwin spent many years developing ideas. He also carried out experimental breeding of pigeons to gain direct evidence that his ideas might work.

At the same time as Darwin was developing his ideas, another scientist, Alfred Wallace, was working on his own theory of evolution in Borneo. In 1858 he sent his ideas to Darwin for peer review before its publication. As Wallace's ideas were so similar to Darwin's, they proposed the theory of evolution through a joint presentation of two scientific papers to the Linnean Society of London on 1st July 1858.

A year later in 1859, Darwin published '*On the Origin of Species*'. It was in this book that he named the theory that he and Wallace had presented independently as the theory of evolution by **natural selection** (see Topic 10.7, Adaptations).

The book was extremely controversial at the time. The theory of evolution conflicted with the religious view that God had created all of the animals and plants on Earth in their current form, and only about six thousand years ago. A further implication of Darwin's theory is that humans are simply a type of animal evolved from apes, which conflicted with the widely held Christian belief that God created 'man' in his own image.

Darwin's theory split the scientific community before his idea became generally agreed. Darwin's theory of evolution is now widely accepted, however, even today, debate with religious groups continues.

Evidence for evolution

Scientists use a number of sources to study the process of evolution. These include:

- palaeontology – the study of fossils and the fossil record
- comparative anatomy – the study of similarities and differences between organisms' anatomy
- comparative biochemistry – similarities and differences between the chemical makeup of organisms.

▲ **Figure 3** *One of the most common fossils found in the UK is an ammonite, an organism that lived in the sea. These organisms became extinct about 65 million years ago. Radioisotope dating is used to determine the age of the rock strata and the fossils found within the layer*

Palaeontology

Fossils are formed when animal and plant remains are preserved in rocks. Over long periods of time, sediment is deposited on the earth to form layers (strata) of rock. Different layers correspond to different geological eras, the most recent layer being found on the top. Within the different rock strata the fossils found are quite different, forming a sequence from oldest to youngest, which shows that organisms have gradually changed over time. This is known as the fossil record.

Evidence provided by the fossil record:

- Fossils of the simplest organisms such as bacteria and simple algae are found in the oldest rocks, whilst fossils of more complex organisms such as vertebrates are found in more recent rocks. This supports the evolutionary theory that simple life forms gradually evolved over an extremely long time period into more complex ones.

- The sequence in which the organisms are found matches their ecological links to each other. For example, plant fossils appear before animal fossils. This is consistent with the fact that animals require plants to survive.

- By studying similarities in the anatomy of fossil organisms, scientists can show how closely related organisms have evolved from the same ancestor. For example zebras and horses, members of the genus *Equus*, are closely related to the rhinoceros of the family Rhinocerotidae. An extensive fossil record of these organisms exists, which spans over 60 million years and links them to the common ancestor *Hyracotherium*. This lineage has been based on structural similarities between their skull (including teeth) and skeleton, in particular the feet (Figure 4).

- Fossils allow relationships between extinct and living (extant) organisms to be investigated.

The fossil record is, however, not complete. For example, many organisms are soft-bodied and decompose quickly before they have a chance to fossilise. The conditions needed for fossils to form are not often present. Many other fossils have been destroyed by the Earth's movements, such as volcanoes, or still lie undiscovered.

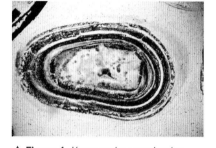

▲ **Figure 4** *You may be surprised to learn that bacteria can become fossils. The cyanobacteria (blue-green algae) have left behind a fossil record. The oldest cyanobacteria-like fossils known are nearly 3.5 billion years old, among the oldest fossils currently known*

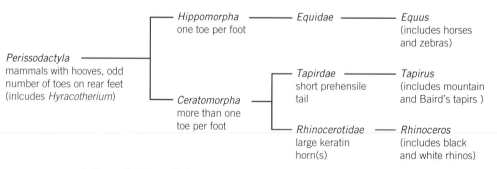

▲ **Figure 5** *This evolutionary tree shows that zebras and horses are closely related to the rhinoceros*

Comparative anatomy

As the fossil record is incomplete, scientists look for other sources of evidence to determine evolutionary relationships. Comparative anatomy is the study of similarities and differences in the anatomy of different living species.

Homologous structures

A **homologous structure** is a structure that appears superficially different (and may perform different functions) in different organisms, but has the same underlying structure. An example is the pentadactyl limb of vertebrates.

Vertebrate limbs are used for a wide variety of functions such as running, jumping, and flying. You would expect the bone structure of these limbs in a flying vertebrate to be very different from that in a walking vertebrate or a swimming vertebrate. However, the basic structures of all vertebrate limbs are actually very similar (Figure 6) – the same bones are adapted to carry out the whole range of different functions. An explanation is that all vertebrates have evolved from a common ancestor, therefore vertebrate limbs have all evolved from the same structure.

The presence of homologous structures provides evidence for **divergent evolution**. This describes how, from a common ancestor, different species have evolved, each with a different set of adaptive features. This type of evolution will occur when closely related species diversify to adapt to new habitats as a result of migration or loss of habitat.

Comparative biochemistry

Comparative biochemistry is the study of similarities and differences in the proteins and other molecules that control life processes. Although these molecules can change over time, some important molecules are

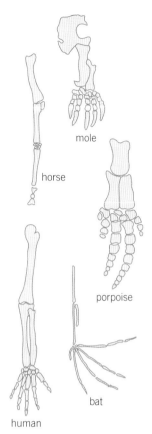

▲ **Figure 6** *These are the pentadactyl limbs from a number of organisms (not to scale). Scientists believe that their function has altered as a result of evolution from a common ancestor. The changes in bone structures of organisms living in the past can be studied through the fossil record*

➕ Evolutionary embryology

Embryology is the study of embryos. It is another source of evidence to show evolutionary relationships. An embryo is an unborn (or unhatched) animal in its earliest phases of development. Embryos of many different animals look very similar and it is often difficult to tell them apart. This shows that the animals develop in a similar way, implying that the processes of embryonic development have a common origin and the animals share common ancestry but have gradually evolved different traits.

Many traits of one type of animal appear in the embryo of another type of animal. For example, fish, and human embryos both have gill slits. In fish these develop into gills, but in humans they disappear before birth.

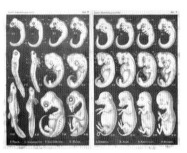

▲ **Figure 7** *Darwin used Haeckel's drawings of embryos as evidence for evolution*

Darwin considered the evidence from embryology to be 'by far the strongest single class of facts in favour of' his theory. He studied a series of drawings produced by biologist Ernst Haeckel that depicted the growth of embryos from various classes of vertebrates. The pictures show that the embryos begin looking virtually identical (suggesting common ancestry), but as they develop, their appearances diverge to take the form of their particular group.

As new organs or structures evolved, these features develop at the end of an organism's embryonic development. As a result, an organism's evolutionary history can be traced in the development of its embryos.

> 1   Explain whether embryos could form part of the fossil record.

Synoptic link

You learnt about ribosomal RNA and its role in protein synthesis in Topic 3.10, Protein synthesis.

Synoptic link

As well as using DNA to demonstrate evolutionary relationships it can also be used for other things. You will learn more examples of this in Topic 21.1, DNA profiling.

▲ **Figure 8** *Humans and chimpanzees have very similar DNA sequences. They have been found to share at least 98% of their DNA. This provides evidence that chimpanzees are humans' closest living relatives*

highly conserved (remain almost unchanged) among species. Slight changes that occur in these molecules can help identify evolutionary links. Two of the most common molecules studied are cytochrome c, a protein involved in respiration, and ribosomal RNA.

The hypothesis of neutral evolution states that most of the variability in the structure of a molecule does not affect its function. This is because most of the variability occurs outside of the molecule's functional regions. Changes that do not affect a molecule's function are called 'neutral'. Since they have no effect on function, their accumulation is not affected by natural selection. As a result, neutral substitutions occur at a fairly regular rate, although that rate is different for different molecules.

To discover how closely two species are related, the molecular sequence of a particular molecule is compared. (Scientists do this by looking at the order of DNA bases, or at the order of amino acids in a protein.) The number of differences that exist are plotted against the rate the molecule undergoes neutral base pair substitutions (which has been determined through studies). From this information scientists can estimate the point at which the two species last shared a common ancestor. Species that are closely related have the more similar DNA and proteins, whereas those that are distantly related have far fewer similarities. Ribosomal RNA has a very slow rate of substitution, so it is commonly used together with fossil information to determine relationships between ancient species.

Summary questions

1   Describe what is shown on a phylogenetic tree.   *(2 marks)*

2   Describe two advantages and two disadvantages of using the fossil record as a source of evidence for evolution.   *(4 marks)*

3   Describe how the work of three scientists was used in the development of the theory of evolution.   *(6 marks)*

4   Explain how comparative biochemistry provides evidence of evolution.   *(3 marks)*

You can tell that a mouse and a bird are different organisms as they have many different characteristics. However, it is more difficult to tell the difference between two individual wood pigeons (*Columba palumbus*). This is because members of the same species share many characteristics.

The differences in characteristics between organisms are called **variations**.

Types of variation

The widest type of variation is between members of different species – these differences are known as **interspecifc variation**. For example, a mouse has four legs, teeth, and fur whereas a bird has two legs, two wings, a beak and feathers.

Every organism in the world is different – even identical twins differ in some ways.

Differences between organisms within a species are called **intraspecific variation**. For example, people vary in height, build, hair colour, and intelligence.

Learning outcomes

Demonstrate knowledge, understanding, and application of:

→ interspecific and intraspecific variation

→ genetic and environmental causes of variation.

▲ **Figure 1** *These organisms all belong to the Plantae kingdom but they are members of different species. The differences between them are examples of interspecific variation*

Causes of variation

Two factors cause variation:

- An organism's genetic material – differences in the genetic material an organism inherits from its parents leads to **genetic variation**.

- The environment in which the organism lives – this causes environmental variation.

▲ **Figure 2** *Dogs show immense variation within their species. For example, this Pug and Great Dane are both members of the species <u>Canis familiaris</u>. The differences between these individuals are examples of intraspecific variation. Due to humans selectively breeding dogs for their particular characteristics, the differences are more extreme than those that would occur naturally*

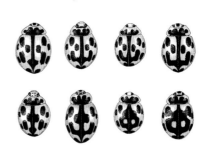

▲ **Figure 3** *These are wingcases of several fourteen-spot ladybirds (<u>Propylea 14-punctata</u>) collected from the same nettle plant. The variation in colouration patterns between the individuals is an example of intraspecific variation*

Genetic causes of variation

Genetic variation is due to the genes (and alleles) an individual possesses. There are several causes for genetic variation being present within a population:

1   **Alleles** (variants) – genes have different alleles (alternative forms). With a gene for a particular characteristic, different alleles produce different effects. For example, the gene for human blood groups has three different alleles (A, B and O). Depending on the parental combination of these alleles (see point 4) four different blood groups can be produced (A, B, AB and O). Individuals in a species population may inherit different alleles of a gene.

2   **Mutations** – changes to the DNA sequence and therefore to genes can lead to changes in the proteins that are coded for. These protein changes can affect physical and metabolic characteristics. If a mutation occurs in somatic (body) cells, just the individual is affected. However, if a mutation occurs in the gametes it may be passed on to the organism's offspring. Both can result in variation.

3   **Meiosis** – gametes (sex cells – ovum and sperm) are produced by the process of meiosis in organisms that reproduce sexually. Each gamete receives half the genetic content of a parent cell. Before the nucleus divides and chromatids of a chromosome separate, the genetic material inherited from the two parents is 'mixed up' by **independent assortment** and **crossing over**. This leads to the gametes of an individual showing variation.

4   Sexual reproduction – the offspring produced from two individuals inherits genes (alleles) from each of the parents. Each individual produced therefore differs from the parents.

5   Chance – many different gametes are produced from the parental genome. During sexual reproduction it is a result of chance as to which two combine (often referred to as random fertilisation). The individuals produced therefore also differ from their siblings as each contains a unique combination of genetic material.

Points 3, 4 and 5 are all aspects of sexual reproduction. As a result there is much greater variation in organisms that reproduce sexually than asexually. Asexual reproduction results in the production of clones (individuals that are genetically identical to their parents). Genetic variation can only be increased in these organisms as a result of mutation.

An example of a characteristic that is determined purely by genetic variation is your blood group. The genes passed onto you from your parents determine if your blood group will be type A, B, AB or O.

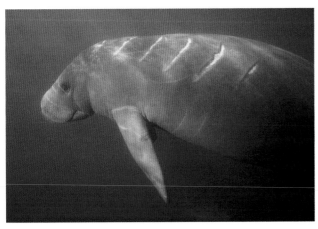

▲ **Figure 4** *These scars on the manatee's back are caused by damage from a boat propeller. They are an example of environmental variation*

Environmental causes of variation

All organisms are affected by the environment in which they live, although plants may be affected to a greater degree than animals due to their lack of mobility. For example, two rose bushes are planted in different positions in a garden. The one that has greater access to the sun will generally grow larger than one in a shadier position. As the plant cannot move to gain sunlight, it is more affected by the environment than an animal, which could move to another area to look for food or shelter.

An example of a characteristic that is determined purely by environmental variation is the presence (or absence) of any scars on your body. They will have occurred as a result of an accident or disease and have no genetic origin. Scars cannot be inherited from a parent.

▲ **Figure 5** *Hydrangeas (*<u>Hydrangea</u>* spp.) produce blue flowers in acidic soils and pink flowers in alkaline soils. This is an example of environmental variation*

Environmental and genetic causes

In most cases variation is caused by a combination of both environmental and genetic factors.

If you have very tall parents, you have most likely inherited the genes to also grow to a tall height. However, if you eat a very poor diet or suffer from disease you may only grow to below average height.

Another example of a characteristic that shows both environmental and genetic causes is your skin colour. This is determined by how much of the pigment, melanin, it contains. The more melanin present in your skin, the darker your skin is. Your skin colour at birth is determined purely by genetics – however, when you expose your skin to sunlight you produce more melanin to protect your skin from harmful UV rays. This results in your skin turning darker.

As many characteristics are caused by a combination of both genetic and environmental causes, it can be very difficult to investigate and draw conclusions about the causes of a variation in any particular case (however, see the Application on studying variation in identical twins). This is often referred to as the 'nature versus nurture' argument. For example, many studies have investigated the primary cause of variation in intelligence – genetics or environment? To date, no definitive conclusion has been reached.

Study tip

Examples of characteristics entirely due to the environment and without genetic influence are few.

For example, scarring as discussed – some people have skin that forms cheloid scars that are very obvious, and others have skin that heals easily with minimal scarring. So even scarring has genetic aspects.

However, the scars themselves cannot be inherited – so have only an environmental cause.

Studying variation in identical twins

Many studies have been carried out on identical twins to determine how much of a characteristic is a result of genetic variation, and how much is a result of the environment in which a person lives.

Identical twins are produced when an egg splits after fertilisation. At this point each twin contains identical genetic material, therefore they show no genetic variation.

If the twins are brought up in different environments, the results of environment on variation can clearly be seen. Even within the same environment, as the twins grow they will show some variation. The characteristics in which they show most variation must be influenced more greatly by the environment than by genes. Those in which they show least variation are controlled more by genes than environment.

One of the most famous case studies on identical twins is the 'Minnesota Study of Twins Reared Apart'. This study looked at the lives of several pairs of identical twins. One pair was known as the 'Jim twins'. Identical twins Jim Lewis and Jim Springer were four weeks old when they were separated, and adopted into different families. They were later reunited, aged 39. At this point, the similarities the twins shared amazed researchers at the University of Minnesota. Many physical characteristics were shared; both twins:

- were 6 feet tall
- had a body mass of 82 kg (13 stone)

- were fingernail biters
- suffered from migraine headaches.

This is perhaps not too surprising. It was also discovered that the twins shared a number of other astonishing similarities. Both twins:

- had owned a dog named Toy
- had been married twice (where both first wives were called Linda, and the second wives both called Betty)
- smoked the same brand of cigarettes
- had studied carpentry and mechanical drawing.

Of course, like other identical twins Jim Lewis and Jim Springer were not identical copies of each other. The two men styled their hair differently – one preferred the medium of speech to communicate, while the other preferred the written word.

1 Why are identical twins used in variation studies?
2 Why do differences between identical twins increase as they age?
3 ⚙ Look at these data. What can you determine about the genetic and environmental causes of these characteristics?

Twin	Height	Eye colour	Ear piercing	Body mass
A	1.79 m	brown	yes	100 kg
B	1.81 m	brown	no	85 kg

Summary questions

1 State the difference between interspecific and intraspecific variation. *(1 mark)*

2 **a** Name two human characteristics with variation caused solely by the environment. *(1 mark)*
   b Name two human characteristics with variation caused solely by genetics. *(1 mark)*

3 Explain some of the causes of variation of human hair. *(3 marks)*

4 Explain why genetic variation is more common in organisms that reproduce sexually. *(3 marks)*

10.6 Representing variation graphically

Specification reference: 4.2.2

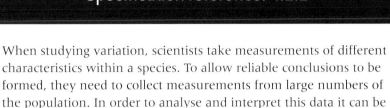

When studying variation, scientists take measurements of different characteristics within a species. To allow reliable conclusions to be formed, they need to collect measurements from large numbers of the population. In order to analyse and interpret this data it can be represented graphically. This allows any patterns to be seen clearly.

Based on the data collected, characteristics can be sorted into those that show **discontinuous variation** and those that show **continuous variation**.

Discontinuous variation

A characteristic that can only result in certain values is said to show discontinuous variation (or discrete variation). There can be no in-between values. Variation determined purely by genetic factors falls into this category. An animal's sex is an example of discontinuous variation as there are only two possible functional values – male or female.

An example of discontinuous variation in microorganisms is the shape of bacteria. They can be spherical (cocci), rods (bacilli), spiral (spirilla), comma (vibrios) or corkscrew shaped (spirochaetes).

Discontinuous variation is normally represented using a bar chart, but a pie chart may also be used. Human blood groups also show discontinuous variation. Like most other characteristics that show discontinuous variation, it is controlled by a single gene, the ABO gene.

Continuous variation

A characteristic that can take any value within a range is said to show continuous variation. There is a graduation in values from one extreme to the other of a characteristic – this is known as a continuum. The height and mass of plants and animals are examples of such characteristics.

Characteristics that show continuous variation are not controlled by a single gene but a number of genes (polygenes). They are also often influenced by environmental factors.

Data on characteristics that show continuous variation are collected in a frequency table (Table 1). These data are then plotted onto a histogram (Figure 1). Normally a curve is then drawn onto the graph to show the trend.

Normal distribution curves

When continuous variation data are plotted onto a graph, they usually result in the production of a bell-shaped curve known as a **normal**

▼ **Table 1** *Frequency of heights (measured to the nearest 2 cm)*

Height/cm	Frequency
140	0
144	1
148	23
152	90
156	261
160	393
164	458
168	413
172	177
176	63
180	17
184	4
188	1
190	0
192	0

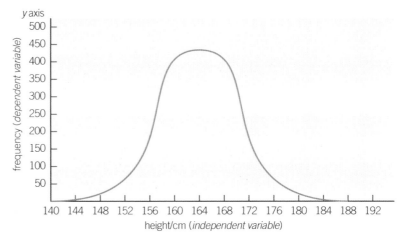

▲ **Figure 1** *This graph shows the frequency against height for a sample of humans. The height of the human population ranges from the shortest person in the world to the tallest person. A person's height can take any value in between. This is an example of continuous variation*

▲ **Figure 2** *The surface area of the leaves on this tree show continuous variation – depending on their position on the tree they have all grown to slightly different sizes*

distribution curve (Figure 3). The data is said to be normally distributed.

Characteristics of a normal distribution:

● The mean, mode, and median are the same.

● The distribution has a characteristic 'bell shape', which is symmetrical about the mean.

● 50% of values are less than the mean and 50% are greater than the mean.

● Most values lie close to the mean value – the number of individuals at the extremes are low.

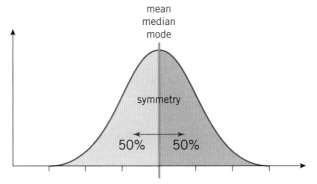

▲ **Figure 3** *A normal distribution curve*

Standard deviation

The standard deviation is a measure of how spread out the data is. The greater the standard deviation is, the greater the spread of the data. In terms of variation, a characteristic which has a high standard deviation has a large amount of variation.

When you calculate the standard deviation of data that display a normal distribution you will generally find that:

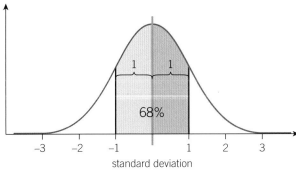

68% of values are within
1 standard deviation of the mean

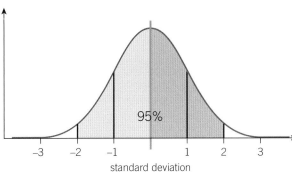

95% of values are within
2 standard deviations of the mean

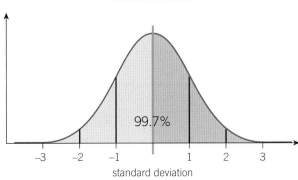

99.7% of values are within
3 standard deviations of the mean

▲ **Figure 5** *The petiole (stalk) length of ivy (*Hedera* spp.) leaves show continuous variation*

▲ **Figure 4** *Graphs showing spread of data on a normal distribution*

 Worked example: Petiole length variation in ivy leaves

Standard deviation (denoted with the Greek letter σ) is a measurement of the spread of data.

It is calculated using the following formula:

$$\sigma = \sqrt{\frac{\sum (x - \overline{x})^2}{n - 1}}$$

$\sum$ = the sum (total) of

x = value measured

$\overline{x}$ = mean value

n = total number of values in the sample

Follow the worked example below to work out the standard deviation of the length of petioles (the stalk attaching the leaf to the stem) of a sample of 10 ivy leaves. The data collected was:

Sample Number	1	2	3	4	5	6	7	8	9	10
Petiole length / mm	28	30	17	31	35	45	46	67	33	57

1 Calculate the mean value $\overline{x}$

$$\overline{x} = \frac{\text{sum of the individual measurements}}{\text{number in the sample}}$$

$$= \frac{389}{10} = 38.9 \, \text{mm}$$

2 Subtract the mean value from each measured value: $x - \overline{x}$

For example, the first measurement was 28 mm.

Measured value (x)– mean value ($\bar{x}$)
= 28 – 38.9 = –10.9 mm

For the other measurements you would calculate: –8.9 mm, –21.9 mm, –7.9 mm, –3.9 mm, 6.1 mm, 7.1 mm, 28.1 mm, –5.9 mm, 18.1 mm

3   Square each of these values $(x - \bar{x})^2$

For example –10.9^2 = 118.81

For the other measurements you would calculate: 79.2 mm, 479.6 mm, 62.4 mm, 15.2 mm, 37.2 mm, 50.4 mm, 789.6 mm, 34.8 mm, 327.6 mm

4   Sum each of these values $\Sigma \, (x - \bar{x})^2$

118.8 + 79.2 + 479.6 + 62.4 + 15.2 + 37.2 + 50.4 + 789.6 + 34.8 + 327.6 = 1994.8 mm

5   Divide this value by the sample size minus one $\dfrac{\Sigma(x - \bar{x})^2}{n-1}$

$$\frac{\Sigma(x - \bar{x})^2}{n-1} = \frac{1994.8}{10-1} = \frac{1994.8}{9} = 221.7$$

6   Find the square root of this value $\sqrt{\dfrac{\Sigma \, (x - \bar{x})^2}{n-1}}$

$$\sigma = \sqrt{\frac{\Sigma \, (x - \bar{x})^2}{n-1}} = \sqrt{221.7} = 14.9$$

Flagella length variation in *Salmonella*

Salmonella bacteria have a number of flagella to enable them to move. Five bacteria were chosen at random, and their longest flagellum measured. This is the data that was collected:

Bacterium	1	2	3	4	5
Longest flagellum (µm)	3.0	2.5	1.8	2.0	2.7

1   Calculate the mean value for the length of flagella in *Salmonella* bacteria.
2   Using the formula given in the worked example, calculate the standard deviation for the length of flagella in *Salmonella* bacteria. State your answer to two decimal places.
3   State the range of flagella lengths that 68% of the *Salmonella* population will have.
4   State and explain what type of variation is shown by the length of flagella in *Salmonella*.

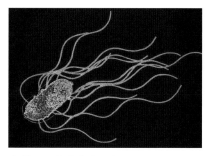

▲ **Figure 6** *This is a* Salmonella *bacteria* (Salmonella *sp.*). *The length of their flagella shows continuous variation*

Other statistical tests

Several statistical tests can be used by scientists to determine the significance of data collected. These tests can be used in a number of situations, for example when comparing variation within populations, or when comparing the effects of abiotic and biotic factors on organisms (Chapter 11, Biodiversity). These include:

- Student's *t* test – this is used to compare the means of data values of two populations
- Spearman's rank correlation coefficient – this is used to consider the relationship of between two sets of data.

Study tip

You will not be expected to learn the formulae for statistical tests, but you should have an understanding of how these tests are calculated, of the circumstances in which to use them, and how to interpret the results.

Student's *t* test

Student's *t* test is used to compare the mean values of two sets of data. To use this test the data collected must be normally distributed and enough data should be collected to calculate a reliable mean. Different sample sizes may be used.

Study tip

A significant difference at *p* = 0.05 means that if the null hypothesis were correct (i.e., the samples or treatments do not differ) then we would expect to get a t value as great as this on exactly 5% of occasions. You can therefore be reasonably confident that the samples do differ from one another, but there is still nearly a 5% chance of this conclusion being wrong

If the calculated t value exceeds the tabulated value for *p* = 0.01, then there is a 99% chance of the means being significantly different (and a 99.9% chance if the calculated t value exceeds the tabulated value for *p* = 0.001). By convention, a difference between means at the 95% level is 'significant', a difference at 99% level is 'highly significant' and a difference at the 99.9% level is 'very highly significant'.

 ## Worked example: Comparing mean petiole length in ivy grown in the light and shade

Student's *t* test is is calculated using the following formula:

$$t = \frac{(\bar{x}_1 - \bar{x}_2)}{\sqrt{\left(\frac{\sigma_1^2}{n_1}\right) + \left(\frac{\sigma_2^2}{n_2}\right)}}$$

$\bar{x}_1, \bar{x}_2$ = mean of populations 1 and 2

$\sigma_1, \sigma_2,$ = standard deviation of populations 1 and 2

n_1, n_2 = total number of values in samples 1 and 2

A sample of ten ivy leaves was collected from either side of a tall tree stump. Those collected from the south-facing side of the trunk were referred to as the 'light' leaves and those on the north-facing side of the stump, the 'shade' leaves. A group of students wanted to see if there was a significant difference in the size of the leaf's petiole, depending on whether ivy is grown in the light or the shade.

Before calculating Student's *t* test, the students had to produce a **null hypothesis**. This is a prediction that there is no significant difference between specified populations, and so any observed difference would be due to chance variation in the sample.

The data the students collected is summarised below:

Sample taken	Number in sample	Mean petiole length/mm	Standard deviation/mm
Light	10	38.9	14.9
Shade	10	52.8	15.1

The students then used the Student's *t* test to determine if there is statistical significance between the petiole length of ivy grown in the light, compared to those grown in the shade.

1 State the null hypothesis:

There will be no difference in the length of ivy petiole length of leaves growing in the light, compared with those in the shade.

2 Subtract the mean petiole length of sample two from sample one:

$\bar{x}_1 - \bar{x}_2 = 38.9 - 52.8 = (-)13.9$

Note: ignore minus signs.

3 For both populations, square the standard deviation and divide by the number in the sample:

Population 1: $\frac{\sigma_1^2}{n_1} = \frac{14.9^2}{10} = 22.201$

Population 2: $\frac{\sigma_2^2}{n_2} = \frac{15.1^2}{10} = 22.801$

4 Sum these values:

$\frac{\sigma_1^2}{n_1} + \frac{\sigma_2^2}{n_2} = 22.201 + 22.801 = 45.002$

5 Square root this value

$$\sqrt{\frac{\sigma_1^2}{n_1} + \frac{\sigma_2^2}{n_2}} = \sqrt{45.002} = 6.71$$

6 Calculate Student's t test

$$t = \frac{(\bar{x}_1 - \bar{x}_2)}{\sqrt{\frac{\sigma_1^2}{n_1} + \frac{\sigma_2^2}{n_2}}} \quad t = \frac{13.9}{6.71} = 2.07$$

To understand what this value means, you must look it up in the Student's t test significance tables (see appendix). First, calculate a quantity known as the 'degrees of freedom' (df) using the formula:

$df = (n_1 + n_2) - 2$ where n_1 = population 1, n_2 = population 2

In this example: $df = (n_1 + n_2) - 2$
$= (10 + 10) - 2 = 18$

Then look at the corresponding probability values. For the data to be considered significantly different from chance alone, the probability (p) must be 5% (0.05) or less.

At $df = 18$, the value of 2.07 falls between 5% and 10%.

The null hypothesis should be accepted, as we cannot be more that 95% confident that the results are not down to chance. We therefore cannot conclude that there is a significant difference between the petiole length of ivy grown in the light and in the shade.

Study tip

The correlation test is used to see if two different variables are correlated in a linear fashion in the context of a scatter-graph. There are several different types of statistical test that can be used, the OCR GCE Biology specifications will only cover Spearman's rank correlation coefficient. You can find a full table of values in the appendix.

Spearman's rank correlation coefficient

If two sets of data are related they are said to be correlated. Two sets of data can show:

- no correlation – no relationship between the data (Figure 9)
- positive correlation – as one set of data increases in value, the other set of data also increases in value (Figure 7)
- negative correlation – as one set of data increases in value, the other set of data decreases in value (Figure 8).

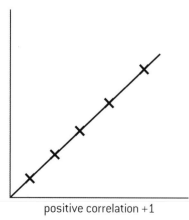

positive correlation +1

▲ **Figure 7** *Graph showing positive correlation*

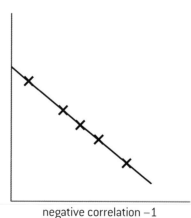

negative correlation −1

▲ **Figure 8** *Graph showing negative correlation*

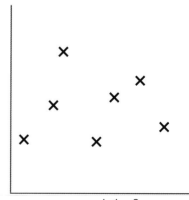

no correlation 0

▲ **Figure 9** *Graph showing no correlation*

Worked example – Using Spearman's rank correlation coefficient to compare ivy leaves' petiole length and leaf width

The correlation coefficient is calculated using the following formula:

$$r_s = 1 - \frac{6\sum d^2}{n(n^2 - 1)}$$

where:

r_s = correlation coefficient $\sum$ = the sum (total) of

d = difference in ranks n = number of pairs of data

The group of students next wanted to find out if the petiole length of the ivy was related to the width of an ivy leaf. They took a sample of 10 ivy leaves from the north facing side of the stump. The data they collected is shown below.

Sample Number	1	2	3	4	5	6	7	8	9	10
Petiole length / mm	28	58	57	59	27	59	44	54	79	63
Leaf width / mm	38	66	64	66	30	65	48	54	78	62

The data for the two variables should be rank ordered, from lowest to highest. Using a table can help to make manipulating the data more straightforward.

Where identical values exist, the 'average rank' should be used. So, if two equal values appear at rank 5, both are assigned the rank 5.5 (between ranks 5 and 6).

petiole length / mm	leaf width / mm	Rank: petiole length	Rank: leaf width	Rank difference d	d^2
28	38	2	2	0	0
58	66	6	8.5	−2.5	6.25
57	64	5	6	−1	1
59	66	7.5	8.5	−1	1
27	30	1	1	0	0
59	65	7.5	7	0.5	0.25
44	48	3	3	0	0
54	54	4	4	0	0
79	78	10	10	0	0
63	62	9	5	4	16
					$\sum d^2 = 24.5$

Substituting values from the table:

$$r_s = 1 - \frac{6\sum d^2}{n(n^2 - 1)} = 1 - \frac{(6 \times 24.5)}{10 \times (10^2 - 1)} = 1 - \left(\frac{147}{990}\right) = 0.852$$

where:

An r_s value of +1 shows a perfect positive correlation.

An r_s value of −1 shows a perfect negative correlation

An r_s value of 0 shows no correlation

Therefore in this example, petiole length and leaf width show an excellent correlation.

To work out the statistical strength of the correlation, the value should be looked up in the correlation coefficient critical value tables. Some tables refer to the number of data pairs (n); others ask you to calculate the degrees of freedom (df). The tables you will be using for your Spearman's rank correlation coefficient use n.

Synoptic link

You will learn how to calculate degrees of freedom in Topic 20.4, Phenotypic ratios.

Then look at the probability values for this number of data pairs. As before, for the data to be considered significantly different from chance alone, the probability must be 5% (0.05) or less – a certainty of 95% or more.

	$p = 0.1$	$p = 0.05$	$p = 0.02$	$p = 0.01$
	10%	5%	2%	1%
n				
1	–	–	–	–
2	–	–	–	–
3	–	–	–	–
4	1.0000	–	–	–
5	0.9000	1.0000	1.0000	–
6	0.8286	0.8857	0.9429	1.0000
7	0.7143	0.7857	0.8929	0.9286
8	0.6429	0.7381	0.8333	0.8810
9	0.6000	0.7000	0.7833	0.8333
10	0.5636	0.6485	0.7455	0.7939
11	0.5364	0.6182	0.7091	0.7545
12	0.5035	0.5874	0.6783	0.7273

If p=0.01 then this correlation has only a 1% probability of having occurred by random chance. As the correlation is positive, we can conclude that the greater the petiole length, the greater the leaf width.

Summary questions

1 Sort the following list into those characteristics which show continuous variation and those which show discontinuous variation: (2 marks)
   a the presence in humans of lobed or lobeless ears
   b the size of an *E.coli* bacterium
   c the height of a group of seedlings, planted for a germination experiment
   d the number of spots present on a ladybird.

2 Describe the differences between the genetic and environmental control of characteristics that show discontinuous and continuous variation.   (4 marks)

3 Explain why a mean value should not normally be calculated for a characteristic showing discontinuous variation.   (2 marks)

4 Describe the pattern of variation that would be seen if the body mass of all wild rabbits was measured.   (4 marks)

5 The following data was collected from a student's fieldwork study:

Diameter of rose bush stem / mm	1	2	3	5	8	10	11	14
Number of thorns per unit length	8	11	9	12	12	27	23	30

   a Calculate the Spearman's ranked correlation coefficient for this set of data   (6 marks)
   b Evaluate the strength of the correlation calculated in part (a)   (3 marks)

You should be familiar with the concept that organisms are adapted to the environment in which they live. Organisms can also be adapted to protect themselves from predators or attract a mate.

What are adaptations?

Adaptations are characteristics that increase an organism's chance of survival and reproduction in its environment. Adaptations can be divided into three groups:

- anatomical adaptations – physical features (internal and external)
- behavioural adaptations – the way an organism acts. These can be inherited or learnt from their parents.
- physiological adaptations – processes that take place inside an organism.

Many adaptations fall into more than one category. For example, the courtship behaviour of a peacock requires it to lift its huge, colourful tail to attract the peahen. This is an example of both a behavioural and anatomical adaptation.

Anatomical adaptations

Some examples of anatomical adaptations:

- Body covering – animals have a number of different body coverings such as hair, scales, spines, feathers, and shells. These can: help the organism to fly, such as feathers on birds – help it to stay warm, such as the thick hair on polar bears – provide protection, such as a snail's shell. Thick waxy layers on plants prevent water loss and spikes can deter herbivores and protect the tissues from sun damage.
- Camouflage – the outer colour of an animal allows it to blend into its environment, making it harder for predators to spot it. For example, the snowshoe hare is white in winter to match the snow, and turns brown in summer to blend in with the soil and rock environment in which it lives.
- Teeth – the shape and type of teeth present in an animal's jaw are related to its diet. Herbivores, such as sheep, have continuously growing molars for chewing tough grass and plants. Carnivores, such as tigers, have sharp large canines to kill prey and tear meat.
- Mimicry – copying another animal's appearance or sounds allows a harmless organism to fool predators into thinking it is poisonous or dangerous. For example, the harmless hoverfly mimics the markings of a wasp to deter predators.

Learning outcomes

Demonstrate knowledge, understanding, and application of:

→ the different types of adaptations of organisms to their environment

→ why organisms from different taxonomic groups may show similar anatomical features.

▲ **Figure 1** *Otters (*Lutra *spp.) have webbed paws. This allows them to swim as well as walk. This is an example of an anatomical adaptation. It increases their chance of survival as they can live and hunt on land and in the water (×10 magnification)*

▲ **Figure 2** *The harmless milk snake (*Lampropeltis triangulum*, top) mimics the markings of the deadly coral snake (*Micrurus alleni*, bottom)*

259

Synoptic link

You learnt about transpiration in Topic 9.3, Transpiration and about plant adaptations to prevent water loss in Topic 9.5, Plant adaptations to water availability.

Marram grass

Marram grass (*Ammophila* spp.) is commonly found on sand dunes around the UK. It is a xerophyte, a plant that has adapted to live in an environment with little water. Its adaptations reduce the rate of **transpiration** and include:

- curled (or rolled) leaves to minimise the surface area of moist tissue exposed to the air, and protect the leaves from the wind

  - hairs on the inside surface of the leaves to trap moist air close to the leaf, reducing the diffusion gradient

  - stomata sunk into pits, which make them less likely to open and lose water

  - a thick waxy cuticle on the leaves and stems, reducing water loss through evaporation.

Behavioural adaptations

Some examples of behavioural adaptations:

- Survival behaviours – for example, an opossum plays dead and a rabbit freezes when they think they have been seen.

- Courtship – many animals exhibit elaborate courtship behaviours to attract a mate. For example, scorpions perform a dance to attract a partner. This increases the organism's chance of reproducing.

- Seasonal behaviours – these adaptations enable organisms to cope with changes in their environment. They include:

  - migration – animals move from one region to another, and then back again when environmental conditions are more favourable. This may be for a better climate or a source of food

  - hibernation – a period of inactivity in which an animal's body temperature, heart rate and breathing rate slow down to conserve energy, reducing the animal's requirement for food. For example, brown bears hibernate during the winter.

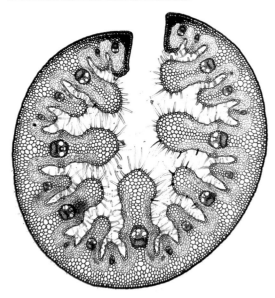

▲ **Figure 3** *A light micrograph of a curled leaf of marram grass plant (*Ammophila arenaria*). You can see the hairs on the inside surface of the leaf, slowing air movement and reducing water loss from the leaf (× 10 magnification)*

Generally, behavioural adaptations fall into two main categories:

- Innate (or instinctive) behaviour – the ability to do this is inherited through genes. For example, the behaviour of spiders to build webs and woodlice to avoid light is innate. This allows the organism to survive in the habitat in which it lives.

- Learned behaviour – these adaptations are learnt from experience or from observing other animals. An example of learned behaviour is the use of tools. For example, sea otters use stones to hammer shells off rocks, and then to crack the hard shells open.

▲ **Figure 4** *This chimpanzee is using a twig/grass to get termites out of a termite mound*

However, many behavioural adaptations are a combination of both innate and learned behaviours.

Physiological adaptations

Some examples of physiological adaptations:

- Poison production – many reptiles produce venom to kill their prey and many plants produce poisons in their leaves to protect themselves from being eaten.

- Antibiotic production – some bacteria produce antibiotics to kill other species of bacteria in the surrounding area.

- Water holding – the water-holding frog (*Cyclorana platycephala*) can store water in its body. This allows it to survive in the desert for more than a year without access to water. Many cacti and other desert plants can hold large amounts of water in their tissues.

Many other examples are less unusual, and include reflexes, blinking and temperature regulation.

Anatomical adaptations provide evidence for convergent evolution

Analogous structures

Although the tail fins of a whale and a fish perform the same role, when you look at them in detail their structures are very different. These are known as **analogous structures** – they have adapted to perform the same function but have a different genetic origin.

Convergent evolution takes place when unrelated species begin to share similar traits. These similarities evolve because the organisms adapt to similar environments or other selection pressures (for an explanation of the term 'selection pressure', Topic 10.8, Changing population characteristics). The organisms live in a similar way to each other. Using our example of whales and fish, their similar characteristics have evolved over time to allow the organisms to move efficiently through water.

Marsupials in Australia and placental mammals in the Americas are an example of convergent evolution. Species in each continent resemble each other because they have adapted to fill similar niches.

In placental mammals, a placenta connects the embryo to its mother's circulatory system in the uterus. This nourishes the embryo, allowing it to reach a high level of maturity before birth. Marsupials also start life in the uterus, but then leave and enter the marsupium (pouch) while they are still embryos. They complete their development here by suckling milk.

These two subclasses of mammals separated from a common ancestor more than 100 million years ago. Each lineage then evolved independently. Despite this large temporal and geographical separation, marsupials in Australia and placental mammals in North America have produced varieties of species that bear a strong resemblance in overall shape, type of locomotion and feeding techniques. This is because they have adapted to similar climates and food supplies. However, these organisms have very different methods of reproduction. This is the feature that accurately reflects their distinct evolutionary relationships.

▲ **Figure 5** *Aestivation is the name given to a period of inactivity in hot, dry places. The land snails aestivate to cope with periods of extreme dry heat. This is a combination of physiological and behavioural adaptations*

Synoptic link

You will learn more about the adaptations of ectotherms and endotherms to temperature in Topics 15.2 and 15.3, respectively.

▲ **Figure 7** *(top) Flying phalanger* ([Petaurus sp.]) *– a marsupial mammal and (bottom) flying squirrel (family Pteromyini) – a placental mammal*

Examples include:

- marsupial and placental mice – both are small, agile climbers that live in dense ground cover and forage at night for small food items. The two mice are very similar in size and body shape

- flying phalangers and flying squirrels (Figure 7) – both are gliders that eat insects and plants. Their skin is stretched between their forelimbs and hind limbs to provide a large surface area for gliding from one tree to the next

- marsupial and placental moles (Figure 8) – both burrow through soft soil to find worms and grubs. They have a streamlined body shape and modified forelimbs for digging. They also have velvety fur, which allows smooth movement through the soil. However, they differ in fur colour – the marsupial mole ranges in colour from white to orange whereas the placental mole is grey.

Convergent evolution can also been seen in some plant species. For example, aloe and agave appear very similar as they have both adapted to survive in the desert. However, these species developed entirely separately from each other. Aloe are sometimes referred to as 'old world', having evolved in sub-Saharan Africa. Agave, by comparison, are 'new world', having evolved in Mexico and the southern United States.

▲ **FIGURE 8** *(top) Marsuipal mole (family Notoryctidae) found in Australia and (bottom) placental mole (family Talpidae) found in North America*

aloe

agave

▲ **Figure 9** *Aloe (Aloe spp.) and agave (Agave spp.) have a similar appearance. They provide an example of convergent evolution amongst plant species*

➕ Classification of giant pandas

Classification aims to place every organism into a particular taxonomical group. However, there are a number of organisms that do not fit easily into a group. An example of this is the giant panda (*Ailuropoda melanoleuca* – Figure 10, top).

Père Armand David, a catholic priest, was the first westerner to see a giant panda. He discovered the panda in 1869, and based on its appearance he concluded that it was related to a bear (family *Ursidae* – Figure 10,

bottom). He gave it a name that included the word *ursus* (the Latin word for bear).

A few years later, Alphonse Milne-Edwards, a French scientist, inspected the remains of a giant panda. He concluded that its anatomical structure was closer to the red panda (Figure 10, middle), a member of the raccoon family. He renamed the giant panda, and classified it into its own category. Many people disagreed with Milne-Edwards' conclusion because of its size. Red pandas have a mass of between 3 and 7 kg – the largest raccoons have a mass of around 30 kg. By comparison, the giant pandas can exceed 100 kg body mass.

The debate over the classification of the giant panda has continued for several decades.

Similarities to a red panda:

- Both eat bamboo and grip bamboo in the same manner.
- Both have similar snouts, teeth and paws.

Similarities to a bear:

- Both are a very similar shape and size.
- Both have shaggy fur.
- Both walk and climb in a similar manner.

Giant pandas and red pandas may have developed similar ways of eating bamboo separately as a result of convergent evolution. Equally, convergent evolution could explain their resemblance to bears.

1 Describe the adaptations of a giant panda.
2 Explain how you would classify a giant panda. Give reasons for your classification.
3 Suggest how recent biological techniques could be used to help classify the giant panda.

In the 1950s, the first molecular-level analysis of the giant panda occurred. Biologists used an immunological method to assess the closeness of bears to pandas. Through studying blood serum, they concluded that the 'serological affinities of the giant panda are with the bears rather than with the raccoons'. The giant panda is therefore a true bear and part of the Ursidae family, although it differentiated early in history from other bears.

Despite the shared name, habitat type, and diet, as well as a unique enlarged bone called the pseudo thumb (which helps them grip bamboo shoots), the giant panda and red panda are only distantly related. Molecular studies place the red panda in its own family – Ailuridae.

Summary questions

1 Classify the following adaptations into anatomical, behavioural or physiological adaptations:
melanin production; camouflage; migration; sharp canine teeth; production of toxins; courtship dance *(3 marks)*

2 State the difference between analogous and homologous structures. *(1 mark)*

3 Which of the following is an example of convergent evolution? *(1 mark)*
Explain your answer.
a Insect wing and bird wing
b Bat wing and human arm

4 Select either a cactus or a hedgehog. For your chosen organism, state and explain how it is adapted to survive successfully in its habitat. *(4 marks)*

5 State and explain how marsupial moles and placental moles provide evidence for convergent evolution. *(4 marks)*

▲ **Figure 10** *The giant panda (*<u>Ailuropoda melanoleuca</u> *– top) shares characteristics with both the red panda (*<u>Ailurus</u> <u>fulgens</u> *– middle) and bears (family Ursidae – for example this brown bear* <u>Ursus</u> <u>arctos</u> *– bottom). As a result, scientists have argued about how to classify giant pandas*

10.8 Changing population characteristics

Specification reference: 4.2.2

You have already learnt about the theory of evolution as a result of natural selection in Topic 10.4 Evidence for evolution. This process takes place over many, many generations; it generally takes several thousand years for a species to evolve. However, evolution is a dynamic process and is always occurring.

Natural selection

All organisms are exposed to **selection pressures**. These are factors that affect the organism's chances of survival or reproductive success (the ability to produce fertile offspring).

Organisms that are best adapted to their environment are more likely to survive and reproduce. As a result of natural selection these adaptations will become more common in the population. Organisms that are poorly adapted are less likely to survive and reproduce. Therefore their characteristics are not passed on to the next generation. As a result, less of the population will display these characteristics.

Natural selection follows a number of steps:

1   Organisms within a species show variation in their characteristics that are caused by differences in their genes (genetic variation). For example, they may have different alleles of a gene for a particular characteristic. New alleles can arise by mutation.

2   Organisms whose characteristics are best adapted to a selection pressure such as predation, competition (for mates and resources) or disease, have an increased chance of surviving and successfully reproducing. Less well-adapted organisms die or fail to reproduce. This process is known as 'survival of the fittest'.

3   Successful organisms pass the allele encoding the *advantageous characteristic* onto their offspring. Conversely, organisms that possess the non-advantageous allele are less likely to successfully pass it on.

4   This process is repeated for every generation. Over time, the proportion of individuals with the advantageous adaptation increases. Therefore the frequency of the allele that codes for this particular characteristic increases in the population's gene pool.

5   Over very long periods of time, many, many generations and often involving multiple genes, this process can lead to the evolution of a new species.

Modern examples of evolution

Antibiotic-resistant bacteria

Methicillin-resistant *Staphylococcus aureus* (MRSA) has developed resistance to many antibiotics. Bacteria reproduce very rapidly and so evolve in a relatively short time. When bacteria replicate, their DNA can be altered and this usually results in the bacteria dying. However, a mutation in some *S. aureus* arose that provided resistance to methicillin.

When the bacteria were exposed to this antibiotic, resistant individuals survived and reproduced, passing the allele for resistance on to their offspring. Non-resistant individuals died. Over time the number of resistant individuals in the population increased.

Peppered moths

Dramatic changes in the moth's environment in the 19th century caused changes in allele frequency in peppered moths (*Biston betularia*). Before the industrial revolution, most peppered moths in Britain were pale coloured. This provided camouflage against light-coloured tree bark, increasing their chance of survival. Those that were dark were easily spotted by birds and eaten. The different colourings are due to different alleles.

During the industrial revolution many trees became darker – partly due to being covered in soot, and partly due to the loss of lichen cover caused by increased atmospheric pollutants. The dark moths were now better adapted, as they were more highly camouflaged. More dark peppered moths survived and reproduced, increasing the frequency of dark moths (and the 'dark' allele) in the population. After a few years the number of dark peppered moths close to industrial towns and cities became much higher than pale peppered moths.

Since the Clean Air Act of 1956 steps have been taken to improve air quality in towns and cities, and to reduce the levels of pollution released from factories. The bark on the vast majority of trees in the UK is once again lighter coloured, and therefore the frequency of the pale allele in the moth gene pool has increased.

Sheep blowflies

Sheep blowflies (*Lucilia cuprina*) lay their eggs in faecal matter around a sheep's tail – the larvae then hatch and cause sores. This condition is known as 'flystrike', and if left untreated is normally fatal.

In the 1950s in Australia, the pesticide diazinon (an organophosphate pesticide) was used to kill the blow flies and prevent the condition. Within six years, blowflies had developed a high level of resistance to diazinon. Individual insects with resistance survived exposure to the insecticide, and passed on this characteristic through their alleles, allowing a resistant population to evolve.

To investigate how this evolution occurred so quickly, scientists extracted DNA from a sample of 70-year-old blowflies kept at

Synoptic link

You will learn more about the development of drug resistance in microorganisms in Topic 12.7, Preventing and treating diseases.

▲ **Figure 1** *Before the industrial revolution, pale peppered moths (*Biston betularia*) were better adapted to the environment. During the industrial revolution dark peppered moths were now better adapted – therefore the frequency of this characteristic increased in the population*

▲ **Figure 2** *Scientists have discovered that sheep blowflies (*Lucilia* spp.) have an inbuilt natural resistance to some organophosphate insecticides. This allowed their rapid evolution to become resistant to the pesticide diazinon*

the Australian National Insect Collection. Two Australian sheep blowflies were studied, *Lucilia cuprina* and the closely related *Lucilia sericata*. The researchers compared the blowflies' resistance genes before and after the introduction of the pesticide. Diazinon resistance was not found in the DNA of the 70-year-old flies, whereas it is present in the modern species. However, when they performed the same investigation with malathion (another organophosphate pesticide), they found resistance alleles in both the old and modern blowflies, showing there was pre-existing resistance to this chemical.

The scientists concluded that pre-adaptation contributed to the development of diazinon-resistance. Pre-adaptation is when an organism's existing trait is advantageous for a new situation. The alteration in the DNA that caused the pre-existing resistance allowed the flies to rapidly develop resistance to organophosphate chemicals in general, and ultimately a specific diazinon-resistance allele.

The existence of pre-adaptation in an organism may help researchers predict potential insecticide resistance in the future.

Flavobacterium

Most evolution occurs as a negative result of selection pressures. However, some organisms have evolved due to opportunities that have arisen in their environment. For example, scientists have found a new strain of *Flavobacterium* living in waste water from factories that produce nylon 6. Nylon 6 is used to make objects like toothbrushes and violin strings. This strain of bacteria has evolved to digest nylon and is therefore beneficial to humans as they help to clear up factory waste.

These bacteria use enzymes to digest the nylon known as nylonases. They are unlike any enzymes found in other strains of *Flavobacterium*, and they do not help the bacteria to digest any other known material. It is beneficial to the bacteria as it provides them with another source of nutrients.

Most scientists believe that the gene mutation that occurred to produce these enzymes was a result of a gene duplication, combined with a frameshift mutation (an insertion or deletion of DNA bases that causes the genetic code to be read incorrectly).

Anolis lizards

When a few individuals of a species colonise a new area, their offspring initially experience a loss in genetic variation, often resulting in individuals that are physically and genetically different from their source population. This is known as the **founder effect**.

A 14-year experiment (led by Kolbe, a biologist at the University of Rhode Island) was carried out to study

evolution. Pairs of *Anolis sagrei* were released across 14 small Caribbean islands that had no previous lizard populations. During the experiment, the lizard populations each became adapted to their respective environments through changes in their body shape driven by the flora in their environment. Several new species of lizards evolved.

1 State what is meant by the founder effect.
2 Explain why these particular islands were an ideal location for the experiment.

To determine how much of the evolution of the new species was due to the founder effect, and how much resulted from natural selection, Kolbe randomly selected pairs of Anolis lizards from the island of Iron Cay. He then released these organisms onto seven smaller islands that had no lizard population. Each island had the same types of insects, birds and short scrub vegetation but differed from Iron Cay, which is covered in forest.

Forest Anolis lizards have long hind limbs, which allow them to move quickly across thick branches, whereas short limbs give scrub-living lizards stability to walk along narrow perches. The scientists therefore predicted that the lizards in their experiment would develop shorter hind limbs than those of the lizards on Iron Cay.

After one year, the researchers noticed that the offspring of the experimental lizards had less genetic variability than the Iron Cay lizards – the founder effect. There were also significant differences in hind-limb length among the lizards on the islands. As the founder effect is a random process independent of the environment, there was no pattern to the length of the lizards' hind limbs. Over the next few years the lizards' hind limbs on all the experimental islands got shorter, making them better suited for their environment – natural selection.

▲ **Figure 3** *A male Anolis lizard (Anolis sagrei) displaying its eye-catching dewlap. When enlarged it makes the lizard appear much bigger than it really is. This mechanism is used to ward off predators and to attract females during the mating season*

Kolbe concluded that both processes were evident during the experimental period.

3 Explain why scientists thought the lizards would develop short hind limbs.
4 Explain how the scientists showed whether the evolution was mainly a result of the founder effect or natural selection.
5 Some scientists were surprised that all the new populations of species survived as the presence of only a few individuals leads to inbreeding. Explain why this is disadvantageous.

Summary questions

1 State three selection pressures that may be experienced by a plant species. *(1 mark)*

2 Describe the process of natural selection. *(3 marks)*

3 DDT is a chemical insecticide that was used to kill mosquitoes to prevent the spread of malaria. Several years after its introduction large populations of mosquitoes became DDT resistant. Explain how this occurred. *(4 marks)*

4 Using examples, state and explain the positive and negative effects on humans of recent examples of evolution in some species. *(6 marks)*

Practice questions

1   As part of a sample collected from the Indian Ocean, scientists identified an organism as *Hydrophis spiralis*. Which of the following statements is/are correct about this organism?

Statement 1: The organism belongs to the genus *Hydrophis*

Statement 2: The organism belongs to the species *spiralis*

Statement 3: The organism belongs to the genus *spiralis*

A   1, 2 and 3 are correct

B   1 and 2 are correct

C   2 and 3 are correct

D   Only 1 is correct   (*1 mark*)

2   Figure 1 shows an electron micrograph of an invertebrate known as 'water bear'.

▲ **Figure 1**

a   Complete the following passage about the classification of water bears using the most appropriate terms.

The water bear, *Eschiniscus trisetosus* is a member of the genus.................and the family *Echiniscidae*. This family belongs to the.....................Eschiniscoidea, which forms part of the class *Heterotardigrada*. Water bears, also known as tardigrades, are classified into a...... of their own called the *Tardigrada*. Tardigrades form part of the kingdom................ Within the domain........... .   (*5 marks*)

b   State the meaning of the term phylogeny and explain how phylogeny is related to classification.   (*3 marks*)

c   Water bears are extremely common in many habitats, including household gardens. However, they were not discovered until approximately 300 years ago.

Suggest reasons why they were not known before this time.   (*2 marks*)

OCR June 2013 F212/01

3   Living organisms can be classified into five kingdoms, based on certain key characteristics.

a   Table 1 shows some of the characteristics of the five kingdoms.

Copy and complete the table

kingdom	membrane-bound organelles	cell wall	type(s) of nutrition
prokaryote	absent	present – made of peptidoglycan	
	present	sometimes present – composition varies	heterotrophic and autotrophic
Fungi		present – made of chitin	heterotrophic
	present		autotrophic
animal		absent	heterotrophic

(*6 marks*)

b   An unknown species is discovered. Its cells contain many nuclei scattered throughout the cytoplasm of thread-like structures.

Suggest the kingdom to which this species belongs   (*1 mark*)

c   Living organisms can also be classified into three groups called **domains.**

Outline the features of this system of classification compared with the five kingdom system.   (*3 marks*)

OCR F212/01 2012

4 Adaptations are characteristics which increase an organism's chance of survival and reproduction in an environment.

a State the difference between a behavioural and a physiological adaptation. *(1 mark)*

b Marram grass is commonly found on sand dunes around the coast of Great Britain.

State and explain three anatomical adaptations which enable this plant to survive in an environment with little access to water. *(3 marks)*

c Anatomical adaptations provide evidence for convergent evolution. Explain what is meant by the term 'convergent evolution'. *(2 marks)*

OCR F212/01 2013

5 Bats are the only mammals that can truly fly. Many species of bat hunt flying insects at night. Bats are able to use sound waves (echolocation) in order to help them find their prey in the dark.

a Suggest how the ability to use echolocation may have evolved from an ancestor that did not have that ability *(4 marks)*

The pipistrelle is the most common species of bat in Europe. It was originally thought that all pipistrelles belonged to the same species, *Pipistrellus pipistrellus*. However, in the 1990s, it was decided that there were two species: the common pipistrelle, *Pipistrellus pipistrellus* and the soprano pipistrelle, *Pipistrellus pygmaeus*.

Data for both species are provided in Table 2

species	mean body mass (g)	mean wingspan (m)	range of echolocation (kHz)	colour
Common pipistrelle	5.5	0.22	42–47	medium to dark brown
Soprano pipistrelle	5.5	0.21	52–60	medium to dark brown

b **(i)** Name the genus to which the soprano pipistrelle belongs *(1 mark)*

(ii) Using the data in Table 2, suggest why pipistrelles were originally classified as one species. *(1 mark)*

(iii) Describe how it is possible to confirm, over a longer period of time, whether two organisms belong to different species or the same species. *(2 marks)*

c The soprano pipistrelle has an echolocation call that is 'high pitched' (between 52 and 60 kHz). The common pipistrelle has an echolocation call that is 'low pitched' (between 42 and 47 kHz).

Variation within and between species can be a result of genetic or environmental factors. Whatever the causes of variation, the type of variation displayed can occur in two different forms.

Using the pipistrelle as an example, describe the key features of both **forms** of variation. *(7 marks)*

OCR F212/01 2012

You may be familiar with the term **biodiversity** – the variety of living organisms present in an area. Biodiversity includes plants, animals, fungi, and other living things. In fact, it includes everything from gigantic redwood trees to single-celled algae.

The importance of biodiversity

Biodiversity is essential in maintaining a balanced ecosystem for all organisms. All species are interconnected – they depend on one another. For example, trees provide homes for animals. Animals eat plants, which in turn need fertile soil to grow. Fungi and other microorganisms help decompose dead plants and animals, returning nutrients to the soil. In regions of reduced biodiversity, these connections may not all be present, which eventually harms all species in the ecosystem.

We rely on balanced ecosystems as they provide us with the food, oxygen and other materials we need to survive. Unfortunately, many human activities, such as farming and clearing land for housing, can lead to a reduction in biodiversity.

▲ **Figure 1** *Coral reefs are amongst the most biodiverse ecosystems on the planet*

Measuring biodiversity

Tropical, moist regions (that are warm all year round) have the most biodiversity. The UK's temperate climate (warm summers and cold winters) has less biodiversity. Very cold areas such as the Arctic, or very dry areas such as deserts, have the least biodiversity. Generally, the closer a region is to the Equator (the line of latitude of the Earth, halfway between the North Pole and South Pole), the greater the biodiversity. For example, over 40 000 plant species live in the Amazon rainforest, whereas less than 3000 live in Northern Canada.

Measuring biodiversity plays an important role in conservation. It informs scientists of the species that are present, thus providing a baseline for the level of biodiversity in an area. From this information, the effect of any changes to an environment can be measured. These may include the effect of human activity, disease or climate change, for example.

▲ **Figure 2** *There is very little biodiversity at the top of a high mountain*

Before a major project is undertaken, such as building a new road or the creation of a new nature reserve, an Environmental Impact Assessment (EIA) is undertaken. This assessment attempts to predict the positive and negative effects of a project on the biodiversity in that area.

Biodiversity can be studied at different levels:

- habitat biodiversity
- species biodiversity
- genetic biodiversity.

Habitat biodiversity

Habitat biodiversity refers to the number of different habitats found within an area. Each habitat can support a number of different species. Therefore in general, the greater the habitat biodiversity, the greater the species biodiversity will be within that area.

The UK is home to large number of habitat types, including meadow, woodland, streams, and sand dunes. It has a large habitat biodiversity. By contrast Antarctica, covered almost entirely by an ice sheet, has a very low habitat biodiversity and very few species live in this region.

On a smaller scale, countryside that is habitat rich, perhaps with a river, woodland, hedgerows and wild grassland, will be more species rich than farmed countryside with large ploughed fields making up a single uniform habitat.

Species biodiversity

Species biodiversity has two different components:

- species richness – the number of different species living in a particular area, and
- species evenness – a comparison of the numbers of individuals of each species living in a community. (The community is all the populations of living organisms in a particular habitat.)

Therefore an area can differ in its species biodiversity even if it has the same number of species. For example, a cornfield and a grass meadow may both contain 20 species. However, in the cornfield, corn will make up 95% of the community with the remaining 5% made up of other organisms including weed plants, insects, mice, and birds. In the grass meadow the species will be more balanced in their populations.

Genetic biodiversity

Genetic biodiversity refers to the variety of genes that make up a species. Humans have about 25 000 genes, but some species of flowering plants have as many as 400 000 genes. Many of these genes are the same for all individuals within a species. However, for many genes, different versions (alleles) exist. This leads to genetic biodiversity within a species (you will learn more about genetic biodiversity in Topic 11.5, Calculating genetic biodiversity).

Genetic biodiversity within a species can lead to quite different characteristics being exhibited. For example, some genes are the same for all breeds of dog – these genes define the organism as a dog. Some of the genes have many alleles – they code for the wide variation in characteristics seen between different breeds of dog, for example coat colour and length.

Greater genetic biodiversity within a species allows for better adaptation to a changing environment, and is more likely to result in individuals who are resistant to disease.

▲ **Figure 3** *These two butterflies appear to belong to different species; however, both are examples of the Gaudy Commodore butterfly (*Precis octavia*). Genetic biodiversity within this species leads to different wing patterns and colours. The actual colours displayed depend on the season in which the butterflies are born*

Study tip

There are many key terms in this topic – make sure you are clear on their meaning. Why don't you try making your own biodiversity glossary?

Summary questions

1 State the difference between species richness and species evenness. *(2 marks)*

2 Compare the biodiversity of an arid desert and a temperate coastline. *(3 marks)*

3 Suggest why greater genetic biodiversity increases a species' chances of long-term survival. *(4 marks)*

You can use a variety of techniques to measure and compare the biodiversity of different habitats. However, it is often impossible to count or measure all of the organisms present in an area, so sampling techniques are used.

What is sampling?

Sampling means taking measurements of a limited number of individual organisms present in a particular area.

Sampling can be used to estimate the *number* of organisms in an area without having to count them all. The number of individuals of a species present in an area is known as the *abundance* of the organism.

Sampling can also be used to measure a *particular characteristic* of an organism. For example, you cannot reliably determine the height of wheat by measuring one wheat plant in a farmer's field. However, if you measure the height of a number of plants and then calculate an average, your result is likely to be close to the average height of the entire crop.

After measuring a sample, you can use the results of the sample to make generalisations or estimates about the number of organisms, distribution of species or measured characteristic throughout the entire habitat.

Sampling can be done in two ways – random and non-random.

Random sampling

Random sampling means selecting individuals by chance. In a random sample, each individual in the population has an equal likelihood of selection, rather like picking names out of a hat.

To decide which organisms to study, random number tables or computers can be used. You have no involvement in deciding which organisms to investigate. For example, to take a random sample at a grass verge you could follow these steps:

1 Mark out a grid on the grass using two tape measures laid at right angles.

2 Use random numbers to determine the *x* coordinate and the *y* coordinate on your grid.

3 Take a sample at each of the coordinate pairs generated.

Non-random sampling

Non-random sampling is an alternative sampling method where the sample is not chosen at random. It can be divided into three main techniques:

● **Opportunistic** – this is the weakest form of sampling as it may not be representative of the population. Opportunistic sampling uses organisms that are conveniently available.

▲ **Figure 1** *This scientist is taking random soil samples in a field*

- **Stratified** – some populations can be divided into a number of strata (sub-groups) based on a particular characteristic. For instance, the population might be separated into males and females. A random sample is then taken from each of these strata proportional to its size.

- **Systematic** – in systematic sampling different areas within an overall habitat are identified, which are then sampled separately. For example, systematic sampling may be used to study how plant species change as you move inland from the sea. Systematic sampling is often carried out using a line or a belt transect. A **line transect** involves marking a line along the ground between two poles and taking samples at specified points, this can include describing all of the organisms which touch the line or distances of samples from the line. A **belt transect** provides more information; two parallel lines are marked, and samples are taken of the area between the two lines.

▲ **Figure 2** *These students are carrying out sampling by combining two different methods. The quadrat seen here is being used along a line transect. This is known as an interrupted belt transect. This is an example of systematic sampling*

Reliability

A sample is never entirely representative of the organisms present in a habitat. This may be due to the following:

- Sampling bias – the selection process may be biased. This may be by accident, or may occur deliberately. For example, you may choose to sample a particular area that has more flowers because it looks interesting. The effects of sampling bias can be reduced using random sampling, where human involvement in choosing the samples is removed.

- Chance – the organisms selected may, by chance, not be representative of the whole population. For example, a sample of five worms collected in a trap may be the five longest in the habitat. Chance can never be completely removed from the process, but its effect can be minimised by using a large sample size. The greater the number of individuals studied, the lower the probability that chance will influence the result. Therefore the larger the sample size, the more reliable the result.

Summary questions 🧪

1 State the difference between random and non-random sampling. *(1 mark)*

2 ⚙ Describe how you can increase the likelihood of a sample being a reliable representation of the population as a whole. *(4 marks)*

3 State and explain which type of sampling you would use to study:
   a how organisms differ throughout the length of a stream *(2 marks)*
   b the distribution of organisms on a school field. *(2 marks)*

You can use many different techniques to sample the living organisms present in a habitat and the environment in which they live, as you saw in the previous topic. The technique you choose is dependent on the information you require. At each sampling point you would normally use more than one technique, so that a range of data can be collected.

Sampling animals

The following techniques can be used to collect living animals for study later. Remember, all living organisms must be handled carefully and for as short a time period as possible. As soon as any sample animals have been identified, counted and measured if required, they must be released back into the habitat at the point they were collected.

- A pooter is used to catch small insects. By sucking on a mouthpiece, insects are drawn into the holding chamber via the inlet tube. A filter before the mouthpiece prevents them from being sucked into the mouth.

- Sweep nets are used to catch insects in areas of long grass.

- Pitfall traps are used to catch small, crawling invertebrates such as beetles, spiders and slugs. A hole is dug in the ground, which insects fall into. It must be deep enough that they cannot crawl out and covered with a roof-structure propped above so that the trap does not fill with rainwater. The traps are normally left overnight, so that nocturnal species are also sampled.

▲ **Figure 1** *This student is using a pooter to collect insects from a tree*

▲ **Figure 2** *This is a pitfall trap. The glass perspex cover prevents rain entering the trap and potentially causing any trapped insects to drown*

- Tree beating is used to take samples of the invertebrates living in a tree or bush. A large white cloth is stretched out under the tree. The tree is shaken or beaten to dislodge the invertebrates. The animals will fall onto the sheet where they can be collected and studied.

- Kick sampling is used to study the organisms living in a river. The river bank and bed is 'kicked' for a period of time to disturb the substrate. A net is held just downstream for a set period of time in order to capture any organisms released into the flowing water.

Sampling plants

Plants are normally sampled using a **quadrat**, which can also be used to pinpoint an area in which the sample of plants should be collected. Quadrats can also be used to sample slow-moving animals such as limpets, barnacles, mussels, and sea anemones.

There are two main types of quadrat:

- Point quadrat – this consists of a frame containing a horizontal bar. At set intervals along the bar, long pins can be pushed through the bar to reach the ground. Each species of plant the pin touches is recorded (Figure 3).

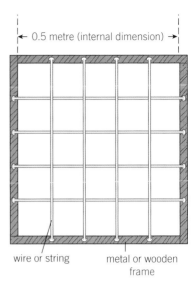

← 0.5 metre (internal dimension) →

wire or string   metal or wooden frame

▲ **Figure 4** *A frame quadrat*

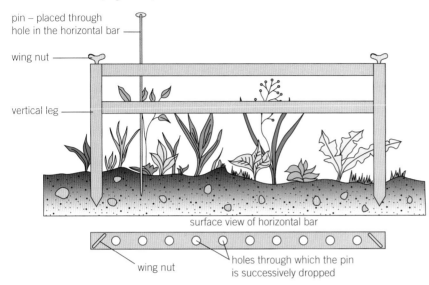

pin – placed through hole in the horizontal bar

wing nut

vertical leg

surface view of horizontal bar

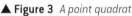

wing nut   holes through which the pin is successively dropped

▲ **Figure 3** *A point quadrat*

- Frame quadrat – this consists of a square frame divided into a grid of equal sections. The type and number of species within each section of the quadrat is recorded (Figure 4). Further details are given below.

To collect the most valid representative sample of an area, quadrats should be used following a random sampling technique (as discussed in Topic 11.2, Types of sampling). To study how the presence and distribution of organisms across an area of land varies, the quadrats can be placed systematically along a line or belt transect.

Measuring species richness

As you learnt in Topic 11.1, Biodiversity species richness is a measure of the number of different species living in a specific area. You should use a combination of the techniques described above to try to identify all the species present in a habitat. A list should be compiled of each species identified. The total number of species can then be calculated.

To enable scientists to accurately identify organisms, identification keys are often used. These may contain images to identify the organism, or a series of questions, which classify an organism into a particular species based on the presence of a number of identifiable characteristics.

Measuring species evenness

As you learnt in Topic 11.1, Biodiversity, species evenness refers to how close in numbers the populations of each species in an environment are. For example, 50 organisms are found living under

Synoptic link

You learned about how organisms are classified in Topic 10.1, Classification and 10.2, The five kingdoms.

a decaying log. Of these, 20 are woodlice, 15 are spiders, and 15 are centipedes – the community is quite evenly distributed between species. However, if the 50 insects comprised just 45 woodlice and 5 spiders, the community would be described as uneven.

Using frame quadrats

A frame quadrat is used to sample the population of plants living in a habitat. There are three main ways of doing this:

- Density – if individual large plants can be seen clearly, count the number of them in a 1m by 1m square quadrat. This will give you the density per square metre. This is an absolute measure, not an estimate as the following two methods.

- Frequency – this is used where individual members of a species are hard to count, like grass or moss. Using the small grids within a quadrat, count the number of squares a particular species is present in. For example, if clover is present in 65 out of 100 squares, the frequency of its occurrence is 65%. (Each square represents 1%.) Another commonly used quadrat contains 25 squares – in this case, each square represents 4% of the study area. Therefore if, during a sampling of grassland, eight quadrat squares contained buttercups, the frequency of occurrence would be 32%.

- Percentage cover – this is used for speed as lots of data can be collected quickly. It is useful when a particular species is abundant or difficult to count. It is an estimate by eye of the area within a quadrat that a particular plant species covers.

For each approach, samples should be taken at a number of different points. The larger the number of samples taken, the more reliable your results. You should then calculate the mean of the individual quadrat results to get an average value for a particular organism per m² (To calculate the mean value, sum the individual quadrat results, then divide by the number of samples taken). To work out the total population of an organism in an area that has been sampled, multiply the mean value per m² by the total area.

Synoptic link

You will learn how to use the Lincoln index to estimate population size in Topic 23.5, Measuring the distribution and abundance of organisms.

Estimating animal population size

As animals are constantly moving through a habitat and others may be hidden, it can be difficult to accurately determine their population size. A technique known as capture-mark-release-recapture is often used to estimate a population size. This involves capturing as many individuals of a species in an area as possible. The organisms are marked and then released back into the community. Time is allowed for the organisms to redistribute themselves throughout the habitat before another sample of animals is collected. By comparing the number of marked individuals with the number of unmarked individuals in the second sample, scientists can estimate population size. The greater the number of marked individuals recaptured, the smaller the population.

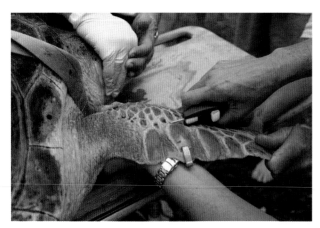

▲ **Figure 5** *An electronic tag is being fitted to this green sea turtle (Chelonia mydas). This is as part of a study to estimate population size*

The species evenness in an area can then be calculated by comparing the total number of each organism present. Populations of plants or animals that are similar in size or density represent an even community and hence a high species evenness. Species evenness can also be expressed as a ratio between the numbers of each organism present.

Measuring abiotic factors

Abiotic factors are the non-living conditions in a habitat. They have a direct effect on the living organisms that reside there. Examples are the amount of light and water available. To enable them to draw conclusions about the organisms present and the conditions they need for survival, scientists normally measure these conditions at every sampling point.

Table 1 summarises the ways in which common abiotic factors can be measured.

▼ Table 1

Abiotic factor	Sensor used	Example unit of measurement
wind speed	anemometer	$m\,s^{-1}$
light intensity	light meter	lx
relative humidity	humidity sensor	$mg\,dm^{-3}$
pH	pH probe	pH
temperature	temperature probe	°C
oxygen content in water	dissolved oxygen probe	$mg\,dm^{-3}$

Many abiotic factors can be measured quickly and accurately using a range of sensors, which are advantageous for a number of reasons:

- Rapid changes can be detected.
- Human error in taking a reading is reduced.
- A high degree of precision can often be achieved.
- Data can be stored and tracked on a computer.

 Belt transect in a National Park

A group of students was asked to investigate the impact of human land use in a National Park on the species of plants which were found there. The students decided to carry out a belt transect of open countryside, across a main walking path, in mid-Dartmoor.

Part of the data the students collected is shown. Data was collected on the percentage cover of six species, and the maximum height of vegetation at each position was noted.

Species present	Quadrat position along line, from starting point / m									
	0	1	2	3	4	5	6	7	8	9
Grasses	25	5	10	85	85	80	75	40	45	0
Heathers	10	15	0	0	0	0	0	10	10	35
Mosses	5	0	0	0	0	0	0	0	0	10
Gorse	20	75	90	10	0	0	0	30	0	25
Bracken	30	0	0	0	0	0	0	0	40	20
Bare ground	5	0	0	5	15	20	25	20	5	0
Maximum vegetation height / cm	70	65	45	10	10	5	5	30	70	65

The students produced the following graph of this data:

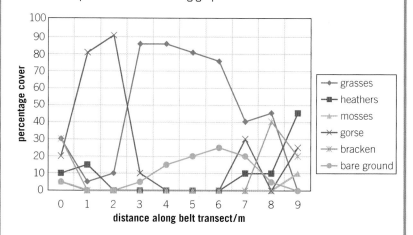

Questions

1 Describe how you would use a transect to obtain the data which the students collected.
2 Explain why random sampling would not be appropriate to develop this data.
3 Plot a graph showing how the maximum vegetation height varied with distance along the belt transect.
4 Using the data gathered by the students, suggest how land use in this region affects the vegetation present.
5 A range of abiotic factors may also affect the species in the region studied. Suggest and explain how you would investigate the effect of one of these factors on the vegetation present.
6 Evaluate the quality of the data collected by the students.

Summary questions

1 State which piece of equipment you would use to collect sample data on the following:
   a number of beetles on an oak tree trunk
   b number of moths in an area of woodland
   c pH of soil along a line transect
   d number of plant species present on a school playing field. (*4 marks*)

2 Describe the advantages of using a temperature probe over a thermometer. (*2 marks*)

3 Discuss the advantages and disadvantages of the different ways to measure species evenness. (*4 marks*)

Ecologists, such as those working for the Environment Agency, often perform calculations using specific formulae to determine the biodiversity of an area. One such calculation is a measure of the species diversity. The diversity of the organisms present in an area is normally proportional to the stability of the ecosystem, so the greater the species diversity the greater the stability. The most stable communities have large numbers of fairly evenly distributed species, in good-sized populations.

Pollution often reduces biodiversity. As a result of harsh conditions, a few species tend to dominate. If corrective steps are taken to improve environmental conditions, biodiversity levels usually increase. Monitoring biodiversity is therefore a useful tool in successful conservation and environmental management.

How to calculate biodiversity

The simplest way to measure biodiversity is to count up the number of species present – the species richness. However, this measure does not take into account the number of individuals present. Therefore in a meadow containing two daisies and 1000 buttercups, the daisies have as much influence on the richness of the area as 1000 buttercups. A community dominated by one or two species is considered to be less diverse than one in which several different species have a similar abundance.

Simpson's Index of Diversity (D) is a better measure of biodiversity as it takes into account both species richness, and species evenness.

It is calculated using the formula:

$$D = 1 - \sum \left(\frac{n}{N}\right)^2$$

where:

Σ = sum of (total)

N = the total number of organisms of all species and

n = the total number of organisms of a particular species.

When using a technique such as Simpson's Index of Diversity, scientists normally have to estimate population size using a variety of sampling techniques, such as using a quadrat to estimate the population of a plant species in an area.

Simpson's Index of Diversity always results in a value between 0 and 1, where 0 represents no diversity and a value of 1 represents infinite diversity. The higher the value of Simpson's Index of Diversity, the more diverse the habitat.

Learning outcomes

Demonstrate knowledge, understanding, and application of:

→ the use and interpretation of Simpson's Index of Diversity (D) to calculate the biodiversity of a habitat.

Synoptic link

You learnt about species richness and species evenness in Topic 11.1, Biodiversity.

Study tip

Be careful not to confuse species diversity with species richness. Species diversity takes into account both the number of different species present (species richness) and the number of individuals of each species.

Biodiversity values

What do low and high biodiversity values tell us about a habitat?

▼ **Table 1** *Typical habitat features for environments with low and high biodiversity*

Habitat features	Low biodiversity	High biodiversity
number of successful species	relatively few	a large number
nature of the environment	stressful and/or extreme with relatively few ecological niches	relatively benign/not stressful, with more ecological niches
adaptation of species to environment	relatively few species live in the habitat, often with very specific adaptations for the environment	many species live in the habitat, often with few specific adaptations to the environment
type of food webs	relatively simple	complex
effect of a change to the environment on ecosystem as a whole	major effects on the ecosystem	often relatively small effect

▲ **Figure 1** *A habitat with a low biodiversity value*

Although some habitats of low biodiversity are unable to support a large species diversity, those organisms that are present in the habitat can be highly adapted to the extreme environment of the habitat. These organisms may not survive elsewhere. It is therefore important to conserve some habitats with low biodiversity, as well as those with high biodiversity, in order to conserve rare species that may be too specialised to survive elsewhere.

 Worked example: Calculating Simpson's Index of Diversity

Individuals of the species living in a pond habitat were identified and counted (Table 2). They were sampled by sweeping a net through the pond.

▼ **Table 2** *Species identified in a pond*

Species	Number
water boatman	4
water strider	6
dragonfly larvae	3
mayfly larvae	8
caddisfly larvae	2

Calculate Simpson's Index of Diversity for this habitat.

Use the following formula:

$$D = 1 - \sum \left(\frac{n}{N}\right)^2$$

1   Calculate total number of organisms (N)

$N = 4 + 6 + 3 + 8 + 2 = 23$

2   Calculate $\left(\frac{n}{N}\right)^2$ for each organism. For example, for the water boatman:

n = number of organisms = 4

N = total number of organisms = 23

$\left(\frac{n}{N}\right)^2 = \left(\frac{4}{23}\right)^2 = 0.03$

3   Sum $\left(\frac{n}{N}\right)^2$ for all organisms

$\sum\left(\frac{n}{N}\right)^2 = 0.03 + 0.07 + 0.02 + 0.12 + 0.01 = 0.25$

▲ **Figure 2** *Ecologist studying biodiversity in a pond habitat*

4   Calculate Simpson's Index of Diversity

$D = 1 - \sum\left(\frac{n}{N}\right)^2 = 1 - 0.25 = 0.75$

This indicates that the pond habitat has a relatively high biodiversity.

Summary questions

1   Describe the conditions likely to be found in an area of high biodiversity. *(3 marks)*

▲ **Figure 3** *Pollution reduces biodiversity in a habitat. These trees have been killed by acid rain*

2   The organisms present in two pond habitats were sampled. The following values of Simpson's Index of Diversity were calculated:

Pond A: $D = 0.27$ Pond B: $D = 0.63$

a   Which pond habitat is the more biodiverse? *(1 mark)*

b   Which pond habitat is most likely to be polluted? Explain your answer. *(3 marks)*

3   A sample was taken of wildflowers growing in a meadow. Calculate Simpson's Index of Diversity for this habitat. *(4 marks)*

Species	Number
Bird's-foot trefoil	2
Crested dog's-tail	5
Meadow buttercup	9
Oxeye daisy	7
Rough hawkbit	2
Smaller cat's-tail	3

11.5 Calculating genetic biodiversity

Specification reference: 4.2.1

Maintaining genetic biodiversity is essential to the survival of a species. In isolated populations, such as those present within a captive breeding programme, genetic biodiversity is often reduced. This means that the individuals may suffer from a range of problems associated with in-breeding.

Scientists can calculate the genetic biodiversity of a population of a species (sometimes referred to as the gene pool) to monitor the health of the population and ensure its long-term survival.

The importance of genetic biodiversity

Within a species, *individuals* have very little variation within their DNA.

All members of the species share the same genes. However, they may have different versions of some of these genes. The different 'versions' of genes are called alleles. The differences in the alleles among individuals of a species creates genetic biodiversity within the species, or within a population of the species. The more alleles present in a population, the more genetically biodiverse the population.

Species that contain greater genetic biodiversity are likely to be able to adapt to changes in their environment, and hence are less likely to become extinct. This is because there are likely to be some organisms within the population that carry an advantageous allele, which enables them to survive in the altered conditions. For example, when a potentially fatal new disease is introduced to a population, all organisms will be killed unless individuals carry resistance to the disease. Those organisms are likely to survive the disease, and therefore be able to reproduce – leading to the survival of the species.

Factors that affect genetic biodiversity

For genetic biodiversity to increase, the number of possible alleles in a population must also increase. This can occur through:

- **mutation(s)** in the DNA of an organism, creating a new allele.
- interbreeding between different populations. When an individual migrates from one population and breeds with a member of another population, alleles are transferred between the two populations. This is known as **gene flow**.

In order for genetic biodiversity to decrease, the number of possible alleles in a population must also decrease. This can occur through:

- selective breeding (also known as artificial selection), where only a few individuals within a population are selected for their advantageous characteristics and bred. For example, the breeding of pedigree animals or of human food crops

- captive breeding programmes in zoos and conservation centres, where only a small number of *captive* individuals of a species are available for breeding. Often the wild population is endangered or extinct

- rare breeds, where selective breeding has been used historically to produce a breed of domestic animal or plant with characteristics which then become less popular or unfashionable, so the numbers of the breed fall catastrophically. When only a small number of individuals of a breed remain and are available for breeding, and all of these animals will have been selected for the specific breed traits, the genetic diversity of the remaining population will be low. This can cause serious problems when trying to restore numbers yet maintain breed characteristics, for example, a Gloucester Old Spot pig must have at least one spot on the body to be accepted into the registry of this rare breed

- artificial cloning (asexual reproduction), for example using cuttings to clone a farmed plant

- **natural selection.** As a result, species will evolve to contain primarily the alleles which code for advantageous characteristics. Over time, alleles coding for less advantageous characteristics will be lost from a population, or only remain in a few individuals.

- **genetic bottlenecks**, where few individuals within a population survive an event or change (e.g., disease, environmental change or habitat destruction), thus reducing the 'gene pool'. Only the alleles of the surviving members of the population are available to be passed on to offspring

- the **founder effect**, where a small number of individuals create a new colony, geographically isolated from the original. The gene pool for this new population is small.

- genetic drift, due to the random nature of alleles being passed on from parents to their offspring, the frequency of occurrence of an allele will vary. In some cases, the existence of a particular allele can disappear from a population altogether. Genetic drift is more pronounced in populations with a low genetic biodiversity.

▲ **Figure 1** *Gloucester Old Spot pig*

Synoptic link

An example of the founder effect is given in Topic 10.8, Changing population characteristics. You will learn more about genetic bottlenecks and the founder effect in Chapter 20, Patterns of inheritance and variation.

Synoptic link

You will find out about how polymorphic genes, and in particular your blood type, are inherited in Chapter 20, Patterns of inheritance and variation.

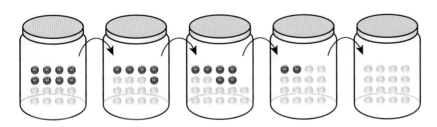

◀ **Figure 2** *The effect of genetic drift on genetic biodiversity can be shown using the 'marbles in a jar' analogy. At each generation, the random nature of alleles being passed on leads to a change in the frequency of the alleles present in the population. In this case, by the fifth generation the purple allele is removed from the population altogether*

Measuring genetic biodiversity

One way in which scientists quantify genetic biodiversity is by measuring polymorphism. Polymorphic genes have more than one

Summary questions

1 Describe how genetic biodiversity in a population can increase.   (*2 marks*)

2 Explain why it is advantageous for a species to be genetically biodiverse.   (*3 marks*)

3 ⚙ A scientist was studying two species of *Drosophila* (flies). DNA was extracted from each species and 25 gene loci compared.
For species A, 12 of the loci studied were polymorphic. For species B, 15 loci were polymorphic.

Use the data collected to explain which of the species was more genetically diverse.   (*4 marks*)

allele. For example, different alleles exist for the immunoglobulin gene, which plays a role in determining human blood type – this is therefore defined as a polymorphic gene. The three alleles are:

● I^A – resulting in the production of antigen A

● I^B – resulting in the production of antigen B

● I^O – resulting in the production of neither antigen

Most genes are not polymorphic. These genes are said to be monomorphic – a single allele exists for this gene. This ensures that the basic structure of individuals within a species remains consistent. The proportion of genes that are polymorphic can be measured using the formula:

$$\text{proportion of polymorphic gene loci} = \frac{\text{number of polymorphic gene loci}}{\text{total number of loci}}$$

(The locus (plural – loci) of a gene refers to the position of the gene on a chromosome.)

The greater the proportion of polymorphic gene loci, the greater the genetic biodiversity within the population.

🖩 Worked example: Measuring genetic biodiversity

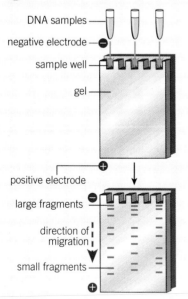

Gel electrophoresis is a technique used to separate fragments of DNA, based on their size. In this technique restriction enzymes are used to cut DNA into smaller pieces, which are then placed in a gel. The gel is placed between positive and negative electrodes, which cause the negatively charged DNA to move towards the positive side. The smaller the fragment of DNA, the further the movement through the gel. The pattern produced, known as a banding pattern, can be used to compare DNA samples from different individuals.

The following section of data was collected from the gel electrophoresis of five genetic loci within 20 individuals in an ibex (mountain goat) population. The five loci studied were labelled V, W, X, Y, and Z.

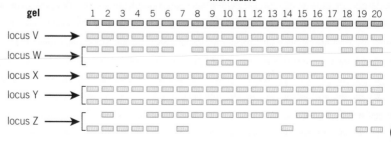

From this sample, the scientist could tell that:

- Loci W Y and Z were polymorphic
- Loci V and X were not polymorphic

Calculate the proportion of polymorphic gene loci for this mountain ibex population.

1 Count the number of polymorphic gene loci: 3 (W, Y and Z)

2 Count the total number of loci: 5 (V, W, X, Y and Z)

3 Calculate the proportion of polymorphic gene loci:

proportion of polymorphic gene loci

$$= \frac{\text{number of polymorphic gene loci}}{\text{total number of loci}}$$
$$= \frac{3}{5}$$
$$= 0.6$$

4 The proportion of polymorphic gene loci is often expressed
as a percentage:

percentage of polymorphic gene loci

$$= \text{proportion of polymorphic gene loci} \times 100$$
$$= 0.6 \times 100$$
$$= 60\%$$

11.6 Factors affecting biodiversity

Specification reference: 4.2.1

Learning outcomes

Demonstrate knowledge, understanding, and application of:

→ the factors affecting biodiversity, including human population growth, agriculture and climate change.

▲ **Figure 1** *This photo shows a remnant of rainforest surrounded by farmland near Iguacu National Park in Brazil. Rainforests are being destroyed at a rapid pace. Almost 90% of West Africa's rainforest has been destroyed and the island of Madagascar has lost two thirds of its original rainforest since humans arrived 2000 years ago. Removal of tropical rainforests causes the greatest loss of global biodiversity – even though rainforests cover less than 10% of the Earth's surface, they contain approximately 80% of the world's documented species*

Maintaining biodiversity is essential for preserving a balanced ecosystem for all organisms. As species are interconnected within an ecosystem, the removal of one species can have a profound effect on others. For example, it could lead to a loss of another species' food source or shelter.

As part of the human population you rely on biodiversity for many of the materials you need to survive, such as food, wood, and oxygen. However, humans are the leading cause of loss of biodiversity.

Human influence on biodiversity

The human population is growing at a dramatic rate. There are now over seven billion people living in the world, over double the number alive in the 1960s and over seven times more than in 1800. This increasing growth rate is linked to improvements in medicine, hygiene, housing, and infrastructure, which enable people to live for longer.

To create enough space for housing, industry, and farming to support the increasing population, humans are severely disrupting the ecology of many areas. The main problems are occurring as a result of:

- deforestation – the permanent removal of large areas of forest to provide wood for building and fuel (known as logging), and to create space for roads, building and agriculture.

- agriculture – an increasing amount of land has to be farmed in order to feed the growing population. This has resulted in large amounts of land being cleared and in many cases planted with a single crop (**monoculture**).

- climate change – there is much evidence that the release of carbon dioxide and other pollutants into the atmosphere from the burning of fossil fuels is increasing global temperatures.

Other forms of pollution result from industry and agriculture, such as the chemical pollution of waterways. The improper disposal of waste and packaging is a form of environmental pollution called littering.

Deforestation

Deforestation can occur naturally, for example as a result of forest fires caused by lightning or extreme heat and dry weather. However, most deforestation now occurs deliberately as a result of human action. Some areas of forest have also been destroyed indirectly by humans through acid rain, which forms as a result of pollutants being released into the atmosphere.

Deforestation affects biodiversity in a number of ways. For example:

- It directly reduces the number of trees present in an area.

- If only a specific type of tree is felled, the species diversity is reduced. For example, rosewood is often extracted from rainforests (it is used in the manufacture of furniture and guitars), but less useable trees may be left intact.

- It reduces the number of animal species present in an area as it destroys their habitat, including their food source and home. This in turn reduces the number of other animal species that are present, by reducing or removing their food source.

- Animals are forced to migrate to other areas to ensure their survival. This may result in the biodiversity of neighbouring areas increasing.

▲ **Figure 2** *This land is being cleared to make space for housing. As well as reducing biodiversity, burning the trees increases carbon dioxide levels in the atmosphere*

In some areas forests are now being replaced. Although this helps to restore biodiversity, generally only a few commercially viable tree species are planted. Therefore, biodiversity is still significantly reduced from its original level.

Agriculture

In general, farmers will only grow a few different species of crop plants, or rear just a few species of animals. Farmers often select the species based on characteristics that give a high yield (high levels of production), for example, wheat that produces the most grain or dairy cows that produce the most milk. The selection of only a few species greatly reduces the biodiversity of the area.

In order to be economically viable, once the farmers have selected their desired species, a number of techniques are used to produce as many of the desired species as possible, maximising food production. Unfortunately many of these techniques lead to a reduction in biodiversity, for example:

- Deforestation – to increase the area of land available for growing crops or rearing animals.

- Removal of hedgerows – as a result of mechanisation, farmers remove hedgerows to enable them to use large machinery to help them plant, fertilise, and harvest crops. It also frees up extra land for crop growing. This reduces the number of plant species present in an area and destroys the habitat of animals such as blackbirds, hedgehogs, mice and many invertebrates.

- Use of chemicals such as pesticides and herbicides. Pesticides are used to kill pests that would eat the crops or live on the animals. This reduces species diversity directly as it destroys the pest species (normally insects), and indirectly by destroying the food source of other organisms.

- Herbicides are used to kill weeds. A weed is any plant growing in an area where it is not wanted. Weeds are destroyed as they compete with the cultivated plants for light, minerals, and

▲ **Figure 3** *This hedgerow supports a large diversity of species. As well as destroying the habitats of many organisms, removing hedges or trees causes soil erosion. Hedges act as natural windbreaks. Once they are removed, when the fields are left bare during winter the soil can be blown or washed away*

water. By destroying weeds, plant diversity is reduced directly, and animal diversity may also be reduced by the removal of an important food source.

- Monoculture – many farms specialise in the production of only one crop, with many acres of land being used for the growth of one species. This has an enormous local effect in lowering biodiversity as only one species of plant is present. As relatively few animal species will be supported by only one type of plant, this results in low overall biodiversity levels. The growth of vast oil palm plantations is one of the leading causes of rainforest deforestation, leading to a loss of habitat for critically endangered species like the rhino.

▲ **Figure 4** *This rice field covers a huge area of land. It is an example of monoculture – only rice is being grown*

Climate change

In 2007, the Intergovernmental Panel on Climate Change (IPCC) released a report summarising scientists' current understanding of climate change. The report took six years to produce and involved over 2500 scientific personnel in its production. Some of the key findings include the following:

- The warming trend over the last 50 years (about 0.13° C per decade) is nearly twice that for the previous 100 years.

- The average amount of water vapour in the atmosphere has increased since the 1980s over land and ocean. The increase is broadly consistent with the extra water vapour that warmer air can hold.

- Since 1961, the average temperature of the global ocean down to depths of 3 km has increased. The ocean has been absorbing more than 80% of the heat added to the climate system, causing seawater to expand and contributing to sea-level rise.

- The global average sea level rose by an average of 1.8 mm per year from 1961 to 2003. There is high confidence that the rate of observed sea level rise increased from the 19th to the 20th century.

- Average Arctic temperatures have increased at almost twice the global average rate in the past 100 years.

- Mountain glaciers and snow cover have declined on average in both hemispheres. Widespread decreases in glaciers and ice caps have contributed to sea-level rise.

- Long-term upward trends in the amount of precipitation have been observed over many regions from 1900 to 2005.

To enable our understanding of climate change to develop, significant quantities of data have been developed charting changes to the Earth's climate over time. This has required an enormous international co-operative effort over many years. It is only on the basis of reliable, irrefutable evidence that decisions of an international significance can take place. Decisions made now may have far-reaching consequences for the populations of individual countries or continents today, as well as far-reaching global implications for the future.

The need to produce reliable data for issues of this scale is paramount. Despite the weight of evidence for climate change, some scientists still believe that a causal link between human activity and climate change is yet to be established.

Global warming refers to a rise in the Earth's mean surface temperature. The Earth's climate has shown fluctuations in temperature throughout its history, so it is not possible to say for certain that humans are directly causing global warming. However, carbon dioxide levels in the atmosphere have significantly increased since the industrial revolution, trapping more thermal energy in the atmosphere. Therefore most scientists believe that human activities are contributing to global warming.

If global warming continues biodiversity will be affected. For example:

- The melting of the polar ice caps could lead to the extinction of the few plant and animal species living in these regions. Some species of animals present in the Arctic are migrating further and further north to find favourable conditions as their habitat shrinks. Increasing global temperatures would allow temperate plant and animal species to live further north than currently.

- Rising sea levels from melting ice caps and the thermal expansion of oceans could flood low-lying land, reducing the available terrestrial habitats. Saltwater would flow further up rivers, reducing the habitats of freshwater plants and animals living in the river and surrounding areas.

- Higher temperatures and less rainfall would result in some plant species failing to survive, leading to drought-resistant species (**xerophytes**, Figure 5) becoming more dominant. The loss of non-drought-resistant species of plants would lead to the loss of some animal species dependent on them as a food source. These would be replaced by other species that feed on the xerophytes.

- Insect life cycles and populations will change as they adapt to climate change. Insects are key pollinators of many plants, so if the range of an insect changes, it could affect the lives of the plants it leaves behind, causing extinction. And as insects carry many plant and animal pathogens, if tropical insects spread, this in turn could lead to the spread of tropical diseases towards the poles.

If climate change is slow, species may have time to adapt (for example by eating a different food source) or to migrate to new areas. This will lead to a loss of native species, but in turn other species may move into the area – so biodiversity would not necessarily be lost. The species mix would simply change.

▲ **Figure 5** *These Joshua trees ([Yucca](#) [brevifolia](#)) are an example of a xerophyte. A xerophyte is a species of plant that has adapted to survive in an environment with little water, such as a desert or an ice- or snow-covered region in the Alps or the Arctic*

Synoptic link

You learned about xerophyte adaptations in Topic 9.5, Plant adaptations to water availability.

▲ **Figure 6** *The small red-eyed damselfly ([Erythromma](#) [viridulum](#)). Changing species distribution provides evidence for climate change. Since 1980, 34 of the 37 British species of dragonfly and damselfly have expanded their range northwards by an average of 74 km. This is evidence that the UK's climate is growing warmer*

Loss of biodiversity in the UK

▲ **Figure 7** *Heathland*

▲ **Figure 8** *Woodland*

Scientists have estimated that the present worldwide rate of extinction is between 100 and 1000 times greater than at any other point in evolutionary history. This is primarily the result of the increase in the world human population. This has resulted in large areas of land being cleared worldwide, to meet the demand for food. Twelve to fifteen million hectares of forest are lost worldwide each year – the equivalent of 36 football fields per minute. These highly diverse habitats are replaced with agricultural land, which has far lower levels of biodiversity.

Conservation agencies have estimated the percentage of various habitats that have been lost in the UK since 1900. Their findings are summarised in Table 1.

1 State the key reason why the habitats stated in Table 1 have been lost over the past century.

2 Compare the proportion of chalk grassland that has been lost since 1900 with lowland mixed woodland.

3 There are approximately 1500 hectares of hay meadow in the UK. What was the equivalent figure in 1900?

4 Farmers can receive financial subsidies for farming areas of land in a traditional, sustainable manner. State and explain how this may affect species diversity in these areas.

▼ **Table 1**

Habitat	Habitat loss since 1900 (%)	Main reason for habitat loss
Hay meadow	95	Conversion to highly productive grass and silage
Chalk grassland	80	Conversion to highly productive grass and silage
Lowland fens and wetlands	50	Drainage and reclamation of land for agriculture
Limestone pavements in England	45	Removal for sale as rockery stone
Lowland heaths on acid soils	40	Conversion to grasslands and commercial forests
Lowland mixed woodland	40	Conversion to commercial conifer plantations and farmland
Hedgerows	30	To make larger fields to accommodate farm machinery

Summary questions

1 State and explain how the following factors reduce species diversity:
   a monoculture (*2 marks*)
   b building of roads (*2 marks*)
   c use of pesticides. (*2 marks*)

2 Explain why there is a reduction in species diversity when an area of forest is cleared to create additional land for the grazing of cattle. (*2 marks*)

3 Suggest and explain ways in which climate change can affect the biodiversity in an area. (*3 marks*)

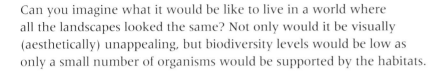

11.7 Reasons for maintaining biodiversity

Specification reference: 4.2.1

Can you imagine what it would be like to live in a world where all the landscapes looked the same? Not only would it be visually (aesthetically) unappealing, but biodiversity levels would be low as only a small number of organisms would be supported by the habitats.

Reasons for maintaining biodiversity

It is important to maintain biodiversity for a number of reasons. These reasons can be broadly arranged into three groups – aesthetic, economic and ecological, although some reasons fall across more than one category.

Aesthetic reasons

- The presence of different plants and animals in our environment enriches our lives. For example, you might like to relax on a beach, walk in your local woodland or park or visit a rainforest.

- The natural world provides inspiration for people such as musicians and writers, who in turn provide pleasure for many others through music and books.

- Studies have shown that patients recover more rapidly from stress and injury when they are supported by plants and a relatively natural environment.

Economic reasons

If biodiversity in an ecosystem is maintained, levels of long-term productivity are higher.

- Soil erosion and desertification may occur as a result of deforestation. These reduce a country's ability to grow crops and feed its people, which can lead to resource- and economic-dependence on other nations.

- It is important to conserve all organisms that we use to make things. Non-sustainable removal of resources, such as hardwood timber, will eventually lead to the collapse of industry in an area. Once all or enough of the raw material has been lost, it does not become economically viable to continue the industry. Note that even when 'sustainable' methods are used – for example replanting forest areas – the new areas will not be as biodiverse as the established habitats they replace.

- Large-scale habitat and biodiversity losses mean that species with potential economic importance may become extinct before they are even discovered. For example, undiscovered species in tropical rainforests may be chemically or medically useful. A number of marine species use a chemical-based defence mechanism. These are rich potential sources of new and economically important medicines.

Learning outcomes

Demonstrate knowledge, understanding, and application of:

→ the ecological, economic and aesthetic reasons for maintaining biodiversity.

▲ **Figure 1** *Most people would agree that natural, biodiverse areas are more attractive than acres of cultivated crops and concrete. Visiting these areas provides space for relaxation and exercise, both essential for a healthy life. Protecting these landscapes is therefore essential for human well-being*

- Continuous monoculture results in soil depletion – a reduction in the diversity of soil nutrients. It happens because the crop takes the same nutrients out of the soil year after year and is then harvested, not left for the nutrients to be recycled. This depletion of soil nutrients makes the ecosystem more fragile. The crops it can support will be weaker, increasing vulnerability to opportunistic insects, plant competitors, and microorganisms. The farmer will become increasingly dependent on expensive pesticides, herbicides, and fertilisers in order to maintain productivity.

- High biodiversity provides protection against abiotic stresses (including extreme weather and natural disasters) and disease. When biodiversity is not maintained, a change in conditions or a disease can destroy entire crops. The Irish potato famine of the 1840s was a direct consequence of the reliance on only two varieties of potato. When a new disease spread to the area (the oomycete *Phytophthora infestans*), neither species contained alleles for genetic resistance, so the entire crop was destroyed. This led to widespread famine and the deaths of around 1 million people.

- Areas rich in biodiversity provide a pleasing, attractive environment that people can enjoy. Highly biodiverse areas can promote tourism in the region, with its associated economic advantages.

- The greater the diversity in an ecosystem, the greater the potential for the manufacture of different products in the future. These products may be beneficial to humans. For example, it may make food production more financially viable or provide cures or treatment for disease.

- Plant varieties are needed for cross breeding, which can lead to better characteristics such as disease resistance or increased yield. The wild relatives of cultivated crop plants provide an invaluable reservoir of genetic material to aid the production of new varieties of crops. Also, through genetic engineering, scientists aim to use genes from wild plants and animals to make crop plants and animals more efficient, thus reducing the land required to feed more people. If these wild varieties are lost, the crop plants may themselves also become more vulnerable to extinction. This is also important ecologically.

Ecological reasons

- All organisms are interdependent on others for their survival. The removal of one species may have a significant effect on others, for example a food source or a place to live may be lost. For example, decomposers break down dead plant and animal remains, releasing nutrients into the soil, which plants later use for healthy growth. Plants rely on bees for pollination – this is important for both wild plant species and commercially produced crops. Fruit farmers use bees to pollinate their crops; a decrease in the wild bee population would decrease crop yields.

- Some species play a key role in maintaining the structure of an ecological community. These are known as **keystone species**. They have a disproportionately large effect on their environment relative to their abundance (in terms of their biomass or productivity). They affect many other organisms in an ecosystem and help to determine the species richness and evenness in the community. When a keystone species is removed the habitat is drastically changed. All other species are affected and some may disappear altogether. It is therefore essential to protect keystone species to maintain biodiversity. (See the Application for examples of keystone species.)

Human activity versus biodiversity

We have discussed the negative impact humans have on biodiversity, such as deforestation and clearing land for monoculture. However, human activity also plays an important role in increasing biodiversity. In many countries, including the UK, the natural habitat is created by human intervention and the management of land. For example, farming, grazing, planting of hedges, meadows, and forest management have changed the landscapes, the habitats and the ecology over thousands of years. Even the wildest of habitats, such as Dartmoor and the Scottish mountains, are a result of farmers and landowners managing the ecosystems.

One example is sheep grazing on downlands. This enables rare species like the Glanville fritillary (an orange patterned butterfly) to survive. By maintaining the grass at low levels it allows the plantains that the caterpillars feed on to thrive and therefore maintains biodiversity. Research has also shown that after annual controlled burning of gorse and heather in the New Forest (an area of lowland heath), biodiversity soars. If left to its own devices, bracken and pioneer tree species such as pine and silver birch would start to dominate. Areas of lowland heath worldwide are now rarer than rainforest and provide habitats for rare UK bird and reptile species such as the nightjar and sand lizard.

Keystone species

Sea stars, American alligators and prairie dogs are all examples of key stone species:

- Like many keystone species, sea stars are predators. They maintain a balanced ecosystem by limiting the population of other species. Sea stars eat mussels and sea urchins, which have no other natural predators. If the sea star is removed from the ecosystem, the mussels undergo a population explosion, reducing the number of other species present in an area (such as barnacles and limpets) as they compete for space and other resources. Similarly, if sea urchins are not eaten, their growing population crowds coral reefs, preventing other species from occupying the same area.

▲ **Figure 2** *Sea stars (*Pisaster ochraceus*) feeding on mussels. Sea stars are keystone species in the intertidal zones of the Pacific ocean*

- Alligators make burrows for nesting and to stay warm. When they abandon their burrow, fresh water fills the space, which is used by other species during the dry season for breeding and drinking. Alligators are predators, which also contributes to the maintenance of biodiversity in these habitats.

- It is estimated that up to 200 species rely on prairie-dog colonies, primarily due to their tunnelling activities. Prairie colonies provide a food source and burrows for other animals such as snakes. Their tunnelling aerates the soil, which, combined with their droppings, leads to a redistribution of nutrients. It also channels rainwater into the water table. These processes help to maintain a biodiverse range of plant life in the region. So essential is the prairie dog to its habitat that its loss would lead to a change in the ecosystem itself.

▲ **Figure 3** *American alligators (*Aligator mississippiensis*) are keystone species in the Everglade wetlands*

▲ **Figure 4** *Prairie dogs (*Cynomys *spp.) are a keystone species for the prairies – their existence adds to a diversity of life*

1 Define the term 'keystone species'.
2 Explain why keystone species are often predators.
3 The purple coneflower is a plant species found on the North American prairies. Extracts from the plant have anti-bacterial properties, which have been used for centuries to treat fevers and infections. Explain how a reduction in the population of prairie dogs could affect the number of purple coneflowers in this habitat.

Summary questions

1 State the differences between aesthetic, economic, and ecological arguments for maintaining biodiversity. (*1 mark*)

2 Suggest two ethical reasons why we should maintain biodiversity. (*3 marks*)

3 The Irish potato famine of the 1840s had a devastating effect on the population.
   a Explain how a lack of agricultural biodiversity led to this disaster. (*2 marks*)
   b Suggest and explain how a similar famine could be prevented from occurring in the future. (*2 marks*)

11.8 Methods of maintaining biodiversity

Specification reference: 4.2.1

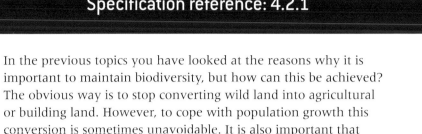

In the previous topics you have looked at the reasons why it is important to maintain biodiversity, but how can this be achieved? The obvious way is to stop converting wild land into agricultural or building land. However, to cope with population growth this conversion is sometimes unavoidable. It is also important that scientists try to repair damage that has already been done, and increase biodiversity.

Maintaining biodiversity

Conservation is the name given to the preservation and careful management of the environment and of natural resources. By conserving the natural habitat in an area, organisms' chances of survival are maintained, allowing them to reproduce. As a consequence species and genetic diversity can be safeguarded.

There are many different ways in which scientists try to conserve biodiversity. They can be divided into two main categories:

- *in situ* **conservation** – within the natural habitat
- *ex situ* **conservation** – out of the natural habitat.

Scientists are currently trying to conserve a number of species to prevent their extinction. Species are classified, for the purposes of conservation, according to their abundance in the wild:

- extinct – no organisms of the species exist anywhere in the world
- extinct in the wild – organisms of the species only exist in captivity
- endangered – a species that is in danger of extinction
- vulnerable – a species that is considered likely to become endangered in the near future.

Non-threatened and categories of least concern follow below. Many conservation techniques focus on increasing the numbers of organisms from species that are classified as endangered.

Scientists also promote the practice of **sustainable development** – economic development that meets the needs of people today, without limiting the ability of future generations to meet their needs.

In situ conservation

In situ conservation takes place inside an organism's natural habitat. This maintains not only the genetic diversity of species, but also the evolutionary adaptations that enable a species to adapt continually to changing environmental conditions, such as changes in pest populations or climate. By allowing the endangered species to interact with other species, it also preserves the interdependent relationships present in

Learning outcomes

Demonstrate knowledge, understanding, and application of:

→ *in situ* and *ex situ* methods of maintaining biodiversity

→ international and local conservation agreements made to protect species and habitats.

▲ **Figure 1** *The giant panda (*Ailuropoda melanoleuca*) is an example of an endangered species. Its numbers have been severely reduced by loss of habitat and poaching. Current estimates suggest there are less than 2000 giant pandas living in the wild*

Synoptic link

You will find more about sustainable development in Topic 24.5, Sustainability and about the ecology of environmentally sensitive regions in Topic 24.9, Environmentally sensitive ecosystems.

▲ **Figure 2** *This is Wistman's Wood on Dartmoor. It is a SSSI – a Site of Special Scientific Interest. The UK has over 4000 conservation areas where habitats are protected, covering around 8% of the nation's land*

▲ **Figure 3** *This white rhinoceros is having its horn removed in Umhlametsi Private Nature Reserve, South Africa as an anti-poaching measure. The horn is highly desirable for its use in ornaments and some traditional medicines*

▲ **Figure 4** *In Britain, the rhododendron has taken over large areas, virtually eliminating some native plants. Invasive plants and animals are the second greatest threat to biodiversity after habitat loss*

a habitat, therefore interlinked species may also be preserved. *In situ* conservation is generally cheaper than *ex situ* conservation.

Marine (saltwater), aquatic (freshwater) and terrestrial (land) nature reserves are examples of areas that have been specifically designated for the conservation of wildlife.

Wildlife reserves

Once an area has been designated as a wildlife reserve, active management is required. Active management techniques may include:

- controlled grazing – only allowing livestock to graze a particular area of land for a certain period of time to allow species time to recover, or keeping a controlled number of animals in a habitat to maintain it (see below)

- restricting human access – for example, not allowing people to visit a beach during the seal reproductive season, or by providing paths which must be followed to prevent plants being trampled

- controlling poaching – this includes creating defences to prevent access, issuing fines, or more drastic steps such as the removal of rhino horns

- feeding animals – this technique can help to ensure more organisms survive to reproductive age

- reintroduction of species – adding species to areas that have become locally extinct, or whose numbers have decreased significantly

- culling or removal of invasive species – an invasive species is an organism that is not native to an area and has negative effects on the economy, environment, or health. These organisms compete with native species for resources

- halting succession – **succession** is a natural process in which early colonising species are replaced over time until a stable mature population is achieved. For example, as a result of natural succession any piece of land left alone for long enough in the UK will develop into woodland. The only way to protect some habitats such as heath-, down- or moorland from becoming woodland is through controlled grazing. In different parts of the country ponies, deer, sheep, and cows eat tree seedlings as they appear, preventing succession from heathland to scrubland to woodland. This is an important role played by humans in maintaining some of our most beautiful habitats for future generations.

Marine conservation zones

Marine conservation zones are less well established than terrestrial ones. Lundy Island is currently the only statutory marine reserve in England, but there are many other protected areas.

Marine reserves are vital in preserving species-rich areas such as coral reefs, which are being devastated by non-sustainable fishing methods. The purpose of the marine reserve is not to prevent fisherman from visiting the entire area, but to create areas of refuge within which populations can build up and repopulate adjacent areas. Large areas of sea are required for marine reserves as the target species often move large distances, or breed in geographically different areas.

Ex situ conservation

Ex situ conservation involves the removal of organisms from their natural habitat. It is normally used in addition to *in situ* measures, ensuring the survival of a species.

Botanic gardens

Plant species can be grown successfully in botanic gardens. Here the species are actively managed to provide them with the best resources to grow, such as the provision of soil nutrients, sufficient watering, and the removal or prevention of pests.

There are roughly 1500 botanic gardens worldwide, holding 35 000 plant species. Although this is a significant number (more than 10% of the world's flora), the majority of species are not conserved. Many wild relatives of selectively bred crop species are under-represented amongst the conserved species. These wild species are a potential source of genes, conferring resistance to diseases, pests, and parasites.

Seed banks

A **seed bank** is an example of a gene bank – a store of genetic material. Seeds are carefully stored so that new plants may be grown in the future. They are dried and stored at temperatures of −20 °C to maintain their viability, by slowing down the rate at which they lose their ability to germinate. Almost all temperate seeds, and many tropical seeds, can be stored in this way. Scientists expect that they will remain viable for centuries, providing a back-up against the extinction of wild plants. The Svalbard 'Doomsday Vault' in Norway stores seeds in the permafrost and already houses around 800 000 species. It will eventually have 3 million different types of seeds and aims to provide a back-up against the extinction of plants in the wild by storing seeds for future reintroduction and research, for breeding and for genetic engineering in the future.

Seed banks don't work for all plants. Some seeds die when dried and frozen, and sadly the seeds of most tropical rainforest trees fall into this category.

Captive breeding programmes

Captive breeding programmes produce offspring of species in a human-controlled environment. These are often run and managed by zoos and aquatic centres. For example, The National Marine Aquarium in South West England is playing an important role in the conservation of sea horse species. Several species are now solely represented by animals in captivity.

Scientists working on captive breeding programmes aim to create a stable, healthy population of a species, and then gradually reintroduce the species back into its natural habitat. The Arabian Oryx is an example of a species that was extinct in the wild before its reintroduction.

Synoptic link

You will find more about succession in Topic 23.4, Succession.

▲ **Figure 5** *The Millennium Seed Bank Project at Kew Gardens contains over a billion seeds from over 34 000 species in underground frozen vaults. It is the world's largest collection of seeds and aims to provide a back-up against the extinction of plants in the wild by storing seeds for future use*

▲ **Figure 6** *These critically endangered Western lowland gorillas (*Gorilla gorilla gorilla*) are part of a New York Zoo captive breeding programme. Although far more invertebrates than vertebrates face extinction, most captive breeding programmes focus on vertebrates as people find it easier to relate to, and have sympathy with vertebrates. This generates financial support for their conservation and extends public education to wider issues*

Captive breeding programmes provide the animals with shelter, an abundant supply of nutritious food, an absence of predators and veterinary treatment. Suitable breeding partners or semen (which can be used to artificially inseminate females) can be imported from other zoos if not available within the zoo's own population.

Maintaining genetic diversity within a captive breeding population can be difficult. As only a small number of breeding partners are available, problems related to inbreeding can occur. To overcome this, an international catalogue is maintained, detailing genealogical data on individuals. Mating can thus be arranged to ensure that genetic diversity is maximised. Techniques such as artificial insemination, embryo transfer and long-term cryogenic storage of embryos allow new genetic lines to be introduced without having to transport the adults to new locations, and do not require the animals' cooperation.

Some organisms born in captivity may not be suitable for release in the wild. These are some of the reasons:

- Diseases – there may be a loss of resistance to local diseases in captive-bred populations. Also, new diseases might exist in the wild, to which captive animals have yet to develop resistance.

- Behaviour – some behaviour is innate, but much has to be learned through copying or experience. In an early case of reintroduction, a number of monkeys starved because they had no concept of having to search for food – they had become domesticated. Now food is hidden in cages, rather than just supplied, so that the animals learn to look for it.

- Genetic races – the genetic make-up of captive animals can become so different from the original population that the two populations cannot interbreed.

- Habitat – in many cases the natural habitat must first be restored to allow captive populations to be reintroduced. If only a small suitable habitat exists it is likely that there are already as many individuals as the habitat can support. The introduction of new individuals can lead to stress and tension as individuals fight for limited territory and resources such as food.

Conservation agreements

To conserve biodiversity successfully, local, and international cooperation is required to ensure habitats and individual species are preserved. Animals do not respect a country's boundaries. Therefore, to increase the chances of a species' survival, cross-border protections should be offered.

International Union for the Conservation of Nature

Intergovernmental organisations, such as the International Union for the Conservation of Nature (IUCN), assist in securing agreements between nations. At least once a year the IUCN publishes the Red List, detailing the current conservation status of threatened animals. Countries can then work together to conserve these species.

The IUCN was also involved in the establishment of the Convention on International Trade in Endangered Species (CITES). This treaty regulates the international trade of wild plant and animal specimens and their products. As the trade in wild animals and plants crosses borders between countries, the effort to regulate it requires international cooperation to safeguard certain species from over-exploitation. Today, more than 35 000 species of animals and plants are protected by this treaty.

The Rio Convention

In 1992, an historic meeting of 172 nations was held in Rio de Janeiro, which became known as the Earth Summit. The summit resulted in some new agreements between nations in the Rio Convention:

- The Convention on Biological Diversity (CBD) requires countries to develop national strategies for sustainable development, thus ensuring the maintenance of biodiversity.

- The United Nations Framework Convention on Climate Change (UNFCCC) is an agreement between nations to take steps to stabilise greenhouse gas concentrations within the atmosphere.

- The United Nations Convention to Combat Desertification (UNCCD) aims to prevent the transformation of fertile land into desert and reduce the effects of drought through programmes of international cooperation.

Each convention contributes to maintaining biodiversity. They are intrinsically linked, operating in many ecosystems and addressing interdependent issues.

Countryside stewardship scheme

Many conservation schemes are set up at a more local level. An example is the Countryside Stewardship Scheme in England. The scheme, which operated from 1991–2014, offered governmental payments to farmers and other land managers to enhance and conserve the English landscape. Its general aim was to make conservation a part of normal farming and land management practice. Specific aims of the scheme included:

- sustaining the beauty and diversity of the landscape

- improving, extending and creating wildlife habitats

- restoring neglected land and conserving archaeological and historic features

- improving opportunities for countryside enjoyment.

This scheme has now been replaced by the Environmental Stewardship Scheme, which operates similarly.

▲ **Figure 7** *A number of non-governmental organisations exist whose aim is to promote conservation and sustainability. One example is the World Wide Fund for Nature (WWF)*

Summary questions

1 Describe three methods of *ex situ* conservation. (*3 marks*)

2 Discuss the advantages and disadvantages of captive breeding programmes. (*4 marks*)

3 State and explain four techniques used in the active management of wildlife reserves. (*4 marks*)

4 Explain why local and international agreements can help to preserve biodiversity. (*4 marks*)

Practice questions

1   Which of the following can be used to measure the biodiversity of an area?

    A   Student's *t*-test

    B   Simpson's Index

    C   Correlation coefficient

    D   Standard deviation   (*1 mark*)

2   Which of the following statements are true with respect to what you would expect to find in an area of high biodiversity?

    Statement 1: A large number of successful species

    Statement 2: Complex food chains

    Statement 3: An extreme environment

    A   1, 2 and 3 are correct

    B   Only 1 and 2 are correct

    C   Only 2 and 3 are correct

    D   Only 1 is correct   (*1 mark*)

3   The table below contains a number of terms ecologists use when studying biodiversity. Complete the table using the appropriate term or description.   (*5 marks*)

Term	Description
a.	The variety of living organisms present in an area.
Genetic biodiversity	b.
c.	A sample produced without bias; each individual has an equal likelihood of selection.
Abiotic factor	d.
e.	A sampling technique where different areas within a habitat are identified and sampled separately; for example, by using a line transect.

4   When studying biodiversity, scientists often take samples of a habitat.

    a   (i)   State what is meant by the term sampling.   (*1 mark*)

       (ii)   State two reasons why sampling is carried out.   (*1 mark*)

    b   Describe the difference between random and non-random sampling, stating one advantage of each technique.   (*3 marks*)

    c   Explain how you could sample an area of grassland, to study the organisms present.   (*6 marks*)

5   a   The black poplar tree was once a common tree throughout southern Britain. Its numbers have decreased by 94% since 1942 and it is in danger of becoming extinct in the wild. There are thought to be approximately 2500 black poplars surviving in Britain today. Use the information above to calculate the original number of black poplar trees in 1942. Show your working.   (*2 marks*)

    b   Species such as the black poplar contribute to biodiversity in the UK.

       Suggest three reasons why the conservation of the black poplar is important.   (*3 marks*)

    c   Botanic gardens are important in the conservation of plant species.

       (i)   State why the conservation of a species in a botanic garden is described as *ex situ*.   (*1 mark*)

       (ii)   Many botanic gardens use seed banks as a method of plant conservation.

          Outline the advantages of using a seed bank, as opposed to adult plants, in order to conserve an endangered plant species.   (*4 marks*)

       (iii)   Suggest why it is important to ensure that, for each species, the seeds in a seed bank have been collected from several different sites in the wild.

           (*3 marks*) OCR F212 2012

6   On a biology field trip, a pair of students collected some data about plant species in an area of ash woodland. Their results are shown in the table.

species	Number of individuals (*n*)	*n/N*	(*n/N*)^2
Dog's mercury	40		
Wild strawberry	13	0.13	0.0169
Common avens	43		
Wood sorrel	4		
	N =		
			$\sum (n/N)^2$ =
			$1 - [\sum (n/N)^2]$ =

    a   (i)   Use the information in the table to work out Simpson's index of diversity (*D*) for the area of woodland sampled using the formula: $D = 1 - (\sum (n/N)^2)$

          Copy and complete the table.

(ii) Simpson's index of diversity takes into account both species richness and species evenness. In a school exercise book a student wrote the following definitions:

Species richness is a measure of the amount of species in an area

Species evenness shows how many individuals there are of a species in an area.

The teacher did not award a mark for either of these statements. Suggest how each statement could be improved. *(2 marks)*

(iii) if the value for Simpson's Index of Diversity is high, this indicates that the biodiversity of the habitat is high.

Outline the **implications** for a habitat if the Simpson's Index of Diversity is **low**. *(2 marks)*

b when collecting data on the field trip, the students placed quadrats in 15 locations and calculated a mean number of plants for each species.

Suggest two **other** steps they could have taken to ensure that their value for Simpson's Index of Diversity was as accurate as possible. *(2 marks)*

OCR F212/01 2013

7 Scientists have identified approximately 1.8 million different species. The number of species that actually exist is likely to be significantly higher than 1.8 million.

a Suggest two reasons why the number of species identified is likely to be lower than the actual number of species present on Earth. *(2 marks)*

b Many organisations, such as the International Union for the Conservation of Nature (IUCN), gather annual data about the number of species that are known to exist and to what extent they are considered to be endangered.

Figure 1 shows the total number of species assessed by the IUCN over a 10 year period and the number of those species assessed that are considered to be threatened with extinction.

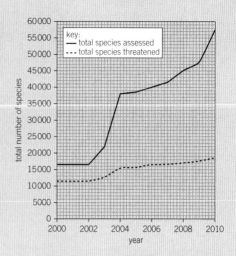

(i) Using the graph, **compare** the changes in the total number of species assessed with the changes in the total number of threatened species over the 10 year period. *(3 marks)*

(ii) Using the graph, calculate the percentage of species assessed that were threatened with extinction in **2010**. Show your working. Give your answer to the **nearest whole number**. *(2 marks)*

(iii) Suggest explanations for the shape of the two curves between **2005** and **2010**. *(2 marks)*

c A study of the biodiversity of an area considers not only the total number of species but also the relative number of individuals within each species.

State **one** relative factor that could be taken into account when describing the biodiversity of an area. *(1 mark)*

d In any attempt to protect global diversity, cooperation between countries is important.

Two examples of such international cooperation are:

- Convention on International trade in Endangered Species (CITES)

- Rio Convention on Biological Diversity.

Other than the conservation of biodiversity, state **two** aims for each of these conventions. *(4 marks)*

OCR F212/01 2013

12 COMMUNICABLE DISEASES
12.1 Animal and plant pathogens
Specification reference: 4.1.1

Learning outcomes

Demonstrate knowledge, understanding, and application of:

→ the different types of pathogen that can cause communicable diseases in plants and animals.

Communicable diseases are caused by infective organisms known as **pathogens**. Pathogens include bacteria, viruses, fungi, and protoctista. Each has particular characteristics that affect the way they are spread and the ways we can attempt to prevent or cure the diseases they cause.

A communicable disease can be passed from one organism to another. In animals they are most commonly spread from one individual of a species to another, but they can also be spread between species. Communicable diseases in plants are spread directly from plant to plant. **Vectors**, which carry pathogens from one organism to another, are involved in the spread of a number of important plant and animal diseases. Common vectors include water and insects.

Globally, around 13 million people a year die as a result of communicable diseases. That is 23% of all deaths – non-communicable diseases cause around 68% of deaths, and injuries cause the rest. Communicable diseases are also a major problem in domestic and wild animals, and in the plants on which life on Earth depends.

Types of pathogens

Bacteria

There are probably more bacteria than any other type of organism. A small proportion of these bacteria are pathogens, causing communicable diseases.

Bacteria are prokaryotes, so they have a cell structure that is very different from the eukaryotic organisms they infect. They do not have a membrane-bound nucleus or organelles.

Bacteria can be classified in two main ways:

Synoptic link

You learnt about prokaryotic cell structure in Topic 2.6, Prokaryotic and eukaryotic cells.

- By their basic shapes – they may be rod shaped (bacilli), spherical (cocci), comma shaped (vibrios), spiralled (spirilla), and corkscrew (spirochaetes).

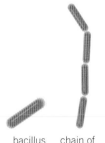

| bacillus (rod) | chain of bacilli (known as streptobacilli) | coccus (spherical) | pair of cocci | chain of cocci (streptococci) | cluster of cocci (staphylococci) | vibrio (comma) | spirillum (spiral) | spirochaete (corkscrew) |

▲ **Figure 1** *The main types of bacteria*

- By their cell walls – the two main types of bacterial cell walls have different structures and react differently with a process called Gram staining. Following staining **Gram positive bacteria** look purple-blue under the light microscope, for example methicillin-resistant *Staphylococcus aureus* (MRSA). **Gram negative bacteria** appear red, for example the gut bacteria *Escherichia coli* (*E.coli*). This is useful because the type of cell wall affects how bacteria react to different **antibiotics** (a compound that kills or inhibits the growth of bacteria).

Synoptic link

You learnt about the structure of proteins and nucleic acids in Topic 3.1 Biological elements.

Viruses

Viruses are non-living infectious agents. At 0.02–0.3 µm in diameter, they are around 50 times smaller in length than the average bacterium. The basic structure of a virus is some genetic material (DNA or RNA) surrounded by protein. Viruses invade living cells, where the genetic material of the virus takes over the biochemistry of the host cell to make more viruses. Viruses reproduce rapidly and evolve by developing adaptations to their host, which makes them very successful pathogens. All naturally occurring viruses are pathogenic. They cause disease in every other type of organism. There are even viruses that attack bacteria, known as **bacteriophages**. They take over the bacterial cells and use them to replicate, destroying the bacteria at the same time. People now use bacteriophages both to identify and treat some diseases, and they are very important in scientific research. Medical scientists consider viruses to be the ultimate **parasites**.

protein coat
nucleic acid strand
the human immunodeficiency virus (HIV)
Herpes simplex virus causes cold sores
tobacco mosaic virus infects tobacco plants
bacteriophages are viruses which infect bacteria

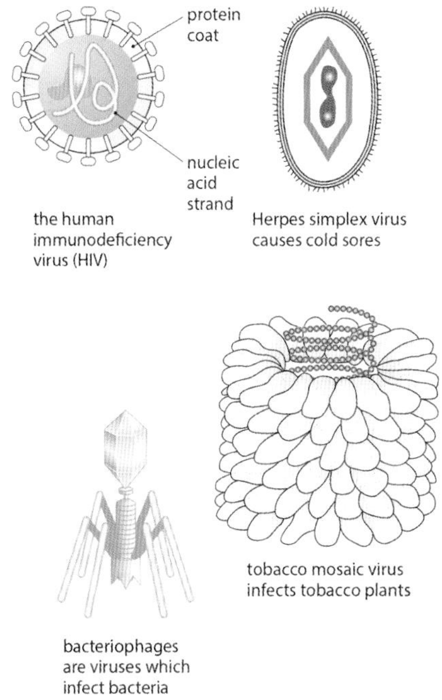

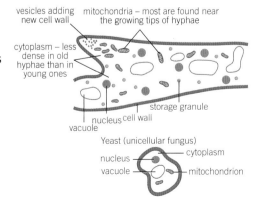

▲ **Figure 2** *The structure of viruses*

Protoctista (protista)

The protoctista (now widely known as protista) are a group of eukaryotic organisms with a wide variety of feeding methods. They include single-celled organisms and cells grouped into colonies. A small percentage of protoctista act as pathogens, causing devastating communicable diseases in both animals and plants. The protists which cause disease are parasitic – they use people or animals as their host organism. Pathogenic protists may need a vector to transfer them to their hosts – malaria and sleeping sickness are examples – or they may enter the body directly through polluted water – amoebic dysentery and *Giardia* are examples of these.

Fungi

Fungal diseases are not a major problem in animals, but they can cause devastation in plants. Fungi are eukaryotic organisms that are often multicellular, although the yeasts which cause human diseases such as thrush are single-celled. Fungi cannot photosynthesise and they digest their food extracellularly before absorbing the nutrients. Many fungi are saprophytes which means they feed on dead and decaying matter. However some fungi are parasitic, feeding on living plants and animals. These are the pathogenic fungi which cause communicable diseases. Because fungal infections often affect the leaves of plants, they stop them photosynthesising and so can quickly kill the plant. When fungi reproduce they produce millions of tiny spores which can spread huge distances, this adaptation

Fungal hyphae (filamentous fungus)
vesicles adding new cell wall
mitochondria – most are found near the growing tips of hyphae
cytoplasm – less dense in old hyphae than in young ones
storage granule
nucleus
cell wall
vacuole
Yeast (unicellular fungus)
cytoplasm
nucleus
vacuole
mitochondrion

▲ **Figure 3** *The structure of fungi*

means they can spread rapidly and widely through crop plants. Fungal diseases of plants cause hardship and even starvation in many countries around the world.

Pathogens – modes of action

Damaging the host tissues directly

Many types of pathogen damage the tissues of their host organism. It is this damage, combined with the way in which the body of the host responds to the damage, that causes the symptoms of disease. Different types of pathogens attack and damage the host tissues in different ways:

- Viruses take over the cell metabolism. The viral genetic material gets into the host cell and is inserted into the host DNA. The virus then uses the host cell to make new viruses which then burst out of the cell, destroying it and then spread to infect other cells

- Some protoctista also take over cells and break them open as the new generation emerge, but they do not take over the genetic material of the cell. They simply digest and use the cell contents as they reproduce. Proctists which cause malaria are an example of this.

- Fungi digest living cells and destroy them. This combined with the response of the body to the damage caused by the fungus gives the symptoms of disease.

Producing toxins which damage host tissues:

- Most bacteria produce toxins that poison or damage the host cells in some way, causing disease. Some bacterial toxins damage the host cells by breaking down the cell membranes, some damage or inactivate enzymes and some interfere with the host cell genetic material so the cells cannot divide. These toxins are a by-product of the normal functioning of the bacteria

- Some fungi produce toxins which affect the host cells and cause disease.

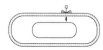

1. attachment of virus to host cell   2. insertion of viral nucleic acid

3. replication of viral nucleic acid   4. synthesis of viral protein

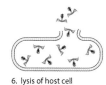

5. assembly of virus particles   6. lysis of host cell

▲ **Figure 4** *This virus is a bacteriophage and it destroys the bacterial cell as the new viruses burst out. Viruses have the same effect in multicellular eukaryotic organisms such as animals and plants*

Summary questions

1 a Explain what is meant by the term 'communicable disease'. (*2 marks*)
  b Produce a pie chart to show the main causes of death worldwide each year. (*4 marks*)

2 Make a table to compare bacteria, viruses, protoctista, and fungi. (*4 marks*)

3 If approximately 13 million people die of communicable diseases each year, give an approximate figure for the numbers who die of non-communicable disease and explain how you arrived at your answer. (*4 marks*)

4 Give four ways in which pathogens can attack the cells of their host organism and cause disease. (*6 marks*)

5 a Explain the difference between the way viruses and protists cause disease. (*4 marks*)
  b Suggest why viruses are described as the ultimate parasite. (*2 marks*)

Plant diseases

Plant diseases threaten people, because when crop plants fail, people suffer. They may starve, economies may struggle and jobs are lost. Plant diseases threaten ecosystems too – entire species can be threatened.

Learning outcomes

Demonstrate knowledge, understanding, and application of:

→ the different types of pathogens that can cause communicable diseases in plants and animals.

The threat to English oak trees

Oak woodlands are a traditional part of the British countryside. These mighty trees (*Quercus robur*) can live for centuries and are home to up to 284 species of insects. As many as 324 different lichens have been identified on a single tree. Future generations, however may not see these trees in the countryside – they are under threat from a new disease. Acute oak decline first appeared in the UK the 1980s, having spread from continental Europe. It causes dark fluid to ooze from the bark, with a rapid decline in the tree and often death.

Scientists still do not know exactly what the cause of this deadly tree disease is – which makes it very difficult to understand how to prevent it spreading. Some of the evidence so far includes:

- The discovery of previously unknown bacteria in the tree which may play a role in the disease
- Evidence of oak jewel beetle activity in infected trees – they may be important in disease development. For example, they may act as a vector or their presence may just be a coincidence

A massive research project is currently underway involving DNA analysis of the microorganisms on infected and healthy oak trees, along with a careful study of the behaviour of the oak jewel beetle. For now, the advice is to try and avoid spreading the disease by careful hygiene procedures for both people who work in oak woodlands and their machinery.

1 Suggest two pieces of evidence which would help to show whether oak jewel beetles are vectors of Acute oak decline or not.

Plant diseases are caused by a range of pathogens. They include:

Disease	Effect on plants
Ring rot – a bacterial disease of potatoes, tomatoes, and aubergines caused by the Gram positive bacterium *Clavibacter michiganensis*. It damages leaves, tubers and fruit. It can destroy up to 80% of the crop and there is no cure. Once bacterial ring rot infects a field it cannot be used to grow potatoes again for at least two years.	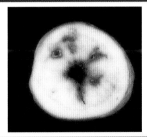 ◀ **Figure 1** *Bacterial ring rot infects a field so it cannot be used to grow potatoes for at least two years*

Disease	Effect on plants
Tobacco mosaic virus (TMV) – a virus that infects tobacco plants and around 150 other species including tomatoes, peppers, cucumbers, petunias and delphiniums. It damages leaves, flowers and fruit, stunting growth and reducing yields, and can lead to an almost total crop loss. Resistant crop strains are available but there is no cure.	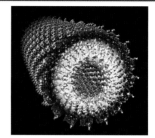 ▲ **Figure 2** *TMV and its effects*
Potato blight (tomato blight, late blight) – caused by the fungus-like protoctist oomycete *Phytophthora infestans*. The hyphae penetrate host cells, destroying leaves, tubers and fruit, causing millions of pounds worth of crop damage each year. There is no cure but resistant strains, careful management and chemical treatments can reduce infection risk.	 ◀ **Figure 3** *Tomato blight caused by the protoctista* Phytophthora infestans
Black sigatoka – a banana disease caused by the fungus *Mycosphaerella fijiensis*, which attacks and destroys the leaves. The hyphae penetrate and digest the cells, turning the leaves black. If plants are infected it can cause a 50% reduction in yield. Resistant strains are being developed – good husbandry and **fungicide** (a chemical that kills fungi) treatment can control the spread of the disease but there is no cure.	 ▲ **Figure 4** *Banana leaf infected by Black sigatoka*

➕ Banana diseases and food security ⚙️

Food security is one of the biggest issues globally. In an ideal world everyone would consistently have a balanced diet provided in a sustainable way. One of the main problems, however, for many people is getting enough to eat. If plant diseases threaten staple crops such as rice, maize, cassava, and bananas then they threaten food security and the survival of the population.

- Bananas are grown in over 130 countries where they are important both as a food crop and economically as a cash crop. They are the 4th most important crop in the developing world after rice, wheat, and maize

- In East Africa bananas (known as plantains) are the staple food for around 50% of the population. People eat around 400 kg of bananas per year

- 90% of the bananas cultivated are produced on small farms and eaten locally. In recent years, as a result of Black Sigatoka there has been a 40% fall in banana yields

- Around 10% of bananas are produced on big plantations for Western supermarkets. These are all from the same clone of a variety called Cavendish so they are genetically very similar. Black Sigatoka is invading these plantations too.

1   There are around 150 million people in East Africa. How many bananas need to be cultivated to feed the 50% of the population for whom bananas are their staple diet?

2   Suggest the effect on Black Sigatoka on the population of East Africa

3   What problems are likely to affect the control of the disease on

   a   small local farms

   b   large plantations of Cavendish bananas

Animal diseases

The diseases that affect animals – and in particular human beings – have a profound effect on human health and wellbeing – and on national economies. Communicable diseases range from mild to fatal. Examples include:

Tuberculosis (TB)

A bacterial disease of humans, cows, pigs, badgers, and deer commonly caused by *Mycobacterium tuberculosis* and *M. bovis*. TB damages and destroys lung tissue and suppresses the immune system, so the body is less able to fight off other diseases. Worldwide in 2012 around 8.6 million people had TB of which 1.3 million died. The global rise of HIV/AIDS has had a big impact on the numbers of people also suffering from diseases such as TB, because people affected by HIV/AIDS are much more likely to develop TB infections. In people TB is both curable (by antibiotics) and preventable (by improving living standards and vaccination).

TB, cows, and badgers

TB affects animal populations. In 2013, almost 33 000 UK cattle were destroyed because they were infected with bovine TB. There is clear evidence that TB is passed from wild animals, such as badgers or possums, to cattle, and vice versa. This presents a problem as cattle can be tested and culled, but it is very difficult to prevent them becoming re-infected from wildlife, particularly when they are out at pasture.

Scientists are still unsure how this wildlife infection can best be controlled. One method is to cull the wildlife source – in countries where this has been done, TB rates in cattle have fallen substantially, however it must be carried out carefully and thoroughly or it can lead to greater disease spread as animals are dispersed.

Some people, however, feel culling is not an acceptable approach and vaccination of either cattle or the wild animals is a better route. The test for TB cannot currently distinguish between an infected animal and a vaccinated animal, so current EU law bans cattle vaccines. Research is continuing on an improved version of both the vaccine and test. Vaccinating a population of wild animals is not an easy task, and it is as yet an unproven method to control the spread of disease.

The problem of TB in animals will not be solved easily, but the research and the debate continues.

1 Why do you think it is so difficult to prevent re-infection of cattle when they are out grazing in the fields?

2 Suggest why EU law bans the vaccination of cattle against TB

3 Why do you think that people are against culling badgers, when thousands of cattle infected with TB are slaughtered each year?

4 What might be the difficulties of vaccinating a wild population of animals?

5 Investigate the impact of TB on infected cows and badgers

- Bacterial meningitis – a bacterial infection (commonly *Streptococcus pneumoniae* or *Neisseria meningitidis*) of the meninges of the brain (protective membranes on the surface of the brain), which can spread into the rest of the body causing septicaemia (blood poisoning) and rapid death. It mainly affects very young children and teenagers aged 15–19. They have different symptoms but in both, a blotchy red/purple rash that does not disappear when a glass is pressed against it is a symptom of septicaemia and immediate medical treatment is needed. About 10% of people infected will die. Up to 25% of those who recover have some permanent damage. Antibiotics will cure the disease if delivered early. Vaccines can protect against some forms of bacterial meningitis.

- HIV/AIDS (acquired immunodeficiency syndrome) – caused by HIV (human immunodeficiency virus), which targets T helper cells in the immune system of the body (see Topic 12.6, The specific immune system). It gradually destroys the immune system so affected people are open to other infections, such as TB and pneumonia, as well as some types of cancer. HIV/AIDS can affect humans and some non-human primates. HIV is a retrovirus with RNA as its genetic material. It contains the enzyme reverse transcriptase, which transcribes the RNA to a single strand of DNA to produce a single strand of DNA in the host cell. This DNA interacts with the genetic material of the host cell. The virus is passed from one person to another in bodily fluids, most commonly through unprotected sex, shared needles, contaminated blood products and from mothers to their babies during pregnancy, birth or breast feeding. In 2012 around 35 million people worldwide were living with HIV infection and about 1.6 million died of the disease. There is as yet no vaccine and no cure, but anti-retroviral drugs slow the progress of the disease to give many years of healthy life. Girls and women are at particularly high risk of HIV/AIDS in many countries. Traditional practices such as female genital mutilation (FGM) increase the infection rate – if the same equipment is used multiple times then this can spread the infection, in addition, women who have undergone FGM are also more vulnerable to infection during intercourse. Sub-Saharan Africa is the region worst affected by HIV/AIDS, with 25 million people living with HIV/AIDS – around 70% of the global total. This disease has massive social and economic consequences as well as the personal impact to each person infected.

- Influenza (flu) – a viral infection (*Orthomyxoviridae* spp.) of the ciliated epithelial cells in the gas exchange system. It kills them, leaving the airways open to secondary infection. Flu can be fatal, especially to young children, old people and people with chronic illnesses. Many of these deaths are from severe secondary bacterial infections such as pneumonia on top of the original viral infection. Flu affects mammals, including humans and pigs, and birds, including chickens. There are three main strains – A, B and C. Strain

Synoptic link

You will learn more about TB in Topic 12.3, The transmission of communicable diseases and Topic 12.7, Preventing and treating disease.

Synoptic link

You learnt about ciliated epithelial cells of the gaseous exchange system in Topics 6.4, Cell specialisation and levels of organisation.

A viruses are the most virulent and they are classified further by the proteins on their surfaces, for example A(H1N1) and A(H3N3). Flu viruses mutate regularly. The change is usually quite small, so having flu one year leaves you with some immunity for the next. Every so often, however there is a major change in the surface antigens and this heralds a flu epidemic or pandemic as there are no antibodies available. Vulnerable groups are given a flu vaccine annually to protect against ever changing strains. There is no cure.

Zoonotic Influenza

A disease which people can catch from animals is known as a zoonosis. Influenza, for example attacks a range of animals including birds and pigs. Sometimes the virus which causes bird flu or swine (pig) flu mutates and becomes capable of infecting people. These new strains can be particularly serious, because few people have any natural immunity to them.

In March 2009 60% of the population of a small town in Mexico became infected with a new disease and two babies died. Some of those infected tested positive for H1N1, a form of flu usually found in pigs, rather than the usual human flu strains.

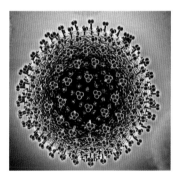

▲ *The H1N1 virus which caused a flu pandemic in 2009*

The outbreak spread to the US, where using DNA analysis techniques, the virus was identified as a new mutant strain of the H1N1 swine flu virus, which had not been seen before in either pigs or people. Three months after it first appeared people were infected with H1N1 flu in 62 countries, and some of them were dying. None of the available flu vaccines were any use against this zoonotic virus.

Within five months almost 3000 people around the world had died.

Fortunately, only six months after swine flu H1N1 was first identified, scientists produced an effective vaccine. In spite of this, recent analyses of the data suggest between 200 000–300 000 people died as a result of H1N1 infection in the 2009 outbreak – and up to 80% of those deaths were in people aged 65 and younger. In a normal seasonal flu outbreak, only around 10% of deaths occur in people who are under 65.

H1N1 is now part of the normal seasonal flu vaccine and scientists remain on the lookout for the next mutation which may enable the flu virus to pass from pigs or birds to people, known as a species jump.

1 H1N1 is a virus not a bacterium. Why does this make it so much more dangerous?
2 Explain how modern DNA technology helps in the case of a zoonotic disease outbreak such as this
3 a How did the age profile of the people who died from H1N1 differ from the normal pattern of flu-related deaths?
  b Suggest reasons for this difference.

- Malaria – caused by the protoctista *Plasmodium* and spread by the bites of infected *Anopheles* mosquitoes (the vector – Topic 12.3). The *Plasmodium* parasite has a complex life cycle with two hosts – mosquitoes and people. They reproduce inside the female mosquito. The female needs to take two blood meals to provide her with protein before she lays her eggs – and this is when *Plasmodium* is passed on to people. It invades the red blood cells, liver, and even the brain. Around 200 million people are reported to have malaria

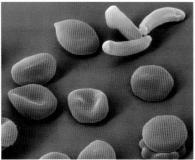

▲ **Figure 5** *Top: Only female* <u>Anopheles</u> *mosquitoes (top) spread malaria – they need two blood meals before they lay their eggs. When infected females feed they transmit the parasite to the human host – bottom: scanning electron microscope showing red blood cells and* <u>Plasmodium falciparum</u> *protozoa. × 500 magnification.*

Synoptic link

You will learn about the principles of DNA sequencing and the development of new DNA sequencing techniques in Chapter 21, Manipulating genomes.

Summary questions

1 Make a table to summarise all the bacterial, viral, protoctist, and fungal diseases described in this section, including the main organisms that are affected. *(6 marks)*

2 Compare and contrast a bacterial disease of plants and of animals. *(6 marks)*

3 Suggest three ways in which animal diseases and three ways in which plant diseases may be spread from one organism to the next. *(6 marks)*

each year, and over 600 000 die. The disease recurs, making people weak and vulnerable to other infections. There is no vaccine against malaria and limited cures, but preventative measures can very effective. The key is to control the vector. *Anopheles* mosquitoes can be destroyed by insecticides and by removing the standing water where they breed. Simple measures such as mosquito nets, window and door screens and long sleeved clothing can prevent them biting people and spreading the disease.

● Ring worm – a fungal disease affecting mammals including cattle, dogs, cats and humans. Different fungi infect different species – in cattle, ring worm is usually caused by *Trichophyton verrucosum*. It causes grey-white, crusty, infectious, circular areas of skin. It is not damaging but looks unsightly and may be itchy. Antifungal creams are an effective cure.

● Athlete's foot – a human fungal disease caused by *Tinia pedia*, a form of human ring worm that grows on and digests the warm, moist skin between the toes. It causes cracking and scaling, which is itchy and may become sore. Antifungal creams are an effective cure.

✚ Identifying pathogens

When an outbreak of a disease occurs in plants or animals, the key to successful control or cure is to identify the pathogens involved. Our ability to do this has increased along with our understanding of the causes of disease and developments in technology:

● Traditionally pathogens were cultured in the laboratory and identified using a microscope.

● Monoclonal antibodies (antibodies made by cells of the immune system that recognise one specific antigen) can be used now to identify pathogenic organisms in both plants and animals (Topic 12.6, The specific immune system).

● DNA sequencing technology means pathogens can be identified precisely, down to a single mutation (see Zoonotic influenza).

Case study: In a transplant ward, four patients developed infections caused by methicillin resistant *Staphylococcus aureus* (MRSA). If the bacterium was being transmitted to patients by a member of staff, this was a serious outbreak. But DNA sequencing at the Sanger Institute gave rapid results – new technology means a bacterial genome can be sequenced in less than 24 hours. Researchers showed that each of the patients had a different strain of MRSA. The cases were not linked, so it was not a hospital-based outbreak requiring staff to be screened or treated.

1 Why is it important to identify the pathogen causing a communicable disease?
2 Suggest the benefits and limitations of culturing pathogens and using a light microscope to identify them.
3 Find out as much as you can about the use of monoclonal antibodies in the detection of plant diseases.

12.3 The transmission of communicable diseases

Specification reference: 4.1.1

For the pathogens that cause communicable diseases to be successful, they have to be transmissible. So how are pathogenic bacteria, viruses, protoctista, and fungi transmitted from one host to another?

Transmission of pathogens between animals

Understanding how diseases are transmitted from one individual to another allows us to work out ways to reduce or prevent it happening. There are two main types of transmission – direct transmission and indirect transmission.

Direct transmission

Here the pathogen is transferred directly from one individual to another by:

Direct contact (contagious diseases):
- kissing or any contact with the body fluids of another person, for example, bacterial meningitis and many sexually transmitted diseases
- direct skin-to-skin contact, for example, ring worm, athlete's foot
- microorganisms from faeces transmitted on the hands, for example, diarrhoeal diseases.

Inoculation:
- through a break in the skin, for example, during sex (HIV/AIDS)
- from an animal bite, for example, rabies
- through a puncture wound or through sharing needles, e.g. septicaemia.

Ingestion:
- taking in contaminated food or drink, or transferring pathogens to the mouth from the hands, for example, amoebic dysentery, diarrhoeal diseases.

Indirect transmission

This is where the pathogen travels from one individual to another indirectly.

Fomites:
- inanimate objects such as bedding, socks, or cosmetics can transfer pathogens, for example, athlete's foot, gas gangrene and *Staphylococcus infections*.

Droplet infection (inhalation):
- Minute droplets of saliva and mucus are expelled from your mouth as you talk, cough or sneeze. If these droplets contain pathogens, when healthy individuals breathe the droplets in they may become infected, for example, influenza, tuberculosis.

Vectors:
- A vector transmits communicable pathogens from one host to another. Vectors are often but not always animals, for example,

⚛ **Preventing the spread of communicable diseases in humans**

Key factors in reducing the spread of communicable diseases in humans include:

- hand washing – regular hand washing is the single most effective way of preventing the spread of many communicable diseases

- improvements in living and working conditions, for example, reducing overcrowding, ensuring good nutrition

- disposal of both bodily and household waste effectively.

1 Explain why hand washing is so effective at preventing disease transmission.

2 Why does improving living standards have such an impact on disease transmission?

3 Effective management of household waste, leaving no empty containers around and keeping drains clear, substantially reduces the incidence of malaria in an area. Explain why this happens.

4 Research and report on other low-tech ways in which the incidence of malaria can be prevented.

mosquitoes transmit malaria, rat fleas transmit bubonic plague, dogs, foxes and bats transmit rabies.

- Water can also act as a vector of disease, for example, diarrhoeal diseases.

Transmission between animals and humans

Some communicable diseases can be passed from animals to people, for example the bird flu strain H1N1 and brucellosis, which is passed from sheep to people. Minimising close contact with animals and washing hands thoroughly following any such contact can reduce infection rates. People can also act as vectors of some animal diseases, sometimes with fatal results, for example foot-and-mouth disease.

Factors affecting the transmission of communicable diseases in animals

The probability of catching a communicable disease is increased by a number of factors:

- overcrowded living and working conditions

- poor nutrition

- a compromised immune system, including (in humans) having HIV/AIDS or needing immunosuppressant drugs after transplant surgery

- (in humans) poor disposal of waste, providing breeding sites for vectors

- climate change – this can introduce new vectors and new diseases, for example increased temperatures promote the spread of malaria as the vector mosquito species is able to survive over a wider area

- Culture and infrastructure – in many countries traditional medical practises can increase transmission

- Socioeconomic factors – for example, a lack of trained health workers and insufficient public warning when there is an outbreak of disease can also affect transmission rates.

Transmission of pathogens between plants

Plants do not move around, cough or sneeze, yet diseases spread rapidly through plant communities, plant pollen and seed, for example move widely. Plants also have a less well developed immune system than humans.

Direct transmission

This involves direct contact of a healthy plant with any part of a diseased plant. Examples are ring rot, tobacco mosaic virus (TMV), tomato and potato blight, and black sigatoka.

Indirect transmission

Soil contamination

Infected plants often leave pathogens (bacteria or viruses) or reproductive spores from protoctista or fungi in the soil. These can infect the next crop. Examples are black sigatoka spores, ring rot bacteria, spores of *P. infestans* and TMV. Some pathogens (often as spores) can survive the composting process so the infection cycle can be completed when contaminated compost is used.

Vectors

- Wind – bacteria, viruses and fungal or oomycete spores may be carried on the wind, e.g. Black sigatoka blown between Caribbean islands, *P. infestans* sporangia form spores which are carried by the wind to other potato crops/tomato plants.

- Water – spores swim in the surface film of water on leaves; raindrop splashes carry pathogens and spores, etc. Examples are spores of *P. infestans* (potato blight) which swim over films of water on the leaves.

- Animals – insects and birds carry pathogens and spores from one plant to another as they feed. Insects such as aphids inoculate pathogens directly into plant tissues.

- Humans – pathogens and spores are transmitted by hands, clothing, fomites, farming practices and by transporting plants and crops around the world. For example, TMV survives for years in tobacco products, ring rot survives on farm machinery, potato sacks, etc.

▲ **Figure 1** *Disease can spread rapidly through monocultures such as these crop plants*

Factors affecting the transmission of communicable diseases in plants

A number of factors are responsible:

- planting varieties of crops that are susceptible to disease
- over-crowding increases the likelihood of contact
- poor mineral nutrition reduces resistance of plants
- damp, warm conditions increase the survival and spread of pathogens and spores
- climate change – increased rainfall and wind promote the spread of diseases; changing conditions allow animal vectors to spread to new areas; drier conditions may reduce the spread of disease.

Summary questions

1 Explain the difference between direct and indirect transmission of communicable pathogens. (*2 marks*)

2 Compare and contrast *direct* transmission of animal and plant diseases. (*4 marks*)

3 Compare and contrast *indirect* transmission of animal and plant diseases. (*4 marks*)

4 Suggest different approaches to control the spread of malaria. (*5 marks*)

 Preventing the spread of communicable diseases in plants

Key factors in reducing the spread of communicable diseases in plants:

- Leave plenty of room between plants to minimise the spread of pathogens.

- Clear fields as thoroughly as possible – remove all traces of plants from the soil at harvesting.

- Rotate crops – the spores or bacteria will eventually die if they do not have access to the host plant.

- Follow strict hygiene practices – measures such as washing hands, washing boots, sterilising storage sacks, washing down machinery, etc.

- Control insect vectors.

1 Discuss ways of controlling communicable plant diseases with reference to the plant disease triangle in Figure 2.

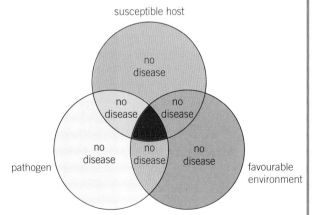

▲ **Figure 2** *Disease in plants is the result of the interaction between a susceptible host, a pathogen and a favourable environment. Modifying any of these factors can reduce the incidence of disease*

12.4 Plant defences against pathogens

Specification reference: 4.1.1

Learning outcomes

Demonstrate knowledge, understanding, and application of:

→ plant defences against pathogens.

Synoptic link

You learnt about cell division and meristems in plants in Topic 6.2, Mitosis and Topic 6.3, Meiosis.

Plants have evolved a number of ways to defend themselves against the pathogens that cause communicable diseases. The waxy cuticle of plant leaves, the bark on trees, and the cellulose cell walls of individual plant cells act as barriers, which prevent pathogens getting in. Unlike animals, plants do not heal diseased tissue – they seal it off and sacrifice it. Because they are continually growing at the meristems, they can then replace the damaged parts.

Recognising an attack

Plants are not passive – they respond rapidly to pathogen attacks. Receptors in the cells respond to molecules from the pathogens, or to chemicals produced when the plant cell wall is attacked. This stimulates the release of signalling molecules that appear to switch on genes in the nucleus. This in turn triggers cellular responses, which include producing defensive chemicals, sending alarm signals to unaffected cells to trigger their defences, and physically strengthening the cell walls.

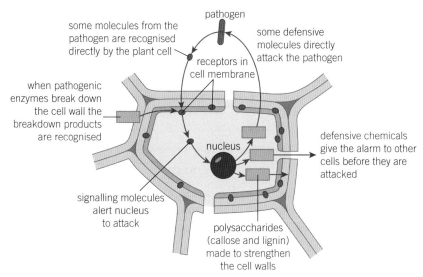

▶ **Figure 1** *The presence of a pathogen stimulates a series of defensive strategies in the plant*

Physical defences

When plants are attacked by pathogens they rapidly set up extra mechanical defences. They produce high levels of a polysaccharide called **callose**, which contains β-1,3 linkages and β-1,6 linkages between the glucose monomers. Scientists still do not fully understand the roles played by callose in the defence mechanisms of the plant but current research suggests that:

- within minutes of an initial attack, callose is synthesised and deposited between the cell walls and the cell membrane in cells next to the infected cells. These callose papillae act as barriers, preventing the pathogens entering the plant cells around the site of infection

Synoptic link

You learnt about α and β linkages in polysaccharides in Topic 3.3, Carbohydrates.

- large amounts of callose continue to be deposited in cell walls after the initial infection. Lignin is added, making the mechanical barrier to invasion even thicker and stronger
- callose blocks sieve plates in the phloem, sealing off the infected part and preventing the spread of pathogens
- callose is deposited in the plasmodesmata between infected cells and their neighbours, sealing them off from the healthy cells and helping to prevent the pathogen spreading.

Chemical defences

Many plants produce powerful chemicals that either repel the insect vectors of disease or kill invading pathogens. Some of these chemicals are so powerful that we extract and use them or synthesise them to help us control insects, fungi and bacteria. Some have strong flavours and are used as herbs and spices. Examples of plant defensive chemicals include:

- insect repellents – for example, pine resin and citronella from lemon grass
- insecticides – for example, pyrethrins – these are made by chrysanthemums and act as insect neurotoxins; and caffeine – toxic to insects and fungi
- antibacterial compounds including antibiotics – for example, phenols – antiseptics made in many different plants; antibacterial gossypol produced by cotton; defensins – plant proteins that disrupt bacterial and fungal cell membranes; lysosomes – organelles containing enzymes that break down bacterial cell walls
- antifungal compounds – for example, phenols – antifungals made in many different plants; antifungal gossypol produced by cotton; caffeine – toxic to fungi and insects; saponins – chemicals in many plant cell membranes that interfere with fungal cell membranes; **chitinases** – enzymes that break down the chitin in fungal cell walls
- anti-oomycetes – for example, **glucanases** – enzymes made by some plants that break down glucans; polymers found in the cell walls of oomycetes (e.g., *P.infestans*)
- general toxins – some plants make chemicals that can be broken down to form cyanide compounds when the plant cell is attacked. Cyanide is toxic to most living things.

▲ **Figure 2** *Castor oil beans (shown here) produce ricin, which is used as a defensive chemical. Just 0.2 milligrams would be fatal if ingested*

Summary questions

1 Describe how the plant response to a pathogen attack is triggered.
(*3 marks*)

2 Make a table or diagram to summarise plant defences against pathogens.
(*4 marks*)

3 Investigate the evidence for one of the roles of callose in plant defence responses. (*6 marks*)

12.5 Non-specific animal defences against pathogens

Specification reference: 4.1.1

Synoptic link

You learnt about the gas exchange system in Topic 7.2, Mammalian gaseous exchange system.

Synoptic link

You learnt about the enzymatic-control of the blood clotting cascade in Topic 4.4, Cofactors, coenzymes, and prosthetic groups.

Mammals (for example humans) have two lines of defence against invasion by pathogens. The primary non-specific defences against pathogens are always present or activated very rapidly. This system defends against all pathogens in the same way. Mammals have a specific immune response, which is specific to each pathogen but is slower to respond (as discussed in Topic 12.6).

Non-specific defences – keeping pathogens out

The body has a number of barriers to the entry of pathogens:

- The skin covers the body and prevents the entry of pathogens. It has a skin flora of healthy microorganisms that outcompete pathogens for space on the body surface. The skin also produces sebum, an oily substance that inhibits the growth of pathogens.

- Many of the body tracts, including the airways of the gas exchange system, are lined by **mucous membranes** that secrete sticky mucus. This traps microorganisms and contains lysozymes, which destroy bacterial and fungal cell walls. Mucus also contains phagocytes, which remove remaining pathogens.

- Lysozymes in tears and urine, and the acid in the stomach, also help to prevent pathogens getting into our bodies.

We also have expulsive reflexes. Coughs and sneezes eject pathogen-laden mucus from the gas exchange system, while vomiting and diarrhoea expel the contents of the gut along with any infective pathogens.

Blood clotting and wound repair

If you cut yourself, the skin is breached and pathogens can enter the body. The blood clots rapidly to seal the wound. When platelets come into contact with collagen in skin or the wall of the damaged blood vessel, they adhere and begin secreting several substances. The most important are:

- thromboplastin, an enzyme that triggers a cascade of reactions resulting in the formation of a blood clot (or thrombus, Figure 1)

- serotonin, which makes the smooth muscle in the walls of the blood vessels contract, so they narrow and reduce the supply of blood to the area.

The clot dries out, forming a hard, tough scab that keeps pathogens out. This is the first stage of wound repair. Epidermal cells below the scab start to grow, sealing the wound permanently, while damaged blood vessels regrow. Collagen fibres are deposited to give the new

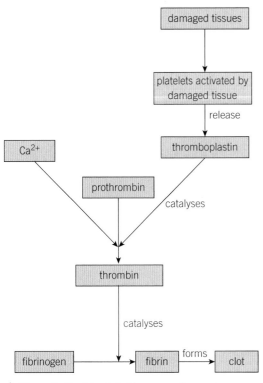

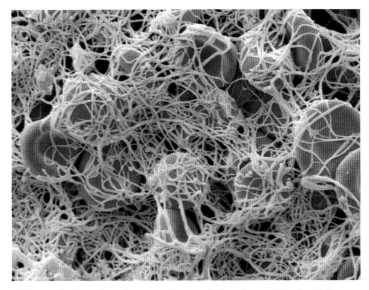

▲ **Figure 1** *The blood clotting cascade – once a clot forms, blood cannot leak out of the body and pathogens cannot get in*

▲ **Figure 2** *Scanning electron micrograph showing red blood cells in a fibrin mesh. × 800 magnification*

tissue strength. Once the new epidermis reaches normal thickness, the scab sloughs off and the wound is healed.

Inflammatory response

The inflammatory response is a localised response to pathogens (or damage or irritants) resulting in **inflammation** at the site of a wound. Inflammation is characterised by pain, heat, redness, and swelling of tissue.

Mast cells are activated in damaged tissue and release chemicals called **histamines** and **cytokines**.

- Histamines make the blood vessels dilate, causing localised heat and redness. The raised temperature helps prevent pathogens reproducing.
- Histamines make blood vessel walls more leaky so blood plasma is forced out, once forced out of the blood it is known as tissue fluid. Tissue fluid causes swelling (oedema) and pain.
- Cytokines attract white blood cells (phagocytes) to the site. They dispose of pathogens by **phagocytosis**.

If an infection is widespread, the inflammatory response can cause a whole-body rash.

Non-specific defences – getting rid of pathogens

If the pathogens get into the body, the next lines of defence are adaptations to prevent them growing or to destroy them.

Synoptic link

You learnt about phagocytosis as a form of bulk transport in Topic 5.4, Active transport.

Synoptic link

You will learn more about the control of body temperature in Chapter 15, Homeostasis.

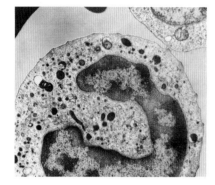

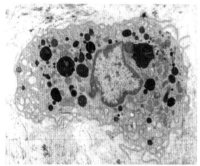

▲ **Figure 3** *Common phagocytes in the body: 70% of white blood cells are neutrophils (top) with their multi-lobed nuclei and rapid action against pathogens. Macrophages (bottom), with their simpler round nuclei, make up 4% of the white blood cells and are involved in both the non-specific defence system and the specific immune system. Coloured transmission electron micrographs, × 2000 magnification*

Fevers

Normal body temperature of around 37 °C is maintained by the hypothalamus in your brain. When a pathogen invades your body, cytokines stimulate your hypothalamus to reset the thermostat and your temperature goes up. This is a useful adaptation because:

- most pathogens reproduce best at or below 37 °C. Higher temperatures inhibit pathogen reproduction
- the specific immune system works faster at higher temperatures.

Phagocytosis

Phagocytes are specialised white cells that engulf and destroy pathogens. There are two main types of phagocytes – neutrophils and macrophages (Figure 2).

Phagocytes build up at the site of an infection and attack pathogens. Sometimes you can see pus in a spot, cut or wound. Pus consists of dead neutrophils and pathogens.

The stages of phagocytosis

1   Pathogens produce chemicals that attract phagocytes.

2   Phagocytes recognise non-human proteins on the pathogen. This is a response not to a specific type of pathogen, but simply a cell or organism that is non-self.

3   The phagocyte engulfs the pathogen and encloses it in a vacuole called a **phagosome.**

4   The phagosome combines with a lysosome to form a **phagolysosome.**

5   Enzymes from the lysosome digest and destroy the pathogen.

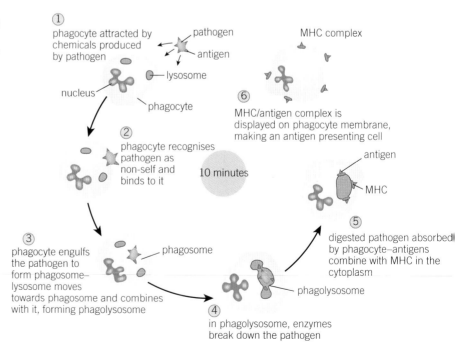

▲ **Figure 4** *Phagocytosis – a key process in both the non-specific and specific defence systems of the body*

It usually takes a human neutrophil under 10 minutes to engulf and destroy a bacterium. Macrophages take longer but they undergo a more complex process. When a macrophage has digested a pathogen, it combines antigens from the pathogen surface membrane with special glycoproteins in the cytoplasm called the **major histocompatibility complex** (**MHC**). The MHC complex moves these pathogen antigens to the macrophage's own surface membrane, becoming an **antigen-presenting cell** (APC). These antigens now stimulate other cells involved in the specific immune system response (see Topic 11.6, Factors affecting biodiversity).

Counting blood cells

In your previous studies you learnt how to examine microscope slides and draw the cells you saw. You also learned how to count the number of cells in a given area of a slide. Both of these skills are very important when looking at blood smears, made by spreading a single drop of blood very thinly across a slide. They are often stained to show up the nuclei of the lymphocytes, making them easier to identify. Identifying the numbers of different types of lymphocytes in a blood smear indicates if a non-specific or specific immune response is taking place.

Helpful chemicals

Phagocytes that have engulfed a pathogen produce chemicals called **cytokines.** Cytokines act as cell-signalling molecules, informing other phagocytes that the body is under attack and stimulating them to move to the site of infection or inflammation. Cytokines can also increase body temperature and stimulate the specific immune system.

Opsonins are chemicals that bind to pathogens and 'tag' them so they can be more easily recognised by phagocytes. Phagocytes have receptors on their cell membranes that bind to common opsonins, and the phagocyte then engulfs the pathogen. There are a number of different opsonins, but antibodies such as immunoglobulin G (IgG) and immunoglobulin M (IgM) have the strongest effect.

> ### Synoptic link
>
> You learnt about smear slides in Topic 2.1, Microscopy. It may help to look back at Topics 6.4, The organisation and specialisation of cells, and 6.5, Stem cells, to remind yourself of the different types of cells in the blood.

Summary questions

1 Make a table to show the main adaptations of the body that prevent the entry of pathogens. *(4 marks)*

2 A woman gets a bad scratch from a bramble. The scratch gets very red and hot and the next day it contains pus. Explain what is happening. *(4 marks)*

3 a Describe the process of phagocytosis. *(6 marks)*
   b Explain how cytokines and opsonins make the process of phagocytosis more effective than it would be without them. *(5 marks)*

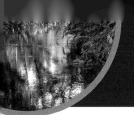

12.6 The specific immune system

Specification reference: 4.1.1

All cells have molecules called **antigens** on their surfaces. The body recognises the difference between *self* antigens on your own cells and *non-self* antigens on the cells of pathogens. Some toxins also act as antigens. Antigens trigger an immune response, which involves the production of polypeptides called **antibodies**.

The **specific immune system** (also known as active or acquired immunity) is slower than the non-specific responses – it can take up to 14 days to respond effectively to a pathogen invasion. However, the immune memory cells mean it reacts very quickly to a second invasion by the same pathogen.

Demonstrate knowledge, understanding, and application of:

➔ modes of action of B and T lymphocytes in the specific immune response

➔ the structure and functions of antibodies

➔ actions of opsonins, agglutinins and anti-toxins

➔ the primary and secondary immune responses

➔ autoimmune diseases.

Antibodies

Antibodies are Y-shaped glycoproteins called **immunoglobulins**, which bind to a specific antigen on the pathogen or toxin that has triggered the immune response. There are millions of different antibodies, and there is a specific antibody for each antigen.

Antibodies are made up of two identical long polypeptide chains called the heavy chains and two much shorter identical chains called the light chains (Figure 1). The chains are held together by disulfide bridges and there are also disulfide bridges within the polypeptide chains holding them in shape.

Antibodies bind to antigens with a protein-based 'lock-and-key' mechanism similar to the complementarity between the active site of an enzyme and its substrate. The binding site is an area of 110 amino acids on both the heavy and the light chains, known as the variable region. It is a different shape on each antibody and gives the antibody its specificity. The rest of the antibody molecule is always the same, so it is called the constant region.

When an antibody binds to an antigen it forms an **antigen–antibody complex.**

The hinge region of the antibody provides the molecule with flexibility, allowing it to bind two separate antigens, one at each of its antigen-binding sites.

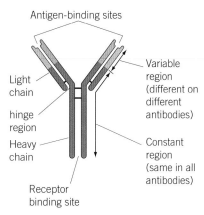

▲ **Figure 1** *All antibodies have the same basic structure*

How antibodies defend the body

1 The antibody of the antigen–antibody complex acts as an opsonin so the complex is easily engulfed and digested by phagocytes.

2 Most pathogens can no longer effectively invade the host cells once they are part of an antigen–antibody complex.

3 Antibodies act as **agglutinins** causing pathogens carrying antigen–antibody complexes to clump together. This helps prevent them spreading through the body and makes it easier for phagocytes to engulf a number of pathogens at the same time.

4 Antibodies can act as **anti-toxins**, binding to the toxins produced by pathogens and making them harmless.

Synoptic link

You learnt about protein structure and glycoproteins in Topic 3.5, Lipids and Topic 3.6, Structure of proteins, and about the active sites of enzymes in Topic 4.1 Enzyme action.

Lymphocytes and the immune response

The specific immune system is based on white blood cells called **lymphocytes**. **B lymphocytes** mature in the **B**one marrow, while **T lymphocytes** mature in the **T**hymus gland.

The main types of T lymphocytes:

- **T helper cells** – these have CD4 receptors on their cell-surface membranes, which bind to the surface antigens on APCs (Topic 12.5). They produce **interleukins,** which are a type of cytokine (cell-signalling molecule). The interleukins made by the T helper cells stimulate the activity of B cells, which increases antibody production, stimulates production of other types of T cells and attracts and stimulates macrophages to ingest pathogens with antigen–antibody complexes.

- **T killer cells** – these destroy the pathogen carrying the antigen. They produce a chemical called **perforin,** which kills the pathogen by making holes in the cell membrane so it is freely permeable.

- **T memory cells** – these live for a long time and are part of the **immunological memory**. If they meet an antigen a second time, they divide rapidly to form a huge number of clones of T killer cells that destroy the pathogen.

- **T regulator cells** – these cells suppress the immune system, acting to control and regulate it. They stop the immune response once a pathogen has been eliminated, and make sure the body recognises self antigens and does not set up an **autoimmune response**. Interleukins are important in this control.

The main types of B lymphocytes:

- **Plasma cells** – these produce antibodies to a particular antigen and release them into the circulation. An active plasma cell only lives for a few days but produces around 2000 antibodies per second while it is alive and active.

- **B effector cells** – these divide to form the plasma cell clones.

- **B memory cells** – these live for a very long time and provide the immunological memory. They are programmed to remember a specific antigen and enable the body to make a very rapid response when a pathogen carrying that antigen is encountered again.

Cell-mediated immunity

In cell-mediated immunity, T lymphocytes respond to the cells of an organism that have been changed in some way, for example by a virus infection, by antigen processing or by mutation (for example cancer cells) and to cells from transplanted tissue. The cell-mediated response is particularly important against viruses and early cancers.

1 In the non-specific defence system, macrophages engulf and digest pathogens in phagocytosis. They process the antigens from the surface of the pathogen to form antigen-presenting cells (APCs).

2   The receptors on some of the T helper cells fit the antigens. These T helper cells become activated and produce interleukins, which stimulate more T cells to divide rapidly by mitosis. They form clones of identical activated T helper cells that all carry the right antigen to bind to a particular pathogen.

3   The cloned T cells may:

- develop into T memory cells, which give a rapid response if this pathogen invades the body again

- produce interleukins that stimulate phagocytosis

- produce interleukins that stimulate B cells to divide

- stimulate the development of a clone of T killer cells that are specific for the presented antigen and then destroy infected cells.

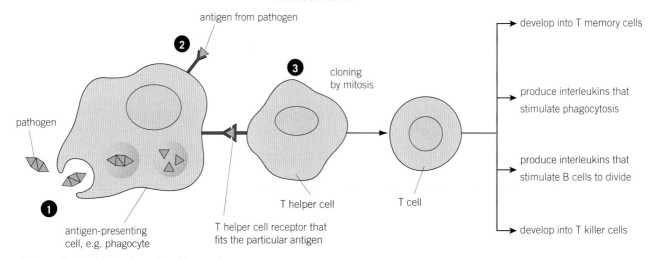

▲ **Figure 2** *Model of cell-mediated immunity*

Humoral immunity

In humoral immunity the body responds to antigens found outside the cells, for example bacteria and fungi, and to APCs. The humoral immune system produces antibodies that are soluble in the blood and tissue fluid and are not attached to cells.

B lymphocytes have antibodies on their cell-surface membrane (immunoglobulin M or IgM) and there are millions of different types of B lymphocytes, each with different antibodies. When a pathogen enters the body it will carry specific antigens, or produce toxins that act as antigens. A B cell with the complementary antibodies will bind to the antigens on the pathogen, or to the free antigens. The B cell engulfs and processes the antigens to become an APC (see Figure 3).

1   Activated T helper cells bind to the B cell APC. This is **clonal selection** – the point at which the B cell with the correct antibody to overcome a particular antigen is selected for cloning.

2   Interleukins produced by the activated T helper cells activate the B cells.

3   The activated B cell divides by mitosis to give clones of plasma cells and B memory cells. This is **clonal expansion.**

4   Cloned plasma cells produce antibodies that fit the antigens on the surface of the pathogen, bind to the antigens and disable them, or act as opsonins or agglutinins. This is the **primary immune response** and it can take days or even weeks to become fully effective against a particular pathogen. This is why we get ill – the symptoms are the result of the way our body reacts when the pathogens are dividing freely, before the primary immune response is fully operational.

5   Some cloned B cells develop into B memory cells. If the body is infected by the same pathogen again, the B memory cells divide rapidly to form plasma cell clones. These produce the right antibody and wipe out the pathogen very quickly, before it can cause the symptoms of disease. This is the **secondary immune response.**

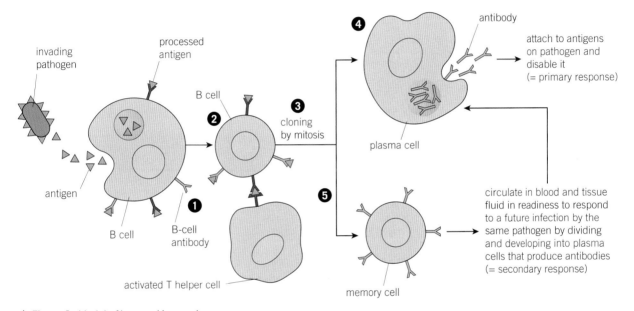

▲ **Figure 3** *Model of humoral immunity*

Autoimmune diseases

Sometimes the immune system stops recognising 'self' cells and starts to attack healthy body tissue. This is termed an **autoimmune disease**. Scientists still do not understand fully why this happens. There appears to be a genetic tendency in some families, sometimes the immune system responds abnormally to a mild pathogen or normal body microorganisms and in some cases the T regulator cells do not work effectively.

There are around 80 different autoimmune diseases that can cause chronic inflammation or the complete breakdown and destruction of healthy tissue. Immunosuppressant drugs, which prevent the immune system working, may be used as treatments but they deprive the body of its natural defences against communicable diseases.

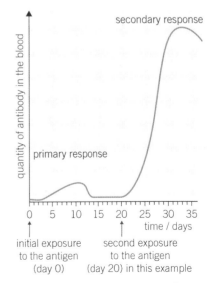

▲ **Figure 4** *Primary and secondary immune responses*

Synoptic link

Look back to the Application on gene therapy in Topic 6.5, Stem cells, for an example of an autoimmune disease.

▼ **Table 1** *Some common autoimmune diseases*

Autoimmune disease	Body part affected	Treatment
Type 1 diabetes	• the insulin-secreting cells of the pancreas	• insulin injections • pancreas transplants • immunosuppressant drugs
Rheumatoid arthritis	• joints—especially in the hands, wrists, ankles and feet	• no cure • anti-inflammatory drugs • steroids • immunosuppressants • pain relief
Lupus	• often affects skin and joints and causes fatigue • can attack any organ in the body including kidneys, liver, lungs or brain	• no cure • anti-inflammatory drugs • steroids • immunosuppressants • various others

Summary questions

1   What are antibodies and how do they work?   (*2 marks*)

2   Compare the main types of T and B lymphocytes – what are their similarities and differences?   (*6 marks*)

3   Discuss the problems that could arise from treating an autoimmune disease with immunosuppressant drugs.   (*3 marks*)

4   The humoral immune system deals well with bacterial and fungal infections but the cell-mediated immune system is more effective at tackling viral infections. Explain the biology behind this statement.   (*6 marks*)

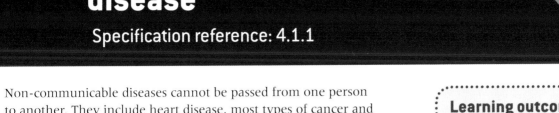

12.7 Preventing and treating disease

Specification reference: 4.1.1

Non-communicable diseases cannot be passed from one person to another. They include heart disease, most types of cancer and many diseases of the nervous, endocrine and digestive systems. Communicable diseases are caused by pathogens and can be passed from person to person.

When you come into contact with a foreign antigen, you need some form of immunity to prevent you getting the disease. There are several ways of achieving this immunity.

Natural immunity

Some forms of immunity occur naturally in the body:

- When you meet a pathogen for the first time, your immune system is activated and antibodies are formed, which results in the destruction of the antigen (Topic 12.6, The specific immune system). The immune system produces T and B memory cells so if you meet a pathogen for a second time, your immune system recognises the antigens and can immediately destroy the pathogen, before it causes disease symptoms. This is known as **natural active immunity**. It is known as active because the body has itself acted to produce antibodies and/or memory cells.

- The immune system of a new-born baby is not mature and it cannot make antibodies for the first couple of months. A system has evolved to protect the baby for those first few months of life. Some antibodies cross the placenta from the mother to her fetus while the baby is in the uterus, so it has some immunity to disease at birth. The first milk a mammalian mother makes is called **colostrum**, which is very high in antibodies. The infant gut allows these glycoproteins to pass into the bloodstream without being digested. So within a few days of birth, a breast-fed baby will have the same level of antibody protection against disease as the mother. This is **natural passive immunity** and it lasts until the immune system of the baby begins to make its own antibodies. The antibodies the baby receives from the mother are likely to be relevant to pathogens in its environment, where the mother acquired them.

Artificial immunity

Some diseases can kill people before their immune system makes the antibodies they need. Medical science can give us immunity to some of these life-threatening diseases without any contact with live pathogens.

Learning outcomes

Demonstrate knowledge, understanding, and application of:

→ the differences between active and passive immunity, and between natural and artificial immunity

→ the principles of vaccination and the role of vaccination programmes in the prevention of epidemics

→ the benefits and risks of using antibiotics to manage bacterial infection

→ possible sources of new medicines.

Artificial passive immunity

For certain potentially fatal diseases, antibodies are formed in one individual (often an animal), extracted and then injected into the bloodstream of another individual. This **artificial passive immunity** gives temporary immunity – it doesn't last long but it can be lifesaving. For example, tetanus is caused by a toxin released by the bacterium *Clostridium tetani*, found in the soil and animal faeces. It causes the muscles to go into spasm so you cannot swallow or breathe. People who might be infected with tetanus (for example after a contaminated cut) will be injected with tetanus antibodies extracted from the blood of horses, preventing the development of the disease but not providing long-term immunity. Rabies is another fatal disease that is treated with a series of injections that give artificial passive immunity.

Artificial active immunity – the principles of vaccination

In **artificial active immunity** the immune system of the body is stimulated to make its own antibodies to a safe form of an antigen (a **vaccine**), which is injected into the bloodstream (vaccination). The antigen is not usually the normal live pathogen, as this could cause the disease and have fatal results. The main steps are as follows:

1   The pathogen is made safe in one of a number of ways so that the antigens are intact but there is no risk of infection. Vaccines may contain:

- killed or inactivated bacteria and viruses, for example, whooping cough (pertussis)
- attenuated (weakened) strains of live bacteria or viruses, for example, rubella, BCG against TB, polio (vaccine taken orally)
- toxin molecules that have been altered and detoxified, for example, diphtheria, tetanus
- isolated antigens extracted from the pathogen, for example, the influenza vaccine
- genetically engineered antigens, for example, the hepatitis B vaccine.

2   Small amounts of the safe antigen, known as the vaccine, are injected into the blood.

3   The primary immune response is triggered by the foreign antigens and your body produces antibodies and memory cells as if you were infected with a live pathogen.

4   If you come into contact with a live pathogen, the secondary immune response is triggered and you destroy the pathogen rapidly before you suffer symptoms of the disease.

The artificial active immunity provided by vaccines may last a year, a few years or a lifetime. Sometimes boosters (repeat vaccinations) are needed to increase the time you are immune to a disease.

Vaccines and the prevention of epidemics

Vaccines are used to give long-term immunity to many diseases. However, they are also used to help prevent epidemics. An **epidemic**

Synoptic link

You will learn about genetic engineering in Topic 21.4, Genetic engineering.

▲ **Figure 2** *In the UK babies are given vaccinations against a range of diseases including pertussis, diphtheria, tetanus, polio, meningitis, measles, mumps, and rubella. As a result, the number of children who die from preventable infections is now very low*

is when a communicable disease spreads rapidly to a lot of people at a local or national level. A **pandemic** is when the same disease spreads rapidly across a number of countries and continents.

At the beginning of an epidemic, mass vaccination can prevent the spread of the pathogen into the wider population. When vaccines are being deployed to prevent epidemics, they often have to be changed regularly to remain effective.

When a significant number of people in the population have been vaccinated, this gives protection to those who do not have immunity. This is known as herd immunity, as there is minimal opportunity for an outbreak to occur.

 Case Study: Influenza

Flu is a disease that has caused epidemics at intervals throughout history. The virus that causes strain A influenza mutates regularly, so the antigens on the surface change too. Some forms of this virus can cross the species barrier from animals such as birds or pigs to people. Although people develop resistance to one strain of flu, the next year the antigens on the surface of the virus may have changed so much that the immune system does not recognise it and many of the same people become ill again.

Every year in the UK older people and anyone who has a compromised immune system are given a flu vaccine. Every year the mixture of flu antigens in the vaccine is different, reflecting the forms of the virus that the World Health Organisation (WHO) predicts will be most common and most likely to cause serious disease.

If a flu epidemic begins, more people are vaccinated to control infection rates. Because people travel freely and frequently, an epidemic can spread rapidly from one country to another. People across the world need to be vaccinated to stop the spread of disease and stock piles of vaccines are in place in case this becomes necessary.

SARS was a new flu-like disease that appeared in 2002 and spread from birds to people. It spread rapidly across countries as people travelled around. However, in spite of a lack of vaccine, careful management of cases by isolation meant the outbreak was contained and quickly closed down with relatively few deaths (Figure 3).

1 Suggest three ways in which an outbreak of 'flu' or 'flu-like illness' might be contained.

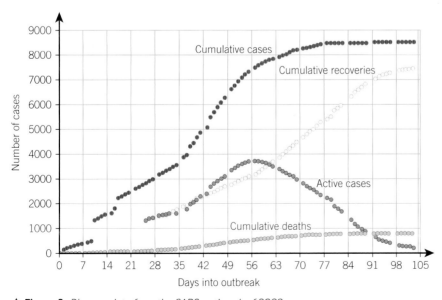

▲ **Figure 3** *Disease data from the SARS outbreak of 2002.*

Some communicable diseases that cause problems at a global level cannot yet be prevented by vaccination. Examples include:

- malaria – *Plasmodium,* the protoctist that causes malaria. It is very evasive – it spends time inside the erythrocytes so it is protected by self antigens from the immune system, and within an infected individual its antigens reshuffle

- HIV, the human immunodeficiency virus that causes AIDS. It enters the macrophages and T helper cells, so it has disabled the immune system itself.

So far scientists have been unable to develop a vaccine for these diseases, which between them affect millions of people globally every year.

Medicines and the management of disease

Medicines can be used to treat communicable and non-communicable diseases. Medicines can be used to treat symptoms and cure them, making people feel better. Common medicines include painkillers, anti-inflammatories and anti-acid medicines (which reduce indigestion).

Medicines that cure people include chemotherapy against some cancers, antibiotics that kill bacteria, and antifungals that kill fungal pathogens.

Sources of medicines

Penicillin was the first widely used, effective, safe antibiotic capable of curing bacterial diseases. It comes from a mould, *Penicillium chrysogenum,* famously discovered by Alexander Fleming in 1928, when he found it growing on his *Staphylococcus* spp. cultures. Fleming saw what the mould did to his bacteria but could not extract enough to test its potential. It needed Howard Florey and Ernst Chain to develop an industrial process for making the new drug, which has since saved millions of lives around the world.

The medicines we use today come from a wide range of sources. Scientists design drugs using complex computer programmes. They can build up 3-dimensional models of key molecules in the body, and of pathogens and their antigen systems. This allows models of potential drug molecules to be built up which are targeted at particular areas of a pathogen. Computers are also used to search through enormous libraries of chemicals, to isolate any with a potentially useful action against a specific group of feature of a pathogen, or against the mutated cells in a cancer.

Analysis of the genomes of pathogens and genes which have been linked to cancer enable scientists to target their novel drugs to attack any vulnerabilities. However, many of the drugs most commonly used in medicine are still either derived from, or based on, bioactive compounds discovered in plants, microorganisms or other forms of life. Table 2 lists some examples.

▼ **Table 2** *Some common medicinal drugs, derived from living organisms*

Drug	Source	Action
Penicillin	commercial extraction originally from mould growing on melons	antibiotic – the first effective treatment against many common bacterial diseases
Docetaxel/paclitaxel	derived originally from yew trees	treatment of breast cancer
Aspirin (acetylsalicylic acid)	based on compounds from sallow (willow) bark	painkiller, anti-coagulant, anti-pyretic (reduces fever) and anti-inflammatory
Prialt	derived from the venom of a cone snail from the oceans around Australia	new pain-killing drug 1000 times more effective than morphine
Vancomycin	derived from a soil fungus	one of our most powerful antibiotics
Digoxin	based on digitoxin, originally extracted from foxgloves	powerful heart drug used to treat atrial fibrillation and heart failure

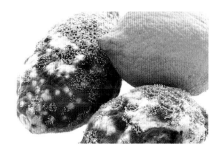

In the 21st century biodiversity is rapidly being lost around the world, including the destruction of rain forests, the loss of coral reefs and loss of habitat for natural ecosystems in countries all around the world. This is at least partly due to human activities. Scientists have not yet explored and identified and analysed a fraction of life on Earth. One of many reasons why it is so important to maintain biodiversity is to make sure we do not destroy a plant, animal or microorganism which could give us the key to a life-saving drug.

Drug design for the future
Pharmacogenetics

Personalised medicine – a combination of drugs that work with your individual combination of genetics and disease – is the direction in which medicine is going. The human genome can be analysed relatively rapidly and cheaply, giving a growing understanding of the genetic basis of many diseases. The science of interweaving knowledge of drug actions with personal genetic material is known as **pharmacogenomics**. We already know that genotypes and drugs interact. For example, in approximately 30% of all breast cancers there is a mutation in the HER2 gene. The activity of this gene can be shut down by specific drugs – trastuzumab (known an Herceptin) and lapatinib. By analsying breast tumours and treating those which have this mutation with the relevant drugs, doctors can reduce the deaths from HER2 breast cancer by up to 50%. In future, this type of treatment, where clinicians looks at the genome of their patients and the genome of the invading pathogen before deciding how to treat them, will become increasingly common.

▲ **Figure 4** *Mould, a cone snail and foxgloves – this diverse range of organisms provide the origins of some important medicines*

Synthetic biology

Another major step forward in drug development is synthetic biology. Using the techniques of genetic engineering, we can develop populations of bacteria to produce much needed drugs that would otherwise be too rare, too expensive or just not available. Synthetic biology enables the use of bacteria as biological factories. Mammals have also been genetically modified to produce much needed therapeutic proteins in their milk. This re-engineering of biological systems for new purposes has great potential in medicine. Nanotechnology is another strand of synthetic biology, where tiny, non-natural particles are used for biological purposes – for example, to deliver drugs to very specific sites within the cells of pathogens or tumours.

Synoptic link

You will learn more about the human genome, synthetic biology, and genetic engineering in Chapter 21, Manipulating genomes and Chapter 22, Cloning and biotechnology.

polymixines make holes in cell membrane altering its permeability

penicillin and cephalosporins – weaken the cell wall so bacterium is more easily damaged by immune reaction

sulfonamides interfere with metabolic reactions

DNA

tetracyclines and streptomycin inhibit protein synthesis

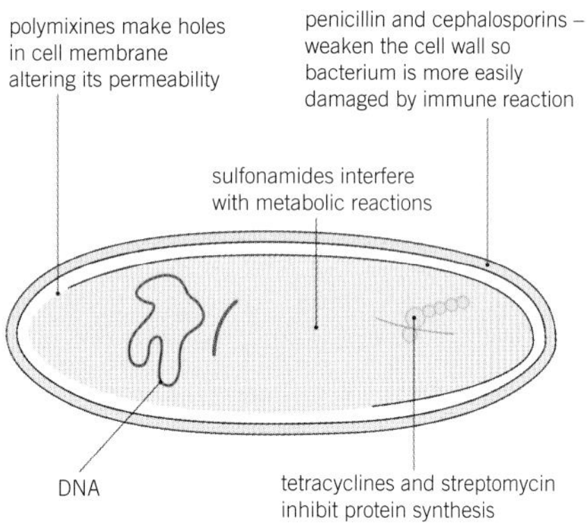

▲ **Figure 5** *Some of the different ways in which antibiotic drugs damage bacteria*

The antibiotic dilemma

At the beginning of the 20th century, 36% of all deaths – and 52% of all childhood deaths – were from communicable diseases. Antibiotics interfere with the metabolism of the bacteria without affecting the metabolism of the human cells – this is called **selective toxicity**. They gave doctors, for the first time, medicines that were effective against bacteria, so antibiotics were understandably, widely used. By the start of this century, the numbers of children dying per year had fallen dramatically and, of those remaining few deaths, only about 7% were due to communicable diseases.

There are many different types of bacteria and a range of antibiotics is used against them, including streptomycin, amoxicillin (very like penicillin), cephalosporins, tetracyclines, sulfonamides, polymixines, ampicillin, and vancomycin. In 2014 up to 1 in 6 of all prescriptions were still for antibiotics. They are often used for relatively minor infections where the immune system of the patient would deal with the infection with no serious difficulty.

Unfortunately, antibiotics are becoming less effective in the treatment of bacterial diseases. Bacteria are becoming resistant to more and more antibiotics. This trend started with penicillin – now there are microorganisms that are resistant to all of the antibiotics we have.

The development of antibiotic resistance

There is an evolutionary race between scientists and bacteria. An antibiotic works because a bacterium has a binding site for the drug, and a metabolic pathway that is affected by the drug. If a random mutation during bacterial reproduction produces a bacterium that is not affected by the antibiotic, that is the one which is best fitted to survive and reproduce, passing on the antibiotic resistance mutation to the daughter cells. Bacteria reproduce very rapidly, so once a mutation occurs it does not take long to grow a big population of **antibiotic-resistant bacteria**.

In a few decades we have reached a stage where increasing numbers of bacterial pathogens are resistant to most or all of our antibiotics.

In some countries, including the US, farmers routinely add antibiotics to animal feed prophylactically to prevent animals losing condition due to infections, and reducing business profits. There are concerns that such routine exposure to the antibiotics accelerates natural selection of antibiotic-resistant strains of both human and animal pathogens. However, in the UK it is illegal to give animals routine antibiotics this way. Evidence suggests that it is the over subscription of antibiotics to people which is the prime cause of the rise in antibiotic resistance.

MRSA and *C. difficile*

Antibiotic-resistant bacteria are a particular problem in hospitals and care homes for older people, where antibiotics are often needed and used. **MRSA (methicillin-resistant *Staphylococcus aureus*)** and *Clostridium difficile* (*C. difficile*) have been high-profile examples of antibiotic-resistant bacteria. They are summarised in Table 3.

Figure 6 ▶

Development of resistance in bacteria to ciprofloxacin, an antibiotic used for treating a wide variety of infections

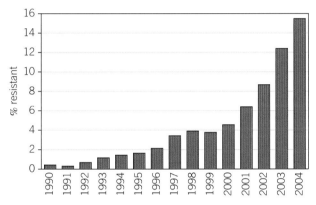

▼ **Table 3**

MRSA	C. difficile
• bacterium carried by up to 30% of the population on their skin or in their nose • in the body it can cause boils, abscesses and potentially fatal septicaemia • was treated effectively with methicillin, a penicillin-like antibiotic but mutation has produced methicillin-resistant strains	• bacterium in the guts of about 5% of the population • produces toxins that damage the lining of the intestines, leading to diarrhoea, bleeding and even death • not a problem for healthy person *but* when commonly-used antibiotics kill off much of the 'helpful' gut bacteria it survives, reproduces and takes hold rapidly

Antibiotic-resistant infections can be reduced in the long term by measures including:

- minimising the use of antibiotics, and ensuring that every course of antibiotics is completed to reduce the risk of resistant individuals surviving and developing into a resistant strain population

- good hygiene in hospitals, care homes and in general – this has a major impact on the spread of all infections, including antibiotic-resistant strains.

Solving the problem

The development of antibiotic-resistant bacteria is one of the biggest health problems of our time – there is a fear that we may return to the days when bacterial infections killed thousands of people each year in the UK alone. Scientists are working on developing new antibiotics using computer modelling and looking at possible sources in a wide variety of places, including soil microorganisms, crocodile blood, fish slime, honey and the deepest abysses of the oceans. But at the moment, bacterial resistance is building faster than new antibiotics can be found. In 2014 it was announced that a new Lottery-funded prize of £10 million named Longitude will be reserved for anyone who can come up with a cost-effective, accurate and easy-to-use test for bacterial infections so that doctors all over the world can use the right antibiotics at the right time, and only when they are needed.

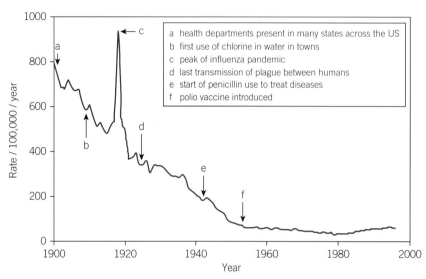

a   health departments present in many states across the US
b   first use of chlorine in water in towns
c   peak of influenza pandemic
d   last transmission of plague between humans
e   start of penicillin use to treat diseases
f   polio vaccine introduced

▲ **Figure 7**   *Graph showing crude death rate per year per 100 000 in the US from 1900–1996*

Summary questions

1   Make a flow diagram to explain how artificial active immunity is stimulated when a vaccination is given.   *(4 marks)*

2   Suggest why some children still die of communicable diseases in the UK even in the 21st century.   *(4 marks)*

3   Using the data from Figure 3.
   a   When did the peak of the SARS outbreak occur?   *(1 mark)*
   b   How many people were infected during the outbreak of the disease   *(1 mark)*
   c   How many people died in the outbreak   *(1 mark)*
   d   Approximately what percentage of the people affected by the SARS virus recovered?   *(3 marks)*

4   Explain how vaccinations may be used:
   a   to prevent rabies after a person has been bitten by a rabid dog, fox or bat   *(2 marks)*
   b   to control epidemics.   *(4 marks)*

5   Using data from Figure 7, discuss the factors that affect mortality from communicable diseases and comment on the effects of using antibiotics to treat bacterial infections on mortality since penicillin was first introduced.   *(6 marks)*

Practice questions

1   Doctors regularly receive advice with regard to the prescribing of antibiotics.

    Which of the following statements is/are likely to form part of this advice.

    Statement 1: avoid the use of broad spectrum antibiotics

    Statement 2: only prescribe antibiotics when patients have a bacterial infection

    Statement 3: use targeted antibiotics whenever possible

    A   1, 2 and 3 are correct

    B   Only 1 and 2 are correct

    C   Only 2 and 3 are correct

    D   Only 1 is correct   (*1 mark*)

2   a   Define the term pathogen.   (*2 marks*)

    b   Describe the physical barriers of plants which act as a defence against infection from pathogens.

                                             (*3 marks*)

    c   Explain why plants cannot acquire immunity.   (*4 marks*)

3   The 2014 Ebola virus outbreak was more correctly called an epidemic. It is often fatal and there is no known cure or vaccine available.

    The diagram below shows the progression of the disease in the first few months of the outbreak.

a   (i)   Describe the trends shown in the graph.
                                             (*4 marks*)

    (ii) Suggest the reasons for the difference in the trends.   (*3 marks*)

b   Describe the structure of a virus and explain why viruses are not considered to be living organisms.   (*4 marks*)

c   Ebola outbreaks usually happen in remote areas of developing countries. They are quite rare and usually on a relatively small scale.

    Discuss why there are no effective drugs or vaccines available to treat Ebola. (*5 marks*)

4   The graph below shows the change in incidence of TB in patients and their HIV status, if known.

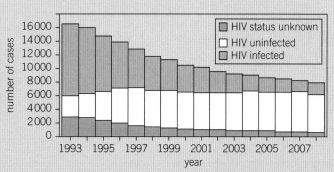

a   Describe and explain the changes from 1993 to 2008.   (*5 marks*)

b   Describe what is meant by the term opportunistic infection.   (*3 marks*)

c   Explain the differences in the way TB and HIV are transmitted.   (*4 marks*)

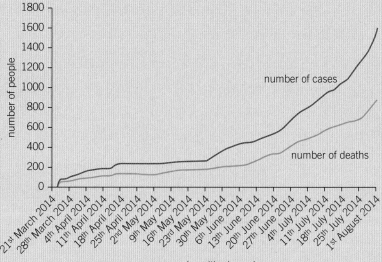

non-logarithmic scale

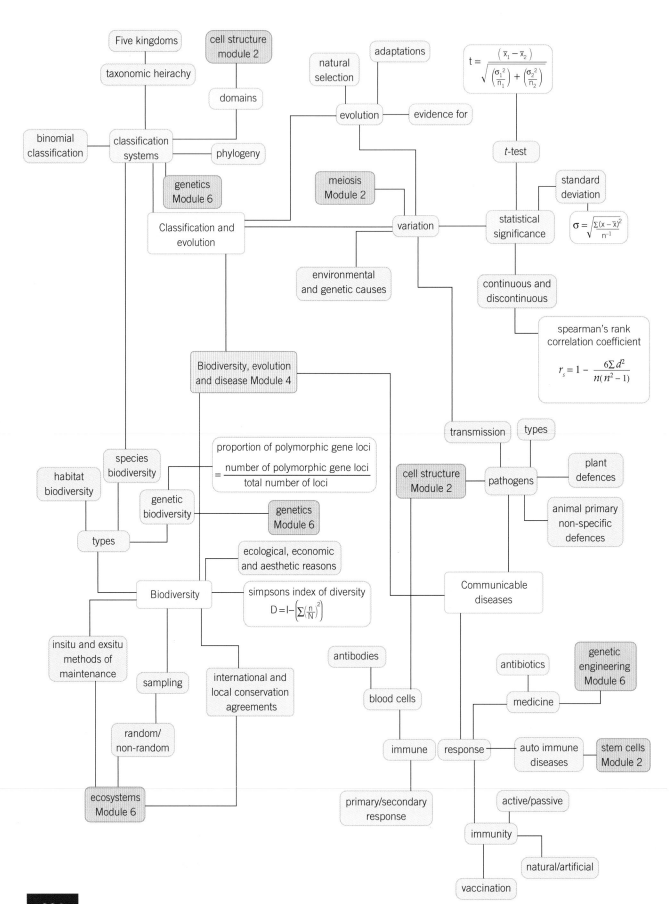

Application

Doctors at a Cambridge hospital were puzzled. A 50 year old Chinese male patient reported seizures, headaches, a changed sense of smell and memory flashbacks, however, tests for communicable diseases such as TB, HIV, syphilis and Lyme disease were all negative. MRI scans showed an abnormal region in his brain, however a biopsy showed only inflammation. Over four years, the symptoms changed and scans showed the abnormal region move across the patient's brain until finally in 2014 doctors operated. They were amazed to remove a 1 cm long tapeworm which was the cause of the symptoms. The patient recovered completely once the worm was removed!

This type of tapeworm was new to the UK. A tissue sample was sent to the Wellcome Trust Sanger Institute for genome sequencing and identified as *Spirometra erinaceieuropae*, a tapeworm species found in China, Japan, South Korea and Thailand. Known human infections are rare – only 300 in the last 60 years. The normal lifecycle of this tapeworm is: Eggs hatch in water → Infects tiny crustaceans → eaten by and infects reptiles e.g., snakes and amphibians e.g., frogs → infects

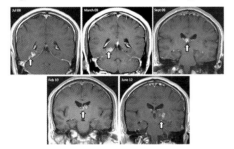

▲ **Figure 1** *The movement of a parasitic tapeworm through a human brain over time*

carnivores e.g., cats, dogs which eat intermediate hosts → proglottids passed out in faeces to start the cycle again.

Sequencing also showed the tapeworm genome was approximately 10 times bigger than any other sequenced tapeworm species. This gives the potential for it to express many different proteins, and therefore invade a wide variety of host animals in its lifecycle. It also showed that the species is not sensitive to a common anti-tapeworm drug, but might be sensitive to an alternative treatment.

1 a What is a communicable disease?
  b The patient described above was tested for TB and HIV/AIDS. What are the similarities and differences between these two diseases and the tapeworm infestation affecting the patient?
  c What is inflammation? Suggest a role for inflammation in the symptoms experienced by the patient during the infestation by the tapeworm?

2 a What methods are available to classify an organism such as *Spirometra erinaceieuropae*, found in an unexpected place?
  b What is natural selection and how is it involved in the process of evolution?
  c The genome of *Spirometra erinaceieuropae* is 10 times larger than any other species of tapeworm sequenced and almost a third of the size of the human genome. How might this large genome be a successful adaptation for this parasite.
  d Using the identification of *Spirometra erinaceieuropae* as one example, discuss the role of DNA sequencing in both classification and medicine.

Extension

The parasite *Spirometra erinaceieuropae* remained in the brain of a patient for over 4 years. It was not destroyed by the specific immune system.

1 Summarise how you would expect the specific immune system to respond to an invasion by another organism.
2 Investigate and describe the adaptations which enable a tapeworm such as *Spirometra*

erinaceieuropae to avoid destruction by the immune system of the host.

3 Identify one biochemical response by a parasitic worm and produce a clear explanation of the way it interacts with the host cells to reduce or moderate the host immune response. Suggest potential uses for these types of responses in the pharmaceutical industry.

1 In 1940 it was observed that about 70% of Europeans had the ability to curl their tongues up at the sides to form a tube shape.

It was suggested that 'tongue rolling' is controlled by one gene.

a **(i)** State the name of the type of variation shown if tongue rolling is only controlled by one gene. (*1 mark*)

It was later observed that it was possible for someone to learn to roll their tongue and that only 70% of identical twins have been found to share the tongue rolling trait.

(ii) Discuss, giving your reasons, whether tongue rolling does actually follow the type of variation named in a (i).
(*4 marks*)

Put your hands together, without giving it any thought, and interlock your fingers.

It has been found that 55% of people place their left thumb on top and that 45% place their right thumb on top and the rest show no preference.

Although the majority of identical twins share this trait, it does not follow a predictable pattern of inheritance.

b Outline the factors that influence which thumb is placed on top. (*3 marks*)

2 Mutation plays an important role in the processes of speciation.

a **(i)** State the definition of a species.
(*2 marks*)

(ii) Explain the role of mutation in the formation of a new species. (*4 marks*)

b **(i)** Describe what is meant by the term 'selection pressure'. (*2 marks*)

(ii) Discuss whether selection pressures are viewed as positive or negative influences on the rate of speciation.
(*3 marks*)

3 Human populations have an enormous impact on the biodiversity of ecosystems.

a Define the term biodiversity. (*2 marks*)

The impact of biodiversity is usually particularly apparent on farmland due to the agricultural practices in use.

b Explain the effects of each of the following practices on biodiversity.
Deforestation
Monoculture
Fertilisers
Culling of predators (*8 marks*)

4 Taiwan is an island with varied geography and landscape leading to the presence of a variety of habitats.

As global temperatures increase it has been predicted that many plants in Taiwan will migrate from lower to higher elevations above sea level.

A scientific study was carried out comparing the average number and Simpson's diversity index of six different species of plant against different elevations.

The results are shown in the graph.

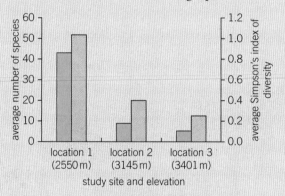

a **(i)** Describe how species diversity changes with elevation. (*4 marks*)

(ii) Explain what the Simpson's index of diversity shows. (*3 marks*)

It has been predicted that three species will migrate upwards to the higher zone. One species will migrate downwards to the lower zone and two species will stay where they are.

b Suggest how this will affect the species diversity at the different elevations.
(*4 marks*)

5 Copy and complete the paragraph, filling in the missing words.

T memory cells live for a long time and are part of the................. .... If they meet an .... for a second time, they undergo .... to form a large .... of T .... cells that destroy the pathogen. *(5 marks)*

From the AS paper 1 practice questions

6 Consider the statement:

'evolution can be summarised as change over time'

a Discuss why this statement is an over simplification as a description of evolution. *(4 marks)*

b Explain, using the finch populations observed by Darwin in the Galapagos islands the process of disruptive selection. *(6 marks)*

From the AS paper 1 practice questions

7 a Describe the difference between conservation and preservation. *(3 marks)*

b Describe the difference between 'in situ' and 'ex situ' conservation. *(4 marks)*

From the AS paper 1 practice questions

8 Which of the following statements is/are correct with respect to phylogenetic trees?

Statement 1: Phylogenetic trees depict the evolutionary relationships among groups of organisms

Statement 2: Species and their most recent common ancestor form a clade within a phylogenetic tree

Statement 3: Phylogenetic trees produced more recently show the relationship between clades and taxonomic groups

A 1, 2 and 3

B Only 1 and 2

C Only 2 and 3

D Only 1 *(1 mark)*

From the AS paper 1 practice questions

9 Variation is a fundamental characteristic of living organisms. Variation can be influenced by both genetics and the environment. The graphs below represent the two main types of variation.

A

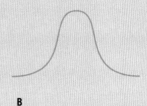

B

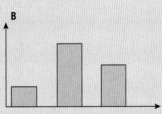

a Name the two types of variation shown in the graphs. *(2 marks)*

b Explain why evolution would not be possible without variation. *(4 marks)*

c Describe how variation arises. *(4 marks)*

d As humans, we have an anthropocentric view of evolution. This means the success of a species is measured by our own standards. Discuss whether our view of evolution is right. *(6 marks)*

From the AS paper 2 practice questions

10 a Describe the process of phagocytosis. *(5 marks)*

b Explain what is meant by the term pathogen. *(2 marks)*

c Explain why antibiotics are not prescribed to treat influenza. *(3 marks)*

d Suggest why antibiotics might still be prescribed to someone with influenza. *(3 marks)*

e Outline the different ways in which pathogens are made safe to use in vaccines. *(4 marks)*

f The diagram shows an antibody.

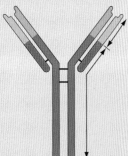

Describe how an antibody, such as the one in the diagram, has the ideal structure to carry out its role in an immune response. *(6 marks)*

OCR F224 2011

From the AS paper 2 practice questions

11 Advances in technology particularly in biochemistry have provided evidence that casts doubt on the way in which some organisms have been grouped in the five kingdom classification system.

DNA sequencing suggests that some organisms are more closely related to organisms belonging to another kingdom than other members of their own kingdom.

a Explain the meaning of the following terms:

(i) Phylogeny *(2 marks)*

(ii) Hierarchy *(2 marks)*

(iii) Taxonomy *(2 marks)*

(iv) Cladistics *(2 marks)*

b Outline the differences between kingdom and domain classification systems. *(5 marks)*

c Explain the meaning of the term 'descent with modification'. *(6 marks)*

From the AS paper 2 practice questions

12 The diagram shows the origin and development of a B lymphocyte in a tissue, X. It also shows the immune response in a lymph node following a vaccination with the measles virus.

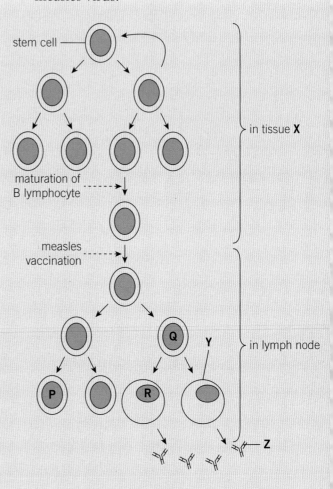

a Name tissue X, cell Y, and molecule Z. *(3 marks)*

b Molecules found on the surface of the measles virus can stimulate an immune response.

State the term given to molecules that stimulate an immune response. *(1 mark)*

c (i) Identify which of the cells P, Q, or R could become a memory cell. *(1 mark)*

(ii) State one function of memory cells. *(1 mark)*

d The graph shows the concentration of molecule Z in the blood following a vaccination for measles at day 0.

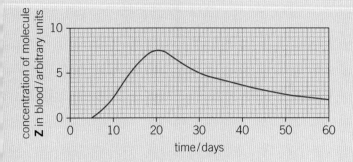

(i) Using the graph, describe how the concentration of molecule Z in the blood changes following vaccination.

(3 marks)

(ii) Copy and complete the table putting a tick (✓) in the appropriate box to indicate the type of immunity provided by this vaccination. *(1 mark)*

natural passive immunity	
natural active immunity	
artificial passive immunity	
artificial active immunity	

OCR 2802 2008

MODULE 5
Communication, homeostasis, and energy

Introduction

It is important that organisms, both plants and animals, are able to respond to stimuli. This is achieved by communication within the body, which may be chemical and/or electrical. Both systems are covered in detail in this module. Communication is also fundamental to homeostasis with control of temperature, blood sugar, and blood water potential being studied as examples. This module also explores the biochemical pathways of photosynthesis and respiration, with an emphasis on the formation and use of ATP as the source of energy for biochemical processes and synthesis of biological molecules.

Neuronal communication introduces you to how electrical systems are used to monitor and respond to any deviation from the body's steady state. This includes looking at action potentials and transmission between neurones at synapses.

Hormonal communication looks at how specific hormones bring about their effects.

Diabetes is used as an example of a defect in a hormonal control system. The kidneys and liver are examined in relation to the removal of toxic products of metabolism.

Homeostasis studies animal responses, which involve nervous, hormonal, and muscular coordination. This is contrasted with plants which use hormones to respond to environmental changes. You will also look at how these processes be exploited commercially.

Energy for biological processes looks in detail at this complex process, including how it is used to drive the production of chemicals, including ATP, and how large organic molecules are synthesised from inorganic molecules.

Respiration studies the series of enzyme controlled reactions which result in energy being transferred to ATP. ATP provides the immediate source of energy for biological processes.

Knowledge and understanding checklist

From your Key Stage 4 study you should be able to answer the following questions. Work through each point, using your Key Stage 4 notes and the support available on Kerboodle.

☐ Describe the relationship between the structure and function of the human nervous system.

☐ Describe the relationship between structure and function in a reflex arc.

☐ Explain the principles of hormonal coordination and control in humans.

☐ Describe how hormones are used in human reproduction.

☐ State a range of hormonal and non-hormonal methods of contraception.

☐ Describe the process of homeostasis.

☐ Describe photosynthesis as the key process for food production and therefore biomass for life.

☐ Explain the process of photosynthesis.

☐ State the factors affecting the rate of photosynthesis.

☐ Explains the importance of cellular respiration.

☐ Describe the processes of aerobic and anaerobic respiration.

☐ Define carbohydrates, proteins, nucleic acids, and lipids as key biological molecules.

Maths skills checklist

All Biologists need to use maths in their studies and field of work. In this module, you will need to use the following maths skills.

☐ **Standard deviation.** You will need to calculate the standard deviation to measure the spread in a set of data.

☐ **Student _t_ test.** You will need to use this test to compare the means of data values of two populations.

☐ **Respiratory quotient.** You will need to use this calculation to work out which substrate is being metabolised.

MyMaths.co.uk
Bringing Maths Alive

▲ **Figure 1** *Complex coordination systems are needed to enable your body to cope with the sort of changes you encounter when moving from a warm, lit house to a dark, frosty outside world*

Synoptic link

You will find out about the nervous system throughout Chapter 13, Neuronal communication, the hormonal system in Chapter 14, Hormonal communication, and coordination in plants in Chapter 16, Plant responses.

When changes occur in an organism's internal or external environment, the organism must respond to these changes in order to survive. Examples of these changes are shown in Table 1.

▼ Table 1

Internal environment	External environment
blood glucose concentration	humidity
internal temperature	external temperature
water potential	light intensity
cell pH	new or sudden sound

Animals and plants respond to these changes in a variety of ways. Animals react through electrical responses (via neurones), and through chemical responses (via hormones). Plant responses are based on a number of chemical communication systems including plant hormones. These communication systems must be coordinated to produce the required response in an organism.

Why coordination is needed

As species have evolved, cells within organisms have become specialised to perform specific functions. As a result organisms need to coordinate the function of different cells and systems to operate effectively. Few body systems can work in isolation (apart from a few exceptions, for example, a heart can continue to beat if placed in the right bathing solution). For example, red blood cells transport oxygen effectively, but have no nucleus. This means that these cells are not able to replicate – a constant supply of red blood cells to the body is maintained by haematopoietic stem cells. In order to contract, muscle cells must constantly respire, and thus require a consistent oxygen supply. As these cells cannot transport oxygen, they are dependent on red blood cells for this function. In plants flowering needs to coordinate with the seasons, and pollinators must coordinate with the plants. In temperate climates light-sensitive chemicals enable plants to coordinate the development of flower buds with the lengthening days that signal the approach of spring and summer.

◀ **Figure 2** *Everything from the fresh spring leaves to the reproductive behaviour of these birds and of the insects with which they feed their young is controlled by chemicals and/or nervous coordination*

Homeostasis

In many relatively large multicellular animals, different organs have different functions in the body. Therefore, the functions of organs must be coordinated in order to maintain a relatively constant internal environment. This is known as **homeostasis**. For example, the digestive organs such as the exocrine pancreas, duodenum, and ileum along with the endocrine pancreas and the liver work together to maintain a constant blood glucose concentration.

Cell signalling

Nervous and hormonal systems coordinate the activities of whole organisms. This coordination relies on communication at a cellular level through cell signalling. This occurs through one cell releasing a chemical which has an effect on another cell, known as a target cell. Through this process, cells can:

- transfer signals locally, for example, between neurones at synapses. Here the signal used is a neurotransmitter.
- transfer signals across large distances, using hormones. For example, the cells of the pituitary gland secrete antidiuretic hormone (ADH), which acts on cells in the kidneys to maintain water balance in the body.

Coordination in plants

Plants do not have a nervous system like animals. However, to survive they still must respond to internal and external changes to their environment. For example, plant stems grow towards a light source to maximise their rate of photosynthesis. This is achieved through the use of plant hormones.

Synoptic link

You will find out about homeostasis in Chapter 15, Homeostasis.

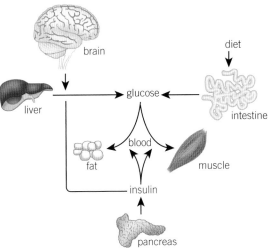

▲ **Figure 3** *To maintain blood glucose concentration, the function of many organs must be coordinated*

Synoptic link

You will find out more about cell signalling in Topic 14.5, Coordinated responses.

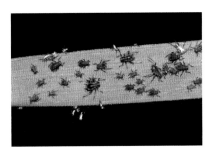

▲ **Figure 4** *When aphids attack certain types of wheat, the wheat plant cells produce chemicals both to alert other cells within the plant to the attack, and to signal to other aphids, putting them off landing on the plant and attacking it further*

Synoptic link

You will find out more about plant hormones in Chapter 16, Plant responses.

Summary questions

1 State one internal factor which causes a response in: *(2 marks)*
   a a plant
   b an animal

2 Describe how cells are able to communicate with one another. *(2 marks)*

3 Using examples, explain how and why coordination is required in a multicellular organism. *(6 marks)*

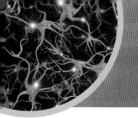

13.2 Neurones

Specification reference: 5.1.3

Learning outcomes

Demonstrate knowledge, understanding, and application of:

→ the structure and functions of sensory, relay, and motor neurones.

Synoptic link

You will find out about hormonal communication in Chapter 14, Hormonal communication.

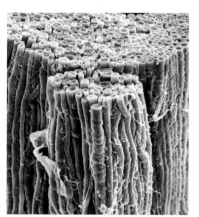

▲ **Figure 1** *Nerve fibres (axons). They are grouped together into bundles called nerves, which transmit electrical impulses to and from the central nervous system approx ×800 magnification*

The nervous system is responsible for detecting changes in the internal and external environment. These changes are known as a **stimulus**. This information then needs to be processed and an appropriate **response** triggered.

Both the nervous system and hormonal system play a role in reacting to stimuli, but they do so in very different ways. In this chapter you will focus on neuronal communication. This is generally a much faster and more targeted response than that produced by hormonal communication.

Neurones

You have already learnt that the nervous system is made up billions of specialised nerve cells called **neurones**. The role of neurones is to transmit electrical impulses rapidly around the body so that the organism can respond to changes in its internal and external environment. There are several different types of neurone found within a mammal. They work together to carry information detected by a sensory receptor to the effector, which in turn carries out the appropriate response.

Structure of a neurone

Mammalian neurones have several key features:

- Cell body – this contains the nucleus surrounded by cytoplasm. Within the cytoplasm there are also large amounts of endoplasmic reticulum and mitochondria which are involved in the production of **neurotransmitters**. These are chemicals which are used to pass signals from one neurone to the next. You will find out more about the important role of neurotransmitters in the nervous system in Topic 13.5, Synapses.

- Dendrons – these are short extensions which come from the cell body. These extensions divide into smaller and smaller branches known as dendrites. They are responsible for transmitting electrical impulses towards the cell body.

- Axons – these are singular, elongated nerve fibres that transmit impulses away from the cell body. These fibres can be very long, for example, those that transmit impulses from the tips of toes and fingers to the spinal cord. The fibre is cylindrical in shape consisting of a very narrow region of cytoplasm (in most cases approximately 1 μm) surrounded by a plasma membrane.

Types of neurone

Neurones can be divided into three groups according to their function. As a result they have slightly different structures:

- Sensory neurones – these neurones transmit impulses from a sensory receptor cell to a relay neurone, motor neurone, or the brain. They have one dendron, which carries the impulse to the cell body, and one axon, which carries the impulse away from the cell body.

- Relay neurones – these neurones transmit impulses between neurones. For example, between sensory neurones and motor neurones. They have many short axons and dendrons.

- Motor neurones – these neurones transmit impulses from a relay neurone or sensory neurone to an effector, such as a muscle or a gland. They have one long axon and many short dendrites.

In most nervous responses the electrical impulse follows the pathway:

Receptor → sensory neurone → relay neurone → motor neurone → effector cell

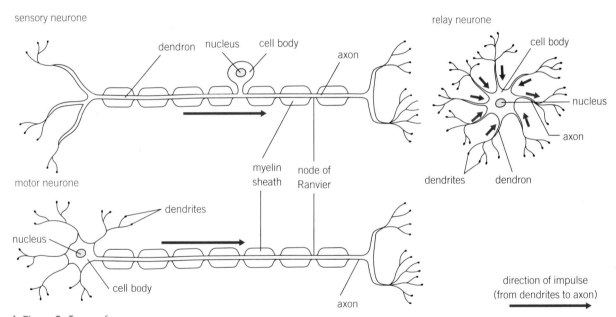

▲ **Figure 2** *Types of neurone*

Myelinated neurones

The axons of some neurones are covered in a **myelin sheath**, made of many layers of plasma membrane. Special cells, called Schwann cells, produce these layers of membrane by growing around the axon many times. Each time they grow around the axon, a double layer of phospholipid bilayer is laid down. When the Schwann cell stops growing there may be more than 20 layers of membrane. The myelin sheath acts as an insulating layer and allows these myelinated neurones to conduct the electrical impulse at a much faster speed than unmyelinated neurones. Myelinated neurones can transmit impulses at up to 100 metres per second. In comparison, non-myelinated neurones can only conduct impulses at approximately 1 metre per second.

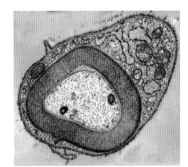

▲ **Figure 3** *Transverse section through an axon showing the myelin sheath created by the Schwann cell's membrane wrapping around the axon many times, approx ×25 000 magnification*

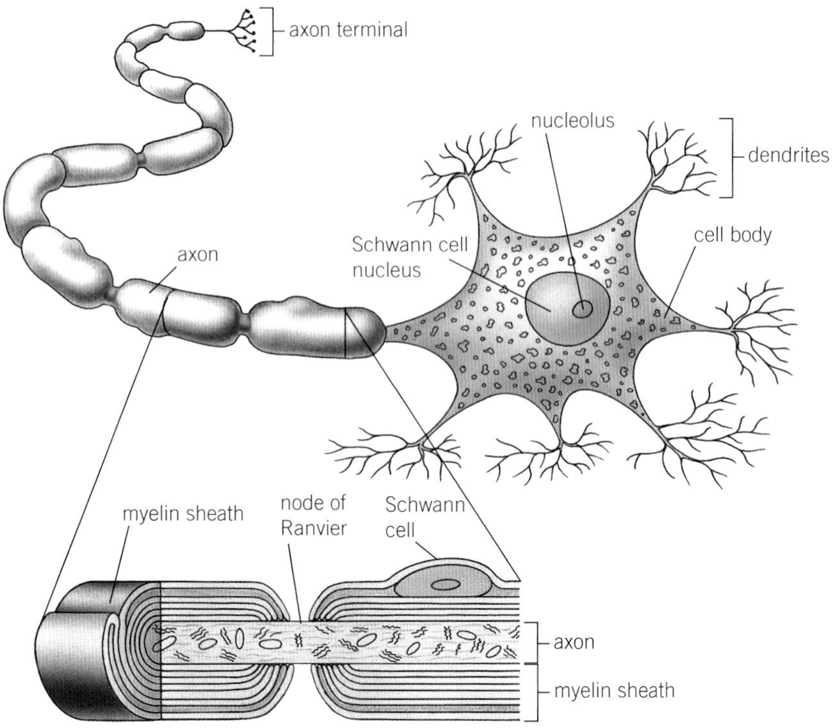

axon terminal

nucleolus

dendrites

Schwann cell
nucleus

cell body

axon

myelin sheath

node of
Ranvier

Schwann
cell

axon

myelin sheath

▲ Figure 4 *Structure of a myelinated motor neurone*

Summary questions

1 State the difference between the function of a motor and a sensory neurone. *(1 mark)*

2 Draw and annotate a diagram of a motor neurone. *(4 marks)*

3 Describe the difference in structure between a myelinated and a non-myelinated neurone and how this affects the speed a nerve impulse is transmitted. *(4 marks)*

Between each adjacent Schwann cell there is a small gap (2–3 μm) known as a node of Ranvier. This creates gaps in the myelin sheath. In humans these occur every 1–3 mm. The myelin sheath is an electrical insulator. In myelinated neurones, the electrical impulse 'jumps' from one node to the next as it travels along the neurone. This allows the impulse to be transmitted much faster. In non-myelinated neurones the impulse does not jump – it transmits continuously along the nerve fibre, so is much slower. You will find out more detail about the transmission of impulses along axons in Topic 13.4, Nervous transmission.

Multiple sclerosis

◀ Figure 5 *Computer artwork comparing a healthy myelinated nerve fibre (bottom) with one from a person with multiple sclerosis (top)*

Multiple sclerosis (MS) is a neurological condition which affects around 100 000 people in the UK. Most people are diagnosed between the ages of 20 and 40.

MS affects nerves in the brain and spinal cord, causing a wide range of symptoms, including problems with muscle movement, balance, and vision.

MS is known to be an autoimmune disease. This is where the immune system mistakenly attacks healthy body tissue. This results in a thinning or complete loss of the myelin sheath and, as the disease advances, results in the breakdown of the axons of neurones. It is not known what triggers this disorder but is thought to be a combination of genetic and environmental factors, such as a viral infection.

1 The population of the UK is around 65 000 000. Calculate the proportion of the population who suffer from MS.
2 State what is meant by an autoimmune disease.
3 Describe the role of myelin in the body.
4 One of the symptoms of MS is the loss of vision, normally in only one eye. Suggest why a damaged myelin sheath could prevent a person from being able to see.

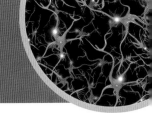

13.3 Sensory receptors
Specification reference: 5.1.3

The body is able to detect changes in its environment using groups of specialised cells known as sensory receptors. These are often located in the sense organs, such as the ear and eye.

Sensory receptors convert the stimulus they detect into a nerve impulse. The information is then passed through the nervous system and on into the central nervous system (CNS) – normally to the brain. The brain coordinates the required response and sends an impulse to an **effector** (normally a muscle or gland) to result in the desired response.

Learning outcomes
Demonstrate knowledge, understanding, and application of:
→ the roles of mammalian sensory receptors in converting different types of stimuli into nerve impulses.

Features of sensory receptors

All sensory receptors have two main features:

- They are specific to a single type of stimulus.
- They act as a transducer – they convert a stimulus into a nerve impulse.

There are four main types of sensory receptor present in an animal, shown in Table 1.

▼ Table 1

Type of sensory receptor	Stimulus	Example of receptor	Example of sense organ
mechanoreceptor	pressure and movement	Pacinian corpuscle (detects pressure)	skin
chemoreceptor	chemicals	olfactory receptor (detects smells)	nose
thermoreceptor	heat	end-bulbs of Krause	tongue
photoreceptors	light	cone cell (detects different light wavelengths)	eye

Synoptic link
Plants also respond to stimuli, but their receptor cells produce chemicals rather than nerve impulses. You will find out more about plant responses in Chapter 16, Plant responses.

Role as a transducer

Sensory receptors detect a range of different stimuli including light, heat, sound, or pressure. The receptor converts the stimulus into a nervous impulse, called a generator potential. For example, a rod cell (found in your eye) responds to light and produces a generator potential.

Pacinian corpuscle

Pacinian corpuscles are specific sensory receptors that detect mechanical pressure. They are located deep within your skin and are most abundant in the fingers and the soles of the feet. They are also found within joints, enabling you to know which joints are changing direction.

▲ Figure 1 *A Pacinian corpuscle. These sensory receptors are found mostly in the skin of the feet, hands, genitals, and nipples. They respond to vibration and deep pressure*

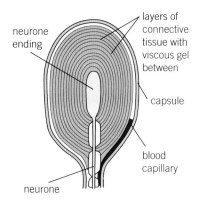

neurone ending

layers of connective tissue with viscous gel between

capsule

blood capillary

neurone

▲ **Figure 2** *The structure of a Pacinian corpuscle*

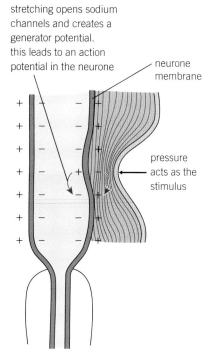

stretching opens sodium channels and creates a generator potential. this leads to an action potential in the neurone

neurone membrane

pressure acts as the stimulus

▲ **Figure 3** *A generator potential is created when mechanical pressure causes a Pacinian corpuscle to change shape*

Figure 2 shows the structure of a Pacinian corpuscle. The end of the sensory neurone is found within the centre of the corpuscle, surrounded by layers of connective tissue. Each layer of tissue is separated by a layer of gel.

Within the membrane of the neurone there are sodium ion channels. These are responsible for transporting sodium ions across the membrane. The neurone ending in a Pacinian corpuscle has a special type of sodium channel called a stretch-mediated sodium channel. When these channels change shape, for example, when they stretch, their permeability to sodium also changes.

You learnt about the structure of different neurones in Topic 13.2, Neurones and you will learn more about the transmission of nerve impulses in Topic 13.4, Nervous transmission.

The following steps explain how a Pacinian corpuscle converts mechanical pressure into a nervous impulse:

1   In its normal state (known as its resting state), the stretch-mediated sodium ion channels in the sensory neurone's membrane are too narrow to allow sodium ions to pass through them. The neurone of the Pacinian corpuscle has a **resting potential**.

2   When pressure is applied to the Pacinian corpuscle, the corpuscle changes shape. This causes the membrane surrounding its neurone to stretch.

3   When the membrane stretches, the sodium ion channels present widen. Sodium ions can now diffuse into the neurone.

4   The influx of positive sodium ions changes the potential of the membrane – it becomes *depolarised*. This results in a generator potential.

5   In turn, the generator potential creates an action potential (a nerve impulse) that passes along the sensory neurone.

The **action potential** will then be transmitted along neurones to the CNS.

Summary questions

1   Describe the role of a sensory receptor in the body. *(2 marks)*

2   State the transformation that takes place in a cone cell. *(1 mark)*

3   Explain how your body detects that your finger has touched a pin. *(6 marks)*

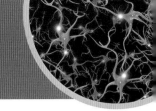

After the sensory receptor has detected a change in the environment, an impulse is sent along the neurone by temporarily changing the voltage (potential difference) across the axon's membrane. As a result, the axon membrane switches between two states – a resting potential and an action potential. You will find out how the impulse travels between neurones in Topic 13.5, Synapses.

Resting potential

When a neurone is not transmitting an impulse, the potential difference across its membrane (difference in charge between the inside and outside of the axon) is known as a resting potential. In this state, the outside of the membrane is more positively charged than the inside of the axon. The membrane is said to be *polarised* as there is a potential difference across it. It is normally about −70 mV.

The resting potential occurs as a result of the movement of sodium and potassium ions across the axon membrane. The phospholipid bilayer prevents these ions from diffusing across the membrane and, therefore, they have to be transported via channel proteins. Some of these channels are gated – they must be opened to allow specific ions to pass through them. Other channels remain open all of the time allowing sodium and potassium ions to simply diffuse through them.

The following events result in the creation of a resting potential:

● Sodium ions (Na⁺) are actively transported *out* of the axon whereas potassium ions (K⁺) are actively transported *into* the axon by a specific intrinsic protein known as the sodium–potassium pump. However, their movement is not equal. For every three sodium ions that are pumped out, two potassium ions are pumped in.

● As a result there are more sodium ions outside the membrane than inside the axon cytoplasm, whereas there are more potassium ions inside the cytoplasm than outside the axon. Therefore, sodium ions diffuse back into the axon down its electrochemical gradient (this is the name given to a concentration gradient of ions), whereas potassium ions diffuse out of the axon.

● However, most of the 'gated' sodium ion channels are closed, preventing the movement of sodium ions, whereas many potassium ion channels are open, thus allowing potassium ions to diffuse out of the axon. Therefore, there are more positively charged ions outside the axon than inside the cell. This creates the resting potential across the membrane of −70 mV, with the inside negative relative to the outside.

Learning outcomes

Demonstrate knowledge, understanding, and application of:

→ the generation and transmission of nerve impulses in mammals.

Synoptic link

Look back at Chapter 5, Plasma membranes to ensure you fully understand the structure of a plasma membrane and the role of intrinsic proteins.

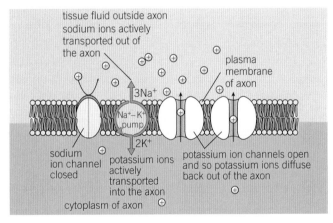

▲ **Figure 1** *Axon membrane during a resting potential*

Action potential

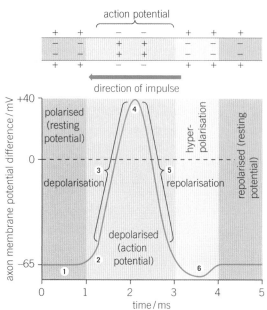

▲ **Figure 2** *Change in potential difference across an axon membrane during an action potential*

When a stimulus is detected by a sensory receptor, the energy of the stimulus temporarily reverses the charges on the axon membrane. As a result the potential difference across the membrane rapidly changes and becomes positively charged at approximately +40 mV. This is known as **depolarisation** – a change in potential difference from negative to positive. As the impulse passes **repolarisation** then occurs – a change in potential difference from positive back to negative. The neurone returns to its resting potential.

An action potential occurs when protein channels in the axon membrane change shape as a result of the change of voltage across its membrane. The change in protein shape results in the channel opening or closing. These channels are known as *voltage-gated ion channels*.

Figure 2 shows the changes in potential difference which occur across the axon membrane during an action potential.

The numbers on the graph correspond to the sequence of events that take place during an action potential:

1 The neurone has a resting potential – it is not transmitting an impulse. Some potassium ion channels are open (mainly those that are not voltage-gated) but sodium voltage-gated ion channels are closed.

2 The energy of the stimulus triggers some sodium voltage-gated ion channels to open, making the membrane more permeable to sodium ions. Sodium ions therefore diffuse into the axon down their electrochemical gradient. This makes the inside of the neurone less negative.

3 This change in charge causes more sodium ion channels to open, allowing more sodium ions to diffuse into the axon. This is an example of *positive feedback*.

4 When the potential difference reaches approximately +40 mV the voltage-gated sodium ion channels close and voltage-gated potassium ion channels open. Sodium ions can no longer enter the axon, but the membrane is now more permeable to potassium ions.

5 Potassium ions diffuse out of the axon down their electrochemical gradient. This reduces the charge, resulting in the inside of the axon becoming more negative than the outside.

6 Initially, lots of potassium ions diffuse out of the axon, resulting in the inside of the axon becoming more negative (relative to the outside) than in its normal resting state. This is known as *hyperpolarisation*. The voltage-gated potassium channels now close. The sodium-potassium pump causes sodium ions to move out of the cell, and potassium ions to move in. The axon returns to its resting potential – it is now repolarised.

Propagation of action potentials

A nerve impulse is an action potential that starts at one end of the neurone and is propagated along the axon to the other end of the neurone.

The initial stimulus causes a change in the sensory receptor which triggers an action potential in the sensory receptor, so the first region of the axon membrane is depolarised. This acts as a stimulus for the depolarisation of the next region of the membrane. The process continues along the length of the axon forming a wave of depolarisation. Once sodium ions are inside the axon, they are attracted by the negative charge ahead and the concentration gradient to diffuse further along inside the axon, triggering the depolarisation of the next section.

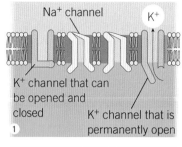

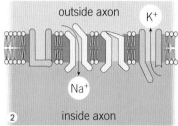

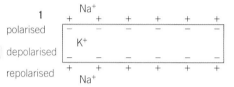

1 At resting potential the concentration of sodium ions outside the axon membrane is high relative to the inside, whereas that of the potassium ions is high inside the membrane relative to the outside. The overall concentration of positive ions is, however, greater on the outside, making this positive compared with the inside. The axon membrane is polarised.

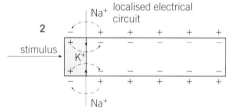

2 A stimulus causes a sudden influx of sodium ions and hence a reversal of charge on the axon membrane. This is the action potential and the membrane is depolarised.

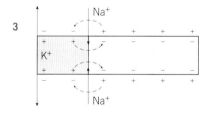

3 The localised electrical circuits established by the influx of sodium ions cause the opening of sodium voltage-gated channels a little further along the axon. The resulting influx of sodium ions in this region causes depolarisation. Behind this new region of depolarisation, the sodium voltage-gated channels close and the potassium ones open. Potassium ions begin to leave the axon along their electrochemical gradient.

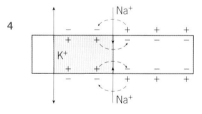

4 The action potential (depolarisation) is propagated in the same way further along the axon. The outward movement of the potassium ions has continued to the extent that the axon membrane behind the action potential has returned to its original charged state (positive outside, negative inside), that is, it has been repolarised.

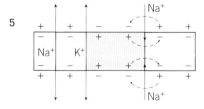

5 Following repolarisation the axon membrane returns to its resting potential in readiness for a new stimulus if it comes.

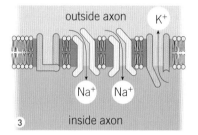

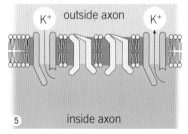

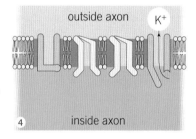

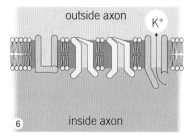

▲ Figure 3 *Propagation of an action potential along a non-myelinated neurone*

▲ Figure 4 *Changes in channel proteins in axon membrane during an action potential*

The region of the membrane which has been depolarised as the action potential passed along now undergoes repolarisation to return to its resting potential.

After an action potential there is a short period of time when the axon cannot be excited again, this is known as the *refractory period*. During this time, the voltage-gated sodium ion channels remain closed, preventing the movement of sodium ions into the axon.

A refractory period is important because it prevents the propagation of an action potential backwards along the axon as well as forwards. The refractory period makes sure action potentials are unidirectional. It also ensures that action potentials do not overlap and occur as discrete impulses.

Saltatory conduction

Myelinated axons transfer electrical impulses much faster than non-myelinated axons. This is because depolarisation of the axon membrane can only occur at the nodes of Ranvier where no myelin is present. Here the sodium ions can pass through the protein channels in the membrane. Longer localised circuits therefore arise between adjacent nodes. The action potential then 'jumps' from one node to another in a process known as saltatory conduction. This is much faster than a wave of depolarisation along the whole length of the axon membrane. Every time channels open and ions move it takes time, so reducing the number of places where this happens speeds up the action potential transmission. Long-term, saltatory conduction is also more energy efficient. Repolarisation uses ATP in the sodium pump, so by reducing the amount of repolarisation needed, saltatory conduction makes the conduction of impulses more efficient.

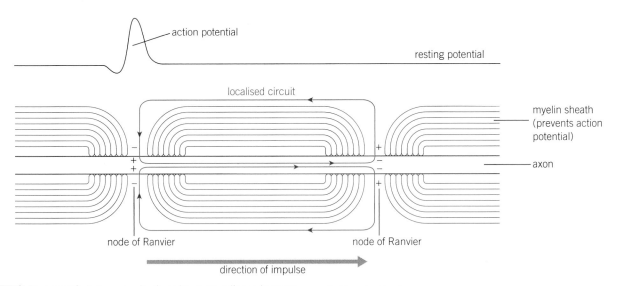

▲ **Figure 5** *Saltatory conduction along a myelinated axon*

Apart from myelination, two other factors affect the speed at which an action potential travels:

- Axon diameter – the bigger the axon diameter, the faster the impulse is transmitted. This is because there is less resistance to the flow of ions in the cytoplasm, compared with those in a smaller axon.

- Temperature – the higher the temperature, the faster the nerve impulse. This is because ions diffuse faster at higher temperatures. However, this generally only occurs up to about 40 °C as higher temperatures cause the proteins (such as the sodium–potassium pump) to become denatured.

All-or-nothing principle

Nerve impulses are said to be all-or-nothing responses. A certain level of stimulus, the *threshold value*, always triggers a response. If this threshold is reached an action potential will always be created. No matter how large the stimulus is, the same sized action potential will always be triggered. If the threshold is not reached, no action potential will be triggered.

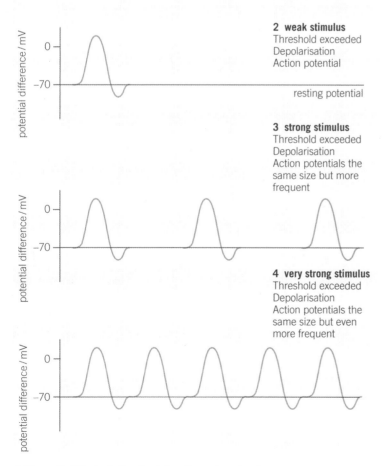

1 very weak stimulus
Threshold not exceeded No depolarisation No action potential

2 weak stimulus
Threshold exceeded
Depolarisation
Action potential

resting potential

3 strong stimulus
Threshold exceeded
Depolarisation
Action potentials the same size but more frequent

4 very strong stimulus
Threshold exceeded
Depolarisation
Action potentials the same size but even more frequent

▲ Figure 6 *Effect of stimulus intensity on impulse frequency*

The size of the stimulus, however, does affect the number of action potentials that are generated in a given time. The larger the stimulus the more frequently the action potentials are generated.

The effect of the size of the stimulus on the frequency of nerve impulses can be seen in Figure 6.

 Measuring action potentials

The presence and frequency of action potentials can be recorded using an oscilloscope. The diagram below shows some sample data collected in this manner, showing two action potentials:

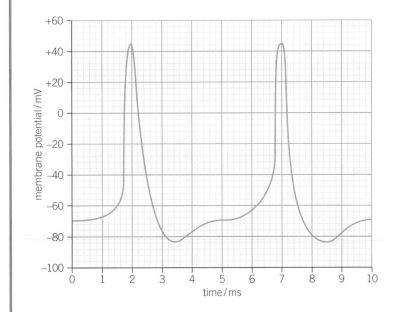

▲ Figure 7

1   State what occurs in the neurone between 1 ms and 2 ms.
2   State and explain how the membrane potential changes between 5.5 ms and 7 ms.
3   Using the data shown in the graph, calculate the frequency of action potentials.

Summary questions

1   State how the body detects the difference between a small and a large stimulus.   (*1 mark*)

2   State the difference between depolarisation, repolarisation, and hyperpolarisation.   (*2 marks*)

3   Describe what would happen if a refractory period did not exist.   (*2 marks*)

4   Describe how the movement of ions establishes the resting potential in an axon.   (*4 marks*)

5   Explain how temperature receptors in the hand generate an action potential in the sensory neurone to tell the body that you are touching a hot object.   (*6 marks*)

13.5 Synapses

Specification reference: 5.1.3

In Topic 13.4, Nervous transmission, you learnt about how impulses travel along each neurone in the form of an action potential. However, to reach the CNS or an effector, the impulse often needs to be passed between several neurones. The junction between two neurones (or a neurone and effector) is called a **synapse**. Impulses are transmitted across the synapse using chemicals called **neurotransmitters**.

Learning outcomes

Demonstrate knowledge, understanding, and application of:

→ the structure and roles of synapses in neurotransmission.

Synapse structure

All synapses have a number of key features:

- Synaptic cleft – the gap which separates the axon of one neurone from the dendrite of the next neurone. It is approximately 20–30 nm across.

- Presynaptic neurone – neurone along which the impulse has arrived.

- Postsynaptic neurone – neurone that receives the neurotransmitter

- Synaptic knob – the swollen end of the presynaptic neurone. It contains many mitochondria and large amounts of endoplasmic reticulum to enable it to manufacture neurotransmitters (in most cases).

- Synaptic vesicles – vesicles containing neurotransmitters. The vesicles fuse with the presynaptic membrane and release their contents into the synaptic cleft.

- Neurotransmitter receptors – receptor molecules which the neurotransmitter binds to in the postsynaptic membrane.

Types of neurotransmitter

Neurotransmitters can be grouped into two categories:

1 Excitatory – these neurotransmitters result in the depolarisation of the postsynaptic neurone. If the threshold is reached in the postsynaptic membrane an action potential is triggered. Acetylcholine is an example of an excitatory neurotransmitter.

2 Inhibitory – these neurotransmitters result in the hyperpolarisation of the postsynaptic membrane. This prevents an action potential being triggered. Gamma-aminobutyric acid (GABA) is an example of an inhibitory neurotransmitter that is found in some synapses in the brain.

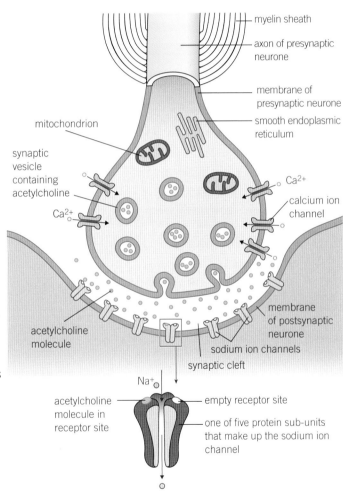

▲ Figure 1 Structure of a synapse. This synapse is known as a cholinergic synapse as the neurotransmitter which passes across the synapse is acetylcholine

Study tip

There are many different neurotransmitters such as dopamine, serotonin, and adrenaline. The one you have to concentrate on is acetylcholine and its role in a cholinergic synapse.

Transmission of impulses across synapses

Synaptic transmission occurs as a result of the following:

- The action potential reaches the end of the presynaptic neurone
- Depolarisation of the presynaptic membrane causes calcium ion channels to open
- Calcium ions diffuse into the presynaptic knob
- This causes synaptic vesicles containing neurotransmitters to fuse with the presynaptic membrane. Neurotransmitter is released into the synaptic cleft by exocytosis
- Neurotransmitter diffuses across the synaptic cleft and binds with its specific receptor molecule on the postsynaptic membrane
- This causes sodium ion channels to open
- Sodium ions diffuse into the postsynaptic neurone
- This triggers an action potential and the impulse is propagated along the postsynaptic neurone.

Once a neurotransmitter has triggered an action potential in the postsynaptic neurone, it is important that it is removed so the stimulus is not maintained, and so another stimulus can arrive at and affect the synapse. Any neurotransmitter left in the synaptic cleft is removed. Acetycholine is broken down by enzymes, which also releases them from the receptors on the postsynaptic membrane. The products are taken back into the presynaptic knob. Removing the neurotransmitter from the synaptic cleft prevents the response from happening again and allows the neurotransmitter to be recycled.

Transmission across cholinergic synapses

Cholinergic synapses use the neurotransmitter acetylcholine. They are common in the CNS of vertebrates and at neuromuscular junctions – where a motor neurone and a muscle cell (an effector) meet. If the neurotransmitter reaches the receptors on a muscle cell, it will cause the muscle to contract. Acetylcholine is released from the vesicles in the presynaptic knob (Figure 2). It then diffuses across the synaptic cleft where it binds with specific receptors in the postsynaptic membrane. This triggers an action potential in the postsynaptic neurone or muscle cell. Once an action potential has been triggered, acetylcholine is hydrolysed by a specific enzyme – acetylcholinesterase. This enzyme is also situated on the postsynaptic membrane. Acetylcholine is hydrolysed to give choline and ethanoic acid. One molecule of acetylcholinesterase can break down around 25 000 molecules of acetylcholine per minute. The breakdown products are taken back into the presynaptic knob to be reformed into acetylcholine, and the postsynaptic membrane is ready to receive another impulse.

1 The arrival of an action potential at the end of the presynaptic neurone causes calcium ion channels to open and calcium ions (Ca^{2+}) enter the synaptic knob.

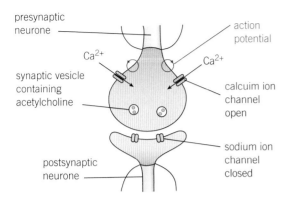

2 The influx of calcium ions into the presynaptic neurone causes synaptic vesicles to fuse with the presynaptic membrane, so releasing acetylcholine into the synaptic cleft.

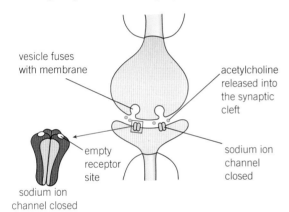

3 Acetylcholine molecules fuse with receptor sites on the sodium ion channel in the membrane of the postsynaptic neurone. This causes the sodium ion channels to open, allowing sodium ions (Na^+) to diffuse in rapidly along a concentration gradient.

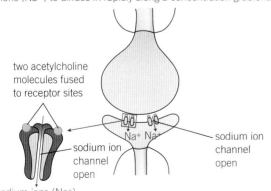

4 The influx of sodium ions generates a new action potential in the postsynaptic neurone.

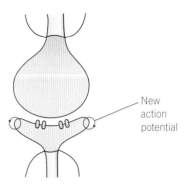

5 Acetylcholinesterase hydrolyses acetylcholine into choline and ethanoic acid (acetyl), which diffuse back across the synaptic cleft into the presynaptic neurone (= recycling). In addition to recycling the choline and ethanoic acid, the breakdown of acetylcholine also prevents it from continuously generating a new action potential in the postsynaptic neurone.

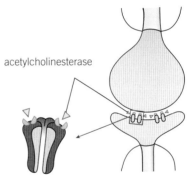

6 ATP released by mitochondria is used to recombine choline and ethanoic acid into acetycholine. This is stored in synaptic vesicles for future use. Sodium ion channels close in the absence of acetylcholine in the receptor sites.

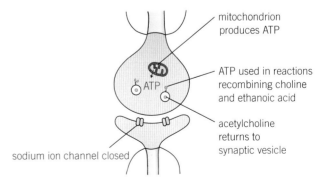

▲ **Figure 2** *Mechanism of transmission across a cholinergic synapse – only essential structures are shown*

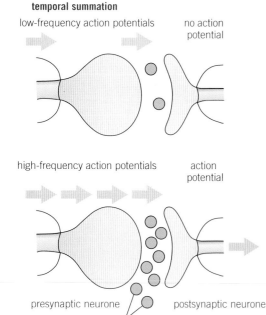

spatial summation

presynaptic neurone A — no action potential

presynaptic neurone B

no action potential

action potential

▲ **Figure 3** *On their own, neurone A and neurone B do not produce enough neurotransmitter to trigger an action potential. However, together there is enough neurotransmitter to reach the threshold and trigger an action potential in the postsynaptic neurone*

temporal summation

low-frequency action potentials — no action potential

high-frequency action potentials — action potential

presynaptic neurone — postsynaptic neurone

neurotransmitter

Role of synapses

In simple models a neurone with a single synapse is shown. In the body, however, one neurone may make thousands of synapses using the dendron, dendrites, and axons. Synapses play an important role in the nervous system:

● They ensure impulses are unidirectional. As the neurotransmitter receptors are only present on the postsynaptic membrane, impulses can only travel from the presynaptic neurone to the postsynaptic neurone.

● They can allow an impulse from one neurone to be transmitted to a number of neurones at multiple synapses. This results in a single stimulus creating a number of simultaneous responses.

● Alternatively, a number of neurones may feed in to the same synapse with a single postsynaptic neurone. This results in stimuli from different receptors interacting to produce a single result.

Summation and control

Each stimulus from a presynaptic neurone causes the release of the same amount of neurotransmitter into the synapse. In some synapses, however, the amount of neurotransmitter from a single impulse is not enough to trigger an action potential in the postsynaptic neurone, as the threshold level is not reached. However, if the amount of neurotransmitter builds up sufficiently to reach the threshold then this will trigger an action potential. This is known as **summation**. There are two ways this can occur:

● *Spatial summation* – this occurs when a number of presynaptic neurones connect to one postsynaptic neurone. Each releases neurotransmitter which builds up to a high enough level in the synapse to trigger an action potential in the single postsynaptic neurone (Figure 3).

● *Temporal summation* – this occurs when a single presynaptic neurone releases neurotransmitter as a result of an action potential several times over a short period. This builds up in the synapse until the quantity is sufficient to trigger an action potential (Figure 4).

◀ Figure 4 *If neurotransmitter is released several times in quick succession from a presynaptic neurone, the level builds up in the synapse and triggers an action potential in the postsynaptic neurone*

Effects of drugs on synapses

Many recreational and medical drugs cause their effects by acting on synapses. This will result in the nervous system being stimulated or inhibited.

Drugs that stimulate the nervous system create more action potentials in postsynaptic neurones, resulting in an enhanced response. For example, if the targeted synapse is with a neurone that transmits an impulse from a sound receptor, the body will perceive a louder sound. These drugs may work by:

- Mimicking the shape of the neurotransmitter – nicotine is the same shape as acetylcholine. It can therefore bind to acetylcholine receptors on the postsynaptic membrane and trigger action potentials in the postsynaptic neurone.

- Stimulating the release of more neurotransmitter. For example, amphetamines.

- Inhibiting the enzyme responsible for breaking down the neurotransmitter in the synapse. For example, nerve gases stop acetylcholine being broken down. This can result in a loss of muscle control.

Drugs that inhibit the nervous system create fewer action potentials in postsynaptic neurones, resulting in a reduced response. For example, if the targeted synapse is with a neurone that transmits an impulse from a sound receptor, the body will perceive a quieter sound. These drugs may work by:

- Blocking receptors – this means the neurotransmitter can no longer bind and activate the receptor. For example, curare blocks acetylcholine receptors at neuromuscular junctions. The muscle cells cannot therefore be stimulated, and the person suffers from paralysis.

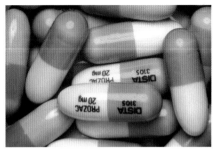

▲ **Figure 5** *Prozac is used to treat depression. It acts on neurotransmitters in synapses in the brain*

- Binding to specific receptors on the post-synaptic membrane of some neurones and changing the shape of the receptor such that binding of the neurotransmitter increases. This therefore increases activity. An example of this is alcohol binding to $GABA_A$ receptors.

1 State the difference between the result of an inhibitory and a stimulatory drug that acts on the nervous system.
2 Explain how amphetamines will affect the nervous system.
3 Explain how alcohol affects the nervous system.
4 Serotonin is a neurotransmitter involved in the regulation of sleep and other emotional states. Low levels of serotonin are found in patients suffering from depression. Prozac is an example of a drug used to treat depression. It works be blocking the reuptake of serotonin into the presynaptic neurone. Using your knowledge of synapses, explain how Prozac causes its effects.

Summary questions

1 State what is meant by a synapse. (*1 mark*)

2 Explain how synapses ensure impulses are only transmitted in one direction. (*2 marks*)

3 Describe one similarity and one difference between temporal and spatial summation. (*2 marks*)

4 Explain in detail how a motor neurone causes a postsynaptic neurone to depolarise. (*6 marks*)

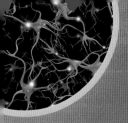

13.6 Organisation of the nervous system

Specification reference: 5.1.5

In the last few topics, you have looked in detail at how nervous impulses are transmitted around the nervous system – how are the billions of neurones in your nervous system organised?

Structural organisation

The mammalian nervous system is organised structurally into two systems:

- **Central nervous system** (CNS) – this consists of your brain and spinal cord.
- **Peripheral nervous system** (PNS) – this consists of all the neurones that connect the CNS to the rest of the body. These are the sensory neurones which carry nerve impulses from the receptors to the CNS, and the motor neurones which carry nerve impulses away from the CNS to the effectors.

Functional organisation

The nervous system is also functionally organised into two systems:

- **Somatic nervous system** – this system is under conscious control – it is used when you voluntarily decide to do something. For example, when you decide to move a muscle to move your arm. The somatic nervous system carries impulses to the body's muscles.
- **Autonomic nervous system** – this system works constantly. It is under subconscious control and is used when the body does something automatically without you deciding to do it – it is involuntary. For example, to cause the heart to beat, or to digest food. The autonomic nervous system carries nerve impulses to glands, smooth muscle (for example, in the walls of the intestine), and cardiac muscle.

The autonomic nervous system is then further divided by function into the sympathetic and parasympathetic nervous system. Generally, if the outcome increases activity it involves the sympathetic nervous system – for example, an increase in heart rate. If the outcome decreases activity it involves the parasympathetic nervous system – for example, a decrease in heart or breathing rate after a period of exercise.

Study tip

Although primarily unconscious, many aspects of the autonomic nervous system can come under conscious control. For example, people can choose to hold their breath or swallow rapidly. When people do not actively choose to control these functions, the autonomic nervous system takes over and controls them. This frees up the conscious areas of the brain – it would be hard to think of much else if you had to concentrate on breathing and keeping your heart beating.

▶ **Table 1** *Examples of the effects of sympathetic and parasympathetic stimulation. Note that most of these result in opposite effects on the body*

Structure	Sympathetic stimulation	Parasympathetic stimulation
salivary glands	saliva production reduced	saliva production increased
lung	bronchial muscle relaxed	bronchial muscle contracted
kidney	decreased urine secretion	increased urine secretion
stomach	peristalsis reduced	gastric juice secreted
small intestine	peristalsis reduced	digestion increased

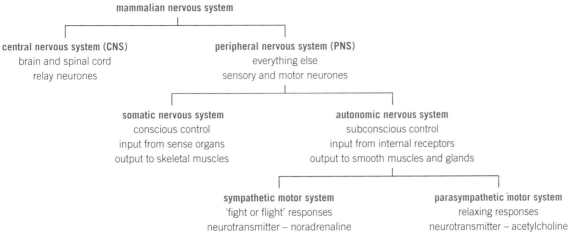

▲ Figure 1 *Summary of the organisation of the mammalian nervous system*

▼ Table 2 *Comparison of autonomic and somatic motor systems, where ACh = acetylcholine and NA = noradrenaline*

Cell bodies in central nervous system			Peripheral nervous system	Neurotransmitter at effector	Effector organs	Effect
somatic nervous system			single neuron from CNS to effector organs / heavily myelinated axon	ACh	skeletal muscle	+ stimulatory
autonomic nervous system	sympathetic		neurons from CNS to effector organs / lightly myelinated preganglionic axons / ganglion / unmyelinated postganglionic axon / noradrenaline / adrenal medulla / blood vessel	NA		+ −
	parasym-pathetic		lightly myelinated preganglionic axon / ganglion / unmyelinated postganglionic axon	ACh	smooth muscle (e.g., in gut), glands, cardiac muscle	stimulatory or inhibitory, depending on neuro-transmitter and receptors on effector organs

Summary questions

1 State the difference between the peripheral and the central nervous system.   (*1 mark*)

2 Sort the following activities into those which are controlled by the somatic nervous system and those which are controlled by the autonomic nervous system
   a pupil dilation   b blood pressure
   c throwing a ball   d walking   (*2 marks*)

3 State and explain one reason why many autonomic functions can also be controlled by the somatic nervous system.   (*3 marks*)

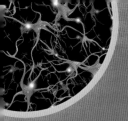

13.7 Structure and function of the brain

Specification reference: 5.1.5

An adult human brain contains approximately 86 billion neurones. The brain is responsible for processing all the information collected by receptor cells about changes in the internal and external environment. It also receives and processes information from the hormonal system through molecules in the blood. It must then produce a coordinated response.

The advantage of having a central control centre for the whole body is that communication between the billions of neurones involved is much faster than if control centres for different functions were distributed around the body. With the exception of reflex actions, all other nervous reactions are processed by the brain. You will find out about reflex actions in Topic 13.8, Reflexes.

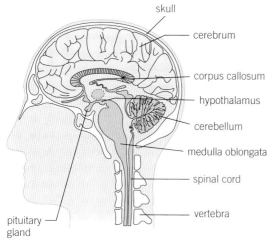

skull
cerebrum
corpus callosum
hypothalamus
cerebellum
medulla oblongata
spinal cord
vertebra
pituitary gland

▲ Figure 1 *Main structures in the brain*

Gross structure

The brain is protected by the skull. It is also surrounded by protective membranes (called meninges). The human brain is extremely complex, but the structures you need to know about are shown in Figure 1.

There are five main areas. They are distinguishable by their shape, colour, or microscopic structure:

● Cerebrum – controls voluntary actions, such as learning, memory, personality, and conscious thought.

● Cerebellum – controls unconscious functions such as posture, balance, and non-voluntary movement.

● Medulla oblongata – used in autonomic control, for example, it controls heart rate and breathing rate.

● Hypothalamus – regulatory centre for temperature and water balance.

● Pituitary gland – stores and releases hormones that regulate many body functions

Different images of the brain

Many different techniques are used to study the brain in order to understand its function. Figures 2, 3, and 4 all show a cross-section through the brain. Can you identify the main structures?

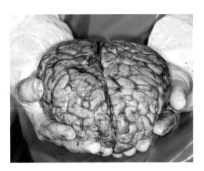

▲ **Figure 2** *Photo of the brain. Images like this have been taken during autopsies. The position of a lesion caused as a result of an accident, tumour, or stroke can be linked to observed changes in a patient's behaviour or capabilities before their death*

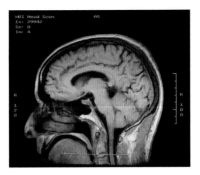

▲ **Figure 3** *Magnetic resonance imaging (MRI) of the brain. MRI is used to investigate the structure of the brain. A specialised version of MRI, called functional magnetic resonance imaging (fMRI), has been developed which allows the brain to be studied during activity. Active areas of the brain can be identified due to increased blood flow*

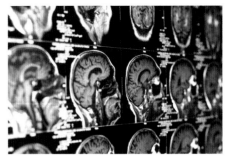

▲ **Figure 4** *Computerised tomography (CT) scan of the brain. A CT scan uses a series of X-rays to create detailed three-dimensional images of the inside of the body*

Cerebrum

The cerebrum receives sensory information, interprets it with respect to that stored from previous experiences, and then sends impulses along motor neurones to effectors to produce an appropriate response. It is responsible for coordinating all of the body's voluntary responses as well as some involuntary ones.

The cerebrum is highly convoluted, which increases its surface area considerably and therefore its capacity for complex activity. It is split into left and right halves known as the cerebral hemispheres. Each hemisphere controls one half of the body, and has discrete areas which perform specific functions – these areas are mirrored in each hemisphere. The outer layer of the cerebral hemispheres is known as the cerebral cortex. It is 2–4 mm thick. The most sophisticated processes such as reasoning and decision-making occur in the frontal and prefrontal lobe of the cerebral cortex.

Each sensory area within the cerebral hemispheres receives information from receptor cells located in sense organs. The size of the sensory area allocated is in proportion to the relative number of receptor cells present in the body part. The information is then passed on to other areas of the brain, known as association areas, to be analysed and acted upon. Impulses come into the motor areas where motor neurones send out impulses, for example, to move skeletal muscles. The size of the motor area allocated is in proportion to the relative number of motor endings in it. The main region which controls movement is the primary motor cortex located at the back of the frontal lobe.

In the base of the brain, impulses from each side of the body cross – therefore the left hemisphere receives impulses from the right-hand side of the body, and the right hemisphere receives impulses from the left-hand side of the body. For example, inputs from the eye pass to the visual area in the occipital lobe. Impulses from the right side of the field of vision in each eye are sent to the visual cortex in the left hemisphere, whereas impulses from the left side of the field of vision are sent to the right hemisphere. Through the integration of these inputs the brain is able to judge distance and perspective.

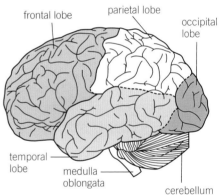

▲ **Figure 5** *The folded structure of the cerebral cortex viewed from the left side*

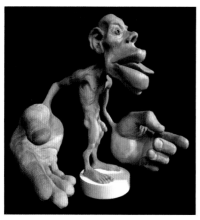

▲ **Figure 6** *A sensory homunculus. Neurobiologists have constructed models of the body in which the body part is made in proportion to the number of sensory inputs received from it. The hands and lips are very sensitive, and therefore they are drawn as particularly large in relation to other body parts*

Synoptic link

You will find out more about controlling heart rate in Topic 14.6, Controlling heart rate.

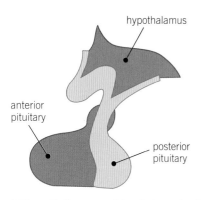

hypothalamus

anterior pituitary

posterior pituitary

▲ **Figure 7** *Structure of the pituitary gland*

Cerebellum

This area of the brain is concerned with the control of muscular movement, body posture, and balance – it does not initiate movement, but coordinates it. Therefore, if this area of the brain is damaged, a person suffers from jerky and uncoordinated movement. The cerebellum receives information from the organs of balance in the ears and information about the tone of muscles and tendons. It then relays this information to the areas of the cerebral cortex that are involved in motor control.

Medulla oblongata

The medulla oblongata contains many important regulatory centres of the autonomic nervous system. These control reflex activities such as ventilation (breathing rate) and heart rate. It also controls activities such as swallowing, peristalsis, and coughing.

Hypothalamus

This is the main controlling region for the autonomic nervous system. It has two centres – one for the parasympathetic and one for the sympathetic nervous system. It has a number of functions, which include:

● controlling complex patterns of behaviour, such as feeding, sleeping, and aggression

● monitoring the composition of blood plasma, such as the concentration of water and blood glucose – therefore it has a very rich blood supply

● producing hormones – it is an endocrine gland, that is, it produces hormones.

Pituitary gland

This is found at the base of the hypothalamus. It is approximately the size of a pea but it controls most of the glands in the body. It is divided into two sections:

● Anterior pituitary (front section) – produces six hormones including follicle-stimulating hormone (FSH), which is involved in reproduction and growth hormones.

● Posterior pituitary (back section) – stores and releases hormones produced by the hypothalamus, such as ADH involved in urine production.

Summary questions

1 State the difference between the function of the anterior pituitary and the posterior pituitary. (*1 mark*)

2 Sounds are interpreted by the auditory area in the temporal lobe. State the pathway followed by a nervous impulse produced by a sound wave. (*3 marks*)

3 A patient displays three symptoms: *asynergia* – a lack of coordination in their motor movement, *adiadochokinesia* – an inability to perform rapid movements, and *ataxic gait* – staggering movements. Suggest and explain which part of the brain may have been damaged to cause these symptoms. (*2 marks*)

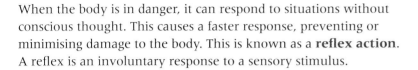

When the body is in danger, it can respond to situations without conscious thought. This causes a faster response, preventing or minimising damage to the body. This is known as a **reflex action**. A reflex is an involuntary response to a sensory stimulus.

Reflex arc

The pathway of neurones involved in a reflex action is known as a reflex arc. Most reflexes follow the same steps between the stimulus and the response:

- Receptor – detects stimulus and creates an action potential in the sensory neurone.
- Sensory neurone – carries impulse to spinal cord.
- Relay neurone – connects the sensory neurone to the motor neurone within the spinal cord or brain.
- Motor neurone – carries impulse to the effector to carry out the appropriate response.

Figure 1 illustrates what happens when you touch a hot candle – this is known as a withdrawal reflex. Before your brain registers that your hand is hot, the muscles in your arm have already pulled your hand away from the danger, minimising damage to your hand.

Learning outcomes

Demonstrate knowledge, understanding, and application of:

→ the reflex actions.

5 motor neurone passes impulses to the muscle

transverse section through spinal cord (magnified five times in relation to man)

7 response hand is moved quickly away from flame

6 effector contracts

1 stimulus heat from candle flame

4 relay neurone passes impulses across the spinal cord

2 thermoreceptor in skin detects heat

3 sensory neurone passes nerve impulses to spinal cord

▲ Figure 1 *Reflex arc involved in the withdrawal of the hand from a heat stimulus*

Spinal cord

The spinal cord is a column of nervous tissues running up the back. It is surrounded by the spine for protection. At intervals along the spinal cord pairs of neurones emerge, as shown in Figure 2.

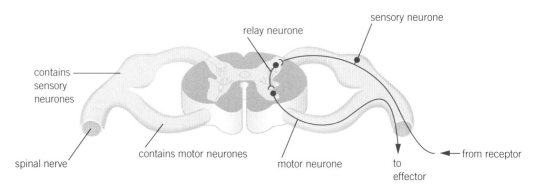

▲ **Figure 2** *Section through a spinal cord showing the neurones in a reflex arc*

Knee-jerk reflex

The knee-jerk reflex is a reflex commonly tested by doctors. It is a spinal reflex – this means that the neural circuit only goes up to the spinal cord, not the brain.

When the leg is tapped just below the kneecap (patella), it stretches the patellar tendon and acts as a stimulus. This stimulus initiates a reflex arc that causes the extensor muscle on top of the thigh to contract. At the same time, a relay neurone inhibits the motor neurone of the flexor muscle, causing it to relax. This contraction, coordinated with the relaxation of the antagonistic flexor hamstring muscle, causes the leg to kick.

After the tap of a hammer, the leg is normally extended once and comes to rest. The absence of this reflex may indicate nervous problems and multiple oscillation of the leg may be a sign of a cerebellar disease.

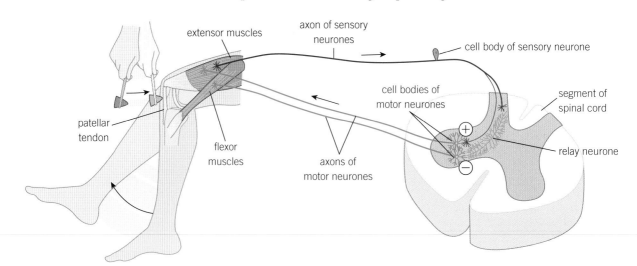

▲ **Figure 3** *The knee-jerk reflex*

This reflex is used by the body to help maintain posture and balance, allowing you to remain balanced with little effort or conscious thought.

Blinking reflex

The blinking reflex is an involuntary blinking of the eyelids (Figure 4). It occurs when the cornea is stimulated, for example, by being touched. Its purpose is to keep the cornea safe from damage due to foreign bodies such as dust or flying insects entering the eye – this type of response is known as the corneal reflex. A blink reflex also occurs when sounds greater than 40–60 dB are heard, or as a result of very bright light. Blinking as a reaction to over-bright light (to protect the lens and retina) is known as the optical reflex. The blinking reflex is a cranial reflex – it occurs in the brain, not the spinal cord.

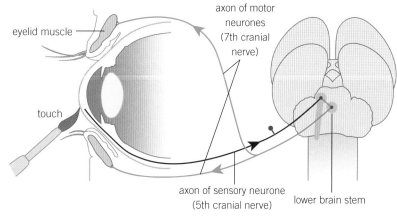

▲ **Figure 4** *The blinking reflex*

When the cornea of the eye is irritated by a foreign body, the stimulus triggers an impulse along a sensory neurone (the fifth cranial nerve). The impulse then passes through a relay neurone in the lower brain stem. Impulses are then sent along branches of the motor neurone (the seventh cranial nerve) to initiate a motor response to close the eyelids. The reflex initiates a consensual response – this means that both eyes are closed in response to the stimulus.

The blinking reflex is very rapid – it occurs in around one tenth of a second.

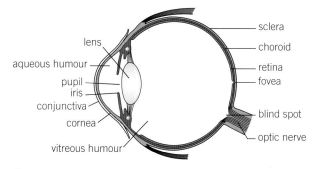

▲ **Figure 5** *Structure of the eye*

Doctors test for the blinking reflex when examining unconscious patients. If this reflex is present, it indicates that the lower brain stem is functioning. This procedure is therefore used as part of an assessment to determine whether or not a patient is brain-dead – if the corneal reflex is present the person cannot be diagnosed as brain-dead.

Measuring reaction time

When a person catches a falling object, at least part of this response is a reflex reaction. Measuring the time taken to catch a falling object can therefore be used to measure a person's reaction time.

To measure the reaction time, a suitable scale is placed onto the ruler which converts the distance dropped by the ruler into a reaction time. One investigation which can be carried out using this approach is to measure the effect of caffeine concentration on a person's reaction time.

1 State two variables which should be controlled when carrying out this investigation.
2 Explain why, in this investigation, only 'part of this response is a reflex action'.
3 Plan an investigation into how the concentration of caffeine affects a person's reaction time.

When carrying out this investigation, a researcher may choose to give a placebo caffeine drink to some of the people being tested. This is a drink labelled as a caffeine drink, but contains no caffeine.

4   Explain why a researcher may choose to give some people being tested a placebo.

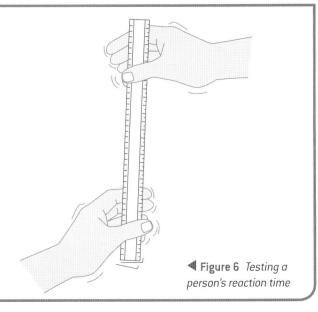

◀ Figure 6 *Testing a person's reaction time*

Survival importance

Reflexes are essential for survival as they avoid the body being harmed, or reduce the severity of any damage. For example, the iris contracts the pupil in bright light to prevent damage to the retina. In dim light, the reverse occurs to enable you to see as much as possible. Reflexes increase your chances of survival by:

● Being involuntary responses – the decision-making regions of the brain are not involved, therefore the brain is able to deal with more complex responses. It prevents the brain from being overloaded with situations in which the response is always the same.

● Not having to be learnt – they are present at birth and therefore provide immediate protection.

● Extremely fast – the reflex arc is very short. It normally only involves one or two synapses, which are the slowest part of nervous transmission.

● Many reflexes are what we would consider everyday actions, such as those which keep us upright (and thus not falling over), and those which control digestion.

Summary questions

1   State the reflex arc which occurs when a doctor tests the knee-jerk reflex.   (*1 mark*)

2   Which of the following actions are reflexes
    a   gagging   b   speaking
    c   jumping   d   pupil dilation.   (*1 mark*)

3   State and explain how a reflex action can improve an organism's chances of survival.   (*2 marks*)

4   ⚙ State and explain the considerations a researcher should take into account when planning an investigation into the effect of drugs, such as caffeine, on a group of human volunteers.   (*4 marks*)

13.9 Voluntary and involuntary muscles

Specification reference: 5.1.5

There are around 650 muscles in the body, making up roughly half of the body's weight. The contraction of many muscle cells causes the body to move (Topic 13.10, Sliding filament theory). However, there are many muscle cells in the body whose contractions you are largely unaware of.

Types of muscle

There are three types of muscle in the body:

- Skeletal muscle – skeletal muscles make up the bulk of body muscle tissue. These are the cells responsible for movement, for example, the biceps and triceps.

- Cardiac muscle – cardiac muscle cells are found only in the heart. These cells are myogenic, meaning they contract without the need for a nervous stimulus, causing the heart to beat in a regular rhythm.

- Involuntary muscle (also known as smooth muscle) – involuntary muscle cells are found in many parts of the body – for example, in the walls of hollow organs such as the stomach and bladder. They are also found in the walls of the blood vessels and the digestive tract, where through peristalsis they move food along the gut.

▼ **Table 1** *Important differences in the structure and function of the different types of muscle*

Type of muscle	Skeletal	Cardiac	Involuntary
Fibre appearance	striated	specialised striated	non-striated
Control	conscious (voluntary)	involuntary	involuntary
Arrangement	regularly arranged so muscle contracts in one direction	cells branch and interconnect resulting in simultaneous contraction	no regular arrangement – different cells can contract in different directions
Contraction speed	rapid	intermediate	slow
Length of contraction	short	intermediate	can remain contracted for a relatively long time
Structure	Muscles showing cross striations are known as striated or striped muscles. Fibres are tubular and multinucleated.	Cardiac muscle does show striations but they are much fainter than those in skeletal muscle. Fibres are branched and uninucleated.	Muscles showing no cross striations are called non-striated or unstriped muscles. Fibres are spindle shaped and uninucleated.

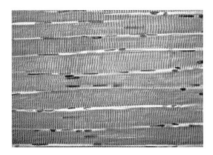

▲ **Figure 1** *Skeletal muscle, ×2 magnification*

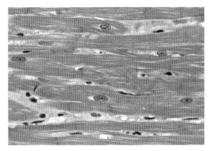

▲ **Figure 2** *Cardiac muscle, approx ×300 magnification*

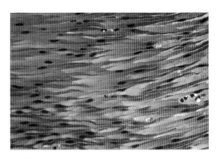

▲ **Figure 3** *Involuntary muscle, approx ×157 magnification*

Structure of skeletal muscle

Muscle fibres

Skeletal muscles are made up of bundles of muscle fibres. These are enclosed within a plasma membrane known as the *sarcolemma*.

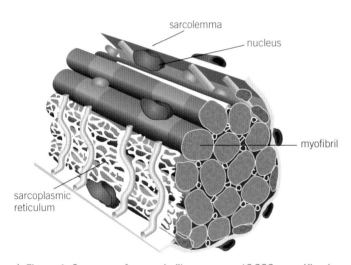

▲ **Figure 4** *Structure of a muscle fibre, approx ×10 000 magnification*

The muscle fibres contain a number of nuclei and are much longer than normal cells, as they are formed as a result of many individual embryonic muscle cells fusing together. This makes the muscle stronger, as the junction between adjacent cells would act as a point of weakness. The shared cytoplasm within a muscle fibre is known as *sarcoplasm*.

Parts of the sarcolemma fold inwards (known as transverse or T tubules) to help spread electrical impulses throughout the sarcoplasm. This ensures that the whole of the fibre receives the impulse to contract at the same time.

Muscle fibres have lots of mitochondria to provide the ATP that is needed for muscle contraction. They also have a modified version of the endoplasmic reticulum, known as the sarcoplasmic reticulum. This extends throughout the muscle fibre and contains calcium ions required for muscle contraction.

Myofibrils

Each muscle fibre contains many **myofibrils**. These are long cylindrical organelles made of protein and specialised for contraction. On their own they provide almost no force but collectively they are very powerful. Myofibrils are lined up in parallel to provide maximum force when they all contract together. Myofibrils are made up of two types of protein filament:

- Actin – the thinner filament. It consists of two strands twisted around each other.
- Myosin – the thicker filament. It consists of long rod-shaped fibres with bulbous heads that project to one side.

Study tip

When thinking about the structure of muscles think of a rope. A rope is made up of lots of strings (muscle fibres), which themselves are made up of lots of threads (myofibrils), acting together to give it its strength.

Myofibrils have alternating light and dark bands – these result in their striped appearance:

- Light bands – these areas appear light as they are the region where the actin and myosin filaments do not overlap. (They are also known as isotopic bands or I-bands.)

- Dark bands – these areas appear dark because of the presence of thick myosin filaments. The edges are particularly dark as the myosin is overlapped with actin. (They are also known as anisotropic bands or A-bands.)

- Z-line – this is a line found at the centre of each light band. The distance between adjacent Z-lines is called a **sarcomere**. The sarcomere is the functional unit of the myofibril. When a muscle contracts the sarcomere shortens.

- H-zone – this is a lighter coloured region found in the centre of each dark band. Only myosin filaments are present at this point. When the muscle contracts the H-zone decreases.

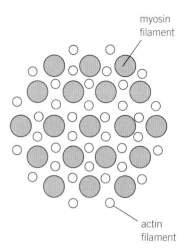

▲ **Figure 5** *Transverse section through a myofibril, approx × 50 000 magnification*

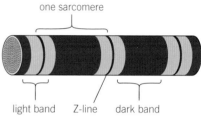

▲ **Figure 6** *The structure of a myofibril*

Drawing a labelled diagram of a sarcomere

Key points you should remember when drawing and labelling a sarcomere are:

- Show two Z-lines to demonstrate your understanding of the length of a sarcomere.
- Ensure there are heads present on the myosin filaments.
- Connect actin filaments to the Z-line.
- Clearly label the light and dark bands.
- Show the position of the H-zone.

It is useful to note the position of the A band, I band, H zone, and Z-line but you do not need to learn them.

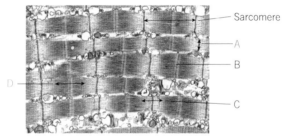

▲ **Figure 7** *TEM of skeletal muscle, ×5 000 magnification*

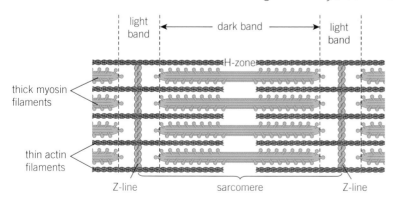

▲ **Figure 8** *The structure of a sarcomere*

Why are there light bands within a sarcomere?

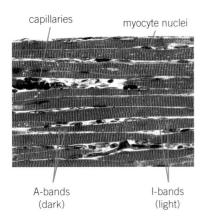

capillaries   myocyte nuclei

A-bands   I-bands
(dark)   (light)

▲ **Figure 9** *Stained skeletal muscle as seen through a microscope, ×570 magnification*

Histology of skeletal muscle

Figure 9 is a stained section of skeletal muscle viewed through a microscope. You should be able to identify the following features:

● Individual muscle fibres – long and thin multinucleated fibres that are crossed with a regular pattern of fine red and white lines.

● The highly structured arrangement of sarcomeres which appear as dark (A-bands) and light (I-bands) bands.

● Streaks of connective and adipose tissue.

● Capillaries running in between the fibres.

➕ Slow-twitch and fast-twitch muscles

There are two types of muscle fibres found in your body. Different muscles in the body have different proportions of each fibre.

Properties of slow-twitch fibres:

● fibres contract slowly

● provide less powerful contractions but over a longer period

● used for endurance activities as they do not tire easily

● gain their energy from aerobic respiration

● rich in myoglobin, a bright red protein which stores oxygen – this makes the fibres appear red

● rich supply of blood vessels and mitochondria.

Slow-twitch fibres are found in large proportions in muscles which help to maintain posture such as those in the back and calf muscles which have to contract continuously to keep the body upright.

Properties of fast-twitch fibres:

● fibres contract very quickly

● produce powerful contractions but only for short periods

● used for short bursts of speed and power as they tire easily

● gain their energy from anaerobic respiration

● pale coloured as they have low levels of myoglobin and blood vessels

● contain more, and thicker, myosin filaments

● store creatine phosphate – a molecule that can rapidly generate ATP from ADP in anaerobic conditions.

Fast-twitch fibres are found in high proportions in muscles which need short bursts of intense activity, such as biceps and eyes.

1 🜨 State and explain how you could tell the difference between areas of fast-twitch and slow-twitch muscle when observing skeletal muscle under the microscope.

2 State and explain any differences in composition which may exist in the skeletal muscles of marathon runners and sprinters.

Summary questions

1 Describe simply the structure of skeletal muscle.   (*3 marks*)

2 🧪 Figure 7 shows an image of skeletal muscle viewed under an electron microscope. Name the structures labelled A–D.   (*4 marks*)

3 Describe the similarities and differences in the structure and function of cardiac and involuntary muscle.   (*4 marks*)

4 The drawings in Figure 10 show myofibrils in transverse section.
   a Describe the difference between taking a transverse and a longitudinal section of muscle.   (*1 mark*)
   b State and explain which section through a sarcomere is represented by each image in Figure 10.   (*4 marks*)

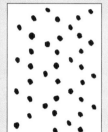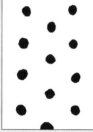

▲ **Figure 10** *Transverse sections of striated muscle*

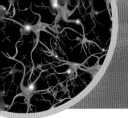

13.10 Sliding filament model

In the previous topic, you looked in detail at the structure of skeletal muscle fibres. In order to contract and cause movement, the actin and myosin filaments within the myofibrils have to slide past each other. Muscle contraction is usually described using the **sliding filament model**.

Sliding filament model

During contraction the myosin filaments pull the actin filaments inwards towards the centre of the sarcomere. This results in:

- the light band becoming narrower
- the Z lines moving closer together, shortening the sarcomere
- the H-zone becoming narrower.

The dark band remains the same width, as the myosin filaments themselves have not shortened, but now overlap the actin filaments by a greater amount.

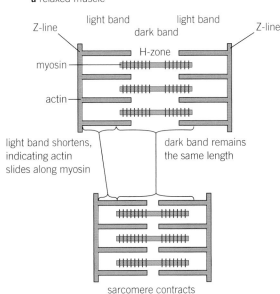

a relaxed muscle

light band shortens, indicating actin slides along myosin

dark band remains the same length

sarcomere contracts

b contracted muscle

▲ Figure 2 *Comparison of a relaxed and contracted sarcomere*

▲ Figure 1 *Electron micrograph of relaxed and contracted sarcomeres*

The simultaneous contraction of lots of sarcomeres means that the myofibrils and muscle fibres contract. This results in enough force to pull on a bone and cause movement. When sarcomeres return to their original length the muscle relaxes.

Structure of myosin

Myosin filaments have globular heads that are hinged which allows them to move back and forwards. On the head is a binding site for each of actin and ATP. The tails of several hundred myosin molecules are aligned together to form the myosin filament.

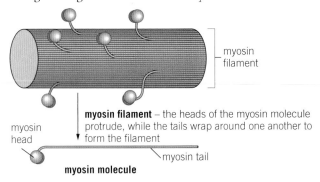

myosin filament

myosin filament – the heads of the myosin molecule protrude, while the tails wrap around one another to form the filament

myosin head

myosin tail

myosin molecule

▶ Figure 3 *Myosin structure*

Structure of actin

Actin filaments have binding sites for myosin heads. These are called actin–myosin binding sites. However, these binding sites are often blocked by the presence of another protein called tropomyosin which is held in place by the protein troponin.

When a muscle is in a resting state (relaxed) the actin–myosin sites are blocked by tropomyosin. The myosin heads can therefore not bind to the actin, and the filaments cannot slide past each other.

When a muscle is stimulated to contract, the myosin heads form bonds with actin filaments known as actin–myosin cross-bridges. The myosin heads then flex (change angle) in unison, pulling the actin filament along the myosin filament. The myosin then detaches from the actin and its head returns to its original angle, using ATP. The myosin then reattaches further along the actin filament and the process occurs again. This is repeated up to 100 times per second.

How muscle contraction occurs

Neuromuscular junction

Muscle contraction is triggered when an action potential arrives at a neuromuscular junction – this is the point where a motor neurone and a skeletal muscle fibre meet (Topics 13.4, Nervous transmission and 13.5, Synapses). There are many neuromuscular junctions along the length of a muscle to ensure that all the muscle fibres contract simultaneously. If only one existed, the muscle fibres would not contract together therefore the contraction of the muscle would not be as powerful. It would also be much slower, as a wave of contraction would have to travel across the muscle to stimulate the individual fibres to contract.

All the muscle fibres supplied by a single motor neurone are known as a motor unit – the fibres act as a single unit. If a strong force is needed, a large number of motor units are stimulated, whereas only a small number are stimulated if a small force is required.

When an action potential reaches the neuromuscular junction, it stimulates calcium ion channels to open. Calcium ions then diffuse from the synapse into the synaptic knob, where they cause synaptic vesicles to fuse with the presynaptic membrane. Acetylcholine is released into the synaptic cleft by exocytosis and diffuses across the synapse. It binds to receptors on the postsynaptic membrane (the sarcolemma), opening sodium ion channels, and resulting in depolarisation.

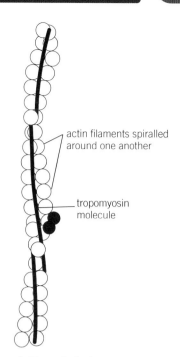

▲ **Figure 4** *Actin structure*

> **Study tip**
>
> Try to visualise the sliding filament model as a rowing boat containing several rowers. When the oars (myosin heads) are dipped into the river (bind to actin filament), the oars change angle (myosin heads flex), then are removed (myosin heads detach). This is then repeated further along the river. The rowers work in unison, and so the boat and water move relative to one another.

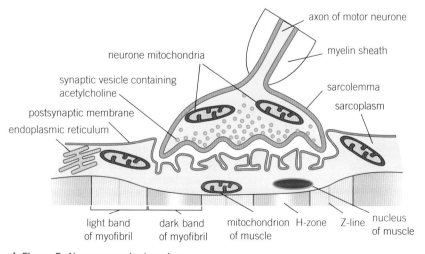

▲ **Figure 5** *Neuromuscular junction*

Acetylcholine is then broken down by acetylcholinesterase into choline and ethanoic acid. This prevents the muscle being overstimulated. Choline and ethanoic acid diffuse back into the neurone, where they are recombined into acetylcholine, using the energy provided by mitochondria.

Sarcoplasm

The depolarisation of the sarcolemma travels deep into the muscle fibre by spreading through the T-tubules. These are in contact with the sarcoplasmic reticulum. The sarcoplasmic reticulum contains stored calcium ions which it actively absorbs from the sarcoplasm.

When the action potential reaches the sarcoplasmic reticulum it stimulates calcium ion channels to open. The calcium ions diffuse down their concentration gradient flooding the sarcoplasm with calcium ions.

The calcium ions bind to troponin causing it to change shape. This pulls on the tropomyosin moving it away from the actin–myosin binding sites on the actin filament. Now that the binding sites have been exposed the myosin head binds to the actin filament forming an actin–myosin cross-bridge.

Once attached to the actin filament the myosin head flexes, pulling the actin filament along. The molecule of ADP bound to the myosin head is released. An ATP molecule can now bind to the myosin head. This causes the head to detach from the actin filament.

The calcium ions present in the sarcoplasm also activate the ATPase activity of the myosin. This hydrolyses the ATP to ADP and phosphate, releasing energy which the myosin head uses to return to its original position.

The myosin head can now attach itself to another actin–myosin binding site further along the actin filament and the cycle is repeated. The cycle continues as long as the muscle remains stimulated. During the period of stimulation many actin–myosin bridges form and break rapidly, pulling the actin filament along. This shortens the sarcomere and causes the muscle to contract.

Figure 6 summarises what takes place in the sarcoplasm.

Energy supply during muscle contraction

Muscle contraction requires large quantities of energy. This is provided by the hydrolosis of ATP into ADP and phosphate. The energy is required for the movement of the myosin heads and to enable the sarcoplasmic reticulum to actively reabsorb calcium ions from the sarcoplasm. The three main ways ATP is generated are described here – many activities use a combination of these processes.

Aerobic respiration

Most of the ATP used by muscle cells is regenerated from ADP during oxidative phosphorylation. This chemical reaction takes place inside the mitochondria which are plentiful in the muscle. However, this can only occur in the presence of oxygen. Aerobic respiration is therefore used for long periods of low-intensity exercise.

1 Tropomyosin molecule prevents myosin head from attaching to the binding site on the actin molecule.

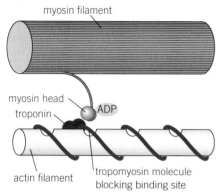

myosin filament

myosin head
troponin
ADP
actin filament
tropomyosin molecule
blocking binding site

2 Calcium ions released from the endoplasmic reticulum cause the tropomyosin molecule to pull away from the binding sites on the actin molecule.

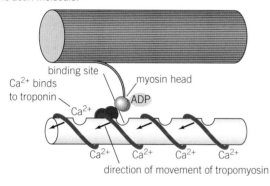

binding site
Ca^{2+} binds
to troponin
myosin head
Ca^{2+}
ADP
Ca^{2+} Ca^{2+} Ca^{2+} Ca^{2+}
direction of movement of tropomyosin

3 Myosin head now attaches to the binding site on the actin filament.

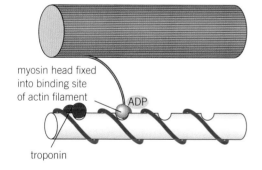

myosin head fixed
into binding site
of actin filament
ADP

troponin

4 Head of myosin changes angle, moving the actin filament along as it does so. The ADP molecule is released.

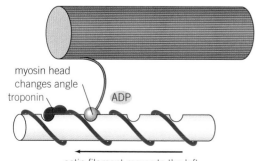

myosin head
changes angle
troponin
ADP

actin filament moves to the left

5 ATP molecule fixes to myosin head, causing it to detach from the actin filament.

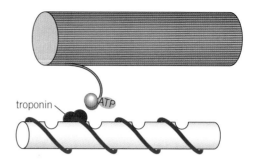

troponin
ATP

6 Hydrolysis of ATP to ADP by myosin provides the energy for the myosin head to resume its normal position.

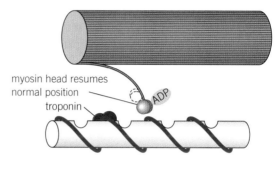

myosin head resumes
normal position
troponin
ADP

7 Head of myosin reattaches to a binding site further along the actin filament and the cycle is repeated.

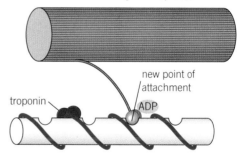

new point of
attachment
troponin
ADP

▲ **Figure 6** *Interaction of myosin and actin during muscle contraction. Only one myosin head is shown for simplicity*

Anaerobic respiration

In a very active muscle, oxygen is used up more quickly than the blood supply can replace it. Therefore, ATP has to be generated anaerobically. ATP is made by glycolysis but, as no oxygen is present, the pyruvate which is also produced is converted into lactate (lactic acid). This can quickly build up in the muscles resulting in muscle fatigue. Anaerobic respiration is used for short periods of high-intensity exercise, such as sprinting.

Creatine phosphate

Another way the body can generate ATP is by using the chemical *creatine phosphate* which is stored in muscle. To form ATP, ADP has to be phosphorylated – a phosphate group has to be added. Creatine phosphate acts as a reserve supply of phosphate, which is available immediately to combine with ADP, reforming ATP. This system generates ATP rapidly, but the store of phosphate is used up quickly. As a result this is used for short bursts of vigorous exercise, such as a tennis serve. When the muscle is relaxed, the creatine phosphate store is replenished using phosphate from ATP.

> **Synoptic link**
>
> You will study in detail the chemical reactions which take place during respiration in Chapter 18, Respiration.

 Monitoring muscle activity with sensors

Sensors can be used to monitor the electrical activity in a muscle. These can be used to measure the strength of a muscle contraction, or to track muscle fatigue levels.

The resultant trace, an electromyogram (EMG), is a record of the electrical activity in a muscle during an activity.

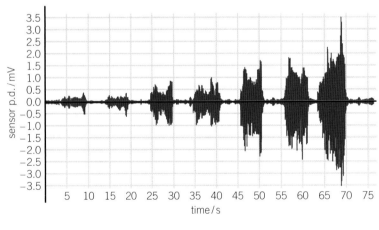

◀ **Figure 7** *EMG sensor trace from a biceps muscle. The subject is applying a gradually increasing force, which results in a larger output trace on the EMG, p.d = partial discharge*

Muscle fatigue is a long-lasting reduction of the ability to contract and exert force. It is normally localised and occurs after prolonged, relatively strong muscle activity. Occasionally, this can be beneficial, through promoting muscle growth (as seen in bodybuilders). However, it is usually harmful – serious injury is most likely to occur when the level of fatigue in a muscle is high.

The detection and classification of muscle fatigue is important in research into human–computer interactions, sport injuries and performance, ergonomics, and prosthetics.

A typical experiment in muscle fatigue research involves a subject performing a set task such as moving a limb in a specified manner. A signal is acquired using sensors attached to the skin, which is recorded and processed to reveal the characteristics of the muscle during that particular exercise.

To gain the signal trace, electromyography uses electrodes which detect the electrical currents created when muscles contract. Changes in this signal are used to identify fatigue, which may include:

- an increase in the mean amplitude of the signal
- a decrease in the frequency of the signal
- disruption to the overall pattern existing within the signal.

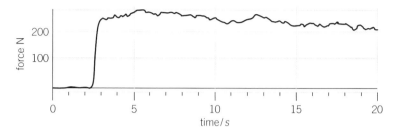

◀ **Figure 8** *At 2.5s the subject grasped a grip strength meter as hard as possible. Notice how the strength of the subject's grip peaked at 5.5s and then gradually declined until recording ended at 20s*

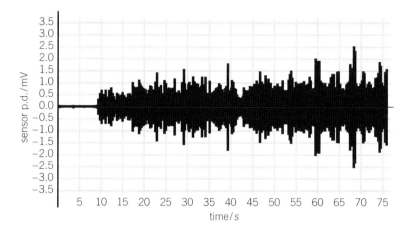

◀ **Figure 9** *EMG sensor trace from a muscle undergoing fatigue. Notice that the electrical activity in the muscle increases over time while, as shown in Figure 8, the maximum force produced by a muscle decreases*

1 Plan an investigation into muscle fatigue, using an EMG to monitor the activity of the muscle.

2 🜂 ⚙️ Suggest how the outcomes from the investigation could be used to identify when the muscle became fatigued.

Summary questions

1 🜂 State two differences in the appearance of a sarcomere in a relaxed and contracted muscle when observed through a microscope. (*2 marks*)

2 Professional sprinters have high levels of creatine phosphate in their muscle cells. Describe why this is advantageous. (*2 marks*)

3 After a person's death, their body can no longer produce ATP. This results in the stiffening of muscles (rigor mortis). Explain why a lack of ATP prevents muscles relaxing. (*3 marks*)

4 Bepridil is a drug that can be used to treat angina, a form of heart disease. It works by partially blocking calcium ion channels. Explain the effect bepridil will have on heart muscle contraction. (*4 marks*)

Practice questions

1 Tension in muscles is created as the myosin filaments pull the actin filaments towards each other as cross bridges form between the filaments. The resting length of muscle fibres determines, along with the frequency of stimulation, the magnitude of this tension.

Elastic proteins, such as titins, present in muscle resist the overstretching of muscle fibres.

a (i) State the name of cross bridges that form between the filaments. (*1 mark*)

(ii) State the term that describes the fact that both the frequency of stimulation and arrival of action potentials determine the size of the response. (*1 mark*)

The diagram shows the changes in tension as a muscles contracts.

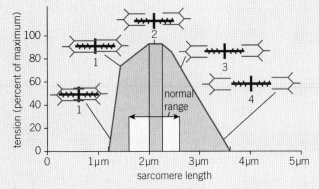

b (i) Describe how muscle tension changes with changing sarcomere length.
(*4 marks*)

(ii) State the optimal resting length of a sarcomere. (*2 marks*)

(iii) Explain, using the diagram, why muscle fibres have a small range of optimal resting lengths. (*5 marks*)

2 Multiple sclerosis (MS) is a condition that affects the nervous system. The immune system of people with MS treats parts of the nervous system as foreign, and launches an immune response damaging neurons. An example of the damage caused is shown in the diagram.

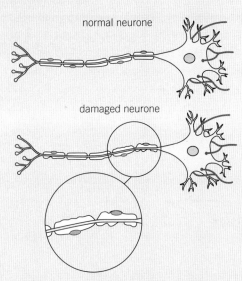

a (i) State the name of the specific part of the nervous system damaged in people with multiple sclerosis. (*1 mark*)

(ii) Describe and explain how this damage would affect the functioning of the nervous system. (*4 marks*)

Some of the symptoms of MS are described here:

temporary loss of vision

loss of balance and co-ordination

incontinence / constipation

temporary loss of sensation in limbs

sensory impairment

b Explain why people with MS may get these symptoms. (*4 marks*)

3 An imbalance in the neurotransmitters in the brain, particularly a shortage of serotonin, is now considered to be one of the causes of depression.

The diagram shows a synapse in the brain which uses serotonin as a neurotransmitter.

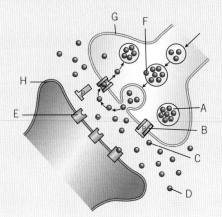

a (i) Name the structures labelled A, F, G, and H. *(4 marks)*

(ii) State the name of the process that results in the release of F into the synaptic cleft. *(1 mark)*

Structure B is a serotonin transporter (SERT) that transports serotonin back into the synaptic knob. Structure P is a drug that is used to treat depression.

b (i) Explain why serotonin needs to be removed from a synapse after it has been released. *(2 marks)*

(ii) Suggest the action of drug P on the synapse in the diagram and explain how this would improve the symptoms of depression. *(4 marks)*

Fluoxetine is an example of a drug used in the treatment of depression. Fluoxetine only inhibits the reuptake of serotonin. It is known as a selective serotonin reuptake inhibitor (SSRI).

Cocaine, a drug of misuse, also inhibits the reuptake of serotonin as well as the neurotransmitters dopamine and noradrenaline.

c Discuss why cocaine is not used to treat depression. *(4 marks)*

4 When patients arrive at hospital with a suspected stroke they are usually given an MRI or CT brain scan. Doctors need to find out as much information as possible to help with the diagnosis.

Strokes are often diagnosed by studying images of the brain produced during these brain scans. New guidelines have suggested that doctors should use MRI scans to diagnose strokes instead of CT scans.

MRI scans are better at detecting stroke damage caused by a lack of blood flow (usually due to a blockage or a blood clot) in the brain compared to CT scans. The majority of strokes are caused by a lack of blood flow in the brain. There is only a short time that treatment used to reverse the damage is effective.

In diffusion MRI, the movement of water in brain tissue is measured. The movement of water is restricted in damaged tissue.

a Suggest why the movement of water is restricted in damaged tissue. *(2 marks)*

CT scans use X-rays to produce multiple images building up a detailed, three-dimensional picture of the brain. An injection of a dye into one of the veins in the arm can be administered during the scan to help improve the clarity of the image.

b Explain how the dye improves the clarity of the images produced. *(2 marks)*

MRI scans use a strong magnetic field and radio waves to produce a detailed picture of the brain. A dye can also be used to improve scan images.

The photos show two brain scans performed on patients after admission to hospital with suspected strokes.

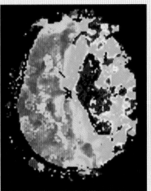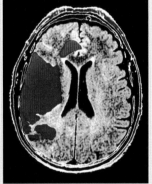

The photo on the left shows a functional magnetic resonance imaging (fMRI) scan of the brain of a 32 year old woman after a massive stroke. This scan shows the amount of blood flow received by areas of the brain. Green and blue areas are receiving normal blood flow, while yellow, red, and black are receiving abnormal blood flow. The lack of blood flow in the right hemisphere is due to a blocked right internal carotid artery. The photo on the right shows a coloured computed tomography (CT) scan of a section through a patient's brain showing internal bleeding (red) due to a stroke.

c Describe the differences between the two images. *(3 marks)*

d Outline the advantages and disadvantages of MRI and CT scans. *(4 marks)*

e Suggest why a CT scan may still be recommended if an MRI scanner is not immediately available for patients that require an emergency injection to break up blood clots. *(2 marks)*

14 HORMONAL COMMUNICATION
14.1 Hormonal communication
Specification reference: 5.1.4

Learning outcomes

Demonstrate knowledge, understanding, and application of:

→ endocrine communication by hormones

→ the structure and functions of the adrenal glands.

In the previous chapter you looked in detail at how the nervous system detects and responds to changes in the internal and external environment. The body has a second system, the endocrine system, which works alongside the neuronal system to react to changes. The endocrine system uses **hormones** to send information about changes in the environment around the body to bring about a designated response.

The endocrine system

Endocrine glands

The endocrine system is made up of **endocrine glands**. An endocrine gland is a group of cells which are specialised to secrete chemicals – these chemicals are known as hormones, and are secreted directly into the bloodstream. Examples of endocrine glands include the pancreas and adrenal glands.

Figure 1 shows the positions of the major endocrine glands in the body and the hormones they secrete. The pituitary gland at the base of the brain makes several hormones, which in turn control the release of other hormones. The close proximity of the pituitary gland to the hypothalamus ensures that the nervous and hormonal responses of the body are closely linked and coordinated.

Pituitary gland – produces growth hormone, which controls growth of bones and muscles; anti-diuretic hormone, which increases reabsorption of water in kidneys; and gonadotrophins, which control development of ovaries and testes.

Thyroid gland – produces thyroxine which controls rate of metabolism and rate that glucose is used up in respiration, and promotes growth.

Adrenal gland – produces adrenaline which increases heart and breathing rate and raises blood sugar level.

Testis – produces testosterone which controls sperm production and secondary sexual characteristics.

Pineal gland – produces melatonin which affects reproductive development and daily cycles.

Thymus – produces thymosin which promotes production and maturation of white blood cells.

Pancreas – produces insulin which converts excess glucose into glycogen in the liver; and glucagon, which converts glycogen back to glucose in the liver.

Ovary – produces oestrogen, which controls ovulation and secondary sexual characteristics; and progesterone, oestrogen, which controls ovulation and secondary sexual characteristics; and progesterone, which prepares the uterus lining for receiving an embryo.

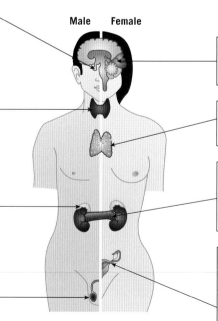

Male   Female

▲ **Figure 1** *Position of the major endocrine glands in the body, hormones are highlighted in green, white boxes are features common to both sexes*

(Remember that, by contrast, exocrine glands, such as those in the digestive system, secrete chemicals through ducts into organs, or to the surface of the body.)

Hormones

Hormones are often referred to as chemical messengers because they carry information from one part of the body to another. They can be steroids, proteins, glycoproteins, polypeptides, amines, or tyrosine derivatives. Although they are chemically different, they share many characteristics.

Hormones are secreted directly into the blood when a gland is stimulated. This can occur as a result of a change in concentration of a particular substance, such as blood glucose concentration. It can also occur as the result of another hormone or a nerve impulse.

Once secreted, the hormones are transported in the blood plasma all over the body. The hormones diffuse out of the blood and bind to specific receptors for that hormone, found on the membranes, or in the cytoplasm of cells in the target organs. These are known as **target cells**. Once bound to their receptors the hormones stimulate the target cells to produce a response. You will find out more about how blood glucose concentration is maintained in Topic 14.3, Regulation of blood glucose concentration.

The type of hormone determines the way it causes its effect on a target cell. For example:

- *Steroid hormones* are lipid-soluble. They pass through the lipid component of the cell membrane and bind to steroid hormone receptors to form a hormone–receptor complex. The receptors may be present in the cytoplasm or the nucleus depending on the hormone. The hormone–receptor complex formed acts as a transcription factor which in turn facilitates or inhibits the transcription of a specific gene. Oestrogen is an example of a hormone which works in this way.

- *Non-steroid hormones* are hydrophilic so cannot pass directly through the cell membrane. Instead they bind to specific receptors on the cell surface membrane of the target cell. This triggers a cascade reaction mediated by chemicals called second messengers. Adrenaline is an example of a hormone which works in this way.

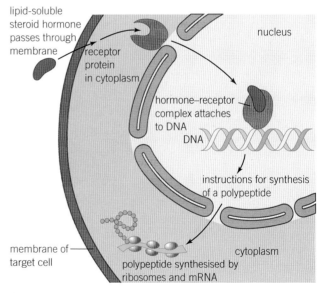

▲ **Figure 2** *Mechanism of action of steroid hormones*

Hormonal versus neuronal communication

As hormones are not released directly onto their target cells, this results in a slower and less specific form of communication than neuronal communication. However, as hormones are not broken down as quickly as neurotransmitters, it can result in a much longer lasting and widespread effect. For example, the hormones insulin and

Synoptic link

You will find out how the second messenger model works in Topic 15.6, The kidney and osmoregulation.

glucagon are responsible for controlling blood glucose concentration. A number of organs are involved in this response.

Table 1 summarises the main differences between the actions of the hormonal and nervous systems.

▼ Table 1 *Comparison of the hormonal and nervous systems*

Hormonal system	Nervous system
communication is by chemicals called hormones	communication is by nerve impulses
transmission is by the blood system	transmission is by neurones
transmission is usually relatively slow	transmission is very rapid
hormones travel to all parts of the body, but only target organs respond	nerve impulses travel to specific parts of the body
response is widespread	response is localised
response is slow	response is rapid
response is often long-lasting	response is short-lived
effect may be permanent and irreversible	effect is temporary and reversible

Adrenal glands

The adrenal glands are two small glands that measure approximately 3 cm in height and 5 cm in length. They are located on top of each kidney and are made up of two distinct parts surrounded by a capsule:

● The adrenal cortex – the outer region of the glands. This produces hormones that are vital to life, such as cortisol and aldosterone.

● The adrenal medulla – the inner region of the glands. This produces non-essential hormones, such as adrenaline which helps the body react to stress.

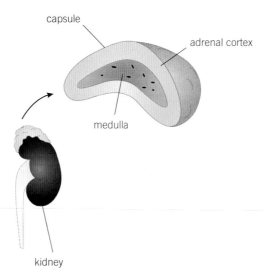

▲ Figure 3 *Structure of adrenal glands*

Adrenal cortex

The production of hormones by the adrenal cortex is itself controlled by hormones released from the pituitary gland in the brain. There are three main types of hormones produced by the adrenal cortex:

- *Glucocorticoids.* These include cortisol which helps regulate metabolism by controlling how the body converts fats, proteins, and carbohydrates to energy. It also helps regulate blood pressure and cardiovascular function in response to stress. Another glucocorticoid hormone released is corticosterone. This works with cortisol to regulate immune response and suppress inflammatory reactions. The release of these hormones is controlled by the hypothalamus.

- *Mineralocorticoids.* The main one produced is aldosterone which helps control blood pressure by maintaining the balance between salt and water concentrations in the blood and body fluids. Its release is mediated by signals triggered by the kidney.

- *Androgens.* Small amounts of male and female sex hormones are released – their impact is relatively small compared with the larger amounts of hormones, such as oestrogen and testosterone, released by the ovaries or testes after puberty, but they are still important, especially in women after the menopause.

Adrenal medulla

The hormones of the adrenal medulla are released when the sympathetic nervous system is stimulated. This occurs when the body is stressed. You can find out more about the fight or flight response in Topic 14.5, Coordinated responses.

The hormones secreted by the adrenal medulla are:

- *Adrenaline.* This increases the heart rate sending blood quickly to the muscles and brain. It also rapidly raises blood glucose concentration levels by converting glycogen to glucose in the liver.

- *Noradrenaline.* This hormone works with adrenaline in response to stress, producing effects such as increased heart rate, widening of pupils, widening of air passages in the lungs, and the narrowing of blood vessels in non-essential organs (resulting in higher blood pressure).

Summary questions

1 Using a named example, explain the function of an endocrine gland.
(2 marks)

2 Describe the pathway triggered by a stimulus in hormonal communication. *(2 marks)*

3 Bright light causes the iris muscles in your eyes to contract, constricting the pupil and preventing damage to the eye. State and explain whether hormonal or neuronal communication would be used in this response. *(2 marks)*

4 A person falls into a fast-flowing river. State and explain the changes that may occur in the body and increase the person's chances of survival in this situation. *(6 marks)*

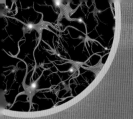

14.2 Structure and function of the pancreas

Specification reference: 5.1.4

The pancreas is found in the upper abdomen, behind the stomach (Figure 1). It plays a major role in controlling blood glucose concentration, and in digestion. It is a glandular organ – its role is to produce and secrete hormones and digestive enzymes.

Function of the pancreas

The pancreas has two main functions in the body, as an:

- exocrine gland – to produce enzymes and release them via a duct into the duodenum
- endocrine gland – to produce hormones and release them into the blood.

Role as an exocrine gland

Most of the pancreas is made up of exocrine glandular tissue. This tissue is responsible for producing digestive enzymes and an alkaline fluid known as pancreatic juice. The enzymes and juice are secreted into ducts which eventually lead to the pancreatic duct. From here they are released into the duodenum, the top part of the small intestine. The pancreas produces three important types of digestive enzymes:

- Amylases – break down starch into simple sugars. For example, pancreatic amylase.

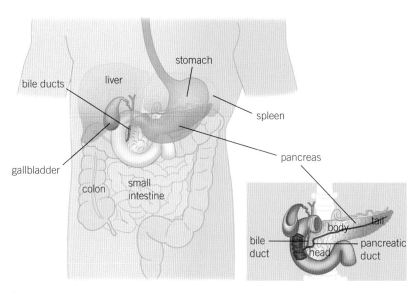

▲ Figure 1 *Position of the pancreas in the body*

- Proteases – break down proteins into amino acids. For example, trypsin.
- Lipases – break down lipids into fatty acids and glycerol. For example, pancreatic lipase.

Role as an endocrine gland

The pancreas is responsible for producing insulin and glucagon. These two hormones play an essential role in controlling blood glucose concentration, which you will read about in Topic 14.3, Regulation of blood glucose concentration. Within the exocrine tissue there are small regions of endocrine tissue called **islets of Langerhans**. The cells of the islets of Langerhans are responsible for producing insulin and glucagon, and secreting these hormones directly into the bloodstream.

Histology of the pancreas

When viewed under a microscope, you can clearly see the differences between endocrine and exocrine pancreatic tissue. The main differences are summarised in Table 1.

▼ Table 1

Structure	Appearance	Shape	Type of tissue	Function
islets of Langerhans	lightly stained	large, spherical clusters	endocrine pancreas	produce and secrete hormones
pancreatic acini (singular – acinus)	darker stained	small, berry-like clusters	exocrine pancreas	produce and secrete digestive enzymes

Islets of Langerhans

Within the islets of Langerhans are different types of cell. They are classified according to the hormone they secrete:

- α (alpha) cells – these produce and secrete glucagon
- β (beta) cells – these produce and secrete insulin

Alpha cells are larger and more numerous than beta cells within an islet.

Using standard staining techniques, it is often very difficult to distinguish between the cell types within an islet of Langerhans. In Figure 2, a differential stain has been used. The β cells of the islets that produce insulin are stained blue, and the α cells that produce glucagon are stained pink.

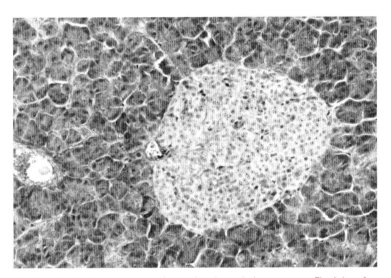

▲ **Figure 2** *Light micrograph of a section through the pancreas. The islet of Langerhans (centre right) is composed of groups of secretory cells. The main secretions from these cells are the hormones insulin and glucagon, which control blood sugar. These cells are endocrine – their secretions go straight into the bloodstream. The cells surrounding the islet are packed into secretory acini (pink) which secrete digestive enzymes. This part of the pancreas is exocrine – the enzymes pass straight out into the gut, via ducts. The structure on the left is a branch of the pancreatic duct, ×195 magnification*

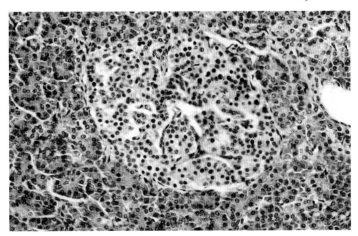

▲ **Figure 3** *Islet of Langerhan cell in human pancreas, ×300 magnification*

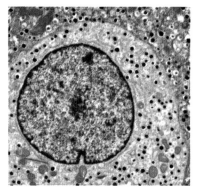

▲ **Figure 4** *A colour transmission electron micrograph of a pancreatic alpha cell. Alpha cells make up 15–20% of the cells in the islets of Langerhans. They produce glucagon, a hormone that regulates blood sugar levels by increasing the amount of glucose available. The glucagon is carried into the bloodstream by granules (dark red). The nucleus (round, blue) and mitochondria (brown) can also be seen, ×5 000 magnification*

Summary questions

1 State the difference between endocrine and exocrine glandular tissue. *(1 mark)*

2 State two features which would enable you to identify the islets of Langerhans through a cross-section of pancreatic material, when viewed under a light microscope. *(2 marks)*

3 ⚙ Figure 6 shows pancreatic tissue which has been stained for insulin. Areas which contain insulin appear brown. Describe the structure and function of the pancreatic tissue shown on this slide. *(6 marks)*

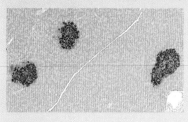

▲ Figure 6

✚ Histology of the pancreas 🅐

The following activity will allow you to view the structures within the pancreas.

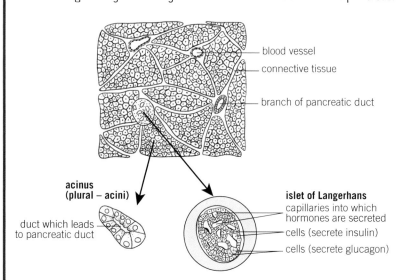

▲ **Figure 5** *Cross-section of pancreatic material, as viewed through a light microscope*

Place a stained slide of pancreatic tissue on the stage of a light microscope.

Begin by viewing under low power. Compare this with Figure 5 – can you identify the following structures?

a Exocrine tissue, which secretes digestive enzymes
b Blood vessels
c Branches of the pancreatic duct.

Select one of the groups of exocrine cells and view under high power. You will be viewing an acinus – can you identify the following features?

d Small group of cells formed in a cluster
e Clearly visible nuclei within these cells
f Central duct (which leads to the pancreatic duct).

Make a scientific drawing of an acinus. Remember to include a scale on your drawing and label all of the key features.

Return to low power and look for small groups of cells which look different to the rest of the material. These are the islets of Langerhans. These groups of cells are likely to appear lighter than the surrounding material.

View an islet of Langerhans under high power. Can you spot the following?

g Capillaries, which transport the secreted hormones
h α cells, which secrete glucagon
i β cells, which secrete insulin.

You may find that your slide has been prepared using a differential stain, which colours the α and β cells differently to enable you to identify these structures more readily.

Make a drawing of an islet of Langerhans. Remember to include a scale on your drawing and label all of the key features.

14.3 Regulation of blood glucose concentration

Specification reference: 5.1.4

During respiration the body uses glucose to produce ATP. To remain healthy it is important that the concentration of glucose in your blood is kept constant. Without control, blood glucose concentration would range from very high levels after a meal, to very low levels several hours later. At these very low levels cells would not have enough glucose for respiration. Blood glucose concentration is kept constant by the action of the two hormones – insulin and glucagon.

Increasing blood glucose concentration

Glucose is a small, soluble molecule that is carried in the blood plasma. Blood glucose is normally maintained at a concentration of around $90\,mg\,cm^{-3}$ of blood. Blood glucose concentration can increase as a result of:

- Diet – when you eat carbohydrate-rich foods such as pasta and rice (which are rich in starch) and sweet foods such as cakes and fruit (which contain high levels of sucrose), the carbohydrates they contain are broken down in the digestive system to release glucose. The glucose released is absorbed into the bloodstream, and the blood glucose concentration rises.

- **Glycogenolysis** – glycogen stored in the liver and muscle cells is broken down into glucose which is released into the bloodstream increasing blood glucose concentration.

- **Gluconeogenesis** – the production of glucose from non-carbohydrate sources. For example, the liver is able to make glucose from glycerol (from lipids) and amino acids. This glucose is released into the bloodstream and causes an increase in blood glucose concentration.

Decreasing blood glucose concentration

Blood glucose concentration can be decreased by:

- Respiration – some of the glucose in the blood is used by cells to release energy. This is required to perform normal body functions. However, during exercise, more glucose is needed as the body needs to generate more energy in order for muscle cells to contract. The higher the level of physical activity, the higher the demand for glucose and the greater the decrease of blood glucose concentration.

- **Glycogenesis** – the production of glycogen. When blood glucose concentration is too high, excess glucose taken in through the diet is converted into glycogen which is stored in the liver.

Role of insulin

Insulin is produced by the β cells of the islets of Langerhans in the pancreas. If the blood glucose concentration is too high, the β cells detect this rise in blood glucose concentration and respond by secreting insulin directly into the bloodstream.

Virtually all body cells have insulin receptors on their cell surface membrane (an exception being red blood cells). When insulin binds to its glycoprotein receptor, it causes a change in the tertiary structure of the glucose transport protein channels. This causes the channels to open allowing more glucose to enter the cell. Insulin also activates enzymes within some cells to convert glucose to glycogen and fat.

Insulin therefore lowers blood glucose concentration by:

- increasing the rate of absorption of glucose by cells, in particular skeletal muscle cells
- increasing the respiratory rate of cells – this increases their need for glucose and causes a higher uptake of glucose from the blood
- increasing the rate of glycogenesis – insulin stimulates the liver to remove glucose from the blood by turning the glucose into glycogen and storing it in the liver and muscle cells
- increasing the rate of glucose to fat conversion
- inhibiting the release of glucagon from the α cells of the islets of Langerhans.

Insulin is broken down by enzymes in the cells of the liver. Therefore, to maintain its effect it has to be constantly secreted. Depending on the food eaten, insulin secretion can begin within minutes of the food entering the body and may continue for several hours after eating.

As blood glucose concentration returns to normal, this is detected by the β cells of the pancreas. When it falls below a set level, the β cells reduce their secretion of insulin. This is an example of *negative feedback*. Negative feedback ensures that, in any control system, changes are reversed and returned back to the set level.

Role of glucagon

Glucagon is produced by the α cells of the islets of Langerhans in the pancreas. If the blood glucose concentration is too low, the α cells detect this fall in blood glucose concentration and respond by secreting glucagon directly into the bloodstream.

Unlike insulin, the only cells in the body which have glucagon receptors are the liver cells and fat cells – therefore these are the only cells that can respond to glucagon.

Glucagon raises blood glucose concentration by:

- glycogenolysis – the liver breaks down its glycogen store into glucose and releases it back into the bloodstream
- reducing the amount of glucose absorbed by the liver cells
- increasing gluconeogenesis – increasing the conversion of amino acids and glycerol into glucose in the liver.

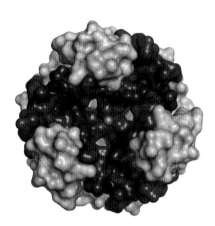

▲ Figure 1 *Computer model showing the structure of a molecule of the hormone insulin. Insulin is a globular protein made up of 51 amino acids which are arranged into two chains*

Synoptic link

You will find out more about the structure and function of the liver in Topic 15.4, Excretion, homeostasis, and the liver.

▲ Figure 2 *Computer model showing the structure of glucagon. The secondary structure of the hormone as a coiled ribbon can be seen and the atoms it is made up of. Atoms are colour-coded spheres (carbon: grey, nitrogen: blue, and oxygen: red)*

As blood glucose concentration returns to normal, this is detected by the α cells of the pancreas. When it rises above a set level, the α cells reduce their secretion of glucagon. This is another example of negative feedback. The feedback causes the corrective measures to be switched off, returning the system to its original (normal) level.

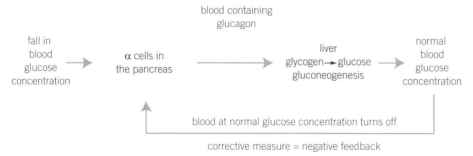

▲ **Figure 3** *Negative feedback in control of blood glucose concentration*

Interaction of insulin and glucagon

Figure 4 shows how insulin and glucagon work together to maintain a constant blood glucose concentration. Insulin and glucagon are antagonistic hormones, that is, they work against each other.

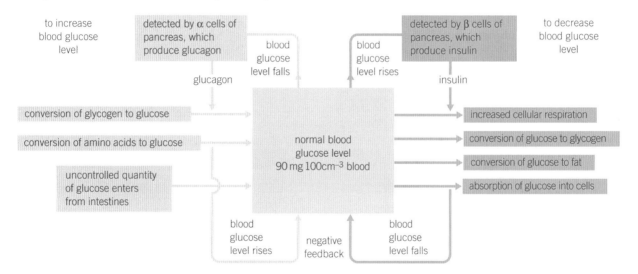

▲ **Figure 4** *Summary of blood glucose concentration regulation*

The system of maintaining blood glucose concentration is said to be self-regulating, as it is the level of glucose in the blood that determines the quantity of insulin and glucagon that is released. Blood glucose concentration is not constant, but fluctuates around a set point as the result of negative feedback. In times of stress adrenaline is released by the body. One of the effects of this hormone is to raise the blood glucose concentration to allow more respiration to occur. You will find out more about the fight and flight response in Topic 14.5, Coordinated responses.

Control of insulin secretion

When blood glucose concentration rises above the set level, this is detected by the β cells in the islets of Langerhans and insulin is released. The mechanism by which this occurs is as follows:

1 At normal blood glucose concentration levels, potassium channels in the plasma membrane of β cells are open and potassium ions diffuse out of the cell. The inside of the cell is at a potential of −70 mV with respect to the outside of the cell.

2 When blood glucose concentration rises, glucose enters the cell by a glucose transporter.

3 The glucose is metabolised inside the mitochondria, resulting in the production of ATP.

4 The ATP binds to potassium channels and causes them to close. They are known as ATP-sensitive potassium channels.

5 As potassium ions can no longer diffuse out of the cell, the potential difference reduces to around −30 mV and depolarisation occurs.

6 Depolarisation causes the voltage-gated calcium channels to open.

7 Calcium ions enter the cell and cause secretory vesicles to release the insulin they contain by exocytosis.

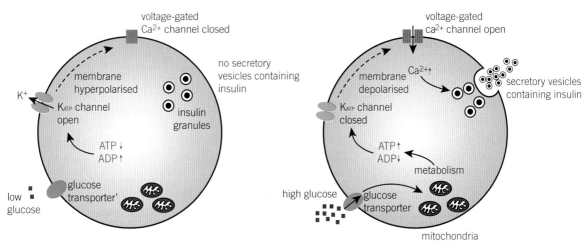

▲ **Figure 5** *Changes which occur in β cells in the islets of Langerhans in the pancreas when stimulated by a high blood glucose concentration*

Summary questions

1 Describe what is meant by negative feedback. *(1 mark)*

2 Describe the role glucagon plays in the control of blood glucose concentration. *(3 marks)*

3 Describe the changes which take place inside a β cell to cause the release of insulin in the presence of high blood glucose concentration. *(4 marks)*

4 Explain how hormones return blood glucose concentration to normal after a meal. *(6 marks)*

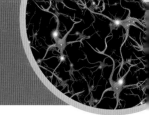

14.4 Diabetes and its control

Specification reference: 5.1.4

To keep blood glucose concentration constant the body relies on the interaction between glucagon and insulin. However, for over 300 million people in the world this system of regulation does not work properly. They suffer from the chronic disease, diabetes mellitus (usually referred to as diabetes). This means they are unable to metabolise carbohydrates properly, in particular glucose.

Types of diabetes

If you suffer from diabetes your pancreas either does not produce enough insulin, or your body cannot effectively respond to the insulin produced. This means that blood glucose concentration remains high. Hyperglycaemia, or raised blood sugar, is a common effect of uncontrolled diabetes. Over time this can lead to serious damage of many body systems, especially the nerves and blood vessels.

There are two main types of diabetes:

- Type 1 diabetes. Patients with type 1 diabetes are unable to produce insulin. The β cells in the islets of Langerhans do not produce insulin. The cause of type 1 diabetes is not known and so, at the moment, the disease cannot be prevented or cured. It is possible, however, to treat the symptoms. Evidence suggests that in many cases the condition arises as a result of an autoimmune response where the body's own immune system attacks the β cells. This condition normally begins in childhood, and people develop symptoms of the disease quickly.

- Type 2 diabetes. Patients with type 2 diabetes cannot effectively use insulin and control their blood sugar levels. This is either because the person's β cells do not produce enough insulin *or* the person's body cells do not respond properly to insulin. This is often because the glycoprotein insulin receptor on the cell membrane does not work properly. The cells lose their responsiveness to insulin, and therefore do not take up enough glucose, leaving it in the bloodstream. Globally, approximately 90% of people with diabetes have type 2 diabetes. This is largely as a result of excess body weight, physical inactivity, and habitual, excessive overeating of (refined) carbohydrates. Symptoms are similar to those of type 1 diabetes, but are often less severe and develop slowly. As a result, the disease is often only diagnosed after complications have already arisen. Risk of type 2 diabetes increases with age. Until recently, this type of diabetes was seen only in adults (normally over the age of 40), but it is now also occurring in children.

Learning outcomes

Demonstrate knowledge, understanding, and application of:

→ the differences between type 1 and type 2 diabetes mellitus

→ the potential treatments for diabetes mellitus.

▼ Table 1 *Symptoms of diabetes*

Common symptoms of diabetes
● High blood glucose concentration
● Glucose present in urine
● Excessive need to urinate (polyuria)
● Excessive thirst (polydipsia)
● Constant hunger
● Weight loss
● Blurred vision
● Tiredness

▲ Figure 1 *People with diabetes have to measure their blood glucose levels regularly. This is normally done using a finger prick test*

Type 2 diabetes

There has been a significant increase in the number of cases of diabetes diagnosed in the UK, rising from 1.4 million in 1996 to around 3 million in 2014. It is estimated that there may be as many as 5 million sufferers of diabetes by 2025, with around 85% having type 2 diabetes.

A clear causal link has been established between obesity and the onset of type 2 diabetes. This information has been used to launch initiatives to promote healthy eating and exercise. These include, for example, the Change4Life campaign. This is an example of where scientific evidence has been used to inform decision-making at a national level.

Discussion

1. To what extent is a government responsible for the health of its citizens?
2. What steps would you take to minimise the risk of diabetes amongst the UK population?
3. Should universal benefits – for example, free healthcare – be available to those whose lifestyle choices cause the onset of a medical condition?

Study tip

Most of the symptoms of diabetes are logical. Think about what would happen if your blood glucose concentration remained high, but the level of glucose in your cells was low.

▲ **Figure 2** *People with type 1 diabetes need regular insulin injections. As insulin is a protein, it cannot be taken by mouth because it will be digested*

Diabetes treatment

Diabetes is not a curable disease, but it can be controlled successfully, allowing sufferers to lead a normal life. Treatment differs for both types of diabetes.

Type 1 diabetes

Type 1 diabetes is controlled by regular injections of insulin and is therefore said to be insulin-dependent.

People with the condition have to regularly test their blood glucose concentration, normally by pricking their finger. The drop of blood is then analysed by a machine, which tells the person their blood glucose concentration. Based on this concentration, the person can work out the dose of insulin they need to inject. The insulin administered increases the amount of glucose absorbed by cells and causes glycogenesis to occur, resulting in a reduction of blood glucose concentration.

If a person with diabetes injects himself or herself with too much insulin, they may experience hypoglycaemia (very low blood glucose concentrations) that can result in unconsciousness. However, too low an insulin dose results in hyperglycaemia, which can also result in unconsciousness and death if left untreated. Careful monitoring and dose regulation is therefore required.

Figure 3 shows how blood glucose concentration and insulin levels vary in a person with type 1 diabetes, and a person without diabetes.

If the person with diabetes injects himself or herself with insulin, there will be a surge of insulin in their blood which will cause their blood glucose level to drop quickly.

Type 2 diabetes

The first line of control in type 2 diabetes is to regulate the person's carbohydrate intake through their diet and matching this to their exercise levels. This often involves increasing exercise levels. Overweight people are also encouraged to lose weight.

In some cases, diet and exercise are not enough to control blood glucose concentration so drugs also have to be used. These can include drugs that stimulate insulin production, drugs that slow down the rate at which the body absorbs glucose from the intestine, and ultimately even insulin injections.

Medically produced insulin

Originally, insulin was obtained from the pancreas of cows and pigs which had been slaughtered for food. This process was difficult and expensive. The insulin extracted could also cause allergic reactions as it differed slightly from human insulin.

In 1955, the structure of human insulin was identified and it is now made by genetically modified bacteria. This has a number of advantages:

- Human insulin is produced in a pure form – this means it is less likely to cause allergic reactions.
- Insulin can be produced in much higher quantities.
- Production costs are much cheaper.
- People's concerns over using animal products in humans, which may be religious or ethical, are overcome.

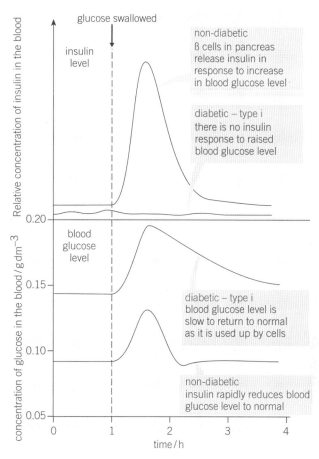

▲ **Figure 3** *Comparison of blood glucose and insulin levels in a person with type 1 diabetes and a person without diabetes after each has swallowed a glucose tablet*

Potential use of stem cells in diabetes treatment

For decades, diabetes researchers have been searching for ways to replace the faulty β cells in the pancreatic islets of diabetic sufferers. Each year, over 1000 people with type 1 diabetes receive a pancreas transplant. After a year, over 80% of these patients have no symptoms of diabetes and do not have to take insulin. However, the demand for transplantable pancreases far outweighs their availability. The risk of having a transplant can also be a greater health risk than the diabetes itself – immunosuppressant drugs are required to ensure the body accepts the transplanted pancreas, which can leave a person susceptible to infection.

Doctors have attempted to cure diabetes by injecting patients with pancreatic β islet cells, but fewer than 8% of cell transplants performed have been successful. The immunosuppressant drugs used to prevent rejection of these cells increases the metabolic demand on insulin-producing cells. Eventually this exhausts their capacity to produce insulin.

Synoptic link

You will find out how bacteria are genetically engineered to produce human insulin in Topic 21.4, Genetic engineering, and about the industrial production of insulin in 22.5, Microorganisms, medicines, and bioremediation.

Synoptic link

Look back at Topic 6.5, Stem cells to remind yourself of the different types of stem cell that exist and the sources of stem cells.

As type 1 diabetes results from the loss of a single cell type, and there is evidence that a relatively small number of islet cells can restore insulin production, the disease is a perfect candidate for stem cell therapy. Totipotent stem cells have the potential to grow into any of the body's cell types. Scientists have been researching the best type of stem cells and the signals required to promote their differentiation into β cells, either directly in the patient or in the laboratory before being transplanted. It is likely that the stem cells used in diabetes treatment would be taken from embryos. To obtain the stem cells, the early embryo has to be destroyed. This means destroying a potential human life. However, the embryos used as a source for these stem cells would usually be destroyed anyway – they are 'spare' embryos from infertility treatments or from terminated pregnancies.

Stem cells lines formed from a small number of embryos can be used to treat many patients – each treatment does not require a separate embryo. An alternative to using embryonic matter is that of using preserved umbilical stem cells.

Stem cells offer many advantages over current therapies:

- donor availability would not be an issue – stem cells could produce an unlimited source of new β cells
- reduced likelihood of rejection problems as embryonic stem cells are generally not rejected by the body (although some evidence contradicts this). Stem cells can also be made by somatic cell nuclear transfer (SCNT)
- people no longer have to inject themselves with insulin.

However, because our ability to control growth and differentiation in stem cells is still limited, a major consideration is whether any precursor or stem-like cells transplanted into the body might induce the formation of tumours as a result of unlimited cell growth.

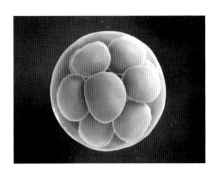

▲ **Figure 4** *A human embryo. Embryonic stem cells could lead to the discovery of new medical treatments (such as the replacement of β cells in diabetes patients) that would alleviate the suffering of many people. However, many people argue against using the cells, as an embryo has to be destroyed, approx ×600 magnification*

Summary questions

1. Copy and complete the table to compare and contrast the differences between the causes of type 1 and type 2 diabetes. *(3 marks)*

	Type 1	Type 2
Cause		
When does it develop		
Period of development		

2. Explain why people with type 1 have to constantly monitor their blood glucose concentration. *(3 marks)*

3. Explain why some who suffer from diabetes mellitus who can produce insulin cannot control their blood glucose concentration. *(3 marks)*

4. ⚙ Discuss the advantages and disadvantages of the different treatments for diabetes. *(6 marks)*

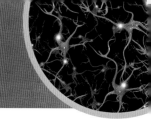

On many occasions the body responds to changes in its internal and external environment through a coordinated response. The nervous and endocrine systems work together to detect and respond appropriately to stimuli. One example of coordination between these two systems is the mammalian 'fight or flight' response.

Fight or flight response

The fight or flight response is an instinct that all mammals possess. When a potentially dangerous situation is detected, the body automatically triggers a series of physical responses. These are intended to help mammals survive by preparing the body to either run or fight for life, hence the name of the response.

Once a threat is detected by the autonomic nervous system, the hypothalamus communicates with the sympathetic nervous system and the adrenal–cortical system. The sympathetic nervous system uses neuronal pathways to initiate body reactions whereas the adrenal–cortical system uses hormones in the bloodstream. The combined effects of these two systems results in the fight or flight response. The overall process is summarised in Figure 1.

The sympathetic nervous system sends out impulses to glands and smooth muscles and tells the adrenal medulla to release adrenaline and noradrenaline into the bloodstream. These 'stress hormones' cause several changes in the body, including an increased heart rate.

The release of other stress hormones which have a longer-term action from the adrenal cortex is controlled by hormones produced by the pituitary gland in the brain. The hypothalamus stimulates the pituitary gland to secrete adrenocorticotropic hormone (ACTH). This travels in the bloodstream to the adrenal cortex, where it activates the release of many hormones that prepare the body to deal with a threat. Look back at Topic 14.1, Hormonal communication to remind yourself of the structure and function of the adrenal glands.

The physiological responses which occur as part of the fight or flight response are summarised in Table 1.

> ### Learning outcomes
> Demonstrate knowledge, understanding, and application of:
> → the coordination of responses by the nervous and endocrine systems.

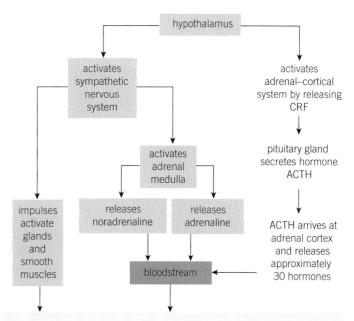

combination of neuronal and hormonal activity results in 'fight or flight' response

▲ Figure 1 *Summary of coordination of 'fight or flight' response*

▼ Table 1 *Fight or flight physiological responses*

Physical response	Purpose
heart rate increases	to pump more oxygenated blood around the body
pupils dilate	to take in as much light as possible for better vision
arterioles in skin constrict	more blood to major muscle groups, brain, heart, and muscles of ventilation
blood glucose level increases	increase respiration to provide energy for muscle contraction
smooth muscle of airways relaxes	to allow more oxygen into lungs
non-essential systems (like digestion) shut down	to focus resources on emergency functions
difficulty focusing on small tasks	brain solely focused only on where threat is coming from

Action of adrenaline

One of adrenaline's main functions during the fight and flight response is to trigger the liver cells to undergo glycogenolysis so that glucose is released into the bloodstream. This allows respiration to increase so more energy is available for muscle contraction.

Adrenaline is a hormone. It is hydrophilic therefore cannot pass through cell membranes. Adrenaline binds with receptors on the surface of a liver cell membrane and triggers a chain reaction inside the cell:

● When adrenaline binds to its receptor, the enzyme *adenylyl cyclase* (which is also present in the cell membrane) is activated.

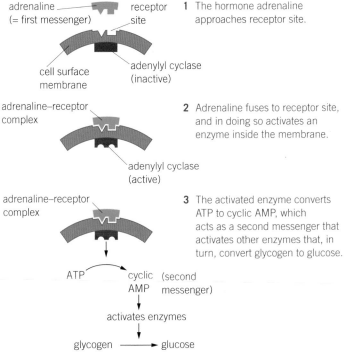

▲ Figure 2 *Second messenger model of hormone action*

- Adenylyl cyclase triggers the conversion of ATP into *cyclic adenosine mono-phosphate* (cAMP) on the inner surface of the cell membrane in the cytoplasm.

- The increase in cAMP levels activates specific enzymes called *protein kinases* which phosphorylate, and hence activate, other enzymes. In this example, enzymes are activated which trigger the conversion of glycogen into glucose.

This model of hormone action is known as the *second messenger model*. The hormone is known as the first messenger (in this example, adrenaline) and cAMP is the second messenger. One hormone molecule can cause many cAMP molecules to be formed. At each stage, the number of molecules involved increases so the process is said to have a cascade effect (Figure 3).

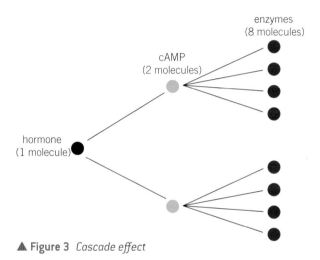

▲ **Figure 3** *Cascade effect*

Summary questions

1   State and explain two physical responses which occur as a result of the 'fight and flight' response.   (*2 marks*)

2   Explain why people often feel cold in times of stress.   (*2 marks*)

3   Explain how the nervous and endocrine systems work together to enable the body to respond to danger.   (*6 marks*)

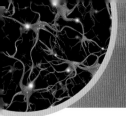

14.6 Controlling heart rate

Specification reference: 5.1.5

Learning outcomes

Demonstrate knowledge, understanding, and application of:

→ the effects of hormones and nervous mechanisms on heart rate.

Synoptic link

Look back at Topic 8.5, The heart to remind yourself of the heart's structure and the intrinsic rhythmicity of the heart.

Synoptic link

Look back at Topic 13.6, Organisation of the nervous system to remind yourself of the two different functional systems of the autonomic nervous system – the parasympathetic and sympathetic nervous system.

Synoptic link

Look back at Topic 7.4, Ventilation and gas exchange in other organisms to remind yourself of transport of carbon dioxide in the blood.

The human heart beats at approximately 70 beats per minute at rest. However, when you exercise, or in times of danger, it is essential that the heart rate increases to provide the extra oxygen required for increased respiration.

Controlling heart rate

Heart rate is involuntary and controlled by the autonomic nervous system.

The medulla oblongata in the brain is responsible for controlling heart rate and making any necessary changes. There are two centres within the medulla oblongata, linked to the sinoatrial node (SAN) in the heart by motor neurones:

- one centre increases heart rate by sending impulses through the sympathetic nervous system, these impulses are transmitted by the accelerator nerve

- one centre decreases heart rate by sending impulses through the parasympathetic nervous system, these impulses are transmitted by the vagus nerve.

Which centre is stimulated depends on the information received by receptors in the blood vessels. There are two types of receptors which provide information that affects heart rate:

- **baroreceptors** (pressure receptors) – these receptors detect changes in blood pressure. For example, if a person's blood pressure is low, the heart rate needs to increase to prevent fainting. Baroreceptors are present in the aorta, vena cava, and carotid arteries.

- **chemoreceptors** (chemical receptors) – these receptors detect changes in the level of particular chemicals in the blood such as carbon dioxide. Chemoreceptors are located in the aorta, the carotid artery (a major artery in the neck that supplies the brain with blood), and the medulla.

Chemoreceptors

Chemoreceptors are sensitive to changes in the pH level of the blood. If the carbon dioxide level in the blood increases, the pH of the blood decreases because carbonic acid is formed when the carbon dioxide interacts with water in the blood. If the chemoreceptors detect a decrease in blood pH, a response is triggered to increase heart rate – blood therefore flows more quickly to the lungs so the carbon dioxide can be exhaled.

This process is summarised in Figure 2

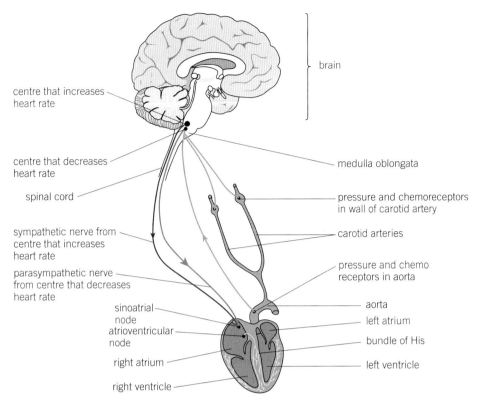

▲ **Figure 1** *Control of heart rate*

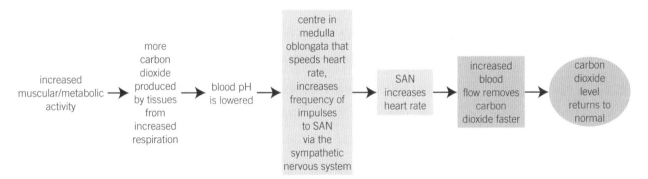

▲ **Figure 2** *Effects of exercise on cardiac output*

When the carbon dioxide level in the blood decreases, the pH of the blood rises. This is detected by the chemoreceptors in the wall of the carotid arteries and the aorta. This results in a reduction in the frequency of the nerve impulses being sent to the medulla oblongata. In turn, this reduces the frequency of impulses being sent to the SAN via the sympathetic nervous system, and thus heart rate decreases back to its normal level.

Baroreceptors

Baroreceptors present in the aorta and carotid artery wall detect changes in pressure. If blood pressure is too high, impulses are sent to the medulla oblongata centre which decreases heart rate. The medulla oblongata sends impulses along parasympathetic neurones to the SAN which decreases the rate at which the heart beats. This reduces blood pressure back to normal.

If blood pressure is too low, impulses are sent to the medulla oblongata centre which increases heart rate. The medulla oblongata sends impulses along sympathetic neurones to the SAN which increases the rate at which the heart beats. This increases blood pressure back to normal.

Hormonal control

Heart rate is also influenced by the presence of hormones. For example, in times of stress adrenaline and noradrenaline are released. These hormones affect the pacemaker region of the heart itself – they speed up your heart rate by increasing the frequency of impulses produced by the SAN. Look back at Topic 14.1, Hormonal control and Topic 14.5, Coordinated responses to remind yourself how adrenaline is released and its importance in the 'fight or flight' response.

 Monitoring heart rate

Heart rate can be determined by taking a pulse. Each time the heart beats a surge of blood travels around the body, which can be felt as a pulse in your blood vessels. The most common places to take a pulse are at the wrist (radial artery) or in the neck (carotid artery). The number of pulses felt per minute is equivalent to the number of times your heart beats per minute (bpm). Many people who exercise regularly and have a keen interest in monitoring their fitness levels wear a pulse monitor.

A group of students wanted to investigate how their heart rate was affected by exercise (Table 1).

1 Devise a practical procedure to collect these data.

The following were collected from a sample of 10 students. The students' pulse rates were measured at rest, during exercise, and two minutes after completing exercise.

2 State why it was important for the students to know the resting pulse rates.

3 a Calculate the mean resting heart rate of the students.
  b Calculate the standard deviation of the students' heart rate during exercise.

$$\sigma = \sqrt{\frac{\sum(x - \bar{x})^2}{n - 1}}$$

4 a Calculate the percentage increase in the average heart rate two minutes after exercise, compared with the resting heart rate. *(2 marks)*
  b Calculate the Spearman's rank correlation coefficient between the students' resting pulse rates, and their pulse rates during exercise. *(6 marks)*
  c Evaluate the strength of the correlation between the two sets of data. *(2 marks)*

Study tip

You can find a full table of values of Spearman's rank correlation coefficient in the appendix.

Study tip

Look back at Topic 10.6, Representing variation graphically, to see a worked example of how to calculate the standard deviation of a set of data, and how to calculate a Spearman's rank correlation coefficient.

▼ Table 1

Student	Resting pulse / bpm	Pulse rate during exercise / bpm	Pulse rate after exercise / bpm
1	57	84	62
2	64	89	66
3	54	84	58
4	78	101	80
5	74	98	75
6	68	92	72
7	65	90	68
8	60	86	66
9	64	95	67
10	72	94	75

Synoptic link

Look back at Topic 10.6, Representing variation graphically to see a worked example of how to calculate the standard deviation of a set of data.

Summary questions

1 Copy and complete the table to summarise the control of heart rate:

Stimulus	Receptor	Nervous system involved	Effect on heart rate
high blood pressure			
low blood pressure			
low blood CO_2 concentration			
high blood CO_2 concentration			

(4 marks)

2 Explain why an athlete's heart rate may increase just before a race has begun. (2 marks)

3 a Explain why blood pH values vary during exercise. (2 marks)
  b Explain why an increase in the pH of blood leads to a decrease in heart rate. (3 marks)

Practice questions

1 Insulin-dependent diabetes most often appears during childhood. It is caused by the autoimmune destruction of β cells in the pancreas. In this autoimmune response, specific white blood cells respond to proteins on the surface of the β cells.

 a (i) Name the hormone that will become deficient due to the autoimmune destruction of the β cells. *(1 mark)*

 (ii) Suggest what happens to the proteins on the cell surface membrane of the β cells to stimulate the autoimmune response. *(1 mark)*

 b In individuals with insulin-dependent diabetes, there is excessive secretion of glucagon. Increased glucagon concentration in the blood results in additional metabolic changes to those cause by damaged β cells. Describe the effects of increased glucagon concentration on the liver. *(3 marks)*

 c The graph shows the changes in blood glucose concentration following a meal in a diabetic individual and in a healthy individual.

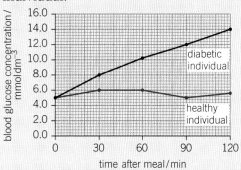

 (i) **Describe** the changes in blood glucose concentration in the **diabetic** individual for the 120 minutes following the meal. *(3 marks)*

 (ii) **Explain** the changes in blood glucose concentration in the **healthy** individual for the 120 minutes following the meal. *(3 marks)*

 OCR Jan 2010

2 One risk factor that increases the risk of developing Type 2 diabetes is obesity.

 The graph shows how prevalence of diabetes and mean body mass changed in one population from 1990 to 2000.

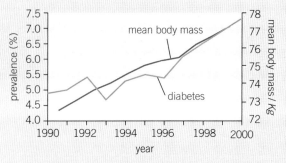

 a Explain why the data in the graph indicates a correlation and does not prove that obesity is a risk factor in diabetes. *(2 marks)*

 b Describe how the change in prevalence of diabetes changes with mean body mass in the graph. *(3 marks)*

 c Explain why the following symptoms may be seen in diabetics. *(5 marks)*

 sudden weight loss, increased need to urinate, dehydration

 d It is believed by some scientists that insulin resistance, the loss of sensitivity of receptors to insulin, is a result of an 'energy surplus' within cells.

 Some drugs used in the treatment of type 2 diabetes have been found to be inhibitors of ATP synthase of protein complexes in the electron transport chains.

 Suggest why this activity would lead to an increase in receptor sensitivity to insulin. *(3 marks)*

3 The pancreas acts as both an endocrine and exocrine gland. The diagram shows the arrangement of some of the different types of cells present in the pancreas.

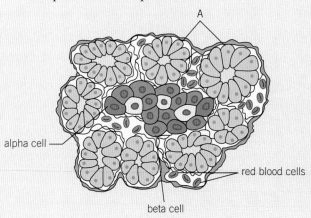

a (i) State the name given to a group of cells that work together to carry out a function. *(1 mark)*

   (ii) State the name of the structure that contains the alpha and beta cells in the pancreas. *(1 mark)*

   (iii) Outline the roles of the alpha and beta cells. *(4 marks)*

b Describe the role of the red blood cells with regard to the function of the alpha and beta cells. *(3 marks)*

c (i) State the name of the cell labelled A. *(1 mark)*

   (ii) State the function of cell A. *(1 mark)*

   (iii) Explain why cells, of which cell A is an example, are arranged in a different way to the alpha and beta cells. *(5 marks)*

4 A student carried out an investigation to study the effect of caffeine on daphnia heart rate. The following procedure was followed:

1 Daphnia, *transparent* aquatic organisms, of *the same species and size* were *placed on ice for 20 minutes* before the experiment.

2 A single daphnia was then placed in a beaker containing the solution under investigation *for 5 minutes*.

3 A small quantity of *cotton wool was placed on a microscope slide*.

4 A few *drops of pond water were placed on the cotton wool* on the microscope slide.

5 The daphnia was then placed on a microscope slide using a piece of filter paper. *A cover slip was not used*.

6 A cavity slide containing cold water was put under the microscope slide.

7 The daphnia was viewed under a *low power lens* of light microscope focused on the heart.

8 The number of heart beats every 30 seconds was then recorded.

9 This was *repeated* seven times for each concentration of caffeine.

10 The same procedure was repeated for solutions with different concentrations of caffeine.

The table shows the results that the student obtained.

Caffeine	1%	0.50%	0.25%	0.10%
1	401	392	308	232
2	400	380	276	228
3	380	360	284	204
4	404	400	268	232
5	360	332	292	248
6	428	368	260	256
7	440	340	280	240
8	368	320	308	216
Average	398	362	285	232

a Describe the significance of the terms in italics at each step in the procedure with regard to the following:

   (i) Reliability *(3 marks)*

   (ii) Experimental errors *(3 marks)*

   (iii) Ethics *(2 marks)*

b Plot a graph using the results in the table *(3 marks)*

c Describe the trends shown in the graph you have drawn for the different concentrations of caffeine. *(3 marks)*

The standard deviation, calculated as the square root of the variance, is used in the *t*-test calculation to compare two sets of data.

d Describe what standard deviation shows. *(1 mark)*

The variance is calculated in the following way:

Subtract the mean from each number in a column and square the difference. Add all of these squared differences together for each column and divide by $n - 1$ (one less than the number of results obtained). t is calculated using the following formula:

$$t = \frac{(\overline{x_1} - \overline{x_2})}{\sqrt{\dfrac{\sigma_1^2}{N_1} + \dfrac{\sigma_2^2}{N_2}}}$$

Where:

X_1 is the mean of the first data set
X_2 is the mean of the second data set

σ_1 is the standard deviation of the first data set
σ_2 is the standard deviation of the second data set
N_1 is the number of elements in the first data set
N_2 is the number of elements in the second data set

e Calculate t for the 1% and 0.5% solutions of caffeine and state whether there is a significant difference between these two sets of results. *(4 marks)*

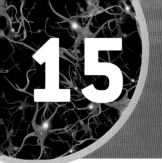

▲ Figure 1 *Mammals maintain the conditions inside their body within very narrow limits wherever they live and whatever they do*

Synoptic link

You learnt about sensory receptors, sensory and motor neurones, and coordination in the brain in Chapter 13, Neuronal communication.

Synoptic link

You learnt about negative feedback in Chapter 14, Hormonal communication.

The enzyme-controlled reactions of life can only take place if the conditions are right. The concentration of chemicals such as glucose and sodium ions must be kept within a narrow range, as must the pH and water balance of the body fluids, and the core temperature of the body. Organisms use both chemical and electrical systems to monitor and respond to any changes from the steady state of the body, and use the information to maintain a dynamic equilibrium.

Receptors and effectors

It is impossible to maintain a living mammal in a completely stable state because everything causes minute changes. Instead, the body maintains a dynamic equilibrium, with small fluctuations over a narrow range of conditions. This is known as **homeostasis**.

Receptors and effectors are vital for the body to maintain this dynamic equilibrium. As you have seen, sensory receptors detect changes in the internal and external environment of an organism. In homeostasis, it is essential to monitor changes in the internal environment, for example, the pH of the blood, core body temperature, and concentrations of urea and sodium ions in the blood.

Information from the sensory receptors is transmitted to the brain and impulses are sent along the motor neurones to the effectors to bring about changes to restore the equilibrium in the body. Effectors are the muscles or glands that react to the motor stimulus to bring about a change in response to a stimulus. Both are vital in a homeostatic system – detecting change is no use without the means to react to that change, but effectors cause chaos unless responding to a need.

Feedback systems

Homeostasis depends on sensory receptors detecting small changes in the body, and effectors working to restore the status quo. These precise control mechanisms in the body are based on feedback systems that enable the maintenance of a relatively steady state around a narrow range of conditions.

Negative feedback systems

Most of the feedback systems in the body involve negative feedback. A small change in one direction is detected by sensory receptors. As a result, effectors work to reverse the change and restore conditions to their base level. Negative feedback systems work to reverse the initial stimulus. You have seen negative feedback in action in the control of blood sugar levels by insulin and glucagon. Negative feedback systems are also important in many other aspects of homeostasis including temperature control and the water balance of the body. The general principles of negative feedback systems are shown in Figure 2.

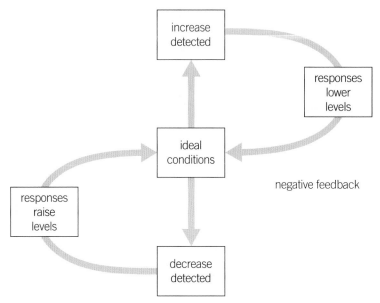

▲ **Figure 2** *The principles of negative feedback*

Positive feedback systems

There are relatively few positive feedback systems in the body. In a positive feedback system, a change in the internal environment of the body is detected by sensory receptors, and effectors are stimulated to reinforce that change and increase the response. One example occurs in the blood clotting cascade. When a blood vessel is damaged, platelets stick to the damaged region and they release factors that initiate clotting and attract more platelets. These platelets also add to the positive feedback cycle and it continues until a clot is formed. Another example of a positive feedback mechanism is seen during childbirth. The head of the baby presses against the cervix, stimulating the production of the hormone oxytocin. Oxytocin stimulates the uterus to contract, pushing the head of the baby even harder against the cervix and triggering the release of more oxytocin. This continues until the baby is born.

> ### Synoptic link
>
> You learnt about the clotting cascade in Topic 12.5, Non-specific animal defences against pathogens.

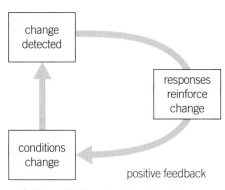

▲ **Figure 3** *The principles of positive feedback*

Summary questions

1 a Suggest three different types of receptors explaining what changes they detect. (*3 marks*)

   b Suggest two different types of effector and give an example of what they do. (*4 marks*)

2 a What is homeostasis? (*2 marks*)

   b Why are both receptors and effectors important in homeostasis? (*3 marks*)

3 Suggest why effective homeostasis depends on negative rather than positive feedback systems. (*6 marks*)

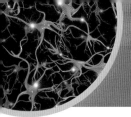
The enzymes controlling the rates of the chemical reactions needed for life are very temperature sensitive. Each enzyme has an optimum temperature at which it works most efficiently. If the temperature gets too high the enzymes are denatured. An important aspect of homeostasis in many animals is the maintenance of a relatively constant core body temperature to maintain optimum enzyme activity. This process is known as **thermoregulation**.

Endotherms and ectotherms

Organisms are constantly heating up and cooling down as a result of their surroundings. These changes depend on a number of physical processes. These include:

- Exothermic chemical reactions.
- Latent heat of evaporation – objects cool down as water evaporates from a surface.
- Radiation – the transmission of electromagnetic waves to and from the air, water, or ground.
  - Convection – the heating and cooling by currents of air or water, warm air or water rises and cooler air or water sinks setting up convection currents around an organism.
  - Conduction – heating as a result of the collision of molecules. Air is not a good conductor of heat but the ground and water are.

In many cases, the balance between heating and cooling determines the core temperature of the organism. Animals can be classified as **ectotherms** or **endotherms** depending on how they maintain and control their body temperature.

gains in heat   losses of heat

waste heat from cell respiration
conduction from surroundings
convection from surroundings
radiation from surroundings

organism

evaporation of water
conduction to surroundings
convection to surroundings
radiation to surroundings

▲ Figure 1 *Ways in which animals warm up and cool down*

Ectotherms

Most animals are ectotherms and use their surroundings to warm their bodies (ectotherm literally means 'outside heat'). Their core body temperature is heavily dependent on their environment. Ectotherms include all the invertebrate animals, along with fish, amphibians, and reptiles.

Many ectotherms living in water do not need to thermoregulate. The high heat capacity of water means that the temperature of their environment does not change much. Ectotherms that live on land have a much bigger problem with temperature regulation. The temperature of the air can vary dramatically both between seasons and even over a 24-hour period from the middle of the day to the end of the night. As a result ectotherms have evolved a range of strategies that enable them to cool down or warm up.

Endotherms

Mammals and birds are endotherms. They rely on their metabolic processes to warm up and they usually maintain a very stable core body temperature regardless of the temperature of the environment (endotherm literally means 'inside heat'). They have adaptations which enable them to maintain their body temperature and to take advantage of warmth from the environment. As a result, endotherms survive in a wide range of environments. Keeping warm in cold conditions and cooling down in hot conditions are both active processes. The metabolic rate of endotherms is around five times higher than ectotherms, so they need to consume more food to meet their metabolic needs than ectotherms of a similar size.

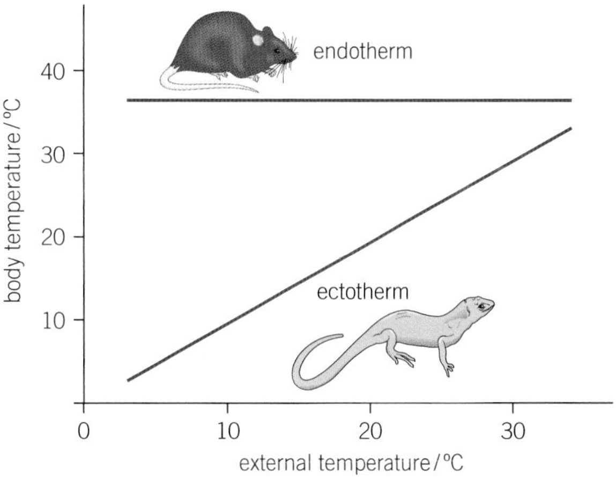

Temperature regulation in ectotherms

Ectotherms cannot control their body temperature using their metabolism – however, they have evolved a range of behavioural responses that enable them to overcome the limitations imposed by the temperature of their surroundings.

▲ Figure 2 *Graph to show simplified effect of changes in the internal and external temperature on ectotherms and endotherms*

Behavioural responses

Ectotherms display a number of behaviours which increase or reduce the radiation they absorb from the Sun. Sometimes they need to warm up to reach a temperature at which their metabolic reactions happen fast enough for them to be active. They may bask in the Sun, orientate their bodies so that the maximum surface area is exposed to the Sun, and even extend areas of their body to increase the surface area exposed to the Sun. For example, lizards often bask for long periods of time to get warm enough to move fast and hunt their prey, and insects such as locusts and butterflies orientate themselves for maximum exposure to the Sun and spread their wings to increase the available surface area to get warm enough to fly.

Ectotherms can increase their body temperature through conduction by pressing their bodies against the warm ground. They also get warmer as a result of exothermic metabolic reactions. Galapagos iguanas will contract their muscles and vibrate increasing cellular metabolism to raise their body temperature. Similarly, moths and butterflies may vibrate their wings to warm their muscles before they take flight.

Ectotherms sometimes need to cool down to prevent their core temperature reaching a point where enzymes begin to denature. To cool down, many of the warming processes are reversed. Ectotherms shelter from the sun by seeking shade, hiding in cracks in rocks, or even digging burrows. They will press their bodies against cool, shady earth or stones, or move into available water or mud. They orientate their bodies so that the minimum surface area is exposed to the sun, and minimise their movements to reduce the metabolic heat generated.

Physiological responses to warming

Much of the thermoregulation by ectotherms is the result of behavioural responses but some of them have physiological responses

▲ Figure 3 *The black pigment of the marine iguanas observed by Darwin on the Galapagos Islands enables them to absorb enough heat to swim and feed in the relatively cold seawater surrounding the famous archipelago*

as well. Dark colours absorb more radiation than light colours. Lizards living in colder climates tend to be darker coloured than lizards living in hotter countries so that they get warmer. Some ectotherms also alter their heart rate to increase or decrease the metabolic rate and sometimes to affect the warming or cooling across the body surfaces.

Ectotherms are always more vulnerable to fluctuations in the environment than endotherms. However, by using a variety of behavioural and physiological strategies many of them can maintain relatively stable core temperatures. They need less food than endotherms as they use less energy regulating their temperatures, and so they can survive in some very difficult habitats where food is in short supply.

Summary questions

1 **a** What is an ectotherm?
(2 marks)
   b Give two examples of ectotherms. (1 mark)

2 Give an example of an ectotherm warming up or cooling down through interaction with the environment by:
   a radiation (2 marks)
   b conduction (2 marks)
   c convection (2 marks)
   d evaporation. (2 marks)

3 Galapagos marine iguanas are unique reptiles because they swim and feed in the sea. Read the following statements and discuss the observations in terms of thermoregulation.
   a Marine iguanas are black in colour. They spend a lot of time on the exposed rocks and alter their position and posture regularly. They need to have a body temperature of around 36 °C before they dive for food.
(5 marks)
   b Dives usually last a few minutes but can last up to 30 minutes. The length of the dive seems to be related to body size. The core temperature of the iguanas can drop by about 10 °C during a dive. The animals are slow and clumsy when they emerge from the sea.
(6 marks)

➕ The Namaqua chameleon – a highly adapted ectotherm

The Namaqua chameleon lives in the Namib desert, one of the most inhospitable hot and waterless environments on Earth. Several observations have been made on this rare and extremely well-adapted ectotherm:

- It is black in the morning. It may even appear black on the side exposed to the sun and pale grey on the other side of the body.
- It orientates its body sideways to the Sun.
- It has an increased heart rate early in the morning when basking.
- It inflates its body in the early morning.
- It presses its body to the desert sand in the morning.
- During the day the chameleon deflates its body.
- The animal becomes a very pale grey.
- It holds itself well away from the desert surface.
- The heart rate slows down.
- The chameleon opens its mouth and pants in the middle of the day.

▲ **Figure 4** *The Namaqua chameleon looks very different depending on whether it is trying to gain or lose heat*

Using each of the adaptations of the Namaqua chameleon described, explain how they help the animal to warm up or cool down.

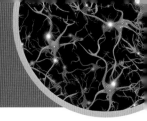

Endotherms can regulate their body temperature within a very narrow range in a wide variety of external conditions. Human beings, like all mammals, have a number of physiological responses that make this thermoregulation possible.

Detecting temperature changes

In any homeostatic system receptors are needed to detect a change in the internal environment. The peripheral temperature receptors are in the skin and detect changes in the surface temperature. Temperature receptors in the hypothalamus detect the temperature of the blood deep in the body. The temperature of the skin is much more likely to be affected by external conditions than the temperature of the hypothalamus. The combination of the two gives the body great sensitivity and allows it to respond not only to actual changes in the temperature of the blood but to pre-empt possible problems that might result from changes in the external environment.

The temperature receptors in the hypothalamus act as the thermostat of the body, controlling the responses that maintain the core temperature in a dynamic equilibrium to within about 1 °C of 37 °C.

Principles of thermoregulation in endotherms

Endotherms use their internal exothermic metabolic activities to keep them warm, and energy-requiring physiological responses to help them cool down. They also have passive ways of heating up and cooling down, to reduce the energy demands on their bodies. Like ectotherms, endotherms have a range of behavioural responses to temperature changes that include basking in the Sun, pressing themselves to warm surfaces, wallowing in water and mud to cool down, and digging burrows to keep warm or cool. Some animals even become dormant through the coldest weather (hibernation) or through the hottest weather (aestivation is a period of prolonged or deep sleep similar to hibernation but occurs in summer or during dry seasons to avoid heat stress rather than cold).

Humans have additional behavioural adaptations to help control body temperature – clothes are worn to stay warm, houses are built, and then heated up or cooled down to maintain the ideal temperature.

In spite of these behavioural responses, endotherms mainly rely on physiological adaptations to maintain a stable core body temperature, regardless of the environmental conditions or the amount of exercise being done. These adaptations include the peripheral temperature receptors, the thermoregulatory centres of the hypothalamus, the skin, and muscles.

Synoptic link

You learnt about the areas of the brain in Topic 13.7, Structure and function of the brain.

▲ Figure 1 Thermoregulation makes it possible for endotherms to survive in many extreme environments

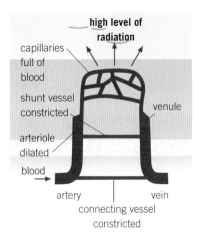

▲ Figure 2 *Vasodilation leading to heat loss by radiation*

▲ Figure 3 *Horses sweat all over when they are very hot. This horse is also panting, so he is cooling by evaporation both from his skin and his breathing passages. The hair on his body is also lying very flat to avoid trapping an insulating layer of air. The dilated arterioles and venules under the skin take blood to and from the capillary network*

Synoptic link

You learnt about surface area : volume ratio in Topic 7.1, Specialised exchange surfaces.

Cooling down

If the core body temperature increases it is important for an animal to cool down. There are a number of rapid responses to a rise in the core temperature that are common to all endotherms. These include:

Vasodilation

The arterioles near the surface of the skin dilate when the temperature rises. The vessels that provide a direct connection between the arterioles and the venules (the arteriovenous shunt vessels) constrict. This forces blood through the capillary networks close to the surface of the skin. The skin flushes, and cools as a result of increased radiation. If the skin is pressed against cool surfaces, then the cooling results from conduction.

Increased sweating

As the core temperature starts to increase, rates of sweating also increase. Sweat spreads out across the surface of the skin. In some mammals, including humans and horses, there are sweat glands all over the body. As the sweat evaporates from the surface of the skin, heat is lost, cooling the blood below the surface. In some animals, the sweat glands are restricted to the less hairy areas of the body such as the paws. These animals often open their mouths and pant when they get hot, again losing heat as the water evaporates. In human beings, around $1\,dm^3$ of sweat is lost by evaporation on a normal day. If the conditions are very hot and dry or the person is exercising very hard, up to $12\,dm^3$ of sweat a day can be lost. Kangaroos and cats often lick their front legs to keep cool in high temperatures.

Reducing the insulating effect of hair or feathers

As the body temperature begins to increase, the erector pili muscles (the hair erector muscles) in the skin relax – as a result, the hair or feathers of the animal lie flat to the skin. This avoids trapping an insulating layer of air. It has little effect in humans.

Endotherms that live in hot climates often have anatomical adaptations as well as the behavioural and physiological adaptations already described. These minimise the effect of high temperatures and maximise the ability of the animal to cool down through the surface area of the body. They include a relatively large surface area : volume (SA : V) ratio to maximise cooling (e.g., include large ears and wrinkly skin), and pale fur or feathers to reflect radiation.

Warming up

If the core temperature falls it is important for an animal to warm up and prevent further cooling. There are a number of rapid responses to a fall in the core temperature that are common to all endotherms.

Vasoconstriction

The arterioles near the surface of the skin constrict. The arteriovenous shunt vessels dilate, so very little blood flows through the capillary networks close to the surface of the skin. The skin looks pale, and very little radiation takes place. The warm blood is kept well below the surface.

Decreased sweating

As the core temperature falls, rates of sweating decrease and sweat production will stop entirely. This greatly reduces cooling by the evaporation of water from the surface of the skin, although some evaporation from the lungs still continues.

Raising the body hair or feathers

As the body temperature falls, the erector pili muscles in the skin contract, pulling the hair or feathers of the animal erect. This traps an insulating layer of air and so reduces cooling through the skin. The effect can be quite dramatic and it is a very effective way to reduce heat loss to the environment in many animals. In humans this has little effect although you can observe the hairs being pulled upright.

Shivering

As the core temperature falls the body may begin to shiver. This is the rapid, involuntary contracting and relaxing of the large voluntary muscles in the body. The metabolic heat from the exothermic reactions warm up the body instead of moving it and is an effective way of raising the core temperature.

Endotherms living in cold climates often have additional anatomical adaptations to help them keep warm. Many have adaptations that minimise their SA : V ratio to reduce cooling (e.g., small ears). Another common adaptation is a thick layer of insulating fat underneath the skin, for example, blubber in whales and seals. Some animals hibernate – they build up fat stores, build a well-insulated shelter, and lower their metabolic rate so they pass the worst of the cold weather in a deep sleep-like state.

Polar bears demonstrate many of the ways in which endotherms can survive in extremely cold conditions. They have small ears and fur on their feet to insulate them from the ice. The fur and skin of polar bears work together. The hairs are hollow so trap a permanent layer of insulating air. The skin underneath is black, so it absorbs warming radiation. They have a thick layer of fat under the skin. Polar bears are so well insulated that their external surfaces are similar in temperature to the snow and ice on which they live. Females dig dens in the snow and remain in them, warm and insulated, for months while they give birth to their cubs, only emerging when the cubs are large enough to survive the cold. Polar bears are so well adapted to life in temperatures down to –50 °C in the Arctic that they can overheat at temperatures over 10 °C.

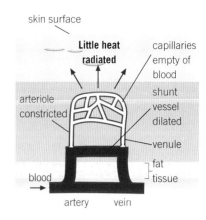

▲ Figure 4 *Vasoconstriction reduces heat loss by radiation*

▲ Figure 5 *The effect of the contraction of the erector pili muscles in robins*

Controlling thermoregulation

The physiological responses of endotherms to changes in the core temperature are the result of complex homeostatic mechanisms involving negative feedback control from the hypothalamus. There are two control centres:

The heat loss centre

This is activated when the temperature of the blood flowing through the hypothalamus increases. It sends impulses through autonomic motor neurones to effectors in the skin and muscles, triggering responses that act to lower the core temperature.

The heat gain centre

This is activated when the temperature of the blood flowing through the hypothalamus decreases. It sends impulses through the autonomic nervous system to effectors in the skin and the muscles, triggering responses that act to raise the core temperature.

The interaction of the sensory receptors, the autonomic nervous system, and the effectors in a sophisticated feedback system enables endotherms to maintain a very stable core body temperature regardless of environmental conditions or activity levels.

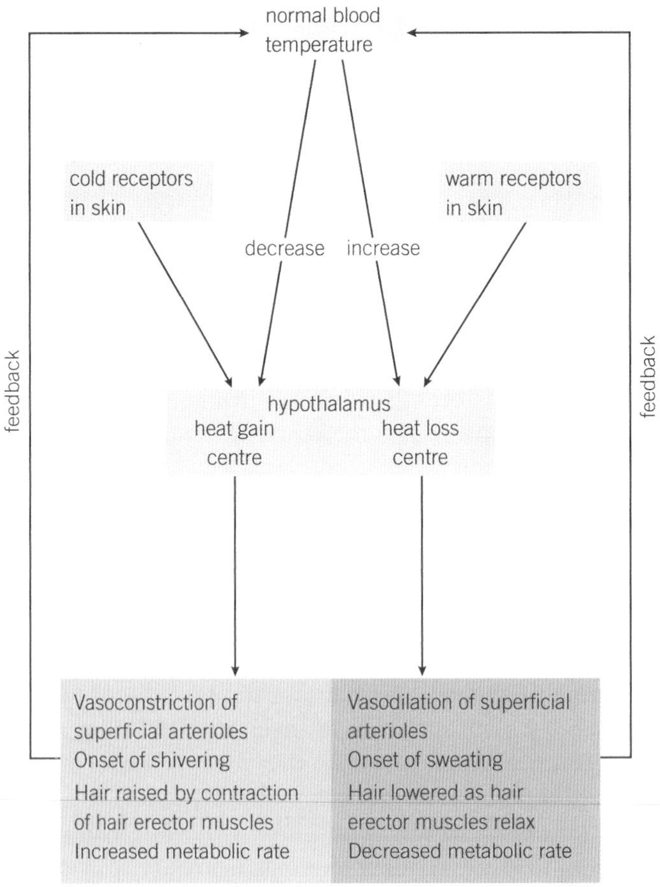

Summary questions

1 Why is the control of the internal temperature so important to both ectotherms and endotherms? (*6 marks*)

2 Explain how the role of evaporation of water in thermoregulation differs between ectotherms and endotherms. (*6 marks*)

3 Explain the difference in the role of the peripheral temperature receptors and the temperature receptors in the hypothalamus in the regulation of the core body temperature in an endotherm. (*3 marks*)

4 Endotherms that live in very hot climates are often pale coloured.
   a Why is this? (*3 marks*)
   b Why might you expect endotherms that live in very cold environments to be dark coloured? (*2 marks*)
   c In fact, very few endotherms that live in very cold environments are dark coloured. Suggest reasons for this. (*4 marks*)

▲ **Figure 6** *Summary of the control of body temperature by the peripheral temperature receptors, the hypothalamus, and the autonomic nervous system*

15.4 Excretion, homeostasis, and the liver

Specification reference: 5.1.2

Many of the chemical reactions of metabolism that take place in the cells of the body produce waste products that are toxic if they are allowed to build up. **Excretion** is the removal of the waste products of metabolism from the body.

Excretion in mammals

The main metabolic waste products in mammals are:

- Carbon dioxide – one of the waste products of cellular respiration which is excreted from the lungs.

- Bile pigments – formed from the breakdown of haemoglobin from old red blood cells in the liver. They are excreted in the bile from the liver into the small intestine via the gall bladder and bile duct. They colour the faeces.

- Nitrogenous waste products (urea) – formed from the breakdown of excess amino acids by the liver. All mammals produce **urea** as their nitrogenous waste. Fish produce ammonia while birds and insects produce uric acid. Urea is excreted by the kidneys in the urine.

The liver

The liver is one of the major body organs involved in homeostasis. It is a reddish-brown organ which makes up about 5% of the total body mass – the largest internal organ of the body. It lies just below the diaphragm and is made up of several lobes. The liver is very fast growing and damaged areas generally regenerate very quickly.

The liver has a very rich blood supply – about $1\,dm^3$ of blood flows through it every minute. Oxygenated blood is supplied to the liver by the hepatic artery and removed from the liver and returned to the heart in the hepatic vein. The liver is also supplied with blood by a second vessel, the **hepatic portal vein**. This carries blood loaded with the products of digestion straight from the intestines to the liver and this is the starting point for many metabolic activities of the liver (Figure 1). Up to 75% of the blood flowing through the liver comes via the hepatic portal vein.

The structure of the liver

The liver carries out many different complex functions but the cells are surprisingly simple and uniform in appearance. Liver cells or **hepatocytes** have large nuclei, prominent Golgi apparatus, and lots of mitochondria, indicating that they are metabolically active cells (Figure 1). They divide and replicate – even if around 65% of the liver is lost, it will regenerate in a matter of months.

The blood from the hepatic artery and the hepatic portal vein is mixed in spaces called sinusoids which are surrounded by hepatocytes. This

Learning outcomes

Demonstrate knowledge, understanding, and application of:

→ the term *excretion* and its importance in maintaining metabolism and homeostasis

→ the structure and mechanisms of action and functions of the mammalian liver

→ the examination and drawing of stained sections to show the histology of liver tissue.

Synoptic link

You learnt about the excretion of carbon dioxide from the body in Topic 7.2, Mammalian gaseous exchange system and Topic 8.4, Transport of oxygen and carbon dioxide in the blood.

Synoptic link

You learnt about macrophages in Topic 12.5, Non-specific animal defences against pathogens.

mixing increases the oxygen content of the blood from the hepatic portal vein, supplying the hepatocytes with enough oxygen for their needs. The sinusoids contain Kupffer cells, which act as the resident macrophages of the liver, ingesting foreign particles and helping to protect against disease. The hepatocytes secrete bile from the breakdown of the blood into spaces called canaliculi, and from these the bile drains into the bile ductules which take it to the gall bladder.

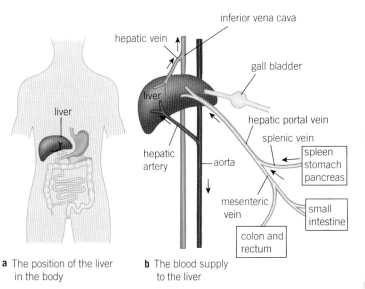

a The position of the liver in the body

b The blood supply to the liver

c The arrangement of the tissues in the liver, giving all of the hepatocytes close contact with the blood for both removing materials and adding substances.

d Single hepatocyte

▲ **Figure 1** *The structure of the liver is related to its variety of functions in the metabolism of carbohydrates, proteins, and fats and in excretion and homeostasis*

The functions of the liver

The liver has many functions – around 500 different metabolic pathways are linked to the liver. Several of these play a major role in homeostasis. These functions will now be explored in more detail.

Carbohydrate metabolism

Synoptic link

You learnt about the interconversion of glucose to glycogen and the storage of glycogen in the liver in Topic 14.3, Regulation of blood glucose concentration.

Hepatocytes are closely involved in the homeostatic control of glucose levels in the blood by their interaction with insulin and glucagon. When blood glucose levels rise, insulin levels rise and stimulate hepatocytes to convert glucose to the storage carbohydrate glycogen. About 100 g of glycogen is stored in the liver. Similarly, when blood sugar levels start to fall, the hepatocytes convert the glycogen back to glucose under the influence of the hormone glucagon.

Deamination of excess amino acids

The liver plays a vital role in protein metabolism where hepatocytes synthesise most of the plasma proteins. Hepatocytes also carry out *transamination* – the conversion of one amino acid into another. This is important because the diet does not always contain the required balance of amino acids but transamination can overcome the problems this might cause.

The most important role of the liver in protein metabolism is in **deamination** – the removal of an amine group from a molecule. The body cannot store either proteins or amino acids. Any excess ingested protein would be excreted and therefore wasted if it were not for the action of the hepatocytes. They deaminate the amino acids, removing the amino group, and converting it first into ammonia which is very toxic and then to urea. Urea is toxic in high concentrations but not in the concentrations normally found in the blood. Urea is excreted by the kidneys (you will learn about the role of the kidneys in excretion and water balance in Topics 15.5 and 15.6). The remainder of the amino acid can then be fed into cellular respiration or converted into lipids for storage.

The ammonia produced in the deamination of proteins is converted into urea in a set of enzyme-controlled reactions known as the **ornithine cycle**. Removing the amino group from amino acids and converting the highly toxic ammonia to the less toxic and more manageable compound urea involves some complex biochemistry. This is simplified and summarised in Figure 3.

Detoxification

The level of toxins in the body always tends to increase. Apart from urea, many other metabolic pathways produce potentially poisonous substances. We also take in a wide variety of toxins by choice such as alcohol and other drugs. The liver is the site where most of these substances are detoxified and made harmless.

One example is the breakdown of hydrogen peroxide, a by-product of various metabolic pathways in the body. Hepatocytes contain the enzyme **catalase**, one of the most active known enzymes, that splits the hydrogen peroxide into oxygen and water. Another example is the way in which liver detoxifies the ethanol – the active drug in alcoholic drinks. Hepatocytes contain the enzyme alcohol dehydrogenase that breaks down the ethanol to ethanal. Ethanal is then converted to ethanoate which may be used to build up fatty acids or used in cellular respiration.

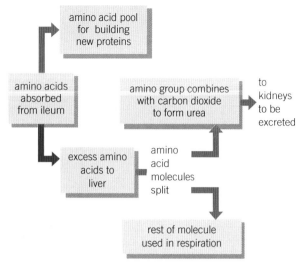

▲ **Figure 2** *The process of deamination*

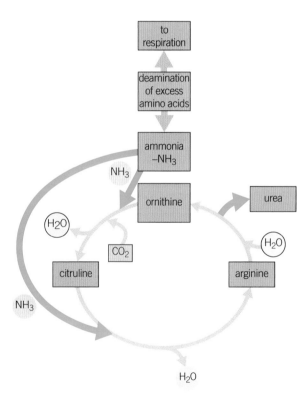

▲ **Figure 3** *The ornithine cycle where ammonia resulting from the breakdown of excess amino acids in the body is converted into useful products and urea*

Synoptic link

You learnt about catalase as an enzyme in Topic 3.7, Types of proteins and Topic 4.2, Factors affecting enzyme activity.

Looking at liver cells 🧪

When liver tissue is stained for use under the light microscope, different stains can be used to show up different things, for example glycogen stored in the hepatocytes.

Using low magnification with a light microscope enables you to see the arrangement of the liver cells, the blood vessels, and the sinusoids.

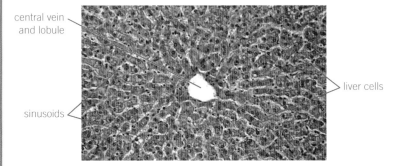

central vein
and lobule

liver cells

sinusoids

▲ **Figure 4** *Light micrograph through a section of human liver ×34 magnification*

Using a high magnification gives more detail of the individual hepatocytes.

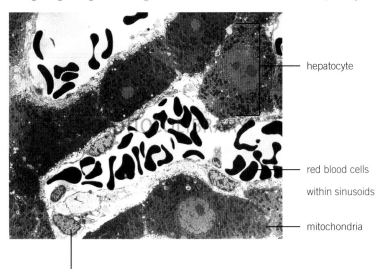

hepatocyte

red blood cells
within sinusoids

mitochondria

Kupffer cells

▲ **Figure 5** *High magnification light micrograph of liver cells ×3 500 magnification*

Make a scientific drawing of some of the cells seen in Figure 5. Annotate your drawings to indicate the role of the cells in the liver.

Synoptic link

You learnt about the light microscope in Topic 2.1, Microscopy.

Cirrhosis of the liver

Cirrhosis is a disease where the normal liver tissue is replaced by fibrous scar tissue. There are lots of different causes including genetic conditions and hepatitis C, however, in the UK the most common cause is drinking excessive amounts of alcohol.

There are three stages of alcoholic liver disease – alcoholic fatty liver disease, alcoholic hepatitis, and liver cirrhosis. In fatty liver, big fat-filled vesicles displace the nuclei of the hepatocytes and the liver gets larger. In alcoholic hepatitis, the patient has fatty liver along with damaged hepatocytes and the sinusoids and hepatic veins become narrowed. In alcoholic cirrhosis the liver tissue is irreversibly damaged. Many hepatocytes die and are replaced with fibrous tissue. The hepatocytes can no longer divide and replace themselves so the liver shrinks and its ability to deal with toxins in the body decreases.

1 In many cases, if an affected person stops drinking alcohol the liver may recover. How is this possible?

2 Many people cannot stop drinking – explain why not and discuss ways of helping those affected.

3 Suggest possible effects on the body when the alcoholic liver damage becomes irreversible.

4 Discuss the pros and cons of giving someone with alcoholic cirrhosis of the liver a liver transplant.

Summary questions

1 Many people use the term excretion to describe defecation. Why is this not entirely accurate?   (6 marks)

2 Hepatocytes make up approximately 70–85% of the liver's mass.
   a What is a hepatocyte?   (2 marks)
   b Hepatocytes have large nuclei, lots of Golgi apparatus, and many mitochondria. What does this tell you about the cells?   (3 marks)

3 a Draw a labelled diagram to show the structure of the liver.   (6 marks)
   b Explain how the structure of the liver is adapted for its functions in the body   (6 marks)

4 a Why do you think the liver is particularly affected by excess drinking?   (6 marks)
   b Why is a build-up of fatty tissue a common symptom of excess drinking?   (4 marks)

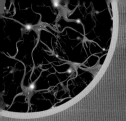

Human kidneys are typical of all mammalian kidneys. They are a pair of reddish-brown organs attached to the back of the abdominal cavity. They are usually surrounded by a thick, protective layer of fat and a layer of fibrous connective tissue. The kidneys play two important homeostatic roles in the body – they are involved in excretion and **osmoregulation**. They filter nitrogenous waste products out of the blood, especially urea. They also help to maintain the water balance and pH of the blood, and hence the tissue fluid that surrounds all the cells.

The anatomy of the kidneys

If you put your hands on your hips, the place where your thumbs are is the approximate position of your kidneys. The kidneys are supplied with blood at arterial pressure by the renal arteries that branch off from the abdominal aorta. Blood that has circulated through the kidneys is removed by the renal vein that drains into the inferior vena cava. About 90–120 cm^3 of blood passes through the kidneys every minute. All of the blood in the body passes through the kidneys about once an hour. The kidneys filter 180 dm^3 of blood a day, producing 1–2 dm^3 of urine. The final volume depends on many different factors. You will learn more about the factors that determine the amount of urine produced in Topic 15.6, The kidney and osmoregulation.

The kidneys are made up of millions of small structures called **nephrons** that act as filtering units. The sterile liquid produced by the kidney tubules is called urine. The urine passes out of the kidneys down tubes called **ureters**. It is collected in the bladder, a muscular sac that can store around 400–600 cm^3 of urine. When the bladder is getting full, the sphincter at the exit to the bladder opens and the urine passes out of the body down the **urethra**.

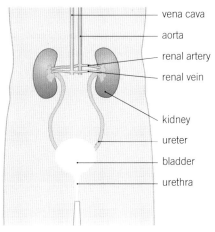

vena cava
aorta
renal artery
renal vein
kidney
ureter
bladder
urethra

▲ Figure 1 *The human urinary system*

Kidney structure

If you slice open a kidney, you will see three main areas – the **cortex**, the **medulla**, and the **pelvis** (Figure 2).

● The cortex is the dark outer layer. This is where the filtering of the blood takes place and it has a very dense capillary network carrying the blood from the renal artery to the nephrons.

● The medulla is lighter in colour – it contains the tubules of the nephrons that form the pyramids of the kidney and the collecting ducts.

● The pelvis (which is Latin for *basin*) of the kidney is the central chamber where the urine collects before passing out down the ureter.

Nephrons – the functional units of the kidney

In the nephrons the blood is filtered and then the majority of the filtered material is returned to the blood, removing nitrogenous wastes and balancing the mineral ions and water. Each nephron is around 3 cm long and there are around 1.5 million nephrons in each kidney. This provides the body with several kilometres of tubules for the reabsorption of water, glucose, salts, and other substances back into the blood.

Structure of the nephron

The main structures and functions of the nephron are as follows:

- **Bowman's capsule** – cup-shaped structure that contains the glomerulus, a tangle of capillaries. More blood goes into the glomerulus than leaves it due to the ultrafiltration processes that take place.

- **Proximal convoluted tubule** – the first, coiled region of the tubule after the Bowman's capsule, found in the cortex of the kidney. This is where many of the substances needed by the body are reabsorbed into the blood

- **Loop of Henle** – a long loop of tubule that creates a region with a very high solute concentration in the tissue fluid deep in the kidney medulla. The descending loop runs down from the cortex through the medulla to a hairpin bend at the bottom of the loop. The ascending limb travels back up through the medulla to the cortex.

- **Distal convoluted tubule** – a second twisted tubule where the fine-tuning of the water balance of the body takes place. The permeability of the walls to water varies in response to the levels of the **antidiuretic hormone (ADH)** in the blood (ADH is explored in more detail in Topic 15.6, The kidney and osmoregulation). Further regulation of the ion balance and pH of the blood also takes place in this tubule.

- **Collecting duct** – the urine passes down the collecting duct through the medulla to the pelvis. More fine-tuning of the water balance takes place – the walls of this part of the tubule are also sensitive to ADH.

The nephron has a network of capillaries around it which finally lead into a venule and then to the renal vein. The blood that leaves the kidney has greatly reduced levels of urea, but the levels of glucose and other substances such as amino acids needed by the body are almost the same as when the blood entered the kidneys (may be slightly less as some glucose will have been used for selective reabsorption). The mineral ion concentration in the blood has also been restored to ideal levels.

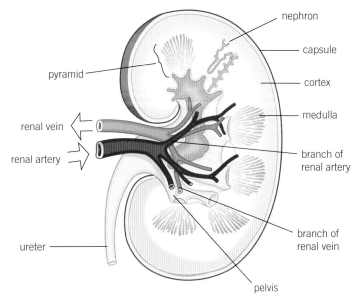

▲ **Figure 2** *Internal structure of a kidney*

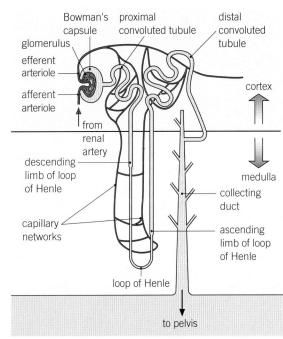

▲ **Figure 3** *The structure of a nephron and its blood supply*

Synoptic link

You learnt about arteries, arterioles, capillaries, venules, and veins in Topic 8.2, The blood vessels, and about the formation of tissue fluid in Topic 8.3, Blood, tissue fluid, and lymph.

Investigating the kidneys

Lamb, pig, or even beef kidneys can be used to look at the external and internal structures of these fascinating organs.

1 The protective layer of fat around kidneys (renal capsule) is impressive – these vital organs are well-cushioned from physical damage in the body (Figure 4).

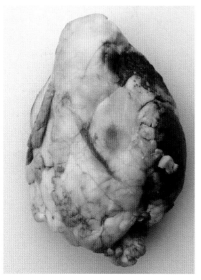

▲ **Figure 4** *Here the kidney is still surrounded by perirenal fat (adipose tissue)*

2 After carefully removing the fat, the external appearance of the kidney can be seen. You can see the colour, the fibrous coat, the ureter, and if you are skilful, you will be able to identify the renal artery and vein as well (depending on whether these have been left on by the butcher).

▲ **Figure 5** *External appearance of the kidney once perirenal fat has been removed*

3 Slice the kidney open carefully using a scalpel or a very sharp knife by making small cuts. Look carefully and identify as many of the internal features as you can (coloured pins are useful for this).

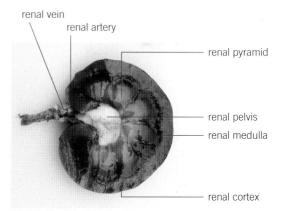

▲ **Figure 6** *Labelled photo of a dissected kidney*

4 Dissecting a kidney cannot show us the histology of the kidneys. To observe the structure of the individual nephrons requires a hand-lens or stained sections of kidney tissue and a microscope. It is very difficult to see an entire tubule but different parts of the tubules can be identified (Figures 7 and 8). To make the nephrons more visible a drop of hydrogen peroxide can be applied onto the cut surface of the kidney (wearing safety glasses and gloves). There will be rapid effervescence (foaming), after this is wiped off, the renal tubules, collecting duct, and loops of Henle should be a little clearer to see as shown by strings of bubbles.

In a section through the cortex you will see Bowman's capsules and glomeruli, as well as sections through proximal and distal convoluted tubules. The lumen of the distal tubules tends to be bigger and more open than those of proximal tubules which can be helpful in identifying them.

In a section through the medulla you will see mainly loops of Henle and collecting duct. In a transverse section you will see the lumens of the tubules – the collecting ducts are larger than the thick ascending loops of Henle while the thin-walled descending limbs are only visible at very high magnifications. In a longitudinal section you will see the parallel tubes – low magnifications give an overall impression whilst higher magnifications enable you to see individual tubules.

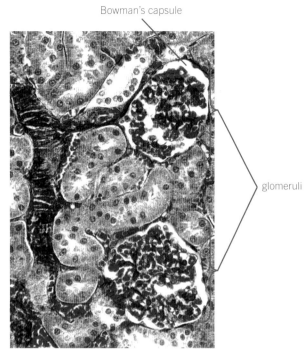

Bowman's capsule

glomeruli

▲ **Figure 7** *Light micrograph of glomeruli in kidney. These are a tightly coiled network of capillaries surrounded by the lumen of the Bowman's capsule. Surrounding this are the tubules (rounded) where reabsorption takes place, × 230 magnification*

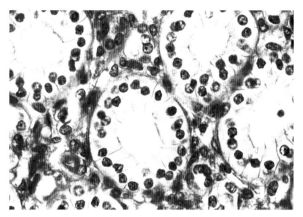

▲ **Figure 8** *Light micrograph showing a section through a normal human kidney, ×275 magnification*

1 Compare the section of a kidney in Figure 6 with the stylised diagram in Figure 2. How does the dissection differ from the diagram? Compare their value in helping to understand the structure of the organ.
2 Which micrograph do you find most useful in developing an understanding of the structure of the kidney. Why?

The functions of the nephrons

Ultrafiltration

The first stage in the removal of nitrogenous waste and osmoregulation of the blood is **ultrafiltration**. Ultrafiltration in the kidney tubules is a specialised form of the process that results in the formation of tissue fluid in the capillary beds of the body and it is the result of the structure of the glomerulus and the cells lining the Bowman's capsule (Figure 9 and 10). The glomerulus is supplied with blood by a relatively wide afferent (incoming) arteriole from the renal artery. The blood leaves through a narrower efferent (outward) arteriole and as a result there is considerable pressure in the capillaries of the glomerulus. This forces the blood out through capillary wall – it acts rather like a sieve. Then the fluid passes through the basement membrane – scientists are increasingly recognising the basement membrane as an important factor in the filtration process. The basement membrane is made up of a network of collagen fibres and other proteins that make up a second 'sieve'. Most of the plasma contents can pass through the basement membrane but the blood cells and many proteins are retained in the capillary because of their size.

> **Synoptic link**
>
> You learnt about collagen and its properties in Topic 3.7, Types of proteins.

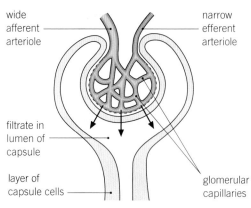

▲ **Figure 9** *Ultrafiltration takes place in the Bowman's capsule as a result of the high blood pressure in the glomerulus*

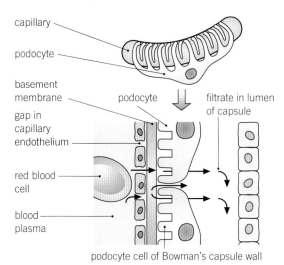

▲ **Figure 10** *The microstructure of the cells in the capillaries of the glomerulus and the podocytes of the Bowman's capsule is important in the ultrafiltration of the blood*

▼ **Table 1** *Comparison of the mean composition of human blood plasma and the ultrafiltrate as it enters the nephron*

Molecule or ion	Approximate concentration / $g\,dm^{-3}$	
	plasma	filtrate
water	900.0	900.0
protein	80.0	0.0
glucose	1.0	1.0
amino acids	0.5	0.5
urea	0.3	0.3
inorganic ions	7.2	7.2

The wall of the Bowman's capsule also involves special cells called *podocytes* that act as an additional filter. They have extensions called *pedicels* that wrap around the capillaries, forming slits that make sure any cells, platelets, or large plasma proteins that have managed to get through the epithelial cells and the basement membrane do not get through into the tubule itself. The filtrate which enters the capsule contains glucose, salt, urea, and many other substances in the same concentrations as they are in the blood plasma (Figure 11 and Table 1). The process is so efficient that up to 20% of the water and solutes are removed from the plasma as it passes through the glomerulus. The volume of blood that is filtered through the kidneys in a given time is known as the glomerular filtration rate.

Reabsorption

Ultrafiltration removes urea, the waste product of protein breakdown, from the blood but it also removes a lot of water along with the glucose, salt, and other substances which are present in the plasma. Many of these substances are needed by the body – for example, glucose is used for cellular respiration and is never normally excreted. The ultrafiltrate is also hypotonic to (less concentrated than) the blood plasma. The main function of the nephron after the Bowman's capsule is to return most of the filtered substances back to the blood.

The proximal convoluted tubule

In the proximal convoluted tubule all of the glucose, amino acids, vitamins, and hormones are moved from the filtrate back into the blood by active transport. Around 85% of the sodium chloride and water is reabsorbed as well – the sodium ions are moved by active transport while the chloride ions and water follow passively down concentration gradients. The cells lining the proximal convoluted tubule have clear adaptations:

- they are covered with microvilli, greatly increasing the surface area over which substances can be reabsorbed

- they have many mitochondria to provide the ATP needed in active transport systems.

Once the substances have been removed from the nephron, they diffuse into the extensive capillary network which surrounds the tubules down steep concentration gradients. These are maintained by the constant flow of blood through the capillaries. The filtrate reaching the loop of Henle at the end of the proximal convoluted tubule is isotonic (at same concentration) with the tissue fluid surrounding the tubule and isotonic with the blood. At this stage over 80% of the glomerular filtrate has been reabsorbed back into the blood. This remains the same regardless of the conditions in the body.

The loop of Henle

The loop of Henle is the section of the kidney tubule that enables mammals to produce urine more concentrated than their own blood. Different areas of the loop have different permeabilities to water and this is central to the way the loop of Henle functions. It acts as a countercurrent multiplier, using energy to produce concentration gradients that result in the movement of substances such as water from one area to another. Cells use ATP to transport ions using active transport and this produces a diffusion gradient in the medulla.

The changes that take place in the descending limb of the loop of Henle depend on the high concentrations of sodium and chloride ions in the tissue fluid of the medulla that are the result of events in the ascending limb of the loop.

- The descending limb leads from the proximal convoluted tubule. This is the region where water moves out of the filtrate down a concentration gradient. The upper part is impermeable to water but the lower part of the descending limb is permeable to water and runs down into the medulla. The concentration of sodium and chloride ions in the tissue fluid of the medulla gets higher and higher moving through from the cortex to the pyramids, as a result of the activity of the ascending limb of the loop of Henle.

  The filtrate entering the descending limb of the loop of Henle is isotonic with the blood. As it travels down the limb, water passes out of the loop into the tissue fluid by osmosis down a concentration gradient. It then moves down a concentration gradient into the blood of the surrounding capillaries (the vasa recta).

  The descending limb is not permeable to sodium and chloride ions, and no active transport takes place in the descending limb. The fluid that reaches the hairpin bend is very concentrated and hypertonic to the blood in the capillaries.

- The first section of the ascending limb of the loop of Henle is very permeable to sodium and chloride ions and they move out of the concentrated solution by diffusion down a concentration gradient. In the second section of the ascending limb, sodium and chloride ions are actively pumped out into the medulla tissue fluid against a concentration gradient. This produces very high sodium and chloride ion concentrations in the medulla tissue. Importantly, the ascending limb of the loop of Henle is impermeable to water, so water cannot follow the chloride and sodium ions down a concentration gradient. This means the fluid left in the ascending limb becomes increasingly dilute, while the tissue fluid of the medulla develops the very high concentration

Synoptic link

You learnt about the process of active transport in Topic 5.4, Active transport.

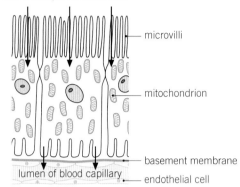

▲ **Figure 11** *The cells of the proximal convoluted tubule show a high level of adaptation to their function*

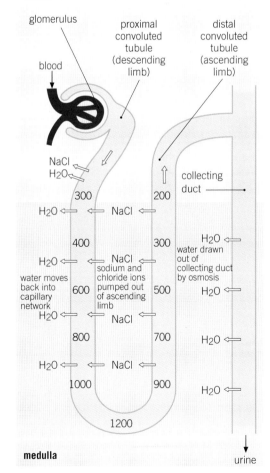

▲ **Figure 12** *The countercurrent multiplier system of the loop of Henle depends on a combination of active transport, diffusion, and osmosis. The higher the number the lower the water potential*

of ions that is essential for the kidney to produce urine that is more concentrated than the blood. This is a key part of the countercurrent multiplier system.

By the time the dilute fluid reaches the top of the ascending limb it is hypotonic to the blood again, and it then enters the distal convoluted tubule and collecting duct.

Distal convoluted tubule

Balancing the water needs of the body takes place in the distal convoluted tubule and the collecting duct. These are the areas where the permeability of the walls of the tubules varies with the levels of ADH. The cells lining the distal convoluted tubule also have many mitochondria so they are adapted to carry out active transport.

If the body lacks salt, sodium ions will be actively pumped out of the distal convoluted tubule with chloride ions following down an electrochemical gradient. Water can also leave the distal tubule, concentrating the urine, if the walls of the tubule are permeable in response to ADH. The distal convoluted tubule also plays a role in balancing the pH of the blood.

The collecting duct

The collecting duct passes down through the concentrated tissue fluid of the renal medulla (Figure 3 and 6). This is the main site where the concentration and volume of the urine produced is determined. Water moves out of the collecting duct by diffusion down a concentration gradient as it passes through the renal medulla. As a result the urine becomes more concentrated. The level of sodium ions in the surrounding fluid increases through the medulla from the cortex to the pelvis. This means water can be removed from the collecting duct all the way along its length, producing very hypertonic urine when the body needs to conserve water. The permeability of the collecting duct to water is controlled by the level of ADH, which determines how much or little water is reabsorbed. You will learn more about the role of the collecting duct in maintaining the water potential of the blood in Topic 15.6, The kidney and osmoregulation.

Study tip

Remember that the amount of reabsorption that occurs in the proximal tubule is always the same – the fine-tuning of the water balance takes place further along the nephron.

How long is your loop of Henle?

The ability of animals to produce very concentrated urine depends on several factors, one is the length of the loop of Henle. As you have seen the loop of Henle develops the concentration gradient across the kidney medulla, meaning water can leave the collecting duct all the way through, concentrating the urine as it goes. Fish have no loop of Henle and cannot produce urine that is more concentrated than their blood. Desert animals tend to have lots of nephrons that have very long loops of Henle that travel deep into the medulla.

▼ Table 2 *Urine concentrations and urine/plasma ratios in mammal species from different habitats*

Mammal	Urine concentration/mOsmol l^{-1}	Urine : plasma ratio
rat	2900	9
kangaroo rat	5500	16
beaver	520	1.7
human	1400	4–5
porpoise	1800	5
camel	2800	8

1 Demonstrate how this data could be presented in a different way.
2 How would you expect the loops of Henle of a camel to compare with those of a beaver? Explain your answer.
3 Why do you think a porpoise has a similar urine : plasma ratio as a human being?
4 The kangaroo rat has long loops of Henle but they are not comparatively as long as some animals such as the camel. The cells lining the loops in the kangaroo rat, however, have a very large number of mitochondria that have a very large number of cristae. Suggest why these observations are important in the development of the urine : plasma ratio seen in the kangaroo rat.

Summary questions

1 Give three examples of how the kidneys are well adapted for their functions in the body (excluding cellular adaptations). (6 marks)

2 a How much blood is filtered over 24 hours through the kidneys? (1 mark)
   b What percentage of this filtrate is lost to the body as urine? (4 marks)

3 a What is ultrafiltration? (4 marks)
   b What would you expect to see in normal glomerular filtrate? (4 marks)
   c Kidney infections can damage the lining of the Bowman's capsule. How might this result in protein appearing in the urine? (6 marks)

4 The normal concentration of urea in the blood is around 2.5–7.1 mmol / litre of blood. The concentration of urea in the urine varies considerably but most people pass 0.43–0.72 moles of urea in the urine over 24 hours.
   a Calculate the mean concentration of urea in the urine over a 24-hour period. (3 marks)
   b Approximately how much more urea is there in urine than in blood? (1 mark)
   c i Suggest three possible factors that might affect the amount of urea in the urine. (3 marks)
     ii Why do these factors not affect the concentration of urea in the blood? (1 mark)

5 Explain the main stages of how the kidney tubules produce urine that is more concentrated than the blood in these different regions
   a The proximal tubule (4 marks)
   b The descending limb of the loop of Henle (5 marks)
   c The ascending limb of the loop of Henle (5 marks)
   d The distal convoluted tubule (6 marks)
   e The collecting duct. (6 marks)

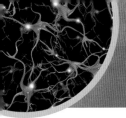
The mammalian kidney has a vital excretory function – it removes urea, the nitrogenous waste product of metabolism from the body. The kidney, however, also plays another important homeostatic role in the body – it is the main organ of *osmoregulation*. This involves controlling the water potential of the blood within very narrow boundaries, regardless of the activities of the body. Eating a salty meal, drinking large volumes of liquid, exercising hard, running a fever, or visiting a very hot climate can all put osmotic stresses on the body. It is very important to keep the water potential of the tissue fluid as stable as possible, because if water moves into or out of the cells by osmosis it can cause damage and even death.

Osmoregulation

Every day the body has to deal with many unpredictable events. The water potential of the blood has to be maintained regardless of the water and solutes taken in as you eat and drink, and the water and mineral salts lost by sweating, in defaecation, and in the urine. Changing the concentration of the urine is crucial in this dynamic equilibrium. The amount of water lost in the urine is controlled by ADH in a negative feedback system. ADH is produced by the hypothalamus and secreted into the posterior pituitary gland, where it is stored. ADH *increases* the permeability of the distal convoluted tubule and, most importantly, the collecting duct to water. You will be concentrating on the effect of ADH on the collecting duct walls.

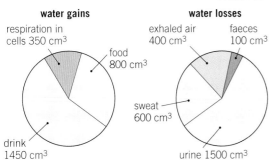

▲ **Figure 1** *Typical water gains and losses over a 24-hour period in an adult human being*

water gains — respiration in cells 350 cm³, food 800 cm³, drink 1450 cm³

water losses — exhaled air 400 cm³, faeces 100 cm³, sweat 600 cm³, urine 1500 cm³

The mechanism of ADH action

ADH is released from the pituitary gland and carried in the blood to the cells of the collecting duct where it has its effect. The hormone does not cross the membrane of the tubule cells – it binds to receptors on the cell membrane and triggers the formation of cyclic AMP (cAMP) as a second messenger inside the cell. A second messenger is a molecule which relays signals received at cell surface receptors to molecules inside the cell. The cAMP causes a cascade of events:

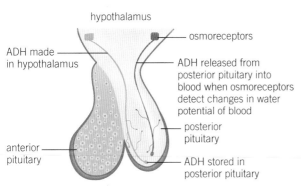

▲ **Figure 2** *ADH is made in the hypothalamus but stored and released from the posterior pituitary gland when it is stimulated by neurones from the osmoreceptors in the hypothalamus*

hypothalamus — osmoreceptors — ADH made in hypothalamus — ADH released from posterior pituitary into blood when osmoreceptors detect changes in water potential of blood — posterior pituitary — anterior pituitary — ADH stored in posterior pituitary

- Vesicles in the cells lining the collecting duct fuse with the cell surface membranes on the side of the cell in contact with the tissue fluid of the medulla.

- The membranes of these vesicles contain protein-based water channels (aquaporins) and when they are inserted into the cell surface membrane, they make it permeable to water.

- This provides a route for water to move out of the tubule cells into the tissue fluid of the medulla and the blood capillaries by osmosis.

The more ADH that is released, the more water channels are inserted into the membranes of the tubule cells. This makes it easy for more water to leave the tubules by diffusion, resulting in the formation of a small amount of very concentrated urine. Water is returned to the capillaries, maintaining the water potential of the blood and therefore the tissue fluid of the body.

When ADH levels fall, the reverse happens. Levels of cAMP fall, then the water channels are removed from the tubule cell membranes and enclosed in vesicles again. The collecting duct becomes impermeable to water once more, so no water can leave. This results in the production of large amounts of very dilute urine, and maintains the water potential of the blood and the tissue fluid.

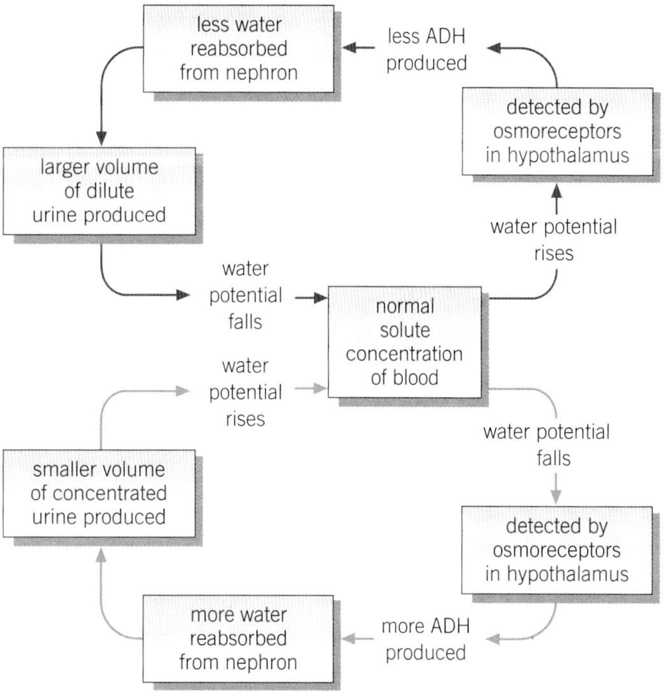

Negative feedback control and ADH

The permeability of the collecting ducts is controlled to match the water requirements of the body very closely. This is brought about by a complex negative feedback system that involves **osmoreceptors** in the hypothalamus of the brain. These osmoreceptors are sensitive to the concentration of inorganic ions in the blood and are linked to the release of ADH.

When water is in short supply

When water is in short supply in the body, the concentration of inorganic ions in the blood rises and the water potential of the blood and tissue fluid becomes more negative. This is detected by the osmoreceptors in the hypothalamus. They send nerve impulses to the posterior pituitary which in turn releases stored ADH into the blood. The ADH is picked up by receptors in the cells of the collecting duct and increases the permeability of the tubules to water. Water leaves the filtrate in the tubules and passes into the blood in the surrounding capillary network. A small volume of concentrated urine is produced (Figure 3).

An excess of water

When large amounts of liquid are taken in, the blood becomes more dilute and its water potential becomes less negative. Again, the change is detected by the osmoreceptors of the hypothalamus. Nerve impulses to the posterior pituitary are reduced or stopped and so the release of ADH by the pituitary is inhibited. Very little reabsorption of water can take place because the walls of the collecting duct remain impermeable to water. In this way the concentration of the blood is maintained – and large amounts of dilute urine are produced (Figure 3).

▲ **Figure 3** *The negative feedback loop that controls the water potential of the blood*

ADH, water balance, and blood pressure

The osmoreceptors in the hypothalamus are not the only sensory receptors that exert control over the release of ADH. It is also stimulated or inhibited by changes in the blood pressure, detected by baroreceptors in the aortic and carotid arteries. These baroreceptors are also involved in the control of the heart rate.

A rise in blood pressure can often be caused by a rise in blood volume. The increase in pressure is detected by the baroreceptors and in turn they prevent the release of ADH. This increases the volume of water lost in the urine, reducing the blood volume and so the blood pressure falls.

If the blood pressure falls it can be a signal that the blood volume has fallen. If the baroreceptors detect a fall in blood pressure there is an increase in the release of ADH from the pituitary, so the kidneys respond to reduce water loss from the body. Water is returned to the blood and a small amount of concentrated urine is produced.

> People with severe diarrhoea or more than 20% blood loss often produce little or no urine. In each case explain why.

Summary questions

1 In cool weather you may produce more urine than on a similar day in hot weather. Suggest a reason for this. *(2 marks)*

2 Draw a diagram or a flow chart to explain how ADH increases the permeability of the cells lining the distal convoluted tubules and the collecting duct to water. *(6 marks)*

3 There is a rare condition called diabetes insipidus where the body does not make ADH (or very rarely, the kidneys do not respond to ADH).
   a What would you expect the symptoms of diabetes insipidus to be? *(4 marks)*

   b Explain what is happening in the kidney tubules in a patient with diabetes insipidus. *(6 marks)*

   c Suggest how you might treat mild diabetes insipidus (when some ADH is made) and severe diabetes insipidus (where no ADH is made). *(3 marks)*

15.7 Urine and diagnosis

Specification reference: 5.1.2

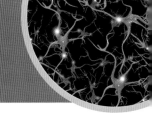

For many centuries people have used urine to try and diagnose diseases by looking at the colour, the smell, and even the taste of it. Urine is still a useful diagnostic tool, although our methods of analysis are more sophisticated.

Urine samples and diagnostic tests

Urine contains water, urea, mineral salts – and much more. It contains the breakdown products of a whole range of chemicals, including hormones and any toxins taken into the body. If you are affected by one of a number of different diseases, new substances will show up in your urine. The presence of glucose in the urine is a well-known symptom of type 1 and type 2 diabetes. If you have muscle damage, large amounts of creatinine will show up in your urine.

Urine and pregnancy testing

The human embryo implants in the uterus, around six days after conception. The site of the developing placenta then begins to produce a chemical called human chorionic gonadotrophin (hCG). Some of this hormone is found in the blood and the urine of the mother. Until the 1960s, the most reliable available pregnancy test was to inject the urine from a pregnant woman into an African clawed toad (*Xenopus laevis*). If she was pregnant, the hCG triggered egg production in the toad within 8–12 hours of the injection. It could not be used until the woman was several weeks pregnant.

Modern pregnancy tests still test for hCG in the urine, but they rely on **monoclonal antibodies**. Some are so sensitive that pregnancy can be detected within hours of implantation.

Making monoclonal antibodies

Monoclonal antibodies are antibodies from a single clone of cells that are produced to target particular cells or chemicals in the body.

A mouse is injected with hCG so it makes the appropriate antibody. The B-cells that make the required antibody are then removed from the spleen of the mouse and fused with a myeloma, a type of cancer cell which divides very rapidly. This new fused cell is known as a hybridoma. Each hybridoma reproduces rapidly, resulting in a clone of millions of 'living factories' making the desired antibody. These monoclonal antibodies are collected, purified and used in a variety of ways.

The main stages in a pregnancy test are as follows:

● the wick is soaked in the first urine passed in the morning – this will have the highest levels of hCG.

● The test contains mobile monoclonal antibodies that have very small coloured beads attached to them. They will only bind to hCG. If the woman is pregnant the hCG in her urine binds to the mobile monoclonal antibodies and forms a *hCG/antibody complex* (complete with coloured bead).

Learning outcomes

Demonstrate knowledge, understanding, and application of:

→ how excretory products can be used in medical diagnosis.

▲ **Figure 1** *The mixture of urine and a toad may not sound very scientific but the combination made a relatively accurate and available pregnancy test for many years*

Synoptic link

You learnt about the specific immune system and antibodies in Topic 12.6, The specific immune system.

▲ **Figure 2** *Pregnancy test*

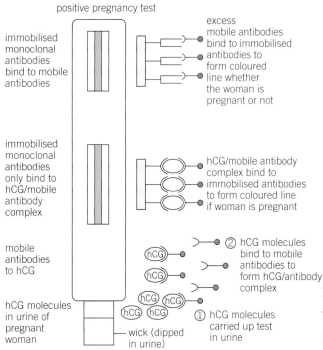

positive pregnancy test

immobilised monoclonal antibodies bind to mobile antibodies

excess mobile antibodies bind to immobilised antibodies to form coloured line whether the woman is pregnant or not

immobilised monoclonal antibodies only bind to hCG/mobile antibody complex

hCG/mobile antibody complex bind to immobilised antibodies to form coloured line if woman is pregnant

mobile antibodies to hCG

② hCG molecules bind to mobile antibodies to form hCG/antibody complex

hCG molecules in urine of pregnant woman

wick (dipped in urine)

① hCG molecules carried up test in urine

▲ **Figure 3** *The presence of hCG in the urine can be picked up using monoclonal antibodies to produce an easy-to-use and very accurate pregnancy test*

- The urine carries on along the test structure until it reaches a window.

- Here there are immobilised monoclonal antibodies arranged in a line or a pattern such as a positive (+) sign that only bind to the hCG/antibody complex. If the woman is pregnant, a coloured line or pattern appears in the first window.

- The urine continues up through the test to a second window.

- Here there is usually a line of immobilised monoclonal antibodies that bind only to the mobile antibodies, regardless of whether they are bound to hCG or not. This coloured line forms regardless of whether the woman is pregnant – it simply indicates that the test is working.

If the woman is pregnant, two coloured patterns appear. If she is not pregnant, only one appears.

Urine and anabolic steroids

Athletes and body builders may try to cheat by using anabolic steroids. Anabolic steroids are drugs that mimic the action of the male sex hormone testosterone and they stimulate the growth of muscles. They are, however, excreted in the urine. By testing the urine using gas chromatography and mass spectrometry, scientists can show that an individual has been using these drugs, which are banned in all sports. The urine sample is vaporised with a known solvent and passed along a tube. The lining of the tube absorbs the gases and is analysed to give a chromatogram that can be read to show the presence of the drugs.

Urine and drug testing

Urine is tested for the presence of many different drugs, including alcohol. Because drugs or metabolites – the breakdown products of drugs – are filtered through the kidneys and stored in the bladder, it is possible to find drug traces in the urine some time after a drug has been used.

If someone is suspected of having taken an illegal drug, they may be asked to provide a urine sample and this will be divided into two. The first sample may be tested by an immunoassay, using monoclonal antibodies to bind to the drug or its breakdown product. If this shows positive, the second sample may be run through a gas chromatograph/ mass spectrometer to confirm the presence of the drug. A wide range of drugs can be detected in the urine in this way as illustrated in Table 1.

▼ Table 1

Substance	Time it persists in the urine
ethanol (alcohol)	6–24 hours
amphetamines	1–3 days
cannabis	22 hours to 30 days depending on use
cocaine	2–5 days

Summary questions

1 Why is urine so useful for diagnostic tests? *(3 marks)*

2 The professional bodies of different sports carry out random urine testing both during training and competition. The winning athletes always have their urine tested.
   a Urine samples are divided into two and only one of them is tested initially. Explain why. *(4 marks)*
   b Why do you think urine tests are carried out at random during training as well as at competitions? *(3 marks)*

3 a Compare the process of pregnancy tests with the process of testing for illegal drugs. *(3 marks)*
   b Some pregnancy tests come with two tests. Explain how a pregnancy test might show a false-negative result. *(3 marks)*

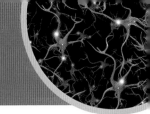

15.8 Kidney failure

Specification reference: 5.1.2

The kidneys play a vital role in homeostasis. If they are damaged and become less efficient or stop working, the effects on the body are serious and may be fatal if they are not treated.

Causes and effects of kidney failure

There are a number of reasons why the kidneys may fail. They include kidney infections, where the structure of the podocytes and the tubules themselves may be damaged or destroyed; raised blood pressure that can damage the structure of the epithelial cells and basement membrane of the Bowman's capsule; and genetic conditions such as polycystic kidney disease where the healthy kidney tissue is replaced by fluid-filled cysts or damaged by pressure from cysts.

If the kidneys are infected or affected by high blood pressure this may cause:

- Protein in the urine – if the basement membrane or podocytes of the Bowman's capsule are damaged, they no longer act as filters and large plasma proteins can pass into the filtrate and are passed out in the urine.
- Blood in the urine – another symptom that the filtering process is no longer working.

If the kidneys fail completely, the concentrations of urea and mineral ions build up in the body. The effects include:

- Loss of electrolyte balance – if the kidneys fail, the body cannot excrete excess sodium, potassium, and chloride ions. This causes osmotic imbalances in the tissues and eventual death.
- Build-up of toxic urea in the blood – if the kidneys fail, the body cannot get rid of urea and it can poison the cells.
- High blood pressure – the kidneys play an important role in controlling the blood pressure by maintaining the water balance of the blood. If the kidneys fail, the blood pressure increases and this can cause a range of health problems including heart problems and strokes.
- Weakened bones as the calcium/phosphorus balance in the blood is lost.
- Pain and stiffness in joints as abnormal proteins build up in the blood.
- Anaemia – the kidneys are involved in the production of a hormone called erythropoietin that stimulates the formation of red blood cells. When the kidneys fail it can reduce the production of red blood cells causing tiredness and lethargy.

Measuring glomerular filtration rate

Kidney problems almost always affect the rate at which blood is filtered in the Bowman's capsules of the nephrons. The glomerular

Learning outcomes

Demonstrate knowledge, understanding, and application of:

→ the effects of kidney failure and its potential treatments.

▼ Table 1 *To show average GFR in healthy adults*

Age (years)	Average eGFR
20–29	116
30–39	107
40–49	99
50–59	93
60–69	85
70+	75

filtration rate (GFR) is widely used as a measure to indicate kidney disease. The rate of filtration is not measured directly – a simple blood test measures the level of creatinine in the blood. Creatinine is a breakdown product of muscles and it is used to give an estimated glomerular filtration rate (eGFR). The units are cm^3/min. If the levels of creatinine in the blood go up, it is a signal that the kidneys are not working properly. Certain factors need to be taken into account in the calculations to work out GFR. For example, GFR decreases steadily with age even if you are healthy, and men usually have more muscle mass and therefore more creatinine than women.

Treating kidney failure with dialysis

As you can see in Table 1, normal GFRs do not fall below 70 even in very elderly people. A GFR of below 60 for more than three months is taken to indicate moderate to severe chronic kidney disease – and if it falls below 15, that is kidney failure. The kidneys are filtering so little blood they are virtually ineffective.

There are two main ways in which kidney failure is treated. In **renal dialysis**, the function of the kidneys is carried out artificially. In a transplant, a new healthy kidney is put into the body to replace the functions of the failed kidneys – an animal can function perfectly well with just one healthy kidney.

There are two main types of dialysis – haemodialysis and peritoneal dialysis.

Haemodialysis

This involves the use of a dialysis machine. It is usually carried out in hospital although sometimes patients will have a machine in their own home. Blood leaves the patient's body from an artery and flows into the dialysis machine, where it flows between partially permeable dialysis membranes. These membranes mimic the basement membrane of the Bowman's capsule. On the other side of the membranes is the dialysis fluid. During dialysis it is vital that patients lose the excess urea and mineral ions that have built up in the blood. It is equally important that they do not lose useful substances such as glucose and some mineral ions.

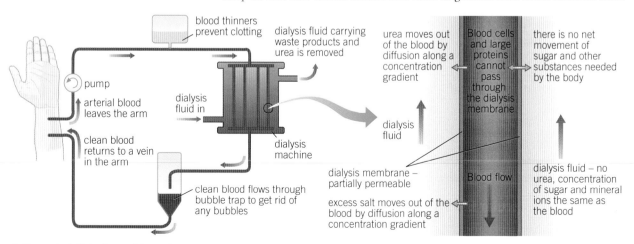

▲ Figure 1 *A dialysis machine relies on simple diffusion to remove waste products from the body*

The loss of these substances is prevented by careful control of the dialysis fluid. It contains normal plasma levels of glucose to ensure there is no net movement of glucose out of the blood. The dialysis fluid also contains normal plasma levels of mineral ions, so any excess mineral ions in the blood move out by diffusion down a concentration gradient into the dialysis fluid, thus restoring the correct electrolyte balance of the blood. The dialysis fluid contains no urea meaning there is a very steep concentration gradient from the blood to the fluid, and as a result, much of the urea leaves the blood. The blood and the dialysis fluid flow in opposite directions to maintain a countercurrent exchange system and maximise the exchange that takes place.

The whole process of dialysis depends on diffusion down concentration gradients – there is no active transport. Dialysis takes about eight hours and has to be repeated regularly. Patients with kidney failure who rely on haemodialysis have to remain attached to a dialysis machine several times a week for many hours. They also need to manage their diets carefully, eating relatively little protein and salt and monitoring their fluid intake to keep their blood chemistry as stable as possible. The only time they can eat and drink what they like is at the beginning of the dialysis process.

Peritoneal dialysis

Peritoneal dialysis is done inside the body – it makes use of the natural dialysis membranes formed by the lining of the abdomen, that is, the peritoneum. It is usually done at home and the patient can carry on with their normal life while it takes place. The dialysis fluid is introduced into the abdomen using a catheter. It is left for several hours for dialysis to take place across the peritoneal membranes, so that urea and excess mineral ions pass out of the blood capillaries, into the tissue fluid, and out across the peritoneal membrane into the dialysis fluid. The fluid is then drained off and discarded, leaving the blood balanced again and the urea and excess minerals removed.

Treating kidney failure by transplant

Long-term dialysis has some serious side effects. The best solution for the patient is a kidney transplant, where a single healthy kidney from a donor is placed within the body. The blood vessels are joined and the ureter of the new kidney is inserted into the bladder. If the transplant is successful, the kidney will function normally for many years.

The main problem with transplanted organs is the risk of rejection. The antigens on the donor organ differ from the antigens on the cells of the recipient and the immune system is likely to recognise this. This can result in rejection and the destruction of the new kidney.

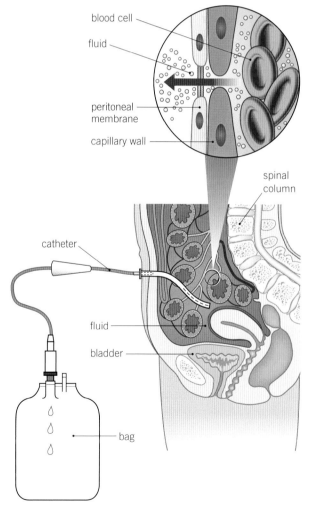

▲ **Figure 2** *Peritoneal dialysis takes place in the body cavity across the peritoneal membranes*

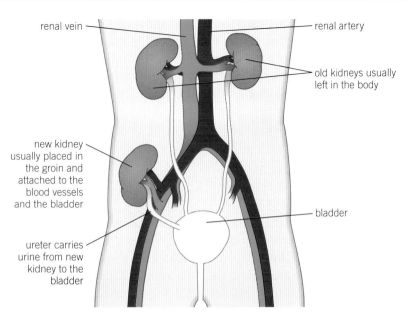

renal vein — renal artery

old kidneys usually left in the body

new kidney usually placed in the groin and attached to the blood vessels and the bladder

bladder

ureter carries urine from new kidney to the bladder

▲ **Figure 3** *A donor kidney takes over the functions of the failed kidneys, which are usually left in the body*

There are a number of ways of reducing the risk of rejection. The match between the antigens of the donor and the recipient is made as close as possible. For example, a donor kidney can be used with a 'tissue type' very similar to the recipient (from people with the same blood group).

The recipient is given drugs to suppress their immune response (immunosuppressant drugs) for the rest of their lives. This helps to prevent the rejection of their new organ. Immunosuppressant drugs are improving all the time and the need for a really close tissue match is becoming less important.

The disadvantage of taking immunosuppressant drugs is that they prevent the patients from responding effectively to infectious diseases. They have to take great care if they become ill in any way. However, most people feel this is a small price to pay for a new, functioning kidney.

Transplanted organs don't last forever with the average transplanted kidney functioning for around 9–10 years, although some have continued working for around 50 years. Once the organ starts to fail the patient has to return to dialysis and wait until another suitable kidney is found.

Dialysis or transplant?

Dialysis is much more readily available than donor organs, so it is there whenever kidneys fail. It enables the patient to lead a relatively normal life. However, patients have to monitor their diet carefully and need regular sessions on the machine. Long-term dialysis is much more expensive than a transplant and can eventually cause damage to the body.

If a patient receives a kidney transplant they are free from the restrictions which come with regular dialysis sessions and dietary monitoring. This is generally the ideal scenario for patients waiting for a transplant.

Synoptic link

You learnt about stem cells in Topic 6.5, Stem cells.

The main source of donor kidneys is from people who die suddenly, often from road accidents, strokes, and heart attacks. In the UK, organs can be taken from people if they carry an organ donor card or are on the online donor register. Alternatively, a relative of someone who has died suddenly can give his or her consent.

Unfortunately for people needing a transplant, there is a shortage of donor kidneys. Many people do not register as donors. In addition, as cars become safer, fewer people die suddenly in traffic accidents. Whilst good, this means there are fewer potential donors. At any one time there are thousands of people undergoing kidney dialysis. Most will not have the opportunity to have a kidney transplant. In 2013–14, 3257 people in the UK had kidney transplants. However, there were still around 6000 people on dialysis waiting for a kidney. There are increasing numbers of live donor transplants, where a family member of someone with a tissue match donates a kidney. In 2013–14, 2143 people received kidneys from donors who had died, and 1114 were from living donors.

In 2011, scientists grew functioning embryonic kidney tissue from stem cells. Going forward, the hope is that whole new kidneys can be grown – perhaps even without the antigens which trigger the immune reaction so that patients don't need to take immunosuppressant drugs.

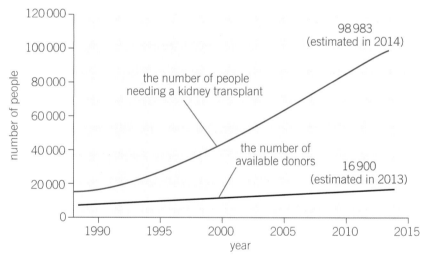

▲ **Figure 4** *This graph shows how the gap between people needing a kidney transplant and the available donors is widening in the US*

Summary questions

1 a Why is kidney failure such a threat to life? (*3 marks*)
  b On what process does dialysis depend? (*1 mark*)

2 Produce a flow chart to describe how a dialysis machine works. (*6 marks*)

3 Sometimes a live donor, usually a close family member, will donate a kidney. These transplants have a higher rate of success than normal transplants from dead, unrelated donors.
  a Suggest two reasons why live transplants from a close family member have a higher success rate than normal transplants. (*2 marks*)
  b Why do you think that live donor transplants are still the minority? (*2 marks*)

4 a Explain the importance of dialysis fluid containing no urea and normal plasma levels of salt, glucose, and minerals. (*4 marks*)
  b Both blood and dialysis fluid are constantly circulated through the dialysis machine. Explain why it is important that the blood and dialysis fluid flow in opposite directions and that there is a constant circulation of dialysis fluid. (*3 marks*)
  c Why can patients with kidney failure eat and drink what they like during the first few hours of dialysis? (*2 marks*)

Practice questions

1 The Cori cycle is the name given to the metabolic pathway that recycles the lactate produced during anaerobic respiration.

The diagram summarises the production of lactate and the Cori cycle.

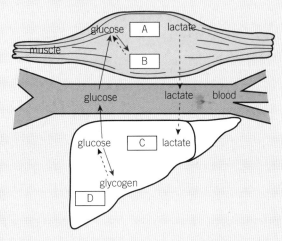

a (i) Name the processes occurring at A and C in the diagram (*2 marks*)

(ii) Name the molecule B. (*1 mark*)

(iii) Name the organ D. (*1 mark*)

b Suggest why the Cori cycle is important for homeostasis. (*4 marks*)

It was believed for many years that lactate was a toxic waste product. In the 1970's this began to be questioned and the phrase '*lactic acid, friend or foe*' was coined.

c Discuss the phrase 'lactic acid, friend or foe'. (*5 marks*)

2 The light micrograph shows a section through a human liver.

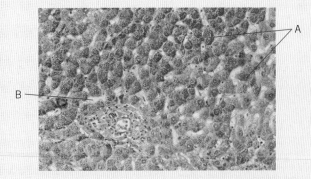

a Name the structures shown by A and B (*2 marks*)

b The diagram outlines the formation of urea in the liver.

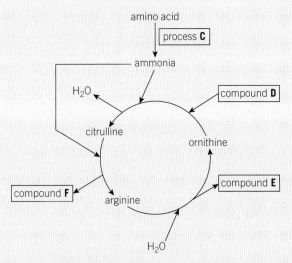

Using the diagram identify:

Process **C**

Compound **D**

Compound **E**

Compound **F** (*4 marks*)

OCR F214 2010 (apart from a)

3 The diagram represent a vertical section through a mammalian kidney.

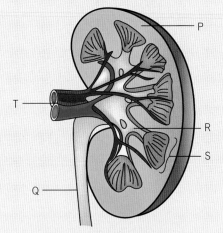

a Name: region P, structure Q, structure R, structure S, and structure T (*5 marks*)

b (i) Draw a labelled diagram of a glomerulus and Bowman's capsule. (*4 marks*)

(ii) Outline the roles of the structures you have drawn in the process of ultrafiltration. (*4 marks*)

c Nephritis is a condition in which the tissue of the glomerulus and proximal convoluted tubule becomes inflamed and damaged.

Suggest and explain two differences in the composition of urine of a person with nephritis when compared to the urine of a person with healthy kidneys. (*4 marks*)

d Caffeine is a mild diuretic. Caffeine prevents the introduction of additional aquaporins into the wall of the collecting duct of the nephron and therefore additional water is not removed from the urine.

Aquaporins are channels in the cell surface membrane that allow water molecules to pass through.

The diagram represents an aquaporin.

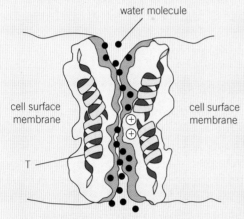

(i) Identify the type of molecule labelled **T** (*1 mark*)

(ii) The aquaporin allows water to travel from the collecting duct into the surrounding tissues but prevents the passage of ions such as sodium ions and potassium ions.

With reference to the diagram, suggest two ways in which the structure of this aquaporin prevents the passage of ions.

(*2 marks*)

OCR F214 2011 (apart from a and b)

4 The graph shows the metabolic rate changes as ambient temperature (temperature of the surroundings) changes.

Non-shivering thermogenesis is stimulated by cold temperatures and leads to an increase in metabolic activity that is not associated with muscle contraction. The increased metabolic activity results in the production of heat.

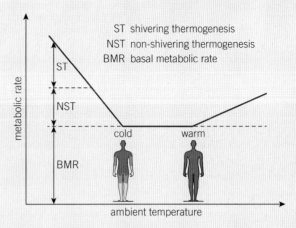

a Describe the change in metabolic rate shown in the graph. (*3 marks*)

b State which part of the involuntary nervous system stimulates thermogenesis. (*1 mark*)

c Explain why an increased metabolic rate leads to an increase in the body temperature. (*3 marks*)

d Explain the change in metabolic rate as the external temperature increases. (*3 marks*)

Recent research has shown that the metabolic activity associated NST occurs mainly in a specialised type of fat tissue called brown adipose tissue (BAT).

A typical brown adipose cell and white adipose cell are shown in the diagram.

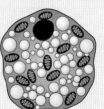

e (i) Describe the visible differences between the two cells in the diagrams. (*3 marks*)

(ii) Suggest how the brown adipose cell is better adapted for the increased heating of the body. (*3 marks*)

f The mitochondria in brown adipose cells contain a protein called uncoupling protein 1 (UCP1) that inhibits the synthesis of ATP.

Explain how this would result in increased warming OR a greater temperature rise. (*2 marks*)

16 PLANT RESPONSES
16.1 Plant hormones and growth in plants
Specification reference: 5.1.5

Learning outcomes

Demonstrate knowledge, understanding, and application of:

→ the roles of plant hormones

→ the experimental evidence for the role of auxins in the control of apical dominance

→ the experimental evidence for the role of gibberellin in the control of stem elongation and seed germination

→ practical investigations into the effect of plant hormones on growth.

People often regard plants as passive objects in the environment that simply grow and sometimes flower. However, plants are dynamic systems, not only photosynthesising and producing food but also responding to their environment in many different ways. They have evolved to cope with abiotic stresses such as lack of water, and they have a range of adaptations to protect them against the attentions of herbivores. They also show directional growth in response to environmental cues such as light and gravity – these are known as **tropisms**.

Chemical coordination

Plants are multicellular organisms living in a complex and ever-changing environment. The key limitations on plants are that they are rooted – they are not mobile, and they do not have a rapidly responding nervous system. They are, however, coordinated organisms that show clear responses to their environment, communication between cells, and even communication between different plants. The timescales of most plant responses are slower than animal responses, but they still respond as a result of complex chemical interactions. Plants have evolved a system of hormones – chemicals that are produced in one region of the plant and transported both through the transport tissues and from cell to cell and have an effect in another part of the plant. Important plant hormones include **auxins**, **gibberellins**, abscisic acid (ABA), and ethene. These chemicals have a wide range of functions within the plant.

▼ Table 1 *Some of the roles of plant hormones*

Hormone	Some of their known roles in plants
auxins	control cell elongation, prevent leaf fall (abscission), maintain apical dominance, involved in tropisms, stimulate the release of ethene, involved in fruit ripening
gibberellin	cause stem elongation, trigger the mobilisation of food stores in a seed at germination, stimulate pollen tube growth in fertilisation
ethene	causes fruit ripening, promotes abscission in deciduous trees
ABA (abscisic acid)	maintains dormancy of seeds and buds, stimulates cold protective responses, for example, antifreeze production, stimulates stomatal closing

440

Plants produce chemicals which signal to other species – for example, to protect themselves from attack by insect pests – and may communicate with other plants. They also produce chemical defences against herbivores. In plant responses, chemicals are essential.

The growth of plants, from the germination of a seed to the long-term growth of a tree, is controlled by plant hormones. You will look at the different chemicals and their roles in isolation, but in fact the growth and form of a plant are the result of the interaction of many different hormonal and environmental factors.

Scientists are still unsure about the details of many plant responses. There are a number of reasons for this. Plant hormones work at very low concentrations, so isolating them and measuring changes in concentrations is not easy. The multiple interactions between the different chemical control systems also make it very difficult for researchers to isolate the role of a single chemical in a specific response.

Outlined are a number of key aspects of plant growth with the role of hormones highlighted.

Plant hormones and seed germination

For a plant to start growing, the seed must germinate.

- When the seed absorbs water, the embryo is activated and begins to produce gibberellins. They in turn stimulate the production of enzymes that break down the food stores found in the seed. The food store is in the cotyledons in dicot seeds and the endosperm in monocot seeds. The embryo plant uses these food stores to produce ATP for building materials so it can grow and break out through the seed coat. Evidence suggests that gibberellins switch on genes which code for amylases and proteases – the digestive enzymes required for germination. There is also evidence suggesting that another plant hormone, ABA, acts as an antagonist to gibberellins (interferes with the action of gibberellin), and that it is the relative levels of both hormones which determine when a seed will germinate.

Experimental evidence

Experimental evidence supporting the role of gibberellins in the germination of seeds includes:

- Mutant varieties of seeds have been bred which lack the gene that enables them to make gibberellins. These seeds do not germinate. If gibberellins are applied to the seeds externally, they then germinate normally.
- If gibberellin biosynthesis inhibitors are applied to seeds, they do not germinate as they cannot make the gibberellins needed for them to break dormancy. If the inhibition is removed, or gibberellins are applied, the seeds germinate.

> **Synoptic link**
>
> You learnt about plant transport systems in Chapter 9, Transport in plants.

▲ **Figure 1** *Plants' responses enable them to grow and flower in almost every land environment*

> **Synoptic link**
>
> You learnt about the role of cotyledons as food stores in the seeds of dicotyledenous plants in Topic 9.1, Transport systems in dicotyledonous plants.

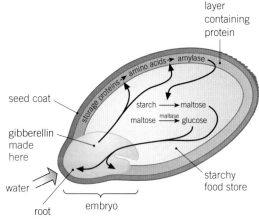

▲ **Figure 2** *The role of gibberellins in germination*

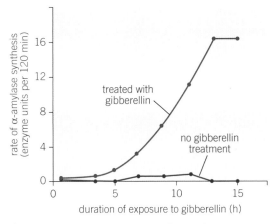

▲ **Figure 3** *The effect of gibberellins on the synthesis of amylase, a starch-digesting enzyme, in isolated tissue from barley seeds*

Plant hormones, growth, and apical dominance

The growth of a plant shoot after a seed has germinated is controlled by a number of plant hormones.

Auxins

Auxins such as indoleacetic acid (IAA) are growth stimulants produced in plants. Small quantities can have powerful effects. They are made in cells at the tip of the roots and shoots, and in the meristems. Auxins can move down the stem and up the root both in the transport tissue and from cell to cell. The effect of the auxin depends on its concentration and any interactions it has with other hormones. Auxins have a number of major effects on plant growth.

● They stimulate the growth of the main, apical shoot. Evidence suggests that auxins affect the plasticity of the cell wall – the presence of auxins means the cell wall stretches more easily. Auxin molecules bind to specific receptor sites in the plant cell membrane, causing a fall in the pH to about 5. This is the optimum pH for the enzymes needed to keep the walls very flexible and plastic. As the cells mature, auxin is destroyed. As the hormone levels fall, the pH rises so the enzymes maintaining plasticity become inactive. As a result, the wall becomes rigid and more fixed in shape and size and the cells can no longer expand and grow.

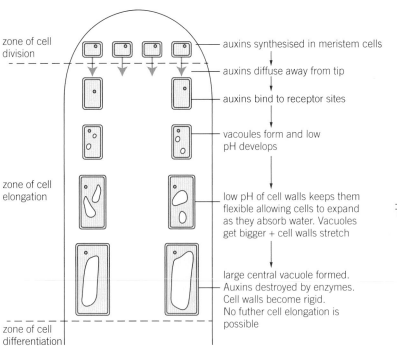

▲ **Figure 4** *The effect of auxin on apical shoot growth*

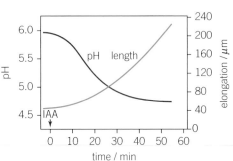

▲ **Figure 5** *Graph to show the effect of an external application of auxin on pH levels in the cell walls and on shoot growth*

High concentrations of auxins suppress the growth of lateral shoots. This results in apical dominance. Growth in the main shoot is stimulated by the auxin produced at the tip so it grows quickly. The lateral shoots are inhibited by the hormone that moves back down the stem, so they do not grow very well. Further down the stem, the auxin concentration is lower and so the lateral shoots grow more strongly. There is a lot of experimental evidence for the role of auxins in apical dominance. For example, if the apical shoot is removed, the auxin-producing cells are removed and so there is no auxin. As a result, the lateral shoots, freed from the dominance of the apical shoot, grow faster. If auxin is applied artificially to the cut apical shoot, apical dominance is reasserted and lateral shoot growth is suppressed.

Low concentrations of auxins promote root growth. Up to a given concentration, the more auxin that reaches the roots, the more they grow. Auxin is produced by the root tips and auxin also reaches the roots in low concentrations from the growing shoots. If the apical shoot is removed, then the amount of auxin reaching the roots is greatly reduced and root growth slows and stops. Replacing the auxin artificially at the cut apical shoot restores the growth of the roots. High auxin concentrations inhibit root growth.

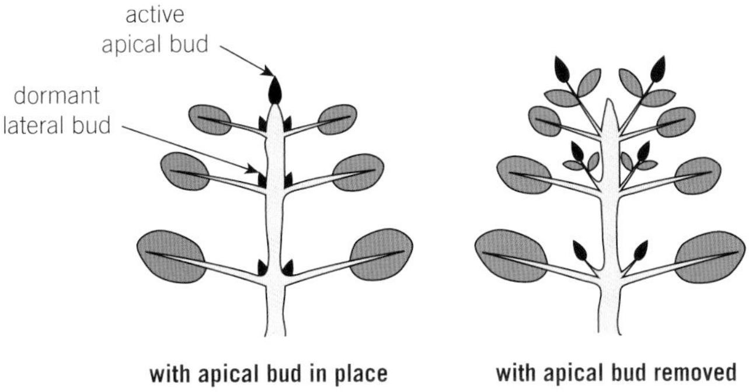

▲ **Figure 6** *Apical dominance affects the form of a plant and can clearly be seen in conifers*

Gibberellins

As you have learnt, gibberellins are involved in the germination of seeds. They are also important in the elongation of plant stems during growth. Gibberellins affect the length of the internodes – the regions between the leaves on a stem. Gibberellins were discovered because they are produced by a fungus from the genus *Gibberella* that affects rice. The infected seedlings grew extremely tall and thin. Scientists investigated the rice and isolated chemicals – gibberellins – which produce the same spindly growth in the plants. It was then discovered that plants themselves produce the same compounds. Plants that have short stems produce few or no gibberellins. There are well over a hundred different naturally produced gibberellins. Scientists have bred many dwarf varieties of plants where the gibberellin synthesis pathway is interrupted. Without gibberellins the plant stems are much shorter. This reduces waste and also makes the plants less vulnerable to damage by weather and harvesting.

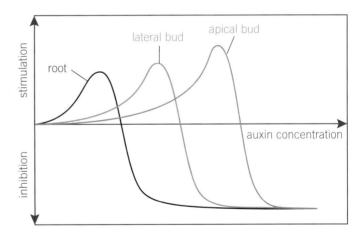

▲ **Figure 7** *Different concentrations of hormones affect different tissues in different ways – the graph shows the effect of auxin on roots, lateral buds, and apical buds*

> **Synoptic link**
>
> You learnt about the use of standard deviation to measure the spread of data in Topic 10.6, Representing variation graphically.

Investigating the effect of hormones on plant growth

There are many different ways to investigate the effect of plant hormones on the growth of the shoots, the roots, and the germination of seeds. These include growing seedlings hydroponically (in nutrient solution rather than soil) in serial dilutions of different hormones, or applying different concentrations of hormones to the cut ends of stems or roots and observing the effects.

In most experiments, it is important to make serial dilutions to observe the effects of different concentrations of the hormones, as they can have different effects on growth at different concentrations.

Experiments investigating the effect of hormones on plant growth usually involve large numbers of plants. When you have completed your measurements, the spread of data from each experimental group should be measured using standard deviation.

1 Suggest an advantage and a disadvantage of using serial dilutions of hormones in nutrient solution to investigate the effect of a hormone on plant growth.

2 An experiment was conducted to investigate using plant hormones to increase the growth of cuttings. It was found that in the first experiment there was a significant increase ($p \leq 0.05$) in fresh mass with potting medium 2, and leaf length with potting medium 1, both at 100 mg/l IAA. Explain the meaning of the term **significant increase** ($p \leq 0.05$) and explain how it can be calculated.

Synoptic link

You learnt about serial dilutions in Topic 4.2, Factors affecting enzyme activity.

Synergism and antagonism

Most plant hormones do not work on their own but by interacting with other substances. In doing so, very fine control over the responses of the plant can be achieved. If different hormones work together, complementing each other and giving a greater response than they would on their own, the interaction is known as synergism. If the substances have opposite effects, for example one promoting growth and one inhibiting it, the balance between them will determine the response of the plant. This is known as antagonism. Our knowledge of plant hormones and the mechanisms by which they have an effect is still far from complete – this is an active and important area of research.

Summary questions

1 Why are chemicals so important in coordinating the growth of plants? *(3 marks)*

2 a Give three examples of plant hormones and for each give one function in the plant. *(3 marks)*
   b Why are the chemicals you have listed in answer 2a described as plant hormones? *(3 marks)*

3 Plant hormones have very different effects on different plant tissues.
   a Give an example with experimental evidence. *(6 marks)*
   b Explain the importance of these multifunction hormones in a plant. *(2 marks)*

4 Explain how the data in Figure 5 appear to confirm our current model of auxin action. *(6 marks)*

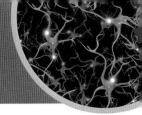

When the environmental conditions around plants change, they have to cope or die. Abiotic stresses include changes in day length, cold and heat, lack of water, excess water, high winds, and changes in salinity. Plants need to be able to cope with these changes. As you have already learnt, plant responses involve both physical and physiological adaptations. They may have very thick cuticles, hairy leaves, sunken stomata or a wilting response in hot, dry or extremely windy conditions, or develop aerenchyma if they grow in an aquatic environment.

Leaf loss in deciduous plants as a response to abiotic stress

Plants that grow in temperate climates experience great environmental changes during the year. For example, the range of daylight hours in parts of northern Scotland ranges from about 6.5 hours midwinter to just under 18.5 hours midsummer. Temperatures vary as well – in England the mean temperature is 3–6°C in winter and 16–21°C in summer. As light and temperature affect the rate of photosynthesis, seasonal changes have a big impact on the amount of photosynthesis possible. The point comes when the amount of glucose required for respiration to maintain the leaves, and to produce chemicals from chlorophyll that might protect them against freezing is greater than the amount of glucose produced by photosynthesis. In addition, a tree that is in leaf is more likely to be damaged or blown over by winter gales. This means deciduous trees in temperate climates lose all of their leaves in winter and remain dormant until the days lengthen and temperatures rise again in spring.

Daylength sensitivity

Scientists have discovered that plants are sensitive to a lack of light in their environment. This is known as photoperiodism. For many years it was assumed that plants responded to the length of daylight, but more recent evidence suggests that it is lack of light that is the trigger for change. Many different plant responses are affected by the photoperiod including the breaking of the dormancy of the leaf buds so they open up, the timing of flowering in a plant and when tubers are formed in preparation for overwintering.

The sensitivity of plants to day length (or dark length) results from a light-sensitive pigment called phytochrome. This exists in two forms – P_r and P_{fr}. Each absorbs a different type of light and the ratio of P_r to P_{fr} changes depending on the levels of light.

Learning outcomes
Demonstrate knowledge, understanding, and application of:
→ the types of plant responses
→ the roles of plant hormones.

Synoptic link
You learnt about the way plants respond to very dry and to watery conditions in Topic 9.5, Plant adaptations to water availability.

Synoptic link
You will learn more about photosynthesis in Chapter 17, Energy for biological processes.

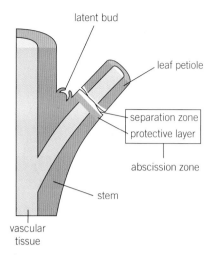

latent bud

leaf petiole

separation zone

protective layer

abscission zone

stem

vascular tissue

▲ Figure 1 *The abscission layer in a leaf stalk*

▲ Figure 2 *The scar where a leaf has fallen from this horse chestnut twig is clear to see, including the dark dots which are the sealed ends of the xylem and phloem vessels*

Synoptic link

You learnt about the mechanism of opening and closing the stomata as a result of turgor changes in Topic 9.3, Transpiration.

Abcission or leaf fall

After a summer of long days, short nights, and warm temperatures, the nights lengthen, days shorten, and temperatures fall as autumn develops. The lengthening of the dark period triggers a number of changes, including abscission or leaf fall and a period of dormancy during the winter months.

The falling light levels result in falling concentrations of auxin. The leaves respond to the falling auxin concentrations by producing the gaseous plant hormone ethene. At the base of the leaf stalk is a region called the abscission zone, made up of two layers of cells sensitive to ethene. Ethene seems to initiate gene switching in these cells resulting in the production of new enzymes. These digest and weaken the cell walls in the outer layer of the abscission zone, known as the separation layer.

The vascular bundles which carry materials into and out of the leaf are sealed off. At the same time fatty material is deposited in the cells on the stem side of the separation layer. This layer forms a protective scar when the leaf falls, preventing the entry of pathogens. Cells deep in the separation zone respond to hormonal cues by retaining water and swelling, putting more strain on the already weakened outer layer. Then further abiotic factors such as low temperatures or strong autumn winds finish the process – the strain is too much and the leaf separates from the plant. A neat, waterproof scar is left behind.

Preventing freezing

Another major abiotic factor which affects plants is a decrease in temperature. If cells freeze, their membranes are disrupted and they will die. Many plants, however, have evolved mechanisms that protect their cells in freezing conditions. The cytoplasm of the plant cells and the sap in the vacuoles contain solutes which lower the freezing point. Some plants produce sugars, polysaccharides, amino acids, and even proteins which act as antifreeze to prevent the cytoplasm from freezing, or protect the cells from damage even if they do freeze.

Most species only produce the chemicals which make them hardy and frost resistant during the winter months. It appears that different genes are suppressed and activated in response to a sustained fall in temperatures along with a reduction in day length, effectively preparing the plants to withstand frosty conditions. A sustained spell of warm weather along with extended day length reverses these changes in the spring.

Stomatal control

As you have already learnt, heat and water availability are major abiotic stresses for plants. One of the major ways in which plants can respond to these stresses is to open the stomata to cool the plant as water evaporates from the cells in the leaves in transpiration, or to close the stomata to conserve water.

The opening and closing of the stomata in response to abiotic stresses is largely under the control of the hormone ABA. The leaf cells appear to release ABA under abiotic stress, causing stomatal closure. However, scientists now think that the roots also provide an early warning of water stresses through ABA. So, for example, when the levels of soil water fall and transpiration is under threat, plant roots produce ABA which is transported to the leaves where it binds to receptors on the plasma membrane of the stomatal guard cells. ABA activates changes in the ionic concentration of the guard cells, reducing the water potential and therefore turgor of the cells. As a result of reduced turgor, the guard cells close the stomata and water loss by transpiration is greatly reduced.

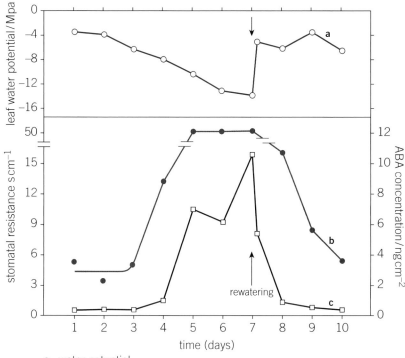

a water potential
b stomatal resistance
c ABA content in corn

▲ **Figure 3** *Changes in water potential, stomatal resistance, and ABA content in corn in response to water stress – stomatal resistance increases as stomata close. All measurements of stomatal resistance in excess of 20 s cm⁻¹ are shown as 50 s cm⁻¹*

Summary questions

1 Why is it so important for plants to be able to respond to their surroundings? (*2 marks*)

2 Why do many trees in temperate climates lose all of their leaves in winter? (*6 marks*)

3 Produce a flow diagram to explain the process of abscission. (*6 marks*)

4 **a** Explain how plant hormones are involved in protecting the plant cells from damage in freezing conditions. (*5 marks*)
  b Give two more adaptations by which plant cells may be protected against damage by freezing during the winter. (*3 marks*)
  c If there is a sudden spell of freezing weather early in the autumn, many plants which can normally survive the winter may be killed. Similarly, if there is a late frost after several weeks of warm spring weather, plants that have already survived a harsh winter can die. Explain the difference in the response of the plants in these circumstances compared with the winter months. (*6 marks*)

5 Explain how the experimental evidence shown in Figure 3 supports the idea of ABA from the roots affecting stomatal opening in times of water stress. (*6 marks*)

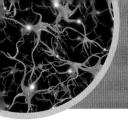

16.3 Plant responses to herbivory
Specification reference: 5.1.5

Herbivores are animals that eat plants. They range from tiny insects to enormous animals such as elephants and rhinos. Herbivory is the process by which herbivores eat plants.

Responses to herbivory

Plants cannot escape animals which want to eat them, so they have evolved a wide range of defences to prevent attack by herbivores or minimise the damage they do.

Physical defences

Common physical defences include thorns, barbs, spikes, spiny leaves, fibrous and inedible tissue, hairy leaves, and even stings to protect themselves and discourage herbivores from eating them.

Chemical defences

Plants have also evolved a wide range of chemical responses to herbivory – the stinging nettle manages to include both physical and chemical defences in its vicious trichomes (stinging hairs), but many other plants produce a cocktail of unpleasant chemicals too. These include:

- **Tannins** – part of a group of compounds called phenols produced by many plants. Tannins can make up to 50% of the dry weight of the leaves. They have a very bitter taste which puts animals off eating the leaves. They are toxic to insects – they bind to the digestive enzymes produced in the saliva and inactivate them. Tea and red wine are both rich in plant tannins.

- **Alkaloids** – a large group of very bitter tasting, nitrogenous compounds found in many plants. Many of them act as drugs, affecting the metabolism of animals that take them in and sometimes poisoning them. Alkaloids include caffeine, nicotine, morphine, and cocaine. Caffeine is toxic to fungi and insects, and the caffeine produced by coffee bush seedlings spreads through the soil and prevents the germination of the seeds of other plants – so caffeine protects the plant both against herbivores and against plant rivals. Nicotine is a toxin produced in the roots of tobacco plants, transported to the leaves and stored in vacuoles to be released when the leaf is eaten.

- **Terpenoids** – a large group of compounds produced by plants which often form essential oils but also often act as toxins to insects and fungi that might attack the plant. Pyrethrin, produced by chrysanthemums, acts as an insect neurotoxin, interfering with the nervous system. Some terpenoids act as insect repellents, for example, citronella produced by lemon grass repels insects.

Learning outcomes

Demonstrate knowledge, understanding, and application of:

→ the types of plant responses.

Synoptic link

You learnt about the responses of plants to attack by pathogens and the defensive chemicals they make in Topic 12.4, Plant defences against pathogens.

▲ **Figure 1** These thorns act to deter herbivores. The fruit, which have evolved to be eaten to spread the seeds, are held on thorn-free stalks

▲ **Figure 2** A stinging nettle magnified 20×, these hollow stinging hairs act like hypodermic needles and inject histamines and other irritant chemicals into animals that try to eat the leaves

448

Pheromones

A pheromone is a chemical made by an organism which affects the social behaviour of other members of the same species. Because plants do not behave socially, they do not rely a lot on pheromones. There are a few instances where they could be regarded as using pheromones to defend themselves:

● If a maple tree is attacked by insects, it releases a pheromone which is absorbed by leaves on other branches. These leaves then make chemicals such as callose to help protect them if they are attacked. Scientists have observed that leaves on the branches of nearby trees also prepare for attack in response to these chemical signals.

● There is some evidence that plants communicate by chemicals produced in the root systems and one plant can 'tell' a neighbour if it is under water stress.

▲ **Figure 3** *Human uses for some of the compounds produced by plants*

However, plants do produce chemicals called volatile organic compounds (VOCs) which act rather like pheromones between themselves and other organisms, particularly insects. They diffuse through the air in and around the plant. Plants use these chemical signals to defend themselves in some amazing ways. They are usually only made when the plant detects attack by an insect pest through chemicals in the saliva of the insect. This may elicit gene switching. For example:

● When cabbages are attacked by the caterpillars of the cabbage white butterfly, they produce a chemical signal which attracts the parasitic wasp *Cotesia glomerata*. This insect lays its eggs in the caterpillars which are then eaten alive, protecting the plant. The signal from the plant also deters any other female cabbage white butterflies from laying their eggs. Scientists estimate up to 90% of cabbage white caterpillars are affected by the parasite. If the cabbage is attacked by the mealy cabbage greenfly, it sends out a different signal which attracts the parasitic wasp *Diaretiella rapae* which only attacks greenfly.

▲ **Figure 4** *Foxgloves produce a chemical, digitalis, which is toxic to mammals – it slows the heart rate. It also contains chemicals which cause vomiting, so after eating a foxglove leaf a small mammal will feel very ill but then vomit, removing the toxins. Digitalis can kill people too, but drugs based on the chemical are also used to treat human heart problems and save lives*

● When apple trees are attacked by spider mites, they produce VOCs which attract predatory mites that come and destroy the apple tree pests.

● Some types of wheat seedling produce VOCs when they have been attacked by aphids and these repel other aphids from the plant.

Sometimes a VOC produced by a plant that has been attacked will not only attract predators of the pest organism – it may also act as a 'pheromone' so that neighbouring plants begin to produce the VOC before they are actually attacked.

Folding in response to touch

Most plants, with the exception of a few insectivorous plants such as the Venus fly trap, move so slowly you cannot follow the movement with the naked eye. It is revealed over hours or days, or by time-lapse photography. There are, however, some exceptions – the sensitive plant *Mimosa pudica* is one of a small number of plants which move at a speed you can see. This plant uses conventional defences against herbivores – it contains a toxic alkaloid and the stem has sharp prickles, but if the leaves are touched, they fold down and collapse. Scientists think this frightens off larger herbivores, and dislodges small insects which have landed on the leaves. The leaf falls in a few seconds, and recovers over 10–12 minutes as a result of potassium ion movement into specific cells, followed by osmotic water movement. The exact causes of the dramatic change in the leaves are still being researched.

✚ *Mimosa pudica* – nerves and muscles in plants

The dramatic leaf-folding in response to touch seen in *Mimosa pudica* seems to be the result of chemical changes in some large, fairly thin-walled cells found at the bases of the leaves and the individual leaflets. These special cells have relatively elastic walls and they surround a vascular bundle to form a thickened region called a pulvinus, which acts rather like a joint (Figure 5). The cells at the top are the flexor region and the cells at the bottom are the extensor region.

When the leaf is touched, scientists think there is an electrochemical change in the cells which results in something rather like the action potential in nerve cells which causes the active movement of potassium ions into the cells on the upper, flexor side of the pulvinus, while potassium ions are similarly moved out of cells on the lower, extensor side. Water follows the potassium ions by osmosis, so turgor in the top cells increases and in the lower cells decreases. There are elastic tissues in the cells that increase this effect. They include actin, one of the proteins found in mammalian muscle cells. As a result the leaflet, or whole leaf, bends down. When the plant recovers the situation is reversed – potassium ions return to their resting levels down concentration and electrochemical gradients and water follows by osmosis. (Figure 6).

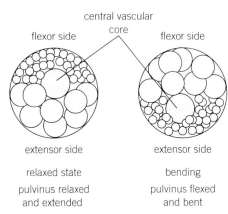

▲ **Figure 5** *The structure of a pulvinus when extended and flexed. The best current model for the cellular basis of the rapid folding response of* <u>Mimosa pudica</u> *to touch*

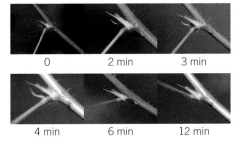

▲ **Figure 6** *The change in angle of pulvinus during recovery*

The initial leaf folding of *Mimosa pudica* takes seconds but recovery takes 10–12 minutes. Both involve similar changes in potassium levels. Suggest why such different rates of change may have evolved.

Summary questions

1 What is herbivory? *(1 mark)*

2 Describe two examples of chemical defences against herbivory by animals, explaining how they protect the plant and how they are used by people. *(6 marks)*

3 a Why are the chemicals sometimes known as plant pheromones not strictly pheromones? *(2 marks)*

   b Why is it so important that these chemicals are volatile? *(2 marks)*

   c Give two examples of how these chemicals may be used to protect a plant against herbivory, including a discussion as to whether your examples are pheromones or not. *(6 marks)*

16.4 Tropisms in plants

Specification reference: 5.1.5

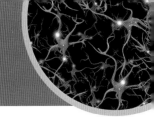

Plant growth responses to stimuli from one direction are known as tropisms. The growth of plants in response to light which comes from one direction only is called **phototropism**; the response to gravity is called **geotropism**; the response to chemicals chemotropism; and to touch thigmotropism. Tropisms involve differential growth of plant cells triggered by chemical messages produced in response to a particular stimulus.

Tropisms as a response to environmental cues

To be able to make the maximum use of the environmental conditions, plants must grow and respond to variations in those conditions. For example, once a seed begins to germinate in the soil, the shoot and root must keep growing in the right direction if the developing plant is to survive. The shoot must grow up towards the light source for photosynthesis to take place. The roots must grow downwards into the soil which will provide support, minerals, and water for the plant. The movements of the root and shoot take place in direct response to environmental stimuli. The direction of the response is related to the direction from which the stimulus comes. These responses are examples of tropisms.

Much of the research on tropisms uses germinating seeds and very young seedlings. They are easy to work with and manipulate and as they are growing and responding rapidly, any changes show up quickly. Changes also tend to affect the whole organism rather than a small part (as with a mature plant) and this makes any tropisms much easier to observe and measure. The seedlings of monocotyledonous plants – usually cereals such as oats and wheat – are most commonly used as the shoot that emerges is a single spike with no apparent leaves known as a coleoptile. It is easier to manipulate and observe than a dicotyledonous shoot. However, coleoptiles are relatively simple plant systems, so it is important to remember that the control of the responses to light in an intact adult plant may be more complex.

Phototropism

The basic model of the way plants respond to light as they grow was based on experiments where shoots were kept entirely in the dark or in full illumination. However, this is rarely the case in real life. Phototropisms are the result of the movement of auxins across the shoot or root if it is exposed to light that is stronger on one side than the other.

If plants are grown in bright, all-round light in normal conditions of gravity they grow more or less straight upwards. In even but low light they will also grow straight upwards – in fact in these conditions they will grow faster and taller than in bright light. If plants, however, are exposed to light which is brighter on one side than another, or

Synoptic link

You learnt about dicotyledonous plants in Topic 9.1, Transport systems in dicotyledonous plants.

to unilateral light that only shines from one side, then the shoots of the plant will grow towards that light and the roots, if exposed, will grow away. Shoots are said to be positively phototropic and roots are negatively phototropic. This response has an obvious survival value for a plant. It helps to ensure that the shoots receive as much all-round light as possible, allowing the maximum amount of photosynthesis to take place. Also, if the roots should emerge from the soil – as they might do after particularly heavy rain, for example – they will rapidly turn back to the soil.

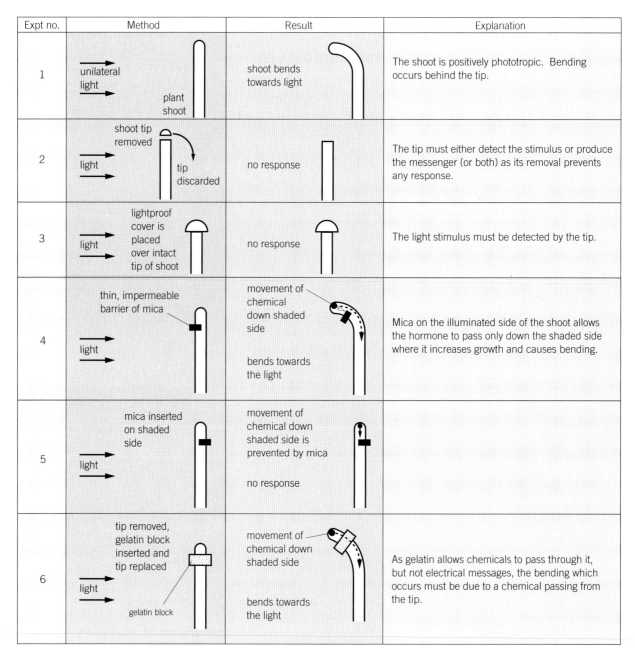

Expt no.	Method	Result	Explanation
1	unilateral light / plant shoot	shoot bends towards light	The shoot is positively phototropic. Bending occurs behind the tip.
2	shoot tip removed / light / tip discarded	no response	The tip must either detect the stimulus or produce the messenger (or both) as its removal prevents any response.
3	lightproof cover is placed over intact tip of shoot / light	no response	The light stimulus must be detected by the tip.
4	thin, impermeable barrier of mica / light	movement of chemical down shaded side / bends towards the light	Mica on the illuminated side of the shoot allows the hormone to pass only down the shaded side where it increases growth and causes bending.
5	mica inserted on shaded side / light	movement of chemical down shaded side is prevented by mica / no response	
6	tip removed, gelatin block inserted and tip replaced / light / gelatin block	movement of chemical down shaded side / bends towards the light	As gelatin allows chemicals to pass through it, but not electrical messages, the bending which occurs must be due to a chemical passing from the tip.

▲ **Figure 1** *Experimental observations such as these by Darwin and Boysen-Jensen are the basis of our understanding of phototropisms. Darwin's experiments helped to show that it is the tips of shoots which are the source of the phototropic response, meanwhile Boysen-Jensen helped show the nature of the 'messenger' in the phototropic response*

The effect of unilateral light

Examples of the response of plants to unilateral light can be seen in any garden or woodland. Where plants are partially shaded the shoots grow towards the light and then grow on straight towards it. This response appears to be the result of the way auxin moves within the plant under the influence of light.

Figure 2 shows that the side of a shoot exposed to light contains less auxin than the side which is not illuminated. It appears that light causes the auxin to move laterally across the shoot, so there is a greater concentration on the unilluminated side. This in turn stimulates cell elongation and growth on the dark side, resulting in observed growth towards the light. Once the shoot is growing directly towards the light, the unilateral stimulus is removed. The transport of auxin stops and the shoot then grows straight towards the light. The original theory was that light destroyed the auxin, but this has been disproved by experiments showing that the levels of auxin in shoots are much the same regardless of whether they have been kept in the dark or under unilateral illumination.

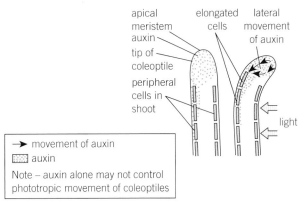

▲ **Figure 2** *In unilateral light, auxin moves laterally across the tip of the shoot away from the light. This stimulates growth on the shady side, so the shoot grows asymmetrically towards the light*

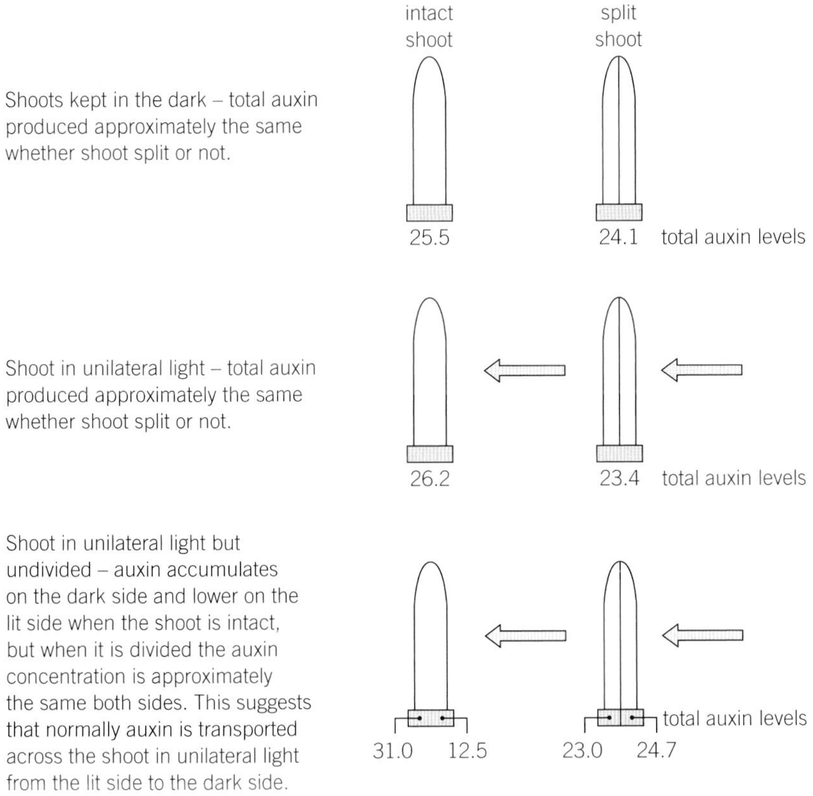

Shoots kept in the dark – total auxin produced approximately the same whether shoot split or not.

intact shoot — 25.5
split shoot — 24.1   total auxin levels

Shoot in unilateral light – total auxin produced approximately the same whether shoot split or not.

26.2   23.4   total auxin levels

Shoot in unilateral light but undivided – auxin accumulates on the dark side and lower on the lit side when the shoot is intact, but when it is divided the auxin concentration is approximately the same both sides. This suggests that normally auxin is transported across the shoot in unilateral light from the lit side to the dark side.

31.0   12.5   23.0   24.7   total auxin levels

▲ **Figure 3** *Experiments such as these with maize coleoptiles help us to determine what happens to auxin levels within a shoot illuminated by unilateral light*

Practical investigations into phototropisms

There are many different ways to investigate phototropisms. Some of them are listed here.

- Germinate and grow seedlings in different conditions of dark, all-round light, and unilateral light. Observe, measure, and record the patterns of growth. Time-lapse photography can give a good record of the changes as they take place.

- Germinate and grow seedlings in unilateral light with different colour filters to see which wavelengths of light trigger the phototropic response.

- Repeat some of the classic experiments (Figure 1) – cover the tips of coleoptiles with foil, remove the tips of some coleoptiles, place auxin-impregnated agar jelly blocks or lanolin on decapitated coleoptiles, place auxin-impregnated agar blocks on one side only of decapitated coleoptiles.

▲ **Figure 4** *Experiments on phototropisms can be done on dicot seedlings as well as coleoptiles – these cress seedlings have been grown in all-round light, the dark, and unilateral light from left to right respectively*

Discuss potential advantages and disadvantages of using new technology such as smart phones and tablets in recording and displaying data in practical investigations into phototropisms.

Growing in the dark

The fact that plants grow more rapidly in the dark than when they are illuminated can at first seem illogical. If a plant, however, is in the dark the biological imperative is to grow upwards rapidly to reach the light to be able to photosynthesise. The seedlings that break through the soil first will not have to compete with other seedlings for light. Evidence suggests that it is gibberellins that are responsible for the extreme elongation of the internodes when a plant is grown in the dark. Once a plant is exposed to the light, a slowing of upwards growth is valuable. Resources can be used for synthesising leaves, strengthening stems, and overall growth. Scientists have demonstrated that levels of gibberellin fall once the stem is exposed to light.

Gardeners sometimes use this response to 'force' growth in plants – early rhubarb is famously grown in dark sheds in Yorkshire. The rapid upward growth which takes place in a plant grown in the dark is known as etiolation. Etiolated plants are thin and pale – because the plant is deprived of light little chlorophyll develops in the leaves.

Geotropisms

Light is not the only thing to which plants are sensitive. Plants are also sensitive to gravity, and the different responses of the roots and shoots are very important in the control of plant growth.

In normal conditions, plants always receive a unilateral gravitational stimulus – gravity always acts downwards. The response of plants to gravity can be seen in the laboratory using seedlings placed on their sides either in all-round light or in the dark. Shoots are usually negatively geotropic (grow away from gravitational pull) and roots are positively geotropic (grow towards gravitational pull). This adaptation ensures that the roots grow down into the soil and the shoots grow up to the light. Geotropisms are also known as gravitropisms.

 Practical investigations into geotropisms

Figure 5 shows two ways in which geotropic responses can be demonstrated. These basic techniques can be adapted to investigate different aspects of the geotropic response.

* The geotropic response can be investigated in shoots and roots using a rotating drum known as a clinostat. The plants can be grown on a slowly rotating clinostat (about four revolutions per hour) so the gravitational stimulus is applied evenly to all sides of the plant – and the root and (in the dark) shoot grow straight.

* Alternatively, the seeds can be placed in petri dishes stuck to the wall of the lab, and the dishes rotated 90° at intervals as the seedlings grow. A geotropic response in the roots can be seen within about two hours.

Investigations into geotropisms are usually carried out with the plants exposed to all round light or in the dark. Suggest a reason for these conditions.

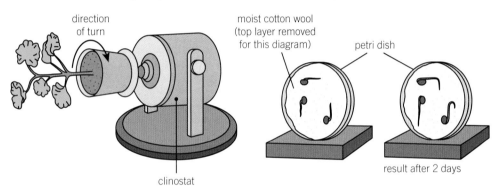

▲ **Figure 5** *Simple investigations can demonstrate geotropisms in shoots (left) and roots (right)*

Summary questions

1 a What is a tropism? *(1 mark)*
  b Explain what is meant by phototropisms and geotropisms. *(2 marks)*
  c Describe the different phototropic and geotropic responses in shoots and roots. *(2 marks)*

2 Produce a flow diagram to explain the events which bring about a phototropic response in a shoot. *(6 marks)*

3 If a block of butter is used in Figure 1 instead of the gelatin block, there is no response in the decapitated shoot. Explain how this informs scientists that the message is water soluble. *(5 marks)*

4 Originally scientists thought geotropisms were the result of auxin movements in response to gravity. Investigate current models of how geotropisms occur and write a brief report. *(6 marks)*

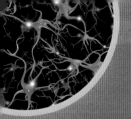

Plant hormones are involved in the control of many different aspects of plant life, from germination of the seeds and growth of the stems to ripening of the fruit and the fall of the leaves. As scientists have unravelled more and more about the role of these fascinating chemicals in the life of plants, people have developed a number of ways of using plant hormones commercially in agriculture and horticulture.

Control of ripening

The gaseous plant hormone ethene is involved in the ripening of climacteric fruits. These are fruits that continue to ripen after they have been harvested. Their ripening is linked to a peak of ethene production triggering a series of chemical reactions including a greatly increased respiration rate. Climacteric fruits include bananas, tomatoes, mangoes, and avocados. Non-climacteric fruit (such as oranges, strawberries, and watermelon) do not produce large amounts of ethene and do not ripen much after picking.

The effect of ethene on climacteric fruit can easily be seen if part of a bunch of green bananas is put in a bag with a single ripe banana. The bunch with the ripe banana will ripen faster than the rest of the bunch, even if the temperature is exactly the same in both cases. Ethene from the ripe banana stimulates the rapid ripening of the green ones.

Ethene is widely used commercially in the production of perfectly ripe climacteric fruit for greengrocers and supermarkets. These fruit are harvested when they are fully formed but long before they are ripe, and then cooled, stored, and transported. The unripe fruit is hard and much less easily damaged during transport around the world than the ripe versions. When the fruit are needed for sale, they are exposed to ethene gas under controlled conditions. This ensures that each batch of fruit ripens at the same rate and are all at the same stage to be put on the shelves for sale to the public. This careful control of ripening prevents a lot of wastage of fruit during transport, and increases the time available for them to be sold.

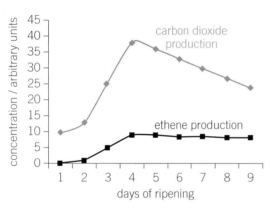

▲ **Figure 1** *Graph to show the effect of ethene on respiration rate of ripening fruit*

Hormone rooting powders and micropropagation

Auxin affects the growth of both shoots and roots. Scientists have discovered that the application of auxin to cut shoots stimulates the production of roots. This makes it much easier to propagate new plants from plant cuttings. A cutting is a small piece of the stem of a plant, usually with some leaves on. If this is placed in compost or soil – or even water – roots may eventually appear and a new plant forms. Dipping the cut stem into hormone rooting powder increases the chances of roots forming, and of successful propagation taking place. This has made it much easier for horticulturists to develop cuttings to sell and for individuals taking their own cuttings.

In both horticulture and agriculture, many plants are now propagated on a large scale by micropropagation, when thousands of new plants are grown from a few cells of the original plant. Plant hormones are essential in this process – they control the production of the mass of new cells and then the differentiation of the clones into tiny new plants.

Hormonal weedkillers

As you have seen, the interactions between the different plant hormones are finely balanced to enable the plant to grow. If this balance is lost it can interrupt the metabolism of the whole plant and may lead to plant death. Sometimes, this is exactly what we want to achieve, and plant hormones can help us. Weeds are plants that grow where they are not wanted. Commercial food crops are vital globally for producing the food people need to eat. Weeds interfere with crop plants, competing for light, space, water, and minerals.

Scientists have developed synthetic auxins which act as very effective weedkillers. Many of the main staple foods around the world are narrow-leaved monocot plants such as rice, maize, and wheat. Most of the weeds are broad-leaved dicots. If synthetic dicot auxins are applied as weedkiller, they are absorbed by the broad-leaved plants and affect their metabolism. The growth rate increases and becomes unsustainable, so they die. The narrow-leaved crop plants are not affected and continue to grow normally, freed from competition. The synthetic auxins used by farmers and gardeners are simple and cheap to produce, have a very low toxicity to mammals, and are selective.

Other uses of plant hormones

There are many different ways in which plant hormones are used commercially as well as those explored here. Examples include:

- Auxins can be used in the production of seedless fruit.

- Ethene is used to promote fruit dropping in plants such as cotton, walnuts, and cherries.

- Cytokinins are used to prevent ageing of ripened fruit and products such as lettuces, and in micropropagation to control tissue development.

- Gibberellins can be used to delay ripening and ageing in fruit, to improve the size and shape of fruits, and in beer brewing to speed up the malting process.

Synoptic link

You will learn more about the use of plant hormones in taking cuttings in Topic 22.1, Natural cloning in plants and the use of plant hormones in micropropagation in Topic 22.2, Artificial cloning in plants.

▲ **Figure 2** *Using rooting powder on cuttings is very simple and effective*

Summary questions

1 How are plant hormones used to control the ripening of fruit? *(4 marks)*

2 Why is it commercially important to be able to control fruit ripening? *(6 marks)*

3 Produce a table to summarise as many commercial uses as you can find for four named plant hormones. *(6 marks)*

4 Look at the graph in Figure 1.
 a Describe the changes in ethene production and carbon dioxide production in the tomato as it ripens. *(6 marks)*
 b Suggest what is happening. *(6 marks)*
 c Why do cool conditions slow the rate of ripening even if ethene is present? *(3 marks)*

Practice questions

1 a Plant responses to environmental changes are co-ordinated by plant growth substances (plant hormones).

Explain why plants need to be able to respond to their environment. *(2 marks)*

b The following investigation was carried out into the effects of plant growth substances on germination:

- A large number of lettuce seeds was divided into eight equal batches

- Each batch of seeds was placed on moist filter paper in a Petri dish and given a different treatment.

The different treatments are shown in the table. Each tick represents one of the eight batches of seeds.

	treatment	concentration of gibberellin (mol dm⁻³)			
		0.00	0.05	0.50	5.00
A	water	✔	✔	✔	✔
B	Abscisic acid	✔	✔	✔	✔

The batches of seeds were left to germinate at 25°C in identical conditions and the percentage germination was calculated. The graph shows the results of this investigation.

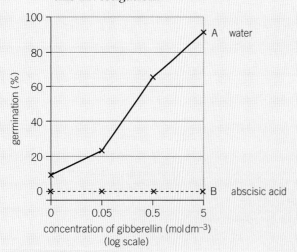

(i) Describe, with reference to the graph, the effects of the plant growth substances on the germination of lettuce seeds. *(4 marks)*

(ii) Explain how the plant hormones have these effects. *(4 marks)*

(iii) Explain why all the lettuce seeds were kept at 25°C. *(2 marks)*

(iv) State **three** variables, **other than temperature**, that needed to be controlled in the investigation. *(3 marks)*

c State two commercial uses of plant growth substances. *(2 marks)*

OCR F215 2010 (apart from 1b(ii))

2 Plants are able to respond to changes in their environment.

a Describe two ways in which hormones may alter a plants growth in response to overcrowding by other plants. *(4 marks)*

OCR F215 2012

b Suggest how hormones alter the growth and morphology, or growth and development of a plant *(4 marks)*

3 The growth and development of a fruit tree is controlled by plant growth regulators. The table shows the stages that occur as the tree grows and develops and indicates the stages at which giberellin is involved (green shading).

	Germination	Growth to maturity	Flowering	Fruit development	Abscission	Seed dormancy
Gibberellin						
Auxin						
Cytokinins						
Ethylene						
ABA						

a In the past plant chemicals such as auxins and gibberellins were referred to as plant hormones. At one stage this was changed to plant growth regulators. Now they are again generally referred to in university plant biology departments as plant hormones.

Explain why plant hormone is a more accurate term than plant growth regulator. (*5 marks*)

b Copy and complete the table and indicate which hormone(s) is/are involved at each stage, using crosses. Gibberellins have already been completed as an example. (*4 marks*)

c Compare and contrast the activity of auxins and cytokinins. (*6 marks*)

4 In an investigation into the effects of water stress, cowpea seeds were sown and the seedlings were thinned to one per pot. The plants were watered normally until mature then watering was completely stopped for 14 days to induce water stress.

The number of leaves and total leaf surface area were measured daily throughout this period.

The results are shown in the graph.

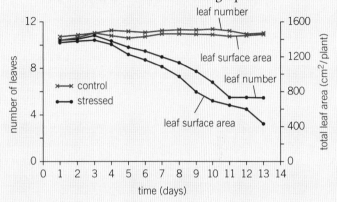

a **(i)** Describe the changes in leaf number and surface area over the 14 day period shown in the graph. (*3 marks*)

(ii) Describe and explain the effects of water stress on the growth and development of plants. (*3 marks*)

b Outline the process of leaf abscission including the roles of plant hormones. (*5 marks*)

c Compare and contrast the response of cowpeas to water stress to the loss of leaves by trees in the UK in autumn. (*5 marks*)

Learning outcomes

Demonstrate knowledge, understanding, and application of:

→ the need for cellular respiration

→ the interrelationship between the processes of photosynthesis and respiration.

Synoptic link

You learnt about active transport in Topic 5.4, Active transport.

Synoptic link

You will learn more about the transfer of energy through ecosystems in Topic 23.2, Biomass transfer through an ecosystem.

Synoptic link

You learnt about glucose as an energy store in Topic 3.3, Carbohydrates, and about ATP as a molecule of energy in Topic 3.11, ATP.

Study tip

A glucose molecule contains more energy than the single metabolic reactions needed to break it down. The energy contained within a glucose molecule is used to synthesise many molecules of ATP, and it is molecules of ATP that drive metabolic reactions.

The need for energy

Living organisms have to be active to survive. Organisms grow, respond to changes in their environment, and deal with threats from other organisms. They have to find or make food, and reproduce. All of this activity depends on metabolic reactions and processes continually taking place in individual cells.

A few examples of these metabolic activities amongst many include:

● active transport, which is essential for the uptake of nitrates by root hair cells, loading sucrose into sieve tube cells, the selective reabsorption of glucose and amino acids in the kidney, and the conduction of nerve impulses

● anabolic reactions, such as the building of polymers like proteins, polysaccharides, and nucleic acids essential for growth and repair

● movement brought about by cilia, flagella, or the contractile filaments in muscle cells.

All of these metabolic activities require energy.

Energy flow through living organisms

The total amount of energy in the universe hasn't changed from the time of the Big Bang, when the universe began, until now. The universe is gradually cooling as it expands because this energy is being dispersed over a larger area. This is inevitable and irreversible. Energy cannot be created or destroyed.

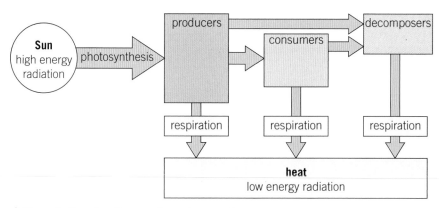

▲ Figure 1 Transfer of energy through ecosystems

Radiation from the Sun is used to fuel the metabolic reactions and processes necessary to keep organisms alive before, eventually, being transferred back to the atmosphere as heat (Figure 1).

Photosynthesis

Organisms make use of the energy in the bonds of organic molecules, such as glucose. These bonds are formed during **photosynthesis** in plants and other photosynthetic organisms (Topic 17.3, Photosynthesis).

Light is trapped by chlorophyll molecules. This energy is used to drive the synthesis of glucose from carbon dioxide and water.

Respiration

All organisms need to respire. **Respiration** is the process by which organic molecules, such as glucose, are broken down into smaller inorganic molecules, like carbon dioxide and water. The energy stored within the bonds of the organic molecules is used to synthesise adenosine triphosphate (ATP).

Two of the most important chemical reactions in the living world are photosynthesis and respiration. Photosynthesis is the reaction behind the production of most of the biomass on the earth. Respiration is the process by which organisms break down biomass to provide the ATP needed to drive the metabolic reactions that take place in cells.

The two reactions are intimately linked (Table 1). The raw materials for one are the products of the other so they are interrelated throughout the living world. The overall reactions for the two are as follows:

Photosynthesis

$$6CO_2 + 6H_2O \leftrightarrow C_6H_{12}O_6 + 6O_2$$

Respiration

$$C_6H_{12}O_6 + 6O_2 \rightarrow 6CO_2 + 6H_2O$$

The importance of carbon–hydrogen bonds

A general rule in biochemistry is that energy is used to break bonds, and energy is released when bonds are formed. The same quantity of energy is involved whether a particular bond is being broken or formed. This is called the bond energy. Whether an overall reaction is **exothermic** (releases energy) or **endothermic** (takes in energy) depends on the total number and strength of bonds that are broken or formed during the reaction.

The atoms in small inorganic molecules like water and carbon dioxide are joined by strong bonds that release a lot of energy when they form but require a lot of energy to break. Organic molecules like glucose and amino acids contain many more bonds than small inorganic molecules. These are weaker bonds compared with those in inorganic molecules and, therefore, release less energy when they form and require less energy to be broken.

In respiration large organic molecules are broken down forming small inorganic molecules. The total energy required to break all the bonds in a complex organic molecule is less than the total energy released in the formation of all the bonds in the smaller inorganic products. The excess energy released by the formation of the bonds is used to synthesise ATP.

> **Synoptic link**
>
> You will learn more about respiration in Chapter 18, Respiration.

▼ **Table 1** *Comparison of respiration and photosynthesis*

	Respiration	Photosynthesis
Reactants	glucose and oxygen	water and carbon dioxide
Products	water and carbon dioxide	glucose and oxygen
Purpose	release energy	trap energy

> **Study tip**
>
> It is important to remember that energy is not produced, created, made, or lost. Say instead that energy is released or absorbed, or transferred as heat.

Organic molecules contain large numbers of carbon–hydrogen bonds, particularly lipids. Carbon and hydrogen share the electrons almost equally in bonds that form between them. This results in a non-polar bond which does not require a lot of energy to break. The carbon and hydrogen released then form strong bonds with oxygen atoms, forming carbon dioxide and water, resulting in the release of large quantities of energy. The reverse happens in photosynthesis when organic molecules are made from small inorganic molecules. The energy required to build these molecules comes from the Sun.

Energy transfer

When thinking about energy transfers, you need to remember that the energy required to form a particular type of bond (e.g. O–H) is equal to the amount of energy released when the bond is broken. When bonds are broken, the bond energy is given a positive sign, e.g. breaking an O–H bond: +464 kJ/mol. When bonds are formed, the bond energy is given a negative sign, e.g. making an O–H bond: −464 kJ/mol.

The overall reaction that takes place in photosynthesis is:

$$6CO_2 + 6H_2O \rightarrow C_6H_{12}O_6 + 6O_2$$

The overall reaction that takes place in respiration is the reverse of this reaction.

The structures of the molecules involved are shown in Figure 2.

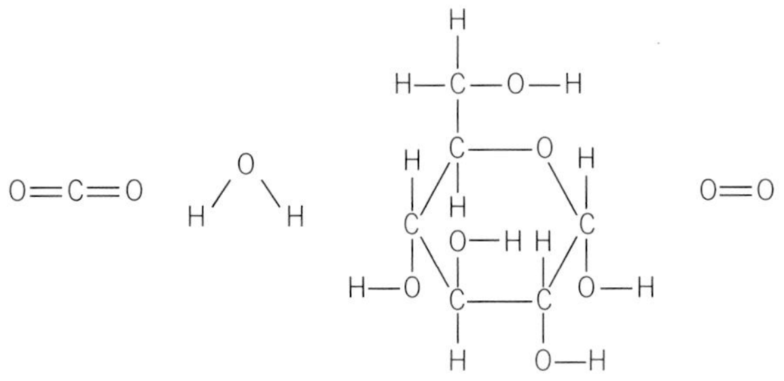

O=C=O

H–O–H (water)

H–C–O–H with glucose ring structure

O=O

▲ Figure 2

1 Copy and complete the table.

Bond	Number of bonds involved in reaction	Bond energy (kJ/mol)	Total (kJ/mol)
C=O		803	
O—H		464	
C—H		414	
C—C		347	
C—O		358	
O=O		495	

2 Using information from the completed table, calculate how much energy is required to make one mole of glucose from six moles of carbon dioxide and six moles of water.

3 State how much energy would be released when one mole of glucose is respired.

4 The breakdown of one glucose molecule results in the production of 38 molecules of ATP. The formation of ATP requires 30.6 kJ/mol. Calculate the percentage of energy released when one mole of glucose is respired.

Summary questions

1 Explain why it is incorrect to say that energy is produced. *(2 marks)*

2 Explain why ATP is not a good energy storage molecule but why organic molecules, like lipids or carbohydrates, are. *(4 marks)*

3 Explain the interrelationship between respiration and photosynthesis in organisms. *(5 marks)*

17.2 ATP synthesis

Specification reference: 5.2.2

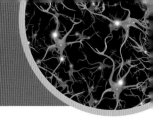

ATP production

As you have already learnt, ATP is the universal energy currency in cells. The bond energy in the ATP molecule is used to drive essential metabolic processes. The production of ATP is therefore fundamental to all forms of life.

In photosynthesis, light provides the energy needed to build organic molecules like glucose. This energy is used to form chemical bonds in ATP, which are then broken to release the energy needed to make bonds as glucose is formed.

In respiration organic molecules, such as glucose, are broken down and the energy released is used to synthesise ATP. This ATP is then used to supply the energy needed to break bonds in the metabolic reactions of the cell.

Synthesis of ATP is therefore a crucial step in both respiration and photosynthesis.

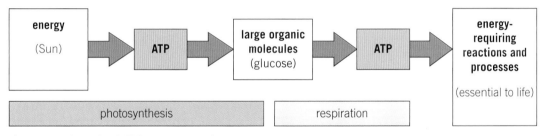

▲ **Figure 1** *The role of ATP in energy transfer*

Chemiosmosis

The ATP produced in both photosynthesis and respiration is synthesised primarily by a process called **chemiosmosis**. Chemiosmosis involves the diffusion of protons from a region of high concentration to a region of low concentration through a partially permeable membrane. The movement of the protons as they flow down their concentration gradient releases energy that is used in the attachment of an inorganic phosphate (P_i) to ADP, forming ATP.

Chemiosmosis depends on the creation of a proton concentration gradient. The energy to do this comes from high-energy electrons (excited electrons).

Excited electrons

Electrons are raised to higher energy levels, or excited, in two ways:

- electrons present in pigment molecules (e.g., **chlorophyll**) are excited by absorbing light from the Sun
- high energy electrons are released when chemical bonds are broken in respiratory substrate molecules (e.g., glucose).

The excited electrons pass into an electron transport chain and are used to generate a proton gradient.

> ### Learning outcomes
> Demonstrate knowledge, understanding, and application of:
> → the chemiosmotic theory.

> ### Synoptic link
> You learnt about ATP–ADP in Topic 3.11, ATP.

Electron transport chain

An electron transport chain is made up of a series of **electron carriers**, each with progressively lower energy levels. As high energy electrons move from one carrier in the chain to another, energy is released. This is used to pump protons across a membrane, creating a concentration difference across the membrane and therefore a proton gradient. The proton gradient is maintained as a result of the impermeability of the membrane to hydrogen ions.

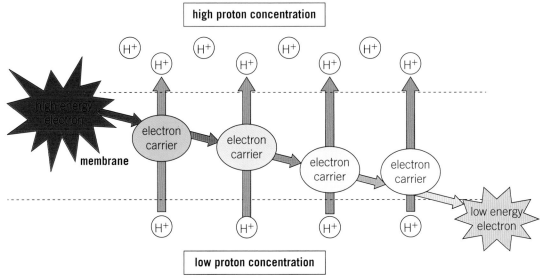

▲ **Figure 2** *The flow of electrons along an electron transport chain releases energy which is used to pump protons across a membrane, resulting in the formation of a proton gradient*

The only way the protons can move back through the membrane down their concentration gradient is through hydrophilic membrane channels linked to the enzyme ATP synthase (catalyses the formation of ATP). The flow of protons through these channels provides the energy used to synthesise ATP (from ADP and P_i).

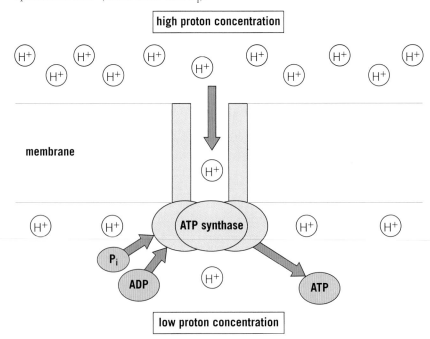

▲ **Figure 3** *The role of ATP synthase in the synthesis of ATP from ADP and P_i*

In photosynthesis, ATP is used to synthesise glucose and other organic molecules. The ATP produced in respiration provides the energy needed by metabolic processes and reactions essential to life.

A simple way of modelling this process is to think of the flow of water through a hydroelectric power station causing turbines to spin, generating electricity. Both chemiosmosis and hydroelectric power generation result in energy in a very useable form.

The processes of oxidative phosphorylation (in respiration) and photophosphorylation (in photosynthesis) are also vital in chemiosmosis. You will learn about these in more detail in Topic 18.4, Oxidative phosphorylation and Topic 17.3, Photosynthesis, respectively.

Synoptic link

You learnt about diffusion and active transport across cell membranes in Chapter 5, Plasma membranes.

Summary questions

1   Explain the importance of ATP to living organisms.   (*3 marks*)

2   Describe the properties of cell membranes necessary for the formation of a proton gradient.   (*5 marks*)

3   Name the type of diffusion which enables protons to move through ATP synthase and explain the role of ATP synthase in the production of ATP.   (*4 marks*)

4   The synthesis and breakdown of ATP is an example of a reversible reaction:

$$ADP + P_i \rightleftharpoons ATP$$

ATPase is often the name given to the enzyme which hydrolyses ATP, producing ADP and P_i. ATPase and ATP synthase are, in fact, the same enzyme. Explain how this is possible.   (*5 marks*)

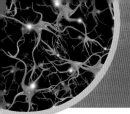

17.3 Photosynthesis

Specification reference: 5.2.1

Synoptic link

You will learn more about how respiration releases energy in Chapter 18, Respiration.

Synoptic link

You learnt about endosymbiosis in Topic 2.6, Prokaryotic and eukaryotic cells.

Photosynthesis is the process by which energy, in the form of light from the Sun, is used to build complex organic molecules, such as glucose. Light energy is transformed into chemical energy trapped in the bonds of the complex organic molecules produced. Organisms that can photosynthesise, like plants and algae, are said to be **autotrophic**.

Heterotrophic organisms, like animals, obtain complex organic molecules by eating other (heterotrophic and/or autotrophic) organisms.

Both autotrophic and heterotrophic organisms then break down complex organic molecules during the process of respiration to release the energy they need to drive metabolic processes.

Photosynthesis can be summarised by the equation:

$$6CO_2 \ + \ 6H_2O \ \rightarrow \ C_6H_{12}O_6 \ + \ 6O_2$$
$$\text{carbon dioxide} \ + \ \text{water} \ \rightarrow \ \text{glucose} \ + \ \text{oxygen}$$

Structure and function of chloroplasts

As you have learnt, photosynthesis takes place in chloroplasts. The network of membranes present within chloroplasts provides a large surface area to maximise the absorption of light essential in the first step of photosynthesis. The membranes form flattened sacs called thylakoids which are stacked to form grana (singular granum) (Figure 1). The grana are joined by membranous channels called lamellae.

Light is absorbed by complexes of pigments, such as chlorophyll, which are embedded within the thylakoid membranes.

The fluid enclosed in the chloroplast is called the stroma and is the site of the many chemical reactions resulting in the formation of complex organic molecules.

Chlorophyll

Pigment molecules absorb specific wavelengths (colours) of light and reflect others. Different pigments absorb and reflect different wavelengths and this is why they have different colours. The primary pigment in photosynthesis is chlorophyll. Chlorophyll absorbs mainly red and blue light and reflects green light. The presence of large quantities of chlorophyll is the reason for the familiar green colour of plants.

Although there are a number of different pigments that absorb light, the primary pigment is chlorophyll a. Other pigments, like chlorophyll b, xanthophylls, and carotenoids absorb different wavelengths of light than those absorbed by chlorophyll a. Different combinations of pigments are the reason for the different shades and colours of leaves.

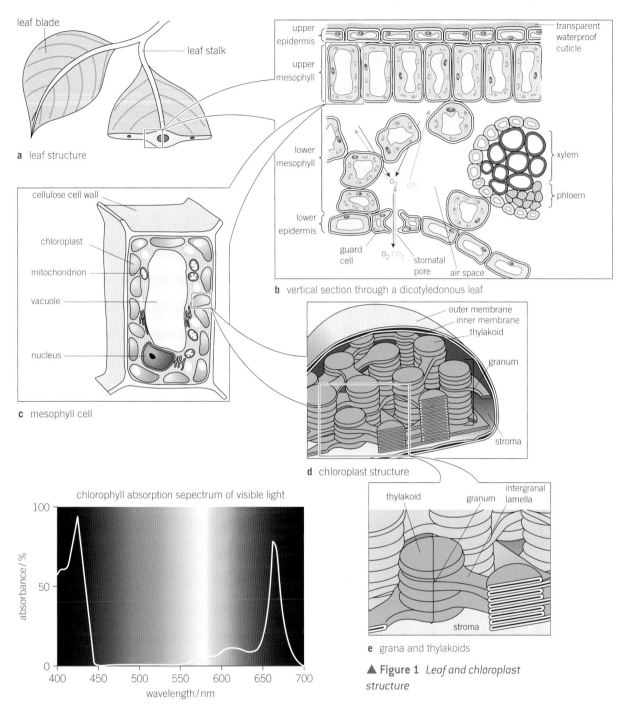

a leaf structure

b vertical section through a dicotyledonous leaf

c mesophyll cell

d chloroplast structure

e grana and thylakoids

▲ **Figure 1** *Leaf and chloroplast structure*

chlorophyll absorption sepectrum of visible light

▲ **Figure 2** *Absorption spectra of chlorophyll showing that red and blue light is absorbed and green light is reflected*

Chlorophyll b, xanthophylls, and carotenoids are embedded in the thylakoid membrane of the chloroplast. These and other proteins and pigments form a light harvesting system (also known as an antennae complex). The role of the system is to absorb, or harvest, light energy of different wavelengths and transfer this energy quickly and efficiently to the reaction centre. Chlorophyll a is located in the reaction centre, which is where the reactions involved in photosynthesis take place.

The light harvesting system and reaction centre are collectively known as a photosystem.

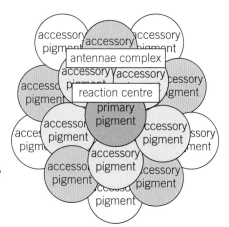

▶ **Figure 3** *The reaction centre is surrounded by an antennae complex, maximising the absorption of light*

Plants use sunscreen as well

Plants need light to photosynthesise but if sunlight is too intense chlorophyll is destroyed. Chlorophyll has to be continuously synthesised during the summer to maintain the level needed to photosynthesise at the required rate. Chlorophyll is not produced when there is little or no sunlight – this is why areas of grass that have been covered turn yellow.

Carotenoids are accessory pigments responsible for the yellow/orange colours seen in plant leaves. Orange carotene and yellow xanthophyll are two examples. These colours are not normally seen because they are masked by the green colour of chlorophyll. Carotenoids are not broken down, unlike chlorophyll, in strong sunlight and are present throughout the growing season.

The shorter days and cooler nights of autumn cause changes in the pigment composition in leaves. Chlorophyll a is no longer produced and leaves turn yellow/orange as we see the carotenoids.

Anthocyanin is a red/purple pigment formed from a reaction between sugars and proteins present in cell sap. It is produced when the concentration of sugars is high. High light intensity also promotes the production of anthocyanins.

Anthocyanin produces the red skin of apples and the purple of black grapes. The colour of the anthocyanin

pigments is pH-dependent, leading to a range of different colours from red to purple.

Anthocyanins act as a sunscreen by absorbing blue-green and ultraviolet light, thereby inhibiting the destruction of chlorophyll.

In their role as pigments they help trees maximise production towards the end of the growing season as the weather changes in autumn.

The red/purple coloration of leaves is also thought to camouflage leaves from herbivores blind to red wavelengths.

1 Suggest explanations for the following observations:
   a Apples are often red on one side and green on the other.
   b Leaves with more vibrant red colours are seen during years when there has been lots of sunlight and dry weather. When it has been raining and overcast there will not be as much red foliage present.
2 Suggest why the production of anthocyanins is temperature-dependent.

Investigating photosynthetic pigments

Chromatography can be used to separate the different pigments in a plant extract. The mobile phase would be the solution containing a mixture of pigments and the stationary phase a thin layer of silica gel applied to glass.

The different solubilities of the pigments in the mobile phase, and their differing interactions with the stationary phase, lead to them moving at different rates. This results in the pigments being separated as they move through the silica gel.

The retention value (R_f) for each pigment can be calculated using the formula:

$$R_f = \frac{\text{distance travelled by component}}{\text{distance travelled by solvent}}$$

Synoptic link

You also learnt about chromatography and how it is used to separate amino acids in solution in Topic 3.6, Structure of proteins.

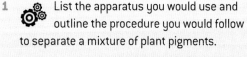

1. List the apparatus you would use and outline the procedure you would follow to separate a mixture of plant pigments.

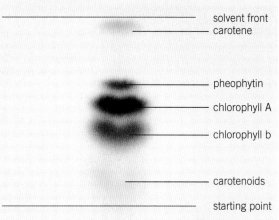

▲ **Figure 4** *Thin layer chromatogram (TLC) of an extract of thylakoid membranes from the leaf of annual meadow grass. A drop of extract was laid at the bottom of the sheet. The sheet was then placed in a beaker of solvent separating out the pigments. Five bands can be seen.*

2. Calculate the R_f value for each pigment.

The two stages of photosynthesis

There are two stages in photosynthesis:

- Light-dependent stage – energy from sunlight is absorbed and used to form ATP (Figure 5). Hydrogen from water is used to reduce coenzyme **NADP** to reduced NADP.

- Light-independent stage – hydrogen from reduced NADP and carbon dioxide is used to build organic molecules, such as glucose. ATP supplies the required energy (Figure 6).

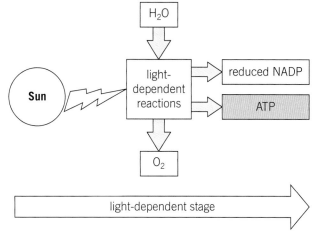

▲ **Figure 5** *Summary of the light-dependent stage of photosynthesis which occurs within and across the thylakoid membranes*

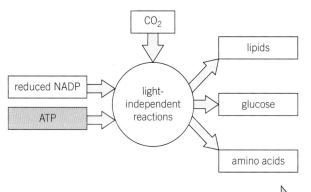

◀ **Figure 6** *Summary of the light-independent stage of photosynthesis which occurs in the stroma*

The light-dependent stage of photosynthesis

Non-cyclic photophosphorylation

Two photosystems are involved in **non-cyclic photophosphorylation**, photosystem II (PSII) followed by photosystem I (PSI). The reaction centre of PSI absorbs light at a higher wavelength (700 nm) than PSII (680 nm). The light absorbed excites electrons at the reaction centres of the photosystems.

The excited electrons are released from the reaction centre of PSII and are passed to an electron transport chain. ATP is produced by the process of chemiosmosis (Topic 17.2, ATP synthesis).

The electrons lost from the reaction centre at PSII are replaced from water molecules broken down using energy from the Sun (Topic 17.4, Factors affecting photosynthesis).

Excited electrons are released from the reaction centre at PSI, passed to another electron transport chain, and ATP is again produced by chemiosmosis. The electrons lost from this reaction centre are replaced by electrons that have just travelled along the first electron transport chain after being released from PSII.

The electrons leaving the electron transport chain following PSI are accepted, along with a hydrogen ion, by the coenzyme NADP, forming reduced NADP. Reduced NADP provides the hydrogen or reducing power in the production of organic molecules, such as glucose, in the light-independent stage which follows.

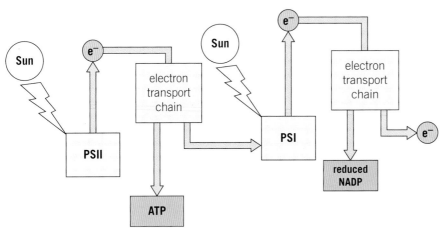

▲ **Figure 7** *Diagram summarising the two electron transport chains involved in cyclic and non-cyclic photophosphorylation, this is often referred to as the Z-scheme*

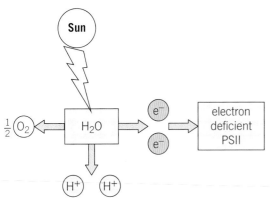

▲ **Figure 8** *Photolysis which is catalysed by the oxygen-evolving complex present in PSII*

Photolysis

Water molecules are split into hydrogen ions, electrons, and oxygen molecules using energy from the Sun in a process called photolysis. The electrons released replace the electrons lost from the reaction centre of PSII. This is why water, along with light and carbon dioxide, is a raw material of photosynthesis.

The oxygen-evolving complex which forms part of PSII is an enzyme that catalyses the breakdown of water. Here water molecules are split into hydrogen ions, electrons, and oxygen molecules using energy from the Sun in a process called photolysis. The electrons released replace the electrons lost from the reaction centre of PSII. This is why water, along with light and carbon dioxide, is a raw material of photosynthesis. The photolysis reaction is summarised as:

$$H_2O \longrightarrow 2\,H^+ + 2e^- + \tfrac{1}{2}O_2$$

Oxygen gas is released as a by-product. The protons are released into the lumen of the thylakoids, increasing the proton concentration across the membrane. As they move back through the membrane down a concentration and electrochemical gradient, they drive the formation of more ATP. Once the hydrogen ions are returned to the stroma, they combine with NADP and an electron from PSI to form reduced NADP. This is used in the light-independent reactions of photosynthesis. This process removes hydrogen ions from the stroma so it helps to maintain the proton gradient across the thylakoid membranes.

Cyclic photophosphorylation

The electrons leaving the electron transport chain after PSI can be returned to PSI, instead of being used to form reduced NADP, leading to **cyclic photophosphorylation**. This means PSI can still lead to the production of ATP without any electrons being supplied from PSII. Reduced NADP is not produced when this happens.

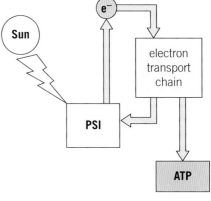

▲ Figure 9 *Cyclic photophosphorylation*

The light-independent stage of photosynthesis

The light-independent stage of photosynthesis takes place in the stroma of chloroplasts and uses carbon dioxide as a raw material. The products from the light-dependent stage – ATP and reduced NADP, are also required. Organic molecules, like glucose, are produced in a series of reactions collectively known as the Calvin cycle.

Calvin cycle

Carbon dioxide enters the intercellular spaces within the spongy mesophyll of leaves by diffusion from the atmosphere through stomata. It diffuses into cells and into the stroma of chloroplasts where it is combined with a five-carbon molecule called **ribulose bisphosphate (RuBP)**. The carbon in carbon dioxide is therefore **fixed**, meaning that it is incorporated into an organic molecule.

The enzyme **ribulose bisphosphate carboxylase (RuBisCO)** catalyses the reaction and an unstable six-carbon intermediate is produced. RuBisCO is the key enzyme in photosynthesis. It is a very inefficient enzyme as it is competitively inhibited by oxygen (see the Photorespiration application box) so a lot of it is needed to carry out photosynthesis successfully. Biologists estimate that it is probably the most abundant enzyme in the world.

The unstable six-carbon compound formed immediately breaks down, forming two three-carbon **glycerate 3-phosphate (GP)** molecules.

Each GP molecule is converted to another three-carbon molecule, **triose phosphate (TP)**, using a hydrogen atom from reduced NADP and energy supplied by ATP, both supplied from the light-dependent reactions of photosynthesis.

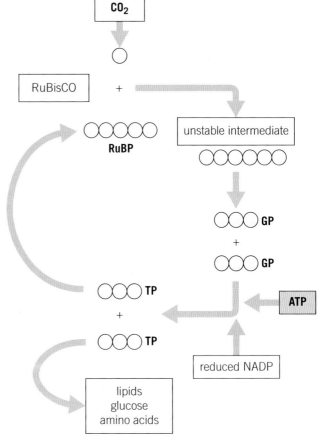

▲ Figure 10 *The Calvin cycle*

Triose phosphate is a carbohydrate, a three-carbon sugar, the majority of which is recycled to regenerate RuBP so that the Calvin cycle can continue. It is the starting point for the synthesis of many complex biological molecules, including other carbohydrates, lipids, proteins, and nucleic acids.

The Calvin cycle can be summarised in three steps:

● *Fixation* – carbon dioxide is fixed (incorporated into an organic molecule) in the first step.

● *Reduction* – GP is reduced to TP by the addition of hydrogen from reduced NADP using energy supplied by ATP.

● *Regeneration* – RuBP is regenerated from the recycled TP.

Regeneration of RuBP

For one glucose molecule to be produced six carbon dioxide molecules have to enter the Calvin cycle, resulting in six full turns of the cycle. This will result in the production of 12 TP molecules, two of which will be removed to make the glucose molecule.

This means that 10 TP molecules are recycled to regenerate six RuBP molecules (used in the six turns of the cycle).

10 × three-carbon TP = 30 carbons 'shuffled' gives 6 × five-carbon RuBP = 30 carbons

Energy is supplied by ATP for the reactions involved in the regeneration of RuBP.

Study tip

Reduced NADP must supply a hydrogen atom to GP in the Calvin cycle, not a hydrogen ion or proton. If there is no electron present a bond will not be formed with carbon.

Synoptic link

You learnt about transpiration in Topic 9.3, Transpiration.

Photorespiration

Stomata need to be open for plants to obtain carbon dioxide for photosynthesis. Water vapour leaves plants through open stomata by the process of transpiration. When the temperature is high and humidity of the atmosphere is low plants can lose too much water. To prevent excess water loss, stomata close.

This prevents the entry of carbon dioxide into the leaves of the plant. The plant will still be photosynthesising and so the carbon dioxide levels fall and the oxygen levels increase.

Oxygen is a competitive inhibitor of the enzyme RuBisCO, leading to the production of phosphoglycolate and reducing the production of GP. This only happens when the concentration of carbon dioxide becomes very low.

Phosphoglycolate is a toxic two-carbon molecule that needs to be removed. It is converted by the plant into other organic molecules and energy from ATP is needed for the conversion.

RuBisCO has a higher affinity for carbon dioxide than oxygen and approximately 25% of the products of the Calvin cycle are lost in photorespiration, reducing the efficiency of photosynthesis.

1 Explain why photorespiration is not something commercial producers would want to encourage.

2 Suggest why plants evolved with such an important enzyme as RuBisCO being inhibited by such a common molecule as oxygen.

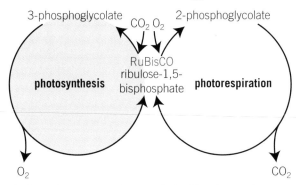

▲ **Figure 11** *Photorespiration – RuBisCO and two substrates*

Summary of photosynthesis

The light-dependent stage, including photolysis, and the light-independent stage are summarised in Figure 12 showing how energy from the Sun in the form of light is used to build the chemical bonds in complex biological molecules.

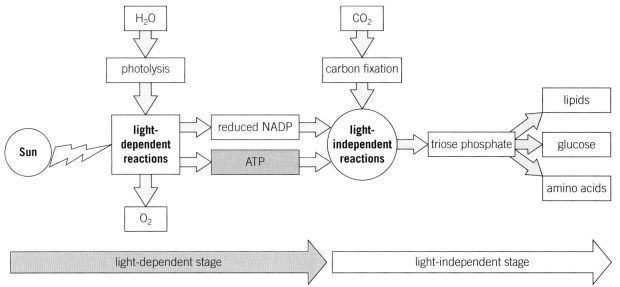

▲ **Figure 12** *Overview of the light-dependent and light-independent stages*

Summary questions

1 Explain the meaning of the term photophosphorylation. *(2 marks)*

2 Look at Figure 13. Explain why the absorption spectra of the pigments present in the thylakoid membranes of chloroplasts (top graph) follows the same pattern as the action spectrum (the rate of photosynthesis at different wavelengths of light, (bottom graph)). *(4 marks)*

3 Explain why photosynthesis stops when plants are exposed to green light only. *(3 marks)*

4 Explain what is meant by the term fixation. *(2 marks)*

5 a The Calvin cycle used to be called the 'dark reaction'. This term is now rarely used.
   Explain why this term is incorrect. *(1 mark)*

   b Explain why the alternate name of the Calvin cycle, the light-independent stage, is also not completely accurate. *(3 marks)*

6 Suggest the possible benefits of cyclic photophosphorylation. *(4 marks)*

7 Describe how RuBP is regenerated from TP in the Calvin cycle. *(4 marks)*

8 The overall reaction for photosynthesis is summarised by the chemical equation:

$$6CO_2 + 6H_2O \rightarrow C_6H_{12}O_6 + 6O_2$$

Outline why this is an oversimplification. *(6 marks)*

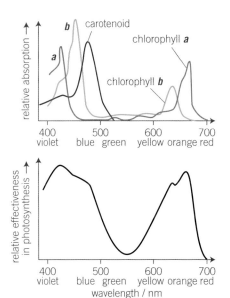

▲ **Figure 13**

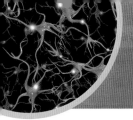
Photosynthesis is a multi-step reaction pathway that takes place in plants. Like any other chemical reaction, it is affected by various environmental factors. Plants are affected by changes in temperature and the availability of raw materials required for photosynthesis.

Plants obtain carbon dioxide through open stomata but this also involves the loss of water by transpiration. However, the loss of water vapour results in the transport of important ions and minerals from the soil to the leaves.

A balance between these different processes has to be maintained and this can be upset by changes in the environment of the plant.

Limiting factors

When one of the factors needed for a plant to photosynthesise is in short supply, it reduces the rate of photosynthesis, and is therefore a **limiting factor**.

The factors that affect the rate of photosynthesis are:

- *Light intensity* – light is needed as an energy source. As light intensity increases, ATP and reduced NADP are produced at a higher rate.

- *Carbon dioxide concentration* – carbon dioxide is needed as a source of carbon, so if all other conditions are met, increasing the carbon dioxide concentration increases the rate of carbon fixation in the Calvin cycle and, therefore, the rate of TP production.

- *Temperature* – affects the rate of enzyme-controlled reactions. As temperature increases, the rate of enzyme activity increases until the point at which the proteins denature. An increase in temperature increases the rates of the enzyme-controlled reactions in photosynthesis, such as carbon fixation. The rate of photorespiration, however, also increases above 25°C meaning higher photosynthetic rates may not be seen at higher temperatures even if enzymes are not actually denatured.

Stomata on plant leaves and other surfaces will close to avoid water loss by transpiration during dry spells when plants undergo water stress. The closure of stomata stops the diffusion of carbon dioxide into the plant, reducing the rate of the light-independent reaction, and eventually stopping photosynthesis.

Although water is required for photosynthesis it is never considered a limiting factor because for water potential to have become low enough to limit the rate of photosynthesis the plant will already have closed its stomata and ceased photosynthesis. Plants, except those with

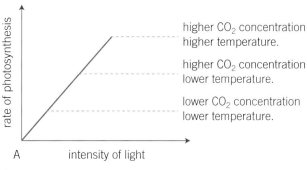

▲ Figure 1 *Following the blue line from A, rate of photosynthesis increases with increasing light intensity meaning light is the limiting factor. This happens until carbon dioxide concentration becomes limiting. Increasing the light intensity will then no longer have an effect until carbon dioxide concentration is increased*

adaptations to tolerate drought conditions, are unlikely to survive these conditions.

The law of limiting factors states that the rate of a physiological process will be limited by the factor which is in shortest supply.

Investigating the factors that affect the rate of photosynthesis

Data loggers are electronic devices that record data over time using sensors. Physical properties are recorded such as light intensity, temperature, pressure, pH (which can be used as a measure of carbon dioxide concentration), and humidity. Readings can be displayed in graphical form or on a spreadsheet.

They are usually equipped with a microprocessor (which inputs digital data) and internal memory for data storage. Data loggers can usually interface with a computer using specialised software.

Readings are taken with high degrees of accuracy and can be taken over long periods of time. They can be set to take many readings in a short period of time or used when there is a risk involved, for example, extreme cold or heat.

The factors affecting rate of photosynthesis can be investigated using a live pond weed, such as *Elodea*. The rate of photosynthesis can be estimated by calculating the rate of oxygen produced, carbon dioxide used, or increase in dry mass of a plant.

Apparatus could be set up as shown in Figure 2. Sodium hydrogen carbonate would be used to provide carbon dioxide. The pond weed should be kept illuminated before use. The apparatus should be left to equilibrate for 10 minutes or so before readings are taken. The oxygen sensor may also need to be calibrated using the oxygen concentration of air (21%).

The software can be set to take readings at desired intervals for the required length of time.

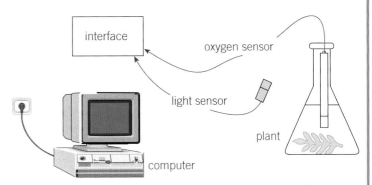

▲ **Figure 2** *The plant is supplied with carbon dioxide from sodium hydrogen carbonate in the solution containing the pond weed*

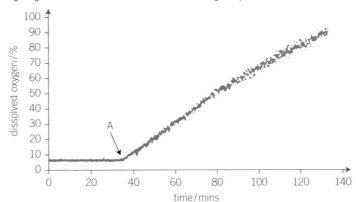

▲ **Figure 3** *Shows an example of a graph produced using the apparatus described*

1 Use the graph to calculate the rate of oxygen production after point A.

2 Suggest what happened at point A.

3 Outline how you could use the apparatus to investigate the effects of changing light intensity, temperature, and carbon dioxide concentration on the rate of photosynthesis.

4 *Elodea* will release bubbles of oxygen when photosynthesising which can be counted and used as an estimate of the rate of photosynthesis.
Evaluate the advantages and disadvantages of this method and of using data loggers in estimating the rate of photosynthesis.

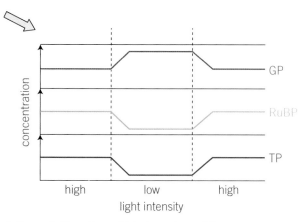

▲ **Figure 4** *The effect of light intensity on GP, TP, and RuBP concentrations*

The effect of reducing light intensity on the Calvin cycle

Reducing light intensity will reduce the rate of the light-dependent stage of photosynthesis. This will reduce the quantity of ATP and reduced NADP produced. ATP and reduced NADP are needed to convert GP to TP. The concentration of GP will therefore increase and the concentration of TP will decrease. As there will be less TP to regenerate RuBP, the concentration of RuBP will also decrease. The reverse will happen when the light intensity is increased (Figure 4).

The effect of carbon dioxide concentration and temperature on the Calvin cycle

All the reactions making up the Calvin cycle are catalysed by enzymes, for example, RuBisCO in carbon fixation. At lower temperatures enzyme and substrate molecules have less kinetic energy resulting in fewer successful collisions and a reduced rate of reaction. This means decreasing temperature results in lower concentrations of GP, TP, and RuBP.

The same effect will be seen at high temperatures as enzymes will be denatured – this is irreversible.

As carbon dioxide is an essential substrate of the Calvin cycle, low concentrations will lead to reduced concentrations of GP (as there is less carbon dioxide to be fixed) and TP. The concentration of RuBP will increase as it is still being formed from TP but not being used to fix carbon dioxide.

Artificial photosynthesis, a win-win solution

The burning of fossil fuels, and respiration, is continually releasing huge quantities of carbon dioxide into the atmosphere. The overall concentration of carbon dioxide, a greenhouse gas, is increasingly leading to more heat from the Sun being trapped in the atmosphere. This is enhanced global warming and is causing the polar ice caps to melt, increasing sea levels, and changing the climate around the world.

Fossil fuels have a limited supply and will eventually run out, leading to fuel shortages, and many parts of the world already suffer from food shortages.

Therefore, we have a surplus of carbon dioxide, which needs removing, and a shortage of fuel and food, both of which are forms of biomass produced by plants using carbon dioxide.

Photosynthesis would appear to offer a solution. It uses carbon dioxide and energy from the Sun to produce carbohydrates. Carbohydrates can be used as both food and fuel.

It is said that 'more energy hits Earth from the Sun in one hour than mankind uses in an entire year', so there is no shortage of energy. We already collect and use energy from the Sun in the form of solar power but the Sun doesn't always shine and at the moment there are no practical ways to store energy for a 'rainy day'. However, the carbohydrate fuel produced by plants can be stored for long periods.

The problem is that the process of photosynthesis, which has taken millions of years to evolve, is still not particularly efficient, so relying on plants is not the answer.

Artificial photosynthesis is seen as a possible solution. By improving on the natural process of photosynthesis carried out by plants, more carbon dioxide could be removed from the atmosphere and more carbohydrate products could be produced which could help with fuel and food shortages.

Suggest what you think would be the basic components of an artificial photosynthetic process.

Different types of photosynthesis

Most plants use the form of photosynthesis that you have learnt about in this chapter. This is referred to as C3 photosynthesis, and it is most efficient in cool, wet climates with average sunshine values.

Plants that live in hot, arid climates like the desert which are exposed to intense sunlight use different types of photosynthesis. Plants which use C4 and crassulacean acid metabolism (CAM) types of photosynthesis use water more efficiently and can photosynthesise at faster rates at higher temperatures and light intensities. Plants are adapted to their different environments to photosynthesise as efficiently as possible.

C4 photosynthesis

Plants that undergo C4 photosynthesis are adapted to high temperatures and limited water supply. They are able to fix carbon dioxide more efficiently and so do not need to have their stomata open for as long as C3 plants meaning there is less water lost by transpiration.

PEP carboxylase present in mesophyll cells, which first fixes carbon dioxide, is not inhibited by oxygen (like RuBisCO), increasing the efficiency of fixation. The four-carbon molecules produced are transported to bundle sheaths, formed from tightly packed cells, deeper inside the plant. These molecules are then decarboxylated, and the carbon dioxide is then fixed by RuBisCO and enters the Calvin cycle. As RuBisCO is shielded from atmospheric oxygen, the waste of resources by photorespiration is reduced. Corn is an example of a C4 plant.

CAM photosynthesis

Other plants use CAM photosynthesis and open their stomata at night, usually closing them during the day, again reducing water loss by transpiration. Carbon dioxide is converted to an acid and stored during the night. During the day the acid is broken back down releasing carbon dioxide to RuBisCO. During very dry spells stomata can remain closed night and day. The carbon dioxide released from respiration is used in photosynthesis and the oxygen released by photosynthesis is used for respiration. Cacti are CAM plants.

1   Some plants drop their leaves and twigs and become dormant during dry spells. Describe the way in which a cactus survives dry spells and the advantages of this method.
2   Suggest why CAM plants can only keep their stomata closed night and day for short periods.

Summary questions

1   Describe what is meant by a limiting factor.   (*3 marks*)

2   Suggest why the rate of oxygen production is only an estimate of the rate of photosynthesis.   (*2 marks*)

3   Discuss, using what you have learnt in this chapter, how an understanding of the effect of limiting factors on the rate of photosynthesis is used to design more efficient glasshouses.   (*5 marks*)

Study tip

Always refer to light 'intensity' rather than 'level' or 'amount'. When referring to rate of photosynthesis it is often a good idea to refer to the rate of the Calvin cycle or carbon fixation as an example.

Practice questions

1 Photosynthesis is dependent on a supply of water, light energy, and carbon dioxide. Restricting the supply of any one of these factors reduces the rate of photosynthesis regardless of the availability of the other factors. A lower rate of photosynthesis results in a lower rate of growth and this has important implications for commercial plant growers.

a Outline the problems that commercial plant growers might have when using glasshouses. (*3 marks*)

The graph shows the changes in photosynthesis and respiration rates over a range of temperatures for tomato plants grown in a glasshouse. Tomatoes need to have reached a certain size and sweetness before they can be sold.

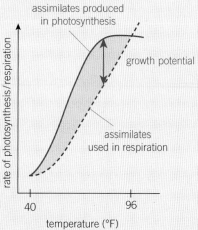

b (i) Describe, and explain, the changes in rates of photosynthesis and respiration shown in the diagram. (*5 marks*)

(ii) Discuss how these changes would affect the saleability of the tomatoes grown at different temperatures. (*4 marks*)

(iii) Comment on why there might be a difference between maximum rate of photosynthesis and optimum economic rate of photosynthesis. (*3 marks*)

2 It is known that C4-type plants carry out a much more efficient type of photosynthesis than C3 plants particularly under conditions of high light intensity and temperature.

The graph shows the changes in rate of photosynthesis for a C3 and C4 plant at different light intensities.

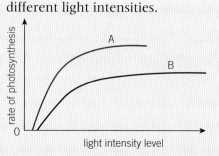

a State which curve in the graph is the C3 plant. (*1 mark*)

Plants are often adapted to grow in bright sunshine or shade, even leaves on the same tree can develop differently depending on the conditions they are exposed to.

The graph shows the changes in rate of photosynthesis in sun and shade leaves at different light intensities.

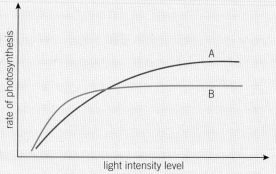

b (i) Suggest how leaves on the same plant can be exposed to different conditions that might affect leaf development. (*2 marks*)

(ii) State which curve in the diagram is most adapted to photosynthesis in direct sunlight. (*1 mark*)

(iii) Describe, making reference to cells, leaves and whole plants, how plants can be adapted to different light conditions. (*5 marks*)

3 The leaves of flowering plants have the ability to develop differently, depending on environmental conditions such as the amount of sun or shade a leaf receives.

A student carried out an investigation into sun and shade leaves from different parts of the same plant. Their observations and results are shown in the table.

Type of leaf	Number of leaves studied	Mean no. of stomata per mm² on lower surface	Mean thickness of leaf (µm)	Cuticle
sun	55	170	208	thick
shade	8	92	93	thin

a Calculate the percentage difference in the mean thickness of the sun leaves compared to the shade leaves.

Show your working. *(2 marks)*

b Suggest and explain one benefit of the greater mean number of stomata per mm² on the lower surface of the sun leaves. *(2 marks)*

c Describe two ways in which the student could improve her investigation. *(2 marks)*

OCR F214 2011

4 One way to determine the rate of photosynthesis is to measure the uptake of carbon dioxide.

a Discuss why measuring carbon dioxide uptake may or may not give a better indication of photosynthetic activity than measuring oxygen production. *(2 marks)*

b The graph shows the relationship between light intensity and the relative carbon dioxide uptake and production in a plant.

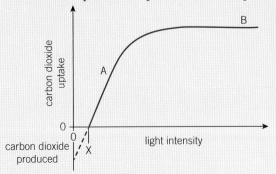

(i) State the factor limiting the rate of photosynthesis at **A** on the graph. *(1 mark)*

(ii) Suggest one factor that may limit the rate of photosynthesis at **B**. *(1 mark)*

(iii) Carbon dioxide is given off by the plant when the light intensity is lower than **X**.

Name the process that **produces** carbon dioxide in the plant. *(1 mark)*

(iv) With reference to the graph, explain the biochemical processes that are occurring in the plant:

- as light intensity increases from 0 (zero) to **X**.
- at light intensity **X**.
- at light intensities greater than **X**. *(3 marks)*

c **(i)** Name the products of the light-dependent stage of photosynthesis. *(3 marks)*

(ii) Paraquat is a weed killer. It binds with electrons in photosystem I.

Suggest how paraquat results in the death of a plant. *(2 marks)*

OCR F214 2012

Cellular respiration

Glucose is a hexose (six-carbon sugar) produced during photosynthesis. It is a complex molecule containing energy absorbed from sunlight 'trapped' within its carbon hydrogen bonds. Respiration is essentially the reverse of photosynthesis.

The carbon framework of glucose is broken down and the carbon-hydrogen bonds broken. The energy released is then used in the synthesis of ATP by chemiosmosis. ATP, the universal energy currency, is constantly synthesised and used in energy-requiring reactions and processes.

Respiration is a complex multi-step reaction pathway (Figure 1). You will be considering respiration in eukaryotic cells. A similar process takes place in prokaryotic cells but they do not have mitochondria so many of the reactions take place on cell membranes. To make the biochemistry of respiration clearer to understand you will look at it in stages, but it is important to remember that in the cell the process is continuous. The first stage of respiration is glycolysis.

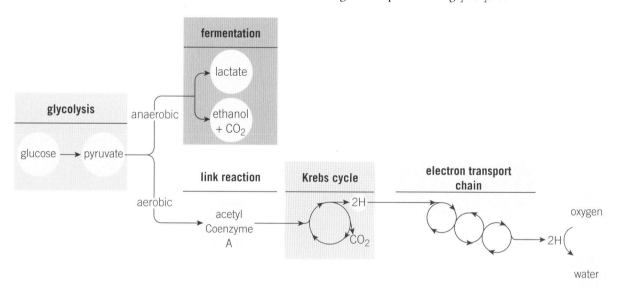

▲ Figure 1 *Summary of respiration*

Glycolysis

Glycolysis occurs in the cytoplasm of the cell. It does not require oxygen – it is an **anaerobic** process. Glucose, a six-carbon sugar, is split into two smaller, three-carbon pyruvate molecules. ATP and reduced nicotinamide adenine dinucleotide (NAD) are also produced. Glycolysis, summarised here, actually involves 10 reaction steps involving many enzymes.

The main steps in glycolysis are:

1 *Phosphorylation* – the first step of glycolysis requires two molecules of ATP. Two phosphates, released from the two ATP molecules, are attached to a glucose molecule forming **hexose bisphosphate.**

2 *Lysis* – this destabilises the molecule causing it to split into two **triose phosphate** molecules.

3 *Phosphorylation* – another phosphate group is added to each triose phosphate forming two triose bisphosphate molecules. These phosphate groups come from free inorganic phosphate (P_i) ions present in the cytoplasm.

4 *Dehydrogenation and formation of ATP* – the two triose bisphosphate molecules are then

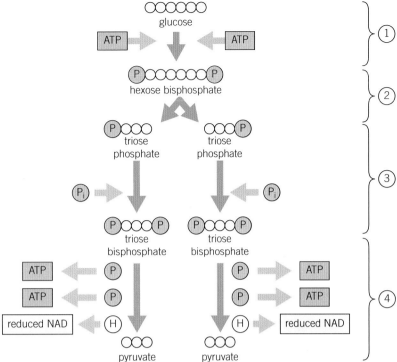

▲ **Figure 2** *Summary of glycolysis*

oxidised by the removal of hydrogen atoms (**dehydrogenation**) to form two **pyruvate** molecules. NAD coenzymes accept the removed hydrogens – they are reduced, forming two reduced NAD molecules.

At the same time, four ATP molecules are produced using phosphates from the triose bisphosphate molecules.

This is an example of substrate level phosphorylation – the formation of ATP without the involvement of an electron transport chain. ATP is formed by the transfer of a phosphate group from a phosphorylated intermediate (in this case triose bisphosphate) to ADP.

Two ATP molecules are used to prime the process at the beginning, and four ATP molecules are produced, so the overall net ATP yield from glycolysis is two molecules of ATP.

The reduced NAD is used in a later stage to synthesise more ATP.

Synoptic link

It may be useful to look back to Topic 4.4, Cofactors, coenzymes, and prosthetic groups and Topic 3.11, ATP.

Study tip

Hexose bisphosphate, triose phosphate, and pyruvate are the only compounds that you have to be able to recall.

Summary questions

1 Describe the processes of dehydrogenation and phosphorylation. (*3 marks*)

2 Explain how NAD acts as a coenzyme in glycolysis. (*3 marks*)

3 Explain the meaning of substrate-level phosphorylation. (*3 marks*)

4 ⚙️ Outline the importance of dehydrogenation and phosphorylation in glycolysis. (*5 marks*)

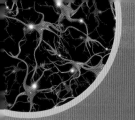

18.2 Linking glycolysis and the Krebs cycle

Specification reference: 5.2.2

As you have learnt, glycolysis takes place in the cytoplasm of the cell. In eukaryotic cells the remaining aerobic (oxygen-requiring) reactions of cellular respiration take place inside the mitochondria.

outer mitochondrial membrane separates the contents of the mitochondrion from the rest of the cell, creating a cellular compartment with ideal conditions for aerobic respiration

matrix contains enzymes for the Krebs cycle and the link reaction, also contains mitochondrial DNA

intermembrane space Proteins are pumped into this space by the electron transport chain. The space is small so the concentration builds up quickly

inner mitochondrial membrane contains electron transport chains and ATP synthase

cristae are projections of the inner membrane which increase the surface area available for oxidative phosphorylation

▲ Figure 1 *In the presence of oxygen, aerobic respiration occurs in the mitochondria of eukaryotic cells*

Oxidative decarboxylation (the link reaction)

The first step in aerobic respiration is oxidative decarboxylation. This is sometimes referred to as the link reaction, because it is the step that links anaerobic glycolysis, occurring in the cytoplasm, to the aerobic steps of respiration, occurring in the mitochondria.

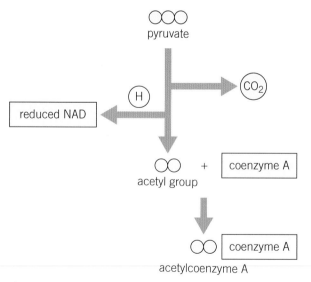

▲ Figure 2 *Summary of oxidative decarboxylation, linking the anaerobic and aerobic stages of cellular respiration in eukaryotic cells*

In eukaryotic cells, pyruvate enters the **mitochondrial matrix** by active transport via specific carrier proteins. Pyruvate then undergoes oxidative decarboxylation – carbon dioxide is removed (**decarboxylation**) along with hydrogen (oxidation). The hydrogen atoms removed are accepted by NAD. NAD is reduced to form **NADH** (reduced NAD). The resulting two-carbon acetyl group is bound by **coenzyme A** forming **acetylcoenzyme A (acetyl CoA)**. The International Union of Pure and Applied Chemistry (IUPAC) name for the acetyl group is the ethanoyl group, but the terms acetyl and acetylcoenzyme A are so widely known and used by biologists that the traditional names are retained.

Acetyl CoA delivers the acetyl group to the next stage of aerobic respiration, known as the Krebs cycle. The reduced NAD is used in oxidative phosphorylation to synthesise ATP (Topic 18.4, Oxidative phosphorylation).

Acetyl groups are now all that is left of the original glucose molecules. The carbon dioxide produced will either diffuse away and be removed from the organism as a metabolic waste or, in autotrophic organisms, it may be used as a raw material in photosynthesis.

Summary questions

1 Explain why the removal of carbon dioxide in the link reaction is called oxidative. *(2 marks)*

2 Name one organic compound and one inorganic compound produced in the link reaction. *(2 marks)*

3 Copy and complete the equation. *(3 marks)*

_____ + CoA + NAD → _____ + CO_2 + _____

4 Suggest why glycolysis occurs in the cytoplasm but not the mitochondrial matrix. *(4 marks)*

Synoptic link

You learnt about coenzymes in Topic 4.4, Cofactors, coenzymes, and prosthetic groups.

Study tip

Remember that each glucose molecule produces two pyruvate molecules which both go through the subsequent stages.

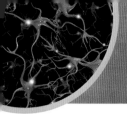

The Krebs cycle also takes place in the mitochondrial matrix and each complete cycle results in the breakdown of an acetyl group. Acetyl groups are all that remain of the glucose that entered glycolysis. It is another complex multi-step pathway, summarised here.

As in the previous stages, the Krebs cycle involves decarboxylation, dehydrogenation, and substrate-level phosphorylation. The hydrogen atoms released are picked up by the coenzymes NAD and flavin adenine dinucleotide (**FAD**). Carbon dioxide is a by-product of these reactions and the ATP produced is available for use by energy-requiring processes within the cell.

The reduced NAD and reduced FAD produced are used in the final, oxygen-requiring step of aerobic respiration (Topic 18.4, Oxidative phosphorylation) to produce large quantities of ATP by chemiosmosis.

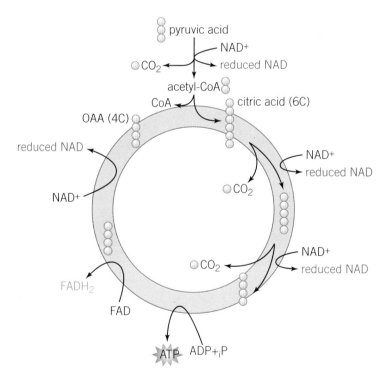

▲ **Figure 1** *The Krebs cycle*

1   Acetyl CoA delivers an acetyl group to the Krebs cycle. The two-carbon acetyl group combines with four-carbon **oxaloacetate** to form six-carbon **citrate**.

2   The citrate molecule undergoes decarboxylation and dehydrogenation producing one reduced NAD and carbon dioxide. A five-carbon compound is formed.

3   The five-carbon compound undergoes further decarboxylation and dehydrogenation reactions, eventually regenerating oxaloacetate,

and so the cycle continues. More carbon dioxide, two more reduced NADs, and one reduced FAD are produced. ATP is also produced by substrate-level phosphorylation.

The importance of coenzymes in respiration

Respiration is a complex multi-step reaction pathway. Coenzymes are required to transfer protons, electrons, and functional groups between many of these enzyme-catalysed reactions.

Redox reactions have an important role in respiration and without coenzymes transferring electrons and protons between these reactions many respiratory enzymes would be unable to function.

NAD and FAD are both coenzymes that accept protons and electrons released during the breakdown of glucose in respiration. The differences between these two enzymes are:

- NAD takes part in all stages of cellular respiration but FAD only accepts hydrogens in the Krebs cycle
- NAD accepts one hydrogen and FAD accepts two hydrogens
- reduced NAD is oxidised at the start of the electron transport chain releasing protons and electrons while reduced FAD is oxidised further along the chain
- reduced NAD results in the synthesis of three ATP molecules but reduced FAD results in the synthesis of only two ATP molecules.

You will see reduced NAD represented in a number of ways – for example, NADH, NADH + H$^+$, or NADH$_2^+$. The reason for this is that NAD is actually charged so is more accurately represented as NAD$^+$. When NAD$^+$ is reduced it accepts two protons and an electron pair (from a C–H bond) forming NADH + H$^+$. NADH, or reduced NAD, then transfers the proton and electron pair to a subsequent reaction.

Coenzymes are usually derived from vitamins. This is why, although coenzymes are mostly recycled, vitamins are an essential micronutrient.

Study tip

Oxaloacetate and citrate are the only names of Krebs cycle intermediate compounds that you need to remember.

The number of ATP molecules produced as a result of reduced NAD and reduced FAD can vary.

Summary questions

1 Compare the structures of ATP and NAD. (3 marks)

2 ATP can be described as a coenzyme. Explain why. (2 marks)

3 Draw a simple diagram summarising the breakdown of glucose to carbon dioxide and reduced coenzymes. (4 marks)

4 Calculate the number of ATP molecules produced by substrate-level phosphorylation after two rounds of the Krebs cycle. (2 marks)

5 The Krebs cycle does not use oxygen at any point. Suggest why the Krebs cycle is termed aerobic. (4 marks)

6 Suggest a reason for the involvement of FAD rather than NAD in only one specific step of the Krebs cycle. (6 marks)

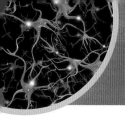

18.4 Oxidative phosphorylation

Specification reference: 5.2.2

Learning outcomes

Demonstrate knowledge, understanding, and application of:

→ the process and site of oxidative phosphorylation.

Synoptic link

You learnt about chemiosmosis in Topic 17.2, ATP synthesis.

Oxidative phosphorylation

The hydrogen atoms that have been collected by the coenzymes NAD and FAD are delivered to electron transport chains present in the membranes of the cristae of the mitochondria.

The hydrogen atoms dissociate into hydrogen ions and electrons. The high energy electrons are used in the synthesis of ATP by chemiosmosis. Energy is released during redox reactions as the electrons reduce and oxidise electron carriers as they flow along the electron transport chain. This energy is used to create a proton gradient leading to the diffusion of protons through ATP synthase resulting in the synthesis of ATP.

At the end of the electron transport chain the electrons combine with hydrogen ions and oxygen to form water. Oxygen is the final electron acceptor and the electron chain cannot operate unless oxygen is present. Respiration which involves the complete breakdown of glucose is therefore an aerobic process.

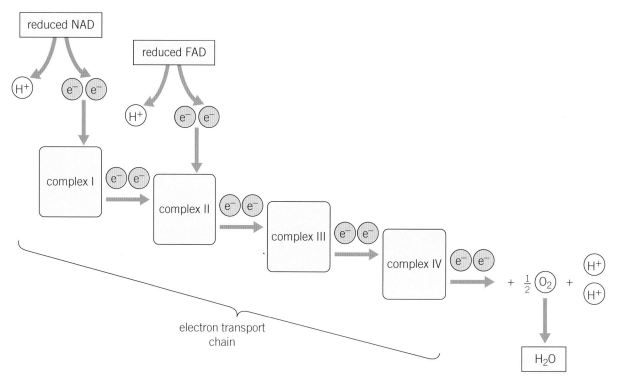

▲ Figure 1 *Electron transport chain in respiration*

Synoptic link

You learnt about coenzymes in Topic 4.4, Cofactors, coenzymes.

The phosphorylation of ADP to form ATP is dependent on electrons moving along electron transport chains. This requires the presence of oxygen and is known as oxidative phosphorylation.

The hydrogens released from NAD and FAD could combine directly with oxygen, releasing energy from the formation of bonds during the production of water. However, this energy could not be used to synthesise ATP. Heat released in the exothermic reaction would simply raise the temperature of the cell.

Substrate level phosphorylation

Substrate level phosphorylation is the production of ATP involving the transfer of a phosphate group from a short-lived, highly reactive intermediate such as creatine phosphate. This is different from oxidative phosphorylation which couples the flow of protons down the electrochemical gradient through ATP synthase to the phosphorylation of ADP to produce ATP.

Summary questions

1   Explain why hydrogens have to be actively pumped across the membrane from the matrix and return to the matrix by diffusion through ATP synthase. (*4 marks*)

2   Explain why the electrons released from reduced FAD lead to the synthesis of less ATP than the electrons released from reduced NAD. (*4 marks*)

3   Cyanide is a respiratory poison. It attaches to the iron in the haem group of cytochrome c oxidase in complex IV of the electron transport chain.
    Suggest an explanation for the toxicity of cyanide. (*4 marks*)

4   ⚙ Explain, with reasons, whether you agree with the following statements. (*6 marks*)
    * ATP synthase is not actually part of the electron transport chain.
    * Oxygen is required for the transfer of electrons along the electron transport chain.
    * Hydrogen ions return to the matrix by facilitated diffusion.

Study tip

Oxidative phosphorylation is a process that occurs along the electron transport chain, which involves a series of membrane-bound enzyme complexes.

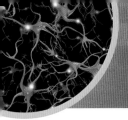

Aerobic respiration was not possible when life began, as there was no oxygen present in the atmosphere of the Earth at that time. It is a relatively new process in evolutionary terms but as it is a far more efficient process than **anaerobic respiration** it was rapidly selected for. Aerobic respiration produces around 38 molecules of ATP per glucose molecule whereas fermentation (a form of anaerobic respiration) only produces two molecules of ATP (net).

Anaerobic respiration in eukaryotic organisms

Eukaryotic cells respire aerobically if enough oxygen is available. Anaerobic respiration, resulting in the synthesis of smaller quantities of ATP, occurs in the absence of oxygen and is also used when oxygen cannot be supplied fast enough to respiring cells. The use of this less efficient process to produce ATP is a temporary 'emergency' measure to keep vital processes functioning.

Organisms fall into different categories determined by their dependence on oxygen or not:

● **obligate anaerobes** – cannot survive in the presence of oxygen. Almost all obligate anaerobes are prokaryotes, for example, *Clostridium* (bacteria that cause food poisoning), although there are some fungi as well.

● **facultative anaerobes** – synthesise ATP by aerobic respiration if oxygen is present, but can switch to anaerobic respiration in the absence of oxygen, for example, yeast.

● **obligate aerobes** – can only synthesise ATP in the presence of oxygen, for example, mammals. The individual cells of some organisms, such as muscle cells in mammals, can be described as facultative anaerobes because they can supplement ATP supplies by employing anaerobic respiration in addition to aerobic respiration when the oxygen concentration is low. However, this is only for short periods and oxygen is eventually required. The shortfall of oxygen during the period of anaerobic respiration produces compounds that have to be broken down when oxygen becomes available again, so the organism as a whole is an obligate aerobe.

Fermentation

Fermentation (a form of anaerobic respiration) is the process by which complex organic compounds are broken down into simpler inorganic compounds without the use of oxygen or the involvement of an electron transport chain. The organic compounds, such as glucose, are not fully broken down so fermentation produces much less ATP than aerobic respiration. The small quantity of ATP produced is synthesised by substrate-level phosphorylation alone.

The end products of fermentation differ depending on the organism. **Alcoholic fermentation** occurs in yeast and some plant root cells.

Here the end products are ethanol (an alcohol) and carbon dioxide. **Lactate fermentation** results in the production of lactate and is carried out in animal cells.

When there is no oxygen to act as the final electron acceptor at the end of the electron transport chain in oxidative phosphorylation, the flow of electrons stops. This means the synthesis of ATP by chemiosmosis also stops.

As the flow of electrons along the electron transport chain has stopped, the reduced NAD and reduced FAD are no longer able to be oxidised because there is nowhere for the electrons to go. This means NAD and FAD cannot be regenerated and so the decarboxylation and oxidation of pyruvate and the Krebs cycle comes to a halt as there are no coenzymes available to accept the hydrogens being removed.

Glycolysis would also come to halt due to the lack of NAD if it were not for the process of fermentation.

Lactate fermentation in mammals

In mammals, pyruvate can act as a hydrogen acceptor taking the hydrogen from reduced NAD, catalysed by the enzyme **lactate dehydrogenase**. The pyruvate is converted to lactate (lactic acid) and NAD is regenerated. This can be used to keep glycolysis going so a small quantity of ATP is still synthesised. In mammals in particular, anaerobic respiration in the muscles is often supported by ATP from aerobic respiration, which is still being produced as fast as oxygen can be delivered in other parts of the body.

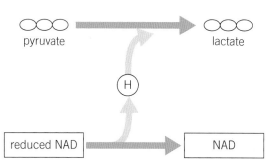

▲ Figure 1 *Summary of lactate fermentation in mammals*

Lactic acid is converted back to glucose in the liver but oxygen is needed to complete this process. This is the reason for the oxygen debt (and the need to breathe heavily) after exercise.

Lactate fermentation cannot occur indefinitely for two main reasons:

- the reduced quantity of ATP produced would not be enough to maintain vital processes for a long period of time

- the accumulation of lactic acid causes a fall in pH leading to proteins denaturing. Respiratory enzymes and muscle filaments are made from proteins and will cease to function at low pH.

Lactic acid is removed from muscles and taken to the liver in the bloodstream. One of the main aims when improving physical fitness is to increase the blood supply and flow, through muscles. This increases the rate of lactic acid removal allowing the intensity and duration of exercise to be increased.

Alcoholic fermentation in yeast (and many plants)

Alcoholic fermentation is not a reversible process like lactate fermentation. Pyruvate is first converted to ethanal, catalysed by the enzyme pyruvate decarboxylase. Ethanal can then accept a hydrogen atom from reduced NAD, becoming ethanol. The regenerated NAD can then continue to act as a coenzyme and glycolysis can continue.

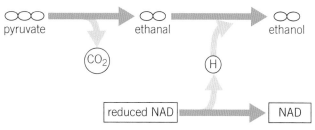

▲ Figure 2 *Summary of alcoholic fermentation in yeast and plant cells*

This is not a short-term process and can continue indefinitely in the absence of oxygen. Ethanol is a toxic waste product to yeast cells and they are unable survive if the ethanol accumulates above approximately 15%. This is allowed to happen during the production of alcohol in brewing or wine making.

Investigation into respiration rates in yeast

The apparatus shown could be used to measure the rate of carbon dioxide production of a yeast suspension. This will be equivalent to the rate of anaerobic respiration or alcoholic fermentation of the yeast cells.

The glucose in solution provides a respiratory substrate. The flask is sealed during the experiment to ensure anaerobic conditions.

As the yeast respires carbon dioxide is released increasing the volume of gas in the flask. As the volume of gas in the tube increases the pressure will increase causing the coloured liquid to move along the capillary tube. The distance moved by the liquid together with diameter of the tube can be used to calculate the increase in volume of gas (carbon dioxide) in the flask over a certain time. This is a measure of the rate of respiration.

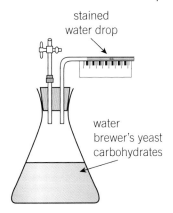

◀ **Figure 3** *Diagram showing the apparatus used to measure the anaerobic respiration in yeast*

Data logging
Respiration is not 100% efficient and energy is lost as heat when organisms respire.

When yeast respires it produces heat which will increase the temperature of a solution containing yeast. Sensors can be used to measure changes in temperature.

A student carried out an investigation into respiration in yeast using a data logger to measure the changes in carbon dioxide concentration as a measure of the rate of respiration.

The student set up the apparatus to be used as shown in Figure 4.

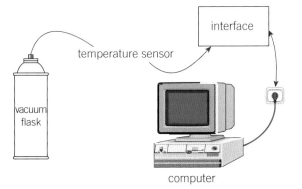

▲ **Figure 4**

The student placed a solution containing yeast and glucose in the flask and inserted a carbon dioxide sensor.

The solution was covered with a layer of liquid paraffin.

The software was set up to record readings every 50 seconds for 1 600 seconds.

The readings were displayed in graphical form shown in Figure 5.

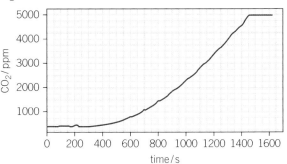

▲ **Figure 5**

1 Explain why the student did the following:
   Carried out the investigation in a vacuum flask
   Covered the solution containing the yeast with liquid paraffin
2 Calculate, using Figure 5, the fastest rate of respiration observed during the investigation.
3 Suggest why the graph eventually reached a plateau.
4 Describe how the apparatus could be adapted to investigate aerobic respiration.

Small-scale and large-scale adaptations to low oxygen environments

Many animals live in or around water and spend time underwater to hunt for food. These animals are adapted in a variety of ways to survive periods of anaerobic respiration while they cannot breathe air. Many bacteria also live in low oxygen environments. There are many adaptations that have evolved in different organisms to overcome the problems of a temporary or permanent lack of oxygen:

Bacterial adaptations

Different groups of bacteria have evolved to use nitrate ions, sulphate ions, and carbon dioxide as final electron acceptors in anaerobic respiration. This enables them to live in very low, or zero, oxygen environments.

Anaerobic bacteria present in the digestive systems of animals play an essential role in the breakdown of food and absorption of minerals. Methanogens are a type of bacteria found in the digestive system of ruminants, such as cows. They digest cellulose from grass cell walls into products that can be further digested, absorbed and used by the ruminants. The final electron acceptor in the respiratory pathway of these bacteria is carbon dioxide, and methane and water are produced. The methane builds up and eventually has to be released – it has been estimated that a cow produces around 500 L of methane per day.

Mammalian adaptations

Marine mammals that dive for long periods, such as seals and whales, have a range of different types of adaptations for surviving when they cannot take in more oxygen:

1   *Biochemical adaptations* include greater concentrations of haemoglobin and myoglobin than land mammals, particularly in the muscles used in swimming. This maximises their oxygen stores, delaying the onset of anaerobic metabolism. Whales have a higher tolerance to lactic acid than human beings, so they can respire anaerobically much longer without suffering tissue damage. They also have a greater tolerance of high carbon dioxide levels – they have very effective blood buffering systems that prevent a catastrophic rise in pH.

▲ **Figure 6** *Sperm whale (*Physeter macrocephalus*) tail*

2   *Physiological adaptations* in many diving mammals include a modified circulatory system. When they dive they show peripheral vasoconstriction, so blood is shunted to the brain, heart, and muscles. The heart slows by up to 85% – this is known as bradycardia and reduces the energy demand of the heart muscle. Whales also exchange 80–95% of the air in the lungs when they breathe – in humans, that figure is around 15%. In some species dives can last up to two hours, so the adaptations are very effective.

3   *Physical adaptations* include streamlining to reduce drag due to friction from water while swimming, therefore reducing the energy demand during a dive. The limbs of marine mammals are 'fin-shaped' to maximise the efficient use of energy in propulsion.

1   The lungs of whales are proportionally no larger than humans but some whales can stay under water for two hours. Suggest how the lungs might be adapted to enable these long dives and why larger lungs would be a disadvantage to a whale.

2   Summarise the adaptations of whales for making long underwater dives.

Synoptic link

It may be useful to look back at Topic 3.11, ATP.

Study tip

Lactate fermentation is important because it regenerates NAD which allows glycolysis to continue. Lactate, or lactic acid, is not a waste product, it is recycled as glucose.

Summary questions

1 Explain why yeast cells are described as facultative anaerobes. (*2 marks*)

2 Describe why alcoholic fermentation can be described as having more in common with aerobic respiration than with fermentation. (*3 marks*)

3 Explain why the build-up of lactic acid eventually stops muscle contraction which we experience as fatigue. (*4 marks*)

4 Glycolysis, the anaerobic stage of respiration, is the only source of ATP in red blood cells. Cardiac muscle is adapted to reduce the chances of anaerobic respiration ever being needed.
Outline the benefits to red blood cells and cardiac muscle of the different types of respiration they undertake. (*6 marks*)

18.6 Respiratory substrates

Specification reference: 5.2.2

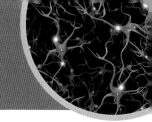

Different respiratory substrates

Glucose is not the only organic molecule that is broken down to release energy for the synthesis of ATP. There are many other **respiratory substrates**. Triglycerides are hydrolysed to fatty acids, which enter the Krebs cycle via acetyl CoA and glycerol.

Glycerol is first converted to pyruvate before undergoing oxidative decarboxylation, producing an acetyl group which is picked up by coenzyme A, forming acetyl CoA. The fatty acids in a triglyceride molecule can lead to the formation of as many as 50 acetyl CoA molecules, resulting in the synthesis of up to 500 ATP molecules.

Gram for gram, lipids store and release about twice as much energy as carbohydrates. Alcohol contains more energy than carbohydrates but less than lipids. Proteins are roughly equivalent to carbohydrates.

Proteins first have to be hydrolysed to amino acids and then the amino acids have to be deaminated (removal of amine groups) before they enter the respiratory pathway, usually via pyruvate. These steps require ATP, reducing the net production of ATP.

The **respiratory quotient (RQ)** of a substrate is calculated by dividing the volume of carbon dioxide released by the volume of oxygen taken in during respiration of that particular substrate. This is measured using a simple piece of apparatus called a respirometer (Figure 2).

$$RQ = \frac{CO_2 \text{ produced}}{O_2 \text{ consumed}}$$

It takes six oxygen molecules to completely respire one molecule of glucose and this results in the production of six molecules of carbon dioxide (and six molecules of water). This results in an RQ of 1.0.

Lipids contain a greater proportion of carbon–hydrogen bonds than carbohydrates which is why they produce so much more ATP in respiration. Due to the greater number of carbon–hydrogen bonds, lipids require relatively more oxygen to break them down and release relatively less carbon dioxide. This results in RQs of less than one for lipids. The structure of amino acids leads to RQs somewhere between carbohydrates and lipids.

- carbohydrates = 1.0
- protein = 0.9
- lipids = 0.7

So, by measuring the volume of oxygen taken in and carbon dioxide released, and calculating RQ, the type of substrate being used for respiration at that point can be roughly determined.

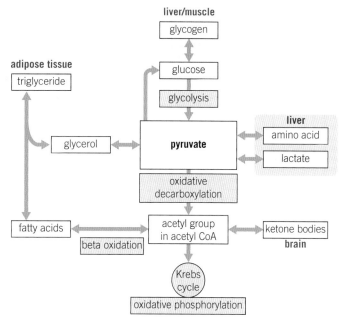

▲ **Figure 1** *The diagram shows how different biological molecules, and fragments of these molecules, can be 'fed' into the respiratory pathway at different points. Some tissues, or organs, are better adapted to use certain substrates. For example, the brain is well adapted to use ketone bodies as a respiratory substrate*

During normal activity, the RQ is in the range of 0.8 to 0.9, showing that carbohydrates and lipids (and probably some proteins) are being use as respiratory substrates.

During anaerobic respiration, the RQ increases above 1.0, although this not easy to measure as the point at which anaerobic respiration begins is not easy to pinpoint.

Calculating the respiratory quotient

The oxygen taken in and carbon dioxide produced by a desert locust was measured experimentally under different conditions:

Conditions	at rest	in flight
Oxygen absorbed / au	10.5	160.0
Carbon dioxide produced / au	10.0	113.6

1 Calculate the RQ of the insect at rest and in flight.
2 What does this suggest about the substrate being respired at rest and in flight?

Low carbohydrate diets

Many people choose low carbohydrate diets when they want to lose weight – and in particular to lose some body fat. The diets can work – but the science suggests that you need to think carefully before cutting out the molecules that are most commonly used as fuel in your body. Here are some of the facts about low carbohydrate diets for you to consider:

1 Triglycerides are hydrolysed into fatty acids and glycerol. The fatty acids are broken down in the mitochondria to give many two-carbon acetyl groups that combine with coenzyme A molecules and enter the Krebs cycle.
2 Triglycerides cannot act as the only respiratory substrate. Carbohydrates are needed to keep the Krebs cycle going so that acetyl groups from the breakdown of fatty acids can be 'fed in'. If carbohydrates are in short supply the body will make them using a process called gluconeogenesis. This process often uses glycerol, but it may also use pyruvate from glycolysis.
3 Oxaloacetate from the Krebs cycle can be used to make glucose when carbohydrate levels are low. Reducing the number of oxaloacetate molecules in the Krebs cycle reduces the rate at which the acetyl groups produced during the breakdown of lipids can be fed into the cycle and produce ATP.
4 Oxaloacetate can be replaced by the conversion of pyruvate from carbohydrate breakdown in the mitochondria. Pyruvate is also synthesised using glycerol from the breakdown of lipids. However, the breakdown of a lipid molecule provides a relatively small quantity of glycerol and so a relatively small

amount of pyruvate. This means carbohydrates are still needed to ensure the continued respiration of fat.
5 Proteins can be hydrolysed into amino acids which are then deaminated in the liver. The remaining keto acids can be converted into glucose molecules. Lean muscle is the protein of choice in this process, so a low carbohydrate diet can lead to the breakdown of muscle tissue. The liver and kidneys also have to remove the nitrogenous waste.
6 If the level of acetyl CoA increases because it is not being taken into the Krebs cycle, the liver starts converting it into ketone bodies. Brain cells normally require glucose as an energy source. They cannot use fatty acids as a respiratory substrate but they can use ketone bodies.
7 When the body is producing more ketone bodies than usual, it is said to be in ketosis. This can lead to a dangerous condition known as ketoacidosis. Ketoacidosis is the result of an accumulation of ketone bodies which cause the pH level of the blood to drop to dangerous, or even fatal, levels. This condition is often seen in alcoholics, untreated diabetes, and during starvation. It is often diagnosed by the fruity smell of propanone (acetone), a breakdown product of ketone bodies, on the breath of an affected person.

1 The term 'fats burn in the flame of carbohydrates' was coined more than a century ago. Discuss the accuracy of this statement.
2 Evaluate the benefits and drawbacks of a low carbohydrate diet.

Practical investigations into the factors affecting rate of respiration using respirometers

A student carried out an experiment to investigate the effect of temperature on the rate of respiration in soaked (germinating) pea seeds and dry (dormant) pea seeds.

A respirometer was used, shown in Figure 2. The potassium hydroxide solution in this apparatus absorbs carbon dioxide. If the apparatus is kept at a constant temperature, any changes in the volume of air in the respirometer will be due to oxygen uptake. The student set up three respirometers, A, B and C in water baths at two different temperatures. The respirometers were left for 10 minutes to equilibrate.

The contents of each respirometer are shown in Table 1.

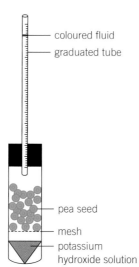

coloured fluid
graduated tube

pea seed
mesh
potassium
hydroxide solution

▲ **Figure 2**

▼ **Table 1**

Temperature [°C]	Respirometer	Contents
15	A	30 soaked pea seeds
	B	glass beads + 30 dry pea seeds
	C	glass beads
25	A	30 soaked pea seeds
	B	glass beads + 30 dry pea seeds
	C	glass beads

At each temperature, respirometer C, which contained only glass beads, was a control. After the student had left each respirometer to equilibrate, a small volume of coloured fluid was introduced into each graduated tube.

The respirometers were then left in the appropriate water baths for 20 minutes and maintained at the correct temperature. During this time, the coloured fluid in the graduated tube moved. The level of the coloured fluid in each respirometer was recorded at the start of the experiment and after 20 minutes. The results are summarised in Table 2.

▼ **Table 2**

Temperature °C	Respirometer	Reading at start [cm³]	Reading after 20 minutes [cm³]	Difference [cm³]	Corrected difference [cm³]	Rate of oxygen uptake [cm³ min⁻¹]
15	A	0.93	0.74	0.19	0.16	0.008
	B	0.93	0.86	0.07	0.04	0.002
	C	0.91	0.88	0.03		
25	A	0.94	0.63	0.31	0.27	
	B	0.93	0.84	0.09	0.05	0.003
	C	0.95	0.91	0.04		

1 Copy and complete the table.
2 Suggest why, at each temperature, respirometer B contained some glass beads.
3 Suggest how the student determined the quantity of glass beads to place in respirometer B at each temperature.
4 Explain why there is an increased rate of respiration in soaked seeds at 25°C compared with soaked seeds at 15°C.
5 Suggest a reason for the difference in the rate of respiration between soaked and dry pea seeds.

Study tip

A respirometer is not the same as a spirometer. A spirometer measures volume changes during breathing but a respirometer measures the change in volume of oxygen or carbon dioxide.

Summary questions

1 Outline the respiration pathway of a triglyceride. *(4 marks)*

2 Describe the difference between a respirometer and a spirometer. *(3 marks)*

3 Consider the three respiratory substrates shown here.
Calculate the percentage of hydrogen in each molecule and use these to compare the relative energy values of the substrates. *(6 marks)*

A CH₂OH

B

C CH₃(CH₂)₁₆COOH

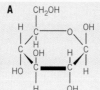

Practice questions

1 a (i) State the meaning of the term phosphorylation. (*1 mark*)

   (ii) Outline the similarities and differences between oxidative phosphorylation and substrate level phosphorylation. (*3 marks*)

b NADH and FADH transfer electrons to the electron transport chain.

> 10 protons are pumped due to the electrons from NADH
>
> 6 protons are pumped due to the electrons from FADH
>
> 4 protons are needed by ATP synthase to make one ATP molecule

   (i) Calculate the number of ATP molecules produced due to each reduced coenzyme. Show your working. (*3 marks*)

   (ii) Calculate the total number of ATP molecules produced due to oxidative phosphorylation. Show your working in your answer. (*3 marks*)

2 a The diagram represents the first stage of respiration.

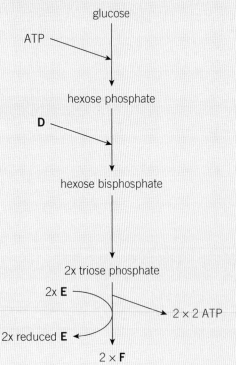

(i) Name the stage represented by the figure. (*1 mark*)

(ii) State precisely where in the cell this stage takes place. (*1 mark*)

(iii) Identify the compounds **D**, **E**, and **F**. (*3 marks*)

b In anaerobic conditions, compound **F** does not proceed to oxidative decarboxylation.

Describe the fate of compound **F** during anaerobic respiration in an animal cell **and** explain the importance of this reaction. (*5 marks*)

c The diagram shows a common seal, *Phoca vitulina*, an aquatic mammal.

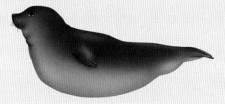

The seal comes to the surface of the water to obtain air and it can then stay underwater for over 20 minutes.

The figure shows a seal at the surface of the water and the following diagram shows the same animal when submerging again.

Suggest how the seal is adapted to respire for such a long time underwater. (*3 marks*)

OCR F214 2010

3 The diagram represents a molecule of ATP.

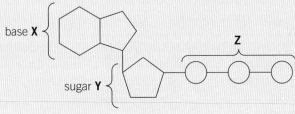

a (i) Name the parts of the ATP molecule labelled **X**, **Y**, and **Z**. (*3 marks*)

   (ii) With reference to the figure, describe and explain the role of ATP in the cell. (*3 marks*)

b The electron micrograph shows a mitochondrion from an animal cell (×31 400 magnification).

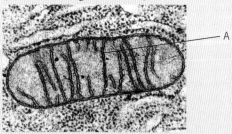

          — A

(i) Name the structure labelled **A**.

(*1 mark*)

(ii) Name the specific process that is carried out by structure **A** in the mitochondrion. (*1 mark*)

c Some animals conserve energy by entering a state of torpor (a short period of dormancy), in which they allow their body temperature to fall below normal for a number of hours.

In an investigation into torpor in the Siberian hamster, *Phodopus sungorus*, the animal's respiratory quotient (RQ) was measured before and during the period of torpor.

The respiratory quotient is determined by the following equation:

$$RQ = \frac{\text{volume of carbon dioxide produced}}{\text{volume of oxygen consumed}}$$
in the same time

RQ values for different respiratory substrates have been determined and are shown in the table.

substrate	RQ
carbohydrate	1.0
lipid	0.7
protein	0.9

Initially, the RQ value determined for the hamster was 0.95, but as the period of torpor progressed, its RQ value decreased to 0.75.

What do these values suggest about the substrates being respired by the hamster during the period of investigation? (*3 marks*)

OCR F214 2010

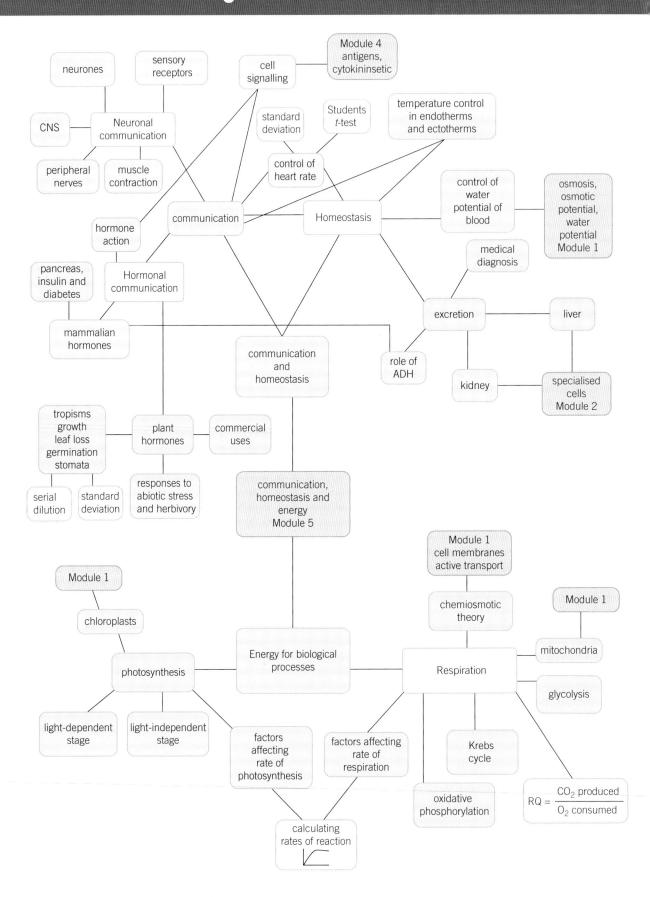

Application

A major drawback of many brain imaging techniques is that individuals need to keep at least their heads completely still. In quantitative electroencephalograms (QEEGs) sensors are attached to the outside of the skull to measure the activity of the brain as people carry out different actions. It allows scientists to build up brain maps indicating which areas are used in different activities and skills. QEEGs are not as spatially accurate as fMRI can be, however, combined with other types of brain imaging they are increasing our knowledge of how the brain works.

Recent work using QEEGs to look at the changes in the brain in children affected by autism has produced some interesting findings, and a new form of therapy. They have used QEEG to show the patterns of different brain waves in autistic brains.

While QEEGs show that the activity in some regions is particularly high, the overall rate of brain activity in people affected by autism has been shown to be lower than that in unaffected people. In fact, levels of brain activity are highest in anxious people, and lowest in people with traumatic brain injuries, but autistic patients are not much above them.

A whole range of therapies is used to help people with autism cope with everyday life. A new tool is the use of neurofeedback training, which uses information from

QEEGs to help children and adults 'retrain' their brains and control the levels of activity in different regions. There is growing evidence that for some people affected by autism this can enable them to function far more effectively and interact successfully with the people and the world around them.

1   a   Far less is understod about the brain than, for example, the heart or kidney.
        Suggest why our understanding of the brain lags behind some other organs.
    b   How have we found out what happens in the brain?
2   a   What is a feedback system? Explain how they work and give examples from the various control and communication systems in the body.
    b   Investigate what is meant by a neurofeedback system and discuss how this might be used to help people retrain their brains to work in different ways.
3   When QEEGs and other recordings of brain activity are taken it is often noted whether the eyes are open or closed.
    a   Summarise how information from open eyes reaches the brain.
    b   Suggest why it is important to record whether the eyes are open or closed.

Extension

Either investigate the main methods used for investigating the brain and make a table or poster to summarise the technology and the information it gives about the structure and function of the brain

OR investigate our current understanding of autism, including the areas and activity of the brain most

affected, the impact on functioning and examples of therapies used to help affected individuals, including at least one both drug-based intervention and neuro-feedback based on QEEGs.

Module 5 practice questions

1 Respiratory inhibitors acting on different structures within a mitochondrion were used to determine the sequence of events in respiration.

Basal respiration or basal metabolic rate (BMR) is the minimum rate of metabolism needed to maintain basic body functions.

a Name two essential functions that require energy. (2 marks)

Proton leak respiration is a measure of the energy (released by electron transfer) lost as protons leak through the membrane at points other than ATP synthase.

b Explain why isolated mitochondria still undergo a slow rate respiration in the absence of ADP. (2 marks)

ATP linked respiration is a measure of the proportion of the energy released in respiration that is used in the production of ATP in mitochondria.

The reserve capacity is the quantity of additional ATP that can be produced by oxidative phosphorylation when there is an increase in energy demand.

The human mitochondrial genome exclusively encodes 13 of the essential subunits of the ETC and ATP synthase.

c (i) State the name of the molecules coded for by the mitochondrial genome. (1 mark)

(ii) Explain the term genome. (2 marks)

(iii) Suggest why the mitochondrial reserve capacity reduces with age. (2 marks)

The diagram shows the changes in mitochondrial respiration after the addition of different inhibitors.

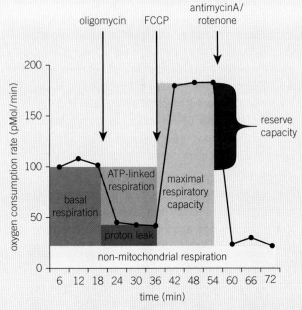

Oligomycin inhibits the final protein carrier in the electron transport chain.

d Describe the effect of Oligomycin. (2 marks)

Carbonyl cyanide-p-trifluoromethoxyphenyl-hydrazon (FCCP) is a protonophore, a lipid soluble molecule that transports ions across phospholipid bilayers.

e (i) Explain, with reference to the structure of the phospholipid bilayer, why protons can only pass through cell membranes by facilitated diffusion or active transport. (4 marks)

(ii) Describe the effect of FCCP. (3 marks)

f Name the process(es) involved and describe how ATP is produced in non-mitochondrial respiration. (5 marks)

g Describe how the following are calculated using the diagram. *(5 marks)*

Basal respiration

ATP-linked respiration

Proton leak respiration

Maximal respiratory capacity

Mitochondrial reserve capacity

Antimycin A and rotenone inhibit two more protein carriers in the electron transport chain which stops the flow of electrons.

h State what can be calculated after Antimycin A and rotenone have been added. *(2 marks)*

2 A short time after the death of a human, or animal, the joints within the organism become locked in place. This process is called rigor mortis and begins with the muscles partially contracting.

The graph shows how the amount of ATP changes in the time after death.

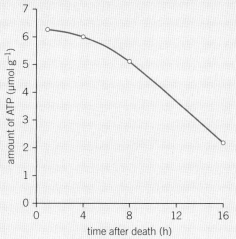

a (i) Describe the changes in ATP shown in the graph. *(4 marks)*

(ii) Explain why the amount of ATP falls after death. *(2 marks)*

(iii) Suggest the reason for the onset of rigor mortis. *(3 marks)*

(iv) Suggest why a small amount of muscle contraction can occur after death. *(3 marks)*

The graph shows how the pH changes in muscle tissue after death.

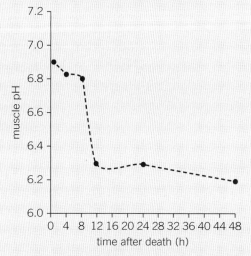

b Suggest a reason for the change in pH in muscle tissue after death. *(2 marks)*

Depending on various factors such as the temperature of the environment, rigor mortis can last up to about three days.

c (i) Suggest an explanation for why rigor mortis is only temporary. *(4 marks)*

(ii) Suggest, with explanations, which factors affect how long rigor mortis lasts. *(4 marks)*

The diagram shows a part of the sliding filament theory.

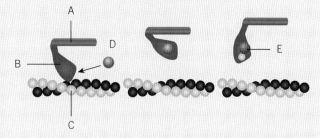

d (i) State the names of the structures A, B, and C *(3 marks)*

(ii) State the name of the molecule labelled D. *(1 mark)*

(iii) Roughly draw the diagram, and indicate, using an arrow, in which direction structure A moves relative to structure C. *(1 mark)*

(iv) Suggest, using your knowledge of the structure of proteins, the role of the ion labelled E in muscle contraction.
(*3 marks*)

(v) Outline how myosin acts as a molecular ratchet in the sliding filament theory. (*3 marks*)

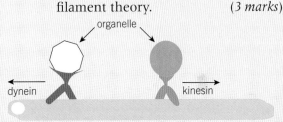

Kinesin is a motor protein that has a structure very similar to myosin. Dynein is another motor protein with a very different structure to myosin.

e Describe what determines the 3D structure of a protein and how this differs in kinesin and dynein. (*3 marks*)

Both dynein and kinesin have regions that binds to microtubules, a region that binds and hydrolyses ATP and a region that binds to an organelle.

The energy released during the hydrolysis of ATP changes their three-dimensional structures resulting in the motor proteins moving along microtubules and transporting organelles within cells. Dynein and kinesin move in opposite directions.

f (i) Compare and contrast the ways in which dyneins, kinesins, and myosin lead to movement. (*4 marks*)

(ii) State two other processes, other than the movement of organelles, in which microtubules also have an important role. (*2 marks*)

3 The chemical structure of the hormone thyroxine, which is lipid soluble, is shown in the diagram.

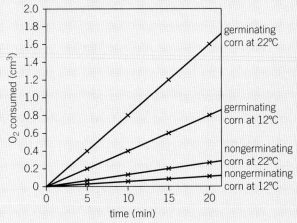

a Explain why secondary messengers are not required for thyroxine to produce a response within target cells. (*4 marks*)

b Suggest why hormones, like thyroxine, travel in the blood stream bound to proteins. (*2 marks*)

4 The oxygen consumption of germinating and non-germinating corn seeds was measured over a period of 20 minutes.

The results are shown in the graph.

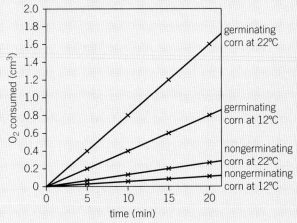

a Describe the relationship between temperature and consumption of oxygen. (*3 marks*)

b Calculate the rate of oxygen consumption for germinating and non-germinating corn at 22°C. (*2 marks*)

c Discuss whether or not non-germinating seeds respire. (*4 marks*)

5 a The rate of photosynthesis is determined mainly by environmental limiting factors. These are light intensity, the availability of carbon dioxide, and temperature. Water supply has indirect effects by influencing the availability of carbon dioxide.

(i) Define the term limiting factor. *(1 mark)*

(ii) Explain how water shortage could have an indirect effect on photosynthesis by influencing the availability of carbon dioxide. *(2 marks)*

b Some plants, such as wood sorrel, and Oxalis acetosella, nearly always grow in shade where light intensity is commonly a limiting factor for photosynthesis. They are known as shade plants.

Plants that live in open habitats, for example the daisy, Bellis perennis, are called sun plants.

The graph shows the net rate of photosynthesis of sun and shade plants in response to increasing light intensity. The net rate of photosynthesis is defined as:

mass of CO_2 fixed in photosynthesis minus mass of CO_2 produced in respiration, per unit time

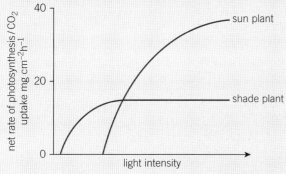

Using the graph, describe the responses of sun and shade plants to increasing light intensity. *(2 marks)*

c Leaves of wood sorrel have been shown to have very low respiration rates per unit leaf area.

The diagram shows sections through the leaves of typical sun and shade plants.

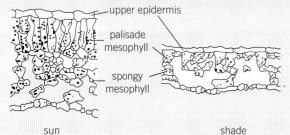

sun   shade

(i) With reference to the diagram, suggest why the leaves of wood sorrel have low rates of respiration. *(1 mark)*

(ii) Explain why a low rate of respiration in leaves is an adaptation to low light intensities. *(3 marks)*

d A large number of seedlings of common orache, Atriplex patula, were grown for several weeks in different environmental conditions as follows:

● group 1 – high light intensities

● group 2 – intermediate light intensities

● group 3 – low light intensities.

When fully grown, the net rate of photosynthesis at different light intensities was measured for each group. The results obtained are shown in the graph.

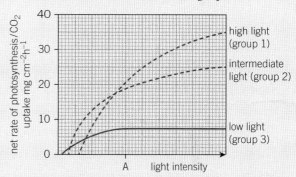

(i) State the net rate of photosynthesis for plants in groups 1 and 3 at light intensity A on the graph. *(2 marks)*

(ii) Suggest how adaptation to light intensity is controlled in A. patula. *(3 marks)*

OCR 2804 2008

MODULE 6
Genetics, evolution, and ecosystems

Chapters in this module

Demonstrate knowledge, understanding, and application of:

Introduction

In this module you will learn how genes regulate and control cell function. This will lead on to the study of heredity, evolution, and DNA manipulation. Finally, you will consider the impacts of human activities on the natural environment and biodiversity.

Genetics of living systems introduces how the genetic control of metabolic reactions determines an organism's growth, development, and function. This also includes looking at the effects of gene mutations on protein function.

Patterns of inheritance and variation allows you to study how genetic and environmental factors contribute to variation within a population. Over a prolonged period, organisms can change so significantly that new species have formed, whereas others have become extinct. Humans can use artificial selection to produce similar changes in plants and animals.

Manipulating genomes has many potential benefits such as the treatment of disease but the implications of genetic techniques are of public debate. You will find out how genomes are sequence as well as how DNA profiling is used in forensics and to determine the risk of certain diseases.

Cloning and biotechnology explores how farmers and growers exploit natural vegetative propagation in the production of uniform crops, as well as the role of scientists in the production of artificial plant and animal clones. The use of microorganisms in biotechnology to produce food drugs, and other products is also studied.

Ecosystems are dynamic and influenced by both biotic and abiotic factors. You will study the comple interactions which occur between organisms and their environment as well as finding out how materials are passed on and recycled.

Populations and sustainability investigates the factors that determine population size and the economic, social, and ethical reasons why ecosystems may need to be managed. This includes looking at how biological resources can be used sustainably to support an increasing human population.

Knowledge and understanding checklist

From your Key Stage 4 study you should be able to answer the following questions. Work through each point, using your Key Stage 4 notes and the support available on Kerboodle.

☐ Describe the process of mitosis in growth, including the cell cycle.

☐ Explain how some abiotic and biotic factors affect communities.

☐ Describe the importance of interdependence and competition in a community.

☐ Explain the role of microorganisms in the cycling of materials through an ecosystem.

☐ Describe human interactions within ecosystems and explain their impact on biodiversity.

☐ Explain some of the benefits and challenges of maintaining local and global biodiversity.

☐ Explain the following terms – genome, gamete, chromosome, gene, allele/variant, dominant, recessive, homozygous, heterozygous, genotype, and phenotype.

☐ Explain single gene inheritance and predict the results of single gene crosses.

☐ State that there is usually extensive genetic variation within a population of a species.

☐ Explain the impact of the selective breeding of food plants and domesticated animals.

☐ Describe the main steps in the process of genetic engineering.

Maths skills checklist

In this module, you will need to use the following maths skills. You can find support for these skills on Kerboodle and through MyMaths.

☐ **Ratios.** You will need to use phenotypic ratios to identify linkage and epistasis.

☐ **Chi-squared test.** You will need to use the chi-squared test to determine the significance of the difference between observed and expected results.

☐ **Hardy–Weinberg principle.** You will need to use the Hardy–Weinberg principle to calculate allele frequencies in populations.

☐ **Correlation coefficient.** You will need to use this test to consider the relationship between two sets of data. This will determine if and how the data are correlated.

MyMaths.co.uk
Bringing Maths Alive

19 GENETICS OF LIVING SYSTEMS
19.1 Mutations and variation
Specification reference: 6.1.1

Synoptic link

Look back at Topic 3.9, DNA replication and the genetic code to remind you of why the genetic code is degenerate.

Gene mutations (when DNA goes wrong)

A **mutation** is a change in the sequence of bases in DNA. Protein synthesis can be disrupted if the mutation occurs within a gene. The change in sequence is caused by the **substitution**, **deletion**, or **insertion** of one or more nucleotides (or base pairs) within a gene. If only one nucleotide is affected it is called a point mutation.

The substitution of a single nucleotide changes the codon in which it occurs. If the new codon codes for a different amino acid this will lead to a change in the primary structure of the protein. The degenerate nature of the genetic code may mean however that the new codon still codes for the same amino acid leading to no change in the protein synthesised.

The position and involvement of the amino acid in R group interactions within the protein will determine the impact of the new amino acid on the function of the protein. For example, if the protein is an enzyme and the amino acid plays an important role within the active site, then the protein may no longer act as a biological catalyst.

The insertion or deletion of a nucleotide, or nucleotides, leads to a frameshift mutation (Figure 1). The triplet code means that sequences of bases are transcribed (or read) consecutively in non-overlapping groups of three. This is the reading frame of a sequence of bases. Each group of three bases corresponds to one amino acid. The addition or deletion of a nucleotide moves, or shifts, the reading frame of the sequence of bases. This will change every successive codon from the point of mutation.

The same effect is seen however many nucleotides are added or deleted, unless the number of nucleotides changed is a multiple of three. Multiples of three correspond to full codons and therefore the reading frame will not be changed – but the protein formed will still be affected as a new amino acid is added.

normal DNA sequence:

AGT CGA TAG
codon 1 codon 2 codon 3

point mutations:

base substitution (A substituted instead of C in codon 2):

AGT **A**GA TAG
codon 1 codon 2 codon 3

frameshift mutations:

insertion (T inserted before G in codon 1):

A**T**G TCG ATA
codon 1 codon 2 codon 3

deletion (G deleted from codon 1):

ATC GAT AGC
codon 1 codon 2 codon 3

▲ **Figure 1** *The diagram shows the effect of different point mutations on codons and sequences of codons*

Effects of different mutations

- No effect – there is no effect on the phenotype of an organism because normally functioning proteins are still synthesised.

- Damaging – the phenotype of an organism is affected in a negative way because proteins are no longer synthesised or proteins synthesised are non-functional. This can interfere with one, or more, essential processes.

- Beneficial – very rarely a protein is synthesised that results in a new and useful characteristic in the phenotype. For example, a mutation in a protein present in the cell surface membranes of human cells means that the human immunodeficiency virus (HIV) cannot bind and enter these cells. People with this mutation are immune to infection from HIV.

Causes of mutations

Mutations can occur spontaneously, often during DNA replication, but the rate of mutation is increased by **mutagens**. A mutagen is a chemical, physical, or biological agent which causes mutations (Table 1).

The loss of a purine base (depurination) or a pyrimidine base (depyrimidination) often occurs spontaneously. The absence of a base can lead to the insertion of an incorrect base through complementary base pairing during DNA replication.

Free radicals, which are oxidising agents, can affect the structures of nucleotides and also disrupt base pairing during DNA replication. Antioxidants, such as vitamins A, C, and E (found in fruit and vegetables), are known anticarcinogens because of their ability to negate the effects of free radicals.

▼ Table 1 *Summary of some of the main mutagens and what they do*

Physical mutagens	ionizing radiations such as X-rays	break one or both DNA strands – some breaks can be repaired but mutations can occur in the process
Chemical mutagens	deaminating agents	chemically alter bases in DNA such as converting cytosine to uracil in DNA, changing the base sequence
Biological agents	alkylating agents	methyl or ethyl groups are attached to bases resulting in the incorrect pairing of bases during replication
	base analogs	incorporated into DNA in place of the usual base during replication, changing the base sequence
	viruses	viral DNA may insert itself into a genome, changing the base sequence

Effects of mutation

Silent mutations

The vast majority of mutations are silent (or neutral) which means they do not change any proteins, or the activity of any proteins synthesised. Therefore, they have no effect on the phenotype of an organism. They can occur in the non-coding regions of DNA (introns) or code for the same amino acid due to the degenerate nature of the genetic code. They may also result in changes to the primary structure but do not change the overall structure or function of the proteins synthesised.

Nonsense mutations

Nonsense mutations result in a codon becoming a stop codon instead of coding for an amino acid. The result is a shortened protein being synthesised which is normally non-functional. These mutations normally have negative or harmful effects on phenotypes.

Missense mutations

Missense mutations result in the incorporation of an incorrect amino acid (or amino acids) into the primary structure when the protein is synthesised. The result depends on the role the amino acid plays in the structure and therefore function of the protein synthesised. The mutation could be silent, beneficial, or harmful. A conservative mutation occurs when the amino acid change leads to an amino acid being coded for which has similar properties to the original, this means the effect of the

mutation is less severe. In contrast, a non-conservative mutation is when the new amino acid coded for has different properties to the original, this is more likely to have an effect on protein structure, and may cause disease.

1 Different mutations have a range of effects on proteins synthesised. These are categorised as follows:
Amorph – mutation that results in the loss of function of a protein
Hypomorph – mutation that results in a reduction of function of a protein
Hypermorph – mutation that results in a gain in function of a protein.
Suggest whether amorphic mutations result in dominant or recessive alleles. Give the reason(s) for your answer.

▼ **Table 2** *Point mutations can be silent, nonsense, or missense depending on their effect on the primary structure of the protein coded for by the gene*

Level where effect occurs	No mutations	Point mutations			
		Silent	Nonsense	Missense	
				conservative	non-conservative
DNA level	TTC	TTT	ATC	TCC	TGC
mRNA level	AAG	AAA	UAG	AGG	ACG
protein level	**Lys**	Lys	**STOP**	**Arg**	**Thr**

Beneficial mutations

The ability to digest lactose, the sugar present in milk, is thought to be the result of a relatively recent mutation. The majority of mammals in the world become lactose intolerant after they cease to suckle. The ability to digest lactose is found primarily in European populations who are more likely to farm cattle. The ability to drink milk and process lactose as an adult helps prevent diseases such as osteoporosis – this could also have prevented individuals with the mutation from starving during famines. This mutation appears to have arisen spontaneously more than once – it has also been found in people in East Africa.

Sickle-cell anaemia

Sickle-cell anaemia is a blood disorder where erythrocytes develop abnormally. The disorder is cause by a mutation in the gene coding for haemoglobin. There is a substitution of just one base. Thymine replaces adenine, making the sixth amino acid valine instead of glutamic acid on the beta haemoglobin chain.

Glutamic acid is a hydrophilic amino acid but valine is a hydrophobic amino acid. When the partial pressure of oxygen is low, and haemoglobin is dissociated from oxygen, the hydrophobic valine amino acids bind to hydrophobic regions on adjacent haemoglobin molecules.

The aggregation of haemoglobin molecules deforms the shape of erythrocytes causing them to become sickle shaped. The erythrocytes are less flexible and have difficulty moving through capillaries resulting in reduced oxygen delivery to tissues, which causes anaemia.

In homozygous individuals, there are two copies of the mutant alleles present. Heterozygous individuals only get mild symptoms of the condition, but are resistant to malaria. Hundreds of millions of people get malaria each year with over a million fatal cases.

1   Evaluate the benefits of being heterozygous for sickle cell anaemia.

2   Explain how the change of one amino acid in haemoglobin could reduce the oxygen-carrying ability of the blood.

Chromosome mutations

Gene mutations occur in single genes or sections of DNA whereas chromosome mutations affect the whole chromosome or number of chromosomes within a cell. They can also be caused by mutagens and normally occur during meiosis. As with gene mutations, the mutations can be silent but often lead to developmental difficulties.

Changes in chromosome structure include:

* **Deletion** – a section of chromosome breaks off and is lost within the cell.

* Duplication – sections get duplicated on a chromosome.

* **Translocation** – a section of one chromosome breaks off and joins another non-homologous chromosome.

* Inversion – a section of chromosome breaks off, is reversed, and then joins back onto the chromosome.

Synoptic link

To remind yourself of the process of meiosis, look back at Topic 6.3, Meiosis.

Summary questions

1   The development of lactose tolerance is thought to have spread over approximately 20 000 years, which in evolutionary terms is very quick.

    Explain why the percentage of adults with the ability to digest lactose increased at such a rate.

2   Outline why the majority of mutations do not have an influence phenotype.   *(2 marks)*

3   Discuss why beneficial mutations are rare and suggest a process that beneficial mutations underpin.   *(4 marks)*

19.2 Control of gene expression

Specification reference: 6.1.1

Gene regulation

Enzymes which are necessary for reactions present in metabolic pathways like respiration are constantly required, and the genes that code for these are called housekeeping genes. Protein-based hormones (required for the growth and development of an organism or enzymes) are only required by certain cells at certain times to carry out a short-lived response. They are coded for by tissue-specific genes.

The entire genome of an organism is present in every prokaryotic cell, or eukaryotic cell that contains a nucleus. This includes genes not required by that cell so the expression of genes and the rate of synthesis of protein products like enzymes and hormones has to be regulated. Genes can be turned on or off, and the rate of product synthesis increased or decreased depending on demand.

Bacteria are able to respond to changes in the environment because of gene regulation. Expressing genes only when the products are needed also prevents vital resources being wasted.

Gene regulation is fundamentally the same in both prokaryotes and eukaryotes. However, the stimuli that cause changes in gene expression and the responses produced are more complex in eukaryotes. Multicellular organisms not only have to respond to changes in the external environment but also the internal environment. Gene regulation is required for cells to specialise and work in a coordinated way.

There are a number of different ways in which genes are regulated, categorised by the level at which they operate:

● Transcriptional – genes can be turned on or off

● Post-transcriptional – mRNA can be modified which regulates translation and the types of proteins produced

● Translational – translation can be stopped or started

● Post-translational – proteins can be modified after synthesis which changes their functions.

Transcriptional control

There are a number of mechanisms that can affect the transcription of genes.

Chromatin remodelling

As you have already learnt, DNA is a very long molecule and has to be wound around proteins called histones in eukaryotic cells, in order to be packed into the nucleus of a cell. The resulting DNA/protein complex is called a chromatin.

Heterochromatin is tightly wound DNA causing chromosomes to be visible during cell division whereas **euchromatin** is loosely wound DNA present during interphase. The transcription of genes is not possible when DNA is tightly wound because RNA polymerase cannot access the genes. The genes in euchromatin, however, can be freely transcribed. Protein synthesis does not occur during cell division but during interphase between cell divisions. This is a simple form of regulation that ensures the proteins necessary for cell division are synthesised in time. It also prevents the complex and energy-consuming process of protein synthesis from occurring when cells are actually dividing.

Histone modification

DNA coils around histones because they are positively charged and DNA is negatively charged. Histones can be modified to increase or decrease the degree of packing (or condensation).

The addition of acetyl groups (**acetylation**) or phosphate groups (phosphorylation) reduces the positive charge on the histones (making them more negative) and this causes DNA to coil less tightly, allowing certain genes to be transcribed. The addition of methyl groups (**methylation**) makes the histones more hydrophobic so they bind more tightly to each other causing DNA to coil more tightly and preventing transcription of genes.

Epigenetics is a term that is increasingly used to describe this control of gene expression by the modification of DNA. It is sometimes used to include all of the different ways in which gene expression is regulated.

> ### Synoptic link
> You learnt about the structure of DNA in Topic 3.9, DNA replication and the genetic code and in Topic 6.1, Cell cycle.

Lac operon

An **operon** is a group of genes that are under the control of the same regulatory mechanism and are expressed at the same time. Operons are far more common in prokaryotes than eukaryotes owing to the smaller and simpler structure of their genomes. They are also a very efficient way of saving resources because if certain gene products are not needed, then all of the genes involved in their production can be switched off.

Glucose is easier to metabolise and is the preferred respiratory substrate of *Escherichia coli* and many other bacteria. If glucose is in short supply, lactose can be used as a respiratory substrate. Different enzymes are needed to metabolise lactose.

The **lac operon** is a group of three genes, lacZ, lacY, and lacA, involved in the metabolism of lactose. They are **structural genes** as they code for three enzymes (β-galactosidase, lactose permease, and transacetylase) and they are transcribed onto a single long molecule of mRNA. A **regulatory gene**, lacI, is located near to the operon and codes for a **repressor protein** that prevents the transcription of the structural genes in the absence of lactose (Figure 1).

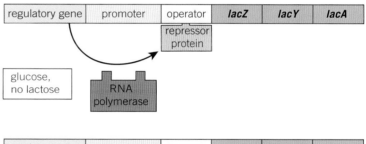

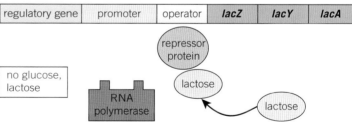

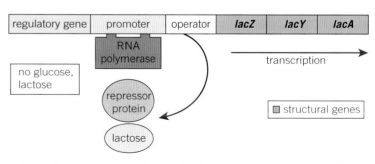

▲ Figure 1 *A repressor protein coded for by a regulatory gene binds to the operator and prevents the binding of RNA polymerase, preventing transcription of the structural genes*

The repressor protein is constantly produced and binds to an area called the operator, which is also close to the structural genes. The binding of this protein prevents RNA polymerase binding to DNA and beginning transcription. This is called down regulation. The section of DNA that is the binding site for RNA polymerase is called the promoter.

When lactose is present, it binds to the repressor protein causing it to change shape so it can no longer bind to the operator. As a result RNA polymerase can bind to the promoter, the three structural genes are transcribed, and the enzymes are synthesised.

Role of cyclic AMP

The binding of RNA polymerase still only results in a relatively slow rate of transcription that needs to be increased or up-regulated to produce the required quantity of enzymes to metabolise lactose efficiently. This is achieved by the binding of another protein, cAMP receptor protein (CRP), that is only possible when CRP is bound to cAMP (a secondary messenger that you are already familiar with).

The transport of glucose into an *E. coli* cell decreases the levels of cAMP, reducing the transcription of the genes responsible for the metabolism of lactose. If both glucose and lactose are present then it will still be glucose, the preferred respiratory substrate, that is metabolised.

Post-transcriptional/pre-translational control

RNA processing

The product of transcription is a precursor molecule, **pre-mRNA**. This is modified forming **mature mRNA** before it can bind to a ribosome and code for the synthesis of the required protein.

A cap (a modified nucleotide) is added to the 5′ end and a tail (a long chain of adenine nucleotides) is added to the 3′ end. These both help to stabilise mRNA and delay degradation in the cytoplasm. The cap also aids binding of mRNA to ribosomes. Splicing also occurs where the RNA is cut at specific points – the introns (non-coding DNA) are removed and the exons (coding DNA) are joined together. Both processes occur within the nucleus.

RNA editing

The nucleotide sequence of some mRNA molecules can also be changed through base addition, deletion, or substitution. These have the same effect as point mutations and result in the synthesis of different proteins which may have different functions. This increases the range of proteins that can be produced from a single mRNA molecule or gene.

Translational control

These following mechanisms regulate the process of protein synthesis:

- degradation of mRNA – the more resistant the molecule the longer it will last in the cytoplasm, that is, a greater quantity of protein synthesised.

Synoptic link

You first encountered cAMP when looking at its role in the mechanism of ADH action in Topic 15.6, The kidney and osmoregulation.

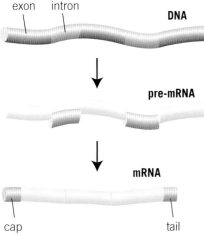

exon intron

DNA

pre-mRNA

mRNA

cap   tail

▲ **Figure 2** *This diagram summarises the modification of a pre-mRNA molecule*

- binding of inhibitory proteins to mRNA prevents it binding to ribosomes and the synthesis of proteins.

- activation of initiation factors which aid the binding of mRNA to ribosomes (the eggs of many organisms produce large quantities of mRNA which is not required until after fertilisation, at which point initiation factors are activated).

Protein kinases

Protein kinases are enzymes that catalyse the addition of phosphate groups to proteins. The addition of a phosphate group changes the tertiary structure and so the function of a protein. Many enzymes are activated by phosphorylation. Protein kinases are therefore important regulators of cell activity. Protein kinases are themselves often activated by the secondary messenger cAMP.

Post-translational control

Post-translational control involves modifications to the proteins that have been synthesised. This includes the following:

- addition of non-protein groups such as carbohydrate chains, lipids, or phosphates

- modifying amino acids and the formation of bonds such as disulfide bridges

- folding or shortening of proteins.

- modification by cAMP – for example, in the lac operon cAMP binds to the cAMP receptor protein increasing the rate of transcription of the structural genes.

> **Study tip**
>
> It is important to appreciate the difference between structural genes and regulatory genes.

Summary questions

1 The lac operon if often referred to as being 'leaky' meaning that it is still transcribed to a limited extent even in the absence of lactose.
   a Using your knowledge of how the lac operon works, explain why this is necessary. *(3 marks)*
   b Suggest the functions of β-galactosidase and lactose permease synthesised by the lac operon. *(3 marks)*

2 Another example of gene regulation in prokaryotes is the trp operon. This operon codes for the production of tryptophan, an essential amino acid for the bacterium *E. coli*. When tryptophan is available in the environment the structural genes in the trp operon are not expressed.

   Suggest a mechanism for the genetic regulation of this operon. *(5 marks)*

3 Using your knowledge of enzymes, explain how enzyme cofactors could play a role in gene regulation. *(4 marks)*

19.3 Body plans

Specification reference: 6.1.1

▲ **Figure 2** *False-colour scanning electron micrograph of a mutant fruit fly, Drosophila melanogaster, with four wings. The normal fly has one pair of wings. Fruit flies have been used for many years in genetic research because it is easy to raise in large numbers in the laboratory, reproduces rapidly, and many of its mutations are easy to see under a low-powered microscope, approx ×9.8 magnification*

Synoptic link

You learnt about evolution in Chapter 10, Classification and evolution.

Body plans

Living organisms come in all shapes and sizes from tulips to mosquitoes to humans. It is the same small group of genes, however, that control the growth and development of these vastly different living forms.

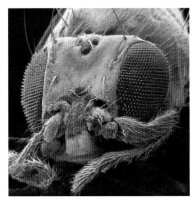

▲ **Figure 1** *Coloured scanning electron micrograph of the head of a mutant fruit fly* Drosophila melanogaster *with leg antennae, approx ×58 magnification*

The regulation of the pattern of anatomical development is called *morphogenesis*.

These genes were discovered by scientists investigating strange mutations observed in fruit flies such as legs on the head in place of antennae or extra pairs of wings. Fruit flies are small flies belonging to the genus *Drosophila* that feed and reproduce on rotting fruit. They are small, easy to keep, and have a short life cycle so have always been a popular choice for use in genetic studies.

Homeobox genes

Homeobox genes are a group of genes which all contain a homeobox. The homeobox is a section of DNA 180 base pairs long coding for a part of the protein 60 amino acids long that is highly conserved (very similar) in plants, animals, and fungi. This part of the protein, a *homeodomain*, binds to DNA and switches other genes on or off. Therefore, homeobox genes are regulatory genes.

The common ancestor of the mouse and human is thought to have lived about 60 million years ago. Mutations have been accumulating ever since and evolution has led to two very different organisms. Many of the homeobox genes present in the mouse and human, however, still have identical nucleotide sequences.

▲ **Figure 3** *This two-headed calf is an example of what can go wrong when there is a mutation in a regulatory Hox gene. Although this calf has two heads, it only has one brain and so the heads react simultaneously*

Pax6 is one of the homeobox genes. When mutated it causes a form of blindness (due to underdevelopment of the retina) in humans. Mice and fruit flies also have this gene and disruption of the gene causes blindness in these organisms as well. These findings suggest that Pax6 is a gene involved in the development of eyes in all three species.

Hox genes

Hox genes (often used interchangeably with homeobox genes) are one group of homeobox genes that are only present in animals. They are responsible for the correct positioning of body parts. In animals the Hox genes are found in gene clusters – mammals have four clusters on different chromosomes.

The order in which the genes appear along the chromosome is the order in which their effects are expressed in the organism. Human beings have 39 Hox genes in total that are all believed to have arisen from one ancient homeobox gene by duplication and accumulated mutations over time.

The layout of living organisms

Body plans are usually represented as cross-sections through the organism showing the fundamental arrangement of tissue layers. *Diploblastic* animals have two primary tissue layers and *triploblastic* animals have three primary tissue layers.

A common feature of animals is that they are segmented, that is, the rings of a worm or the less obvious back bone of vertebrates. These segments have multiplied over time and are specialised to perform different functions. Hox genes in the head control the development of mouthparts and Hox genes in the thorax control the development of wings, limbs, or ribs.

The individual vertebrae and associated structures have all developed from segments in the embryo called somites. The somites are directed by Hox genes to develop in a particular way depending on their position in the sequence.

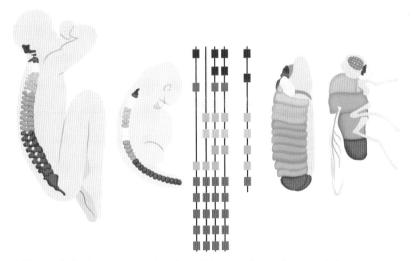

▲ Figure 4 *Body layout from head to tail in both flies and human beings is controlled by Hox genes. The fly has a set of eight Hox genes but human beings have four comparable sets*

Symmetry

The body shape of most animals shows symmetry.

- Radial symmetry is seen in diploblastic animals like jellyfish. They have no left or right sides, only a top and a bottom.

- Bilateral symmetry which is seen in most animals means the organisms have both left and right sides and a head and tail rather than just a top and bottom.

- Asymmetry is seen in sponges which have no lines of symmetry.

Mitosis and apoptosis

Mitosis, which results in cell division and proliferation, and apoptosis (Figure 6), which is programmed cell death, are both essential in shaping organisms.

The role of mitosis is to increase the number of cells leading to growth. The role of apoptosis is not so immediately obvious. Consider how a sculptor works to give shape to a block of wood or stone. The shape is revealed as material is removed bit by bit. This is one of the ways in which apoptosis shapes different body parts – by removing unwanted cells and tissues. Cells undergoing apoptosis can also release chemical signals which stimulate mitosis and cell proliferation leading to the remodelling of tissues. A sculptor working with clay will often add and remove clay during the reshaping process. Hox genes regulate both mitosis and apoptosis.

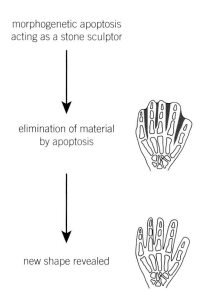

morphogenetic apoptosis acting as a stone sculptor

elimination of material by apoptosis

new shape revealed

▲ **Figure 5** *The development of the hand is an example of morphogenetic apoptosis. The red area shows how apoptosis is used to sculpt the individual fingers*

Synoptic link

You learnt about mitosis in Topic 6.2, Mitosis.

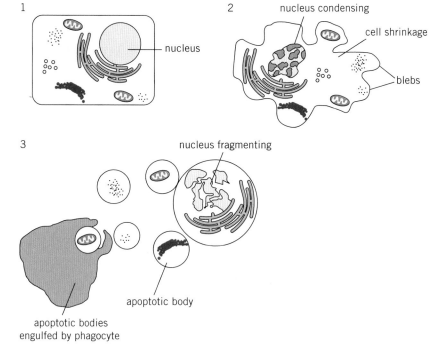

▲ **Figure 6** *The process of apoptosis*

Factors affecting the expression of regulatory genes

The expression of regulatory genes can be influenced by the environment, both internal and external. Stress can be defined as the condition produced when the homeostatic balance within an organism is upset. This can be due to external factors such as a change in temperature or intensity of light. Internal factors can change due to the release of hormones or psychological stress.

These factors will have a greater impact during the growth and development of an organism.

Drugs can also affect the activity of regulatory genes. An example of this was the drug thalidomide. Thalidomide was given to pregnant women in the 1950s and 1960s to treat morning sickness. It was later discovered that it prevented the normal expression of a particular Hox gene. This resulted in the birth of babies with shortened limbs.

Thalidomide is currently used in the treatment of some forms of cancer. The property of this drug that has previously caused problems during pregnancy is being exploited to stop the development of some tumours. It is believed that thalidomide prevents the formation of networks of capillaries which are necessary for some tumours to grow and develop.

▲ **Figure 7** *A girl with birth defects caused by the drug thalidomide. This drug was given to pregnant women in the 1950s and 1960s to treat morning sickness, until it was discovered to cause severe malformations in babies*

 ## Ontology doesn't mimic phylogeny

The theory of recapitulation states that as organisms develop from fertilised egg to embryo they repeat the evolutionary process that they have been through. In biology, this can be summarised by the phrase ontology (the development of an organism) mimics phylogeny (the evolutionary history of an organism).

This theory is not accepted by modern biologists but a modern theory states that 'oncology (the study of cancer) recapitulates ontology'. This refers to the discovery that genes originally expressed in the development of the embryo are expressed again by cancerous cells.

> Use your knowledge of Hox genes to suggest their possible roles in the development of certain tumours.

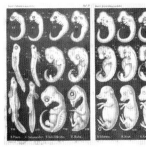

▲ **Figure 8** *The drawing above was made in the 19th century as evidence supporting this theory.*

Summary questions

1 Explain, with reference to their body shape, why human beings are referred to as bilaterally symmetrical but jellyfish are radially symmetrical. *(3 marks)*

2 The Hox gene Pax6 is necessary for the normal development of the retina in humans. A mutation in this gene can lead to blindness. Pax6 mutations also cause blindness in mice and fruit flies. Describe how scientists could have tested the idea that Pax6 plays a role in eye development in all three species. *(5 marks)*

3 Consider the statement.

   All Hox genes are homeobox genes but not all homeobox genes are Hox genes.

   Discuss the validity of this statement. *(5 marks)*

Practice questions

1 The common groundsel, *Senecio vulgaris*, is a weed that is often found in large numbers on cultivated land. It was the first plant species to develop resistance to triazine herbicides. This resistance is the result of a gene mutation in the chloroplast DNA.

Since its first appearance, triazine resistance has spread very rapidly in groundsel populations.

a Explain the rapid spread of herbicide resistance in a weed such as groundsel.

(5 marks)

b DNA was extracted from the chloroplasts of triazine-susceptible and triazine-resistant groundsel plants. Equivalent lengths of DNA, including the site of the mutation, were isolated from each extract and then treated with the restriction enzyme MaeI prior to electrophoresis.

The resulting electrophoresis gel, after staining the DNA, is shown in the diagram.

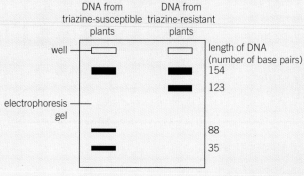

(i) State, giving a reason, whether the mutation giving resistance to triazine is a deletion, a substitution or an addition. *(2 marks)*

(ii) Explain the difference in banding pattern between DNA from triazine-susceptible and triazine-resistant plants. *(2 marks)*

(iii) Suggest one way in which the mutation could give resistance to triazine. *(2 marks)*

OCR 2805/02 2010

2 a The diagram shows a family's history of Huntington's disease.

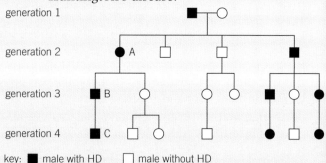

Explain how the diagram provides evidence that Huntington's disease results from the inheritance of an autosomal dominant allele. *(3 marks)*

b Genetic screening for Huntington's disease can be carried out, using a process similar to genetic fingerprinting, to find the length of a repeated triplet, or 'stutter', in an allele. After treatment with a restriction enzyme, fragments of DNA of different lengths are separated by gel electrophoresis.

(i) Describe the role of a restriction enzyme in this technique. *(4 marks)*

(ii) Explain why fragments of DNA of different lengths can be separated by gel electrophoresis *(5 marks)*

c DNA from individuals A, B and C from the family shown in the diagram was analysed.

The resulting banding patterns are shown in the diagram.

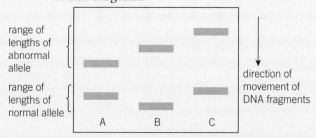

Explain why the DNA of the following bands in the diagram are not the same length for:

(i) the three normal alleles *(1 mark)*

(ii) the three abnormal alleles. *(1 mark)*

OCR 2805/02 2008

3 Homeobox genes show astonishing similarity across widely different species of animal, from fruit flies, which are insects, to mice and humans, which are mammals. The sequences of these genes have remained relatively unchanged throughout evolutionary history and the same genes control embryonic development in flies and mammals.

a State what is meant by a homeobox gene.
(2 marks)

b Homeobox genes show 'astonishing similarity across widely different species of animal'.

Explain why there has been very little change by mutation in these genes.
(2 marks)

c Frogs reproduce by laying eggs in water, each egg develops into a tadpole, which has external gills to extract oxygen from the water, and a tail to help it swim. The tadpole gradually changes into an adult frog as it grows. During this time its gills and tail disappear.

List two cellular processes that must occur during the development of a tadpole into a frog.
(2 marks)

d Name another kingdom of organisms, other than animals, that have similar homeotic genes.
(1 mark)

OCR F215 2012

4 An enhancer of a regulatory gene responsible for limb development in mammals, called the sonic hedgehog gene, is located in the intron of a neighbouring gene. This regulatory gene was first investigated using Drosophila – genetically modified flies. Without this gene the fruit flies grew small projections (known as denticles) all over their bodies.

Point mutations in the enhancer sequence for this gene result in polydactyly, extra fingers or toes.

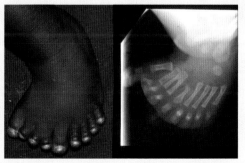

The photograph and X-ray show a child's foot with polydactyl – a deformity in which more than the usual number of digits are present. The condition is genetic in origin, and in most cases causes no harm. The extra digits are often underdeveloped (those at lower left in this case), and if removed surgically soon after birth cause no long term complications.

a Explain how an enhancer works.
(2 marks)

b State the meaning of the term intron.
(1 mark)

c Describe what is meant by a point mutation.
(2 marks)

d Explain why the sonic hedgehog gene is an example of a Hox gene.
(2 marks)

e The sonic hedgehog gene is usually expressed in cells close to tissue that eventually develops into the small fingers or toes.

Suggest how different rates of transcription of this gene leads to the formation of fingers with different sizes and shapes.
(3 marks)

20 PATTERNS OF INHERITANCE AND VARIATION

20.1 Variation and inheritance

Specification reference: 6.1.2

Members of different species are, usually, clearly different from each other and even members of the same species are rarely identical so variation is an important feature of living organisms. Variation arises as a result of mutations – changes to the genetic code which are random and constantly taking place. Variation is essential for the process of natural selection – and therefore evolution.

You have already learnt about how organisms within a species vary. It is important to remember that variation can occur both as a result of environmental variation and genetic variation. In the majority of cases both play a role in determining an organism's characteristics – examples of this include chlorosis in plants and the body mass of an animal.

Chlorosis

Most plants are genetically coded to produce large quantities of chlorophyll, the green pigment that is vital for photosynthesis and gives leaves their green colour. Some plants, however, suffer from a condition known as chlorosis, when the leaves look pale or yellow. This occurs because the cells are not producing the normal amount of chlorophyll. This lack of chlorophyll reduces the ability of the plant to make food by photosynthesis.

Most plants which show chlorosis have normal genes coding for chlorophyll production. The change in their phenotype is the result of environmental factors. There are many different environmental factors which cause chlorosis, each having a different effect on the physiology of the plant but causing the same change in phenotype. Examples include:

● Lack of light – for example, when a toy or gardening tool is left on a lawn. In the absence of light, plants will turn off their chlorophyll production to conserve resources. In this case, chlorosis only occurs where the plant gets no light.

● Mineral deficiencies – for example, a lack of iron or magnesium. Iron is needed as a cofactor by some of the enzymes that make chlorophyll, and magnesium is found at the heart of the chlorophyll molecule. If either of these elements are lacking in the soil, a plant simply cannot make chlorophyll and gradually all the leaves will become yellow.

● Virus infections – when viruses infect plants, they interfere with the metabolism of cells. A common symptom is yellowing in the infected tissues as they can no longer support the synthesis of chlorophyll.

▲ Figure 1 Normal leaf (left), leaf with chlorosis (right)

In summary, even though genetic factors in a plant are likely to code for green leaves, the environment plays a key role in the final leaf appearance.

Animal body mass

Within a species, the body mass of individual animals varies. An organism's body mass is determined by a combination of both genetic and environmental factors. In the majority of cases dramatic variations in size such as obesity and being severely underweight are a result of environmental factors. For example, the amount (and quality) of foods eaten, the quantity of exercise which the organism gets, or the presence of disease can affect the body mass. Being extremely overweight or underweight can result in significant health problems for an animal.

Occasionally obesity can be a result of the genetic make-up of an organism. The obese mouse in Figure 3 has a mutation on chromosome 7. This mutation causes the pattern of fat deposition in its body to be altered. Scientific studies have shown that this gene acts in conjunction with other genes that regulate the energy balance of the body, and as a result mice possessing the mutation grow 35–50% fatter by middle age than a normal mouse would.

Creating genetic variation

Genetic variation is created by the versions of genes you inherit from your parents. For most genes there are a number of different possible alleles or variants. The individual mixture of alleles an organism inherits influences the characteristics they will display. This combination is determined by sexual reproduction involving meiosis (the formation of gametes), and the random fusion of gametes at fertilisation. This results in the vast genetic variation seen between individuals of the same species.

For most genes in your body two alleles are inherited (one from each parent). These alleles may be the same or different versions of the gene. The combination of alleles an organism inherits for a characteristic is known as their **genotype** – this is the genetic make-up of an organism in respect of that gene. The observable characteristics of an organism are known as its **phenotype** (you will learn more about two special cases of inheritance known as codominance and sex linkage in Topic 20.2, Monogenetic inheritance). The actual characteristics that an organism displays are also often influenced by the environment.

▲ **Figure 2** *This severely obese cat has an extremely large body mass as a result of overfeeding and lack of exercise*

▲ **Figure 3** *Left – obese mouse with a mutated chromosome 7. Right – normal mouse with normal chromosome 7*

Synoptic link

You learnt about genetic and environmental variation and examples of these in Topic 10.5, Types of variation. In this topic you also learnt about the detail of how sexual reproduction and meiosis lead to genetic variation.

Study tip

Remember all individuals of the same species have the same genes but not necessarily the same alleles of these genes. Many people refer to alleles and genes interchangeably – make sure you understand the difference between these words, and use the terms correctly.

▲ **Figure 4** *Doctors monitor the height of children at routine developmental check-ups. The genotype of this child may determine that they have the potential to grow to 1.9 m, but the child's final height is influenced greatly by their diet. For example, a shortage of calcium when the child is young can restrict bone development, therefore preventing them reaching their maximum potential height*

▲ **Figure 5** *Dimples in your cheeks are caused by a dominant allele. By simply looking at this man you cannot tell if he is heterozygous or homozygous for this dominant allele*

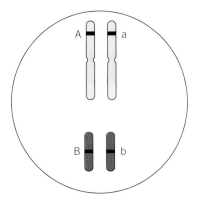

▲ **Figure 6** *In sexually reproducing organisms, most chromosomes occur in pairs known as homologous chromosomes (where one of each pair comes from each parent). The allele of a particular gene occurs in the same position on each of the homologous chromosomes. This is known as its locus. The diagram shows a pair of homologous chromosomes from a pea plant*

Synoptic link

To remind yourself how meiosis creates variation, for example through crossing over and independent assortment, look back at Topic 6.3, Meiosis.

Any changes the environment makes to a person's phenotype are not inherited – these are referred to as modifications. Only mutations (changes to the DNA) in the gametes can be passed on to the offspring.

It is not always possible to determine an organism's genotype from its phenotype owing to the dominance of particular alleles. A **dominant allele** is the version of the gene that will always be expressed if present in an organism. This means an individual showing the dominant characteristic in their phenotype could have one or two copies of the dominant gene – you can't tell from their appearance. A **recessive allele,** however, will only be expressed if two copies of this allele are present in an organism. This means if an individual has a recessive phenotype, you also know their genotype – they must have two alleles coding for the recessive phenotype.

A number of key terms are used to describe an organism's genotype for a particular characteristic:

● **Homozygous** – they have two identical alleles for a characteristic. The organism could be *homozygous dominant* (contain two alleles for the dominant phenotype) or *homozygous recessive* (contain two alleles for the recessive phenotype).

● **Heterozygous** – they have two different alleles for a characteristic. In this case the allele for the dominant phenotype will be expressed.

Continuous and discontinuous variation

The variation of a characteristic displayed within a species can be divided into two groups – those which show continuous variation, and those which show discontinuous variation. Remember – in discontinuous variation individuals fall into distinct groups (for example, blood groups) and normally only one gene is involved and the environment has little, if any, effect (Figure 7). In continuous variation there are two extremes, with every degree of variation possible in between (Figure 8). Examples would include your height or weight.

Many genes will be involved, and the environment has a large effect.

Table 1 summarises some of the important differences between continuous and discontinuous variation.

▼ Table 1

	Continuous variation	Discontinuous variation
definition	a characteristic that can take any value within a range	a characteristic that can only appear in specific (discrete) values
cause of variation	genetic and environmental	mostly genetic
genetic control	polygenes – controlled by a number of genes	one or two genes
examples	leaf surface area animal mass skin colour	blood group albinism round and wrinkled pea shape

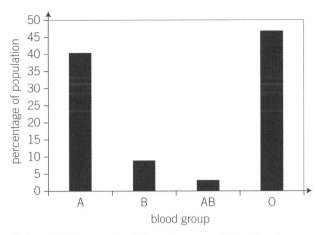

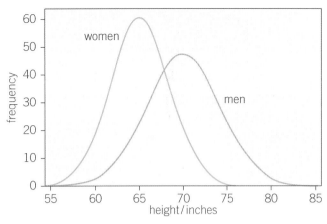

▲ **Figure 7** *An example of discontinuous variation showing individuals falling into distinct categories*

▲ **Figure 8** *Graph showing the range of heights of a large sample of men and women. It should be noted that sex is a discontinuous characteristic but within each category (male or female) there is continuous variation (height)*

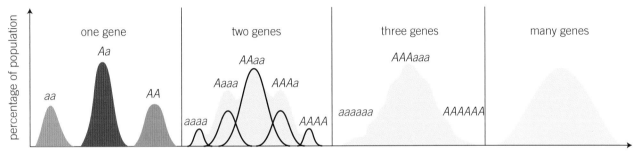

▲ **Figure 9** *Series of graphs showing how the type of variation changes as the number of genes responsible for the production of a characteristic increases*

Summary questions

1 State the difference between a homozygous and heterozygous genotype. *(1 mark)*

2 Explain the difference between the phenotype and genotype of an oak tree. *(2 marks)*

3 Using named examples, state and explain the difference between continuous and discontinuous variation. *(6 marks)*

Synoptic link

Look back at Topic 10.6, Representing variation graphically for more examples of continuous and discontinuous variation.

20.2 Monogenic inheritance
Specification reference: 6.1.2

To explain how characteristics are inherited you need to be able to show how genes are passed on from one generation to the next. This is normally shown using a genetic cross. Most commonly the inheritance of a single gene is shown, this is known as **monogenic inheritance**. The basic laws by which characteristics are inherited were established by Gregor Mendel, a scientist and monk working in the 19th century.

Performing a genetic cross

There are a number of key steps you should follow when analysing a genetic cross. This ensures that your diagram explains fully what is happening to the genes of an organism during fertilisation (and helps you to avoid making errors):

Step 1 – State the phenotype of both the parents.

Step 2 – State the genotype of both parents. To do this, assign a letter code to represent the alleles of the gene being studied. A capital letter should be used to represent the dominant allele and its lower case form to represent the recessive allele. For example, if studying the inheritance of an animal's fur colour, you may choose B to represent brown fur (dominant) and b to represent white fur (recessive).

Step 3 – State the gametes of each parent. It is common practice to circle the letters, for example, Ⓖ.

Step 4 – Use a Punnett Square to show the results of the random fusion of gametes during fertilisation. Remember to label the gametes on the edges of the square.

Step 5 – State the proportion of each genotype which are produced among the offspring. This can be in the form of a percentage, ratio, or 'x out of y offspring ...'.

Step 6 – State the corresponding phenotype for each of the possible genotypes. It must be clear that you know which phenotype results from each genotype.

Homozygous genetic cross

Mendel carried out many of his famous experiments on pea plants. Pea pods come in two colours – green and yellow. The cross in Figure 1 shows what happens when a homozygous green pea pod plant is crossed with a homozygous yellow pea pod plant. The allele for green pea pods is dominant. Organisms that contain homozygous alleles for a particular gene are known as true breeding or pure breeding individuals. Therefore in this experiment Mendel was studying what happened when two true breeding individuals are crossed.

All of the offspring are heterozygous. This means that all plants will have green pods as this is the dominant allele. These offspring are known as the F_1 generation.

Heterozygous genetic cross

The cross in Figure 2 shows what happens if you take two of the heterozygous offspring from the first generation and cross them together.

G = allele for green pods
g = allele for yellow pods

parental phenotypes
parental genotypes

green pods
GG
(arbitrarily assumed as male♂)

yellow pods
gg
(arbitrarily assumed as female♀)

meiosis   meiosis

gametes   G   G   g   g

offspring (F_1) genotypes

♀gametes	♂ gametes	
	G	G
g	Gg	Gg
g	Gg	Gg

offspring (F_1) phenotypes   all plants have green pods (Gg)

▲ Figure 1

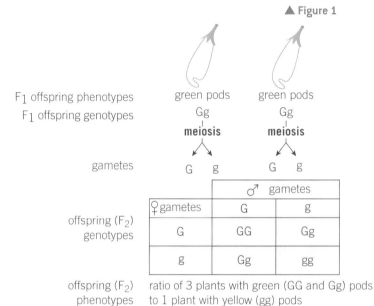

F_1 offspring phenotypes   green pods   green pods
F_1 offspring genotypes   Gg   Gg

meiosis   meiosis

gametes   G   g   G   g

offspring (F_2) genotypes

♀gametes	♂ gametes	
	G	g
G	GG	Gg
g	Gg	gg

offspring (F_2) phenotypes   ratio of 3 plants with green (GG and Gg) pods to 1 plant with yellow (gg) pods

▲ Figure 2

Study tip

Potential outcomes from genetic crosses can be expressed in different forms which are mathematically equivalent:

Probability	Ratio	Percentage
4 in 4	4:0	100%
3 in 4	3:1	75%
2 in 4 (or 1 in 2)	2:2 (or 1:1)	50%
1 in 4	1:3	25%
0 in 4	0:4	0%

The offspring produced from this cross are known as the F_2 generation. Offspring will be produced in a ratio of three pea plants with green pods to one pea plant with yellow pods.

Codominance

Codominance occurs when two different alleles occur for a gene – both of which are equally dominant. As a result both alleles of the gene are expressed in the phenotype of the organism if present.

One example of this condition is the colour of snapdragon flowers. Two equally dominant alleles exist, each of which codes for the colour of the flower:

An allele that codes for red flowers – the allele codes for the production of an enzyme which catalyses the production of red pigment from a colourless precursor.

An allele that codes for white flowers – the allele codes for an altered version of the enzyme which does not catalyse the production of the pigment, therefore the flowers are white.

▲ Figure 3 *Red snapdragon flowers are produced by a plant which is homozygous for the allele coding for the enzyme that catalyses the production of red pigment*

▲ **Figure 4** *White snapdragon flowers are produced by a plant which is homozygous for the allele coding for the enzyme which does not catalyse the production of pigment*

▲ **Figure 5** *Pink snapdragon flowers are produced by a plant which is heterozygous for both alleles*

In this example of codominance, three colours of flower can be produced:

1 Red flowers – the plant is homozygous for the allele coding for the production of red pigment.

2 White flowers – the plant is homozygous for the allele coding for no pigment production.

3 Pink flowers – the plant is heterozygous. The single allele present which codes for red pigmentation produces enough pigment to produce pink flowers.

The genetic cross in Figure 6 shows how pink flowers are produced. Two of the pink flowers produced in the F_1 generation are then crossed in the second cross.

When studying codominance, upper and lower case letters are not used to represent the alleles, as this would imply one allele is dominant and the other recessive. Instead a letter is chosen to represent the gene, in this example C for colour of flowers. The different alleles are then represented using a second letter which is shown as a superscript. In this example C^R is used to represent the allele coding for red flowers and C^W for the allele coding for white flowers.

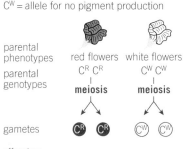

C^R = allele for red pigment production
C^W = allele for no pigment production

▲ **Figure 6** *Production of pink snapdragon flowers*

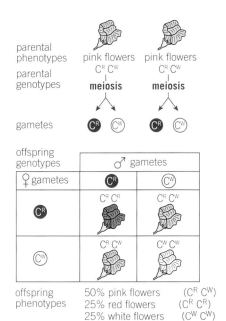

▲ **Figure 7** *Genetic cross between two snapdragons displaying pink flowers*

Multiple alleles

In the previous examples, you have learnt about characteristics coded by a gene with two alleles. Some genes have more than two versions, they have **multiple alleles**. However, as an organism carries only two versions of the gene (one on each of the homologous chromosomes) only two alleles can be present in an individual.

Your blood group is determined by a gene with multiple alleles. The immunoglobulin gene (Gene I) codes for the production of different antigens present on the surface of red blood cells. There are three alleles of this gene:

- I^A – results in the production of antigen A
- I^B – results in the production of antigen B
- I^O – results in the production of neither antigen

I^A and I^B are codominant whereas I^O is recessive to both of the other alleles. Different combinations of these alleles result in the four blood groups:

- Blood group A – $I^A I^A$ or $I^A I^O$
- Blood group B – $I^B I^B$ or $I^B I^O$
- Blood group AB – $I^A I^B$
- Blood group O – $I^O I^O$

There are many different possible crosses. Figure 8 shows how parents of blood group A and B can reproduce to produce children who may display any of the four blood groups.

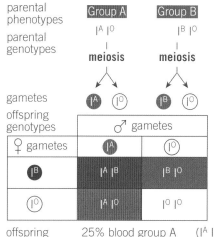

▲ Figure 8

Determining sex

In humans and other mammals, as well as many other species, sex is genetically determined. Humans have 23 pairs of chromosomes of varying sizes and shapes. In 22 of the pairs, both members of the pair are the same but the 23rd pair, known as the sex chromosomes, are different. Human females have two X chromosomes, whereas a male has an X and a Y chromosome. The X chromosome is large and contains many genes not involved in sexual development. The Y chromosome is very small, containing almost no genetic information, but it does carry a gene that causes the embryo to develop as a male.

Therefore the sex of the offspring will be determined by whether the sperm fertilising the egg contains a Y chromosome or an X.

Sex linkage

Some characteristics are determined by genes carried on the sex chromosomes – these genes are called **sex linked.** As the Y chromosome is much smaller than the X chromosome, there are a number of genes in the X chromosome that males have only one copy of. This means that any characteristic caused by a recessive allele on the section of the X chromosome, which is missing in the Y chromosome, occurs more frequently in males. This is because many females will also have a dominant allele present in their cells.

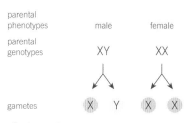

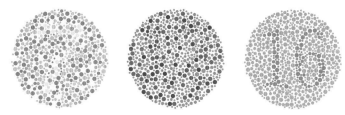

▲ Figure 10 *This is a test used to determine whether a person is colour blind. Red green colour blindness is a sex-linked disorder. It is caused by a recessive allele carried on the X chromosome. It is therefore much more common in males*

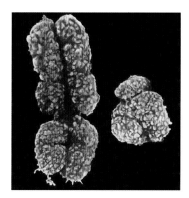

▲ Figure 9 *An X and Y human chromosome viewed through a scanning electron microscope*

▲ Figure 11 *Sex inheritance in humans*

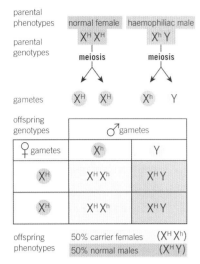

parental phenotypes

parental genotypes

gametes

offspring genotypes

offspring phenotypes 50% carrier females $(X^H X^h)$
50% normal males $(X^H Y)$

▲ **Figure 12** *Inheritance of the haemophiliac allele from a haemophiliac male*

H = allele for production of clotting protein (rapid blood clotting)
h = allele for non-production of clotting protein (slow blood clotting)

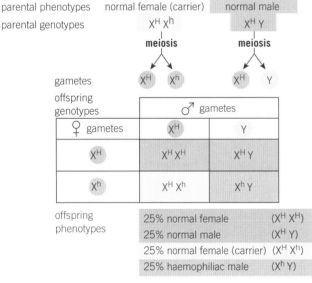

offspring phenotypes 25% normal female $(X^H X^H)$
25% normal male $(X^H Y)$
25% normal female (carrier) $(X^H X^h)$
25% haemophiliac male $(X^h Y)$

▲ **Figure 13** *Inheritance of haemophilia from a normal female (carrier)*

Haemophilia

Haemophilia is an example of a sex-linked genetic disorder. Patients with haemophilia have blood which clots extremely slowly due to the absence of a protein blood-clotting factor (in the majority of cases this is factor VIII). As a result injury can result in prolonged bleeding which, if left untreated, is potentially fatal.

If a male inherits the recessive allele that codes for haemophilia (on their X chromosome) they cannot have a corresponding dominant allele on their Y chromosome, and so develop the condition. As a result the vast majority of haemophilia sufferers are males. Females who are heterozygous for the haemophilia coding gene are known as **carriers**. They do not suffer from the disorder, however they may pass on the allele to their children. This can result in the birth of a son who suffers from haemophilia.

When showing the inheritance of a sex-linked condition the alleles are shown linked to the sex chromosome they are found on. In this example, haemophilia is linked to the X chromosome, therefore:

X^H is used to represent the dominant 'healthy' allele.

X^h is used to represent the recessive allele coding for haemophilia (through the non-production of a blood-clotting protein). This is often known as the faulty allele.

Y is used to represent the Y chromosome – it has no allele attached to it as it does not carry the gene which produces the specific blood-clotting protein (Figure 12).

If a carrier female and a normal male have children, then in theory half of the male offspring produced will have the disorder (Figure 13). Half of the female offspring will be carriers. As male offspring only inherit an X chromosome from their mother, sons can only inherit the condition from their mother.

However, an affected male can pass on the faulty allele to his daughters, resulting in them becoming carriers of the disorder (Figure 12).

Summary questions

1 What is special about the phenotypes formed as a result of a gene which has codominant alleles? *(2 marks)*

2 Using a genetic cross show the blood groups of the potential offspring of a father with blood group AB and a mother with blood group O. *(6 marks)*

3 Some individuals cannot distinguish between the colours red and green. This form of colour blindness is a sex-linked condition, linked to the X chromosome. The allele for red-green colour blindness is recessive to the normal allele. Show how colour blindness can be present in the son of two parents who do not display colour blindness. *(6 marks)*

In the previous topic you learnt how genes were inherited through monogenic inheritance. In reality, thousands of genes are inherited during fertilisation. Dihybrid crosses are used to show the inheritance of two genes and this is known as **dihybrid inheritance**.

Dihybrid cross

A dihybrid cross is used to show the inheritance of two different characteristics, caused by two genes, which may be located on different pairs of homologous chromosomes. Each of these genes can have two or more alleles.

A dihybrid cross is set out in a very similar format to the one used when studying a monohybrid cross – however, four alleles (two for each characteristic) are shown at each stage instead of two.

A classic example is the inheritance of seed phenotype in pea plants. The seeds a pea plant produces can be produced in two different colours – yellow or green. They are also produced in two different shapes – round or wrinkled.

The following codes can be used to represent the alleles:

Y – allele coding for yellow seeds (this is the dominant allele)

y – allele coding for green seeds (this is the recessive allele)

R – allele coding for round seeds (this is the dominant allele)

r – allele coding for wrinkled seeds (this is the recessive allele)

In the dihybrid cross in Figure 2, a true breeding homozygous pea plant with yellow round seeds is crossed with a true breeding homozygous pea plant with green wrinkled seeds.

All of the offspring produced in the F_1 generation will have a heterozygous genotype – YyRr. Therefore, they will all produce yellow round seeds as they will all have inherited one of each of the dominant alleles from the yellow round seeded parent.

In the second example in Figure 2, two of the pea plants grown from the seeds of the F_1 generation are now crossed.

> ## Learning outcomes
> Demonstrate knowledge, understanding, and application of:
> → genetic diagrams to show patterns of inheritance including dihybrid inheritance.

▲ **Figure 1** *Pea seeds can be round or wrinkled. Round seeds are caused by the dominant allele*

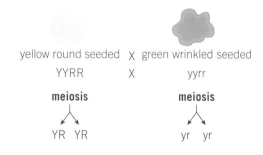

| parental phenotype | yellow round seeded | X | green wrinkled seeded |
| parental genotype | YYRR | X | yyrr |

meiosis   meiosis

gametes   YR   YR   yr   yr

offspring F_1 genotypes

	♀ gametes	
♂ gametes	YR	YR
yr	YyRr	YyRr
yr	YyRr	YyRr

offspring F_1 phenotypes   all yellow round seeded

▲ **Figure 2**

F₁ offspring phenotype   yellow round seeded   X   yellow round seeded

F₁ offspring genotype   YyRr   X   YyRr

Meiosis   **Meiosis**

gametes   YR   Yr   yR   yr   YR   Yr   yR   yr

offspring (F₂) genotypes

♂ gametes	♀ gametes			
	YR	Yr	yR	yr
YR	YYRR	YYRr	YyRR	YyRr
Yr	YYRr	YYrr	YyRr	Yyrr
yR	YyRR	YyRr	yyRR	yyRr
yr	YyRr	Yyrr	yyRr	yyrr

offspring F₂ phenotypes   9 yellow round seeded : 3 yellow wrinkled seeded: 3 green round seeded: 1 green wrinkled

(YYRR, YYRr, YyRR, YyRr)   (YYrr, Yyrr)   (yyRR, yyRr)   (yyrr)

▲ Figure 3

Synoptic Link

Look back at Topic 6.3, Meiosis to remind yourself of the different stages of meiosis.

When the phenotypes of the 16 possible combinations of alleles are identified (as shown in Figure 3), it is found that the expected ratio in the F₂ generation is nine pea plants producing yellow round seeds, to three pea plants producing yellow wrinkled seeds to three pea plants producing green round seeds, to one pea plant producing green wrinkled seeds.

It is worth remembering that this is the expected ratio of the four different phenotypes. As with all genetic crosses the actual ratio of offspring produced can differ from the expected. This may because:

- The fertilisation of gametes is a random process so in a small sample a few chance events can lead to a skewed ratio.
- The genes being studied are both on the same chromosome. These are known as linked genes. If no crossing over occurs the alleles for the two characteristics will always be inherited together.

Summary questions

1 State the difference between monogenic inheritance and dihybrid inheritance. *(1 mark)*

2 State and explain two reasons why the offspring produced from a particular genetic cross may differ from the expected ratio. *(2 marks)*

3 Work through a dihybrid genetic cross between a pea plant which produces green wrinkled seeds and a heterozygous pea plant which produces yellow round seeds to determine the expected ratio of the F₁ offspring. *(6 marks)*

20.4 Phenotypic ratios
Specification reference: 6.1.2

The ratios of phenotypes that you would expect to see in the offspring produced from a dihybrid cross can be easily calculated as long as you know which alleles are dominant and which are recessive. The actual numbers may vary from those expected to some extent because the process is random but the differences should not be large. The larger the sample the closer the numbers will be to the expected ratio.

Linkage

The ratios observed in many dihybrid crosses differ significantly from those expected. This is often due to linkage meaning that the genes are located on the same chromosome. In Topic 20.2 you met the idea of sex linkage, when a particular gene is located on the sex chromosomes. The effects of sex linkage can be seen in conditions such as haemophilia and red-green colour blindness.

When the genes that are linked are found on one of the other pairs of chromosomes it is called **autosomal linkage**. **Linked genes** are inherited as one unit – there is no independent assortment during meiosis unless the alleles are separated by chiasmata. They tend to be inherited together.

Linked genes cannot undergo the normal random 'shuffling' of alleles during meiosis and the expected ratios will not be produced in the offspring. The linked genes are inherited effectively as a single unit. Body colour and wing length, for example, are linked characteristics in fruit flies. The allele B is responsible for a brown body and is dominant to b which results in a black body. The allele V is responsible for long wings and is dominant to the allele v which results in short wings.

A heterozygous brown bodied, long winged fly (BbVv) was crossed with a homozygous black bodied, short winged fly (bbvv). The expected and observed phenotypic ratios from this cross are shown in Figure 1.

The homozygous parent can only produce (bv) gametes and the heterozygous parent produces mainly (BV) and (bv) gametes as the alleles are linked. The heterozygous parent will produce a few (bV) gametes and (Bv) gametes due to crossing over, which results in the separation of some of the linked genes. This is the reason for the small number of brown flies with short wings and black flies with long wings. These are called **recombinant** offspring (they have different combinations of alleles than either parent). The closer the genes are on a chromosome the less likely they are to be separated during crossing over and the fewer recombinant offspring produced.

Learning outcomes

Demonstrate knowledge, understanding, and application of:

→ the use of phenotypic ratios to identify linkage (autosomal and sex linkage) and epistasis

→ using the chi-squared (χ^2) test to determine the significance of the difference between observed and expected results.

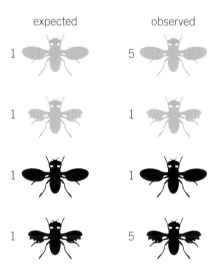

▲ **Figure 1** *A test cross involving linked genes does not produce the expected ratio of phenotypes in the offspring. There is a larger proportion of brown flies with long wings and black flies with short wings*

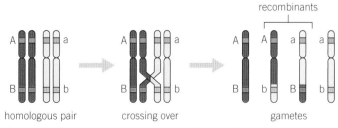

▲ **Figure 2** *During crossing over in prophase I of meiosis, non-sister chromatids of a homologous pair of chromosomes may break and reform. As these chromatids break at the same point, any gene loci below the point of the break will be exchanged as a result of recombination creating new gene combinations*

The **recombination frequency** is a measure of the amount of crossing over that has happened in meiosis.

$$\text{Recombination frequency} = \frac{\text{number of recombinant offspring}}{\text{total number of offspring}}$$

A recombination frequency of 50% indicates that there is no linkage and the genes are on separate chromosomes. Less than 50% indicates that there is gene linkage and the random process of independent assortment has been hindered. As the degree of crossing over reduces, the recombination frequency also gets smaller. The degree of crossing over is determined by how close the genes are on a chromosome. The closer they are the less likely they will be separated during crossing over and vice versa.

The recombination frequencies for a number of characteristics coded for by genes on the same chromosome can be used to map the genes on the chromosome. A recombination frequency of 1% relates to a distance of one map unit on a chromosome.

Chi-squared test

The observed results from a genetic cross will almost always differ to some extent from the expected results and this will be due to chance. If you toss a coin 10 times you would be unlikely to get five heads and five tails. The observed ratio of heads to tails will probably be quite different from the expected ratio. This does not mean there is anything wrong with the coin. If the same coin were tossed a thousand times you would see less relative difference between the expected and observed ratios. The number of observations made, therefore, determines how chance affects the results.

It is important when making comparisons between observed and expected results that it is known whether any differences are due to chance or if there is a reason for the differences (they are significant).

The **chi-squared (χ^2) test** is a statistical test that measures the size of the difference between the results you actually get (observe) and those you expected to get. It helps you determine whether differences in the expected and observed results are significant or not, by comparing the sizes of the differences and the numbers of observations.

The chi-squared test is conventionally used to test the null hypothesis. The null hypothesis is that there is no significant difference between what we expect and what we observe – in other words any differences we do see are due to chance. Calculated chi squared values are used to find the probability of the difference being due to chance alone.

The chi-squared formula is:

$$\chi^2 = \sum \frac{(O - E)^2}{E}$$

Where:

χ^2 = the test statistic

$\sum$ = the sum of

O = observed frequencies

E = expected frequencies

Large chi-squared values mean there is a statistically significant difference between the observed and expected results and the probability that these differences are due to chance is low. There must be a reason, other than chance, for the unexpected results.

The number of categories being compared in an investigation affects the size of the chi-squared value calculated. The degrees of freedom is the number of comparisons being made and is calculated as $n-1$, where n is the number of categories or possible outcomes (phenotypes in the case of phenotypic ratios) present in the analysis. For example, if you were looking at yellow and green peas there would be two categories and therefore one degree of freedom.

Study tip

A full table of chi-squared values can be found in the appendix.

▼ **Table 1** *Table of chi-squared values and the probability of each occurring at different degrees of freedom (df)*

df	p values								df
	0.99	0.95	0.90	0.50	0.10	0.05	0.01	0.001	
1	0.0001	0.0039	0.016	0.46	2.71	3.84	6.63	10.83	1
2	0.02	0.10	0.21	1.39	4.60	5.99	9.21	13.82	2
3	0.12	0.35	0.58	2.37	6.25	7.81	11.35	16.27	3
4	0.30	0.71	1.06	3.36	7.78	9.49	13.28	18.46	4
5	0.55	1.41	1.61	4.35	9.24	11.07	15.09	20.52	5

▦ = null hypothesis accepted

▬ = null hypothesis rejected – another factor involved

If the calculated χ^2 value is less than the critical value found in a table at 5% significance (p=0.05) we do not have sufficiently strong evidence to reject our null hypothesis. Therefore, we accept the null hypothesis – there is no significant difference between what we observed and what we expected. However if the calculated χ^2 value is greater than the critical value we reject the null hypothesis – some other factor, outside our original expectation, is likely to be causing a significant difference between expectation and observation.

The minimum χ^2 value that gives a 5% probability is called the critical value. The critical value increases as the degrees of freedom increase.

If χ^2 is less than the critical value there is no significant difference

If χ^2 is greater than or equal to the critical value there is a significant difference

🖩 Worked example: Calculating the chi-squared value

A monohybrid cross involving two plants heterozygous for purple flowers was carried out.

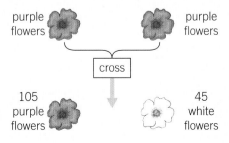

purple flowers   purple flowers

cross

105 purple flowers   45 white flowers

→

In this case, the null hypothesis is that any differences in the observed numbers of purple and white phenotypes from the expected numbers are due to chance.

Step 1: The probability of getting each expected phenotype is calculated.

Purple is dominant to white, so there should be a 3:1 ratio in the offspring.

The chi-squared value is calculated based on this 3:1 hypothesis.

Step 2: The expected number of each phenotype based on the probabilities and total number of offspring is calculated by multiplying the expected proportion by the total:

There are 150 offspring so $150 \times \dfrac{3}{4} = 112.5$

$$150 \times \dfrac{1}{4} = 37.5$$

Step 3: Next, put your calculated and observed values into a table:

Phenotype	Observed	Expected
purple	105	$\dfrac{3}{4} \times 150 = 112.5$
white	45	$\dfrac{1}{4} \times 150 = 37.5$
total	150	

Step 4: Then the chi-squared value can be calculated:

$$\chi^2 = \Sigma \frac{(O - E)^2}{E}$$

$$\chi^2 = \frac{(105 - 112.5)^2}{112.5} + \frac{(45 - 37.5)^2}{37.5}$$

$$\chi^2 = \frac{56.25}{112.5} + \frac{56.25}{37.5}$$

$$\chi^2 = 0.5 + 1.5 = 2.0$$

Step 5: The chi-square value is determined from the chi-squared distribution table at the 0.05 probability level (p) for the correct degrees of freedom. Remember this is $n - 1$. In this case there are only two phenotypes so $n = 2$ and the degrees of freedom is $2 - 1 = 1$.

The table value is 3.84.

Step 6: Because 2.0 < 3.84, the null hypothesis is accepted. The difference between the predicted and actual cross results is not significant (at the 5% probability level). Any difference between the expected results and the actual results is due to chance.

If the χ^2 value had been greater than 3.84, it would indicate that the differences in the observed results were due to another factor such as gene linkage or a new or different way of inheriting alleles.

Study tip

You do not need to learn the chi-squared formula but you need to know how apply it.

Corn and the chi-squared (χ^2) test

There will almost always be differences between expected and observed results because of the random nature of the processes involved. Statistical tests like the chi-squared test are performed to determine whether these differences are due to chance alone or caused by some other factor that may not have been considered.

Maize plants have been used for many years to study genetic crosses. An ear of corn contains around 500 kernels, or seeds. The seeds are produced as the result of the cross-fertilisation of two maize plants. The colour of the seeds is controlled by one gene with a dominant (P, purple) allele and a recessive (p, yellow) allele. Another gene determines the texture of the seeds and there is again a dominant (R, round) allele and recessive (r, wrinkled) allele.

◀ Figure 3

A genetic study was carried out to determine if these two genes are linked.

Two maize plants that were heterozygous for both colour and texture were cross-fertilised and an ear of corn produced as a result of this cross was analysed.

The results are shown in Table 2.

1 State the ratio of phenotypes that you expect from this cross if the genes are not linked.
2 Copy and complete the table.
3 Show how you would use the chi-squared distribution table, shown in Table 1, to determine the critical value for these results.
4 State, giving your reasons, the conclusions you would make from your statistical analysis of these results.

▼ **Table 2** *Shows the number of each of the different types of kernel produced from a cross between two maize plants which are both heterozygous for colour and texture*

Grain phenotype	Observed number	Observed ratio	Expected ratio	Expected number	(observed − expected)^2 / expected
purple and smooth	216			$381 \times \dfrac{9}{16} = 214$	
purple and shrunken	79			$381 \times \dfrac{3}{16} = 71$	
yellow and smooth	65				
yellow and shrunken	21				
				chi-squared value:	

Epistasis

Epistasis is the interaction of genes at different loci. Gene regulation is a form of epistasis with regulatory genes controlling the activity of structural genes, for example, the lac operon. Gene interaction also occurs in biochemical pathways involving only structural genes.

It was originally thought that all genes were expressed independently, and therefore their effects on the phenotype seen. Now it is known that many genes interact *epistatically*. It is the results of these interactions that we see in the phenotypes of living organisms. The characteristics of plants and animals that show continuous variation involve multiple genes and epistasis occurs frequently.

Synoptic Link

You learnt about the lac operon in Topic 19.2, Control of gene expression.

Consider the biochemical pathway shown in Figure 3, where four enzymes a, b, c, and d are required to produce a pigment responsible for the colour of a flower petal. Four genes a, b, c, and d need to be expressed to produce these enzymes. If one of these genes is not expressed then one step will be missing, and the petal will not have the expected colour.

The lack of the enzyme normally produced when gene a, b, or c is expressed means the intermediate molecule necessary for the next reaction in the sequence is not produced. This results in a lack of substrate for the next enzyme in the pathway and so the expression of this gene will not be observed in the phenotype. The gene is effectively 'masked' by the lack of expression of the previous gene in the pathway.

If enzyme d is not produced then precursor **D** will not be converted to a pigment. This means it is often hard to observe the expression of genes **a**, **b**, and **c** in the phenotype. The disabled gene **d** is likely to mask their expression.

A gene that is affected by another gene is said to be hypostatic and a gene that affects the expression of another gene is said to be epistatic.

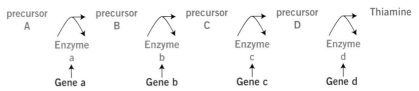

▲ **Figure 3** *The genes a, b, and c code for enzymes that catalyse the production of the substrate needed for the next reaction in the sequence. Gene d produces an enzyme that catalyses the production of the pigment from precursor D*

Dominant and recessive epistasis

An epistatic gene may influence the activity of other genes as result of the presence of dominant or recessive alleles.

In the example previously, if the presence of two recessive alleles at a gene locus led to the lack of an enzyme then it would be called recessive epistasis.

Dominant epistasis occurs if a dominant allele results in a gene having an effect on another gene. This would happen if an epistatic gene (not present in this pathway) coded for an enzyme that modified one of the precursor molecules in the pathway. The next enzyme in the pathway would then lack a suitable substrate molecule and so the pigment would again not be produced. All of the genes in the sequence would be effectively 'masked'.

Labrador colours

The colour of Labrador dogs is produced as a result of the pigment melanin being deposited in the skin and fur. One gene codes for the production of the pigment and has the alleles B (dominant, black pigment produced) and b (recessive, brown pigment produced). A second gene codes for where the pigment is deposited and, again, has two alleles E (dominant, pigment deposited in the skin and fur) and e (recessive, pigment deposited in the skin only).

The colour of a Labrador varies depending on which alleles are present at each locus. The genes are not expressed independently and so this is an example of epistasis. The gene at the E locus is epistatic to the hypostatic gene at the B locus.

The different phenotypes and genotypes are shown in Table 3.

▼ Table 3

B locus genotype	E locus genotype	Fur colour	Skin colour	Overall colour	Phenotype
BB/Be	EE/Ee	black	black	black	
BB/Be	ee	yellow	black	brown	
bb	EE/Ee	brown	brown	yellow	
bb	ee	yellow	brown	yellow	

There are only three registered colours, black, brown (chocolate), and yellow (golden). The yellow coat is an example of recessive epistasis and it ranges from deep gold to pale blond.

Summary questions

1 Explain what the chi-squared (χ^2) test is used to measure.   (*3 marks*)

2 Horse coat colour is an example of epistasis. Two genes are involved. The different colours and the genotypes responsible are summarised here.

   G___ produces a grey horse

   gg E_ produces a black horse

   gg ee produces a chestnut horse.

   Explain, giving your reasons, which form of epistasis this represents.
   (*5 marks*)

3 A biologist test crosses a plant that is heterozygous for the alleles Xx and Yy in order to see how far apart the gene loci are on a chromosome. The offspring of this cross contains 5.2% recombinant individuals.
   a State the distance between these alleles.   (*1 mark*)
   b Describe how the biologist could determine where a third gene with the alleles Zz is on the chromosome relative to Xx and Yy.   (*4 marks*)
   c Suggest why this sort of investigation is best carried out with genes that are relatively close to each other.   (*2 marks*)

20.5 Evolution

Specification reference: 6.1.2

Evolution, the change in inherited characteristics of a group of organisms over time, occurs due to changes in the frequency of different alleles within a population.

Population genetics

Population genetics investigates how allele frequencies within populations change over time. The sum total of all the genes in a population at any given time is known as the **gene pool**. The gene pool of a population includes millions of genes, but you will look at the variation in the different alleles of a single gene within the gene pool. The relative frequency of a particular allele in a population is the **allele frequency**.

The frequency with which an allele occurs in a population is not linked to whether it codes for a dominant or a recessive characteristic, and it is not fixed. It can change over time in response to changing conditions. Evolution involves a long-term change in the allele frequencies of a population, for example, alleles for antibiotic resistance have increased in many bacteria populations over time. Biologists have developed ways of determining allele frequencies and use them in models to determine whether evolution is taking place.

Calculating allele frequency

Imagine a population of 100 diploid organisms that can all breed successfully. You are going to look at a gene that has two possible alleles, A and a. The frequency of allele A in the population is represented by the letter p. The frequency of allele a in the population is represented by q. If every individual in your population of 100 is a heterozygote (Aa), then the frequency of each allele is 100/200 or 0.5 (50%) so $p + q = 1$

In a diploid breeding population with two potential alleles, the frequency of the dominant allele plus the frequency of the recessive allele will *always* equal 1. This simple formula is very important when using the Hardy–Weinberg principle.

The Hardy–Weinberg principle

The Hardy–Weinberg principle models the mathematical relationship between the frequencies of alleles and genotypes in a theoretical population that is stable and not evolving. The Hardy–Weinberg principle states: **in a stable population with no disturbing factors, the allele frequencies will remain constant from one generation to the next and there will be no evolution**. A completely stable population is not common in the real world, but this is still a useful tool. The Hardy–Weinberg principle provides a simple model of a theoretical stable population that allows us to measure and study evolutionary changes when they occur.

The Hardy–Weinberg principle is expressed as:

$$p^2 + 2pq + q^2 = 1$$

where p^2 = frequency of homozygous dominant genotype in the population

   $2pq$ = frequency of heterozygous genotype in the population

   q^2 = frequency of homozygous recessive genotype in the population

How do you use this information? Recessive phenotypes are often easy to observe. As a result you can find the frequency of the recessive genotype and use it to measure the equivalent allele frequency.

Frequency of the recessive genotype = q^2

So, the frequency of the recessive allele is $\sqrt{q^2} = q$

You can then use this to find p because you know $p + q = 1$

Finally, you can substitute these values back into the equation of the Hardy–Weinberg principle to find the frequencies of the three different genotypes.

Hardy–Weinberg worked example

The peppered moth, *Biston betularia*, comes in two forms, light coloured and dark coloured. The light colour is inherited through a recessive allele. Students investigated a population in an area of woodland and found that 48 of the 50 peppered moths they captured were light in colour.

This gives the frequency of the homozygous recessive genotype (q^2) that results in a light colouration as 48/50, or 0.96 (96%). Now you can calculate the value of q, the frequency of the allele in the population.

$q^2 = 0.96$

so $q = \sqrt{0.96} =$ **0.98 (98%)** (2 s.f.)

You know that $p + q = 1$, so $p + 0.98 = 1$, so $p = 1 - 0.98 =$ **0.02 (2%)**

Now substitute these values into the equation for the Hardy–Weinberg principle to work out the frequency of the homozygous dominant genotype and the heterozygous genotype in this population of *Biston betularia*.

Frequency of homozygous dominant genotype (p^2) = **$0.02^2 = 0.0004$ (0.04%)**

Frequency of the heterozygous genotype (**$2pq$**) = **$2 \times 0.02 \times 0.98 = 0.039$ (3.9%)** (2 s.f.)

This gives you the frequencies for the three main genotypes of the *Biston betularia* population in the woodland studied. Around 96% of the moths are homozygous recessive and therefore light coloured,   3.9% are heterozygous and so dark coloured, and 0.04% are homozygous dominant and dark in colour.

Remember allele frequencies must add up to 1 and population percentages to 100% (allowing for rounding numbers up or down).

Disturbing the equilibrium

The Hardy–Weinberg principle assumes a theoretical breeding population of diploid organisms that is large and isolated, with random mating, no mutations, and no selection pressure of any type. In a natural environment these conditions virtually never occur. Species are continuously changing. In the peppered moths of the worked example, the light alleles were dominant historically but the allele frequencies changed dramatically after the Industrial Revolution, when the dark alleles gave individuals an advantage. Now the allele frequencies have changed again as cities and woodlands have become cleaner again. These changes in allele frequencies can be illustrated using the Hardy–Weinberg principle and upsetting the equilibrium may eventually result in evolution.

Factors affecting evolution

There are a number of factors that lead to changes in the frequency of alleles within a population and so they affect the rate of evolution:

- Mutation is necessary for the existence of different alleles in the first place, and the formation of new alleles leads to genetic variation.

- Sexual selection leads to an increase in frequency of alleles which code for characteristics that improve mating success.

- Gene flow is the movement of alleles between populations. Immigration and emigration result in changes of allele frequency within a population.

- Genetic drift occurs in small populations. This is a change in allele frequency due to the random nature of mutation. The appearance of a new allele will have a greater impact (is more likely to increase in number) in a smaller population than in a much larger population where there is a greater number of alleles present in the gene pool.

- Natural selection leads to an increase in the number of individuals that have characteristics that improve their chances of survival. Reproduction rates of these individuals will increase as will the frequency of the alleles coding for the characteristics. This is how changes in the environment can lead to evolution.

The impact of small populations

The gene pool of a large population ensures lots of genetic diversity owing to the presence of many different genes and alleles. Genetic diversity leads to variation within a population which is essential in the process of natural selection. Selection pressures such as changes in the environment, the presence of new diseases, prey, competitors, or even human influences lead to evolution. The population can adapt to change over time.

Small populations with limited genetic diversity cannot adapt to change as easily and are more likely to become extinct. A new strain of pathogen could wipe out a whole population.

The size of a population can be affected by many factors. Factors which limit or decrease the size of a population are called limiting factors. There are two types of limiting factors:

1 Density-dependent factors are dependent on population size and include competition, predation, parasitism, and communicable disease.

2 Density-independent factors affect populations of all sizes in the same way including – climate change, natural disasters, seasonal change, and human activities (for example, deforestation).

Large reductions in population size which last for at least one generation are called population bottlenecks (Figure 1). The gene pool, along with genetic diversity, is greatly reduced and the effects will be seen in future generations. It takes thousands of years for genetic diversity to develop in a population through the slow accumulation of mutations.

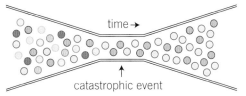

▲ **Figure 1** *A natural disaster or epidemic can drastically reduce a population. The gene pool will be greatly reduced and the remaining individuals may not be representative of the original population as some rarer alleles may not have been present in any of the survivors. The 'founder effect' and genetic drift will influence genetic variation as the population grows again*

Northern elephant seals were almost hunted to extinction in the 19th century. There were probably only about 20 seals left by the time hunting stopped. They now have a population of about 30 000 but show much less genetic diversity than southern elephant seals that did not experience a genetic bottleneck.

Cheetahs are thought to have experienced an initial population bottleneck about 10 000 years ago with other bottlenecks happening more recently. The species now shows low genetic diversity. Cheetahs face the same threats as many other African animals such as habitat loss and poaching, but while the population sizes of other animals are increasing thanks to the efforts of conservationists, cheetahs are not recovering as quickly. They are, in fact, close to extinction.

▲ **Figure 2** *Cheetah mother and cubs, Acinonyx jubatus, Masai Mara Reserve, Kenya*

The reduced genetic diversity of cheetahs means that they share around 99% of their alleles with other members of the species, more than we share with members of our own family. Mammals usually share about 80% of their alleles with other members of a species. As a result they are showing problems of inbreeding including reduced fertility.

Humans and chimpanzees split from a common ancestor about six million years ago. A small group of chimpanzees are likely to show more genetic diversity than all the humans alive today. It is believed that humans have experienced at least one genetic bottleneck, reducing our genetic diversity, as we have evolved into our present form.

A positive aspect of a genetic bottleneck is that a beneficial mutation will have a much greater impact and lead to the quicker development of a new species. This is thought to have a played a role in the evolution of early humans.

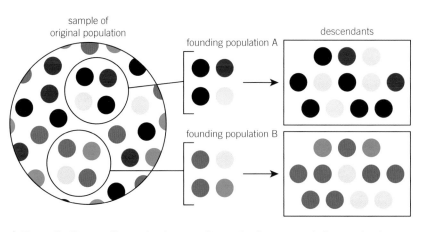

sample of original population

founding population A

descendants

founding population B

▲ **Figure 3** *Diagram illustrating how small samples from a population can lead to populations with very different, and reduced, gene pools*

Founder effect

Small populations can arise due to the establishment of new colonies by a few isolated individuals, leading to the **founder effect**. The founder effect is an extreme example of genetic drift.

These small populations have much smaller gene pools than the original population and display less genetic variation. If carried to the new population, the frequency of any alleles that were rare in the original population will be much higher in the new, smaller population and so they will have a much bigger impact during natural selection.

The Afrikaner population in South Africa is descended mainly from a few Dutch settlers. The population today has an unusually high frequency of the allele that causes Huntington's disease. It is thought that just one of the original settlers carried the disease-causing allele.

The Amish people of America have descended from 200 Germans who settled in Pennsylvania in the 18th century. They rarely marry and have children outside their own religion and are therefore a closed community. The Amish have unusually high frequencies of alleles that cause the normally rare genetic disorder Ellis–van Creveld syndrome. People with the syndrome are short, they often have polydactyly (extra fingers or toes), abnormalities of nails and teeth, and a hole between the two upper chambers of the heart. Ellis–van Creveld syndrome is an example of founder effect caused by one couple, Samuel King and his wife, who settled in the area in 1744.

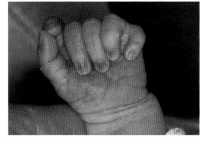

▲ **Figure 4** *Close-up of a baby's hand showing an extra finger. This condition is called polydactylism*

Evolutionary forces

The traits or characteristics of all living organisms show variation within populations. The distribution of the different variants will take the form of a bell-shaped curve if plotted on a graph. This is known in statistics as a **normal distribution**.

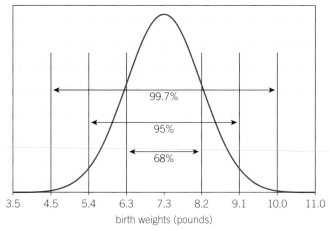

▲ **Figure 5** *Birth weights of baby girls in Europe*

Stabilising selection

Taking the birth weight of babies as an example, babies with an average birth weight will be the most common and therefore form the peak of the graph. Babies with very low birth weight are more prone to infections and very large babies result in difficult births. Both of these extremes in weight reduce the survival chances of babies so the numbers of survivals of very small or very large babies remains low forming the tails on Figure 5.

This is natural selection, or survival of the fittest, at work. Babies with average birth weights are more likely to survive and reproduce than underweight or overweight babies. It is an example of **stabilising selection** because the norm or average is selected for (positive selection) and the extremes are selected against (negative selection). Stabilising selection therefore results in a reduction in the frequency of alleles at the extremes, and an increase in the frequency of 'average' alleles.

Directional selection

Directional selection occurs when there is change in the environment and the normal (most common) phenotype is no longer the most advantageous. Organisms which are less common and have more extreme phenotypes are positively selected. The allele frequency then shifts towards the extreme phenotypes and evolution occurs.

The changes seen in peppered moths during the industrial revolution are a good example of directional selection. During this period of time a lot of smoke was released from factories, which killed lichens growing on barks of trees, and the soot made the bark black. Peppered moths were originally light coloured meaning they were camouflaged by the lichen from predation by birds. There were always a few darker moths present, due to variation, but these were quickly eaten and the allele frequency maintained.

When the lichens died and the trees became black the situation was reversed. The light-coloured moths were very visible and were eaten and the darker moths were camouflaged. Over time the allele frequency shifted due to natural selection and the majority of the peppered moths had the darker colour. The allele frequency had been shifted towards an extreme (less common) phenotype.

As pollution has decreased again the allele frequency of the lighter coloured moths has increased.

Disruptive selection

In **disruptive selection** the extremes are selected for and the norm selected against. The finches observed by Darwin in the Galapagos Islands had been subjected to disruptive selection. This is opposite to stabilising selection when the norm is positively selected.

▲ Figure 6 *Light and dark-coloured peppered moths*

▲ Figure 8 *Top: lazuli bird with bright, blue plumage, middle: lazuli bird with intermediate plumage, and bottom: lazuli bird with dull, brown plumage*

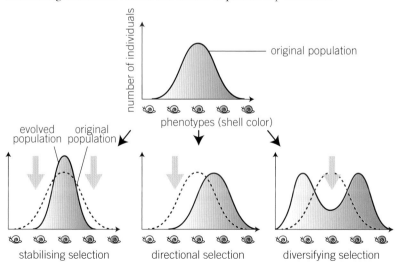

▲ Figure 7 *Graphs showing the different forms of selection. The arrows indicate a selection pressure*

Although examples of disruptive selection are relatively rare, a well-documented example involves feather colour in male lazuli buntings (*Passerina amoena*), birds which are native to North America. The feather colour of young males can range from bright blue to dull brown.

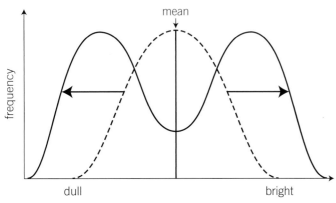

▲ **Figure 9** *The distribution of phenotypes as a result of disruptive selection pressures on lazuli buntings*

There are limited nesting sites in their habitat and so there is a lot of competition between male birds to establish territories and attract female birds. Dull, brown males are seen as non-threatening and bright, blue males too threatening by adult males. Both the brown and blue birds are therefore left alone but birds of intermediate colour are attacked by adult birds and so fail to mate or establish territories.

The extremes are selected for and the distribution of phenotypes shows two peaks as in Figure 9.

Synoptic link

You learnt about natural selection in Topic 10.4, Evidence for evolution.

Summary questions

1   Explain why evolution does not occur within single organisms but groups of organisms. *(3 marks)*

2   Around the world, humans choose their partners for a wide variety of reasons. Explain why this might affect any conclusions about human evolution drawn using the Hardy–Weinberg principle. *(3 marks)*

3   In cats, the short-haired allele L is dominant to the long-haired allele l. In a population of feral cats, 10 out of 90 animals had long hair. Give the expected frequency for the homozygous recessive, homozygous dominant, and heterozygote genotypes in this population of cats. *(6 marks)*

4   Explain why the allele frequency is changing so quickly in Figure 10. *(4 marks)*

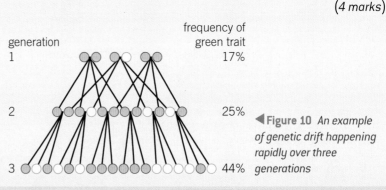

◀ **Figure 10** *An example of genetic drift happening rapidly over three generations*

5   Eukaryotic organisms have large quantities of non-coding DNA whereas most prokaryotic organisms have very little.

Suggest, with reference to the different forms of reproduction in eukaryotes and prokaryotes, why eukaryotes may have evolved to have more non-coding DNA. *(4 marks)*

20.6 Speciation and artificial selection

Speciation is the formation of new species through the process of evolution. The organisms belonging to the new species will no longer be able to interbreed to produce fertile offspring with organisms belonging to the original species. A number of events happen leading to speciation:

- Members of a population become isolated and no longer interbreed with the rest of the population resulting in no gene flow between the two groups.

- Alleles within the groups continue to undergo random mutations. The environment of each group may be different or change (resulting in different selection pressures) so different characteristics will be selected for and against.

- The accumulation of mutations and changes in allele frequencies over many generations eventually lead to large changes in phenotype. The members of the different populations become so different that they are no longer able to interbreed (to produce fertile offspring). They are now reproductively isolated and are different species.

Allopatric speciation

Allopatric speciation is the more common form of speciation and happens when some members of a population are separated from the rest of the group by a physical barrier such as a river or the sea – they are geographically isolated. The environments of the different groups will often be different and so will the selection pressures resulting in different physical adaptations. Separation of a small group will often result in the founder effect leading to genetic drift further enhancing the differences between the populations.

A famous example of allopatric speciation is the finches inhabiting the Galapagos Islands located in the Pacific Ocean off the coast of South America. For about two million years, small groups of finches, from an original population on the mainland, have flown to, and been stranded on, different islands. The finches, separated from finches on other islands and the mainland by the sea, have formed new colonies on the different islands.

The finches have evolved and adapted to the different environments, particularly food sources, present on the islands and are an example of *adaptive radiation* – where rapid organism diversification takes place. As the finches are unable to breed with each other, new species have evolved with unique beaks adapted to the type of food available. Some species

Learning outcomes

Demonstrate knowledge, understanding, and application of:

→ the role of isolating mechanisms in the evolution of new species

→ the principles of artificial selection and its uses

→ the ethical considerations surrounding the use of artificial selection.

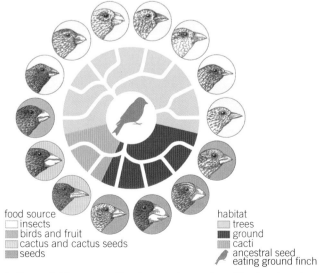

food source
☐ insects
▨ birds and fruit
▨ cactus and cactus seeds
▨ seeds

habitat
☐ trees
■ ground
▨ cacti
✦ ancestral seed eating ground finch

▲ **Figure 1** *The range of beaks seen in species of finches found in the Galapagos Islands*

▲ **Figure 2** *Snapping shrimp*

have large, blunt beaks that can crack nuts, some have long, thin beaks for getting to the nectar in flowers, and some have medium-sized beaks which are ideal for catching insects.

The honeycreepers (family Drepanidinae) of the islands of Hawaii are birds that are an even larger example of adaptive radiation. A single ancestor species has led to the evolution of at least 54 species that have filled every available niche in the different islands.

Panama is a narrow strip of land (isthmus) that joins North and South America and separates the Atlantic and Pacific oceans, and was formed about three million years ago. This was due to the movement of tectonic plates and resulted in the separation of the organisms that had originally occupied the same habitat when the two oceans were joined.

There were originally about 15 species of snapping shrimp present, now there are 15 species present on one side of the isthmus and 15 different species present on the other side. Although the shrimp from either side appear to be identical if males and females are mixed they will snap at each other rather than mate.

In 1995 15 iguanas, *Iguana iguana*, survived a hurricane in the Caribbean on a raft of uprooted trees. They eventually reached the Caribbean island of Anguilla. These iguanas were the first of their species to reach the island. If these iguanas are successful in colonising the island it could be the start of an allopatric speciation. Of course, it could take thousands, if not millions, of years before this is known.

Sympatric speciation

Sympatric speciation occurs within populations that share the same habitat. It happens less frequently than allopatric speciation and is more common in plants than animals. It can occur when members of two different species interbreed and form fertile offspring – this often happens in plants. The hybrid formed, which is a new species, will have a different number of chromosomes to either parent and may no longer be able to interbreed with members of either parent population. This stops gene flow and reproductively isolates the hybrid organisms. Examples of sympatric speciation include fungus-farming ants and blind mole rats.

Fungus-farming ants cultivate the growth of fungi, which is their source of nutrition, by supplying organic material to keep the fungi growing. Parasitic ants have been found in one colony of these industrious ants. Instead of helping in the growth of this fungus these parasitic ants spend their time eating the fungi and reproducing. The parasitic ants are sometimes ignored and at other times attacked and killed. Genetic analysis has shown that although genetically different from the fungus-farming ants the parasitic ants are, in fact, their descendants. They are not a species of ant that has evolved in geographic isolation but within the same habitat as a result of a change in behaviour. It is believed that the genetic division of the original species of ant only happened 37 000 years ago, not long in evolutionary terms.

Blind mole rats live in a small area of northern Israel that is part igneous basalt rock and part chalk bedrock. The different types of soils formed above the bedrock support a different range of plants. Blind mole rats found in both types of soil are sometimes only separated by a few metres of the loose soil. Mole rats will only interbreed with mole rats living in the same type of soil. DNA analysis has shown that the lack of gene flow between the two species is already resulting in genetic differences even though members of the different groups often come into contact with each other as there is no physical barrier. Over time the genetic differences could accumulate to the point that the mole rats from different soil types will no longer be able to interbreed and they will be separate species.

Plants cross with plants of different species forming hybrids much more frequently than animals. The indiscriminate release of large numbers of pollen grains by plants is one reason for this. The hybrids are reproductively isolated from each parent species but could still be present in the same habitat. The evolution of modern wheat has involved at least two hybridisation events and the formation of new species along the way.

Disruptive selection, mating preferences, and other behavioural differences can all result in individuals or small groups becoming reproductively isolated. They will, however, still be living in the same habitat so gene flow, even if reduced, often interferes with the process of speciation.

Reproductive barriers

Barriers to successful interbreeding can form within populations before or after fertilisation has occurred. Prezygotic reproductive barriers prevent fertilisation and the formation of a zygote. Postzygotic reproductive barriers, often produced as a result of hybridisation, reduce the viability or reproductive potential of offspring.

Artificial selection

Populations are usually **polymorphic** (display more than one distinct phenotype) for most characteristics. The allele coding for the most common, or normal, characteristic is called the *wild type* allele. Other forms of that allele, resulting from mutations, are called mutants.

Artificial selection (or **selective breeding**) is fundamentally the same as natural selection except for the nature of the selection pressure applied. Instead of changes in the environment leading to survival of the fittest, it is the selection for breeding of plants or animals with desirable characteristics by farmers or breeders.

Farmers have been selectively breeding plants and animals since before genes were discovered or the theory of evolution was proposed. Individuals with the desired characteristics are selected and interbred.

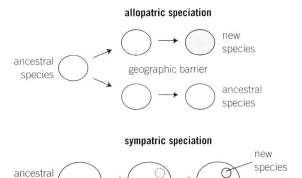

▲ **Figure 3** *Diagram showing the formation of a new species both with and without a geographical barrier*

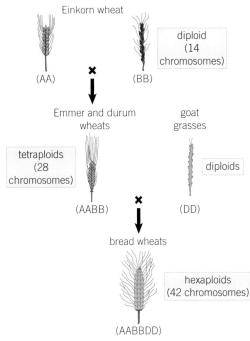

▲ **Figure 4** *The evolution of modern wheat including two hybridisation events which would have both led to sympatric speciation*

Offspring from this cross showing the best examples of the desired traits are then selected to breed. This breeding of closely related individuals is called **inbreeding**. The process is repeated over many generations resulting in changes to the frequency of alleles within the population and eventually speciation.

Brassica oleracea is a wild mustard which has been selectively bred for many centuries producing a number of common vegetables.

Problems caused by inbreeding

Limiting the gene pool and so decreasing genetic diversity reduces the chances of a population of inbred organisms evolving and adapting to changes in their environment.

Many genetic disorders are caused by recessive alleles, for example, cystic fibrosis, a condition where the digestive system and lungs become clogged with mucus. Recessive alleles are not uncommon in most populations but two recessive alleles are needed before they are expressed and most individuals will be heterozygous.

Organisms that are closely related are genetically similar and are likely to have the same recessive alleles. The breeding of closely related organisms therefore results in offspring which have a greater chance of being homozygous for these recessive traits and being affected by genetic disorders. Over time this reduces the ability of these organisms to survive and reproduce. This results in the organisms being less biologically fit – in other words, less likely to survive and produce two surviving offspring to replace themselves.

 Pedigree dogs and the ethics of artificial selection

Domesticated dogs are all members of the same species, *Canis familiaris*. They are another, sometimes controversial, example of how selective breeding has created a variety of different-looking individuals from one wild species. The wild species was the grey wolf and the process began between 18 000 and 32 000 years ago when humans were still leading a hunter-gatherer lifestyle.

It is thought that wolves starting 'hanging around' human hunting parties because of the availability of scraps of leftover food. Over time, the more social wolves would have become integrated into the human groups and so the process of selective breeding began. The wolves that became integrated were eventually used to help catch animals during the hunts or served a protective role like modern-day guard dogs.

Hunter-gatherers eventually started forming settled communities and began the practice of farming. The evolving wolves would have had new roles such as

herding animals. Different traits were selected for depending on whether they were used for hunting, fighting, herding, or even as status symbols, and so a range of dogs with different characteristics evolved.

Many of those characteristics have been exaggerated by continued selective breeding and are most obvious in pedigree dogs seen today. Rather than being selected for the role they performed which was not dissimilar to natural selection, they began to be selected for their looks which took no account of any impact on their health. Interbreeding with wild wolves would have been common which can make tracing the evolution of dogs difficult.

Dachshunds were selected for small size and short legs so that they could follow prey such as foxes and badgers into burrows. Great Danes were selected for large size and strength for hunting and fighting.

The breeding of pedigree dogs is restricted to the descendants of dogs that were registered by the Kennel club in 1873 after the different types or breeds of dog had been developed by breeders and the standard characteristics of each breed identified. With the limited gene pool and lack of outbreeding it is inevitable that unwanted traits are selected for also. Big dogs often have hip or heart problems. The skull of the King Charles spaniel is too small to accommodate the brain comfortably leading to pain and discomfort. Bulldogs usually have breathing difficulties due to the shape of their noses. Dachshunds often have back problems and suffer from epilepsy. Diseased dogs have effectively been deliberately interbred. There are now moves to change some of the breed descriptions to prevent some of the worst examples of this practice.

Selective breeding has been used for centuries in farming to improve the quality and yield of crops and animal produce.

> Discuss the ethical considerations of using selective breeding to produce aesthetically pleasing show dogs and high-yielding plants and animals for agricultural use.

Gene banks

Seed banks keep samples of seeds from both wild type and domesticated varieties. They are an important genetic resource. **Gene banks** store biological samples, other than seeds, such as sperm or eggs. They are usually frozen.

Owing to the problems caused by inbreeding, alleles from gene banks are used to increase genetic diversity in a process called outbreeding. Breeding unrelated or distantly related varieties is also a form of outbreeding. This reduces the occurrence of homozygous recessives and increases the potential to adapt to environmental change.

Synoptic link

You learnt about seed banks in Topic 11.8, Methods of maintaining biodiversity.

Summary questions

1 Describe, with examples, the difference between pre-zygotic and post-zygotic reproductive barriers. *(4 marks)*

2 A small population of adders in Sweden underwent inbreeding depression when farming activities isolated them from other adder populations. The numbers of stillborn and deformed offspring increased as compared to the original population. Researchers introduced adders from other population and the isolated population recovered and produced a higher proportion of viable offspring.
   a Name the type of breeding carried out by the researchers. *(1 mark)*
   b Explain how this type of breeding reduces the problems caused by inbreeding. *(3 marks)*

3 Discuss why variation within a species has to be present for speciation to occur. *(5 marks)*

4 Allopatric speciation is considered by most biologists to be the most common way in which new species evolve.

   Outline the differences between sympatric speciation and allopatric speciation and suggest why some biologists think sympatric speciation is a rare event. *(5 marks)*

Practice questions

1 a The presence or absence of red pigmentation in the outer scales of onion bulbs is controlled by two genes, A/a and B/b.

- The dominant allele, A, codes for the production of a red anthocyanin pigment.
- Onion bulbs homozygous for the recessive allele, a, produce no pigment and are white.
- The dominant allele, B, inhibits the expression of allele A.
- The recessive allele, b, allows anthocyanin production.

(i) Describe the effect of allele B on allele A. *(3 marks)*

(ii) Two onion plants with the genotypes AABB and aabb were cross-pollinated and the resulting F1 generation interbred to give an F2 generation.

Draw a genetic diagram of this cross to show:

- the phenotypes of the parent plants
- the gametes
- the genotypes and phenotypes of the F1 and F2 generations
- the ratio of phenotypes expected in the F2 generation. *(8 marks)*

b Most red-scaled onion bulbs produce a colourless substance which makes them resistant to a fungal infection called 'smudge'. Most white onion bulbs are susceptible to 'smudge'. Suggest why:

(i) Resistance to 'smudge' is almost always inherited together with red pigmentation *(2 marks)*

(ii) Some white onion bulbs are resistant to 'smudge'. *(2 marks)*

OCR 2805/02 2008

2 A pure-breeding variety of tomato plant, variety A, produced red fruit which remained green at their bases even when ripe.

Plants of variety A were crossed with another pure-breeding variety, B, with orange fruit which have no green bases when ripe.

The F1 generation plants all had red fruit with green bases.

a Describe the interaction of the alleles,

(i) At the locus G/g, controlling green-based or not green-based fruit *(1 mark)*

(ii) At the locus R/r, controlling red or orange fruit. *(1 mark)*

b Using the symbols given in (a), state the genotype of variety B. *(1 mark)*

c Plants from the F1 generation were test crossed (backcrossed) to variety B. The ratio of phenotypes expected in a dihybrid test cross such as this is 1 : 1 : 1 : 1.

Using the symbols given in (a), draw a genetic diagram of the test cross to show that the expected ratio of offspring phenotypes is 1 : 1 : 1 : 1. *(4 marks)*

d Two hundred randomly chosen offspring from the test cross described in c had the following phenotypes:

- red fruit with green bases 55
- red fruit with no green bases 45
- orange fruit with green bases 43
- orange fruit with no green bases 57

The χ^2 (chi-squared) test was performed on these data, giving a calculated value for χ^2 of 3.2.

(i) State the number of degrees of freedom applicable to these data. *(1 mark)*

Distribution of χ^2 values

degrees of freedom	probability p				
	0.10	0.05	0.02	0.01	0.001
1	2.71	3.84	5.41	6.63	10.83
2	4.60	5.99	7.82	9.21	13.82
3	6.25	7.81	9.84	11.34	16.27
4	7.78	9.49	11.67	13.28	18.46

(ii) Use the calculated value of χ^2 and the table of probabilities provided in the table to find the probability of the results of the test cross departing significantly by chance from the expected ratio. *(1 mark)*

(iii) State what conclusions may be drawn from the probability found in (d)(ii). *(3 marks)*

e Experiments have shown that loci G/g and R/r are on the same chromosome of the tomato plant genome. The two loci are 44 map units apart.

Explain how the results of the test cross shown in (d) could occur when the two loci are on the same chromosome.

(3 marks)

OCR 2805/02 2008

3 Bread wheat is a hexaploid (6n) plant, with three sets of paired chromosomes.

The likely origin of hexaploid bread wheat from diploid wild grasses is shown in the diagram.

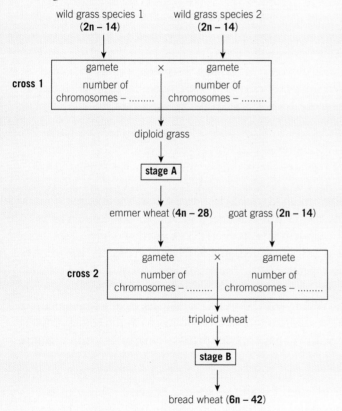

a (i) Copy the diagram and write into the spaces the numbers of chromosomes in the gametes involved in cross 1 and cross 2. *(2 marks)*

(ii) With reference to the diagram, explain what happened at stages A and B to give emmer wheat and bread wheat respectively. *(2 marks)*

b For many years, plant breeders have selectively bred wheat varieties with progressively higher yields. However, bread wheat cannot be interbred with diploid species of grass to establish a variety of wheat with new traits.

Explain why bread wheat cannot be interbred with diploid species of grass to establish new varieties of wheat. *(2 marks)*

c The International Maize and Wheat Improvement Centre (CIMMYT) in Mexico has re-created cross 2 and stage B of the diagram by interbreeding different varieties of wild emmer wheat and goat grass.

In this way more than 1000 varieties of hexaploid 'synthetic wheat' have been produced which can then be interbred with bread wheat.

Explain the need for CIMMYT to maintain seed banks of emmer wheat and goat grass. *(8 marks)*

OCR 2805/02 2008

21 MANIPULATING GENOMES
21.1 DNA profiling
Specification reference: 6.1.3

▲ **Figure 1** *All three of these girls have identical DNA because they were formed from a single fertilised egg – and this included both coding DNA and introns. However, the DNA of the great majority of human beings is unique to each individual*

Every person has a unique combination of DNA in the chromosomes of their cells, unless they are an identical twin or triplet. Your DNA is more similar to your family members' than to other people's, and more similar to other people's than to a gorilla's, a butterfly's or a banana's. It is this combination of individual uniqueness yet similarity to other family or species members that makes DNA so useful in solving crimes, predicting disease, and classifying organisms.

The human genome

The **genome** of an organism is all of the genetic material it contains – for eukaryotes including ourselves, that is the DNA in the nucleus and the mitochondria combined. The chromosomes are made up of hundreds of millions of DNA base pairs, but your genes, the 20–25 000 regions of the DNA that code for proteins, only make up about 2% of your total DNA. They are called exons. The large non-coding regions of DNA that are removed from messenger (m)RNA before it is translated into a polypeptide chain are known as **introns**. Scientists are still investigating the role of non-coding DNA.

Within introns, telomeres, and centromeres there are short sequences of DNA that are repeated many times. This is known as satellite DNA. In a region known as a minisatellite, a sequence of 20–50 base pairs will be repeated from 50 to several hundred times. These occur at more than 1000 locations in the human genome and are also known as variable number tandem repeats (VNTRs). A microsatellite is a smaller region of just 2–4 bases repeated only 5–15 times. They are also known as short tandem repeats (STRs). These satellites always appear in the same positions on the chromosomes, but the number of repeats of each mini- or microsatellite varies between individuals, as different lengths of repeats are inherited from both parents. So just as in the coding DNA, only identical twins will have an identical satellite pattern, although the more closely related you are to someone, the more likely you are to have similar patterns. These patterns in the non-coding DNA were discovered by Professor Sir Alec Jeffreys and his team at Leicester University in 1984. Producing an image of the patterns in the DNA of an individual is known as **DNA profiling** and is a technique employed by scientists to assist in the identification of individuals or familial relationships.

Producing a DNA profile

The process of producing a DNA profile has five main stages:

1 Extracting the DNA

The DNA must be extracted from a tissue sample. When DNA profiling was first discovered, relatively large samples were needed – about 1 μg of

DNA, equivalent to the DNA from the nuclei of about 10 000 human cells. Now, using a technique called the **polymerase chain reaction** (PCR), the tiniest fragment of tissue can give scientists enough DNA to develop a profile.

2 Digesting the sample

The strands of DNA are cut into small fragments using special enzymes called **restriction endonucleases**. Different restriction endonucleases cut DNA at a specific nucleotide sequence, known as a restriction site or recognition site. All restriction endonucleases make two cuts, once through each strand of the DNA double helix. There are many different restriction endonucleases – the recognition sequences and cut sites of three examples are given in Table 1.

Restriction endonucleases give scientists the ability to cut the DNA strands at defined points in the introns. They use a mixture of restriction enzymes that leave the repeating units or satellites intact, so the fragments at the end of the process include a mixture of intact mini- and microsatellite regions.

▼ **Table 1** *Three restriction enzymes and the nucleotide sequences they recognise. The black triangles show where they cut*

Enzyme	Recognition sequence
Sau3A1	5'...▼GATC...3' 3'...CTAG▲...5'
Not1	5'...GC▼GGCCGC...3' 3'...CGCCGG▲CG...5'
Alu1	5'...AG▼CT...3' 3'...TC▲GA...5'

3 Separating the DNA fragments

To produce a DNA profile, the cut fragments of DNA need to be separated to form a clear and recognisable pattern. This is done using **electrophoresis**, a technique that utilises the way charged particles move through a gel medium under the influence of an electric current. The gel is then immersed in alkali in order to separate the DNA double strands into single strands. The single-stranded DNA fragments are then transferred onto a membrane by Southern blotting. (See the Application box for more detail).

4 Hybridisation

Radioactive or fluorescent DNA probes are now added in excess to the DNA fragments on the membrane. DNA probes are short DNA or RNA sequences complementary to a known DNA sequence. They bind to the complementary strands of DNA under particular conditions of pH and temperature. This is called hybridisation. DNA probes identify the microsatellite

❶

extraction
DNA is extracted from the sample

❷

digestion
restriction endo-nucleases cut the DNA into fragments

❸
gel plate

large fragments

small fragments

direction of movement

separation
fragments are separated using gel electrophoresis

❹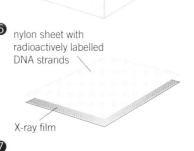
nylon membrane

gel plate

separation (cont.)
DNA fragments are transferred from the gel to nylon membrane in a process known as Southern blotting

❺
DNA probes

hybridisation
DNA probes are added to label the fragments. These radioactive probes attach to specific fragments

❻ nylon sheet with radioactively labelled DNA strands

X-ray film

development
membrane with radioactively labelled DNA fragments is placed onto an X-ray film

❼

development (cont.)
development of the X-ray film reveals dark bands where the radioactive or fluorescent DNA probes have attached

▲ **Figure 2** *Summary of DNA profiling*

regions that are more varied than the larger minisatellite regions. The excess probes are washed off.

5   Seeing the evidence

If radioactive labels were added to the DNA probes, X-ray images are taken of the paper/membrane. If fluorescent labels were added to the DNA probes, the paper/membrane is placed under UV light so the fluorescent tags glow. This is the method most commonly used today. The fragments give a pattern of bars – the DNA profile – which is unique to every individual except identical siblings.

▲ **Figure 3** *DNA profile of a child and its parents. How many fragments were inherited from each parent?*

Separation of nucleic acid fragments by electrophoresis

DNA fragments are put into wells in agarose gel strips (Figure 4 and 5), which also contain a buffering solution to maintain a constant pH. In one or more wells (usually the first and last), DNA fragments of known length are used to provide a reference for fragment sizing.

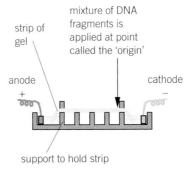

▲ **Figure 4** *Gel electrophoresis apparatus*

▲ **Figure 5** *Preparing a gel plate for electrophoresis*

When an electric current is passed through the electrophoresis plate, the DNA fragments in the wells at the cathode end move through the gel towards the positive anode at the other end. This is due to the negatively charged phosphate groups in the DNA fragments. The rate of movement depends on the mass or length of the DNA fragments – the gel has a mesh-like structure that resists the movement of molecules. Smaller fragments can move through the gel mesh more easily than larger fragments. Therefore, over a period of time, the smaller fragments move further than the larger fragments. When the faster smallest fragments reach the anode end of the gel, the electric current is switched off.

The gel is then placed in an alkaline buffer solution to denature the DNA fragments. The two DNA strands of each fragment separate, exposing the bases.

In a technique called Southern blotting (named after its inventor, Edwin Southern), these strands are transferred to a nitrocellulose paper or a nylon membrane, which is placed over the gel. The membrane is covered with several sheets of dry absorbent paper, drawing the alkaline solution containing the DNA through the membrane by capillary action (Figure 6).

The single-stranded fragments of DNA are transferred to the membrane, as they are unable to pass through it. They are transferred in precisely the same relative positions as they had on the gel. They are then fixed in place using UV light or heated at 80°C.

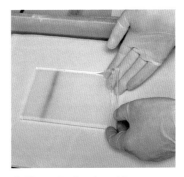

▲ **Figure 6** *Southern blot apparatus*

Gel electrophoresis can also be used to separate proteins. Explain why.

Polymerase chain reaction (PCR)

DNA profiling is often used in solving crimes and only very tiny amounts of DNA may be available. The PCR is a version of the natural process by which DNA is replicated, and allows scientists to produce a lot of DNA from the tiniest original sample.

The DNA sample to be amplified, an excess of the four nucleotide bases A, T, C, and G (in the form of deoxynucleoside triphosphates), small primer DNA sequences, and the enzyme DNA polymerase are mixed in a vial that is placed in a PCR machine (also called a thermal cycler). The temperature within the PCR machine is carefully controlled and changes rapidly at programmed intervals, triggering different stages of the process (Figure 7). The reaction can be repeated many times by the PCR machine, which cycles through the programmed temperature settings. About 30 repeats gives around one billion copies of the original DNA sample – more than enough to carry out DNA profiling.

Step 1 Separating the strands:
The temperature in the PCR machine is increased to 90–95 °C for 30 seconds, this denatures the DNA by breaking the hydrogen bonds holding the DNA strands together so they separate.

Step 2 Annealing of the primers:
The temperature is decreased to 55–60 °C and the primers bind (anneal) to the ends of the DNA strands. They are needed for the replication of the strands to occur.

Step 3 Synthesis of DNA:
The temperature is increased again to 72–75 °C for at least 1 minute, this is the optimum temperature for DNA polymerase to work best. DNA polymerase adds bases to the primer, building up complementary strands of DNA and so producing double-stranded DNA identical to the original sequence. The enzyme Taq polymerase is used, which is obtained from thermophilic bacteria found in hot springs.

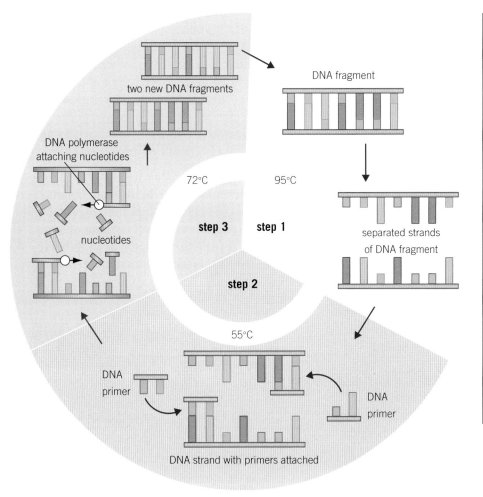

▲ **Figure 7** *The main stages of the polymerase chain reaction*

The uses of DNA profiling

DNA profiling has many uses. Its best known use is in the field of forensic science, especially criminal investigations. PCR and DNA profiling is performed on traces of DNA left at the crime scene. These DNA traces can be obtained from blood, semen, saliva, hair roots, and skin cells. The DNA profile is compared to that of a sample taken from a suspect, or can be identified from a criminal DNA database. DNA profiling is an extremely useful tool in providing evidence for either the guilt or innocence of a suspect.

DNA profiling is also used to prove paternity of a child when it is in doubt. It is used in immigration cases to prove or disprove family relationships. Identifying the species to which an organism belongs can also now be done by DNA profiling, which is much more accurate than any of the older methods. It is also increasingly used to demonstrate the evolutionary relationships between different species.

Another valuable use of DNA profiling is in identifying individuals who are at risk of developing particular diseases. Certain non-coding microsatellites, or the repeating patterns they make, have been found to be associated with an increased risk/incidence of particular diseases, including various cancers and heart disease. These specific gene markers can be identified and observed in DNA profiles.

The information that scientists can obtain from DNA profiling is often used together with the more detailed information obtained from DNA sequencing (Topic 21.2) to make more confident risk assessments for different diseases.

 ## Pitfalls of profiling

One of the earliest recorded cases of mistaken DNA identity occurred in the UK in 2000. Raymond Easton was in the advanced stages of Parkinson's disease – he could hardly dress himself yet he was arrested and charged with a burglary that happened over 200 miles from his home. The arrest was based solely on DNA evidence.

Four years earlier, Raymond had been involved in a family dispute that had got out of hand. He received a police caution and his DNA was taken and kept on file. DNA from the 2000 burglary scene appeared to match Raymond's profile.

Raymond protested his innocence and had a strong alibi for the time of the burglary, so eventually a more rigorous DNA test, looking at satellites in 10 loci rather than the original six, was carried out. None of the additional satellites matched Raymond's DNA and so the charges were dropped. This mis-identification had been caused by an extremely improbable, but not impossible, coincidental DNA profile match.

1. In a UK court of law, until recently DNA was taken as evidence if 11 loci and a sex marker matched. Recently this was put up to 17 loci. Suggest reasons why 11 loci were acceptable and why the number has now risen to 17.
2. DNA evidence at this level is not absolute. The probability that full siblings will have the same DNA profile for these 10 loci is quoted by the Forensic Science Service as 1 in 10 000, and of first cousins, 1 in 100 million. Suggest explanations for these observations.

Summary questions

1. What is an intron?   (3 marks)

2. State the purpose of the polymerase chain reaction and explain how it has advanced DNA profiling.   (4 marks)

3. Discuss the benefits and limitations of DNA profiling.   (6 marks)

4. Produce a flow diagram to show the main stages of the modern DNA profiling process.   (6 marks)

DNA sequencing is the process of determining the precise order of nucleotides within a DNA molecule. This knowledge is invaluable in various scientific applications, from diagnostics to biotechnology, and its uses are explored further in the next topic. In recent years, scientists have not only discovered how to read the information in the genome – they have developed ways of reading it fast.

The beginning of DNA sequencing

DNA sequencing was just an aspiration for scientists until Frederick Sanger and his team developed some techniques for sequencing nucleic acids from viruses and then bacteria. The technique involved radioactive labelling of bases and gel electrophoresis on a single gel. The processes were carried out manually, so it took a long time, but eventually, in the 1970s, the technique now known as Sanger sequencing enabled Sanger and his team to read sequences of 500–800 bases at a time. The first entire genome that they sequenced was just over 5000 bases long and belonged to phiX174, a virus that attacks bacteria. They went on to sequence many other genomes, including the 16 000 base pairs of human mitochondrial DNA. In 1980, Frederick Sanger was awarded the Nobel Prize for his work on sequencing DNA – this was his second Nobel Prize, his first was in 1958 for determining the sequence of the amino acids in insulin. These DNA sequencing techniques are continually being refined. One such development was the swapping of radioactive labels for coloured fluorescent tags, which led to scaling up and automation of the process. This in turn led to the capillary sequencing version of the Sanger sequencing method that was used during the Human Genome Project (HGP), and similar techniques that are used today.

The human genome – pushing the boundaries

In 1990, the HGP was established. It was a massive international project in which scientists from a number of countries worked to map the entire human genome, making the data freely available to scientists all over the world. The early work involved sequencing the DNA of smaller, simpler organisms to refine and develop the techniques. In 1995, after 18 months of work, scientists completed the 1.8 million base pair genome of the bacterium *Haemophilus influenza*. By 1998, the UK team at the Sanger Centre and a US team at Washington University had sequenced the genome of *Caenorhabditis elegans* (*C. elegans*), a nematode worm widely used in scientific experiments, before applying the technique to the three billion base pairs of the human genome itself.

The aim was to complete the HGP in 15 years but the automation of sequencing techniques and the development of more powerful, faster computers meant that the first draft of the human genome was

Learning outcomes

Demonstrate knowledge, understanding, and application of:

→ the principles of DNA sequencing

→ the development of new DNA sequencing techniques.

Synoptic link

You learnt about the structure of proteins and nucleic acids in Chapter 3, Biological molecules and about the ultrastructure of mitochondria in Topic 2.4, Eukaryotic cell structure.

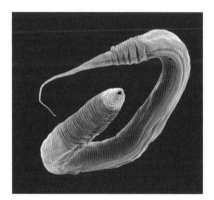

▲ Figure 1 *From* C. elegans *to a human being may seem a giant leap – but our genomes are surprisingly similar. Scanning electron micrograph, approximately ×650 magnification*

▲ **Figure 2** *The contribution of Fred Sanger to our knowledge and understanding of genomes earned him his second Nobel Prize*

Synoptic link

You learnt about the structure of DNA in Topic 3.8, Nucleic acids.

ready in 2000, and the first complete human genome sequence was published in 2003, two years ahead of schedule and under budget.

Principles of DNA sequencing

Sequencing the genome of an organism involves a number of different processes. The DNA is chopped into fragments and each fragment is sequenced. The process involves terminator bases, modified versions of the four nucleotide bases, adenine (A), thymine (T), cytosine (C), and guanine (G), which stop DNA synthesis when they are included. An A terminator will stop DNA synthesis at the location that an A base would be added, a C terminator where a C base would go, and so on. The terminator bases are also given coloured fluorescent tags – A is green, G is yellow, T is red and C is blue. The description of the sequencing process (capillary method) explained here is a simplified version of a technique, has largely been overtaken by much more complex methods – but the basic principles remain the same:

1   The DNA for sequencing is mixed with a primer, DNA polymerase, an excess of normal nucleotides (containing bases A, T, C, and G) and terminator bases.

2   The mixture is placed in a thermal cycler – a piece of equipment as used for PCR (Topic 21.1, DNA profiling) that rapidly changes temperature at programmed intervals in repeated cycles – at 96°C the double-stranded DNA separates into single strands, at 50°C the primers anneal to the DNA strand.

3   At 60°C DNA polymerase starts to build up new DNA strands by adding nucleotides with the complementary base to the single-strand DNA template.

4   Each time a terminator base is incorporated instead of a normal nucleotide, the synthesis of DNA is terminated as no more bases can be added. As the chain-terminating bases are present in lower amounts and are added at random, this results in many DNA fragments of different lengths depending on where the chain terminating bases have been added during the process. After many cycles, all of the possible DNA chains will be produced with the reaction stopped at every base. The DNA fragments are separated according to their length by capillary sequencing, which works like gel electrophoresis in minute capillary tubes. The fluorescent markers on the terminator bases are used to identify the final base on each fragment. Lasers detect the different colours and thus the order of the sequence.

5   The order of bases in the capillary tubes shows the sequence of the new, complementary strand of DNA which has been made. This is used to build up the sequence of the original DNA strand.

The data from the sequencing process is fed into a computer that reassembles the genomes by comparing all the fragments and finding the areas of overlap between them. Once a genome is assembled, scientists want to identify the genes or parts of the genome that code for specific characteristics. Medical researchers want to identify regions that are linked with particular diseases. Many genomes are freely available online – anyone who chooses to can have a look at them.

Stages 1–2: DNA strand chopped up, mixed with primer, bases, DNA polymerase + terminator bases

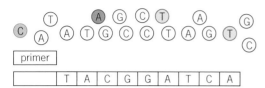

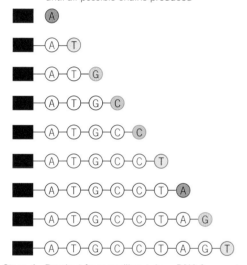

Stage 3: Each time a terminator base is added a strand terminates until all possible chains produced

Stage 4: Readout from capillary tubes: DNA fragments separated by electrophasis in capillary tubes by mass and lasers detect the colours and the sequence

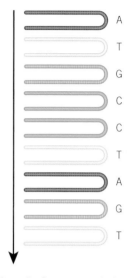

Stage 5: Computer analysis of all data to give original DNA sequence

▲ **Figure 3** *The main steps in the DNA sequencing process*

Next-generation sequencing

Early on, working out the base sequence of even short strands of DNA was difficult and time-consuming for scientists using the original Sanger sequencing method. DNA sequencing technologies have become faster and more automated as they have been developed. Recently, technological developments have led to new, automated, high-throughput sequencing processes. Instead of using a gel or capillaries, the sequencing reaction takes place on a plastic slide known as a flow cell. Millions of fragments of DNA are attached to the slide and replicated in situ using PCR to form clusters of identical DNA fragments. The sequencing process still uses the principle of adding a coloured terminator base to stop the reaction so an image can be taken. As all of the clusters are being sequenced and imaged at the same time, the technique is known as 'massively parallel sequencing' and sometimes referred to as 'next-generation sequencing'.

The process of massively parallel sequencing is integrated with state-of-the-art computer technology and is constantly being refined and developed. These high-throughput methods are extremely efficient and very fast – the 3 billion base pairs of the human genome can be sequenced in days and those of a bacterium in less than 24 hours. High-throughput sequencing also means that the cost has fallen, so more genomes can be sequenced. These techniques are being used by projects such as the 100 000 Genomes Project. They open up the range of questions that scientists can ask and enable us to use the information from the genome in many new and different ways (21.3, Using DNA sequencing).

▲ Figure 4 *The Wellcome Trust Sanger Institute is the biggest DNA sequencing facility in Europe – on this site, the genomes of organisms from bacteria and viruses to humans and even human cancers are sequenced continuously in a highly automated process*

Summary questions

1 Produce a flow chart to summarise the main stages of DNA sequencing. *(5 marks)*

2 a What is the difference in the time it takes to sequence the genetic material of a bacterium today compared to the first complete bacterial genome in 1995? *(1 mark)*
   b Explain the reasons for the difference in these times. *(2 marks)*

3 a What are terminator bases? *(1 mark)*
   b Explain why terminator bases are so important in both the Sanger method of sequencing and in the more modern high-throughput sequencing methods. *(6 marks)*

The development of DNA profiling and DNA sequencing has led to the development of new areas of bioscience that help us analyse, understand, and make use of all the data generated.

Computational biology and bioinformatics

People often use the terms **computational biology** and **bioinformatics** interchangeably. In fact they describe different aspects of the application of computer technology to biology.

Bioinformatics is the development of the software and computing tools needed to organise and analyse raw biological data, including the development of algorithms, mathematical models, and statistical tests that help us to make sense of the enormous quantities of data being generated.

Computational biology then uses this data to build theoretical models of biological systems, which can be used to predict what will happen in different circumstances. Computational biology is the study of biology using computational techniques, especially in the analysis of huge amounts of biodata. For example, it is important in the analysis of the data from sequencing the billions of base pairs in DNA, for working out the 3D structures of proteins, and for understanding molecular pathways such as gene regulation. It helps us to use the information from DNA sequencing – for example in identifying genes linked to specific diseases in populations and in determining the evolutionary relationships between organisms.

Genome-wide comparisons

As whole genome sequencing has become increasingly automated, it has become cheaper and faster, leading to some amazing advances in biology. The field of genetics that applies DNA sequencing methods and computational biology to analyse the structure and function of genomes is called genomics.

Analysing the human genome

Since the first complete draft of the human genome was published in 2003 (Topic 21.2, DNA sequencing and analysis), tens of thousands of human genomes have been sequenced as part of research projects such as the 10 000 Genomes Project UK10K, and most recently the 100 000 Genomes Project.

Computers can analyse and compare the genomes of many individuals, revealing patterns in the DNA we inherit and the diseases to which we are vulnerable. This has enormous implications for health management and the field of medicine in the future. Genomics is changing the face of epidemiology. However, scientists increasingly

Learning outcomes

Demonstrate knowledge, understanding, and application of:

→ how gene sequencing has allowed for:

- genome-wide comparisons between individuals and between species
- the sequences of amino acids in polypeptides to be predicted
- the development of synthetic biology.

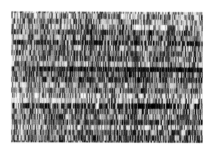

▲ Figure 1 *The impact of computers on the study of biology is growing all the time. This photo shows data from a gel electrophoresis experiment where a grapevine genome was sequenced*

Synoptic link

You learnt about epidemiology in Topic 12.7, Preventing and treating disease.

recognise, with the exception of a few relatively rare genetic diseases caused by changes in a single gene, that our genes work together with the environment to affect our physical characteristics, our physiology, and our likelihood of developing certain diseases.

Analysing the genomes of pathogens

Sequencing the genomes of pathogens including bacteria, viruses, fungi, and protoctista has become fast and relatively cheap. This enables:

- Doctors to find out the source of an infection, for example bird flu or MRSA in hospitals.

- Doctors to identify antibiotic-resistant strains of bacteria, ensuring antibiotics are only used when they will be effective and helping prevent the spread of antibiotic resistance. For example, the bacteria that cause tuberculosis (TB) are difficult to culture, slow growing, and some strains are resistant to most antibiotics. Whole genome analysis makes it easier to track the spread of transmission and to plan suitable treatment options. This has enormous implications for successful treatment of this potentially fatal disease, especially as TB is spreading fast around the world again, linked to the spread of HIV/AIDS.

- Scientists to track the progress of an outbreak of a potentially serious disease and monitor potential epidemics, for example flu each winter, Ebola virus in 2014/15.

- Scientists to identify regions in the genome of pathogens that may be useful targets in the development of new drugs and to identify genetic markers for use in vaccines.

Identifying species (DNA barcoding)

Using traditional methods of observation, it can be very difficult to determine which species an organism belongs to or if a new species has been discovered. Genome analysis provides scientists with another tool to aid in species identification, by comparison to a standard sequence for the species. The challenge for scientists is to produce stock sequences for all the different species.

One useful technique is to identify particular sections of the genome that are common to all species but vary between them, so comparisons can be made – this technique is referred to as DNA barcoding. In the International Barcode of Life (iBOL) project, scientists identify species using relatively short sections of DNA from a conserved region of the genome. For animals, the region chosen is a 648 base-pair section of the mitochondrial DNA in the gene cytochrome c oxidase, that codes for an enzyme involved in cellular respiration. This section is small enough to be sequenced quickly and cheaply, yet varies enough to give clear differences between species. In land plants, that region of the DNA does not evolve quickly enough to show clear differences between species, but two regions in the DNA of the chloroplasts have been identified that can be used in a similar way to identify species.

Synoptic link

You learnt about how organisms are classified and grouped in Topic 10.1, Classification.

The barcoding system is not perfect – so far scientists have not come up with suitable regions for fungi and bacteria, and they may not be able to do so – but DNA sequencing is nevertheless having a big impact on classification.

Searching for evolutionary relationships

Genome sequencing has given scientists a powerful tool to help them understand the evolutionary relationships between organisms. DNA sequences of different organisms can be compared – because the basic mutation rate of DNA can be calculated scientists can calculate how long ago two species diverged from a common ancestor. DNA sequencing enables scientists to build up evolutionary trees with an accuracy they have never had before.

Genomics and proteomics

Proteomics is the study and amino acid sequencing of an organism's entire protein complement. Traditionally, scientists thought that each gene codes for a particular protein, but we now know that there are 20–25 000 coding genes in the human DNA but a very different number of unique proteins. Estimates range from somewhere between 250 000 and 1 000 000 different proteins to only 17–18 000 so there is still a lot of work to be done. More scientific evidence is emerging that highlights the complexity of the relationship between the genotype and the phenotype of an individual. The DNA sequence of the genome should, in theory, enable you to predict the sequence of the amino acids in all of the proteins it produces. The evidence is that the sequence of the amino acids is not always what would be predicted from the genome sequence alone. Some genes can code for many different proteins.

Spliceosomes

The mRNA transcribed from the DNA in the nucleus includes both the exons and introns. Before it lines up on the ribosomes to be translated, this 'pre-mRNA' is modified in a number of ways. The introns are removed, and in some cases, some of the exons are removed as well. Then the exons to be translated are joined together by enzyme complexes known as spliceosomes to give the mature functional mRNA. The spliceosomes may join the same exons in a variety of ways. As a result, a single gene may produce several versions of functional mRNA, which in turn would code for different arrangements of amino acids, giving different proteins and resulting in several different phenotypes.

Protein modification

Some proteins are modified by other proteins after they are synthesised. A protein that is coded for by a gene may remain intact or it may be shortened or lengthened to give a variety of other proteins.

The study of proteomics is constantly giving us increasing knowledge of the extremely complex relationship between the genotype and the phenotype.

> **Synoptic link**
>
> You learnt about pre-mRNA in Topic 19.2, Control of gene expression.

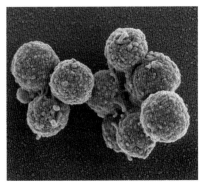

▲ **Figure 2** <u>Mycoplasma</u> <u>mycoides</u>, *the first organism to be controlled by synthetic DNA, was created at the University of California, USA in 2010. Scanning electron micrograph, ×21 000 magnification*

Synoptic link

You will learn more about microorganisms in industrial contexts in Topic 22.4, Microorganisms and biotechnology, Topic 22.6, Culturing microorganisms in the laboratory, and Topic 22.7, Culturing microorganisms on an industrial scale.

Synthetic biology

The ability to sequence the genome of organisms and understand how each sequence is translated into amino acids, along with the ever-increasing ability of computers to store, manipulate, and analyse the data, has led to the development of the new field of science called **synthetic biology**. Synthetic biology is defined by the Royal Society as 'an emerging area of research that can broadly be described as the design and construction of novel artificial biological pathways, organisms or devices, or the redesign of existing natural biological systems.'

Synthetic biology includes many different techniques. These include:

- genetic engineering – this may involve a single change in a biological pathway or relatively major genetic modification of an entire organism (further detail is given in the next topic)

- use of biological systems or parts of biological systems in industrial contexts, for example, the use of fixed or immobilised enzymes and the production of drugs from microorganisms

- the synthesis of new genes to replace faulty genes, for example, in developing treatments for cystic fibrosis (CF), scientists have attempted to synthesise functional genes in the laboratory and use them to replace the faulty genes in the cells of people affected by CF

- the synthesis of an entire new organism. In 2010, scientists announced that they had created an artificial genome for a bacterium and successfully replaced the original genome with this new, functioning genome.

✚ Synthetic life

- Scientists have developed some new nucleotide bases (not adenine, thymine, cytosine, or guanine) which, in a test tube, can be incorporated into a strand of DNA by special enzymes. The bases fit together well – they are not held by hydrogen bonds like the natural bases.

- In 2014, scientists introduced a small section of DNA made with these synthetic bases into bacteria. They found that this unique DNA, including the synthetic nucleotide bases, was replicated time after time as long as they supplied the bacteria with the synthetic bases.

- If these bases can be incorporated into the main DNA of an organism, and then transcribed into RNA, synthetic biologists will have synthetically expanded the genetic code for the very first time.

Suggest a possible practical application for the synthetic expansion of the genetic code.

Infection outbreak – DNA sequencing and clinical intervention

In 2012, there was an outbreak of MRSA (methicillin-resistant *Staphylococcus aureus*) in the Special Care Baby Unit (SCBU) at a UK hospital. The hospital infection control team identified 12 patients carrying MRSA. Researchers at the Wellcome Trust Sanger Institute used DNA sequencing to show that all the bacteria were closely related and this was a hospital-based outbreak.

The sequencing also showed that a number of people living in the community who developed MRSA at the same time all had the same strain as the hospital outbreak. In every case it was found they had a recent link to the hospital.

Two months later, another baby developed MRSA in the same SCBU. Immediate DNA sequencing showed that it was the same strain as the previous outbreak. This suggested that someone working in the hospital was unknowingly carrying MRSA.

Over 150 healthcare workers were screened – and one staff member was found to be carrying MRSA. DNA sequencing confirmed that it was the strain linked to the outbreak. The healthcare worker went through a process to eradicate the MRSA – and the risk of any further infections was removed.

The use of DNA sequencing was critical in identifying that the infections were connected and that a member of staff was a carrier. Without it, this would have been seen as a new outbreak, and many more people could have been infected.

This was the first time DNA sequencing has led to an immediate and successful clinical intervention – but it will certainly not be the last.

▲ **Figure 3** *Babies born early or with health problems are very vulnerable to infection. Sequencing the genome of pathogens so that effective treatment can be introduced as fast as possible and outbreaks halted is a big step forward*

Scientists predict that advances in science and technology may, in a few years, lead to every hospital having its own sequencing machines and hand held devices. These hand held devices would be capable of sequencing the genome, identifying a pathogen within a few minutes from just a drop of blood. Suggest some of the benefits of such developments, and some of the factors that will limit their arrival in our hospitals and local surgeries.

Summary questions

1   Explain the impact of computational biology and bioinformatics on the usefulness of DNA sequencing to scientists.   *(4 marks)*

2   a   Explain how the ability to sequence the genome can be used to identify the source of an outbreak of infectious disease and how this is helpful.   *(4 marks)*

    b   Discuss how DNA sequencing has changed the ways in which we identify species and our understanding of evolutionary relationships.   *(6 marks)*

3   'One-gene-one-polypeptide' is an outdated concept. Discuss how our model of the link between the genotype and phenotype is changing.   *(9 marks)*

21.4 Genetic engineering

Specification reference: 6.1.3

DNA sequencing and proteomics provide us with a detailed understanding of an organism's genetic instructions. Advances in these technologies and molecular biotechnology techniques means it is now possible to manipulate an organism's genome to achieve a desired outcome. This manipulation of the genome is called genetic engineering. The basic principles of genetic engineering involve isolating a gene for a desirable characteristic in one organism and placing it into another organism, using a suitable vector. The two organisms between which the genes are transferred may be the same, similar, or very different species. An organism that carries a gene from another organism is termed 'transgenic' and is often called a genetically modified organism (GMO).

Isolating the desired gene

The first stage of successful genetic modification is to isolate the desirable gene.

The most common technique uses enzymes called restriction endonucleases to cut the required gene from the DNA of an organism. As you learnt in Topic 21.1, DNA profiling, each type of endonuclease is restricted to breaking the DNA strands at specific base sequences within the molecule. Some make a clean, blunt-ended cut in the DNA. However, many restriction endonucleases cut the two DNA strands unevenly, leaving one of the strands of the DNA fragment a few bases longer than the other strand (Figure 2). These regions with unpaired, exposed bases are called sticky ends. The sticky ends make it much easier to insert the desired gene into the DNA of a different organism.

Another technique involves isolating the mRNA for the desired gene and using the enzyme reverse transcriptase to produce a single strand of complementary DNA. The advantage of this technique is that it makes it easier to identify the desired gene, as a particular cell will make some very specific types of mRNA. For example, β cells of the pancreas make insulin, so produce lots of insulin mRNA molecules.

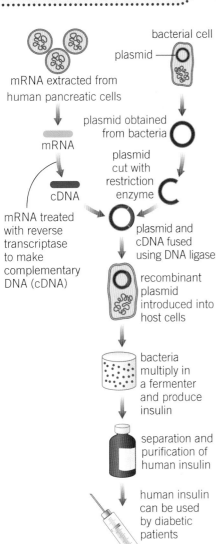

bacterial cell

plasmid

mRNA extracted from human pancreatic cells

mRNA

cDNA

mRNA treated with reverse transcriptase to make complementary DNA (cDNA)

plasmid obtained from bacteria

plasmid cut with restriction enzyme

plasmid and cDNA fused using DNA ligase

recombinant plasmid introduced into host cells

bacteria multiply in a fermenter and produce insulin

separation and purification of human insulin

human insulin can be used by diabetic patients

▲ Figure 1 *The production of human insulin by genetically engineered bacteria was an early success*

a HpaI restriction endonuclease has a recognition site GTTAAC, which produces a straight cut and therefore blunt ends

b HindIII restriction endonuclease has the recognition site AAGCTT, which produces a staggered cut and therefore 'sticky ends'

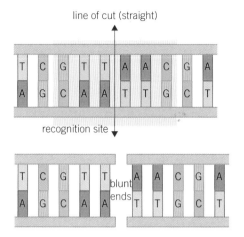

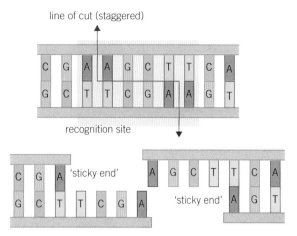

▲ **Figure 2** *Restriction endonucleases are used to isolate genes to be inserted into another organism*

The formation of recombinant DNA

The DNA isolated by restriction endonucleases must be inserted into a vector that can carry it into the host cell.

Vectors

The most commonly used vectors in genetic engineering are bacterial plasmids – small circular molecules of DNA separate from the chromosomal DNA that can replicate independently. Once a plasmid gets into a new host cell it can combine with the host DNA to form what is called **recombinant** DNA. Plasmids are particularly effective in the formation of genetically engineered bacteria used, for example, to make human proteins.

The plasmids that are used as vectors are often chosen because they contain what is known as a marker gene. For example they may have been engineered to have a gene for antibiotic resistance. This gene enables scientists to determine that the bacteria have taken up the plasmid, by growing the bacteria in media containing the antibiotic.

To insert a DNA fragment into a plasmid, first it must be cut open. The same restriction endonuclease as used to isolate the DNA fragment is used to cut the plasmid. This results in the plasmid having complementary sticky ends to the sticky ends of the DNA fragment. Once the complementary bases of the two sticky ends are lined up, the enzyme DNA ligase forms phosphodiester bonds between the sugar and the phosphate groups on the two strands of DNA, joining them together (Figure 2).

The plasmids used as vectors are usually given a second marker gene, which is used to show that the plasmid contains the recombinant gene. This marker gene is itself often placed in the plasmid by genetic engineering methods. The plasmid is then cut by a restriction

> **Synoptic link**
>
> You learnt about bacterial plasmids in Topic 2.6, Prokaryotic and eukaryotic cells.

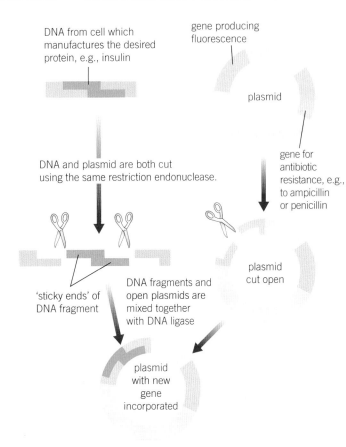

DNA from cell which manufactures the desired protein, e.g., insulin

gene producing fluorescence

plasmid

gene for antibiotic resistance, e.g., to ampicillin or penicillin

DNA and plasmid are both cut using the same restriction endonuclease.

plasmid cut open

'sticky ends' of DNA fragment

DNA fragments and open plasmids are mixed together with DNA ligase

plasmid with new gene incorporated

▲ **Figure 3** *Engineering a desired gene into a plasmid vector*

enzyme within this marker gene to insert the desired gene. If the DNA fragment is inserted successfully, the marker gene will not function. In the early days of genetic engineering, these marker genes were often for antibiotic resistance. There have, however, been many concerns about antibiotic resistance in genetically engineered organisms. As a result, genes producing fluorescence or an enzyme that causes a colour change in a particular medium are now more widely used as marker genes. If a bacterium does not fluoresce, or change the colour of the medium, then it has been engineered successfully and can be grown on (Figure 3).

Transferring the vector

The plasmid with the recombinant DNA must be transferred into the host cell in a process called transformation.

One method is to culture the bacterial cells and plasmids in a calcium-rich solution and increase the temperature. This causes the bacterial membrane to become permeable and the plasmids can enter.

Another method of transformation is **electroporation**. A small electrical current is applied to the bacteria. This makes the membranes very porous and so the plasmids move into the cells.

Electroporation can also be used to get DNA fragments directly into eukaryotic cells. The new DNA will pass through the cell membrane and the nuclear membrane to fuse with the nuclear DNA. Although this technique is effective, the power of the electric current has to be carefully controlled or the membrane is permanently damaged or destroyed, which in turn destroys the whole cell. It is less useful in whole organisms.

Electrofusion

Synoptic link

You learnt about antibodies in Topic 12.6, The specific immune system.

Another way of producing genetically modified (GM) cells is electrofusion. In electrofusion, tiny electric currents are applied to the membranes of two different cells. This fuses the cell and nuclear membranes of the two different cells together to form a hybrid or polyploid cell, containing DNA from both. It is used successfully to produce GM plants.

Electrofusion is used differently in animal cells, which do not fuse as easily and effectively as plant cells. Their membranes have different properties and polyploid animal cells – especially polyploid mammalian cells – do not usually survive in the body of a living organism. However, electrofusion is important in the production of monoclonal antibodies. A monoclonal antibody is produced by a combination of a cell producing one single type of antibody with a tumour cell, which means it divides rapidly in culture. Monoclonal antibodies are now used to identify pathogens in both animals and plants, and in the treatment of a number of diseases including some forms of cancer.

Engineering in different organisms

The techniques of genetic engineering vary between different types of organisms but the principles are the same. It is much easier to carry out genetic modification of prokaryotes than eukaryotes, and among the eukaryotes, plants are easier to work with than animals.

Engineering prokaryotes

Bacteria and other microorganisms have been genetically modified to produce many different substances that are useful to people. These include hormones, for example insulin and human growth hormone, clotting factors for haemophiliacs, antibiotics, pure vaccines, and many of the enzymes used in industry.

Engineering plants

One method of genetically modifying plants uses *Agrobacterium tumefaciens*, a bacterium that causes tumours in healthy plants. A desired gene – for example, for pesticide production, herbicide-resistance, drought-resistance, or higher yield – is placed in the Ti plasmid of *A. tumefaciens* along with a marker gene, for example, for antibiotic resistance or fluorescence. This is then carried directly into the plant cell DNA. The transgenic plant cells form a callus, which is a mass of GM plant cells, each of which can be grown into a new transgenic plant.

Transgenic plant cells can also be produced by electrofusion. The cells produced have chromosomes from both of the original cells and so are polyploid. The cells that are fused may be from similar species, or very different ones. The main stages in this process involve removal of the plant cell wall by cellulases, electrofusion to form a new polyploid cell, the use of plant hormones to stimulate the growth of a new cell wall, followed by callus formation and the production of many cloned, transgenic plants.

Engineering animals

It is much harder to engineer the DNA of eukaryotic animals, especially mammals, than it is to modify bacteria or plants. This is partly because animal cell membranes are less easy to manipulate than plant cell membranes. However, it is an important technique both to enable animals to produce some medically important proteins and to try and cure human genetic diseases such as CF and Huntington's disease.

Synoptic link

You will learn more about cloning plants in Topic 22.1, Natural cloning in plants.

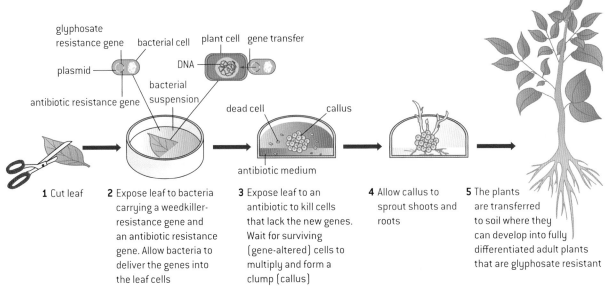

glyphosate
resistance gene   bacterial cell   plant cell gene transfer

plasmid   DNA

antibiotic resistance gene   bacterial suspension

dead cell   callus

antibiotic medium

1 Cut leaf

2 Expose leaf to bacteria carrying a weedkiller-resistance gene and an antibiotic resistance gene. Allow bacteria to deliver the genes into the leaf cells

3 Expose leaf to an antibiotic to kill cells that lack the new genes. Wait for surviving (gene-altered) cells to multiply and form a clump (callus)

4 Allow callus to sprout shoots and roots

5 The plants are transferred to soil where they can develop into fully differentiated adult plants that are glyphosate resistant

▶ **Figure 4** *Top – genetic engineering in plants. Right – A clone of a genetically modified tobacco plant growing in nutrient agar in a petri dish*

Summary questions

1 What is genetic engineering? *(3 marks)*

2 Explain the difference between the way restriction endonucleases and reverse transcriptases produce genes ready for insertion into another organism. *(6 marks)*

3 Describe how a gene is inserted into a bacterial plasmid, which is then taken up by bacteria, and how scientists ensure that they can identify bacteria that have been successfully transformed. *(6 marks)*

4 Produce flow diagrams to show:
  a The process of genetic engineering a bacterium. *(6 marks)*
  b The process of engineering a plant. *(6 marks)*

21.5 Gene technology and ethics

Specification reference: 6.1.3

The rapid development of gene technology in recent years has made many amazing scientific advances in the fields of genetic engineering and biotechnology possible – but it has also raised a number of ethical issues.

All scientists have a responsibility to consider the moral and social values, or ethics, of their work. These ethical considerations are important for many reasons including the protection of human rights, human health and safety, animal welfare, and the protection of the environment. Ethical lapses in scientific research can not only cause harm but can also damage the public's trust in scientists and their research. This in turn can have significant implications on the advancement of knowledge and understanding. Clear and open two-way communication between scientists and the public is vital for building trust and regulating research.

Genetic manipulation of microorganisms

Microorganisms, particularly bacteria, have been genetically modified or engineered to produce many different substances that are useful to people, including insulin and vaccines. These substances can be produced in very large quantities in this way.

GM microorganisms are also used to store a living record of the DNA of another organism in DNA libraries. DNA sequencing projects, such as the HGP (Topic 21.2, DNA sequencing and analysis), enable scientists to build a collection of sequenced DNA fragments from one organism that is then stored (and propagated) in microorganisms (usually bacteria or yeast) through the process of genetic engineering (Topic 21.4, Genetic engineering). These libraries serve as a source of DNA fragments for further genetic engineering applications or for further study of their function.

GM microorganisms are a widely used tool in research for developing novel medical treatments and industrial processes, as well as the development of gene technology itself. Genetically engineered pathogens, however, are not widely used in these applications for the obvious health and safety of the researchers and the wider public. In the few cases that genetic modification of pathogens is carried out, this is usually for the purposes of medical and epidemiological research and is strictly regulated. There is, of course, the concern that genetic engineering of pathogens could be used for the purposes of biological warfare. Attempts to modify the genomes of pathogens to be more virulent, or to be resistant to all known treatments, raises serious ethical concerns and is a largely prohibited area of research except in specialised military research facilities.

Learning outcomes

Demonstrate knowledge, understanding, and application of:

→ the ethical issues relating to the genetic modification of organisms

→ the principles of, and potential for, gene therapy in medicine.

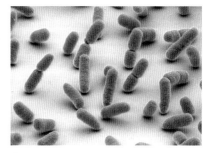

▲ Figure 1 *Microorganisms such as these gut bacteria (*Escherichia coli *) are widely used in scientific research into genetic modification. Scanning electron micrograph, ×7500 magnification*

Synoptic link

You will learn more about the use of genetically modified microorganisms and the enzymes they produce in Chapter 22, Cloning and Biotechnology.

Ethical concerns

Initially, some people were uncomfortable with inserting human genes into microorganisms but the pure human medicines, antibiotics, and enzymes produced are now seen as overwhelmingly beneficial. They have been used safely for many years now. As a result, there is relatively little ethical debate about the use of GM microorganisms except for the manipulation of pathogens in biological warfare.

GM plants

Some people think the genetic modification of plants will help feed the ever-growing human population, and overcome environmental issues including excess carbon dioxide production and pollution. Others have major concerns about the process.

Insect resistance in GM soya beans

Soya beans are a major world crop – around 250 million tonnes are produced each year and over half of the plants are from GM strains. In one such modification, scientists have inserted a gene into soya beans so that they produce the Bt protein. The Bt protein is toxic to many of the pest insects that attack the plant and is widely used as a pesticide by organic farmers. One increasingly widely used strain of soya beans has been engineered to be resistant to a common weed killer and to contain Bt protein. This means farmers can spray to get rid of weeds, making all the resources of light, water, and minerals available to the beans, and they do not need to use pesticides. These plants should enable farmers to grow a much higher-yield crop of soya beans with less labour and less expense.

▲ **Figure 2** *Scientists are currently developing vaccines in transgenic plants to many different diseases such as malaria. Tomato, tobacco, and banana plants are among those used*

Benefits and risks of GM crops

Some of the potential benefits and possible problems of GM plant crops are summarised in Table 1. Also note that whenever antibiotic resistance is used as a marker gene to create GM crops (Topic 21.4, Genetic engineering) there is a perceived risk that this resistance could spread to wild populations of plants and into bacteria.

▼ **Table 1** *Pros and cons of GM crops*

Genetically engineered characteristic	Perceived pros of GM crops	Perceived cons of GM crops
Pest resistance	Pest-resistant GM crop varieties reduce the amount of pesticide spraying, protecting the environment and helping poor farmers. Increased yield.	Non-pest insects and insect-eating predators might be damaged by the toxins in the GM plants. Insect pests may become resistant to pesticides in GM crops.
Disease resistance	Crop varieties resistant to common plant diseases can be produced, reducing crop losses/increasing yield.	Transferred genes might spread to wild populations and cause problems, e.g., superweeds.
Herbicide resistance	Herbicides can be used to reduce competing weeds and increase yield.	Biodiversity could be reduced if herbicides are overused to destroy weeds. Fear of superweeds.

▼ Table 1 *Continued*

Genetically engineered characteristic	Perceived pros of GM crops	Perceived cons of GM crops
Extended shelf-life	The extended shelf-life of some GM crops reduces food waste.	Extended shelf-life may reduce the commercial value and demand for the crop.
Growing conditions	Crops can grow in a wider range of conditions/survive adverse conditions, e.g., flood resistance or drought resistance.	
Nutritional value	Nutritional value of crops can be increased, e.g., enhanced levels of vitamins.	People may be allergic to the different proteins made in GM crops.
Medical uses	Plants could be used to produce human medicines and vaccines.	

Patenting and technology transfer

One of the major concerns about GM crops is that people in less economically developed countries will be prevented from using them by patents and issues of technology transfer. When someone discovers a new technique or invents something, they can apply for a legal patent, which means that no-one else can use it without payment. The people who most need the benefits of, for instance, drought- or flood-resistant crops, high yields, and added nutritional value may therefore be unable to afford the GM seed. They also rely on harvesting seed from one year to plant the next – something that patenting may make impossible.

These concerns are based on evidence. The company that developed the herbicide-resistant and pesticide-producing soya beans, have patented them so farmers can buy the beans from them and grow them to use or sell them for food or processing *only* in the year they are bought. They cannot save the seed to grow again the next year – and in 2013 this was upheld in the US Supreme Court.

Some organisations, however, such as the International Rice Research Institute (IRRI), work to develop engineered rice specifically to support farmers in less economically developed countries with whom they share the technological developments without patent constraints on seed harvesting. For example, they have engineered flood-resistant 'scuba' rice, which gives 70–80% of maximum potential yield even if submerged for 2–3 weeks by flooding.

GM animals

It is much harder to produce GM vertebrates, especially mammals, but scientists are researching the use of microinjections – tiny particles of gold covered in DNA – and modified viruses to carry new genes into animal DNA. Such techniques are used with a number of goals in mind, including the transfer of disease resistance from one animal to another, or to modify physiology in farmed animals.

Some examples of GM animals:

● Swine fever-resistant pigs – in 2013 UK scientists successfully inserted a gene from wild African pigs into the early embryos of a European pig strain giving them immunity to otherwise fatal African swine fever.

● Faster-growing salmon – in the USA, GM Atlantic salmon have received genes from faster-growing Chinook salmon. The genes cause them to produce growth hormones all year round. They grow to full adult size in half the time of conventional salmon, making them a very efficient food source.

Pharming

One of the biggest uses of genetic engineering so far in animals is in the production of human medicines – known as **pharming**. There are two aspects to this field of gene technology:

● Creating animal models – the addition or removal of genes so that animals develop certain diseases, acting as models for the development of new therapies, for example, knockout mice have genes deleted so they are more likely to develop cancer.

● Creating human proteins – the introduction of a human gene coding for a medically required protein. Animals are sometimes used because bacteria cannot produce all of the complex proteins made by eukaryotic cells. The human gene can be introduced into the genetic material of a fertilised cow, sheep, or goat egg, along with a promoter sequence so the gene is expressed only in the mammary glands. The fertilised, transgenic female embryo is then returned to the mother. A transgenic animal is born and when it matures and gives birth, it produces milk. The milk will contain the desired human protein and can be harvested.

Ethical issues

There are many potential benefits to people and indeed to animal health of genetic engineering but the process also raises some ethical questions, which include:

● Should animals be genetically engineered to act as models of human disease?

● Is it right to put human genes into animals?

● Is it acceptable to put genes from another species into an animal without being certain it will not cause harm?

● Does genetically modifying animals reduce them to commodities?

● Is welfare compromised during the production of genetically engineered animals?

▲ **Figure 3** *These transgenic sheep have a human gene in their DNA. This gene codes for the protein alpha-1-antitrypsin (AAT). Hereditary deficiency of AAT leads to emphysema. The gene activates AAT production in the sheep's milk which can be isolated and used for therapy*

Gene therapy in humans

Some human diseases such as CF, haemophilia, and severe combined immunodeficiency (SCIDS) are the result of faulty (mutant) genes. Scientists are looking at different ways of replacing the faulty allele with a healthy one. They can remove the desired alleles from healthy cells or synthesise healthy alleles in the laboratory.

Somatic cell gene therapy

This involves replacing the mutant allele with a healthy allele in the affected somatic (body) cells. The potential for helping people with a wide range of diseases is enormous. Until recently there were few success stories as there are problems in getting the healthy alleles into the affected cells, getting the engineered plasmids into the nucleus of the cells, and finally difficulties in starting and maintaining expression of the healthy allele. Viral vectors are often used.

In recent years, **somatic cell gene therapy** is beginning to show signs of fulfilling its potential. Successful treatments have been reported for diseases including retinal disease (people have regained some vision), immune diseases, leukaemias, myelomas, and haemophilia. The first gene therapy has recently (2012) been approved by the European Medicines Agency for lipoprotein lipase deficiency, which can cause severe pancreatitis. However, somatic cell gene therapy is only a temporary solution for the treated individual. The healthy allele will be passed on every time a cell divides by mitosis but somatic cells have a limited life, and are replaced from stem cells, which will have the faulty allele. In addition, a treated individual will still pass the faulty allele on to any children they have.

Germ line cell gene therapy

The alternative to treating the somatic (body) cells of people already affected by a disease is to insert a healthy allele into the germ cells – usually the eggs – or into an embryo immediately after fertilisation (as part of in vitro fertilisation (IVF) treatment). The individual would be born healthy with the normal allele in place – and would pass it on to their own offspring. This is called **germ line cell gene therapy**.

Such therapy has been successfully done with animal embryos but is illegal for human embryos in most countries as a result of various ethical and medical concerns. These concerns include the fact that the potential impact on an individual of an intervention on the germ cells is unknown. Also, the human rights of the unborn individual could be said to be violated because it is, of course, done without consent and once done the process is irrevocable. Another major ethical concern is that the technology might eventually be used to enable people to choose desirable or cosmetic characteristics of their offspring.

Summary questions

1   Suggest why there is relatively little debate about the ethics of genetically engineering microorganisms.
(3 marks)

2   Draw a table to compare somatic cell gene therapy and germ line cell therapy.
(6 marks)

3   a   Suggest potential benefits of genetically engineering plants *(6 marks)*

    b   Discuss the ethical issues *(6 marks)*

4   Describe the process of genetically modifying animals to produce human proteins and discuss some of the ethical issues this raises.
(6 marks)

Practice questions

1 The diagram is a simplified representation of the apparatus used to separate fragments of DNA in genetic engineering.

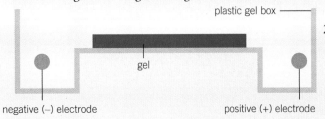

plastic gel box

gel

negative (−) electrode   positive (+) electrode

  a (i) State the name of the process. *(1 mark)*

      (ii) Copy the diagram and indicate where the sample of DNA fragments would be placed and the direction the fragments would move when an electric field is turned on. *(2 marks)*

      (iii) State the reason why the fragments are placed in the position you have indicated. *(1 mark)*

      (iv) Explain the reason for using a gel in the separation process. *(2 marks)*

  b Outline how you could identify the position of a fragment you wanted to locate. *(4 marks)*

The fragment obtained, which usually contains a particular gene under investigation, then undergoes another process called the polymerase chain reaction.

  c Outline the reason for carrying out the polymerase chain reaction. *(2 marks)*

The diagram shows one step in the polymerase chain reaction.

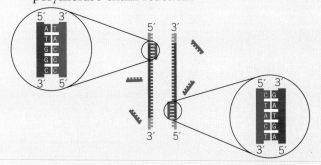

  d (i) Describe what is happening in the diagram. *(2 marks)*

      (ii) Explain the reason for the step you have described. *(2 marks)*

  e (i) Explain why fragments of DNA are denatured during the polymerase chain reaction. *(3 marks)*

      (ii) Compare the denaturation of DNA with protein denaturation. *(5 marks)*

2 It has been discovered that genes make up only about 3% of the human genome and over 50% of the genome consists of repetitive nucleotide sequences. Only recently has it been possible to start to process the data produced from the sequencing of these repetitive DNA sequences. Bioinformatics is a field of study that has developed methods and software programs to handle the large quantities of data involved.

Epigenetic refers to processes that alter the activity of genes without changing nucleotide sequences.

  a (i) State two techniques that might be used to supply the data used in bioinformatics. *(2 marks)*

      (ii) Discuss why the evidence for epigenetics has only been discovered as bioinformatics has developed as a field of study. *(4 marks)*

Synthetic biology refers to the design and synthesis of biological components that do not exist in the natural world or the re-design of existing biological systems.

The guiding principles of synthetic biologists have been stated as below:

- public benefit
- responsible stewardship
- intellectual freedom and responsibility
- democratic deliberation
- justice and fairness

  b Discuss what you believe to be the relative importance of each of these principles. *(7 marks)*

3 DNA profiling is used to identify organisms on the basis of their nucleotide sequences. There are long sections of repetitive DNA, called satellite DNA, in the genomes of most organisms. Different individual have different numbers of repeating segments in their satellite DNA.

A sample of DNA to be sequenced is first amplified, then fragmented and analysed using gel electrophoresis.

a (i) Explain why gel electrophoresis is used to compare the DNA fragments in the samples. *(2 marks)*

(ii) State the name of the process used to amplify a DNA sample. *(1 mark)*

(iii) State the name of the enzymes used to produce the fragments of DNA and explain their specificity to certain sequences of nucleotides. *(3 marks)*

(iv) Explain why the degree of match required differs for the samples of DNA being compared in paternity testing and forensic investigation. *(3 marks)*

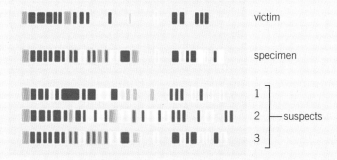

b (i) Using the information in the diagram, identify the most likely suspect. *(1 mark)*

(ii) DNA profiles can also be used to determine if an organism is homozygous or heterozygous at particular loci.

Suggest how this is possible. *(3 marks)*

4 The graph shows the changes in the area of land growing genetically modified crops in different parts of the world.

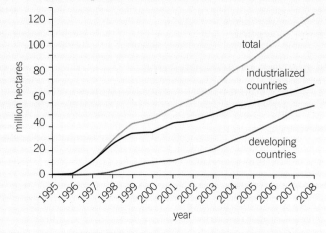

a (i) Describe the changes in the areas of land used for growing GM crops in different parts of the world. *(3 marks)*

(ii) Suggest the reason(s) for the different trends. *(3 marks)*

b Discuss the ethical issues relevant to the genetic modification of plants. *(6 marks)*

5 a Outline the differences between somatic and germ line gene therapy. *(4 marks)*

Some diseases are a result of defective mitochondria such as diabetes, nerve degeneration, and some forms of blindness. It has been proposed that mitochondrial diseases could be cured during the process of IVF.

b (i) Explain why tissues with a high metabolic demand such as muscle and nervous tissue are often affected by mitochondrial disease. *(2 marks)*

(ii) Suggest how the process of IVF could be modified to cure mitochondrial disease. *(3 marks)*

(iii) Discuss whether the manipulation of mitochondria to cure genetic disease overcomes the arguments against germline gene therapy. *(5 marks)*

22 CLONING AND BIOTECHNOLOGY
22.1 Natural cloning in plants
Specification reference: 6.2.1

Synoptic link

You learnt about mitosis in Topic 6.2, Mitosis.

▲ **Figure 1** *Blueberries grow on low bushes – a single clone of plants may stretch for around half a mile*

▲ **Figure 2** *Each baby spider plant is a natural clone*

Asexual reproduction is a form of **cloning** and it results in offspring produced by mitosis and known as **clones.** Clones are usually genetically identical to both the parent organism and to each other.

Natural cloning

Vegetative propagation, or natural cloning, occurs in many species of flowering plants. A structure forms which develops into a fully differentiated new plant, which is genetically identical to the parent. The new plant may be propagated from the stem, leaf, bud, or root of the parent, depending on the type of plant, and it eventually becomes independent from its parent, for example, strawberries and spider plants.

Vegetative propagation often involves perennating organs, which enables plants to survive adverse conditions. These contain stored food from photosynthesis and can remain dormant in the soil. They are often not only a means of asexual reproduction, but also a way of surviving from one growing season to the next. Natural plant cloning occurs in:

● bulbs, for example, daffodil. The leaf bases swell with stored food from photosynthesis. Buds form internally which develop into new shoots and new plants in the next growing season.

● Runners, for example, a strawberry or spider plant. A lateral stem grows away from the parent plant and roots develop where the runner touches the ground. A new plant develops – the runner eventually withers away leaving the new individual independent.

● Rhizomes, for example, marram grass. A rhizome is a specialised horizontal stem running underground, often swollen with stored food. Buds develop and form new vertical shoots which become independent plants.

● Stem tubers, for example, potato. The tip of an underground stem becomes swollen with stored food to form a tuber or storage organ. Buds on the storage organ develop to produce new shoots (e.g., the 'eyes' on a potato).

Using natural clones in horticulture

Natural plant cloning is exploited in horticulture by farmers and gardeners to produce new plants. Splitting up bulbs, removing young plants from runners, and cutting up rhizomes all increase plant numbers cheaply, and the new plants have exactly the same genetic characteristics as their parents.

It is also possible to take cuttings of many plants – short sections of stems are taken and planted either directly in the ground (e.g., sugar cane) or in pots, for example, pelargoniums. Rooting hormone is often applied to the base of a cutting to encourage the growth of new roots. Propagation from cuttings has several advantages over using seeds. It is much faster – the time from planting to cropping is much reduced. It also guarantees the quality of the plants. By taking cuttings from good stock, the offspring will be genetically identical and will therefore crop well. The main disadvantage is the lack of genetic variation in the offspring should any new disease or pest appear or if climate change occurs.

Many of the world's most important food crops are propagated by cloning. Bananas, sugar cane, sweet potatoes, and cassava are all propagated from stem cuttings or rhizomes. Coffee and tea bushes are also propagated from stem cuttings.

Cloning sugar cane

Sugar cane is an internationally important crop used to make sugar and manufacture biofuels. It is one of the fastest growing crop plants in the world – the stems can grow 4–5 metres in 11 months if conditions are good – and it is usually propagated by cloning. Short lengths of cane about 30 cm long, with three nodes, are cut and buried in a clear field in shallow trenches, covered with a thin layer of soil. Per hectare, 10–25 000 lengths of stem are planted.

 Practical cloning

Many popular houseplants are propagated by taking cuttings. There are a number of points which increase the success rate of most cuttings:

- Use a non-flowering stem
- Make an oblique cut in the stem
- Use hormone rooting powder
- Reduce leaves to two or four
- Keep cutting well watered
- Cover the cutting with a plastic bag for a few days.

▲ Figure 3 Taking cuttings on a commercial scale

1. How does each of the above points increase the likelihood that a cutting will succeed?
2. Why are cuttings useful for investigating the effect of growing conditions on plants?

Synoptic link

You learnt about the use of hormonal rooting powders in Topic 16.5, The commercial use of plant hormones.

Study tip

Make sure you use terms such as cloning, propagation, and taking cuttings correctly.

Summary questions

1. What are perennating organs and how are they involved in cloning and survival? (*5 marks*)

2. Explain the advantages and disadvantages of propagating crop plants by cuttings over using seeds. (*4 marks*)

3. Suggest why it is important to describe clones as genetically identical to their parent rather than simply identical – and why even this may not always be true. (*5 marks*)

People have propagated plants by cloning for centuries, but there is a limit to how many 'natural' clones you can make from one plant. Many plant cells are totipotent – they can differentiate into all of the different types of cells in the plant. Scientists have developed ways of using this property to produce huge numbers of identical clones from one desirable plant.

Micropropagation using tissue culture

Micropropagation is the process of making large numbers of genetically identical offspring from a single parent plant using tissue culture techniques. This is used to produce plants when a desirable plant:

- does not readily produce seeds
- doesn't respond well to natural cloning
- is very rare
- has been genetically modified or selectively bred with difficulty
- is required to be 'pathogen-free' by growers, for example, strawberries, bananas, and potatoes.

There are a number of ways in which plants can be micropropagated. One protocol, based on work done at the Royal Botanic Garden at Kew, uses sodium dichloroisocyanurate, the sterilising tablets used to make emergency drinking water and babies' bottles safe. This keeps the plant tissues sterile without being in a sterile lab so it is extremely useful for scientists in the field working with rare and endangered plant material – and also for use in schools. Other protocols are more suited to industrial micropropagation where large sterilising units are available.

The basic principles of micropropagation and tissue culture are as follows:

- Take a small sample of tissue from the plant you want to clone – the meristem tissue from shoot tips and axial buds is often dissected out in sterile conditions to avoid contamination by fungi and bacteria. This tissue is usually virus-free.
- The sample is sterilised, usually by immersing it in sterilising agents such as bleach, ethanol, or sodium dichloroisocyanurate. The latter does not need to be rinsed off which means the tissue is more likely to remain sterile. The material removed from the plant is called the explant.
- The explant is placed in a sterile culture medium containing a balance of plant hormones (including auxins and cytokinins) which stimulate mitosis. The cells proliferate, forming a mass of identical cells known as a callus.

- The callus is divided up and individual cells or clumps from the callus are transferred to a new culture medium containing a different mixture of hormones and nutrients which stimulates the development of tiny, genetically identical plantlets.

- The plantlets are potted into compost where they grow into small plants.

- The young plants are planted out to grow and produce a crop.

The scale of micropropagation is increasing. It now takes place in bioreactors, effectively making artificial embryo plants to be packaged in artificial seeds.

▲ Figure 1 *Micropropagation*

Advantages and disadvantages of micropropagation

The number of common plants that are largely produced by micropropagation is growing constantly and includes potatoes, sugar cane, bananas, cassava, strawberries, grapes, chrysanthemums, Douglas firs, and orchids. Here are some of the points both for and against this process

Arguments for micropropagation

- Micropropagation allows for the rapid production of large numbers of plants with known genetic make-up which will yield good crops.

- Culturing meristem tissue produces disease-free plants.

- It makes it possible to produce viable numbers of plants after genetic modification of plant cells.

- It provides a way of producing very large numbers of new plants which are seedless and therefore sterile to meet consumer tastes (e.g., bananas and grapes).

- It provides a way of growing plants which are naturally relatively infertile or difficult to grow from seed (e.g., orchids).

- It provides a way of reliably increasing the numbers of rare or endangered plants.

Arguments against micropropagation

- It produces a monoculture – many plants which are genetically identical – so they are all susceptible to the same diseases or changes in growing conditions.

- It is a relatively expensive process and requires skilled workers.

- The explants and plantlets are vulnerable to infection by moulds and other diseases during the production process.

- If the source material is infected with a virus, all of the clones will also be infected.

- In some cases, large numbers of new plants are lost during the process.

▲ Figure 2 *Micropropagation of orchids means they are no longer only available to very wealthy people*

▲ Figure 3 *Bananas are the fourth most important food worldwide – and the ones we eat are all clones*

Yes, we have no bananas…

Bananas are now thought to be one of the oldest crops – and possibly the first to be cloned. A wild banana is full of hard seeds and it is virtually inedible. A mutation made them parthenocarpic which means they produce fruit without fertile seeds – which made them good to eat but also made them sterile. Scientists therefore think that since the dawn of agriculture, people cloned bananas using natural asexual reproduction to propagate the plants producing the seedless, tasty fruit. Sweet bananas are widely eaten in more economically developed countries, whilst plantains (cooking bananas) are a staple food in many less economically developed countries.

In the early 20th century almost all of the sweet bananas eaten were the cultivar Gros Michel. Then fungal Panama disease wiped them out in the major banana growing countries – none of the clones had any resistance and a new cultivar took over. Cavendish bananas, while apparently not as tasty as Gros Michel bananas, are resistant to Panama disease. But Cavendish bananas are also clones. Now another banana disease, Black Sigatoka, is destroying Cavendish plantations, and is also spreading to other cooking varieties of bananas.

New biotechnologies for example genetic engineering and micropropagation offer hope for the future. Genetically engineered strains of bananas with resistance genes from the original wild fruit could be micropropagated and used to restock banana plantains across the whole growing region.

1   Gros Michel bananas have been almost entirely lost. Why?
2   How might new technology enable us to retain the Cavendish strain?
3   Discuss how banana culture has changed over the centuries.

Synoptic link

You learnt about Black Sigatoka in Topic 12.2, Animal and plant diseases.

Summary questions

1   Make a flow chart to show the main stages in the micropropagation of plants.   (7 marks)

2   What do you consider the two main advantages of micropropagation of crop plants? Explain your choices.   (6 marks)

3   a   What is the potential of natural cloning for saving important crops such as the banana against disease? Give arguments for and against the technique.   (6 marks)
     b   What is the potential of micropropagation for saving important crops against disease in contrast to natural cloning? Give arguments for and against the technique.   (6 marks)

22.3 Cloning in animals

Specification reference: 6.2.1

Cloning is a natural part of the reproductive cycle in many plants. Perhaps surprisingly, it is not uncommon in many animal species and even occurs in human beings.

Natural animal cloning

Natural cloning is common in invertebrate animals. Although it is less common in vertebrates, it still occurs in the form of twinning.

Cloning in invertebrates

Natural cloning in invertebrates can take several forms. Some animals, such as starfish, can regenerate entire animals from fragments of the original if they are damaged. Flatworms and sponges fragment and form new identical animals as part of their normal reproductive process, all clones of the original. *Hydra* produce small 'buds' on the side of their body which develop into genetically identical clones. In some insects, females can produce offspring without mating. Scientists are increasingly finding differences between the mother and daughters, however, suggesting that as a result of high mutation rates the offspring are not true clones.

Cloning in vertebrates

The main form of vertebrate cloning is the formation of **monozygotic twins (identical twins)**. The early embryo splits to form two separate embryos. No one is sure of the trigger which causes this to happen. The frequency at which identical twins occur varies between species. For example, domestic cattle rarely if ever produce identical twins naturally, while the incidence in natural human pregnancies is around 3 per 1000. When monozygotic twins are born, although genetically identical, they may look different as a result of differences in their position and nutrition in the uterus.

Some female amphibians and reptiles will produce offspring when no male is available. The offspring are often male rather than female, so they are not clones of their mother, yet all of their genetic material arises from her.

Artificial clones in animals

It is relatively easy to produce artificial clones of some invertebrates – liquidise a sponge or chop up a starfish and new animals will regenerate from most of the fragments. It is much more difficult to produce artificial clones of vertebrates, especially mammals. However, two methods are now used widely in the production of high-quality farm animals and in the development of genetically engineered animals for pharming.

Learning outcomes

Demonstrate knowledge, understanding, and application of:

→ natural clones in animal species

→ how artificial clones in animals can be produced by artificial embryo twinning or by enucleation and somatic cell nuclear transfer (SCNT)

→ the arguments for and against artificial cloning in animals.

▲ Figure 1 *The small Hydra is a clone of the parent and will eventually separate and live independently*

Synoptic link

You learnt about pharming in Topic 21.5, Gene technology and ethics.

Artificial twinning

After an egg is fertilised, it divides to form a ball of cells. Each of these individual cells is **totipotent** – it has the potential to form an entire new animal. As the cells continue to divide, the embryo becomes a hollow ball of cells. Soon after this the embryo can no longer divide successfully.

In natural twinning, an early embryo splits and two foetuses go on to develop from the two halves of the divided embryo. In **artificial twinning** the same thing happens, but the split in the early embryo is produced manually. In fact, the early embryo may be split into more than two pieces and results in a number of identical offspring. Artificial twinning, like embryo transfer which preceded it, is used by the farming community to produce the maximum offspring from particularly good dairy or beef cattle or sheep.

The stages of artificial twinning in cattle can be summarised as follows:

- A cow with desirable traits is treated with hormones so she super-ovulates, releasing more mature ova than normal.
- The ova may be fertilised naturally, or by artificial insemination, by a bull with particularly good traits. The early embryos are gently flushed out of the uterus.
- Alternatively, the mature eggs are removed and fertilised by top-quality bull semen in the lab.
- Usually before or around day six, when the cells are still totipotent, the cells of the early embryo are split to produce several smaller embryos, each capable of growing on to form a healthy full-term calf.
- Each of the split embryos is grown in the lab for a few days to ensure all is well before it is implanted into a surrogate mother. Each embryo is implanted into a different mother as single pregnancies carry fewer risks than twin pregnancies.
- The embryos develop into foetuses and are born normally, so a number of identical cloned animals are produced by different mothers.

In pigs, a number of cloned embryos must be introduced into each mother pig. This is because they naturally produce a litter of piglets, and the body may reject and reabsorb a single foetus.

This technology makes it possible to greatly increase the numbers of offspring produced by the animals with the best genetic stock. Some of the embryos may be frozen. This allows the success of a particular animal to be assessed and, if the stock is good, remaining identical embryos can be implanted and brought to term.

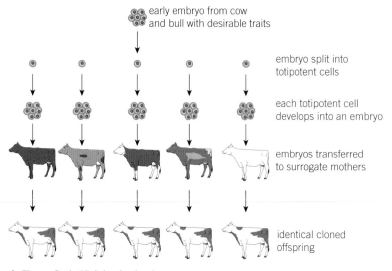

early embryo from cow and bull with desirable traits

embryo split into totipotent cells

each totipotent cell develops into an embryo

embryos transferred to surrogate mothers

identical cloned offspring

▲ Figure 2 *Artificial twinning in cows*

Somatic cell nuclear transfer

Artificial twinning clones an embryo. However, it is now possible to clone an adult animal, by taking the nucleus from an adult somatic (body) cell and transferring it to an **enucleated** egg cell (an oocyte which has had the nucleus removed). A tiny electric shock is used to fuse the egg and nucleus, stimulate the combined cell to divide, and form an embryo that is a clone of the original adult. This process is known as **somatic cell nuclear transfer** (SCNT). The first adult mammal to be cloned in this way was Dolly the sheep in 1996. Since then scientists have cloned a wide range of species including mice, cows, horses, rabbits, cats, and dogs. SCNT is simple in theory, although in practice there are many difficulties so the technique is still not widely used. As you can see in Figure 3, animals of different breeds are often used as the cell donor, the egg donor, and the surrogate mother to make it easier to identify the original animal at each stage.

1 The nucleus is removed from a somatic cell of an adult animal.

2 The nucleus is removed from a mature ovum harvested from a different female animal of the same species (it is enucleated).

3 The nucleus from the adult somatic cell is placed into the enucleated ovum and given a mild electric shock so it fuses and begins to divide. In some cases, the nucleus from the adult cell is not removed – it is simply placed next to the enucleated ovum and the two cells fuse (electrofusion) and begin to divide under the influence of the electric current.

4 The embryo that develops is transferred into the uterus of a third animal, where it develops to term.

5 The new animal is a clone of the animal from which the original somatic cell is derived, although the mitochondrial DNA will come from the egg cell.

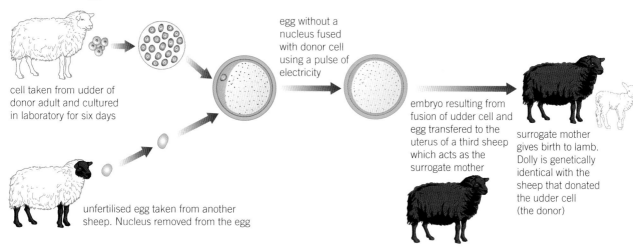

egg without a nucleus fused with donor cell using a pulse of electricity

cell taken from udder of donor adult and cultured in laboratory for six days

embryo resulting from fusion of udder cell and egg transfered to the uterus of a third sheep which acts as the surrogate mother

surrogate mother gives birth to lamb. Dolly is genetically identical with the sheep that donated the udder cell (the donor)

unfertilised egg taken from another sheep. Nucleus removed from the egg

▲ Figure 3 *Somatic cell nuclear transfer (SCNT)*

This process is also known as reproductive cloning, because live animals are the end result. The cloned embryo can then be split to produce several identical clones.

There have been some problems with the animals produced by SCNT – Dolly the sheep had to be put down when she was only six years old

Synoptic link

You learnt about electrofusion in Topic 21.4, Genetic engineering.

because she suffered from arthritis and lung disease, usually seen in much older sheep. However, scientists have improved the technique and whilst concerns about premature ageing in clones produced by SCNT persist, researchers in Japan have produced 581 clones from one original donor mouse, through 25 generations. The mice in each generation were cloned to produce the next generation. Furthermore, they all had babies naturally to prove they functioned normally. All of the mice had normal lifespans. The same team has also produced SCNT clones from the bodies of mice which had been frozen for 16 years.

SCNT can be used in a number of ways. It is used in pharming – the production of animals which have been genetically engineered to produce therapeutic human proteins in their milk. It can also be used to produce genetically modified (GM) animals which grow organs that have the potential to be used in human transplants.

Pros and cons of animal cloning

Animal cloning is still not widespread, although it is increasingly used in agriculture and the world of animal breeding and medicine. A number of arguments are put forward both for and against the process.

Arguments *for* animal cloning

Artificial twinning enables high-yielding farm animals to produce many more offspring than normal reproduction.

Artificial twinning enables the success of a sire (the male animal) at passing on desirable genes to be determined. If the first cloned embryo results in a successful breeding animal, more identical animals can be reared from the remaining frozen clones. The use of meat from animals born to a cloned parent is now permitted in the US.

SCNT enables GM embryos to be replicated and to develop, giving many embryos from one engineering procedure. It is an important process in pharming – the production of therapeutic human proteins in the milk of genetically engineered farm animals, such as sheep and goats.

SCNT enables scientists to clone specific animals, for example, replacing specific pets or cloning top-class race horses. Pet cats and dogs have been cloned in the US at great expense.

SCNT has the potential to enable rare, endangered, or even extinct animals to be reproduced. In theory, the nucleus from dried or frozen tissue could be transferred to the egg of a similar living species and used to produce clones of species that have been dead for a long time.

Arguments *against* animal cloning

SCNT is a very inefficient process – in most animals it takes many eggs to produce a single cloned offspring.

Many cloned animal embryos fail to develop and miscarry or produce malformed offspring.

Many animals produced by cloning have shortened lifespans, although cloned mice have now been developed which live a normal two years.

SCNT has been relatively unsuccessful so far in increasing the populations of rare organisms or allowing extinct species to be brought back to life. For example, scientists have attempted to clone the gaur and the banteng – both extremely rare breeds of wild cattle. One gaur calf was born in 2001 and died within a couple of days. Two banteng calves were born in 2003 – one was deformed and euthanised, the other grew normally but its natural lifespan was halved. The idea of restoring extinct organisms is exciting but scientists are increasingly unconvinced that it will be possible by this method.

Cloning humans

- Scientists have reproduced clones of primates by artificial twinning but it is proving very difficult to produce a SCNT clone of a primate.

- Part of the problem seems to be that the spindle proteins needed for cell division in primate cells are sited very close to the nucleus, so the removal of the nucleus to produce the enucleated primate ovum also destroys the mechanism by which the cell divides. This is not a problem in the ova of many other mammals because the spindle proteins are more dispersed in the cytoplasm.

- In addition, the synchronisation of the stage of the embryo and the state of the reproductive organs of the mother have to be exactly attuned in primates – there seems to be more flexibility in some other mammals.

- In recent years scientists have finally produced embryonic primate stem cell lines by SCNT. This means it may eventually be possible to develop these potentially important therapeutic cells from human beings.

- In most countries there is strict legislation to prevent reproductive cloning of human beings, even if the technical problems of primate cloning are overcome. A modified version of SCNT has the potential, however, to produce human embryonic stem cells from an adult which could produce cells to be used to grow new tissues for that individual patient. Research in this process is strictly controlled so it cannot be used for reproductive cloning – it is known as therapeutic cloning to make it clear that the end result is not to reproduce a person. However, this form of SCNT can potentially make it possible to grow replacement organs which will not trigger an immune response in a patient and which will enable us to cure many currently life-threatening conditions.

> **Synoptic link**
>
> You learnt about spindle formation during cell division in Topic 6.1, Cell cycle.

Some people claim to have produced a cloned human baby, although they have never produced the child and the adult it was cloned from.

1 Explain how this could easily be proved if scientists were given access to the individuals

2 Suggest why these claims seem very unlikely to be true.

Summary questions

1 How is artificial twinning different from natural twinning? *(6 marks)*

2 The evidence suggests that monozygotic twins do not occur naturally in cattle. Suggest ways in which this might be investigated. *(2 marks)*

3 Primate clones have been produced by artificial twinning but not, in 2014, by SCNT.
   a Explain the similarities in the two processes. *(3 marks)*
   b Explain the differences between the two processes. *(6 marks)*

4 Should funding continue for projects attempting to recreate extinct animals by SCNT? Include an evaluation of the process in your answer. *(6 marks)*

22.4 Microorganisms and biotechnology

Specification reference: 6.2.1

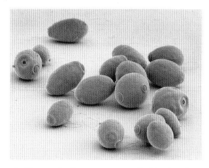

▲ **Figure 1** *Yeasts were probably the first microorganism to be used in biotechnology, and they are still of immense importance worldwide today, scanning electron micrograph approx ×3250 magnification*

The earliest recorded use of microorganisms by people was around 6000 BC when the Sumerians and Babylonians were using yeast to make beer. By 4000 BC the Egyptians were using yeast to make their bread rise. These are all examples of the development and use of biotechnology over several millennia.

Defining biotechnology

Biotechnology involves applying biological organisms or enzymes to the synthesis, breakdown, or transformation of materials in the service of people. It describes a range of processes, from the traditional production of cheese, yoghurt, wine, bread, and beer to the latest molecular technologies using DNA manipulation to produce genetically engineered microorganisms synthesising drugs such as insulin and antibiotics, and the use of biological systems to remove soil and water pollution in processes known as **bioremediation** (Topic 22.5, Microorganisms, medicines, and bioremediation).

The most commonly used organisms in biotechnology processes (bioprocesses) are fungi, particularly the yeasts, and bacteria, which are particularly useful in the newer technologies based around genetic manipulation.

The use of microorganisms

Most biotechnology involves using biological catalysts (enzymes) in a manufacturing process and the most stable, convenient, and effective form of the enzymes is often a whole microorganism. Microorganisms are ideal for a variety of reasons.

● There are no welfare issues to consider – all that is needed is the optimum conditions for growth.

● There is an enormous range of microorganisms capable of carrying out many different chemical syntheses or degradations that can be used.

● Genetic engineering allows us to artificially manipulate microorganisms to carry out synthesis reactions that they would not do naturally, for example, to produce human insulin.

● Microorganisms have a very short life cycle and rapid growth rate. As a result, given the right conditions of food, oxygen, and temperature, huge quantities of microorganisms can be produced in short periods of time.

● The nutrient requirements of microorganisms are often very simple and relatively cheap. Genetic manipulation means we can modify them so that the microorganisms can utilise materials which would otherwise be wasted, making the raw materials for microorganism-controlled syntheses much cheaper than the raw materials needed for most other industrial processes.

- The conditions which most microorganisms need to grow include a relatively low temperature, a supply of oxygen and food, and the removal of waste gases. They provide their own catalysts in the form of enzymes. This makes bioprocesses relatively cheap compared to the high temperatures and pressures and expensive catalysts often needed in non-biological industrial processes.

Indirect food production

Microorganisms are widely used in biotechnological processes to make food such as bread, yogurt, and cheese. The microorganisms have an indirect effect – it is their actions on other food that is important,. When you eat bread you are mainly eating flour, when you eat yoghurt or cheese it is mainly milk.

The advantages of using microorganisms in this way are all of the ones listed previously as advantages of using microorganisms in biotechnology generally. There are few disadvantages to using microorganisms indirectly in the production of human foods. If the conditions are not ideal (e.g., too hot or too cold) the microorganisms do not grow properly and so they do not work efficiently. Conditions that are ideal for the microorganisms can also be ideal for microorganisms that cause the food to go off or cause disease and so the processes have to be sterile. Increasingly the microorganisms used in food production have been genetically engineered, and some people have ethical issues with the use of GM organisms, although this is generally much less the case with microorganisms than with animals and plants.

There are around 900 different types of cheese made around the world. Some are still made by very small-scale, traditional methods and others are produced commercially on a very large scale.

Extra microorganisms

Sometimes microorganisms are used in the same biotechnological process in more than one way. Traditionally bacteria are used in cheesemaking (Table 1) and proteolytic enzymes are also added to the milk to help form curds and whey. Originally these came from rennet, a substance extracted from the stomachs of calves, cows, and pigs containing the enzyme chymosin. In modern cheesemaking, the chymosin used comes mainly from microorganisms – either from fungal sources or GM bacteria.

Suggest three different advantages of using modern sources of chymosin.

▲ Figure 2 *Different types of cheeses*

▼ Table 1 *Examples of microorganisms involved in commercial processes*

Process	Microorganism involved	Steps in commercial process
Baking	Yeast – mixed with sugar and water to respire aerobically. Carbon dioxide produced makes bread rise.	• The active yeast mixture is added to flour and other ingredients. Mixed and left in warm environment to rise. • Dough is knocked back (excess air removed), kneaded, shaped, and left to rise again. • Cooked in a hot oven – the carbon dioxide bubbles expand, so the bread rises more. Yeast cells are killed during cooking.
Brewing	Yeast – respires anaerobically to produce ethanol. Traditional yeasts ferment at 20–28 °C. GM yeasts ferment at lower, and therefore cheaper, temperatures, and clump together (flocculate) and sink at the end of the process leaving the beer very clear.	• Malting – barley germinates producing enzymes that break starch molecules down to sugars which yeast can use. Seeds then killed by slow heating but enzyme activity retained to produce malt. • Mashing – the malt is mixed with hot water (55–65 °C) and enzymes break down starches to produce wort. Hops are added for flavour and antiseptic qualities. The wort is sterilised and cooled. • Fermentation – wort is inoculated with yeast. Temperature maintained for optimum anaerobic respiration (fermentation). Eventually yeast is inhibited by falling pH, build up of ethanol, and lack of oxygen. • Maturation – the beer is conditioned for 4–29 days at temperatures of 2–6 °C in tanks • Finishing – the beer is filtered, pasteurised, and then bottled or canned with the addition of carbon dioxide • The alcohol content varies between about 4% and 9%.

Cheese-making	Bacteria – feed on lactose in milk, changing the texture and taste, and inhibiting the growth of bacteria which make milk go off.	<ul><li>The milk is pasteurised (heated to 95 °C for 20 seconds to kill off most natural bacteria) and homogenised (the fat droplets evenly distributed through the milk).</li><li>It is mixed with bacterial cultures and sometimes chymosin enzyme and kept until the milk separates into solid curds and liquid whey.</li><li>For cottage cheese, the curds are separated from the whey, packaged, and sold.</li><li>For most cheese, the curds are cut and cooked in the whey then strained through draining moulds or cheesecloth. The whey is used for animal feeds.</li><li>The curds are put into steel or wooden drums and may be pressed. They are left to dry, mature, and ripen before eating as the bacteria continue to act for anything from a few weeks to several years.</li></ul>
Yoghurt-making	Bacteria – often *Lactobacillus bulgaricus* (forms ethanal) and *Streptococcus thermophilus* (forms lactic acid). Both produce extracellular polymers that give yoghurt its smooth, thick texture.	<ul><li>Skimmed milk powder is added to milk and the mixture is pasteurised, homogenised, and cooled to about 47 °C.</li><li>The milk is mixed with a 1:1 ratio of *Lactobacillus bulgaricus* and *Streptococcus thermophilus* and incubated at around 45 °C for 4–5 hours.</li><li>At the end of the fermentation, the yoghurt may be put into cartons at a temperature of around 10 °C as plain yoghurt or mixed with previously sterilised fruit.</li><li>Thick-set yoghurts are mixed and ferment in the pot.</li><li>Yoghurt has a shelf-life of about 19 days if stored at 2–3 °C.</li></ul>

Direct food production

People have eaten fungi for thousands of years in the form of a wide variety of mushrooms. In recent times, facing potential protein shortages around the world, scientists are developing more ways of using microorganisms to directly produce protein you can eat. It is known as single-cell protein or SCP.

The best known SCP is Quorn™. This is made of the fungus *Fusarium venetatum*, a single celled fungus that is grown in large fermenters using glucose syrup as a food source (You will learn more about fermenters in Topic 22.7, Culturing microorganisms on an industrial scale). The microorganisms are combined with albumen (egg white) and then compressed and formed into meat substitutes. Quorn™ is not only suitable for vegetarians, it is also a healthy choice as it is high in protein and low in fat. People are very conservative in their food choices and when the new food was launched, no mention was made of the fungi used to produce it. Using the term mycoprotein meant most people did not recognise what it was made of. However a combination of good marketing and a good product meant that people tried Quorn™ and liked it, and it has been internationally successful as a novel protein food.

Other attempts to make proteins from microorganisms have not yet been as successful. Yeasts, algae, and bacteria can be used to grow proteins that match animal proteins found in meat as well as plant proteins. They can be grown on almost anything, are relatively cheap and low in fat, yet none of the alternative protein sources has proved successful so far. People have many reservations about eating food grown on waste. Increasingly single celled proteins are being used to feed animals that we prefer to eat – from fish to cattle. If the world protein shortage continues, however, people may yet turn to eating foods made directly from microorganisms. Table 2 shows some of the advantages and disadvantages of using microorganisms directly to make food for human consumption.

▼ Table 2

Advantages of using microorganisms to produce human food	Disadvantages of using microorganisms to produce human food
microorganisms reproduce fast and produce protein faster than animals and plants	some microorganisms can also produce toxins if the conditions are not maintained at the optimum
microorganisms have a high protein content with little fat	the microorganisms have to be separated from the nutrient broth and processed to make the food
microorganisms can use a wide variety of waste materials including human and animal waste, reducing costs	need sterile conditions that are carefully controlled adding to costs
microorganisms can be genetically modified to produce the protein required	often involve GM organisms and many people have concerns about eating GM food
production of microorganisms is not dependant on weather, breeding cycles etc – it takes place constantly and can be increased or decreased to match demand	the protein has to be purified to ensure it contains no toxin or contaminants
no welfare issues when growing microorganisms	many people dislike the thought of eating microorganisms grown on waste
can be made to taste like anything	has little natural flavour – needs additives

Summary questions

1 What are the main advantages of using microorganisms in biotechnological processes?   (6 marks)

2 Compare the way yeast is used in the process of baking and brewing.   (6 marks)

3 a Why is milk pasteurised before being used commercially to make cheese and yoghurt?   (2 marks)
   b Why is milk homogenised before being used commercially to make cheese and yoghurt?   (2 marks)
   c Give two important differences between the production processes of cheese and yogurt.   (2 marks)

22.5 Microorganisms, medicines, and bioremediation

Specification reference: 6.2.1

Learning outcomes

Demonstrate knowledge, understanding, and application of:

→ the use of microorganisms in biotechnological processes to include penicillin production, insulin production, and bioremediation.

Synoptic link

In Topic 12.7, Preventing and treating disease, you learnt about antibiotics and their importance in the treatment of bacterial diseases. In Topic 14.4, Diabetes and its control, you learnt about the importance of insulin to diabetics.

Since the discovery of penicillin in the 1920s, biotechnology has played a key role in the development of medicines.

Producing penicillin

The first effective antibiotic was penicillin, produced by a mould called *Penicillium notatum*. The yield of penicillin from this mould was very small. Commercial production of the drug in the quantities needed to treat everyone who needed it did not begin until the discovery of *Penicillium chrysogenum* by Mary Hunt on a melon from a market stall.

P. chrysogenum needs relatively high oxygen levels and a rich nutrient medium to grow well. It is sensitive to pH and temperature. This affects the way it is produced commercially. A semi-continuous batch process is used (Topic 22.7, Culturing microorganisms on an industrial scale). In the first stage of the production process the fungus grows. In the second stage it produces penicillin. Finally the drug is extracted from the medium and purified.

- The process uses relatively small fermenters (40–200 dm³) because it is very difficult to maintain high levels of oxygenation in very large bioreactors.
- The mixture is continuously stirred to keep it oxygenated.
- There is a rich nutrient medium.
- The growth medium contains a buffer to maintain pH at around 6.5.
- The bioreactors are maintained at about 25–27 °C.

Making insulin

As you learnt in Topic 21.4, Genetic engineering, biotechnology in the form of genetic engineering is important in the production of human medicines – for example, the production of human insulin. People with type 1 diabetes, and some people with type 2 diabetes, need regular injections of insulin to control their blood sugar levels. In the past, insulin was extracted from the pancreas of animals, usually pigs or cattle, slaughtered for meat. This meant the supply was erratic because it depended on the demand for meat – when fewer animals were killed, less insulin was available but the number of people with diabetes stayed the same. There were a number of other problems. Some people were allergic to the animal insulin as it was often impure, although eventually very pure forms were developed which overcame this problem. The peak activity of animal insulin is several hours after it is injected, which made calculating when to eat meals difficult. For some faith groups, using pig products is not permitted. The development of genetically engineered bacteria which can make

▲ **Figure 1** *Penicillin is produced by the large-scale cultivation of* Penicillium *mould followed by processing and purification of the drug*

human insulin revolutionised the supply from the 1970s onwards. The bacteria are grown in a fermenter and downstream processing results in a constant supply of pure human insulin. You will learn about the bioreactors used to produce human insulin in Topic 22.7, Culturing microorganisms on an industrial scale.

Bioremediation

In bioremediation, microorganisms are used to break down pollutants and contaminants in the soil or in water. There are different approaches to bioremediation:

1   Using natural organisms – many microorganisms naturally break down organic material producing carbon dioxide and water. Soil and water pollutants are often biological, for example, sewage and crude oil. If these naturally occurring microorganisms are supported, they will break down and neutralise many contaminants. For example, in an oil spill, nutrients can be added to the water to encourage microbial growth, and the oil can be dispersed into smaller particles to give the maximum surface area for microbial action.

2   GM organisms – scientists are trying to develop GM bacteria which can break down or accumulate contaminants which they would not naturally encounter, for example, bacteria have been engineered that can remove mercury contamination from water. Mercury is very toxic and accumulates in food chains. The aim is to develop filters containing these bacteria to remove mercury from contaminated sites.

Often, bioremediation takes place on the site of the contamination. Sometimes material is removed for decontamination. In most cases, natural microorganisms outperform GM ones – but as our ability to change the genetic material of microorganisms increases, it may be possible to use bioremediation even more effectively than it is used now.

Synoptic link

You learnt about genetic engineering and the production of insulin in Topic 21.4, Genetic engineering.

▲ **Figure 2** *When huge areas of water are contaminated with oil, as in the Deepwater Horizon spill in 2010, bioremediation by microorganisms is the best hope for the environment*

+ Plants and bioremediation

There are some pollutants which microorganisms cannot, at the moment, remove from the soil. In a number of cases, special plants can be used instead. In the early 1970s, a tree was discovered which produced a blue sap that turned out to contain 26% nickel in its dry mass. Plants which can take up large quantities of metals from the ground are known as hyperaccumulators. The process by which hyperaccumulators take up metals from the ground is known as bioleaching.

Suggest why plants can be used as bioremediators for heavy metal contamination but microorganisms cannot.

Summary questions

1   What is the main difference between the use of fungi to produce penicillin and the use of bacteria to produce human insulin?   (*2 marks*)

2   What is bioremediation? Why it is often carried out on the site of contamination?   (*6 marks*)

3   With your knowledge of genetic engineering and biotechnology so far, produce a flow diagram to show the stages in the production of human insulin.   (*6 marks*)

22.6 Culturing microorganisms in the laboratory

Specification reference: 6.2.1

To investigate microorganisms for the medical diagnosis of disease or for scientific experiments you need to **culture** them. This often involves growing large enough numbers of the microorganisms for us to see them clearly with the naked eye.

Whenever microorganisms are cultured in the laboratory the correct health and safety procedures must be followed because even when the microorganisms are expected to be completely harmless:

● there is always the risk of a mutation taking place making the strain pathogenic

● there may be contamination with pathogenic microorganisms from the environment.

Culturing microorganisms

The microorganisms to be cultured need food as well as the right conditions of temperature, oxygen, and pH. The food provided for microorganisms is known as the nutrient medium. It can be either in liquid form (broth) or in solid form (agar). Nutrients are often added to the agar or the broth to provide a better medium for microbial growth. Some microorganisms need a precise balance of nutrients but often the medium is simply enriched with good protein sources such as blood, yeast extract, or meat. Enriched nutrient media allow samples containing a very small number of organisms to multiply rapidly. The nutrient medium must be kept sterile (free from contamination by microorganisms) until it is ready for use. **Aseptic techniques** are important.

Once the agar or nutrient broth is prepared the bacteria must be added in a process called inoculation.

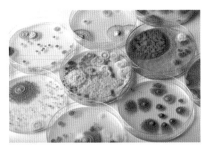

▲ **Figure 1** Culturing microorganisms like these different types of bacteria allows us to see what we are working with

Inoculating broth

1 Make a suspension of the bacteria to be grown.

2 Mix a known volume with the sterile nutrient broth in the flask.

3 Stopper the flask with cotton wool to prevent contamination from the air.

4 Incubate at a suitable temperature, shaking regularly to aerate the broth providing oxygen for the growing bacteria.

Inoculating agar

This also involves a suspension of bacteria but the process is slightly more complicated.

1 The wire inoculating loop must be sterilised by holding it in a Bunsen flame until it glows red hot. It must not be allowed to touch any surfaces as it cools to avoid contamination.

▲ **Figure 2** Cultures grown in broth are usually prepared in flasks or test tubes whilst agar-based cultures are prepared in Petri dishes. The hot sterile liquid agar is poured onto the plates which are immediately resealed and cooled so that they set ready for use

2   Dip the sterilised loop in the bacterial suspension. Remove the lid of the Petri dish and make a zig-zag streak across the surface of the agar. Avoid the loop digging into the agar by holding it almost horizontal. However many streaks are applied, the surface of the agar must be kept intact.

3   Replace the lid of the Petri dish. It should be held down with tape but not sealed completely so oxygen can get in, preventing the growth of anaerobic bacteria. Incubate at a suitable temperature.

▲ **Figure 3** *Inoculating an agar plate with bacteria*

The growth of bacterial colonies

Bacteria can reproduce very rapidly, undergoing asexual reproduction every 20 minutes in optimum conditions. If a single bacterium had unlimited space and nutrients, and if all its offspring continued to divide at the same rate, then at the end of 48 hours there would be 2.2×10^{43} bacteria, weighing 4000 times the weight of the Earth. Fortunately, in a closed system limited nutrients and a build-up of waste products always acts as a brake on reproduction and growth. Logarithmic numbers (logs) are mainly used to represent the bacterial population because the difference in numbers from the initial organism to the billions of descendants is sometimes too great to represent using standard numbers.

There are four stages to this growth curve:

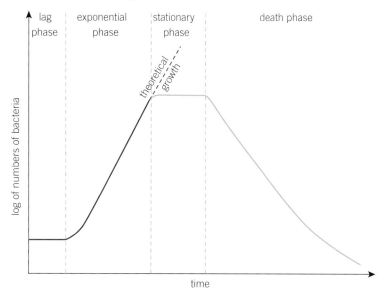

▲ **Figure 4** *The exponential growth theoretically possible for a bacterial population (red dashed line) is prevented by a variety of limiting factors which, fortunately, result in the typical growth curve of a bacterial population in a closed system*

● the lag phase when bacteria are adapting to their new environment. They are growing, synthesising the enzymes they need, and are not yet reproducing at their maximum rate.

● the log or exponential phase is when the rate of bacterial reproduction is close to or at its theoretical maximum.

● the stationary phase occurs when the total growth rate is zero – the number of new cells formed by binary fission is cancelled out by the number of cells dying.

● the decline or death stage comes when reproduction has almost ceased and the death rate of cells is increasing.

There are several limiting factors which prevent exponential growth in a culture of bacteria. These include:

- Nutrients available – initially there is plenty of food, but as the numbers of microorganisms multiply exponentially it is used up. The nutrient level will become insufficient to support further growth and reproduction unless more nutrients are added.
- Oxygen levels – as the population rises, so does the demand for respiratory oxygen so oxygen levels can become limiting.
- Temperature – the enzyme-controlled reactions within microorganisms are affected by the temperature of the culture medium. For most bacteria, a low temperature slows down growth and reproduction, and a higher temperature speeds it up. If the temperature gets too high it will denature the enzymes, killing the microorganisms – even thermophiles have a maximum temperature they can withstand.
- Build-up of waste – as bacterial numbers rise, the build-up of toxic material may inhibit further growth and can even poison and kill the culture.
- Change in pH – as carbon dioxide produced by the respiration of the bacterial cells increases, the pH of the culture falls until a point where the low pH affects enzyme activity and inhibits population growth.

Investigating factors which affect the growth of microorganisms – serial dilutions and bacterial counting

You can investigate the factors which affect the growth of bacterial colonies in a number of ways. For example, you can:

- set up identical colonies in different conditions of temperature
- set up serial dilutions of nutrients or pH, at a set temperature.

It is essential when carrying out these experiments that precautions are taken to ensure aseptic conditions (free from contamination). For example, using sterile equipment and a fresh pipette after each dilution.

To see the effect of the conditions you need to be able to measure the number of microorganisms at the beginning and end of your investigations. One method is to use another application of serial dilutions (Figure 5). The assumption is made that each of the colonies on an agar plate grows from a single, viable microorganism. If two bacterial colonies can be seen after culturing, then there were two living bacteria on the plate, and if 50 patches form there were 50 bacteria on the plate when it was inoculated. However, in most cases when a plate is inoculated a solid mass of microbial growth is present after culturing – you cannot count the individual colonies. This

is overcome by carrying out a serial dilution of the original culture broth until, when you culture a given volume of the broth on an agar plate, you can count the number of colonies. Multiply the number of colonies by the dilution factor to give you a total viable cell count per volume for the original colony. As long as you can count the number of colonies on two or more plates, you can calculate the mean of the number of organisms in a particular culture.

A student made up serial dilutions of a bacterial culture from 10^{-1} to 10^{-10}. A $0.1\,\text{cm}^3$ sample was cultured from each of these dilutions. From the original dilutions the numbers of bacterial colonies counted were:

A Original dilution 10^{-5-} – 500 colonies
B Original dilution 10^{-6} – 52 colonies
C Original dilution 10^{-7} – 4 colonies

Calculate (a) the actual dilution of each sample grown on the plate
(b) the number of bacteria cm^{-3} in the original sample as shown on each plate
(c) the mean number of bacteria cm^{-3} in the original sample.

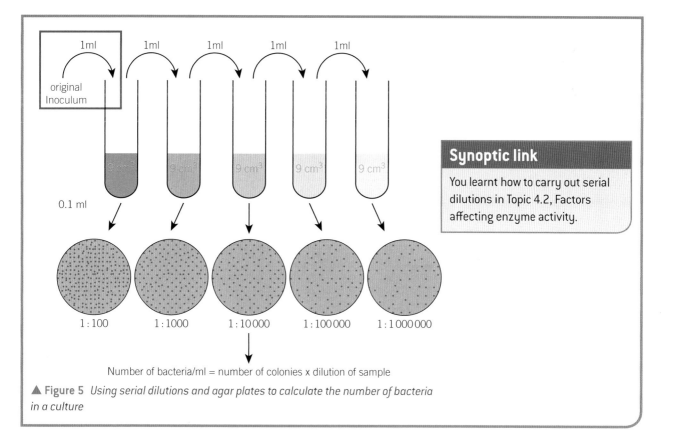

Number of bacteria/ml = number of colonies x dilution of sample

▲ **Figure 5** *Using serial dilutions and agar plates to calculate the number of bacteria in a culture*

Synoptic link

You learnt how to carry out serial dilutions in Topic 4.2, Factors affecting enzyme activity.

Summary questions

1 Compare the processes of culturing bacteria in broth and on agar. (*6 marks*)

2 Why are there such clear differences between the theoretical growth curve of a bacterial colony and the actual growth curve in a closed culture? (*5 marks*)

3 Explain the following statements in terms of factors affecting bacterial growth:
   a Vinegar is a very good preservative. (*2 marks*)
   b Food eventually goes bad in a fridge. (*2 marks*)
   c In the Northern hemisphere, material placed in a compost heap rots down much faster in August than it does in December. (*2 marks*)

4 Make a flow chart to show how you would calculate the affect of a factor on bacterial growth using serial dilutions and agar plating. (*6 marks*)

Learning outcomes
Demonstrate knowledge, understanding, and application of:

→ the importance of manipulating the growing conditions in batch and continuous fermentation in order to maximise the yield of product required.

In any bioprocess the microorganism involved must be able to synthesise or break down the chemical required, work reasonably fast, give a good yield of the product, use relatively cheap nutrients, and not require extreme (and therefore expensive) conditions. It must not produce any poisons that contaminate the product or mutate easily into non-functioning forms.

Primary and secondary metabolites

What is wanted from the microorganisms varies from one bioprocess to another.

Sometimes, you would want as much microorganism as possible, because the microorganism itself is the product to be sold, for example, single-celled protein such as Quorn, or baker's yeast.

Sometimes, substances are wanted which are formed as an essential part of the normal functioning of a microorganism, for example, ethanol (a product of anaerobic respiration in yeast), ethanoic acid, and a range of amino acids and enzymes. They are known as primary metabolites.

In some circumstances, organisms produce substances which are not essential for normal growth, but are still used by the cells. Examples include many pigments, and the toxic chemicals plants produce to protect themselves against attack by herbivores. The organism would not suffer, at least in the short term, without them. These chemicals are known as secondary metabolites, and they are often the required product in a bioprocess, for example, penicillin and many other antibiotics.

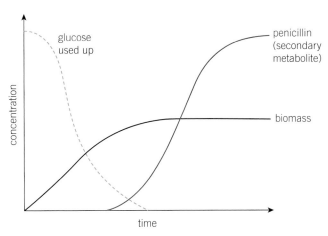

▲ **Figure 1** *Primary metabolites are usually formed in the period of active growth, whilst secondary metabolites tend to be formed during the stationary phase of the life of the culture once the cell mass has reached its maximum. The time at which the culture is harvested will depend on what we actually want from the microorganisms*

Types of bioprocess

Once a microorganism has been chosen, and the ideal size and shape of the bioreactor (reaction vessel) decided, the organisation of the commercial production has to be decided. Two of the main ways of growing microorganisms are **batch fermentation** and **continuous fermentation**.

Batch fermentation

- The microorganisms are inoculated into a fixed volume of medium.
- As growth takes place, nutrients are used up and both new biomass and waste products build up.

- As the culture reaches the stationary phase, overall growth ceases but during this phase the microorganisms often carry out biochemical changes to form the desired end products (such as antibiotics and enzymes).

- The process is stopped before the death phase and the products harvested. The whole system is then cleaned and sterilised and a new batch culture started up.

Continuous culture

- Microorganisms are inoculated into sterile nutrient medium and start to grow.

- Sterile nutrient medium is added continually to the culture once it reaches the exponential point of growth.

- Culture broth is continually removed – the medium, waste products, microorganisms, and product – keeping the culture volume in the bioreactor constant.

Continuous culture enables continuous balanced growth, with levels of nutrients, pH, and metabolic products kept more or less constant.

Both methods of operating a bioreactor can be adjusted to ensure either the maximum production of biomass or the maximum production of the primary or secondary metabolites. Most systems are adapted for maximum yield of metabolites. The majority of industrial processes use batch or semi-continuous cultivation. Continuous cultivation is largely used for the production of single-celled protein and in some waste water treatment processes.

All bioreactors produce a mixture of unused nutrient broth, microorganisms, primary metabolites, possibly secondary metabolites, and waste products. The useful part of the mixture has to be separated out by downstream processing. This is one of the most difficult and expensive parts of the whole bioprocess – the percentage of the total cost of a product which is due to downstream processing costs varies from 15–40%.

Controlling bioreactors

Whether a bioreactor is simply a container containing microbial broth or a complex aseptic fermenter, it is very important to control and manipulate the growing conditions to maximise the yield of product required. Factors which need to be controlled include:

Temperature
If the temperature is too low the microorganisms will not grow quickly enough. If the temperature gets too high, enzymes start to denature and the microorganisms are inhibited or destroyed. Bioreactors often have a heating and/or a cooling system linked to temperature sensors and a negative feedback system to maintain optimum conditions.

Nutrients and oxygen
Oxygen and nutrient medium can be added in controlled amounts to the broth when probes or sample tests indicate that levels are dropping.

▲ Figure 2 *Bioreactors range from around a 10 litre capacity to hundreds of litres and are widely used in the pharmaceutical industry*

Mixing things up

Inside a bioreactor there are large volumes of liquid, which may be quite thick and viscous due to the growth of microorganisms. Simple diffusion is not enough to ensure that all the microorganisms receive enough food and oxygen or that the whole mixture is kept at the right temperature, so most bioreactors have a mixing mechanism and many are stirred continuously.

Asepsis

If a bioprocess is contaminated by microorganisms from the air, or from workers, it can seriously affect the yield. To solve this problem most bioreactors are sealed, aseptic units. If the process involves genetically engineered organisms, it is a legal requirement that they should be contained within the bioreactor and not be released into the environment.

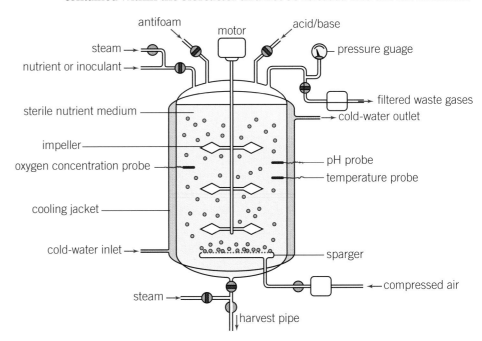

▲ **Figure 3** *The design of bioreactors is moving forward all the time as new and more sensitive ways of controlling conditions during fermentation processes become available, making it easier to get the maximum yield from any process*

Summary questions

1 Bioreactors may run batch or continuous processes. What is the difference? *(6 marks)*

2 Choose three factors which need to be controlled in a bioreactor to give the maximum yield of product and for each explain why it is important and how it might be controlled. *(6 marks)*

3 Look at Figure 1.
   a What is the desired end product of this reaction? *(1 mark)*
   b What tells you that this bioreactor is not producing food? *(3 marks)*
   c Sketch a graph to show what you would expect to see if this process was producing a food material rather than a drug. *(3 marks)*

22.8 Using immobilised enzymes

Specification reference: 6.2.1

Enzymes, mainly from microorganisms, have been used indirectly for thousands of years in biotechnologies including brewing, baking, and making cheese and yoghurt. Many biotechnological processes still use whole microorganisms as their enzyme source. More recently they are also being used in isolation.

Isolated enzymes

Using isolated enzymes instead of whole organisms has some clear advantages.

- Less wasteful – whole microorganisms use up substrate growing and reproducing, producing biomass rather than product. Isolated enzymes do not.

- More efficient – isolated enzymes work at much higher concentrations than is possible when they are part of the whole microorganism.

- More specific – no unwanted enzymes present, so no wasteful side reactions take place.

- Maximise efficiency – isolated enzymes can be given ideal conditions for maximum product formation, which may differ from those needed for the growth of the whole microorganism.

- Less downstream processing – pure product is produced by isolated enzymes. Whole microorganisms give a variety of products in the final broth, making isolation of the desired product more difficult and therefore expensive.

Most of the isolated enzymes used in industrial processes are extracellular enzymes produced by microorganisms. They are generally easier and therefore cheaper to use than intracellular enzymes.

- Extracellular enzymes are secreted, making them easy to isolate and use.

- Each microorganism produces relatively few extracellular enzymes, making it easy to identify and isolate the required enzyme. In comparison, each microorganism produces hundreds of intracellular enzymes which would need extracting from the cell and separating.

- Extracellular enzymes tend to be much more robust than intracellular enzymes. Conditions outside a cell are less tightly controlled than conditions in the cytoplasm, so extracellular enzymes are adapted to cope with greater variations in temperature and pH than intracellular enzymes.

However, in spite of the advantages of using extracellular enzymes, intracellular enzymes are still sometimes used as isolated enzymes in manufacturing processes. This is because there is a bigger range of intracellular enzymes (bullet 2) so in some cases they provide the ideal enzyme for a process. In these cases, the benefits of using a

Learning outcomes

Demonstrate knowledge, understanding, and application of:

→ the uses of immobilised enzymes in biotechnology and the different methods of immobilisation.

active site

▲ Figure 1 *The shape of the active site gives an enzyme great specificity and makes it an invaluable tool to the biotechnologist*

Synoptic link

You learnt about extracellular and intracellular enzymes in Topic 4.1, Enzyme action.

very specific intracellular enzyme outweigh the disadvantages of the more expensive extraction and isolation process and the need for more tightly controlled conditions. Examples of intracellular enzymes used as isolated enzymes in industry include glucose oxidase for food preservation, asparaginase for cancer treatment, and **penicillin acylase** for converting natural penicillin into semi-synthetic drugs which are more effective.

Immobilised enzymes

Isolated enzymes are more efficient than whole organisms, but using free enzymes is often very wasteful. Enzymes are not cheap to produce, but at the end of the process they cannot usually be recovered and so they are simply lost.

Increasingly enzymes used in industrial processes are immobilised – attached to an inert support system over which the substrate passes and is converted to product. This is a case of technology mimicking nature – enzymes in cells are usually bound to membranes to carry out their repeated cycles of catalysis. Because immobilised enzymes are held stationary during the catalytic process, they can be recovered from the reaction mixture and reused time after time. The enzymes do not contaminate the end product, so less downstream processing is needed. These things all make the process more economical.

Advantages of using immobilised enzymes
- Immobilised enzymes can be reused – which is cheaper.
- Easily separated from the reactants and products of the reaction they are catalysing so reduced downstream processing – which is cheaper.
- More reliable – there is a high degree of control over the process as the insoluble support provides a stable microenvironment for the immobilised enzymes.
- Greater temperature tolerance – immobilised enzymes are less easily denatured by heat and work at optimum levels over a much wider range of temperatures, making the bioreactor less expensive to run.
- Ease of manipulation – the catalytic properties of immobilised enzymes can be altered to fit a particular process more easily than those of free enzymes – for example, immobilised **glucose isomerase** can be used continuously for over 1000 hours at temperatures of 60–65 °C. The ability to keep bioreactors running continuously for long periods without emptying and cleaning helps to keep running costs low.

Disadvantages of using immobilised enzymes
- Reduced efficiency – the process of immobilising an enzyme may reduce its activity rate.
- Higher initial costs of materials – immobilised enzymes are more expensive than free enzymes or microorganisms. However, the immobilised enzymes, unlike free enzymes, do not need to be replaced frequently.

- Higher initial costs of bioreactor – the system needed to use immobilised enzymes is different from traditional fermenters so there is an initial investment cost.
- More technical issues – reactors which use immobilised enzymes are more complex than simple fermenters – they have more things which can go wrong.

How are enzymes immobilised?

Enzymes can be immobilised in a number of ways. They may be bound to the surface of insoluble supporting materials either by adsorption onto the surface or by covalent or ionic bonds. They may be entrapped in a matrix, encapsulated in a microcapsule (like proteases for detergent use), or behind a semi-permeable membrane. Each of these methods has advantages and disadvantages as summarised in Table 1.

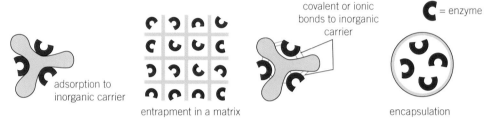

▲ Figure 2 *Four commonly used ways of immobilising enzymes*

▼ Table 1

Method of immobilising the enzymes	Advantages	Disadvantages
surface immobilisation – adsorption to inorganic carriers, e.g., cellulose, silica, carbon nanotubes, and polyacrylamide gel	simple and cheap to do can be used with many different processes enzymes very accessible to substrate and their activity is virtually unchanged	enzymes can be lost from matrix relatively easily
surface immobilisation – covalent or ionic bonding to inorganic carriers covalent bonding, e.g., carriers with amino, hydroxyl, carboxyl groups ionic bonding, e.g., polysaccharides such as cellulose, synthetic polymers	cost varies enzymes strongly bound and therefore unlikely to be lost enzymes very accessible to substrate pH and substrate concentration often have little effect on enzyme activity	cost varies active site of the enzyme may be modified in process, making it less effective
entrapment – in matrix, e.g., polysaccharides, gelatin, activated carbon	widely applicable to different processes	may be expensive can be difficult to entrap diffusion of the substrate to and product from the active site can be slow and hold up the reaction effect of entrapment on enzyme activity very variable, depending on matrix

entrapment – membrane entrapment in microcapsules (encapsulation) or behind a semi-permeable membrane, e.g., polymer-based semi-permeable membranes	relatively simple to do relatively small effect on enzyme activity widely applicable to different processes	relatively expensive diffusion of the substrate to and product from the active site can be slow and hold up the reaction

In some cases whole microorganisms rather than just the enzymes are immobilised. This has many of the same advantages but avoids the time-consuming and expensive process of extracting the pure enzyme and immobilising it before the process starts. On the other hand, the organisms need food, oxygen, and a carefully controlled environment to work at their optimum rate.

Using immobilised enzymes

Immobilised enzymes are very useful when large quantities of product are wanted, because they allow continuous production. Examples include:

- Immobilised penicillin acylase used to make semi-synthetic penicillins from naturally produced penicillins. Many types of bacteria have developed resistance to naturally occurring penicillins so they are no longer very effective drugs. Fortunately, many bacteria are still vulnerable to the semi-synthetic penicillins produced by penicillin acylase so they are very important in treating infections caused by bacteria resistant to the original penicillin. Hundreds of tonnes of these medicines are made every year by immobilised penicillin acylase.

- Immobilised glucose isomerase used to produce fructose from glucose. Fructose is much sweeter than sucrose or glucose and is widely used as a sweetener in the food industries. Glucose is produced from cheap, starch-rich plant material. Glucose isomerise is then used to turn the cheap glucose into very marketable fructose.

- Immobilised **lactase** used to produce lactose-free milk. Some people, and cats, are intolerant of lactose (milk sugar). Immobilised lactase hydrolyses lactose to glucose and galactose, giving lactose-free milk.

- Immobilised aminoacylase used to produce pure samples of L-amino acids used in the production of pharmaceuticals, organic chemicals, cosmetics, and food.

- Immobilised glucoamylase, which can be used to complete the breakdown of starch to glucose syrup. Amylase enzymes break starch down into short chain polymers called dextrins. The final breakdown of dextrins to glucose is catalysed by immobilised glucoamylase.

- Immobilised nitrile hydratase, an enzyme which is playing an increasing role in the plastics industry. Acrylamide is a very important compound which is used in the production of many plastics. It is made by the hydration of acrylonitrile. Traditionally the hydration of acrylonitrile to acrylamide was done using sulphuric acid with a reduced copper catalyst, but the conditions

Study tip

Nitrile hydratase is the enzyme which catalyses the conversion of acrylonitrile to acrylamide in a hydration reaction.

Nitrilases are a group of enzymes that catalyse the hydrolysis of nitriles, for example, acrylonitrile to *carboxylic* acids and ammonia.

needed are extreme and therefore expensive. Furthermore, unwanted by-products form and the yield is poor. Using immobilised nitrile hydratase the conversion takes place under moderate conditions so the process is cheaper and it also gives a 99% yield and no unwanted by-products.

Immobilised enzymes in medicine

Immobilised enzymes and microbial cells are increasingly important both as diagnostic tools in medicine – the manufacture of drugs – for example, the fungus *Rhizopus arrhizus* is immobilised and used in the production of the steroid drug cortisone.

Biosensors are used in accurate monitoring of blood and urine levels of substances such as glucose, urea, amino acids, ethanol, and lactic acid, as well as in the monitoring of waste treatment, water analysis, and the control of complex chemical processes. They are based on an electrochemical sensor in close proximity to an immobilised enzyme membrane. The enzymes react with a specific substrate and the chemicals produced

are detected by the sensor. Because the size of the response is related to the concentration of the substrate, these sensors can be very sensitive and accurate in their measurements and so can be used, for example, by people with diabetes to determine their blood glucose levels and therefore judge the insulin dose they need.

1 Give one clear advantage of using immobilised enzymes in biosensors for medical use and one possible disadvantage, both linked to what you know of the properties of enzymes.
2 Suggest how immobilising the enzyme might help overcome your suggested disadvantage.

Summary questions

1 a What is meant by an immobilised enzyme? (2 marks)
   b What are the main advantages of immobilised enzymes over whole microorganisms? (4 marks)
   c What are the main advantages of immobilised enzymes over free enzymes? (4 marks)

2 Summarise the main ways in which enzymes are immobilised. (4 marks)

3 How can immobilisation:
   a increase the effectiveness of an enzyme (3 marks)
   b decrease the effectiveness of an enzyme? (4 marks)

4 Investigate one reaction catalysed by immobilised enzymes. Give the reaction catalysed, the way the enzyme is immobilised, the importance of the process, and the advantages of using immobilised enzymes over any other way of catalysing the reaction. (6 marks)

Practice questions

1. a Outline the differences between reproductive and therapeutic cloning. *(3 marks)*

   b Evaluate the ethical concerns regarding the reproductive cloning of human beings. *(4 marks)*

   c (i) Describe what is meant by the term somatic cell nuclear transfer (SCNT). *(3 marks)*

   (ii) Suggest why clones produced by SCNT are not exact genetic clones. *(2 marks)*

   d Describe the reasons for embryo splitting. *(3 marks)*

2. Micropropagation is the name of a process used to produce artificially cloned plants.

   a Explain the meaning of the following terms and outline their roles in the micropropagation. *(4 marks)*

   • Explant
   • Callus

   b Describe how plant hormones are used to stimulate the development of cells obtained from the callus. *(5 marks)*

3. The diagram shows a type of fermentation.

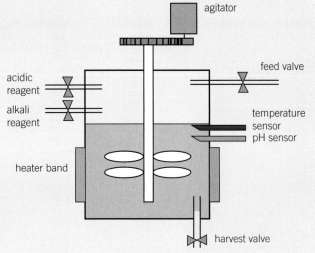

   a State, giving your reasons, the type of fermentation being conducted using the equipment shown in the diagram *(3 marks)*

   b Outline the roles of the valves and sensors shown in the diagram. *(4 marks)*

   c Discuss the advantages and disadvantages of this type of fermentation. *(4 marks)*

4. The cloning of animals is a controversial subject due to the shortened life expectancies and health problems of cloned animals.

   The graph shows the data collected from an investigation into the health problems faced by cloned and conventional cows.

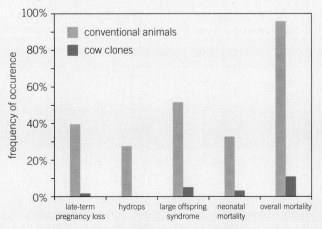

   Calves with large offspring syndrome usually have breathing and cardiovascular problems as well difficult births. Hydrops is an abnormal build-up of fluid in new-born calves and is an indication of more serious health problems.

   A survey carried out by the European Commission investigating European Attitudes Towards Animal Cloning in October 2008 produced the results shown in the graph.

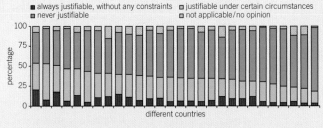

   Discuss, using information from both graphs, the arguments for and against the cloning of animals in the production of food. *(6 marks)*

5. Microorganisms, such as the fungus Fusarium, can be grown and then purified to produce mycoprotein, This mycoprotein can be used as a food source for humans.

   The table compares mycoprotein with beef.

Food	Content per 100g					
	Energy (kJ)	Protein (g)	Carbohydrate (g)	Total fat (g)	Saturated fat	Iron (mg)
mycoprotein	357	12	9	2.9	0.6	0.1
beef	1163	26	0	18.2	7.0	2.6

Use the data in the table to describe and explain the advantages and disadvantages of using microorganisms to produce food for human consumption. *(7 marks)*

OCR F212 2010

6 A number of different methods can be used to determine the number of bacteria in a sample.

a (i) Outline the differences between using a graticule and dilution plating to count bacteria. *(4 marks)*

(ii) State the name of another, more simple technique that uses the same principle as graticule counts. *(1 mark)*

b Suggest why it is necessary to maintain aseptic conditions when working with bacteria. *(3 mark)*

A 1 ml sample containing bacteria was mixed with an equal volume of the dye methylene blue. A drop the resulting mixture was placed on a graticule and observed under a microscope.

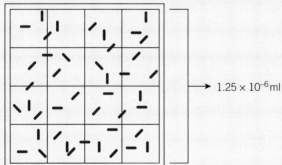

→ 1.25×10^{-6} ml

The graticule held 1.25×10^{-6} ml of sample.

c Calculate the number of bacteria per ml in the original sample. *(3 marks)*

A serial dilution was carried out on a sample containing *E. coli*. Samples from the last three dilutions were plated and incubated. The number of colonies that grew on each plate is shown in the diagram.

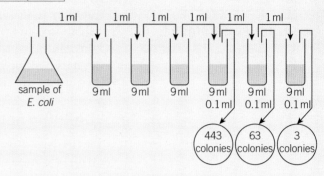

1 ml   1 ml   1 ml   1 ml   1 ml   1 ml

sample of *E. coli*   9 ml   9 ml   9 ml   9 ml 0.1 ml   9 ml 0.1 ml   9 ml 0.1 ml

443 colonies   63 colonies   3 colonies

d Calculate the number of bacteria in the original sample. *(3 marks)*

7 Starch phosphorylase catalyses the conversion of starch and inorganic phosphate into glucose-1-phosphate. It has an important role in the metabolism of starch in plants. It has been used in the production of glucose-1-phosphate for use in the treatment of some heart conditions.

When used for the production of glucose-1-phosphate starch phosphorylase is normally immobilized.

a Outline the different ways that enzymes are immobilized. *(4 marks)*

The diagram shows the changing activities of free and immobilized starch phosphorylase at different temperatures.

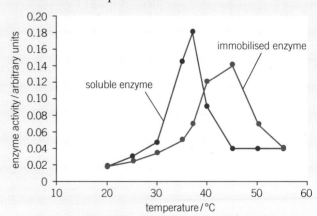

b Explain the reasons for the differences in optimum temperature and maximum activity of the free and immobilized enzymes. *(5 marks)*

▲ **Figure 1** *A biome is a large naturally occurring community of flora (plants) and fauna (animals) occupying a major habitat such as a rainforest*

Ecology is the name given to the study of the relationships between organisms and their environment. It is essential that scientists study the natural world and the vast variety of life that exists, to help us understand the interdependence of living organisms and to help ensure the survival of as much of Earth's biodiversity as possible.

Defining ecosystems

An **ecosystem** is made up of all the living organisms that interact with one another in a defined area, and also the physical factors present in that region. Ecosystems can vary dramatically in size – from a tiny bacterial colony to the entire biosphere of Earth.

The boundaries of a particular ecosystem being studied are defined by the person or team carrying out the study. For example, individual habitats may be studied such as a rock pool or a large oak tree, or small areas of land such as a playing field or a particular stretch of a river.

Factors that affect ecosystems

All ecosystems are dynamic, meaning that they are constantly changing. This is a result of the living organisms present and the environmental conditions.

A large number of factors affect an ecosystem. The factors can be divided into two groups:

- **biotic factors** – the living factors. For example, in a forest ecosystem, the presence of shrews and hedgehogs are biotic factors, as is the size of their populations – the competition between these two animal populations for a food source (e.g., insects) is also a biotic factor.

- **abiotic factors** – the non-living or physical factors. Within the forest ecosystem, abiotic factors include the amount of rainfall received and the yearly temperature range of the ecosystem.

Biotic factors

Biotic factors often refer to the interactions between organisms that are living, or have once lived. These interactions often involve competition, either within a population or between different populations. Examples of things for which animals compete include food, space (territory), and breeding partners.

Abiotic factors

Light

Most plants are directly affected by light availability as light is required for photosynthesis. In general the greater the availability of light, the greater the success of a plant species.

Plants develop strategies to cope with different light intensities. For example, in areas of low light they may have larger leaves. They may also develop photosynthetic pigments that require less light, or reproductive systems that operate only when light availability is at an optimum.

Temperature

The greatest effect of temperature is on the enzymes controlling metabolic reactions. Plants will develop more rapidly in warmer temperatures, as will ectothermic animals. (Endothermic animals control their internal temperature, and so are less affected by the external environment.) Changes in the temperature of an ecosystem, for example, due to the changing seasons, can trigger migration in some animal species, and hibernation in others. In plant species it can trigger leaf-fall, dormancy, and flowering.

Water availability

In most plant and animal populations, a lack of water leads to water stress, which, if severe, will lead to death.

A lack of water will cause most plants to wilt, as water is required to keep cells turgid and so keep the plant upright. It is also required for photosynthesis. Cacti are an example of xerophytes, plants that have developed successful strategies to cope with water stress.

Oxygen availability

In aquatic ecosystems, it is beneficial to have fast-flowing cold water as it contains high concentrations of oxygen. If water becomes too warm, or the flow rate too slow, the resulting drop in oxygen concentration can lead to the suffocation of aquatic organisms.

In waterlogged soil, the air spaces between the soil particles are filled with water. This reduces the oxygen available for plants.

Edaphic (soil) factors

Different soil types have different particle sizes. This has an effect on the organisms that are able to survive in them. There are three main soil types:

- clay – this has fine particles, is easily waterlogged, and forms clumps when wet
- loam – this has different-sized particles, it retains water but does not become waterlogged
- sandy – this has coarse, well-separated particles that allow free draining – sandy soil does not retain water and is easily eroded.

▲ **Figure 2** *Bromeliads, mosses, and small trees growing on the upper branches of a large tree in the cloud forest in Ecuador. This strategy allows small plants to receive sufficient light for photosynthesis*

Synoptic link

You learnt about xerophytes in Topic 9.5, Plant adaptations to water availability.

▲ **Figure 3** *The scimitar Oryx (Oryx dammah) has adapted to survive for 9–10 months without water. Their kidneys are specially adapted to produce the barest minimum of urine. They also sweat very little and so lose hardly any of their overall water intake*

Summary questions

1 Explain what is meant by the term 'dynamic ecosystem'. To which ecosystems does the term apply? *(2 marks)*

2 State two biotic and two abiotic factors in a pond ecosystem. *(4 marks)*

3 Explain why abiotic factors often have a greater effect on plant species than on animal species in an ecosystem. *(4 marks)*

23.2 Biomass transfer through an ecosystem

Learning outcomes

Demonstrate knowledge, understanding, and application of:

→ biomass transfers through ecosystems.

Synoptic link

You learnt about photosynthesis in Chapter 17, Energy for biological processes.

All organisms found within an ecosystem require a source of energy to perform the functions needed to survive. Ultimately, the Sun is the source of energy for almost all ecosystems on Earth. Through the process of photosynthesis, the Sun's light energy is converted into chemical energy in plants and other photosynthetic organisms. This chemical energy is then transferred to other non-photosynthetic organisms as food.

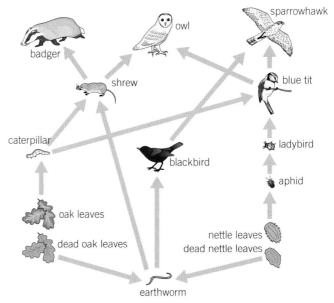

▲ **Figure 1** *Part of a woodland food web. The arrows on the diagram show the direction of energy flow.*

Trophic levels

You should be familiar with food chains and food webs (systems of interlinked food chains). These are diagrams that scientists use to show the transfer of biomass (mass of living material), and therefore energy, through the organisms in an ecosystem. Each stage in the chain is known as a **trophic level**.

The first trophic level is always a **producer** – an organism that converts light energy into chemical energy by the process of photosynthesis. The subsequent trophic levels are all **consumers** – organisms that obtain their energy by feeding on other organisms. The second trophic level is occupied by a primary consumer – an animal that eats a producer.

The following trophic levels are labelled successively – secondary consumer (an animal that eats a primary consumer), tertiary consumer (an animal that eats a secondary consumer), and a quaternary consumer (an animal that eats a tertiary consumer). Food chains rarely have more trophic levels than this as there is not sufficient biomass and stored energy left to support any further organisms.

tertiary consumers

secondary consumers

primary consumers

producers

▲ **Figure 2** *Food chains can be presented diagrammatically as a pyramid of numbers*

Decomposers are also important components of food webs – they break down dead organisms releasing nutrients back into the ecosystem. You will learn more about decomposers in Topic 23.3, Recycling within ecosystems.

Food chains can be presented diagrammatically as a pyramid of numbers, with each level representing the number of organisms at each trophic level (Figure 2). In a pyramid the producers are always placed at the bottom of the diagram with subsequent trophic levels added above.

Measuring biomass

Biomass is the mass of living material present in a particular place or in particular organisms. It is an important measure in the study of food chains and food webs as it can be equated to energy content.

To calculate biomass at each trophic level, you multiply the biomass present in each organism by the total number of organisms in that trophic level. This information can be presented diagrammatically as a pyramid of biomass (Figure 3). This represents the biomass present at a particular moment in time – it does not take into account seasonal changes.

The easiest way to measure biomass is to measure the mass of fresh material present. However, water content must be discounted and the presence of varying amounts of water in different organisms makes this technique unreliable unless very large samples are used. Scientists therefore usually calculate the 'dry mass' of organisms present. This is not without problems. Organisms have to be killed in order to be dried. The organisms are placed in an oven at 80 °C until all water has evaporated – this point is indicated by at least two identical mass readings. To minimise the destruction of organisms (particularly animals) only a small sample is taken. However, this sample may not be representative of the population as a whole.

Biomass is measured in grams per square metre ($g\,m^{-2}$) for areas of land, or grams per cubic metre ($g\,m^{-3}$) for areas of water.

Efficiency of biomass and energy transfer between trophic levels

The biomass in each trophic level is nearly always less than the trophic level below. This is because biomass consists of all the cells and tissues of the organisms present, including the carbohydrates and other carbon compounds the organisms contain. As carbon compounds are a store of energy, biomass can be equated to energy content. When animals eat, only a small proportion of the food they ingest is converted into new tissue. It is only this part of the biomass (and hence energy) which is available for the next trophic level to eat.

The energy available at each trophic level is measured in kilojoules per metre squared per year ($kJ\,m^{-2}\,yr^{-1}$), to allow for changes in photosynthetic production and consumer feeding patterns throughout the year. As biomass is transferred between trophic levels, so the energy contained is

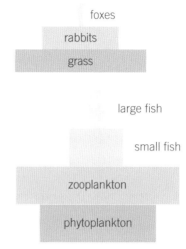

▲ Figure 3 *Pyramids of biomass are virtually always a pyramid shape (top). An exception is a pyramid of biomass for a marine ecosystem (bottom). Phytoplankton are microorganisms that can photosynthesise. The mass of phytoplankton at any given time is often quite small, but they reproduce very quickly so over a period of time there is always more phytoplankton than zooplankton*

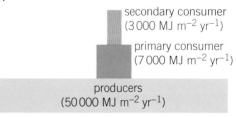

▲ Figure 4 *A pyramid of energy for grassland*

transferred. The efficiency with which biomass or energy is transferred from one trophic level to the next is called the **ecological efficiency**.

The amount of biomass or energy converted to new biomass by each trophic level in a food chain can be represented by a pyramid of energy (Figure 4).

Efficiency at producer level

Producers only convert 1–3% of the sunlight (solar or light energy) they receive into chemical energy and hence biomass. This is because:

- not all of the solar energy available is used for photosynthesis – approximately 90% is reflected, some is transmitted through the leaf, and some is of unusable wavelength
- other factors may limit photosynthesis, such as water availability
- a proportion of the energy is 'lost', as it is used for photosynthetic reactions.

The total solar energy that plants convert to organic matter is called the gross production. However, plants use 20–50% of this energy in respiration. The remaining energy is converted into biomass. This is the energy available to the next trophic level and is known as the net production.

The energy available to the next trophic level can be calculated using the following formula:

Net production = gross production – respiratory losses

Note that this calculation can be applied equally to the biomass or energy production within an organism. The generation of biomass in a producer is referred to as primary production – in a consumer, it is known as secondary production.

 ### Worked example 1: Calculating net production (biomass)

A group of scientists measured the gross production of a grassland area as $60\,g\,m^{-2}\,year^{-1}$. If the respiration loss was $20\,g\,m^{-2}\,year^{-1}$, calculate the annual net production of this area of grassland.

Net production = gross production – respiratory losses

$$= 60 - 20$$
$$= 40\,g\,m^{-2}\,year^{-1}$$

This is an example of primary production.

 ### Worked example 2: Calculating net production (energy)

Sheep in a grassland digest and absorb $15\,000$ $kJ\,m^{-2}\,yr^{-1}$ from the biomass they take in. Of this, $8000\,kJ\,m^{-2}\,yr^{-1}$ is used in respiration. How much energy is available to humans, the next organism in the food chain?

Net production = gross production – respiratory losses

$$= 15\,000 - 8000$$
$$= 7000\,kJ\,m^{-2}\,yr^{-1}$$

This is an example of secondary production.

Efficiency at consumer levels

Consumers at each trophic level convert at most 10% of the biomass in their food to their own organic tissue. This is because:

● not all of the biomass of an organism is eaten, for example, plant roots or animal bones may not be consumed

● some energy is transferred to the environment as metabolic heat, as a result of movement and respiration

● some parts of an organism are eaten but are indigestible – these parts (and their energy content) are egested as faeces

● some energy is lost from the animal in excretory materials such as urine.

Only around 0.001% of the total energy originally present in the incident sunlight is finally embodied as biomass in a tertiary consumer.

You can use the following formula to calculate the efficiency of the energy transfer (approximately equivalent to biomass transfer) between each trophic level of a food chain:

$$\text{Ecological efficiency} = \frac{\text{energy or biomass available after the transfer}}{\text{energy or biomass available before the transfer}} \times 100$$

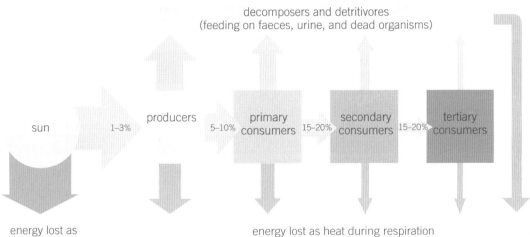

▲ **Figure 5** *This diagram shows the energy flow through different levels of a food chain. The figures stated for percentage energy transfer are approximate – energy transfers vary considerably between different plants, animals, and ecosystems*

 Worked example: Calculating ecological efficiency

Look at Figure 6. This diagram shows the energy available in a food chain in Cayuga Lake.

Calculate the ecological efficiency of the energy transfer between smelt and trout.

$$\text{Ecological efficiency} = \frac{\text{energy or biomass available after the transfer}}{\text{energy or biomass available before the transfer}} \times 100$$

$$= \frac{250}{1250} \times 100 = 20\%$$

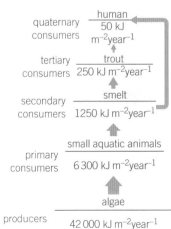

▲ **Figure 6** *This diagram shows the energy available at each trophic level in a food chain found in Cayuga Lake, New York State, USA*

Study tip

Remember that energy cannot be created or destroyed. However, it can be transferred, including to less useful forms such as heat, which is dissipated to the environment. This energy cannot be used to create biomass, and is therefore not available to the next trophic level.

▲ **Figure 7** *Intensively farmed animals are kept in a strictly-controlled environment. The conditions are chosen to ensure as much energy from their food as possible is turned into biomass. For example, small heated cages reduce energy loss through movement and maintaining body temperature*

Summary questions

1. Describe how the biomass of a trophic level is measured.   (*3 marks*)

2. Explain how human activities can manipulate the transfer of biomass through ecosystems.   (*2 marks*)

3. Using Figure 6, calculate the ecological efficiency of energy transfer between algae and small aquatic animals.   (*2 marks*)

4. Explain why biomass decreases at each level in a food chain.   (*4 marks*)

 Worked example: Efficiency of biomass transfer

A region of grassland has a net production of $40\,\mathrm{g\,m^{-2}\,year^{-1}}$. A goat grazes an area of $20\,\mathrm{m} \times 20\,\mathrm{m}$ of this grassland. Assume that the goat consumes all of the biomass in this area.

1. Calculate the total biomass consumed by the goat each year:

    Biomass consumed = mass (per metre squared per year) × area of land

    $$= 40 \times (20 \times 20)$$
    $$= 16\,000\,\mathrm{g}$$
    $$= 16\,\mathrm{kg}$$

2. The mass of the goat increases in this time by $2.4\,\mathrm{kg}$. Calculate the efficiency of biomass transfer between the grass and the goat.

    $$\text{Efficiency of transfer} = \frac{\text{biomass available after transfer}}{\text{biomass available before transfer}} \times 100$$
    $$= \frac{2.4}{16} = 15\%$$

Human activities can manipulate biomass through ecosystems

Human civilisation depends on agriculture. Agriculture involves manipulating the environment to favour plant species that we can eat (crops) and to rear animals for food or their produce. Plants and animals are provided with the abiotic conditions they need to thrive such as adequate watering and warmth (e.g., greenhouse use, stabling of animals). Competition from other species is removed (e.g., the use of chemicals such as pesticides) as well as the threat of predators (e.g., by creating barriers such as fences to exclude wild herbivores or predators).

In a natural ecosystem, humans would occupy the second, third or even fourth trophic level. As you have learnt, at each trophic level there are considerable energy losses, therefore only a tiny proportion of the energy available at the start of the food chain is turned into biomass for consumption at these third and fourth levels.

Agriculture creates very simple food chains. In farming animals or animal produce for human consumption, only three trophic levels are present – producers (animal feed), primary consumers (livestock), and secondary consumers (humans). In cultivating plants for human consumption, there are just two trophic levels – producers (crops) and primary consumers (humans). This means that the minimum energy is lost since there are fewer trophic levels present than in the natural ecosystem. This ensures that as much energy as possible is transferred into biomass that can be eaten by humans.

Monitoring biomass during conservation

Sea urchins (*Strongylocentrotus* spp) are marine invertebrates that feed on kelp (a type of seaweed). In regions where sea urchins are abundant, kelp forest ecosystems can be disrupted. The urchins eat the kelps' holdfasts, these are strong structures which anchor the kelp to the sea bed. The remainder of the plant floats away resulting in an ecosystem known as an 'urchin barren' ecosystem, which contains so little biomass of seaweeds that few species are able to live in this region. The presence or absence of kelp beds therefore has a major influence on the structure of the marine community.

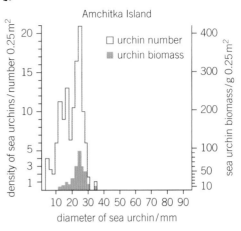

▲ **Figure 8** *Forests of kelp are felled by sea urchins which eat the kelp's holdfasts using their five self-sharpening teeth*

In many areas, sea otters (*Enhydra lutris*) feed on urchins, keeping their levels low and therefore the kelp forests intact. During the 19th century, this ecological balance was destroyed when populations of sea otters were virtually wiped out by excessive hunting for otter fur. As a result, urchin numbers grew rapidly and kelp forests were destroyed. This balance has since been restored by the cessation of the hunting of sea otters, allowing them to again control the abundance of the urchins. In turn, the productive kelp forests have been able to redevelop.

As part of a conservation management project, scientists studied sea urchin populations around two of the Aleutian Islands off the coast of Alaska. Data on sea urchin size, density, and biomass were recorded per 0.25 m² from samples collected from Amchitka Island (which has sea otters) and Sheyma Island (which has no sea otters) The results are shown in Figure 9.

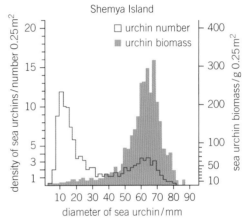

▲ **Figure 9**

1. State which trophic level of this marine community sea otters belong to.

2. a. State the diameter of the largest sea urchin studied on Sheyma Island.

   b. State the diameter of sea urchins with the most frequent biomass on Amchitka Island.

   c. Suggest with reasons which island has the oldest sea urchins.

3. Oil spills can result in the death of sea urchins. Explain how scientists can use the presence of kelp biomass to determine how an area is recovering from oil damage.

▲ Figure 1 *Oyster mushrooms. Fungi are the main decomposers in forests. Only wood-decay fungi have evolved the enzymes necessary to decompose lignin, a chemically complex substance found in wood*

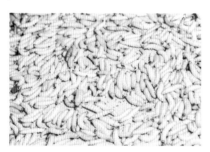

▲ Figure 2 *Maggots are an example of a detritivore that helps to break down animal tissue*

Energy has a linear flow through an ecosystem. It enters the ecosystem from the Sun, and is ultimately transferred to the atmosphere as heat. As long as the Sun continues to supply Earth with energy, life will continue. In contrast, nutrients constantly have to be recycled throughout ecosystems in order for plants and animals to grow. This is because they are used up by living organisms and there is no large external source constantly replenishing nutrients in the way the Sun supplies energy.

Decomposition

Decomposition is a chemical process in which a compound is broken down into smaller molecules, or its constituent elements. Often an essential element, such as nitrogen or carbon, cannot be used directly by an organism in the organic form it is in, in dead or waste matter. This organic material must be processed into inorganic elements and compounds, which are a more usable form, and returned to the environment.

A **decomposer** is an organism that feeds on and breaks down dead plant or animal matter, thus turning organic compounds into inorganic ones (nutrients) available to photosynthetic producers in the ecosystem. Decomposers are primarily microscopic fungi and bacteria, but also include larger fungi such as toadstools and bracket fungi.

Decomposers are saprotrophs because they obtain their energy from dead or waste organic material (saprobiotic nutrition). They digest their food externally by secreting enzymes onto dead organisms or organic waste matter. The enzymes break down complex organic molecules into simpler soluble molecules – the decomposers then absorb these molecules. Through this process, decomposers release stored inorganic compounds and elements back into the environment.

Detritivores are another class of organism involved in decomposition. They help to speed up the decay process by feeding on detritus – dead and decaying material. They break it down into smaller pieces of organic material, which increases the surface area for the decomposers to work on. Examples of detritivores include woodlice that break down wood, and earthworms that help break down dead leaves. Detritivores perform internal digestion.

Recycling nitrogen

Nitrogen is an essential element for making amino acids (and consequently proteins) and nucleic acids in both plants and animals. Animals obtain the nitrogen they need from the food they eat, but plants have to take in nitrogen from their environment.

Nitrogen is abundant in the atmosphere, 78% of air is nitrogen gas (N_2). However, in this form nitrogen cannot be taken up by plants. To be used by living organisms, nitrogen needs to be combined with other elements such as oxygen or hydrogen. Bacteria play a very important role in converting nitrogen into a form useable by plants. Without bacteria, nitrogen would quickly become a limiting factor in ecosystems.

Nitrogen fixation

Nitrogen-fixing bacteria such as *Azotobacter* and *Rhizobium* contain the enzyme nitrogenase, which combines atmospheric nitrogen (N_2) with hydrogen (H_2) to produce ammonia (NH_3) – a form of nitrogen that can be absorbed and used by plants. This process is known as **nitrogen fixation**.

Azotobacter is an example of a free-living soil bacterium. However, many nitrogen-fixing bacteria such as *Rhizobium* live inside root nodules. These are growths on the roots of leguminous plants such as peas, beans, and clover. The bacteria have a symbiotic mutualistic relationship with the plant, as both organisms benefit:

- the plant gains amino acids from *Rhizobium*, which are produced by fixing nitrogen gas in the air into ammonia in the bacteria
- the bacteria gain carbohydrates produced by the plant during photosynthesis, which they use as an energy source.

Other bacteria then convert the ammonia that is produced by nitrogen fixation into other organic compounds that can be absorbed by plants.

✚ Reward and punishment

Recent work suggests that legumes 'select' the *Rhizobium* colonies which provide them with the most nitrates. Careful measurements show that the plants reward the nodules which make lots of nitrates with extra carbohydrates – but the punishment for nodules containing less-productive bacteria is swift and unforgiving. The plant cuts off the supply of carbohydrates and starves the nodule to death. This is a form of natural selection which maximises the benefit to the plant – and to those bacteria which deliver the goods.

Discuss the nitrogen-fixation/carbon-fixation mutualism of legumes and *Rhizobium*, including the pattern of the relationship and its importance to living organisms.

Nitrification

Nitrification is the process by which ammonium compounds in the soil are converted into nitrogen-containing molecules that can be used by plants. Free-living bacteria in the soil called nitrifying bacteria are involved.

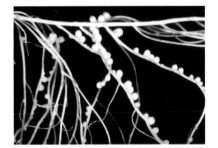

▲ **Figure 3** *Pink nodules of the nitrogen-fixing bacteria* Rhizobium leguminosarum *on the roots of a pea plant. The nodules are the sites where the bacteria fix atmospheric nitrogen, approx ×1.5 magnification*

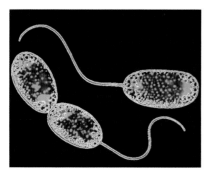

▲ **Figure 4** *Coloured transmission electron micrograph of* <u>Nitrobacter</u> *sp. They oxidise nitrogen-containing molecules for energy, converting nitrites into nitrates. Magnification × 6000*

Nitrification is an oxidation reaction, and so only occurs in well-aerated soil. It takes place in two steps:

1 Nitrifying bacteria (such as *Nitrosomonas*) oxidise ammonium compounds into nitrites (NO_2^-).

2 *Nitrobacter* (another genus of nitrifying bacteria) oxidise nitrites into nitrates (NO_3^-).

Nitrate ions are highly soluble, and are therefore the form in which most nitrogen enters a plant.

Denitrification

In the absence of oxygen, for example, in waterlogged soils, denitrifying bacteria convert nitrates in the soil back to nitrogen gas. This process is known as **denitrification** – it only happens under anaerobic conditions. The bacteria use the nitrates as a source of energy for respiration and nitrogen gas is released.

Ammonification

Ammonification is the name given to the process by which decomposers convert nitrogen-containing molecules in dead organisms, faeces, and urine into ammonium compounds.

Nitrogen cycle

The processes of nitrogen fixation, nitrification, denitrification, and ammonification all form part of the nitrogen cycle. Their place in the cycle can be seen in Figure 5.

▶ **Figure 5** *The nitrogen cycle. This shows how nitrogen is converted into a useable form and then passed on between organisms and the environment. Artificial processes, such as the Haber process used to make ammonia for fertiliser, have an effect on this natural cycle*

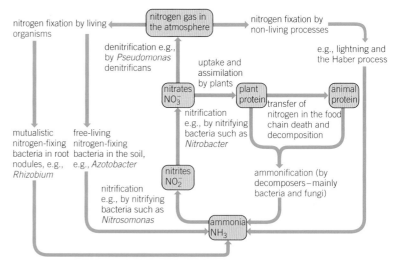

Recycling carbon

Carbon is a component of all the major organic molecules present in living organisms such as fats, carbohydrates, and proteins. The main source of carbon for land-living organisms is the atmosphere. Although carbon dioxide (CO_2) only makes up 0.04% of the atmosphere, there is a constant cycling of carbon between the atmosphere, the land, and living organisms. Figure 6 summarises the key points of the carbon cycle.

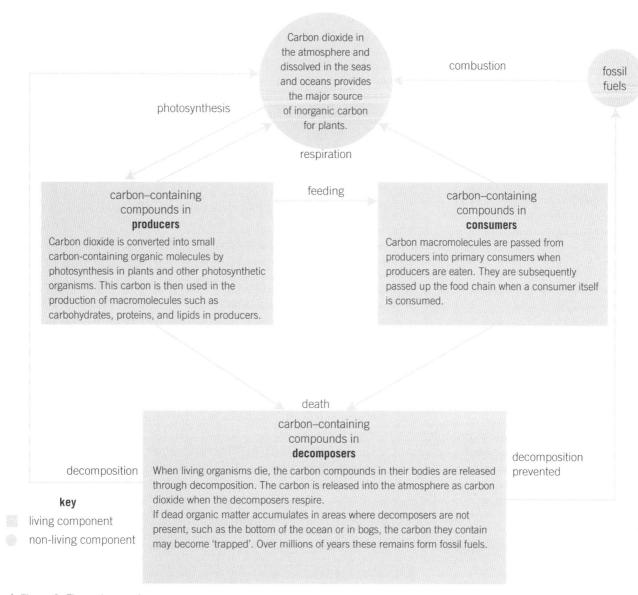

▲ Figure 6 *The carbon cycle*

Fluctuations in atmospheric carbon dioxide

Carbon dioxide levels fluctuate throughout the day. Photosynthesis only takes place in the light, and so during the day photosynthesis removes carbon dioxide from the atmosphere. Respiration, however, is carried out by all living organisms throughout the day and night, releasing carbon dioxide at a relatively constant rate into the atmosphere. Therefore, atmospheric carbon dioxide levels are higher at night than during the day.

Localised carbon dioxide levels also fluctuate seasonally. Carbon dioxide levels are lower on a summer's day than a winter's day, as photosynthesis rates are higher.

▲ Figure 7 *This is limestone. Although the bodies of marine animals decompose quickly, their shells and bones sink to the ocean floor. Over millions of years these are turned into carbon-containing sedimentary rock such as limestone and chalk. Eventually this carbon is returned to the atmosphere as the rock is weathered*

Synoptic link

You learnt about photosynthesis in Chapter 17, Energy for biological processes and about respiration in Chapter 18, Respiration.

Synoptic link

You learnt about climate change and global warming in Topic 11.6, Factors affecting biodiversity.

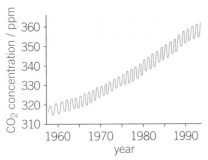

▲ **Figure 8** *Atmospheric levels of carbon dioxide across a 30-year period measured at Mauna Loa, a volcanic mountain in Hawaii. The increase is a result of human activities – primarily burning fossil fuels, and through changed land usage*

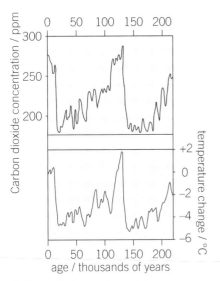

▲ **Figure 9** *Comparison of changes in atmospheric carbon dioxide (from analysis of bubbles in glacial ice) and temperature changes (from oxygen isotope studies)*

Over the past 200 years, global atmospheric carbon dioxide levels have increased significantly. This is mainly due to:

- the combustion of fossil fuels – which has released carbon dioxide back into the atmosphere from carbon that had previously been trapped for millions of years below the Earth's surface
- deforestation – which has removed significant quantities of photosynthesising biomass from Earth. As a result, less carbon dioxide is removed from the atmosphere. In many cases the cleared forest is burnt, therefore releasing more carbon dioxide into the atmosphere.

Increased levels of atmospheric carbon dioxide trap more thermal energy (heat) in the atmosphere – it is called a greenhouse gas for this reason. Its production through human activities is contributing to global warming.

The amount of carbon dioxide dissolved in seas and oceans is affected by the temperature of the water (the higher the temperature, the less gas is dissolved). Therefore, global warming reduces the carbon bank in the oceans and releases more carbon dioxide into the atmosphere – further contributing to the process in a positive feedback loop.

Atmospheric carbon dioxide levels have varied significantly over million-year timescales. To gain information about how the atmosphere has changed over time, samples are taken from deep within a glacier. For example, at a depth of 3.6 km in the Antarctic glacier the ice is 420 000 years old. When the ice formed air bubbles were trapped within the ice – these bubbles reflect the composition of the atmosphere at this point in time. Analysis of the gases present within these bubbles therefore reveals the composition of the atmosphere at this point in history. The graphs in Figure 9 show the variations in carbon dioxide levels which have occurred over time. The temperature of the atmosphere is directly related to the level of carbon dioxide present.

Summary questions

1. State the main differences between a decomposer and a detritivore. *(2 marks)*

2. State and explain three ways in which atmospheric carbon dioxide levels increase. *(3 marks)*

3. Describe what is meant by saprobiotic nutrition. *(2 marks)*

4. ⚙️ Explain how the scientific community have produced evidence that atmospheric carbon dioxide levels have varied naturally over time. *(2 marks)*

5. Explain the role of microorganisms, giving named examples where possible, in the recycling of nitrogen in an ecosystem. *(8 marks)*

23.4 Succession

Specification reference: 6.3.1

As you have learnt, ecosystems are dynamic – they are constantly changing. One process by which ecosystems change over time is called **succession**. Have you noticed how the types of plants present change as you move inland from a beach? In the sand very few species exist – the further you move away from the sea, the more biodiverse the ecosystem becomes. This is an example of succession.

Succession occurs as a result of changes to the environment (the abiotic factors), causing the plant and animal species present to change.

There are two types of succession:

1 Primary succession – this occurs on an area of land that has been newly formed or exposed such as bare rock. There is no soil or organic material present to begin with.

2 Secondary succession – this occurs on areas of land where soil is present, but it contains no plant or animal species. An example would be the bare earth that remains after a forest fire.

Although much of the natural landscape has taken hundreds of years to reach its existing form, primary succession is still taking place. Primary succession occurs when:

● volcanoes erupt, depositing lava – when lava cools and solidifies igneous rock is created

● sand is blown by the wind or deposited by the sea to create new sand dunes

● silt and mud are deposited at river estuaries

● glaciers retreat depositing rubble and exposing rock.

Stages of succession

Succession takes place in a number of steps, each one known as a **seral stage** (or sere). At each seral stage key species can be identified that change the abiotic factors, especially the soil, to make it more suitable for the subsequent existence of other species.

The main seral stages are pioneer community, intermediate community, and climax community, these are summarised in Figure 1.

Learning outcomes

Demonstrate knowledge, understanding, and application of:

→ the process of primary succession in the development of an ecosystem.

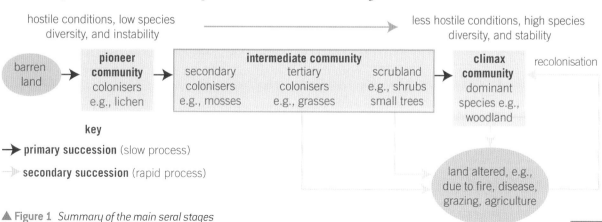

▲ Figure 1 *Summary of the main seral stages*

▲ **Figure 2** *Lichens are an example of a pioneer species. A lichen is not a single organism – it is a stable symbiotic (and mutualistic) association between a fungus and algae and/or cyanobacteria. They are often the first organisms to colonise bare rock. They are able to survive without soil, taking up rainwater and mineral salts through the whole of their body surface*

Pioneer community

Primary succession begins by the colonisation of an inhospitable environment, by organisms known as **pioneer species** (or pioneer colonisers). This represents the first seral stage. These species arrive as spores or seeds carried by the wind from nearby land masses or sometimes by the droppings of birds or animals passing through. Examples of pioneer species include algae and lichen.

Pioneer species have a number of adaptations that enable them to colonise this bare environment, including:

● the ability to produce large quantities of seeds or spores, which are blown by the wind and deposited on the 'new land'
● seeds that germinate rapidly
● the ability to photosynthesise to produce their own energy – light, rainfall, and air (and so carbon dioxide) are often the only abiotic factors present
● tolerance to extreme environments
● the ability to fix nitrogen from the atmosphere, so adding to the mineral content of the soil.

Intermediate community

Over time weathering of the bare rock produces particles that form the basis of a soil. On its own this cannot support other species. However, when organisms of the pioneer species die and decompose small organic products are released into the soil. This organic component of soil is known as **humus**. The soil becomes able to support the growth of new species of plant, known as secondary colonisers, as it contains minerals including nitrates and has an ability to retain some water. These secondary colonisers arrive as spores or seeds. Mosses are an example of a secondary coloniser species. In some cases, pioneer species also provide a food source for consumers, so some animal species will start to colonise the area.

As the environmental conditions continue to improve, new species of plant arrive such as ferns. These are known as tertiary colonisers. These plants have a waxy cuticle that protects them from water loss. These species can survive in conditions without an abundance of water – however, they need to obtain most of their water and mineral salts from the soil.

At each stage the rock continues to be eroded and the mass of organic matter increases. When organisms decompose they contribute to a deeper, more nutrient-rich soil, which retains more water. This makes the abiotic conditions more favourable initially for small flowering plants such as grasses, later shrubs, then finally small trees.

This period of succession is known as the intermediate community and in many cases multiple seral stages evolve during this period until climax conditions are attained.

At each seral stage different plant and animal species are better adapted to the current conditions in the ecosystem. These organisms outcompete many of the species that were previously present and become the dominant species. These are the most abundant species (by mass) present in the ecosystem at a given time.

Climax community

The final seral stage is called the **climax community**. The community is then in a stable state – it will show very little change over time. There are normally a few dominant plant and animal species. Which species make up the climax community depends on the climate. For example, in a temperate climate where the temperatures are mild and there is plenty of water, large trees will normally form the climax community. By comparison in a sub-arctic climate, herbs or shrubs make up the climax community as temperature and water availability are low.

Although biodiversity generally increases as succession takes place, the climax community is often not the most biodiverse. Biodiversity tends to reach a peak in mid-succession. It then tends to decrease due to the dominant species out-competing pioneer and other species, resulting in their elimination. The more successful the dominant species, the less the biodiversity in a given ecosystem.

▲ Figure 3 *Oak woodland is a common climax community in the UK*

Animal succession

Alongside the succession of plant species, animal species undergo similar progression. Primary consumers such as insects and worms are first to colonise an area as they consume and shelter in the mosses and lichens present. They must move in from neighbouring areas so animal succession is usually much slower than plant succession, especially if the 'new land' is geographically isolated, for example, a new volcanic island.

Secondary consumers will arrive once a suitable food source has been established and the existing plant cover will provide them with suitable habitats. Again these species must move in from neighbouring areas. Eventually larger organisms such as mammals and reptiles will colonise the area when the biotic conditions are favourable.

Synoptic link

You learnt about the importance of maintaining biodiversity and the techniques involved in Topic 11.7, Reasons for maintaining biodiversity and Topic 11.8, Methods of maintaining biodiversity.

Deflected succession

Human activities can halt the natural flow of succession and prevent the ecosystem from reaching a climax community. When succession is stopped artificially, the final stage that is formed is known as a **plagioclimax**. Agriculture is one of the main reasons deflected succession occurs. For example:

● grazing and trampling of vegetation by domesticated animals – this results in large areas remaining as grassland

● removing existing vegetation (such as shrub land) to plant crops – the crop becomes the final community

● burning as a means of forest clearance – this often leads to an increase in biodiversity as it provides space and nutrient-rich ash for other species to grow, such as shrubs.

▲ Figure 4 *Mowing your garden prevents succession occurring. The growing shoots of woody plants are cut off by the lawnmower, so larger plants cannot establish themselves. A grassland plagioclimax is formed*

Conservation

Deflected succession is an important conservation technique. To ensure the survival of certain species, it is important to preserve their habitat in its current form. This may require ecological land management to prevent further succession from occurring. An example of this is at the National Trust's Studland Heath nature reserve in Dorset, England.

Studland Heath is home to a number of British reptiles, including the UK's rarest reptile, the smooth snake. This region is heathland – if succession were allowed to occur, woodland would develop as the climax community. Other animal species would then inhabit the area, ultimately leading to the replacement of the smooth snake by other animal species. This would risk the elimination of this reptile species from the UK. So the heathland is managed by the National Trust to ensure that the precious ecosystem is preserved.

A range of conservation techniques are being used:

▲ **Figure 5** *Britain's rarest native reptile, the smooth snake*

- Physical removal of established bracken, and saplings such as birch and pine. Gorse, which is a legume, adds to the nutrient value of the soil. Heathlands are traditionally nutrient-poor areas of land, and so areas of gorse are also removed through physical means (cutting, crushing, and controlled burning). These techniques are used to restore the heathland to its former state.

- Mimicking the controlled grazing to limit the spread of bracken, gorse, and tree saplings. Livestock crush and nibble new growth, therefore limiting the spread of these species. This technique is used to maintain the heathland in its current state.

Through using a combination of these techniques, low nutrient levels are maintained in the soil. This produces a varied vegetation structure, thus supporting a biodiverse heathland community.

> A herd of Red Devon cows are used to graze Studland Heath.
> Explain their role in maintaining the structure of the heathland.

Synoptic link

Look back at 11.8, Methods of maintaining biodiversity, to remind yourself of the various techniques scientists use to maintain biodiversity

 Succession on a sand dune

One of the few examples where all the stages of succession can often be seen clearly in one place is when a series of sand dunes form on a beach. The youngest dunes will be found closest to the shore and the oldest furthest away. Figure 6 shows a simplified version of what occurs:

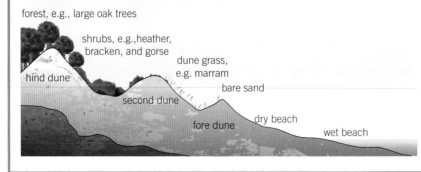

forest, e.g., large oak trees

shrubs, e.g.,heather, bracken, and gorse

dune grass, e.g. marram

bare sand

hind dune

second dune

fore dune

dry beach

wet beach

◀ **Figure 6** *Simplified diagram of succession*

Seeds are blown onto the dunes or washed onto the sand by the sea. At this stage the rooting conditions are poor due to drought, strong winds, and salty sea-water immersion. The presence of a large number of sea shells also makes the environment very alkaline. As the wind blows across the dunes it moves the sand, constantly changing the profile of the dunes and allowing rainwater to soak through rapidly. However, some species such as sea couch grass are able to survive these harsh conditions.

Marram grass becomes the next dominant species. As these plants grow sand is trapped between the roots helping to stabilise the dunes, and the formation of soil begins. Minerals released from decaying pioneer plants create more fertile growing conditions, and the soil becomes less alkaline as pioneer plants grow and trap rainwater. The grasses also reduce salt spray into the hind dunes, and act as a partial wind barrier. Less hardy plants can now grow – over time, these species start to shade out the pioneers. As more species of plants colonise the dunes, the sand disappears and the dunes change colour – from yellow to grey.

Finally taller plants, such as sea buckthorn, or more complex plant species like moorland heathers can grow. Plants from earlier stages are no longer present.

When the water table reaches or nearly reaches the surface, dune slacks can occur. Plants that are specially adapted to be water-tolerant grow here adding to the biodiversity of the sand dunes.

1 State what type of succession occurs on a sand dune.
2 a Name an example of a pioneer species.
  b Suggest and explain two features of the species you named in question 2a.
3 Describe the role marram grass plays in succession.
4 Describe how the abiotic conditions on a sand dune change over time.
5 Explain why fewer pioneer species are present in a climax community of trees or heathers.

Summary questions

1 State the difference between a climax community and a plagioclimax community. (*1 mark*)

2 Describe the difference between primary and secondary succession. (*1 mark*)

3 State and explain the common developments that occur in succession in any ecosystem. (*6 marks*)

4 Evaluate the use of deflected succession as a conservation technique to help maintain the population of an endangered animal species. (*6 marks*)

Synoptic link

You learnt about the adaptations of marram grass in Topic 9.5, Plant adaptations to water availability and Topic 10.7, Adaptations.

23.5 Measuring the distribution and abundance of organisms

Specification reference: 6.3.1

Learning outcomes

Demonstrate knowledge, understanding, and application of:

→ how the distribution and abundance of organisms in an ecosystem can be measured

→ the use of sampling and recording methods to determine the distribution and abundance of organisms in a variety of ecosystems.

Synoptic link

You learnt about sampling techniques in Topic 11.2, Sampling and Topic 11.3, Sampling techniques. This topic reminds you briefly of the techniques, but you should refer to those topics also.

Study tip

Remember that two other types of non-random sampling could also be used – opportunistic and stratified sampling. See Topic 11.2, Sampling.

Synoptic link

You will learn more about how population size changes in Topic 24.1, Population size.

Scientists use a number of techniques to study the distribution and abundance of organisms within an ecosystem. This is a way of measuring and observing the biodiversity present within an ecosystem. These techniques can also be used to study how the organisms present change during succession. Many of these techniques you met in Chapter 11, Biodiversity.

Distribution of organisms

The distribution of organisms refers to *where* individual organisms are found within an ecosystem. The distribution of organisms is usually uneven throughout an ecosystem. Organisms are generally found where abiotic and biotic factors favour them, therefore their survival rate is high as all the resources they need to live are available and predation/pressure from consumers is low.

Measuring distribution

To measure the distribution of organisms within an ecosystem, a line or belt transect is normally used. A line transect involves laying a line or surveyor's tape along the ground and taking samples at regular intervals. A belt transect provides more information – two parallel lines are marked, and samples are taken of the area between these specified points. Belt and line transects are forms of systematic sampling, a type of non-random sampling.

In systematic sampling different areas within an overall habitat are identified, which are then sampled separately. This can have advantages over random sampling as it allows scientists to study how the differing abiotic factors in different areas of the habitat affect the distribution of a species. For example, systematic sampling may be used to study how plant species change as you move inland from the sea. This would therefore be used to study the successional changes that take place along a series of sand dunes.

Abundance of organisms

The abundance of organisms refers to the *number* of individuals of a species present in an area at any given time. This number may fluctuate daily:

● Immigration and births will increase the numbers of individuals.

● Emigration and deaths will decrease the number of individuals.

Measuring abundance

A population is a group of similar organisms living in a given area at a given time. Populations can rarely be counted accurately – for example, some animals elude capture, it may be too time-consuming to count all members of a population, or the counting process could damage the environment. Populations are therefore estimated using sampling techniques.

A sample, however, is never entirely representative of the organisms present in a habitat. To increase its accuracy you should use as large a sample size as possible. The greater the number of individuals studied, the lower the probability that chance will influence the result. You should also use random sampling to reduce the effects of sampling bias.

Measuring plant abundance

To measure the abundance of plants, quadrats are placed randomly in an area. The abundance of the organisms in that area is measured by counting the number of individual plants contained within the quadrat. Using the following formula, the abundance can be estimated:

$$\text{Estimated number in population (m}^{-2}) = \frac{\text{Number of individuals in sample}}{\text{Area of sample (m}^2)}$$

▲ **Figure 1** *This student is conducting a survey using a quadrat. A quadrat is a frame of standard size, often a square divided into equal sections by string, which is used to isolate an area from which a sample is to be taken. Quadrats are often used for measuring the abundance of different species within a given area*

 ## Worked example: Estimating plant abundance

A sample was taken using five quadrats with a combined area of $5\,\text{m}^2$. If the quadrats contained a total of 40 buttercup plants, what is the estimated abundance of buttercups?

Step 1: Identify the equation needed

$$\text{Estimated number in population (m}^{-2}) = \frac{\text{Number of individuals in sample}}{\text{Area of sample (m}^2)}$$

Step 2: Substitute the values into the equation and calculate the answer $= \dfrac{40}{5} = 8\,\text{m}^{-2}$

Meaning eight buttercups per square metre.

Study tip

Individual plants in a quadrat can sometimes be difficult to count so scientists also make estimates using frequency or percentage cover. Look back at Topic 11.3, Sampling techniques to remind yourself of these different techniques.

Measuring animal abundance

Quadrats cannot be used to measure the abundance of animals (unless they are very slow moving, such as barnacles and mussels on a sea shore) so the capture–mark–release–recapture technique is often used to estimate population size.

The technique can be carried out as follows:

1   Capture as many individuals as possible in a sample area.

2   Mark or tag each individual.

3   Release the marked animals back into the sample area and allow time for them to redistribute themselves throughout the habitat.

4   Recapture as many individuals as possible in the original sample area.

5   Record the number of marked and unmarked individuals present in the sample. (Release all individuals back into their habitat.)

6   Use the Lincoln index to estimate the population size:

$$\text{Estimated population size} = \frac{\substack{\text{number of individuals in first sample} \times \\ \text{number of individuals in second sample}}}{\text{number of recaptured marked individuals}}$$

▲ **Figure 2** *The animals sampled can be collected in a number of ways including using pooters, sweep nets, and pitfall traps. (Topic 11.3, Sampling techniques). This student is collecting crawling animals off a tree using a pooter*

Worked example: Capture–mark–release–recapture technique

In a random sample of a school playing field, students captured 12 snails. They marked the underside of each snail's shell with a dot of non-toxic paint.

The students then released the snails back onto the field. A week later, another sample of snails from the same location was taken. This time 15 snails were collected, of which three were found to have a paint dot.

Estimate the total snail population.

Step 1: Sum the total the number of individuals in the first sample = 12

Step 2: Sum the total number of individuals in the second sample = 15

Step 3: Estimate the population size by calculating the Lincoln index:

Estimated population size =

$$\frac{\text{number of individuals in first sample} \times \text{number of individuals in second sample}}{\text{number of recaptured marked individuals}} = \frac{12 \times 15}{3} = 60$$

The estimated total snail population of the playing field is 60 snails.

Once the abundance of all the organisms present in a habitat has been determined, scientists will often mathematically calculate the biodiversity present in a habitat. This can be done using **Simpson's Index of Diversity (D)**:

D = diversity index

N = total number of organisms in the ecosystem

n = number of individuals of each species

$$D = 1 - \Sigma \left(\frac{n}{N}\right)^2$$

Simpson's Index of Diversity always results in a value between 0 and 1, where 0 represents no diversity and a value of 1 represents infinite diversity. The higher the value of Simpson's Index of Diversity, the more diverse the habitat.

Synoptic Link

To remind yourself how to calculate Simpson's Index of Diversity, look back at Topic 11.4, Calculating biodiversity.

Monitoring biodiversity in the Sonoran desert

The Sonoran desert is a region of desert in the south-western part of the United States, including parts of Arizona and California. The environment is harsh – summer temperatures regularly exceed 40 °C. Rainfall is rare, often taking the form of intense, violent storms.

A student carried out a study to test the following hypothesis – 'the greater the abundance of plant species in the Sonoran desert, the greater the abundance of animals'. A selection of the data collected is shown in Table 1.

▼ Table 1

Mean plant abundance / 50 m^{-2}	Mean animal abundance / 50 m^{-2}
0	0.05
1	0.06
2	0.20
3	0.19
4	0.11
5	0.44
6	0.92
7	1.19
8	1.44
9	1.45
10	1.78
11	2.00
12	2.06
13	2.33
14	2.78

1 Suggest how the data could have been collected.
2 Explain how you can tell that the figures stated for animal abundance were a mean of several observations.
3 Plot the data as a scatter diagram.
4 Using your diagram from Q3, state the correlation between the variables, if any.
5 Using the formula $r_s = 1 - \dfrac{6\sum d^2}{n(n^2 - 1)}$, calculate the Spearman's rank correlation coefficient for this data.
6 Use the Spearman's ranked correlation coefficient critical values in the appendix to determine the statistical significance of the correlation, and hence form a conclusion from the data.

Summary questions

1 State the difference between the abundance and distribution of organisms. *(1 mark)*

2 Twenty woodlice were captured in a sample area and marked then released back into the sample area. A week later, 15 woodlice were found in the sample area, of which only two were marked.
   a Suggest which piece of apparatus should be used to collect woodlice. *(1 mark)*
   b Use the Lincoln index to estimate the woodlouse population. *(2 marks)*

3 State and explain one advantage of the following techniques used to measure the distribution of organisms in an ecosystem. *(2 marks)*
   a random sampling   b non-random sampling

4 Discuss the limitations of using the capture–mark–release–recapture technique to estimate population size. *(6 marks)*

Synoptic link

Remind yourself how to calculate the Spearman's rank correlation coefficient by looking back to Topic 10.6, Representing variation graphically.

Study tip

The correlation test is used to see if two different variables are correlated in a linear fashion in the context of a scatter-graph. There are several different types of statistical test that can be used, the OCR GCE Biology specifications will only cover Spearman's rank correlation coefficient. You can find a full table of values in the appendix.

Practice questions

1 Ecological succession describes the change in species within a community over time.

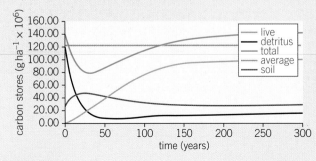

time

| agricultural land is kept in an artificial seral stage | after land is abandoned wild grasses start to grow from wind blown over | over time shrubs start to colonise the grassland | trees become established and climax community develops |

a (i) State, giving your reasons, the type of succession shown in the diagram. *(1 mark)*

(ii) Describe the *ecological role* of the different species involved in the early stages of succession. *(2 marks)*

The graph shows the changing mass of species in a standard example of succession.

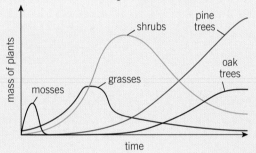

b (i) State the name given to mosses describing their role in this example of succession. *(2 marks)*

(ii) Explain why mosses, grasses, and shrubs all eventually decrease in mass. *(2 marks)*

(iii) State the name given to the community composed primarily of Pine and Oak trees. *(1 mark)*

The quantity of carbon sink changes throughout a period of succession. The graph summarises these changes.

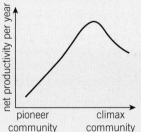

c (i) Describe the changes in carbon content of the components of the ecosystem shown in the diagram. *(4 marks)*

(ii) Suggest how these changes would differ if the succession started without the presence of any soil. *(4 marks)*

As succession progresses the net primary productivity changes as shown in the graph, this is the net rate at which an ecosystem accumulates energy.

d (i) Describe the change in net primary productivity shown in the graph. *(3 marks)*

(ii) Suggest the reasons for the changes you have described in **d** (ii). *(3 marks)*

2 Ecological pyramids summarise the feeding relationships within an ecosystem in terms of number, biomass or energy.

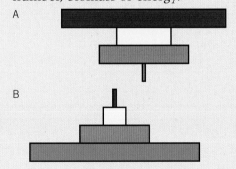

a (i) State the definition of the term trophic level. *(1 mark)*

(ii) Identify which of the lists below represents each of the ecological pyramids shown in the diagram. *(2 marks)*

1 lettuce plant, snail, thrush, sparrowhawk

2 tree, aphid, caterpillar, parasite

(iii) Suggest, giving your reason(s), whether the ecological pyramids in the diagram represent number or biomass. *(2 marks)*

(iv) Describe how you would represent decomposers on pyramid A, or B. *(2 marks)*

3 a Outline the factors that can affect primary productivity. *(5 marks)*

It is usually impossible to count all of the members of a population in a particular habitat so animal populations often have to be estimated.

One method to estimate population size uses the Lincoln Index.

b (i) Outline the principles that underlie the Lincoln Index. *(3 marks)*

(ii) Discuss the limitations of using the Lincoln Index to estimate population size. *(5 marks)*

In different locations of Utah Lake in the US, 24 064 carp were captured, tagged, and released over a 15 day period.

After the tagging was finished, another 10 357 carp at random locations of the lake were caught and the number of tagged fish counted. Only 208 of the recaptured carp had tags.

c Calculate, using the formula, the estimated population number of carp in the lake. *(2 marks)*

$$\frac{\text{Number marked in second sample}}{\text{Total caught in second sample}} = \frac{\text{Number marked in first sample}}{\text{Size of whole population (N)}}$$

d State, giving your reason(s), if the estimate become more accurate if more individuals are captured and marked. *(2 marks)*

e State one adjustment you would have to make to the investigation if you were estimating a population of snails rather than fish. *(1 mark)*

4 Total plant growth within an ecosystem depends on the light intensity, temperature, and the supply of water and inorganic minerals to the ecosystem.

The table shows the net primary production by plants in four different ecosystems.

Ecosystem	Net primary production (kJm^{-2} year^{-1})
temperate grassland	9 240
temperate woodland	11 340
tropical grassland	13 340
tropical rainforest	36 160

a Discuss possible reasons for the differences in net primary production in these ecosystems. *(4 marks)*

b To calculate the net primary production figures in the table in kJm^{-2} year^{-1}, it is necessary to measure the energy content of the primary producers. Outline how the energy content, in kJ, of a primary producer such as grass can be measured in the laboratory. *(2 marks)*

c The efficiency with which consumers convert the food they eat into their own biomass is generally low. The table compares the energy egested, absorbed, and respired in four types of animal.

animal	Percentage of energy consumed that is:			
	egested	absorbed	respired	converted to biomass
grasshopper (herbivorous insect)	63	37	24	13
perch (carnivorous fish)	17	83	61	
cow (herbivorous mammal)	60	40	39	
bobcat (carnivorous mammal)	17	83	77	6

(i) Copy and complete the table to show the percentage of energy consumed that is converted into biomass in the perch and the cow. *(2 marks)*

(ii) Describe and explain, using the data from the table, how the trophic level of a mammal affects the percentage of its food energy that it is able to convert to biomass. *(3 marks)*

(iii) Using the data from the table and your knowledge of energy flow through food chains, suggest which of these four animals could be farmed to provide the maximum amount of food energy in kJm^{-2} year^{-1} for humans. Explain the reason for your choice. *(4 marks)*

OCR F215 2011

24 POPULATIONS AND SUSTAINABILITY
24.1 Population size
Specification reference: 6.3.2

The human population is increasing at a significant rate. The global population has grown from one billion in 1800 to seven billion in 2012, and is predicted to reach 11 billion by the end of this century. Population growth like this cannot be sustained indefinitely as **limiting factors**, such as the availability of food, will prevent the population rising above a certain level. A limiting factor is an environmental resource or constraint that limits population growth.

Population growth curve

If the growth of a new population over time is plotted on a graph, regardless of the organism, most natural populations will show the same characteristics. This is known as a population growth curve.

The graph can be divided into three main phases:

- Phase 1 – a period of *slow growth*. The small numbers of individuals that are initially present reproduce increasing the total population. As the birth rate is higher than the death rate, the population increases in size.

- Phase 2 – a period of *rapid growth*. As the number of breeding individuals increases, the total population multiplies exponentially. No constraints act to limit the population explosion.

- Phase 3 – a *stable state*. Further population growth is prevented by external constraints. During this time the population size fluctuates, but overall its size remains relatively stable. Birth rates and death rates are approximately equal. Slight increases and decreases can be accounted for by fluctuations in limiting factors, such the presence of predators.

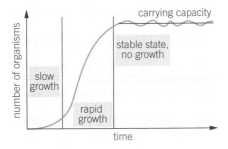

▲ Figure 1 *Growth curve of most natural populations. This is often referred to as a sigmoid population growth curve*

Synoptic link

You learnt about bacterial growth curves in Topic 22.6, Culturing microorganisms in the laboratory.

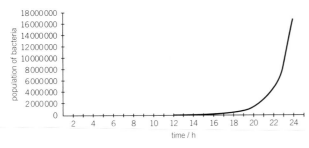

▲ Figure 2a *Exponential growth curve. The size of the population doubles each time a fixed time period elapses. The populations of many organisms follow this trend during an initial expansion of population size*

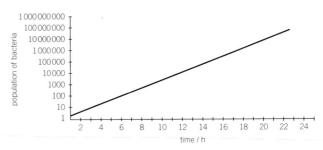

▲ Figure 2b *Exponential growth represented on a logarithmic scale. Where data span several orders of magnitude, it is often more appropriate to plot the data using a logarithmic scale on the relevant axis. In this case, population size covers a large range of values—therefore this data (y-axis) is plotted on a logarithmic scale, whereas time (x-axis) is plotted on a linear scale*

Limiting factors

In theory, if all resources were in plentiful supply a population would continue to grow exponentially. However, this is rarely seen in nature. Instead, a short period of exponential growth occurs when conditions are ideal and the maximum growth rate is achieved.

Limiting factors prevent further growth of a population and in some cases cause it to decline. Examples of limiting factors include competition between the organisms for resources, the build-up of the toxic by-products of metabolism, or disease.

Limiting factors can be divided into abiotic and biotic factors:

- Abiotic factors – these non-living factors include temperature, light, pH, the availability of water or oxygen, and humidity.
- Biotic factors – these living factors include predators, disease, and competition.

The maximum population size that an environment can support is known as its **carrying capacity** (Figure 1), although individual years can show slight increases or decreases in population size. The population size remains stable as the number of births and deaths are approximately equal.

Migration

Another important variable which affects population size is migration:

- Immigration – the movement of individual organisms into a particular area increases population size. For example, millions of Christmas Island red crabs (*Gecarcoidea natalis*) migrate each year from forest to coast to reproduce, dramatically increasing the coastal population of red crabs.
- Emigration – the movement of individual organisms away from a particular area decreases population size. For example, the Norway Lemming (*Lemmus lemmus*) emigrates away from areas of high population density or poor habitat.

Density independent factors

Density independent factors are factors that have an effect on the whole population regardless of its size. These can dramatically change population size. These factors include earthquakes, fires, volcanic eruptions, and storms. In some cases, these factors can remove whole populations of a species from a region.

> **Synoptic link**
>
> Look back at Topic 23.1, Ecosystems to remind yourself of the effect of abiotic and biotic factors on organisms.

> **Study tip**
>
> Competition is often considered to be the most important biotic factor controlling population density. You will find out about competition in more detail in 24.2, Competition.

▲ **Figure 3** *Snow Geese breed in the Arctic Tundra, and winter in farmlands on the American south coast. They make an annual round trip journey of more than 8000 km at speeds of more than 80 km/h, changing the population density of geese in each place dramatically as they arrive and depart*

 ## Human population growth

For many years the human population remained fairly stable, as is the case for other natural populations. The population was kept in check by limiting factors.

> **1** Suggest two limiting factors that historically kept the human population at a constant level.

The development of agriculture, the industrial revolution, and advances in medicine have led to a population explosion. Like other species, the growth of the human population is a result of the imbalance between birth rate and death rate.

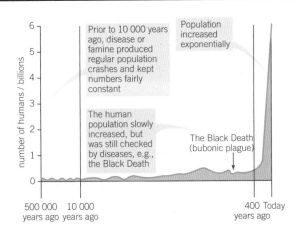

▲ **Figure 4** *Human population growth curve*

Scientists use a measure to calculate how much a population has grown in a certain period of time. This is calculated using the formula:

$$\text{Population growth (\%)} = \frac{\text{population change during the period}}{\text{population at the start of the period}} \times 100$$

If the result is positive, the population has grown. If it is negative, the population has decreased.

> **2** The human population was approximately 3 billion (3×10^9) in 1960. This had risen to 6.8 billion (6.8×10^9) by 2010. Calculate the percentage population growth during this period.

A number of factors affect the birth rate of the human population. In addition to 'natural' limiting factors, economic conditions, cultural and religious backgrounds, and social pressures can limit – or encourage – the birth rate.

Different factors affect the death rate in a population. These include the age profile of the population (in general, the greater the proportion of elderly people, the higher the death rate), the quality of medical care, food availability and quality, and the effects of natural disaster or war.

> **3** Suggest and explain a social, cultural, or religious pressure which may affect the size of a population.

The future size of a population depends upon the number of women of child-bearing age. The age and gender demographic of a population can be represented using an age population pyramid, as shown in Figure 5.

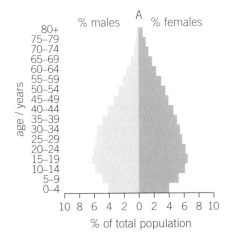

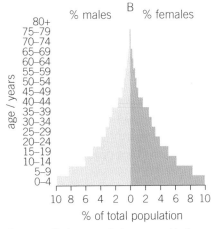

▲ **Figure 5** *Age population pyramids for two different countries*

> **4 a** Explain how the population of each of the countries represented in Figure 5 will change over time.
>
> **b** Sketch or suggest the shape of an age population pyramid for a stable population.

Summary questions

1 State three factors that would cause an increase in population size. *(3 marks)*

2 Describe what happens during a period of exponential population growth. *(2 marks)*

3 Sketch and annotate a graph to show what would happen to the population of duckweed (*Lemna minor*) if a few individuals were introduced into a new pond habitat. *(6 marks)*

In the previous topic, you learnt that population size is influenced by limiting factors. Organisms compete for resources including food, shelter, space, and light. As a result, competition between organisms has a significant effect on the number of organisms present in a particular area.

Types of competition

Competition is an example of a *biotic* limiting factor – it is a result of the interactions between living organisms. There are two main types of competition:

1 **Interspecific competition** – competition between different species.
2 **Intraspecific competition** – competition between members of the same species.

Interspecific competition

Interspecific competition occurs when two or more different species of organism compete for the same resource. This interaction results in a reduction of the resource available to both populations. For example, if both species compete for the same food source, there will usually be less available for organisms of each species. As a result of less food, organisms will have less energy for growth and reproduction, resulting in smaller populations than if only one of the species had been present.

If two species of organism, however, are both competing for the same food source but one is better adapted, the less well adapted species is likely to be *outcompeted*. If conditions remain the same, the less well adapted species will decline in number until it can no longer exist in the habitat alongside the better adapted species. This is known as the *competitive exclusion principle* – where two species are competing for limited resources, the one that uses the resources more effectively will ultimately eliminate the other.

Red and grey squirrels in the UK

An example of interspecific competition is the competition between red and grey squirrels for food and territory in the UK. In the 1870s the grey squirrel, a native of North America, was introduced into the wild in the UK. Its population quickly increased in numbers and resulted in the native red squirrel disappearing from many areas. This is primarily because the grey squirrel can eat a wider range of food than the red squirrel and as it is larger it can store more fat. This increases its chances of survival and therefore its ability to reproduce thus increasing its population. An increasing population of grey squirrels further reduces the food supply available to the red squirrels, reducing their ability to survive and reproduce.

Learning outcomes

Demonstrate knowledge, understanding, and application of:

→ interspecific and intraspecific competition.

▲ **Figure 1** *The bog pondweed (*Potamogeton polygonifolius*, broad leaves) and duckweed (*Lemna minor*, small floating leaves) growing in this pond are both competing for light. This is an example of interspecific competition. In the absence of the pond weed, the duckweed would multiply to cover the whole surface of the pond*

Synoptic link

You learnt about biotic and abiotic factors in Topic 23.1, Ecosystems.

▲ **Figure 2** *The red squirrel (*Sciurus vulgaris*) is native to the UK. The introduction of the grey squirrel (*Sciurus carolinensis*) from North America has led to a significant decline in red squirrel numbers*

Intraspecific competition

Intraspecific competition occurs when members of the same species compete for the same resource. The availability of the resource determines the population size – the greater the availability the larger the population that can be supported. This results in fluctuations in the number of organisms present in a particular population over time.

The effects of intraspecific competition on a population are represented in Figure 3.

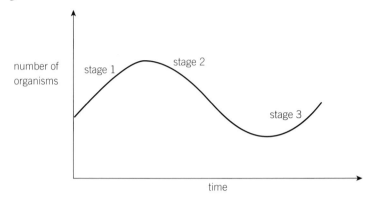

▲ Figure 3 *The effects of intraspecific competition over time*

▲ Figure 4 *Robins compete for breeding territory. Each territory provides adequate food for one family of birds. If food is scarce, territories have to become larger, therefore there are fewer breeding pairs and a smaller population*

- Stage 1 – When a resource is plentiful in a habitat (such as food or space), all organisms have enough of the resource to survive and reproduce. This results in an increase in population size.

- Stage 2 – As a result of the increased population, there are many more individuals that share the food or space available. Resources are now limited; not enough is available for all organisms to survive. The population will decrease in size.

- Stage 3 – Less competition exists as the smaller population means less organisms are competing for the same resources. This means more organisms survive and reproduce, resulting in population growth.

This cycle of events will then repeat.

Summary questions

1   State the difference between interspecific and
    intraspecific competition. *(1 mark)*

2   Describe the cycle of population size changes that occur
    following intraspecific competition. *(3 marks)*

3   Oak tree saplings compete with each other for light, water,
    and minerals.
    a   State the type of competition that exists between oak
        tree saplings. *(1 mark)*
    b   Suggest how the population of oak trees will vary
        over time. *(3 marks)*

In the previous topic, you learnt about the effect of competition on population size. Another major biotic factor that has an influence on population size is the role of **predation**. This is where an organism (the predator) kills and eats another organism (the prey). For example, tigers prey on water buffalo and deer. Predation is a type of interspecific competition, operating between prey and predator species.

Predators have evolved to become highly efficient at capturing prey, for example, through sudden bursts of speed, stealth, and fast reactions. Likewise, prey organisms have evolved to avoid capture through camouflage, mimicry, or defence mechanisms such as spines. Prey organisms have had to evolve alongside their predators (and vice versa) – if evolution had not occurred, one or both of the species may have become extinct.

Predator–prey relationships

The sizes of the predator and prey populations are interlinked. As the population of one organism changes, it causes a change in the size of the other population. This results in fluctuations in the size of both populations.

Predator–prey relationships can be represented on a graph (Figure 2).

In general all predator–prey relationships follow the same pattern. The peaks and troughs in the size of the prey population are mirrored by peaks and troughs in the size of the predator population after a time delay.

- Stage one – An increase in the prey population provides more food for the predators, allowing more to survive and reproduce. This in turn results in an increase in the predator population.

- Stage two – The increased predator population eats more prey organisms, causing a decline in the prey population. The death rate of the prey population is greater than its birth rate.

- Stage three – The reduced prey population can no longer support the large predator population. Intraspecific competition for food increases, resulting in a decrease in the size of the predator population.

- Stage four – Reduced predator numbers result in less of the prey population being killed. More prey organisms survive and reproduce, increasing the prey population – the cycle begins again.

Rarely in the wild is the link between the predator and prey population as simple as this. Other factors will also influence the population size – for example, the availability of the food plants of the prey or the presence of other predators. Fluctuations in numbers may also result from seasonal changes in abiotic factors.

Learning outcomes

Demonstrate knowledge, understanding, and application of:

→ predator–prey relationships

▲ Figure 1 *This predator–prey relationship is often exploited by organic farmers. In a technique known as biological pest control, farmers use the natural predators (in this case ladybirds) to destroy pest populations (aphids) and prevent them damaging crops without the need to use pesticides*

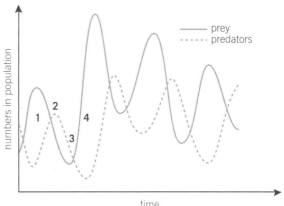

▲ Figure 2 *A general predator–prey graph*

Canadian lynx and snowshoe hare

When a predator feeds on just one type of prey, there is an interdependence between the predator and prey populations. This means that changes in the population of one animal directly affect the population of the other.

The Canadian lynx and snowshoe hare have an interdependent relationship. This relationship has been widely studied – data exist for over 200 years, as records were kept for the number of lynx furs that were traded in Canada. The data collected is shown in Figure 3.

1   Suggest what assumption was made by scientists when estimating population size from the number of furs traded.
2   a   State in which year the hare population was at its highest.
    b   Suggest a reason for this.
3   a   State in which year the lynx population is at its lowest.
    b   Suggest a reason for this.
4   State the approximate time delay between a peak and trough occurring in the snowshoe hare population.
5   Explain the changes in the populations of the lynx and hare that occurred over time.

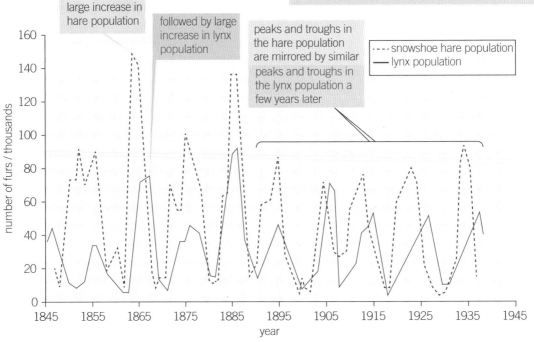

▲ **Figure 3** *The predator–prey relationship illustrated by the number of snowshoe hare and lynx trapped for the Hudson Bay Company in Canada between 1845 and 1940*

Summary questions

1   Describe what would happen to the numbers of a predator population if a fatal disease spread amongst its only prey.   (*1 mark*)

2   Suggest two reasons why the populations of an interlinked predator and prey population don't produce an identical graph to the one shown in Figure 2.   (*4 marks*)

3   In a controlled environment, an experiment was carried out to study the predator–prey relationship between the unicellular organisms *Didinium* and *Paramecium*. A suitable *Paramecium* population prey was established before a small population of *Didinium* was added. Sketch and annotate a graph to show how the populations of these organisms may vary over time. A constant supply of nutrient material was provided for the *Paramecium*.   (*4 marks*)

24.4 Conservation and preservation

You will often hear reports in the media that steps need to be taken to conserve the environment for future generations. This is true not only to preserve the beauty of the natural world, but to try to ensure that the biodiversity of world ecosystems is not lost. There are many reasons for maintaining biodiversity, including economics, preserving genetic variety, and retaining species that might be useful to people. People tend to use the terms **conservation** and **preservation** interchangeably, but in biology they have quite distinct meanings.

Conservation

Conservation means the maintenance of biodiversity through human action or management. This includes maintaining diversity between species, maintaining genetic diversity within a species, and the maintenance of habitats.

Conservation involves the management of ecosystems so that the natural resources in them can be used without running out. This is known as sustainable development. For example, the Forest Stewardship Council ensures that forests are managed so that they provide a sustainable source of timber. Their mission is to promote socially beneficial, environmentally appropriate, and economically viable management of forests across the world. You will find out more about sustainable production in Topic 24.5, Sustainability.

Conservation approaches also include *reclamation* – this is the process of restoring ecosystems that have been damaged or destroyed. For example, a habitat may be destroyed by floods, or as a result of a new building project. Reclamation also involves techniques such as controlled burning of areas of a forest, which can halt succession and increase biodiversity. Conservation is dynamic and needs to adapt to constant change.

Preservation

Preservation is the protection of an area by restricting or banning human interference, so that the ecosystem is kept in its original state. It is most commonly used when preserving ecologically, archaeologically, or palaeontologically sensitive resources, which can easily be damaged or destroyed by disturbances. When lands are preserved, visitation (along with most other activities) is not allowed, except by those who manage and monitor such areas.

Newly discovered caves, called virgin caves, are pristine. These may contain very sensitive geological formations or unique ecosystems – walking from one cave to another can cause irreparable damage, for example, through direct crushing or by the movement of dirt around the cave system. Such damage can be avoided by barring entrance to caves altogether thus preserving these unique habitats. Only through preservation can the integrity of these ecosystems be

▲ **Figure 1** *Controlled grazing of fens is a valuable fen management conservation tool. It helps to maintain open, species-rich fen communities by reducing plant biomass and controlling scrub invasion*

Synoptic link

Look back at Topic 11.8, Methods of maintaining biodiversity to remind yourself of the range of in situ and ex situ techniques that scientists use to conserve biodiversity and the importance of conservation.

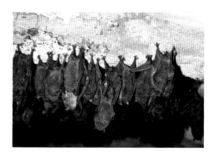

▲ **Figure 2** *Endangered Gray bats (Myotis grisescens) in a cave in Tennessee*

Conservation of gray bats

The gray bat is an example of a species which has been successfully supported through a series of conservation measures. Inhabiting a small number of caves across the south-eastern US, the gray bat became endangered following the human exploitation of its habitat, and through pesticide bioaccumulation. Following the collection of data about bat numbers and particular habitat requirements, a series of measures were introduced to support the remaining population, which is currently estimated to be around one million:

- In 1982, the gray bat was placed on the International Union for Conservation of Nature (IUCN) endangered species list.

- Caves supporting gray bat populations were gated, preventing human access but allowing bats to freely enter and leave the caves.

- The exploitation of land around the caves, which provide the bats with their food sources, was strictly controlled. This included limiting the use of pesticides in these regions.

- A programme of education was launched for those who inhabited the areas surrounding regions which supported a bat population.

1 State one conservation measure and one preservation measure taken to support the population of gray bats.
2 Suggest and explain the advantages and disadvantages of placing the gray bat on the IUCN endangered species list.

guaranteed. However, this can result in no one being able to enjoy the caves and some argue that there is no point in having a resource that cannot be used.

In reality it is objects and buildings that are more commonly preserved, whereas the natural environment is conserved. Examples of preserved habitats include areas set aside in nature reserves and marine conservation zones where human interference is prohibited.

Importance of conservation

Conservation is important for many reasons (look back at Topic 11.7), which can be broadly divided into three categories: economic, social, and ethical reasons.

- Economic – to provide resources that humans need to survive and to provide an income. For example, rainforest species provide medicinal drugs, clothes, and food that can be traded. Other forests are used for the production of timber and paper.

- Social – many people enjoy the natural beauty of wild ecosystems as well as using them for activities which are beneficial to health by providing a means of relaxation and exercise. Examples of these activities include bird watching, walking, cycling, and climbing.

- Ethical – all organisms have a right to exist, and most play an important role within their ecosystem. Many people believe that we should not have the right to decide which organisms can survive, and which we could live without. We also have a moral responsibility for future generations to conserve the wide variety of existing natural ecosystems.

Summary questions

1 State the difference between conservation and preservation. *(1 mark)*

2 There are many reasons why natural habitats may be conserved. State one example of why a habitat may be conserved for:
   a economic reasons *(1 mark)*
   b social reasons *(1 mark)*
   c ethical reasons. *(1 mark)*

3 Suggest and explain one conservation and one preservation technique that could be used to maintain biodiversity in an area. *(4 marks)*

24.5 Sustainability

An increasing global human population results in the need for more resources. To survive at a basic level you require uncontaminated food and water supplies, shelter, clothing, and access to medical care when the need arises. To live in the manner to which you are accustomed, many more resources are required.

To cope with the increased human demand for resources, intensive methods have been used to exploit environmental resources. This can result in the destruction of ecosystems, a reduction in biodiversity, and the depletion of resources.

Sustainable use of resources

The world's natural resources have conflicting demands placed upon them. To conserve natural resources for future generations, sustainable management of the natural environment is necessary. A **sustainable resource** is a renewable resource that is being economically exploited in such a way that it will not diminish or run out.

The aims of sustainability are to:

- preserve the environment
- ensure resources are available for future generations
- allow humans in all societies to live comfortably
- enable less economically developed countries (LEDCs) to develop, through exploiting their natural resources
- create a more even balance in the consumption of these resources between more economically developed countries (MEDCs) and LEDCs.

Alongside the sustainable management of resources, existing resources should be used more efficiently. This helps to prevent finite resources being used up so quickly. For example, many products can be reused – others, such as aluminium cans, can be recycled into new products.

As technology improves, alternatives may be developed that could ease the strain on current finite resources. However, these new resources may take many years to develop, be more costly, and have negative environmental effects of their own.

Sustainable timber production

The sustainable management of forests is possible. This allows for the maintenance of a forest's biodiversity, while sustaining both our supply of wood to meet demands and the economic viability of timber production. The techniques used depend on the scale of timber production.

Small-scale timber production

To produce sustainable timber on a small scale, a technique known as *coppicing* is often used. This is a technique where a tree trunk is cut close to the ground. New shoots form from the cut surface and mature.

Learning outcomes

Demonstrate knowledge, understanding, and application of:

→ how the management of an ecosystem can provide resources in a sustainable way.

▲ Figure 1 The UK government has a campaign to 'Reduce, Reuse, Recycle'. Reduce means lowering your consumption of physical objects and natural resources. Reuse refers to reusing objects in their current form. Recycle means to break an item down into its raw materials, to be used for the manufacture of new items

▲ Figure 2 A hazel coppiced woodland. Some ash trees are left to grow to maturity – the remaining coppice trees are cut to ground level every 7–10 years to provide a constant supply of small wood. The woodland is thus divided into a series of areas cut in different years, ensuring a continuous supply of wood

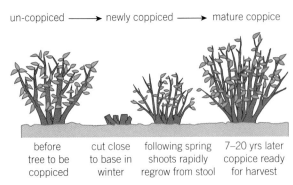

un-coppiced ⟶ newly coppiced ⟶ mature coppice

| before tree to be coppiced | cut close to base in winter | following spring shoots rapidly regrow from stool | 7–20 yrs later coppice ready for harvest |

▲ **Figure 3** *Stages of coppicing in a woodland*

Eventually these shoots are cut and in their place more are produced. These shoots have many uses, including fencing.

In most managed woodlands, rotational coppicing takes place. The woodland is divided into sections and trees are only cut in a particular section until all have been coppiced. Coppicing then begins in another area allowing time for the newly coppiced trees to grow. This process continues until you reach the trees that were first coppiced. These will now have grown to mature-sized trees, and the cycle begins again.

Rotational coppicing maintains biodiversity as the trees never grow enough to block out the light. Hence, succession cannot occur and so more species can survive.

An alternative technique to coppicing which may be used is *pollarding*. The technique is similar to coppicing, but the trunk is cut higher up so deer and other animals cannot eat the new shoots as they appear.

Large-scale timber production

Sustainable timber production on a large scale is based around the technique of felling large areas of forest. The felled trees are destroyed and will not regrow.

To ensure that production is sustainable, timber companies:

- Practise selective cutting, which involves removing only the largest trees.
- Replace trees through replanting rather than waiting for natural regeneration. This also helps to ensure that the biodiversity and mineral and water cycles are maintained.
- Plant trees an optimal distance apart to reduce competition. This results in higher yields as more wood is produced per tree.
- Manage pests and pathogens to maximise yields.
- Ensure that areas of forest remain for indigenous people.

The major disadvantage of this technique is that habitats are destroyed, soil minerals are reduced, and the bare soil which is left is susceptible to erosion. Trees are important for binding soil together, removing water from soil, and maintaining nutrient levels through their role in the carbon and nitrogen cycles.

▲ **Figure 4** *Pollarded alder tree (Alnus glutinosa) at a walled field boundary. This pollarding was recent. Pollarding is a wood-management technique in which all growth is removed from the tree above 2–3 metres at intervals of several years. This provides a regular supply of wood, and has the incidental benefit of prolonging the lifespan of the tree*

Sustainable fishing

As well as the increased demand for fuel and buildings created as a result of population growth, the demand for food is ever-increasing. Fish provide a valuable source of protein within the human diet. However, overfishing has led to the populations of some species of fish decreasing significantly. These fish populations are then unable to regenerate, meaning that they will no longer be able to provide us with a food source in the future.

To try to overcome this problem international agreements are made about the number of fish that can be caught. An example of this is the Common

▲ **Figure 5** *Fishermen sorting out their catch ready for market. Fishing quotas vary depending on the species of fish. Scientists study different species to determine how big their population needs to be for the species to maintain its numbers*

Fisheries Policy in the EU. Fishing quotas provide limits on the numbers of certain species of fish that are allowed to be caught in a particular area. The aim is to maintain a natural population of these species that allows the fish to reproduce sufficiently to maintain their population.

Other techniques that have been used include:

- The use of nets with different mesh sizes. For example, mesh sizes can be made sufficiently large enough that immature fish can escape. Only mature fish are caught, thus allowing breeding to continue.

- Allowing commercial and recreational fishing only at certain times of the year. This protects the breeding season of some fish species and allows the fish levels to increase back to a sustainable level. For example fisherman are only allowed to catch red snapper in the Gulf of Mexico between May and July.

- The introduction of fish farming to maintain the supply of protein food, whilst preventing the loss of wild species. For example, tilapia are among the easiest and most profitable fish to farm due to their diet, tolerance of high stocking densities, and rapid growth. In some regions, the fish are placed in rice fields at planting time where they grow to edible size when the rice is ready for harvest.

> **Study tip**
>
> Be careful with the use of the word 'extinct'. Extinct means no organisms of a particular species are found anywhere in the world. Overfishing can result in the removal of a particular species from an area, but this does not mean the species is extinct.

➕ Overfishing of North Sea cod

As a result of increasing boat numbers and improvements in technology, ever larger numbers of fish have been removed from the sea. In recent years, the number of cod caught in the North Sea has declined as the cod population has fallen.

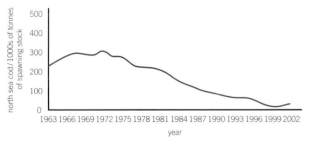

▲ **Figure 6** *The decline of North Sea cod stocks since 1963*

1 State the year the population of cod was at its highest.
2 Suggest why the population of North Sea cod has fallen.

As a result of the near collapse of some fish populations, the European Union introduced fishing quotas to conserve fish stocks.

3 Explain how fishing quotas conserve fish stocks.

These regulations also placed limits on the mesh size of the nets. By increasing the size of the holes in nets, only mature, full-sized fish can be caught. Immature fish can escape.

4 Suggest how increasing mesh size allows the fish population to recover.
5 Evaluate how society uses data on fish populations to inform decisions.

Summary questions

1 State what is meant by a sustainable resource.   (*1 mark*)

2 Suggest two reasons why producing resources in a sustainable way is important.   (*2 marks*)

3 State and explain two ways that large-scale timber production can be managed sustainably.   (*2 marks*)

4 Explain how coppicing is used to produce a sustainable supply of timber.   (*4 marks*)

The Masai Mara National Reserve (MMNR) in southern Kenya is an example of an ecosystem that is actively managed to balance the needs of humans and the need for conservation. The reserve was established as a wildlife sanctuary in 1961, and covers around 1500 km² – it is situated approximately 1500–2000 m above sea level.

Ecosystem

The MMNR is primarily a savannah ecosystem, divided by the main Mara river. The fertile regions close to the river are a combination of rich grasslands and woodland – further from the river are open plains with scattered shrubs and trees. The region is famous for its annual zebra and wildebeest migrations and is home to a wide range of large mammals, including buffalo, elephants, leopards, lions, and black rhinos.

In the past, the region was dominated by the acacia bush. This provided a habitat for the tsetse fly which is a carrier of African trypanosomiasis (sleeping sickness). To attempt to reduce incidences of the disease, government workers and indigenous communities have cleared major tracts of acacia over the last 50 years. Elephants, fire, and cattle grazing have further reduced the presence of acacia and other woody plants.

▲ **Figure 1** *Wildebeest running in the grassland of the Masai Mara*

Farming

Grazing

Traditionally the Masai Mara has been used by local tribes for livestock grazing. In the past, the Masai practised a traditional method of farming known as semi-nomadic farming. Tribes frequently moved depending on climate variation and the presence of tsetse flies. This allowed vegetation time to recover from animal grazing whenever the farmers moved on to another area.

Grazing is now limited to areas on the edge of the reserve, as local tribes are prevented from entering the park. Populations have grown in these marginal areas. Larger herds graze the grassland areas, and more trees are removed for fuel. As the vegetation is removed, the risk of soil erosion increases.

▲ **Figure 2** *The Reserve is named after the local Masai tribes, whose land surrounds the Reserve. Their homes were set up in groups of huts surrounding an enclosed area in which cattle were housed. Each night, the families would bring their livestock into the enclosure in order to prevent predation and cattle theft*

Cultivation

The level of cultivation around the region of the Masai Mara has increased in recent years. As grassland has been converted into cropland, natural vegetation is removed, and nutrients in the soil are used up. Over time this leads to a reliance on fertilisers for effective crop growth.

Ecotourism

The Masai Mara relies on tourism for most of its economic input. Thousands of people travel to the region each year, eager to see for themselves the vast numbers of animals present in their natural habitat. **Ecotourism** is tourism directed towards natural environments, to support conservation efforts and observe wildlife. It is a type of sustainable development which aims to reduce the impact that tourism has on naturally beautiful environments. This is usually seen as a less invasive use of land than agriculture.

The key principles of ecotourism are to:

● ensure that tourism does not exploit the natural environment or local communities

● consult and engage with local communities on planned developments

● ensure that infrastructure improvements benefit local people as well as visitors.

Ecotourism, however, can have negative impacts on the ecosystem. There is evidence that tourist movements such as the repeated use of hiking trails, or the use of mechanised transport, may contribute to soil erosion and other habitat changes.

Conservation and research

The nature reserve plays an important role in the conservation of endangered species. Some of the most popular large mammals have experienced population declines in recent years – beyond those expected from climate or natural variation.

Black rhinos are one of the most endangered animals in Africa, and appear on the IUCN critically endangered list. Despite the trade being illegal, rhino horn is in huge demand, particularly for use in traditional medicine in south-east Asia. People are lured into poaching by the vast sums of money offered to trade in this material.

In 1972, over 100 rhinos lived in the Masai Mara. By 1982, illegal poaching meant that only a handful remained. An active conservation and protection programme was established to encourage a balance between the needs of local communities and those of the wildlife. The programme included the employment of reserve rangers, and the provision of communication equipment, vehicles, and other necessary equipment and infrastructure. These measures have helped to deter poachers from entering the reserve and by the mid-1990s rhino numbers had increased significantly, although it will be some time before population levels return to those seen in the early 1970s.

▲ **Figure 3** *Increasing numbers of tourists who have come to see the wildlife also visit local Masai tribes to observe their way of life and their traditional dances*

▲ **Figure 4** *Most tourists take part in organised safari tours to try and spot the 'big five' – the African lion, African elephant, Cape buffalo, African leopard, and rhinoceros*

> **Synoptic link**
>
> Look back at Topic 11.8, Methods for maintaining biodiversity for more information on the IUCN.

▲ **Figure 5** *The black rhinoceros (Diceros bicornis)*

A number of scientific research projects have been (or are currently being) undertaken in the Masai Mara. These include:

- Michigan State University, studying the behaviour and physiology of the predator spotted hyena.

- Subalusky and Dutton, completing a flow assessment for the Mara River Basin. The aim of this research is to identify the river flows needed to provide for both the basic human needs of the million people who depend on the water, and to sustain the ecosystem in its current form.

- The Mara Predator Project, which catalogues and monitors lion populations throughout the region. The project aims to identify population trends and responses to changes in land management, human settlements, livestock movements, and tourism.

- The Mara-Meru Cheetah Project, which aims to monitor the cheetah population and evaluate the impact of human activity on cheetah behaviour and survival.

Striking a balance

Some human land uses in Masai Mara are incompatible with wildlife survival – increasing wildlife density also threatens pastoral and cultivation lifestyles. A constant balance has to be maintained between the human and animal populations. For example:

- Elephants, in particular, threaten cultivation. Large elephant populations are often responsible for crop trampling and damage to homesteads. Other grazing animals may also eat the crops. To prevent these problems land may be fenced, but this has a negative effect on natural migration.

- Legal hunting is used to cull excess animals. This can successfully maintain population numbers and bring in considerable amounts of money for conservation work. However, numbers must be constantly monitored to ensure that levels are sufficient to maintain the natural balance within the ecosystem.

- Livestock also faces threats from migratory wildlife. For example during the annual wildebeest migration, the wildebeest outcompete cattle for grass. Diseases are introduced to the domesticated animal populations. Equally, the domesticated cattle eat vegetation that could be used by migrating zebras and wildebeest, and diseases can spread from the domestic to the wild animals.

- As the human population expands more homes are required as well as land for cattle and agriculture. Evidence suggests that wildlife density declines significantly as the density of the built environment rises.

Summary questions

1 Describe the ecosystem of the Masai Mara. (*1 mark*)

2 State two ways in which humans use the lands of the Masai Mara (*1 mark*)

3 Explain how and why local Masai tribes have changed their style of farming in recent years. (*4 marks*)

4 The Masai Mara region receives around 300 000 visitors each year. State and explain the positive and negative impacts of this influx of people on the region. (*6 marks*)

24.7 Ecosystem management – Terai region of Nepal

Specification reference: 6.3.2

The southern part of Nepal contains a rich agricultural area known as the Terai region. It stretches along Nepal's southern border with India, with a width of around 25–30 km. The Terai lowlands are defined by a belt of well-watered floodplains stretching from the Indian border in the south to the slopes of the Bhabhar and Siwalik mountain ranges to the north.

The land of the Terai region is fertile, and is the main agricultural region of the country. Alongside farming, people are engaged in a range of trades, industries, and services. As a result of the high population density, natural resources are at risk of being overused.

Learning outcomes

Demonstrate knowledge, understanding, and application of:

→ how ecosystems can be managed to balance the conflict between conservation/preservation and human needs.

Ecosystem

The region is hot and humid in the summer months, and is composed of a fertile alluvial soil which is rich in plant nutrients. The Terai is an area of extreme biodiversity – many subtropical plants are found in this region including pipal and bamboo. There are also large areas of thick forest where animals such as the Bengal tiger, the sloth bear, and the Indian rhinoceros can be found.

Millions of people depend on the Terai forests for their livelihoods. They are also an important source of national income. Primarily as a result of poverty and corruption, large areas of forest have been cleared for agriculture or to sell the timber.

The removal of large parts of the forest has also exacerbated the effects of monsoon flooding, causing severe disruption to communities downstream. If deforestation of the region continues unabated, the communities of the Terai would be left with only small, isolated pockets of forest. This would be devastating not only for the wildlife of the region, but also for the local population who rely on the forests for income through tourism, and through harvesting wood for building products and for burning as fuel.

▲ **Figure 1** *Bengal tigers can be found roaming the forests of Nepal*

▲ **Figure 2** *Royal Chitwan National Park, which is located in the Terai region of Nepal*

Sustainable forest management in Nepal

The aim of sustainable forest management is to provide a livelihood for local people, ensure the conservation of forests, and provide the Nepali state with considerable income for general development. This is being achieved through supportive national legislation, and through the development of local community forestry groups.

The local groups develop their own operational plans, set harvesting rules, set rates and prices for products, and determine how surplus income is distributed or spent. Through the creation of cooperative networks,

▲ **Figure 3** *Map of Nepal showing Terai region*

647

▲ **Figure 4** *Community forestry has contributed to restoring forest resources in Nepal. Forests account for almost 40% of the land in Nepal – however, this area was decreasing at an annual rate of 1.9% during the 1990s. Since 2000 this decline has been reversed, and forest coverage is now increasing*

small forestry businesses can work effectively together – for example, to gain Forestry Stewardship Council (FSC) certification, an international standard which rewards sustainable forestry.

There have been several successes for the community forestry groups:

- Significant improvement in the conservation of the forested regions, both in terms of increased area and improved density.
- Improved soil and water management across the region.
- An increase in the retail price of forestry products, and so a greater economic input to the region.
- Employment and income generation through forest protection, as well as through the production of non-timber forest products.
- Sustainable wood fuel sources, which contribute three-quarters of the local household energy needs.
- Securing the biodiversity of the forested areas.

Promoting sustainable agriculture

The Terai requires a range of management strategies for sustainable land use, to prevent damage of the ecosystem including the further degradation of the Terai forests. These include:

- promoting the production of fruits and vegetables in the hills and mountain regions to avoid further intensification of the Terai
- improving irrigation facilities to enhance crop production
- multiple cropping, where more than one crop is grown on a piece of land each growing season
- the growth of nitrogen-fixing crops such as pulses and legumes to enhance the fertility of the soil
- growing crop varieties resistant to various soil, climatic, and biotic challenges through the use of modern biotechnology and genetic engineering
- improving fertilisation techniques to enhance crop yields – for example, using manure to improve the nutrient content of the soil.

Through the implementation of sustainable forestry and agricultural practices, the Terai region is now being managed in a manner that will secure both the biodiversity of the region, and the economic welfare of its residents for the future.

Summary questions

1 Describe the ecosystem of the Terai region of Nepal.   *(2 marks)*

2 State two ways humans use the Terai region of Nepal.   *(1 mark)*

3 Explain how sustainable forestry and agricultural practices are being used in the Terai region to maintain biodiversity, while also meeting the needs of the local population.   *(4 marks)*

24.8 Ecosystem management – peat bogs

Specification reference: 6.3.2

A peat bog is a region of wet, spongy ground that contains decomposing vegetation. Undisturbed peatland is a 'carbon sink', meaning that it is a store of carbon dioxide. However, once dried, peat can be used as a fuel. As well as releasing thermal energy, burning peat releases carbon dioxide into the atmosphere. It takes many thousands of years for peat bogs to form – the preservation of existing peat bogs is therefore an important component in preventing further climate change.

As well as being used as a fuel, peat is also important for farmers and gardeners, who mix it with soil to improve soil structure and to increase acidity. Peat has very favourable moisture-retaining properties when soil is dry, and prevents excess water killing roots when soil is wet. Although peat can store nutrients, it is not fertile in itself. Commercial peat extraction to supply gardeners and nursery growers is a major threat to this ecosystem.

Ecosystem

Peat forms when plant material is inhibited from fully decaying by acidic and anaerobic conditions. This normally occurs in wet or boggy areas, and therefore peat is mainly composed of wetland vegetation including mosses, sedges, and shrubs.

The plants that grow on peatlands, such as sphagnum mosses (*Sphagnum* spp.), bog cotton or cottonsedge (*Eriophorum angustifolium*), and heathers (typically *Calluna vulgaris*), have adapted to grow and thrive in wet conditions with few nutrients. Bogs also support a wide range of insects such as butterflies, moths, dragonflies, and damselflies. The lack of predators and human disturbance makes some peatlands ideal for birds to nest and bring up their chicks. The abundance of insects, spiders, and frogs, plus nutritious vegetation and berries, provides food for many species. The large areas of open ground provide ideal hunting grounds for birds of prey.

▲ Figure 1 *Peat beds on the Somerset Levels. The peat from here is extracted and sold*

Loss of ecosystem

Lowland raised bogs are one example of a peatland ecosystem. They are a rare and threatened habitat. In the UK, the area of relatively undisturbed lowland raised bog is estimated to have diminished by over 90%, from around 950 km² to only 60 in the last 100 years. It is essential that the remaining areas are conserved to maintain biodiversity – their maintenance will also contribute to flood management, erosion control downstream, and carbon storage.

Learning outcomes

Demonstrate knowledge, understanding, and application of:

→ how ecosystems can be managed to balance the conflict between conservation/preservation and human needs.

Synoptic link

Look back at Topic 23.3, Recycling within ecosystems to remind yourself of how carbon is recycled in the carbon cycle.

▲ **Figure 2** *This peat is drying to be used as fuel. As peat dries out it releases greenhouse gases into the atmosphere*

▲ **Figure 3** *Traditional peat cutting in a bog in south-west Ireland. Slabs of peat are cut from a shallow trench and stacked in small heaps to dry . Hand-cutting peat is a slow, labour-intensive process that can allow the bog to partially recover. It is very different from industrialised, mechanical extraction. The peat companies deep-drain peatlands and strip all vegetation from vast expanses of bog surface, wiping out whole bogs. The Wildlife Trusts are active in attempting to stop this destruction and working to conserve and protect the few remaining areas*

Historically, the greatest decline has occurred through afforestation (the establishment of a forest or stand of trees in an area where there was no forest), peat extraction, and agricultural intensification, including land drainage. These activities have all contributed to the drying out of the bogs.

Conserving lowland bogs

The key to conserving lowland bogs is to maintain or restore appropriate water levels. Steps which are taken to conserve areas of lowland bog include:

- Ensuring that the peat and vegetation of the bog surface is as undisturbed and as wet as possible. Most bogs are surrounded by ditches to allow water to run off, preventing flooding of nearby land. In restoring a bog, ditch blocking may be required for a period of time to raise the water table to the bog surface.

- Removal of seedling trees from the area. Trees have a high water requirement due to transpiration. Therefore, any tree seedling that has the potential to remove water from an area of peatland, or to reduce its ability to support bog vegetation, should be removed to maintain water levels in the area.

- Using controlled grazing to maintain the biodiversity of peatland. Grazing ensures a diverse wetland surface in terms of structure and species composition. This in turn provides a wide range of habitats for many rare insect species.

Continuing intensive land use threatens the existence of much of the remaining peatlands in the UK. Although around 10% of the UK is classified as peat bog of one form or another, around 80% of these areas are in a poor condition. A number of organisations are working to preserve these important ecosystems, including The Wildlife Trusts, Natural England, and the Royal Society for the Protection of Birds (RSPB). It is hoped that, through the conservation work of these and other organisations, an appropriate balance can be met between our need to exploit the land, and the maintenance of a sustainable, biodiverse ecosystem.

Summary questions

1 State what is meant by a peat bog. *(1 mark)*

2 Describe two conservation measures which can be taken to conserve a peat bog ecosystem. *(2 marks)*

3 State and explain why the UK has lost such a large proportion of its peatlands over the past century. *(4 marks)*

Although all ecosystems are, to some extent, fragile, some regions are less resistant to changes than others. These regions are often referred to as *environmentally sensitive ecosystems*. For example, although mass tourism can bring economic vitality to an area, it can also bring attendant changes to an environment that are not always positive, such as the overdevelopment of a coastline. This level of human activity can lead to losses in the biodiversity of the region.

In many environmentally sensitive areas the same types of management techniques are used. These include:

- limiting the areas tourists can visit
- controlling the movement of livestock
- introducing anti-poaching measures
- replanting of forests and native plants
- limiting hunting through quotas and seasonal bans.

This topic looks at examples of some of the world's environmentally sensitive ecosystems, which without control of human activities will be lost.

Learning outcomes

Demonstrate knowledge, understanding, and application of:

→ the effects of human activities on the animal and plant populations in environmentally sensitive ecosystems and how these human activities are controlled.

The Galapagos Islands, Pacific Ocean

The Galapagos is an archipelago of volcanic islands that rise up from the bed of the Pacific Ocean 1000 km west of Ecuador. They are of special interest because they have never been connected to the mainland. The original flora and fauna that reached the islands' shores had to survive a crossing of hundreds of kilometres of ocean. Darwin used evidence he gained from his visits to these islands to develop his theory of evolution by natural selection.

Animals present

The majority of land animals living on the islands are reptiles – there is only one species of land mammal, the Galapagos rice rat. These species arrived here by being washed away from mainland river banks, floating on rafts of vegetation.

Over millions of years these animals, and many of the marine birds that also arrived on the islands, have adapted to their environment in isolation, resulting in many species that are unique to the islands. These include:

- The Galapagos giant tortoise (*Chelonoidis nigra*), which grows to over 150 cm in length.
- The flightless cormorant (*Phalacrocorax harrisi*), whose reduced wings were better for fishing underwater, when flight was not needed to escape mainland predators.

▲ Figure 1 *The Galapagos giant tortoise* (Chelonoidis nigra)

▲ **Figure 2** *The* Scalesia *tree, which dominates the humid zone of the Galapagos Islands. This tree originates from the same family as the diminutive daisy, but grows to heights of over 10 metres*

● The marine iguana (*Amblyrhynchus cristatus*), which contains the advantageous mutation, over a land iguana, of the ability to swim effectively. Unless they are trying to attract a mate, the iguana appears black or dark grey, this allows these ectotherms to bask in the sun and raise raise their body temperatures to approximately 36 °C before swimming in the cold sea where they forage for food. The higher the temperature the longer they can forage.

Plants present

On the larger islands three distinct regions exist, each of which supports particular plant species. These regions are:

● the coastal zone, which contains salt-tolerant species such as mangrove and saltbush

● the arid zone, which contains drought-tolerant species such as cacti and the carob tree

● the humid zone, which contains dense cloud-forest. These trees support populations of mosses and liverworts.

Control of human activities

Until the 19th century the islands were hardly visited by humans. However, as a result of the whaling trade this all changed. The whalers disrupted the islands' fragile ecosystems by allowing domestic animals to roam loose, chopping forests for fires to render down whale fat, and removing tens of thousands of live giant tortoises, whose meat would sustain the whalers on their long sea voyage due to their ability to survive for long periods without food or water. The goats they introduced have also outcompeted giant tortoises on many islands.

The Galapagos National Park was established in 1959. Since then, measures have been taken to protect the living and non-living parts of this unique ecosystem, including:

● introduction of park rangers across the islands

● limiting human access to particular islands, or specific parts of islands

● controlling migration to and from the islands

● strict controls over movement of introduced animals such as pigs (the presence of these were noted by Darwin).

It is hoped that through these and many other measures, the islands' flora and fauna can be protected for future generations.

Antarctica

Antarctica is the coldest, highest, driest, windiest, and emptiest continent. It is almost entirely covered by an ice sheet, which averages 2 km thick. This ice sheet contains around 70% of the world's fresh water.

The average temperature in Antarctica in the winter is below −30 °C. Unlike most parts of the Earth Antarctica has just two seasons, summer and winter – during the summer, many parts of the continent experience 24-hour sunlight – in the winter many parts of the continent experience 24-hour darkness.

Animals present

All endothermic animals living on and around Antarctica rely on thick layers of blubber to insulate them from the cold. These include whales, seals, and penguins. For example, the blubber layer on a Weddell seal (*Leptonychotes weddellii*) can be 10 cm thick.

The emperor penguin (*Aptenodytes forsteri*) is the only warm-blooded animal that remains on the Antarctic continent during the winter. Females lay one egg in June (mid-winter) and leave to spend the winter at sea. The male penguins stay on land, surviving the most extreme winter conditions for up to nine weeks (with no food), keeping their egg warm by balancing it on their feet and covering it with a flap of abdominal skin.

A few invertebrates live on the continent all year. The largest is a wingless midge called *Beligica antarctica*, although at a body size of around 5 mm it can only be viewed properly under a microscope.

Plants present

Plants can only grow in the ice-free regions (around 2% of the continent). Lichens and moss grow in any favourable niche such as in sand, soil, rock, and on the weathered bones and feathers of dead animals. Algae are also able to grow in many sheltered areas.

Control of human activities

With the exception of a few specialised scientific settlements, Antarctica is too cold for people to live. However, in the last 100 years increasing numbers of tourists have been visiting Antarctica. Most visit the Antarctic Peninsula, accessible from Chile, where the climate is mild in comparison with the rest of the continent, and there is much wildlife.

Although strictly controlled in Antarctica (only allowing visits for a few hours to selected area), human activity has had a number of effects on the continent:

- Planet-wide impacts such as global warming (causing some parts of the coastal ice sheet to break up) and ozone depletion, caused by human activities elsewhere.
- Hunting of whales and seals, and fishing of some Antarctic species, has depleted stocks of these organisms.
- Soil contamination, particularly around scientific research stations.
- Discharging of waste into the sea, including human sewage.

The Antarctic treaty was established in 1961 to protect the unique nature of the Antarctic continent. This treaty remains in force indefinitely. Some of its provisions include:

- scientific cooperation between nations
- protection of the Antarctic environment
- conservation of plants and animals
- designation and management of protected areas
- management of tourism.

▲ Figure 3 *There are many different species of penguins in Antarctica, including the large, colourful emperor penguin shown here with its chick*

▲ Figure 4 *These are humpback whales (*Megaptera novaeangliae*) exhaling in calm waters off the Antarctic Peninsula. The International Whaling Commission (IWC) was established in 1949 to protect whales such as the blue and humpback whales whose numbers were diminished by whaling. Indicators show that whale populations are beginning to recover, but such long-lived species with low reproductive rates are incapable of rebuilding their numbers in just a few years*

▲ Figure 5 *Snowdonia National Park*

Snowdonia National Park, Wales

Snowdonia National Park covers 2000 square kilometres of countryside in north-west Wales. It contains the highest mountain range in England and Wales, with four peaks over 1000 metres (including Mount Snowdon at 1085 m). The rugged terrain includes lakes and fast-flowing rivers, and wide tracts of ancient woodland and heath.

Animals present

The rich diversity of habitats in the region provides homes for a wide range of birds. These include:

- Coast and estuary birds such as choughs, cormorants, and oystercatchers.
- Forest birds such as pied flycatcher, redstart, and wood warblers.
- Moorland and mountain birds such as ospreys, buzzards, and sparrowhawks.

There are also over 40 species of land mammal present in Snowdonia including badgers, voles, deer, and hedgehogs.

Plants present

Snowdonia supports an equally diverse range of plant species. For example, if you climb to the top of Mount Snowdon you may come across the Snowdon lily (*Gagea serotina*) and other hardy arctic-alpine plants that have evolved to cope with extreme conditions. Lower down the slopes, the mountain is fringed by woodlands of oak, alder, and wych elm.

Control of human activities

Snowdonia National Park was created in 1951 to conserve the biodiversity of the region. The Park is home to over 25 000 people, many of whom work on the land. It also attracts several million visitors each year. Climbing, walking, cycling, and watersports are some of the most popular activities.

The key purposes of the Park authority are to:

- Conserve and enhance the natural beauty, wildlife, and cultural heritage of the area.
- Promote opportunities for the understanding and enjoyment of the special qualities of the Park.
- Enhance the economic and social well-being of communities within the Park.

▲ Figure 6 *Walkers are asked to keep to designated footpaths in many areas of the park to prevent plants being trampled. The areas in which mountain bikes can be used is even more limited as large tyres with deep treads damage vegetation and top soil, which begins a process of path erosion*

The Dinorwig power station is a pumped-storage hydroelectric power station which was built to help meet the demands of the National Grid during sudden bursts of energy requirement. To preserve the natural beauty of Snowdonia National Park, the power station itself is located deep inside the mountain Elidir Fawr, inside tunnels and caverns. This has minimised the impact to the environment whilst meeting the human demand for energy.

The Lake District, England

The Lake District is England's largest national park, at over 2292 km². The national park contains Scafell Pike, England's highest mountain, and Wastwater, its deepest lake. Terrain within the park includes regions of moorland and fell, and includes 16 lakes dammed by glacial moraines (soil and rock debris) around the end of the last ice age. The dales and fringes of the lakes provide a rich variety of habitats, including areas of ancient woodland.

Animals present

The varied landscapes of the Lake District provide homes for a wide range of wildlife. These include water voles, natterjack toads, and a number of species of bat, through to red deer and birds of prey such as the golden eagle and osprey. A number of native species are under threat including the red squirrel and the vendace, a species of fish that is only found in this region, and which appears on the IUCN endangered list.

Plants present

In the central fells, there are habitats that exist above the tree line. These are rare in the UK and they support a diverse range arctic-alpine plant communities. These include such species as purple saxifrage and alpine cinquefoil. Specialised trees have evolved in these harsh habitats, such as the dwarf juniper and dwarf willow (*Salix herbacea*). Lower regions of the Lake District are home to the sundew (*Drosera rotundifolia*), one of the UK's few carnivorous plants. Insects are trapped within its leaves by a sticky mucilage (which glistens like morning dew, hence its common name) – enzymes are then secreted to digest the insect, releasing its nutrients for the plant.

Control of human activities

The role of the Lake District National Park Authority is like that of Snowdonia – to conserve the region while enabling access for many millions of visitors each year. Through the active management of the countryside, for example, through replanting native tree species, this fragile ecosystem is being secured for generations to come.

▲ **Figure 7** *Wastwater, the deepest of all the lakes in the Lake District, surrounded by mountains*

▲ **Figure 8** *A fly caught in the carnivorous plant,* Drosera rotundifolia, *common name, sundew. The fly is attracted by the glistening droplets at the tips of the tentacles, which the insects then stick to. Proteolytic enzymes secreted by the tentacles digest the insect and the products are absorbed by the leaf*

Summary questions

1   State three reasons why people might want to visit environmentally sensitive areas. *(1 mark)*

2   State one unique feature of:
   a   the Galapagos Islands
   b   Antarctica. *(2 marks)*

3   Using a named region, state and explain how a human activity is controlled to limit its effect on the populations of flora or fauna. *(4 marks)*

4   Environmentally sensitive regions often receive many visitors each year. State and explain the social and environmental impacts on the region of this influx of people. *(6 marks)*

Practice questions

1 The diagram shows two ways in which trees are pruned in order for wood to be continually harvested from the same trees for many years. The health of the trees is not affected in any negative way.

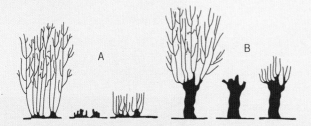

a Name the two types of tree pruning shown in the diagram. (*2 marks*)

b Explain why these forms of timber management are sustainable. (*2 marks*)

c Suggest how these forms of tree pruning help maintain biodiversity. (*3 marks*)

2 Tourism, particularly ecotourism, is a growing industry. The Antarctic Peninsula is even becoming a popular tourist destination. There are many reasons for the increase in tourist numbers such as television making people more aware of extreme environments and more flexible working hours.

a Suggest three more reasons for the increase in tourism to places like Antarctica. (*3 marks*)

The table shows the changes in the number of tourists travelling to Antarctica over the last 20 years.

Year	<1987	1990/ 1991	1992/ 1993	2002/ 2003	2006	2007
Number of visitors	<1000	4698	6500	24 281	>30 000	>40 000

b **(i)** Describe how the numbers of tourists visiting Antarctica have changed over the last 20 years. (*3 marks*)

(ii) Outline the advantages and disadvantages of this change in tourism to Antarctica. (*5 marks*)

(iii) Enhanced global warming has led to melting of the polar ice caps.

Suggest how this has been partly responsible for the increase in tourism. (*1 mark*)

A student carried out an investigation into the perception that people have about the impact of tourism on endangered species such as the blue whale.

The results are shown in the graphs.

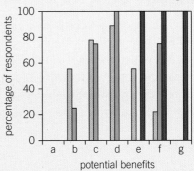

a none
b people learn about whales
c increased appreciation for whales
d people advocate for conservation
e people donate to conservation
f tourism boats carry scientists
g people get involved in research

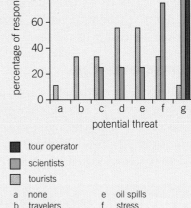

■ tour operator
■ scientists
■ tourists

a none
b travelers
c pollution (other)
d vessel noise
e oil spills
f stress
g collisions

c Discuss any conclusions that could be reached on analysis of these results. (*4 marks*)

The graph shows the changes in numbers of blue whales in the Antarctic over last hundred or so years.

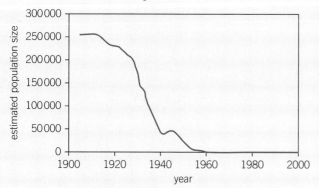

d State, giving your reason(s), whether the outlook for the blue whale population is good or not. (*3 marks*)

3 Peat bogs are large areas of waterlogged land that support a specialised community of plants. Peat bogs take thousands of years to form.

The flow diagram shows the main stages in the formation of a peat bog.

Bull rushes and reeds grow in the shallow water round the margins of a mineral-rich lake.

Dead plant remains accumulate at the margins and trap sediment, which begins to fill the lake.

Different plants now grow, including brown mosses, which form a floating carpet if the water level rises.

New specialised plants grown on the floating brown moss carpet, because it is mineral deficient and acidic.

Sphagnum mosses colonise, increasing the acidity further raising the bog higher, away from sources of minerals.

Plants such as heather, bog cotton, bog asphodel, and carnivorous plants colonise the Sphagnum moss and form a mature peat bog community.

a (i) Name the process summarised in the diagram that changes a lake community into a peat bog community. (*1 mark*)

(ii) Using the flow diagram, list **two** **abiotic** factors that play a role in determining what species of plant can grow in an area. (*2 marks*)

b Most of the minerals in a peant bog are held withing the living plants at all times, **not** in the soil.

- Plants like bog cotton and bog asphodel recycle the minerals they contain,

- The leaves of these plants turn orange as the chlorophyll within them is broken down.

- Minerals such as magnesium ions are transported from the leaves to the plants' roots for storage.

Describe **one** similarity and **two** differences in mineral recycling in a peat bog and in a **deciduous forest**. (*3 marks*)

c In Ireland in 2002, two well-preserved Iron-age bodies were found in peat bogs. Despite having been dead for over two thousand years, the bodies had not decomposed. They had skin, hair, and muscle.

Suggest why these bodies had not decomposed (*2 marks*)

d Suggest two reasons why the large scale removal of peat from bogs for use in gardens is discouraged by conservation groups. (*2 marks*)

OCR 2805/03 2010

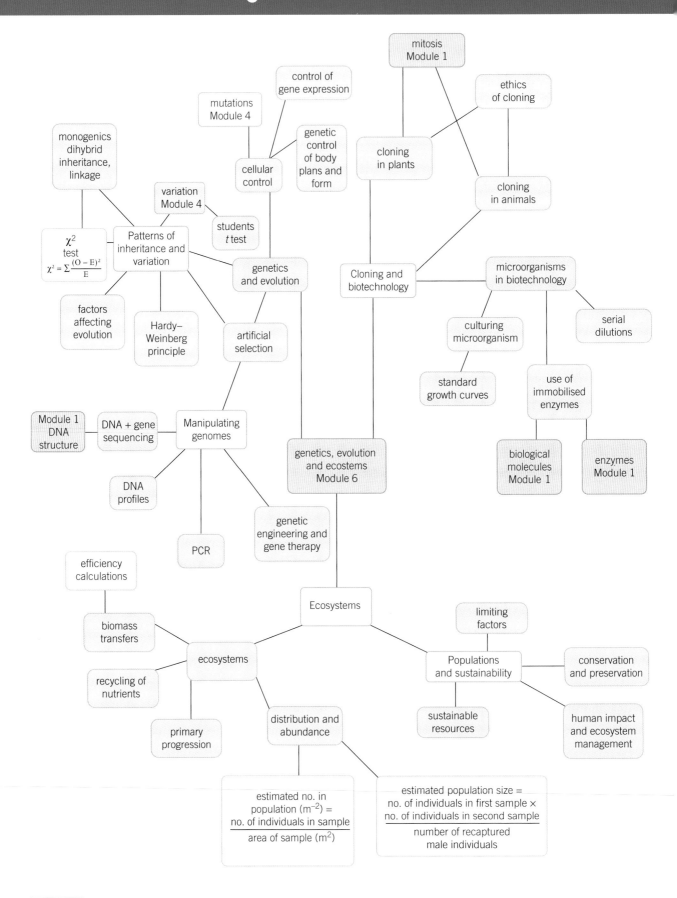

Application

DNA profiling, popularly referred to as DNA fingerprinting, has been used as a form of evidence since 1988 to help prove guilt or innocence in police investigations. The number of areas of DNA used to make the profile has been increased since the early days. Until recently DNA profiling for use in criminal trials in England and Wales involved 11 DNA regions. This has been increased to 17 areas in a new process known as DNA-17.

DNA-17 is a more sensitive test and it can be used (in conjunction with PCR) to get a DNA profile from small quantities of DNA, poor quality samples, and from older cells. This should make it more useful than ever – it will be particularly important in resolving old cases which have been impossible to solve using previous techniques. Advantages include improved discrimination between the profiles of different people (reducing the probability of a chance match between unrelated people), improved sensitivity (so less DNA needed), and improved comparability with the profiles used in the rest of the EU and beyond, where DNA-17 has been the standard for longer.

Possible disadvantages include an increased chance of partial matches from poor samples when more regions are used. Also the test is so sensitive that it is more likely to pick up traces of DNA from innocent parties

1 a Why are DNA profiles used in forensic evidence, rather than, for example, specific proteins?
  b Why is DNA profiling used rather than DNA sequencing in forensic cases and paternity testing?
2 a Summarise the importance of PCR in DNA profiling
  b A limitation in the initial technology meant that a limited number of primers were supplied. As a result, in some individuals with a particular mutation the primers did not bind so poor profiles were produced.
   i Explain what a primer is and what it does
   ii Discuss how a mutation could prevent the binding of the primers
   iii Suggest how this problem could be overcome.
3 a Consider the three advantages of the new test given here and give a scientific explanation for each of them.
   • improved discrimination between the profiles of different people: The more regions of DNA that are analysed to produce a DNA profile, the smaller the chance that two unrelated individuals will have the same pattern of introns. The more regions used in the test, therefore, the less likely it is that errors of identification will be made
   • improved sensitivity
   • improved comparability with the profiles used in the rest of the EU and beyond.
  b Suggest explanations for the disadvantages suggested:
   increased chance of partial matches from poor samples when more regions are used:
   the test is so sensitive that it is more likely to pick up traces of DNA from innocent parties.

SGM (second generation multiplex) target areas

| vWA | TH01 | D8S1179 | FGA | D21S11 | D18S51 |

SGMPlus target areas

| D2S1338 | D16S539 | D19S433 | D3S1358 |

DNA-17 target areas

| D1S1656 | D1S441 | D10S1248 | D12S391 | D22S1045 | SE33 |

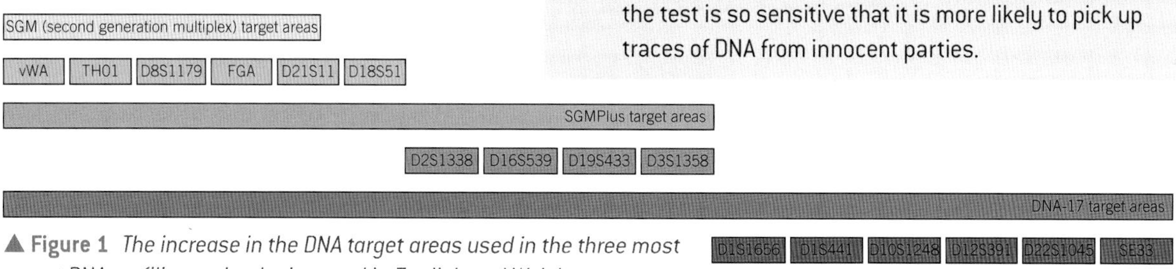

▲ **Figure 1** *The increase in the DNA target areas used in the three most recent DNA profiling technologies used in English and Welsh courts*

Extension

Investigate the story of the development of DNA profiling and the way it was first used to both prove a man innocent and to prove another man, Colin Pitchfork, guilty. Write an account of the story EITHER: for a newspaper with a view to increasing the public understanding of the science of DNA profiling and updating it to mention the introduction of DNA-17

OR: for a popular science magazine, again bringing the story up to date with the way the technology has developed since.

1 Fishing regulations usually require that only fish above a certain size are caught and kept. Smaller fish are thrown back into the sea.

The North East arctic cod have been studied for many years and the average fish size has changed during this period.

The changes are summarised in the table.

Time period	Weight (kg)	Length (cm)
1930s	5.1	85
1970s	4.6	82
2000s	3.2	73

a Describe how the fish have changed over the period of study shown in the diagram.
(*4 marks*)

An investigation was conducted using tank-reared Atlantic silversides. These are small, fast-growing fish that are not usually eaten.

Researchers removed 90% of the fish from the tank every year and allowed the survivors to reproduce.

The fish removed from each tank were chosen in three different ways:

1 The largest fish were removed from one tank.

2 The smallest fish were removed from another tank.

3 The fish were removed randomly from a third tank.

The results are shown in the graph.

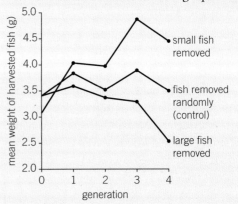

b **(i)** Compare the trends shown in the graph. (*4 marks*)

(ii) Explain the changes in fish size over the four generations shown by the graph. (*5 marks*)

c Suggest whether this is a form of artificial or natural selection. (*2 marks*)

2 a Define the following terms: Genome, Operon, Expression. (*6 marks*)

b Compare the roles of regulatory and structural genes. (*4 marks*)

The diagram summarises the operon responsible for the synthesis of the amino acid tryptophan.

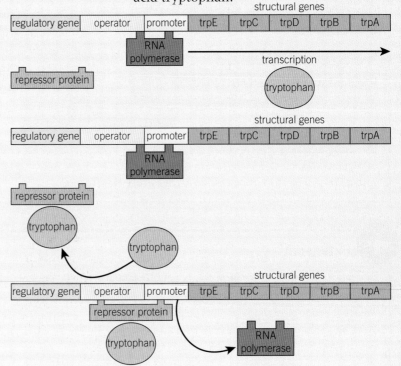

c **(i)** Compare the *structure* of the trp operon to a lac operon. (*5 marks*)

The diagram summarises the action of the repressor protein in the trp operon.

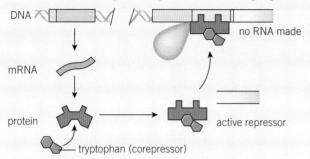

(ii) Suggest how the trp operon is switched on and off. (*4 marks*)

3 Consider the following statements:

Gene expression in prokaryotes is controlled mainly during the process of transcription.

Gene expression in eukaryotes is controlled at the transcriptional, post-transcriptional, translational, and post-translational levels.

Explain, using your knowledge of the structures of eukaryotic and prokaryotic cells, the differences in the control of gene expression. (*7 marks*)

4 **a** Describe the meaning of the following terms: Species evenness, Species richness. (*2 marks*)

The table shows the number of three species found in samples from ponds in the Indiana Dunes.

Lava species	Number of individuals in sample 1	Number of individuals in sample 2
Caddisfly larva	200	20
Dragonfly lava	425	55
Mosquito larva	375	925
Total	1000	1000

b Describe the differences in species evenness and richness between the two sites. (*3 marks*)

The richness and evenness of a community is used to calculate the Simpson's Index of Diversity.

$$D = 1 - \Sigma\left(\frac{n}{N}\right)^2$$

where:

Σ = sum of (total)

N = the total number of organisms of all species and

n = the total number of organisms of a particular species.

Two communities at different sites within the sand dunes were analysed.

The results are shown in table.

Plant species	Number of individuals (n)
Plant species on the Foredune	
marram grass	50
milkweed	10
poison ivy	10
sand cress	4
rose	1
sand cherry	3
Total	N = 78
Plant species on the mature dune	
oak tree	3
hickory tree	1
maple tree	1
beech tree	1
fern	5
moss	3
columbine	3
trillium	3
virginia creeper	4
solomon seal	3
Total	N = 27

c **(i)** Calculate the Simpson diversity index for two local communities. (*2 marks*)

(ii) Suggest reasons for the different diversity indexes. (*3 marks*)

5 It has been estimated that, at the current rate of destruction, tropical rainforests may disappear in less than 150 years.

Outline the ecological, economical, ethical, and aesthetic reasons for the conservation of rainforests. (*7 marks*)

6 There are over 40 000 plants at The Royal Botanic Gardens at Kew. These plants are susceptible to attack by insect pests, such as whitefly, and Trialeurodes spp.

Gardeners at Kew use biological control agents to help prevent pest attack.

The table provides some information regarding the biological control agents of whitefly used at Kew.

Name of control agent	Type of organism	Mode of action	Conditions required
Verticillium lecanii	fungus	infects and kills adults and larvae	high humidity
Encarsia formosa	parasitic wasp	eggs laid on larvae and adult emergence kills pest	low pest density
Macrolophus calliginosus	predatory walking bug	feeds on larvae	high pest density

a Explain, using information from the table, why gardeners at Kew release only one control agent into one greenhouse at any particular time. (*4 marks*)

b Suggest why *E. formosa* would not be effective at high pest densities. (*2 marks*)

c Explain why the gardeners at Kew prefer not to use pesticides. (*4 marks*)

Some researchers investigated the effects of natural predators on another insect pest, Russian wheat aphid *(Diuraphis noxia)*. The results from this study are shown in the graph.

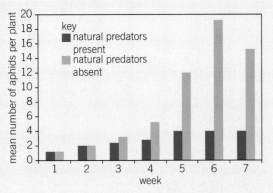

d Describe the results of this study. (*3 marks*)

e Despite the environmental advantages of using biological control agents, most farmers still use pesticides to control pests. Suggest why this is the case. (*3 marks*)

OCR 2805/03 2010

7 a Explain why succession is a dynamic process. (*2 marks*)

b Describe the changes in the following elements of an ecosystem during succession: (*6 marks*)

Species composition, Species diversity, Density and biomass of organisms, Heterotroph population

c In an estuary, a salt marsh often develops on muddy shores between the lowest and the highest tide levels. A salt marsh is characterized by the vegetation that grows in these conditions. Succession can be seen from low tide level through the network of creeks and gulleys up to high tide level and beyond onto higher ground.

Students carried out research on two separate salt marshes, A and B. They measured the elevation range above the mean sea level where five key plant species were found. Their results are shown in the graph.

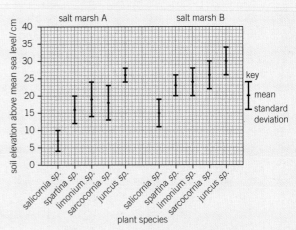

(i) Suggest two environmental features that make a salt marsh an extreme habitat for plants to live in. (*2 marks*)

(ii) State one similarity and one difference in the overall distribution of the five species on the two salt marshes. (*2 marks*)

(iii) Suggest which one of the plant species shown in the graph is a pioneer species. (*1 mark*)

(iv) Suggest how pioneer plants can alter the salt marsh environment. (*2 marks*)

d Where two or more individuals share any resource that is insufficient to satisfy all their requirements fully, then competition results.

  (i) State two species, found in salt marsh A, that are likely to be involved in interspecific competition. (*1 mark*)

  (ii) Name one resource for which these plants will compete. (*1 mark*)

e The students were also interested in measuring how the relative abundance of the five species changed with elevation above the mean sea level in the salt marsh.

Describe how this could be done. (*4 marks*)

OCR 2804 2009 (apart from a and b)

8 a The use of biosensors for the detection of chemicals, such as glucose and organophosphates, has proved to be a reliable alternative to other methods.

Glucose biosensors, first developed in the 1980s, enable diabetics conveniently and easily to monitor their blood glucose. The diagram shows the key components of one type of glucose biosensor.

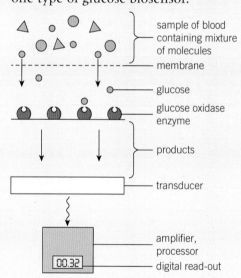

The development of organophosphate biosensors has benefits for health care and environmental monitoring.

With the increasing demand for food production, the use of organophosphates as insecticides to protect crops has increased. However, organophosphates can remain in the environment and are potentially toxic to humans and other animals.

  (i) Explain how water and food supplies may contain organophosphates. (*2 marks*)

  (ii) Many organophosphates are irreversible inhibitors of acetylcholinesterase.

  Explain why this makes them harmful to human health. (*2 marks*)

  (iii) The equation summarises the reaction catalysed by acetylcholinesterase.

$$\text{acetylcholine} + H_2O \xrightarrow{\text{acetylcholinesterase}} \text{choline} + \text{acetate (acetic acid)}$$

  Design a simple biosensor using acetylcholinesterase to detect the presence of a harmful organophosphate in a sample of river water. Include explanations for each element of your design.

  You may wish to draw a labelled diagram help you answer. (*5 marks*)

  (iv) Suggest the advantages of organophosphate biosensors compared with other detection methods. (*3 marks*)

b Using genetic modification, crop plants resistant to the herbicide glyphosate can be produced.

Glyphosate does not act on acetylcholinesterase but inhibits other enzyme systems.

  (i) Explain why a different biosensor to that used in (a)(iii) and (iv) would need to be developed to detect glyphosate in a sample of river water. (*2 marks*)

  (ii) Suggest one advantage and one disadvantage of producing glyphosate-resistant crop plants. (*2 marks*)

c Crop plants can be genetically modified for glyphosate resistance by using a restriction endonuclease and DNA ligase.

Describe the roles of these two enzymes in genetic manipulation. (*4 marks*)

OCR 2805/04 2008

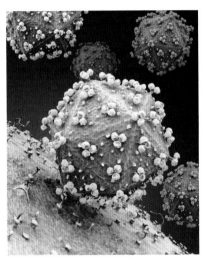

A synoptic concept in biology is an idea or principle that is applied to more than one part of the subject. For example, cell signalling, the basis of communication within an organism is relevant to the action of local signaling, molecules like histamine, hormones in the endocrine system, transmitters in the nervous system, and the identification of cells as foreign by the immune system.

To be a good biologist, it is important that you can identify connections between different topics and apply your knowledge and understanding of each topic to answer novel questions.

In the synoptic biology examination paper, question styles will include short answer questions, practical questions, problem-solving questions, and extended response questions. These questions cover different combinations of topics from different modules, though it is impossible to cover every single one in any given set of questions.

Table 1 shows how just one learning outcome links to many other topics and modules in A Level Biology.

▲ **Figure 1** *The HIV virus is able to infect white blood cells by binding to specific receptor proteins on the cell's membrane*

▼ **Table 1**

Learning outcome	Links with other modules		What you might be asked about...
2.1.5(a) the roles of membranes within cells and at the surface of cells to include the roles of membranes as sites of cell communication (cell signalling)	2.1.2(j)	how the properties of cholesterol molecules relate to their functions in living organisms	• steroid hormones are hydrophobic and can enter cells through phospholipid bilayer • fluidity of cell membranes
	2.1.3(g)	transcription and translation of genes	• the role of the nuclear membrane and the endoplasmic reticulum and Golgi apparatus in the way genes are transcribed and translated • hormones that are water soluble cannot enter cells but bind to membrane receptors and activate cAMP which acts as a secondary messenger and may result in genes being switched on and off, affecting the proteins synthesised.
	2.1.4(a)	the role of enzymes in catalysing reactions that affect metabolism	• hormones that are water soluble cannot enter cells but bind to membrane receptors and activate cAMP which acts as a secondary messenger and activates enzymes

2.1.5(a)	the roles of membranes at the surface of cells	• barrier to the movement of molecules into cells, facilitated diffusion, osmosis with links to turgor in plants and homeostasis in animals, active transport systems, site of receptors for hormones, neurotransmitters, cell recognition
2.1.5(b)	the fluid mosaic model of membrane structure	• hydrophobic core of phospholipid bilayer prevents the entry of water soluble hormones • protein pores, gated channels • osmosis with links to turgor in plants and homeostasis in animals, facilitated diffusion, active transport • surface glycoproteins as part of cell recognition system • infective mechanisms of viruses • immune system • Endo and exocytosis
2.1.5(d)(i)	the movement of molecules across membranes	• steroid hormones can diffuse through phospholipid bilayer • secondary messenger systems • osmosis with links to turgor in plant cells and homeostasis in animals • active transport systems • chemiosmosis
4.1.1(f)	the structure, different roles and modes of action of B and T lymphocytes in the specific immune response	• proteins in cell surface membrane acting as self and non-self-markers
5.1.3(d)	the structure and roles of synapses in neurotransmission	• example of cell signalling • endo and exocytosis
5.1.4(a)	endocrine communication by hormones	• example of cell signalling • secondary messengers
5.1.5(j)	the coordination of responses by the nervous and endocrine systems	• neurotransmitters and hormones involvement in cell signalling
5.1.5(k)	the effects of hormones and nervous mechanisms on heart rate	• neurotransmitters and hormones involvement in cell signalling

Remind yourself of how the heart beat is controlled by the sino-atrial node by looking at Topic 8.5, The heart.

1 The contraction of the heart is **myogenic**, it is **initiated** by a group of muscle cells without stimulation from the nervous system. The natural rhythm of cardiac muscle is about 100–115 contractions per minute.

    a State the name given to the group of cells that initiate cardiac muscle contraction. *(1 mark)*

Answer to part **a**

SA node ✓

For single mark answers, be clear and unambiguous in your answer – and make sure your spelling is correct.

The basis of the resting potential and action potential in neurones was explained in Topic 13.4, Nervous transmission. The same principles can be applied in a number of different places in the living world.

The cells named in **a** are autorhythmic cells. Autorhythmic cell membranes have a **resting potential** like the membranes of neurons. The resting potential in autorhythmic cells, unlike neurons, is unstable. This leads to the membranes continually depolarizing, and subsequently repolarizing. This results in the generation of numerous **action potentials** per minute.

    b (i) Explain how resting membrane potential is maintained. *(3 marks)*

        (ii) State the number of action potentials that would be generated per minute in cardiac muscle isolated from the nervous system. *(1 mark)*

Answer to part **b**

Part **b(i)**

sodium/potassium pumps ✓

sodium ions out of muscle cells and potassium ions into muscle cells ✓

differential permeability of membrane ✓

part **b (ii)** 100–115 ✓

Equal to natural rhythm of cardiac muscle stated in the stem of the question

Check the structure and functions of the sympathetic and parasympathetic nervous systems in Topic 14.1, Hormonal communication and think about how this applies to the question.

The heart is connected to the nervous system by sympathetic and parasympathetic neurons. The average human resting heart rate is 60–80 beats/min.

    c Outline the effect of the **sympathetic** and **parasympathetic** branches of the nervous system on resting heart rate. *(3 marks)*

Answer to part **c**

sympathetic increases heart rate ✓

parasympathetic slows heart rate ✓

idea that parasympathetic is dominant at rest ✓

Atropine is a drug that blocks receptors in neuromuscular junctions of the parasympathetic system.

d Suggest and explain how the heart rate would change when atropine is administered. *(3 marks)*

> You met neuromuscular junctions in Topic 13.5, Synapses. Remind yourself of how they work and which neurotransmitters are involved.

> Give what is asked for in an answer and no more — you don't need a detailed description of a neuromuscular junction for 3 marks, just enough information to make the answer clear.

Answer to part **d**

heart rate would increase ✓

atropine blocks acetylcholine (in parasympathetic neuromuscular junction) from binding ✓

heart rate no longer slowed ✓

Adrenaline, a hormone released from the adrenal glands, also affects the heart rate.

> Look up hormones and the endocrine system in Topic 14.1, Hormonal communication. Make sure you know about the importance of the adrenal glands, the secretion of adrenaline, and its effects on the body.

e (i) State the precise location of the secretion of adrenaline. *(1 mark)*

> Remember that cell signalling usually involves receptors on the cell membranes. Remind yourself of the details in Topic 13.1, Coordination.

(ii) Explain why the effect of adrenaline is a form of cell signalling. *(1 mark)*

(iii) State the effect of adrenaline on the heart rate. *(1 mark)*

Adrenaline is a hydrophilic molecule and therefore is unable to enter cells.

> You need to understand the structure of the cell membrane and understand hydrophilic and hydrophobioc molecules to answer this part of the question, look back at Topic 3.5, Lipids.

(iv) Explain why adrenaline cannot enter cells. *(2 marks)*

Answer to part **e**

Part **e (i)** (adrenal) medulla ✓

Part **e (ii)** molecule released by a cell in the adrenal gland ✓

produces a response / increases rate of contraction ✓

in a cell in the heart ✓

part **e (iii)** increases heart rate ✓

Part **e (iv)** hydrophobic core of phospholipid bilayer ✓

prevents entry of hydrophilic adrenaline molecule ✓

This answer also depends on your understanding of the different types of cell signalling which you learnt about in Topic 14.5, Coordinated responses.

f Outline the way in which adrenaline brings about a response within a cell. *(6 marks)*

Answer to part **f**

adrenaline binds to receptor on cell surface membrane ✓

change in shape of receptor. G protein ✓

activates G protein ✓

G protein activates adenyl cyclase ✓

adenyl cyclase catalyzes formation of cAMP ✓

cAMP activates kinase enzymes ✓

kinase enzymes activate enzymes responsible for response ✓ (max 6)

When you answer a question that carries a lot of marks, make sure you give at least the same number of clear points as there are marks in the question, whether you present your answer as a list or as continuous prose.

You learnt about the role of the adrenal glands and their hormones in Topic 14.1, Hormonal communication.

g Hydrocortisone is a synthetic glucocorticoid derived from cholesterol. It has anti-inflammatory actions due to the suppression of the synthesis of proteins that have roles in the inflammatory process.

Suggest how hydrocortisone brings about its anti-inflammatory response. *(3 marks)*

This links to mechanisms of inflammation with the production of proteins such as histamine. Look back to Topic 12.5, Non-specific animal defences against pathogens.

Answer to part **g**

hydrophobic/lipid soluble due to cholesterol ✓

enter cells through phospholipid bilayer by diffusion ✓

switch genes off/stops transcription ✓

You learnt about cell membrane structure in Topic 5.3, Diffusion.

Check the factors affecting gene expression and how genes are switched on and off in the cell in Topic 3.10, Protein synthesis.

1 Glutamic acid ($C_5H_9NO_4$) is an amino acid. A salt of glutamic acid is known as glutamate. The sodium salt of glutamic acid is called monosodium glutamate (MSG) ($C_5H_8NO_4Na$). MSG occurs naturally in most foods, particularly high protein foods like milk, cheese, meat, and fish. Tomatoes and mushrooms also have high levels of glutamate.

The human body can synthesise glutamic acid and it is an essential molecule in a number of metabolic processes. Children metabolise glutamate in the same way as adults and human breast milk actually contains ten times more glutamate than cow's milk.

Glutamic acid also acts a neurotransmitter.

MSG is a commonly used as a flavour enhancer. It is added to Chinese food, canned vegetables, soups, and processed meats and many other processed foods.

MSG used to be extracted from protein-rich plants such as seaweed.

This was time-consuming process and MSG is now produced by an industrial fermentation process using bacteria.

MSG used to be blamed for causing "Chinese Restaurant Syndrome" after a doctor reported that he suffered from burning sensations along the back of his neck, chest tightness, nausea, and sweating whenever he had eaten at a Chinese restaurant.

Although a small percentage of people are sensitive to monosodium glutamate, double-blind studies have failed to show that monosodium glutamate actually causes any of the symptoms attributed to "Chinese restaurant syndrome".

Chinese food often contains a wide variety of ingredients such as seafood, peanuts, spices and herbs.

The diagram shows a molecule of MSG.

a (i) State the name of the group of molecules that contains glutamic acid. *(1 mark)*

Glutamic acid can be used to build another type of molecule.

(ii) State the name of the group of molecules that contains this larger molecule. *(1 mark)*

(iii) State the name of the bonds formed during the synthesis of these larger molecules. *(1 mark)*

b Outline the role of a neurotransmitter. *(3 marks)*

c Explain why the production of MSG is now an example of biotechnology. *(3 marks)*

The graph shows the results of an investigation into the growth of bacteria and the production of glutamic acid by batch and fed-batch fermentations.

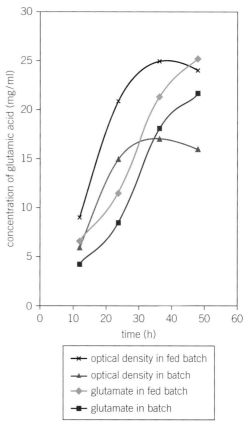

Bacterial growth was estimated by measuring the optical density (OD) of the culture using a spectrophotometer.

d (i) Describe the difference in growth rate and glutamic acid production in the graph. *(4 marks)*

(ii) Compare the processes involved during batch and fed-batch fermentation. *(3 marks)*

(iii) State the overall name given to the processes involved in the production of pure glutamic acid from the slurry removed from the fermentation vessel. *(1 mark)*

e (i) Outline how a double-blind trial would be conducted. *(4 marks)*

(ii) Discuss the arguments for and against the use of monosodium glutamate in food preparation. *(5 marks)*

(iii) Suggest why tomatoes and mushrooms are often used in cooking to flavour food. *(2 marks)*

2 The diagram shows some of the metabolic pathways that take place in a plant cell.

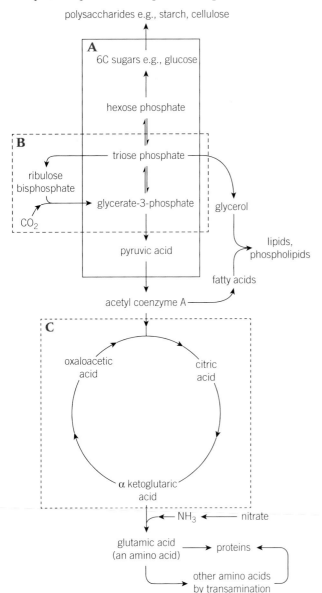

a State the names of the processes labelled A, B and C. *(3 marks)*

b Select an example from the diagram to explain what is meant by an anabolic reaction. *(2 marks)*

c Describe, using information from the diagram, how plants make the chemical components necessary to build cell walls, cell membranes and cytoplasm. *(9 marks)*

d Plants also manufacture chemicals as a protection against herbivory.

Explain why this is a compromise for the plants. *(2 marks)*

e Outline others ways that plants are adapted for adverse conditions. *(6 marks)*

OCR 2806/01 2009 (apart from (d) and (e))

3 Invertebrates, such as cockroaches and squid do not have myelin sheaths around their axons.

Giant axons that are present in cockroaches, earthworms and squid can be up to 1.0 mm in diameter.

Giant axons are essential in parts of the nervous system that control escape reflexes.

a Describe the type of behaviour that includes the escape reflex. *(2 marks)*

The diagram compares the transmission speeds of a mammal with three invertebrate organisms.

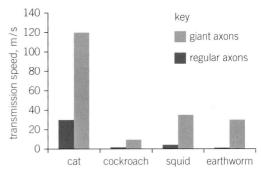

b (i) Describe the variation in the transmission speed of impulses shown in the diagram. *(4 marks)*

(ii) Explain effects of axon diameter and myelination on the speed of impulse transmission. *(4 marks)*

The diagram shows the changes in membrane potential during an action potential.

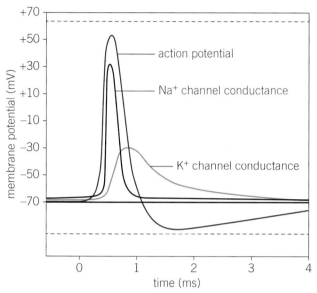

c (i) Explain, using information from the diagram, how an action potential is propagated. *(5 marks)*

(ii) Explain how the differential permeability of neuronal membranes is essential in maintaining resting potential. *(3 marks)*

Tetrodotoxin is a chemical produced by the puffer fish and a number of other animals. It blocks nerve impulses by blocking voltage gated sodium ion channels and is a very powerful toxin.

Some species of puffer fish are considered a delicacy in Japan and Korea and chefs in these countries have to be specially trained to prepare the fish so that the poison is removed.

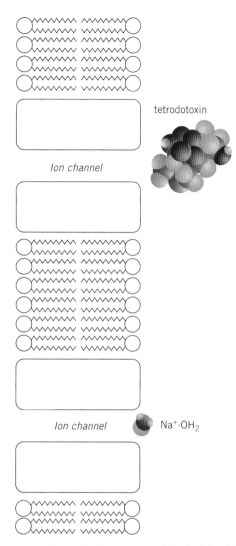

Tetrodotoxin is very specific in blocking the sodium ion voltage gated and the movement of potassium ions is not affected.

Victims of tetrodotoxin poisoning do not die as the result of a heart attack but respiratory paralysis when the diaphragm no longer contracts. The function of cardiac muscle is not affected.

d (i) Explain why tetrodotoxin can block sodium ion voltage gated channels but not potassium ion voltage gated channels. *(2 marks)*

(ii) Explain how paralysis of the diaphragm would lead to death. *(3 marks)*

(iii) Suggest why cardiac muscle is not affected by tetrodotoxin. *(1 mark)*

Pufferfish have undergone a mutation in the gene coding for the protein that forms the sodium ion voltage gated channel.

e (i) Explain the term mutation. (*2 marks*)

(ii) Explain how this mutation means that pufferfish are not susceptible to poisoning by tetrodotoxin. (*4 marks*)

(iii) Describe the process by which this mutation became common in pufferfish. (*4 marks*)

Tetrodotoxin is found in many organisms, including the blue-ringed octopus and some species of poisonous frogs. Toxins are usually very specific to a particular species but in this case tetrodotoxin is produced by bacteria which these organisms accumulate in their diets.

f State the name given to the process shown by these distantly related organisms all evolving to be unaffected by tetrodotoxin. (*1 mark*)

4 A short time after the death of a human being, or animal, the joints within the organism become locked in place. This process is called rigor mortis and begins with the muscles partially contracting. The graphs show how the concentration of ATP change in the hours after death and the time it takes for fully-fixed rigor mortis to develop.

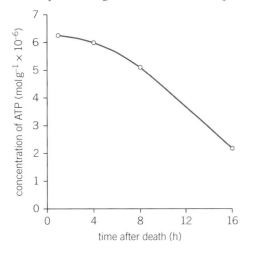

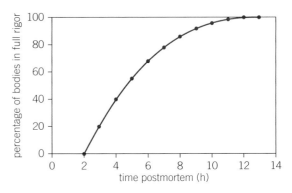

a (i) Explain why the levels of ATP level fall after death. (*2 marks*)

(ii) Describe the changes in ATP level and rigor mortis shown in the graph. (*4 marks*)

(iii) Suggest the reason for the onset of rigor mortis. (*3 marks*)

(iv) Suggest why a small amount of muscle contraction can occur after death. (*3 marks*)

The graph shows how the pH changes in muscle tissue after death.

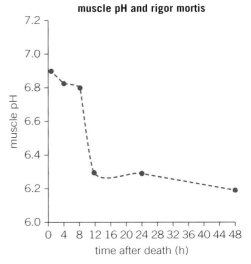

b Suggest a reason for the change in pH in muscle tissue after death. (*2 marks*)

Depending on various factors such as the temperature of the environment, rigor mortis can last up to about three days.

c Suggest an explanation for why rigor mortis is only temporary. (*4 marks*)

The diagram shows a part of the sliding filament theory.

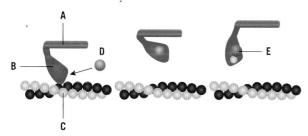

d   (i)   State the names of the structures A, B and C.   (*3 marks*)

(ii) State the name of the molecule labelled D.   (*1 mark*)

(iii) Indicate, using an arrow, in which direction structure A moves relative to structure C.   (*1 mark*)

(iv) Suggest, using your knowledge of the structure of proteins, the role of the ion labelled E in muscle contraction.   (*3 marks*)

(v)   Outline how myosin acts as a molecular ratchet in the sliding filament theory.   (*3 marks*)

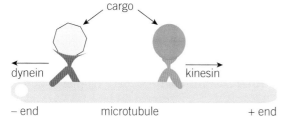

Kinesin is a motor protein that has a structure very similar to myosin. Dynein is another motor protein with a very different structure to myosin.

e   Describe what determines the 3D structure of a protein and how this differs in kinesin and dynein.   (*3 marks*)

Both dynein and kinesin have regions that binds to microtubules, a region that binds and hydrolyses ATP and a region that binds to an organelle.

The energy released during the hydrolysis of ATP changes their three-dimensional structures resulting in the motor proteins moving along microtubules and transporting organelles within cells. Dynein and kinesin move in opposite directions.

f   (i)   Compare and contrast the ways in which dyneins, kinesins and myosin lead to movement.   (*4 marks*)

(ii) State two other processes, other than the movement of organelles, in which microtubules also have an important role.   (*2 marks*)

Glossary

aminoacylase enzyme used to produce pure samples of L-amino acids.

abiotic factors non-living conditions in a habitat.

abscission the fall of leaves.

acetyl coenzyme A molecule that enters the Krebs cycle from glycolysis through a linking reaction when coenzyme A combines with an acetyl group.

acetylation addition of acetyl group.

action potential the change in the potential difference across the neurone membrane of the axon when stimulated (approximately +40 mV).

activation energy the energy required to initiate a reaction.

active site area of an enzyme with a shape complementary to a specific substrate, allowing the enzyme to bind a substrate with specificity.

active transport movement of particles across a plasma membrane against a concentration gradient. Energy is required.

adenosine diphosphate (ADP) a nucleotide composed of a nitrogenous base (adenine), a pentose sugar and two phosphate groups. Formed by the hydrolysis of ATP, releasing a phosphate ion and energy.

adenosine triphosphate (ATP) a nucleotide composed of a nitrogenous base (adenine), a pentose sugar and three phosphate groups. The universal energy currency for cells.

agglutinins chemicals (antibodies) that cause pathogens to clump together so they are easier for phagocytes to engulf and digest.

alcoholic fermentation fermentation that results in the production of ethanol.

alkaloids bitter-tasting compounds found in plant leaves that may affect the metabolism of animals or insects eating them or poison them.

allele version of a gene.

allele frequency the relative frequency of a particular allele in a population at a given time.

alleles different versions of the same gene.

allopatric speciation speciation that occurs as a result of a physical barrier between populations.

amino acids monomer used to build polypeptides and thus proteins.

ammonification conversion of nitrogen compounds in dead organic matter or waste into ammonium compounds by decomposers.

anabolic steroids steroid drugs used illegally by some athletes and bodybuilders to increase muscle mass.

anabolism (anabolic) reactions of metabolism that construct molecules from smaller units. These reactions require energy from the hydrolysis of ATP.

anaerobic respiration respiration in the absence of oxygen.

analogous structures structures that have adapted to perform the same function but have a different origin.

anaphase third stage of mitosis when chromatids are separated to opposite poles of the cell.

antibiotic-resistant bacteria bacteria that undergo mutation to become resistant to an antibiotic and then survive to increase in number.

antibiotics a chemical or compound that kills or inhibits the growth of bacteria.

antibodies y-shaped glycoproteins made by B cells of the immune system in response to the presence of an antigen.

antidiuretic hormone (ADH) hormone that increases the permeability of the distal convoluted tubule and the collecting duct to water.

antigen identifying chemical on the surface of a cell that triggers an immune response.

antigen–antibody complex the complex formed when an antibody binds to an antigen.

antigen-presenting cell (APC) a cell that displays foreign antigens complexed with major histocompatibility complexes on their surfaces.

antisense strand the strand of DNA that runs 3′ to 5′ and is complementary to the sense strand. It acts as a template strand during transcription.

anti-toxins chemicals (antibodies) that bind to toxins produced by pathogens so they no longer have an effect.

apical dominance the growth and dominance of the main shoot as a result of the suppression of lateral shoots by auxin.

apoplast the cell walls and intercellular spaces of plant cells.

apoplast route movement of substances through the cell walls and cell spaces by diffusion and into cytoplasm by active transport.

apoptosis programmed and controlled cell death important in controlling the body form and in the removal of damaged or diseased cells.

arrhythmia an abnormal rhythm of the heart.

artefacts objects or structures seen through a microscope that have been created during the processing of the specimen.

artificial active immunity immunity which results from exposure to a safe form of a pathogen, for example, by vaccination.

artificial passive immunity immunity which results from the administration of antibodies from another animal against a dangerous pathogen.

artificial selection see selective breeding.

artificial twinning the process of producing monozygotic twins artificially.

aseptic techniques techniques used to culture microorganisms in sterile conditions so they are not contaminated with unwanted microorgnisms.

asexual reproduction the production of genetically identical offspring from a single parent.

assimilates the products of photosynthesis that are transported around a plant, e.g., sucrose.

atrial fibrillation an abnormal rhythm of the heart when the atria beat very fast and incompletely.

atrio-ventricular node (AVN) stimulates the ventricles to contract after imposing a slight delay to ensure atrial contraction is complete.

autoimmune disease a condition or illness resulting from an autoimmune response.

autoimmune response response when the immune system acts against its own cells and destroys healthy tissue in the body.

autonomic nervous system part of the nervous system that is under subconscious control.

autosomal linkage genes present on the same, non-sex chromosome.

autotrophic organisms that synthesis complex organic molecules from inorganic molecules via photosynthesis.

auxins plant hormones that control cell elongation, prevent leaf fall, maintain apical dominance, produce tropic responses, and stimulate the use of ethene in fruit ripening.

B effector cells B lymphocytes that divide to form plasma cell clones.

B lymphocytes (B cells) lymphocytes which mature in the bone marrow and that are involved in the production of antibodies.

B memory cells B lymphocytes that live a long time and provide immunological memory of the antibody needed against a specific antigen.

baroreceptors receptors which detect changes in pressure.

batch fermentation an industrial fermentation that runs for a set time.

belt transect two parallel lines are marked along the ground and samples are taken of the area at specified points.

Benedict's reagent an alkaline solution of copper(II)sulfate used in the chemical tests for reducing sugars and non-reducing sugars. A brick-red precipitate indicates a positive result.

beta pleated sheet sheet-like secondary structure of proteins.

binomial nomenclature the scientific naming of a species with a Latin name made of two parts – the first indicating the genus and the second the species.

biodiversity the variety of living organisms present in an area.

bioinformatics the development of the software and computing tools needed to analyse and organise raw biological data.

biomass mass of living material.

bioremediation the use of microorganisms to break down pollutants and contaminants in the soil or water.

biotic factors the living components of an ecosystem.

biuret test the chemical test for proteins; peptide bonds form violet coloured complexes with copper ions in alkaline solutions.

Bohr effect the effect of carbon dioxide concentration on the uptake and release of oxygen by haemoglobin.

Bowman's capsule cup-shaped structure that contains the glomerulusand is the site of ultrafiltration in the kidney.

bradycardia a slow heart rhythm of below 60 beats per minute.

breathing rate the number of breaths (inhalation and exhalation) taken per minute.

bulk transport a form of active transport where large molecules or whole bacterial cells are moved into or out of a cell by endocytosis or exocytosis.

bundle of His conducting tissue composed of purkyne fibres that passes through the septum of the heart.

callose a polysaccharide containing β 1-3 linkages and β 1-6 linkages between the glucose monomers that is important in the plant response to infection.

Calvin cycle the cyclical light independent reactions of photosynthesis.

carbaminohaemoglobin the compound formed when carbon dioxide combines with haemoglobin.

carbohydrates organic polymers composed of the elements carbon, hydrogen and oxygen, usually in the ratio $C_x(H_2O)_y$. Also known as saccharides or sugars.

carbonic anhydrase enzyme which catalyses the reversible reaction between carbon dioxide and water to form carbonic acid.

cardiac cycle the events of a single heartbeat, composed of diastole and systole.

carrier a person who has one copy of a recessive allele coding for a genetically inherited condition.

carrier proteins membrane proteins that play a part in the transport of substances through a membrane.

carrying capacity the maximum population size that an environment can support.

cartilage strong, flexible connective tissue found in many areas of the bodies of humans and other animals.

catabolism (catabolic) reactions of metabolism that break molecules down into smaller units. These reactions release energy.

catalase an enzyme that catalyses the breakdown of hydrogen peroxide.

cell cycle the highly ordered sequence of events that takes place in a cell, resulting in division of the nucleus and the formation of two genetically identical daughter cells.

cell signalling a complex system of intercellular communication.

cellulose a polysaccharide formed from beta glucose molecules where alternate beta glucose molecules are turned upside down. It is unable to coil or form branches but makes hydrogen bonds with other cellulose molecules to produce strong and insoluble fibres. Major component of plant cell walls.

cell wall a strong but flexible layer that surrounds some cell-types.

central nervous system (CNS) consists of the brain and spinal cord.

centrioles component of the cytoskeleton of most eukaryotic cells, composed of microtubules.

centromere region at which two chromatids are held together.

channel proteins membrane proteins that provide a hydrophilic channel through a membrane.

checkpoints control mechanisms of the cell cycle.

chemiosmosis the synthesis of ATP driven by a flow of protons across a membrane.

chemoreceptors receptors which detect chemical changes.

chi-square formula formula used to determine the significance of the difference between observed and expected count data.

chiasmata sections of DNA, which became entangled during crossing over, break and rejoin during anaphase 1 of meiosis sometimes resulting in an exchange of DNA between bivalent chromosomes, forming recombinant chromatids and providing genetic variation.

chloride shift the movement of bicarbonate ions into the red blood cells as hydrogen ions move out to maintain the electrochemical equilibrium.

chlorophyll green pigment that captures light in photosynthesis.

chloroplasts organelles that are responsible for photosynthesis in plant cells. Contain chlorophyll pigments, which are the site of the light reactions of photosynthesis.

chromatids two identical copies of DNA (a chromosome) held together at a centromere.

chromatin uncondensed DNA in a complex with histones.

chromosomes structures of condensed and coiled DNA in the form of chromatin. Chromosomes become visible under the light microscope when cells are preparing to divide.

circulatory system the transport system of an animal.

citrate six carbon molecule formed in Krebs cycle by the combination of oxaloacetate and acetyl coenzyme A.

climax community final stage in succession, where the community is said to be in a stable state.

clonal expansion the mass proliferation of antibody-producing cells by clonal selection.

clonal selection the theory that exposure to a specific antigen selectively stimulates the proliferation of the cell with the appropriate antibody to form numerous clones of these specific antibody-forming cells (clonal expansion).

clones the offspring produced as a result of cloning.

cloning a way of producing offspring by asexual reproduction.

closed circulatory system a circulatory system where the blood is enclosed in blood vessels and does not come into direct contact with the cells of the body beyond the blood vessels.

Clostridium difficile (C. difficile) a species of Gram positive bacteria that is resistant to most antibiotics.

codominance when different alleles of a gene are equally dominant and both are expressed in the phenotype.

codon a three-base sequence of DNA or RNA that codes for an amino acid.

coenzyme A coenzyme with important roles in the oxidation of pyruvate in Krebs cycle and in the synthesis and oxidation of fatty acids.

cofactors non-protein components necessary for the effective functioning of an enzyme.

cohesion-tension theory the best current model explaining the movement of water through a plant during transpiration.

collecting duct final part of the tubule that passes through the renal medulla and the place where hypertonic urine is produced if needed. The permeability of the walls is affected by ADH levels and it is the main site of water balancing.

communicable diseases diseases that can be passed from one organism to another, of the same or different species.

community all the populations of living organisms in a particular habitat.

companion cells the active cells found next to sieve tube elements that supply the phloem vessels with all of their metabolic needs.

competitive inhibitor an inhibitor that competes with substrate to bind to active site on an enzyme.

complementary base pairing specific hydrogen bonding between nucleic acid bases. Adenine (A) binds to thymine (T) or uracil (U) and cytosine (C) binds to guanine (G).

compound light microscope a light microscope which uses two lenses to magnify an object; the objective lens, which is placed near to the specimen and an eyepiece lens, through which the specimen is viewed.

computational biology the study of biology using computational techniques to analyse large amounts of data.

condensation reaction a reaction between two molecules resulting in the formation of a larger molecule and the release of a water molecule. The opposite reaction to a hydrolysis reaction.

conservation the maintenance of biodiversity.

consumer organism that obtains its energy by feeding on another organism.

continuous fermentation an industrial fermentation where culture broth is removed continuously and more nutrient medium added.

continuous variation a characteristic that can take any value within a range, e.g. height.

contrast staining or treating specific cell components so they are visible compared to untreated components.

convergent evolution organisms evolve similarities because the organisms adapt to similar environments or other selection pressures.

correlation coefficient statistical test used to consider the relationship between two sets of data.

cortex the dark outer layer of the kidney containing the Bowman's capsules and glomeruli.

countercurrent exchange system a system for exchanging materials or heat when the two different components flow in opposite directions past each other.

counterstain application of second stain with a contrasting colour to sample for microscopy.

cristae fold of inner mitochondrial membranes, increases the surface area where reactions of the electron transfer chain can take place.

crossing over see chiasmata.

culture growing living matter in vitro, for example, microorganisms in specifically prepared nutrient medium.

cyclic AMP cyclic adenosine monophosphate, a molecule that acts as an important second messenger in many biological systems.

cyclic photophosphorylation synthesis of ATP involving only photosystem I.

cytokines cell-signalling molecules produced by mast cells in damaged tissues that attract phagocytes to the site of infection or inflammation.

cytokinesis cell division stage in the mitotic phase of the cell cycle that results in the production of two identical daughter cells.

cytolysis the bursting of an animal cell caused by increasing hydrostatic pressure as water enters by osmosis.

cytoplasm internal fluid of cells, composed of cytosol (water, salts and organic molecules), organelles and cytoskeleton.

cytoskeleton a network of fibres in the cytoplasm of a eukaryotic cell.

deamination the removal of the amino group from amino acids.

decarboxylation removal of carbon dioxide.

deciduous plants plants that lose all of their leaves for part of the year.

decomposer organism that breaks down dead organisms releasing nutrients back into the ecosystem.

decomposition chemical reaction in which a compound is broken down into simpler compounds or into its constituent elements.

dehydrogenation the removal of a hydrogen atom.

deletion a mutations where one or more nucleotides are deleted and lost from the DNA strand.

denatured (denaturation) change in tertiary structure of a protein or enzyme, resulting in loss of normal function.

denitrification conversion of nitrates to nitrogen gas.

deoxyribonucleic acid (DNA) the molecule responsible for the storage of genetic information.

depolarisation a change in potential difference from negative to positive across the membrane of a neurone.

detoxification removal or breakdown of toxins.

detritivore organism which speeds up decay by breaking down detritus into smaller pieces.

diabetes mellitus medical condition which affects a person's ability to control their blood glucose concentration.

diastole the stage of the cardiac cycle in which the heart relaxes and the atria and then the ventricles fill with blood.

dicotyledonous plants (dicots) plants that produce seeds containing two cotyledons, which act as food stores for the developing embryo and form the first leaves when the seed germinates.

differential staining using specific stains to distinguish different types of cell.

differentiation the process of a cell becoming differentiated. Involves the selective expression of genes in a cell's genome.

digenic inheritance a characteristic controlled by two genes.

dihybrid inheritance A characteristic inherited on two genes.

diploid normal chromosome number; two chromosomes of each type – one inherited from each parent.

directional selection natural selection that favours one extreme phenotype.

disaccharide a molecule comprising two monosaccharides, joined together by a glycosidic bond.

discontinuous variation a characteristic that can only result in certain discrete values, for example, blood type.

disruptive selection natural selection that favours both extremes of a given phenotype.

distal convoluted tubule the second twisted section of the nephron where the permeability of the walls varies in response to ADH levels in the blood.

divergent evolution species diverge over time into two different species, resulting in a new species becoming less like the original one.

DNA helicase enzyme that catalyses the unwinding and separating of strands in DNA replication.

DNA polymerase enzyme that catalyses the formation of phosphodiester bonds between adjacent nucleotides in DNA replication.

DNA profiling producing an image of the patterns in the non-coding DNA of an individual.

DNA replication the semi-conservative process of the production of identical copies of DNA molecules.

DNA sequencing working out the sequence of bases in a strand of DNA.

dominant allele version of the gene that will always be expressed if present.

dominant species the most abundant species in an ecosystem.

double circulatory system a circulatory system where the blood travels twice through the heart for each complete circulation of the body. In the first circulation blood is pumped by the heart to the lungs. In the second circulation oxygenated blood is pumped by the heart to the brain and body to supply cells with oxygen.

ecological efficiency efficiency with which energy or biomass is transferred from one trophic level to the next.

ecosystem all the interacting living organisms and the non living conditions in an area.

ecotourism tourism directed towards natural environments, intended to support conservation efforts.

ectopic heartbeat extra heartbeats that are out of the normal rhythm.

ectotherms animals that use their surroundings to warm their bodies so their core temperature is heavily dependent on the environment.

effector muscle or gland which carries out body's response to a stimulus.

elastic recoil the ability to return to original shape and size following stretching. Particularly of the alveoli of the lungs and of the arteries.

electrocardiogram (ECG) a technique for measuring tiny changes in the electrical conductivity of the skin that result from the electrical activity of the heart. This produces a trace which can be used to analyse the health of the heart.

electron carriers proteins that accept and release electrons.

electron microscopy microscopy using a microscope that employs a beam of electrons to illuminate the specimen. As electrons have a much smaller wavelength than light they produce images with higher resolutions than light microscopes.

electrophoresis a type of chromatography that relies on the way charged particles move through a gel under the influence of an electric current. It is used to separate nucleic acid fragments or proteins.

electroporation the use of a very tiny electric current to transfer genetically engineered plasmids into bacteria or to get DNA fragments directly into eukaryotic cells.

emulsion test laboratory test for lipids using ethanol – a white emulsion indicates the presence of a lipid.

end-product inhibition the product of a reaction inhibits the enzyme required for the reaction.

endocrine glands group of specialised cells which secrete hormones.

endocytosis the bulk transport of materials into cells via invagination of the cell-surface membrane forming a vesicle.

endosymbiosis the widely-accepted theoretical process by which eukaryotic cells evolved from prokaryotic cells.

endothermic reactions that absorb energy.

endotherms animals that rely on their metabolic processes to warm their bodies and maintain their core temperature.

enucleated with the nucleus removed.

enzyme–product complex complex formed as a result of an enzyme-catalysed reaction, when a substrate is converted to a product or products while bound to the active site of an enzyme.

enzymes biological catalysts that interact with substrate molecules to facilitate chemical reactions. Usually globular proteins.

enzyme-substrate complex complex formed when a substrate is bound to the active site of an enzyme.

epidemic when a communicable disease spreads rapidly to a lot of people at a local or national level.

epigenetics external control of genetic regulation.

epistasis the effect of one gene on the expression of another gene.

euchromatin loosely packed DNA.

eukaryotes multicellular eukaryotic organisms like animals, plants and fungi and single-celled protoctista.

eukaryotic cells cells with a nucleus and other membrane-bound organelles.

ex situ **conservation** conservation methods out of the natural habitat.

exchange surfaces surfaces over which materials are exchanged from one area to another.

excretion the removal of the waste products of metabolism from the body.

exocytosis the bulk transport of materials out of cells. Vesicles containing the material fuse with the cell-surface membrane and the contents are released to the outside of the cell.

exoskeleton an external skeleton of some organisms, e.g. insects.

exothermic reactions that release energy.

expiratory reserve volume the extra amount of air that can be forced out of the lungs over and above the normal exhalation (tidal volume).

facilitated diffusion diffusion across a plasma membrane through protein channels.

facultative anaerobes organisms that can respire anaerobically or aerobically.

FAD coenzyme that acts as a hydrogen acceptor in Krebs cycle.

fatty acids long chain carboxylic acids used in the formation of triglycerides.

fermentation anaerobic respiration without the involvement of an electron transport chain.

fibrous proteins long, insoluble, structural proteins.

fluid-mosaic model model of the structure of a cell membrane in which phospholipids within the phospholipid bilayer are free to move and proteins of various shapes and sizes are embedded in various positions.

forensics the application of science to the law, commonly in solving crimes.

fossils the remains or impression of a prehistoric plant or animal preserved in rock.

founder effect when a few individuals of a species colonise a new area, their offspring initially experience a loss in genetic variation, and rare alleles can become much more common in the population.

fungi biological kingdom containing yeasts, moulds and mushrooms.

gametes haploid sex cells produced by meiosis in organisms that reproduce sexually.

gaseous exchange system the complex systems in which the respiratory gases oxygen and carbon dioxide are exchanged in an organism.

gene a section of DNA that contains the complete sequence of bases (codons) to code for a protein.

gene banks store of genetic material.

gene flow when alleles are transferred from one population to another by interbreeding.

gene pool sum total of all the genes in a population at a given time.

genetic bottleneck when large numbers of a population die prior to reproducing, leading to reduced genetic biodiversity within the population.

genetic code the sequences of bases in DNA are the 'instructions' for the sequences of amino acids in the production of proteins.

genetic drift random change of allele frequency.

genetic variation a variety of different combinations of alleles in a population.

genome all of the genetic material of an organism.

genotype genetic makeup of an organism.

geotropism the growth response of plants to gravity.

germ line cell gene therapy inserting a healthy allele into the germ cells or into a very early embryo.

gibberellins plant hormones that cause stem elongation, trigger the mobilisation of food stores in a seed at germination and stimulate pollen tube growth in fertilisation.

gills the gaseous exchange organs of fish, comprised of gill plates, gill filaments and gill lamellae.

globular proteins spherical, water-soluble proteins.

Glomerular filtration rate (GFR) a test used to estimate the volume of blood filtered by the glomeruli each minute, used to indicate a loss of function in the kidneys.

glucoamylase enzyme used to convert dextrins to glucose.

gluconeogenesis production of glucose from non-carbohydrate sources.

glucose a monosaccharide with the chemical formula $C_6H_{12}O_6$. One of the main products of photosynthesis in plants.

glucose isomerase enzyme used to produce fructose from glucose.

glycerate-3-phosphate compound formed in Calvin cycle after carbon fixation.

glycerol alcohol found in triglycerides.

glycogen a branched polysaccharide formed from alpha glucose molecules. A chemical energy store in animal cells.

glycogenesis production of glycogen from glucose.

glycogenolysis process in which glycogen stored in the liver and muscle cells is broken down into glucose.

glycolipids cell-surface membrane lipids with attached carbohydrate molecules of varying lengths and shapes.

glycoproteins extrinsic membrane proteins with attached carbohydrate molecules of varying lengths and shapes.

glycosidic bond a covalent bond between two monosaccharides.

goblet cells differentiated cells specialised to secrete mucus.

Golgi apparatus organelle in most eukaryotic cells formed from an interconnected network of flattened, membrane-enclosed sacs, or cisternae. Plays a role in modifying and packaging proteins into vesicles.

gram negative bacteria bacteria with cell walls that stain red with Gram stain.

gram positive bacteria bacteria with cell walls that stain purple-blue with Gram stain.

granum (plural grana) a structure inside chloroplasts composed of a stack of several thylakoids. Contains chlorophyll pigments, where light reactions occur during photosynthesis.

guard cells cells that can open and close the stomatal pores, controlling gaseous exchange and water loss in plants.

habitat biodiversity the number of different habitats found within an area.

haemoglobin the red, oxygen-carrying pigment of red blood cells.

haemoglobinic acid the compound formed when haemoglobin accepts free hydrogen ions in its role as a buffer in the blood.

haemolymph the transport medium or 'blood' in insects.

haploid half the normal chromosome number; one chromosome of each type.

Hardy-Weinberg equation formula used to calculate the frequency of alleles in a population.

hepatocytes liver cells.

herbivory the process of animals eating plants.

heterochromatin tightly packed DNA.

heterotrophic organisms that acquire nutrients by the ingestion of other organisms.

heterozygous two different alleles for a characteristic.

hexose bisphosphate the compound that results from the phosphorylation of glucose in glycolysis.

hexose monosaccharide a monosaccharide composed of six carbons.

high throughput sequencing new methods of sequencing DNA that are automated, very rapid and much cheaper than the original techniques.

histamines chemicals produced by mast cells in damaged tissues that make the blood vessels dilate (causing redness and heat) and the blood vessel walls leaky (causing swelling and pain).

histones proteins that form a complex with DNA called chromatin.

homeobox genes (Hox genes) genes responsible for the development of body plans.

homeodomain a conserved motif of 60 amino acids found in all homeobox proteins. It is the part of the protein that binds to DNA allowing the protein to act as a transcriptional regulator.

homeostasis the maintenance of a stable equilibrium in the conditions inside the body.

homologous chromosomes matching pair of chromosomes, one inherited from each parent.

homologous structure a structure which appears superficially different but has the same underlying structure.

homozygous two identical alleles for a characteristic.

hormone chemical messengers which travel around the body in the blood stream.

hox genes homeobox genes.

humus organic component of soil formed by the decomposition of leaves and other plant material by soil microorganisms.

hydrolysis reaction the breakdown of a molecule into two smaller molecules requiring the addition of a water molecule. The opposite reaction to a condensation reaction.

hydrophilic the physical property of a molecule that is attracted to water.

hydrophobic the physical property of a molecule that is repelled by water.

hydrophytes plants with adaptations that enable them to survive in very wet habitats or submerged or at the surface of water.

hydrostatic pressure the pressure created by water in an enclosed system.

hypothalamus the region of the brain above the pituitary gland that contains osmoreceptors involved in osmoregulation.

immobilised enzymes enzymes that are attached to an inert support system over which the substrate passes and is converted to product.

immune response a biological response that protects the body by recognising and responding to antigens and by destroying substances carrying non-self antigens.

immunoglobulins γ-shaped glycoproteins that form antibodies.

***in situ* conservation** conservation methods within the natural habitat.

inbreeding breeding between closely related organisms.

inbreeding depression reduced biological fitness due to inbreeding.

independent assortment the arrangement of each homologous chromosome pair (bivalent) in metaphase 1 and metaphase 2 of meiosis is independent of each other and results in genetic variation.

induced-fit hypothesis modified lock and key explanation for enzyme action; the active site of the enzyme is modified in shape by binding to the substrate.

inflammation biological response of vascular tissues to pathogens, damaged cells, or irritants, resulting in pain, heat, redness and swelling.

inhibitor a factor that prevents or reduces the rate of an enzyme-catalysed reaction.

inner mitochondrial membrane the inner most of the two mitochondrial membranes. Separates the mitochondrial matrix from the intermembrane space. Is the site where the electron transport chain takes place.

insertion a mutation where one or more extra nucleotides are inserted into a DNA strand.

inspiratory reserve volume the maximum volume of air that can be breathed in over and above a normal inhalation (tidal volume).

insulin a globular protein hormone involved in the regulation of blood glucose concentration.

intercostal muscles the muscles between the ribs that pull the ribs upwards during inhalation (internal intercostal muscles) and downwards during forced exhalation (external intercostal muscles).

interleukins a type of cytokine produced by T helper cells.

interphase growth period of the cell cycle, between cell divisions (mitotic phase). Consists of stages G_1, S and G_2.

interspecific competition competition between organisms of different species.

interspecific variation the differences between organisms of different species.

intraspecific competition competition between organisms of the same species.

intraspecific variation the differences between organisms of the same species.

introns regions of non-coding DNA or RNA.

iodine test a chemical test for the presence of starch using a potassium iodide solution. A colour change to purple/black indicates a positive result.

ion an atom or molecule with an overall electric charge because the total number of electrons is not equal to the total number of protons. See anion and cation.

ionic bond a chemical bond that involves the donating of an electron from one atom to another, forming positive and negative ions held together by the attraction of the opposite charges.

islets of Langerhans specialised cells within the pancreas responsible for producing insulin and glucagon.

keystone species species which are essential for maintaining biodiversity – they have a disproportionately large effect on their environment relative to their abundance.

kingdom the second biggest and broadest taxonomic group.

lac operon operon responsible for the metabolism of lactose.

lactase enzyme used to hydrolyse lactose to glucose and galactose to produce lactose-free milk.

lactose a disaccharide made up of a galactose and glucose monosaccharide.

lactate dehydrogenase enzyme used in the conversion of pyruvate to lactate.

lactate fermentation fermentation that results in the production of lactate.

lamellae membranous channels which join grana together in a chloroplast.

laser scanning confocal microscope a microscope that employs a laser beam and a pin-hole aperture to produce an image with a very high resolution.

light harvesting system a group of protein and chlorophyll molecules found in the thylakoid membranes of the chloroplasts in a plant cell.

light microscope an instrument that uses visible light and glass lenses to enable the user to see objects magnified many times.

limiting factor factor which limits the rate of a process.

line breeding form of inbreeding using less closely related organisms.

line transect a line is marked along the ground and samples are taken at specified points.

linked genes genes present on the same chromosome.

lipids non-polar macromolecules containing the elements carbon, hydrogen and oxygen. Commonly known as fats (solid at room temperature) and oils (liquid at room temperature).

loop of Henle a long loop of nephron that creates a steep concentration gradient across the medulla.

lung surfactant chemical mixture containing phospholipids and both hydrophilic and hydrophobic proteins, which coats the surfaces of the alveoli and prevents them collapsing after every breath.

lymph modified tissue fluid that is collected in the lymph system.

lymphocytes white blood cells that make up the specific immune system.

lysosomes specialised vesicles containing hydrolytic enzymes for the breakdown of waste materials within a cell.

macromolecules large complex molecules with a large molecular weight.

maltose two glucose molecules linked by a 1, 4 glycosidic bond.

mass transport system a transport system where substances are transported in a mass of fluid.

mature mRNA mRNA after the removal of introns and any other post-transcriptional changes.

medulla the lighter inner layer of the kidney made up of the loops of Henle.

meiosis form of cell division where the nucleus divides twice (meiosis I and meiosis II) resulting in a halving of the chromosome number and producing four haploid cells from one diploid cell.

membrane a selectively-permeable barrier surrounding all cells and forming compartments within eukaryotic cells.

membrane proteins protein components of cell-surface membranes.

meristematic tissue (meristems) tissue found at regions of growth in plants. Contains stem cells.

messenger (m)RNA short strand of RNA produced by transcription from the DNA template strand. It has a base sequence complementary to the DNA from which it is transcribed, except it has uracil (U) in place of thymine (T).

metaphase second stage of mitosis when chromosomes line up at the metaphase plate.

methylation addition of methyl group.

micropropagation the process of making very large numbers of genetically identical offspring from a single parent plant using tissue culture techniques.

mitochondrial DNA DNA present within the matrix of mitochondria.

mitochondrial matrix the part of the mitochondria enclosed by the inner mitochondrial membrane which contains enzymes for the Krebs cycle and the link reaction.

mitosis nuclear division stage in the mitotic phase of the cell cycle.

mitotic phase period of cell division of the cell cycle. Consists of the stages mitosis and cytokinesis.

monoclonal antibodies antibodies from a single clone of cells that are produced to target particular cells or chemicals in the body.

monoculture the cultivation of a single crop in a given area.

monogenic inheritance a characteristic inherited on a single gene.

monomers individual molecules that make up a polymer.

monosaccharide a single sugar molecule.

monozygotic twins twins formed from a single fertilized egg.

mRNA see messenger (m)RNA.

MRSA (methicillin-resistant *Staphylococcus aureus*) a mutated strain of the bacterium *Staphylococcus aureus* that is resistant to the antibiotic, methicillin.

mucous membranes membranous linings of body tracts that secrete a sticky mucus.

multiple alleles a gene with more than two possible alleles.

multipotent a stem cell that can only differentiate into a range of cell types within a certain type of tissue.

mutagens chemical or physical agent which causes mutation.

mutation a change in the genetic material which may affect the phenotype of the organism.

myelin sheath membrane rich in lipid which surrounds the axon of some neurones, speeding up impulse transmission.

myofibril long cylindrical organelles found in muscle which are made of protein and specialised for contraction.

myogenic muscle which has its own intrinsic rhythm.

NAD a coenzyme found in all living cells involved in cellular respiration.

NADP coenzyme which acts as final electron acceptor in photosynthesis.

natural active immunity immunity which results from the response of the body to the invasion of a pathogen.

natural passive immunity the immunity given to an infant mammal by the mother through the placenta and the colostrum.

natural selection the process by which organisms best suited to their environment survive and reproduce, passing on their characteristics to their offspring through their genes.

nephrons tubules that make up the main functional structures of the kidneys.

neurone specialised cell which transmits impulses in the form of action potentials.

neurotransmitter chemical involved in communication across a synapse between adjacent neurones or a neurone and muscle cell.

nitrification conversion of ammonium compounds into nitrites and nitrates.

nitrile hydratase enzyme used to convert acrylonitrile to acrylamide for use in the plastics industry.

nitrogen fixation conversion of nitrogen gas to ammonium compounds.

non-competitive inhibitor an inhibitor that binds to an enzyme at an allosteric site.

non-cyclic photophosphorylation the synthesis of ATP and reduced NADP involving photosystems I and II.

non-random sampling an alternative sampling method to random sampling, where the sample is not chosen at random. It can be opportunistic, stratified or systematic.

normal distribution a distribution of continuous data where the mean, median, and mode have the same value, there is symmetry around the mean with most data points being close to the mean and fewer data points further away from the mean. When plotted produces a bell-shaped or normal distribution curve.

nucleic acids large polymers formed from nucleotides. Contain the elements carbon, hydrogen, nitrogen , phosphorus, and oxygen.

nucleotides the monomers used to form nucleic acids. Made up of a pentose monosaccharide, a phosphate group and a nitrogenous base.

obligate aerobes organisms that can only respire aerobically.

obligate anaerobes organisms that cannot live in environments containing oxygen.

oncotic pressure the tendency of water to move into the blood by osmosis as a result of the plasma proteins.

open circulatory system a circulatory system with a heart but few vessels to contain the transport medium.

operculum the bony flap covering the gills of bony fish. Part of the mechanism that maintains a constant flow of water over the gas exchange surfaces.

operon group of genes expressed together.

opportunistic sampling sampling using the organisms that are conveniently available. The weakest form of sampling as it may not be representative of the population.

opsonins chemicals that bind to pathogens and tag them so they are recognised more easily by phagocytes, e.g. antibodies.

organelle membrane-bound compartments with varying functions inside eukaryotic cells.

ornithine cycle a series of enzyme controlled reactions in the liver converting ammonia formed by deamination of amino acids into urea.

osmoreceptors sensory receptors that respond to changes in the water potential of the blood.

osmoregulation the balancing and control of the water potential of the blood.

osmosis diffusion of water through a partially permeable membrane down a water potential gradient. A passive process.

outbreeding breeding of distantly related organisms.

outer mitochondrial membrane the membrane that separates the contents of the mitochondrion from the rest of the cell, creating a cellular compartment with ideal conditions for aerobic respiration.

oxaloacetate four carbon molecule present at the beginning of Krebs cycle that combines with acetyl coenzyme A to form citrate.

oxidation removal of electrons or hydrogen.

oxygen dissociation curve graph showing the relationship between oxygen and haemoglobin at different partial pressures of oxygen.

oxygenated blood blood that has passed through the gas exchange organs (e.g. lungs) and is high in oxygen.

pandemic when a communicable disease spreads rapidly to a lot of people across a number of countries.

partially permeable membrane that allows some substances to cross but not others.

passive transport transport that is a passive process (does not require energy) and does not use energy from cellular respiration.

pathogens microorganisms that cause disease.

pelvis the central chamber of the kidney where urine collects before passing out down the ureter.

penicillin the first widely used, safe antibiotic, derived from a mould, *Penicillium notatum*.

penicillin acylase enzymes used to make semi-synthetic penicillins from naturally produced penicillins.

pentose monosaccharide a monosaccharide composed of five carbons.

peptide bond bond formed between two amino acids.

peptides chains of two or more amino acid molecules.

peripheral nervous system (PNS) consists of all the neurones that connect the CNS to the rest of the body.

phagocytosis process by which white blood cells called phagocytes recognise non-self cells, engulf them digest them within a vesicle called a phagolysosome.

phagosome the vesicle in which a *pathogen* or damaged cell is engulfed by a phagocyte.

pharming the use of genetically modified animals to produce pharmaceuticals.

phenotype observable characteristics of an organism.

phloem plant transport tissue that carries the products of photosynthesis (assimilates) to all cells of the plant.

phosphodiester bonds covalent bonds formed between the phosphate group of one nucleotide and the hydroxyl (OH) group of another.

phospholipid bilayer arrangement of phospholipids found in cell membranes; the hydrophilic phosphate heads form both the inner and outer surface of a membrane, sandwiching the fatty acid tails to form a hydrophobic core.

phospholipids modified triglycerides, where one fatty acid has been replaced with a phosphate group.

phosphorylation the addition of phosphate group to a molecule.

photosynthesis synthesis of complex organic molecules using light.

photosytem protein complexes involved in the absorption of light and electron transfers in photosynthesis.

phototropism the growth response of plants to unilateral light.

phylogeny the evolutionary relationships between organisms.

pigment molecules that absorb specific wavelengths of light.

pinocytosis endocytosis of liquid materials.

pioneer species the first organisms to colonise an area.

plagioclimax stage in succession where artificial or natural factors prevent the natural climax community from forming.

plasma the main component of blood, a yellow fluid containing many dissolved substances and carrying the blood cells.

plasma cells B lymphocytes that produce about 2000 antibodies to a particular antigen every second and release them into the circulation.

plasma membrane all the membranes of cells, which have the same basic structure described by the fluid-mosaic model.

pluripotent a stem cell that can differentiate into any type of cell, but not form a whole organism.

polymerase chain reaction a process by which a small sample of DNA can be amplified using specific enzymes and temperature changes.

polymers long-chain molecules composed of linked (bonded) multiple individual molecules (monomers) in a repeating pattern.

polymorphic allele a gene with more than two possible alleles.

polypeptide chains of three or more amino acids.

polysaccharide a polymer made up of many sugar monomers (monosaccharides).

posterior pituitary gland the posterior part of the pituitary gland in the brain where ADH is stored ready for release into the blood.

predation the capturing of prey in order to sustain life.

preservation protection of an area by restricting or banning human use – so that the ecosystem is kept exactly as it is.

primary immune response the relatively slow production of a small number of the correct antibodies the first time a pathogen is encountered.

primary or pre mRNA the mRNA transcribed from the DNA before any post-transcriptional regulation to remove introns etc.

producer organism that converts light energy into chemical energy.

prokaryotes single-celled prokaryotic organisms from the kingdom Prokaryotae.

prokaryotic cells cells with no membrane-bound nucleus or organelles.

prophase first stage of mitosis when chromatin condenses to form visible chromosomes and the nuclear envelope breaks down.

prosthetic group non-protein component of a conjugated protein.

proteases enzymes that catalyse the breakdown of proteins and peptides into amino acids.

proteins one or more polypeptides arranged as a complex macromolecule.

protista biological kingdom containing unicellular eukaryotes.

proximal convoluted tubule the first twisted section of the nephron after the Bowman's capsule where many substances are reabsorbed into the blood.

purines double-ringed, nitrogenous bases that form part of a nucleotide.

purkyne fibres tissue that conducts the wave of excitation to the apex of the heart.

pyrimidines single-ringed, nitrogenous bases that form part of a nucleotide.

pyruvate the three carbon product of glycolysis that feeds into Krebs cycle in the presence of oxygen.

quaternary structure the association of two or more protein subunits.

random sampling sampling where each individual in the population has an equal likelihood of selection.

receptors extrinsic glycoproteins that bind chemical signals, triggering a response by the cell.

recessive allele version of a gene that will only be expressed if two copies of this allele are present in an organism.

recombinant new combination of alleles / DNA from two sources.

recombinant chromatids chromatids with a combination of DNA from both homologous chromosomes, formed by crossing over and chiasmata in meiosis.

recombination frequency proportion of recombinant offspring resulting from a cross.

reducing sugars saccharides (sugars) that donate electrons resulting in the reduction (gain of electrons) of another molecule.

reduction division cell division resulting in the production of haploid cells from a diploid cell; meiosis.

reflex action involuntary response to a sensory stimulus.

regulatory gene a gene that codes for proteins involved with DNA regulation.

renal dialysis a process where the functions of the kidney are carried out artificially to maintain the salt and water balance of the blood.

repolarisation a change in potential difference from positive back to negative across the membrane of a neurone.

repressor protein protein that binds to operator affecting the rate of transcription.

respiration breakdown of complex organic molecules linked to the synthesis of ATP.

respiratory quotient ratio of carbon dioxide produced to oxygen used in respiration.

respiratory substrates organic molecules broken down in respiration.

residual volume the volume of air that is left in the lungs after forced exhalation. It cannot be measured directly.

resolution the shortest distance between two objects that are still seen as separate objects.

response the way a body reacts to a stimulus.

resting potential the potential difference across the membrane of the axon of a neurone at rest (normally about −65mV).

restriction endonucleases enzymes that chop a strand of DNA into small pieces.

R-groups variable groups on amino acids.

ribonucleic acid (RNA) Polynucleotide molecules involved in the copying and transfer of genetic information from DNA. The monomers are nucleotides consisting of a ribose sugar and one of four bases; uracil (U), cytosine (C), adenine (A), and guanine (G).

ribose the pentose monosaccharide present in RNA molecules.

ribosomal (r)RNA form of RNA that makes up the ribosome.

ribulose bisphosphate five carbon molecule at beginning of Calvin cycle.

ribulose bisphosphate carboxylase (RuBisCo) key enzyme involved in the first step of carbon fixation in photosynthesis.

RNA polymerase enzyme that catalyses the formation of phosphodiester bonds between adjacent RNA nucleotides.

root hair cells cells found just behind the growing tip of a plant root that have long hair-like extensions that greatly increase the surface area available for the absorption of water and minerals from the soil.

root pressure the active pumping of minerals into the xylem by root cells that produces a movement of water into the xylem by osmosis.

saprophytic/saprotrophic organisms that acquire nutrients by absorption – mainly of decaying material.

sarcomere functional unit of myofibril.

scanning electron microscopy (SEM) an electron microscope in which a beam of electrons is sent across the surface of a specimen and the reflected electrons are focused to produce a three-dimensional image of the specimen surface.

secondary immune response the relatively fast production of very large quantities of the correct antibodies the second time a pathogen is encountered as a result of immunological memory – the second stage of a specific immune response.

seed bank a store of genetic material from plants in the form of seeds.

selection pressure factors that affect an organism's chance of survival or reproductive success.

selective breeding selection of individuals for breeding with desirable characteristics.

selective toxicity the ability to interfere with the metabolism of a pathogen without affecting the cells of the host.

selective reabsorption the reabsorption of selected substances ie those needed by the body in the kidney tubules.

selectively permeable plasma membrane with protein channels that allow specific substances to cross only.

semi-conservative replication DNA replication results in one old strand and one new strand present in each daughter DNA molecule.

sense strand the strand of DNA that runs 5' to 3' and contains the genetic code for a protein.

sensory receptor specialised cell which detects a stimulus.

seral stages the steps in succession.

sex linked genes genes carried on the sex chromosomes.

sliding filament model movement of actin and myosin filaments in relation to each other to cause contraction.

sieve plates areas between the cells of the phloem where the walls become perforated giving many gaps and a sieve-like appearance that allows the phloem contents to flow through.

sieve tube elements the main cells of the phloem that have a greatly reduced living content and sieve plates between the cells.

Simpson's Index of Diversity (D) a measure of biodiversity between 0 and 1 that takes into account both species richness and species evenness.

single circulatory system a circulatory system where the blood flows through the heart and is pumped out to travel all around the body before returning to the heart.

sinks (in plants) regions of a plant that require assimilates to supply their metabolic needs, e.g. roots, fruits.

sino-atrial node (SAN) region of the heart that initiates a wave of excitation that triggers the contraction of the heart.

somatic cell gene therapy replacing a faulty gene with a healthy allele in affected somatic cells.

somatic cell nuclear transfer a method of producing a clone from an adult animal by transferring the nucleus from an adult cell to an enucleated egg cell and stimulating development.

somatic nervous system part of the nervous system that is under conscious control.

smooth endoplasmic reticulum endoplasmic reticulum lacking ribosomes; the site of lipid and carbohydrate synthesis, and storage.

sources (in plants) regions of a plant that produce assimilates (e.g. glucose) by photosynthesis or from storage materials, e.g. leaves, storage organs.

Spearman's rank correlation coefficient a specific type of correlation test that compares the ranked orders of two datasets in order to consider their relationships.

specialised having particular structure to serve a specific function.

speciation the formation of new species.

species the smallest and most specific taxonomic group.

specific immunity also known as active immunity or acquired immunity – the immune system 'remembers' an antigen after an initial response leading to an enhanced response to subsequent encounters.

spiracles small openings along the thorax and abdomen of an insect that open and close to control the amount of air moving in and out of the gas exchange system and the level of water loss from the exchange surfaces.

stabilising selection natural selection that favours average phenotypes.

stage graticule a slide with a scale in micrometres (µm) etched into it. Used to measure the size of a sample under a light microscope.

stains (staining) dyes used in microscopy sample preparation to increase contrast or identify specific components.

starch a polysaccharide formed from alpha glucose molecules either joined to form amylose or amylopectin.

stem cells undifferentiated cells with the potential to differentiate into a variety of the specialised cell types of the organism.

stimulus detectable change in external or internal environment of an organism.

stomata pores in the surface of a leaf or stem that may be opened and closed by guard cells.

stratified sampling sampling where populations are divided into sub-groups (strata) based on a particular characteristic. A random sample is then taken from each of these strata proportional to its size.

stroma fluid interior of chloroplasts.

structural genes genes that code for structural proteins or enzymes not involved in DNA regulation.

Student's *t* test statistical test used to compare the means of data values of two populations.

substrate a substance used, or acted on, by another process or substance. For example a reactant in an enzyme-catalysed reaction.

substitution a mutation where one or more nucleotides are substituted for another in a DNA strand.

substrate level phosphorylation synthesis of ATP by transfer of phosphate group from another molecule.

succession the progressive replacement of one dominant type of species or community by another in an ecosystem, until a stable climax community is established.

sucrose a disaccharide made up of a fructose and glucose monosaccharides.

summation build up of neurotransmitter in a synapse to sufficient levels to trigger an action potential.

sustainable development economic development that meets the needs of people today, without limiting the ability of future generations to meet their needs.

sustainable resource a renewable resource which is being economically exploited in such a way that it will not diminish or run out.

sympatric speciation speciation that occurs when there is no physical barrier between populations.

symplast the continuous cytoplasm of living plant cells connected through the plasmodesmata.

symplast route movement of water and solutes through the cytoplasm of the cells via plasmodesmata by diffusion (passive).

synapse the junction (small gap) between two neurones, or a neurone and an effector.

synthetic biology the design and construction of novel biological pathways, organisms or devices, or the redesign of existing natural biological systems.

systematic sampling different areas of a habitat are identified and sampled separately. Often carried out using a line or belt transect.

systole the stage of the cardiac cycle in which the atria contract, followed by the ventricles, forcing blood out of the right side of the heart to the lungs and the left side of the heart to the body.

T helper cells T lymphocytes with CD4 receptors on their cell-surface membranes, which bind to antigens on antigen-presenting cells and produce interleukins, a type of cytokine.

T killer cells T lymphocytes that destroy pathogens carrying a specific antigen with perforin.

T lymphocytes lymphocytes which mature in the thymus gland and that both stimulate the B lymphocytes and directly kill pathogens.

T memory cells T lymphocytes that live a long time and are part of the immunological memory.

T regulator cells T lymphocytes that suppress and control the immune system, stopping the response once a pathogen has been destroyed and preventing an autoimmune response.

tachycardia a fast heart rhythm of over 100 beats per minute at rest.

tannins bitter tasting chemicals produced to prevent animals eating plant leaves; toxic to many insects.

target cells specific cells which hormones act on.

taxonomic group the hierarchical groups of classification – domain, kingdom, phylum, class, order, family, genus, species.

telophase fourth stage of mitosis when chromosomes assemble at the poles and the nuclear envelope reforms.

temperature coefficient (Q_{10}) a measure of how much the rate of a reaction increases with a 10 °C temperature increase.

template strand the antisense strand of DNA that acts as template during transcription so that the complementary RNA strand formed carries the same code for a protein as the DNA sense strand.

terpenoids chemicals found in plant leaves that may act as toxins to insects or fungi attacking the leaves.

tertiary structure further folding of the secondary structure of proteins involving interactions between R-groups.

test cross a cross used to determine genotype, involving a backcross with a homozygous recessive parent.

thermoregulation the maintenance of a relatively constant core temperature.

thin layer chromatography a technique for separating different pigments through the rate at which they move across an inert surface carried by a solvent.

thylakoid series of membranous compartments in a chloroplast that contain chlorophyll and molecules needed for light-dependent reaction.

tidal volume the volume of air which moves into and out of the lungs with each resting breath.

tissue a collection of differentiated cells that have a specialised function or functions in an organism.

tissue fluid the solution surrounding the cells of multicellular animals.

tonoplast membrane forming a vacuole in a plant cell.

total lung capacity the sum of the vital capacity and the residual volume.

totipotent a stem cell that can differentiate into any type of cell and form a whole organism.

trachea the main airway, supported by incomplete rings of cartilage, which carries warm moist air down from the nasal cavity into the chest.

tracheal fluid fluid found at the ends of the tracheoles in insects that helps control the surface area available for gas exchange and water loss.

transcription the process of copying sections of DNA base sequence to produce smaller molecules of mRNA, which can be transported out of the nucleus via the nuclear pores to the site of protein synthesis.

transcription factors proteins that affect the rate of transcription.

transfer (t)RNA form of RNA that carries an amino acid specific to its anticodon to the correct position along mRNA during translation.

translation the process by which the complementary code carried by mRNA is decoded by tRNA into a sequence of amino acids. This occurs at a ribosome.

translocation the movement of organic solutes around a plant in the phloem.

transmission electron microscopy (TEM) an electron microscope in which a beam of electrons is transmitted through a specimen and focused to produce an image.

transpiration the loss of water vapour from the stems and leaves of a plant as a result of evaporation from cell surfaces inside the leaf and diffusion down a concentration gradient out through the stomata.

transpiration stream the movement of water through a plant from the roots until it is lost by evaporation from the leaves.

transport system the system that transports required substances around the body of an organism.

triglyceride a lipid composed of one glycerol molecule and three fatty acids.

triose phosphate a molecule that is an intermediate in both photosynthesis and respiration and acts as a starting material for the synthesis of carbohydrates, lipids and amino acids.

triplet code the genetic code is a sequence of three nucleic acid bases, called a codon. Each codon codes for one amino acid.

trophic level stage in a food chain.

tropism a growth response by a plant in response to a unidirectional stimulus.

turgor the pressure exerted by the cell-surface membrane against the cell wall in a plant cell.

ultrafiltration the process by which blood plasma is filtered through the walls of the Bowman's capsule under pressure.

ultrastructure the ultrastructure of a cell is those features which can be seen by using an electron microscope.

undifferentiated an unspecialised cell originating from mitosis or meiosis.

urea nitrogenous waste produced from the deamination of excess amino acids in the liver.

ureters tubes carrying urine from the kidneys to the bladder.

urethra tube carrying urine from the bladder to the outside of the body.

vaccine a safe form of an antigen, which is injected into the bloodstream to provide artificial active immunity against a pathogen bearing the antigen.

vascular bundle the vascular system of herbaceous dicots, made up of xylem and phloem tissue.

vascular system a system of transport vessels in animals or plants.

vector a living or non-living factor that transmits a pathogen from one organism to another, e.g. malaria mosquito.

vector (in genetic modification) a means of inserting DNA from one organisms into the cells of another organism.

vegetative propagation the artificial production of natural clones for use in horticulture and agriculture.

ventilation rate is the total volume of air inhaled in one minute. Ventilation rate = tidal volume × breathing rate (per minute).

vesicles membranous sacs used to transport materials in the cell.

vital capacity volume of air that can be breathed in when the strongest possible exhalation is followed by the deepest possible intake of breath.

V_{max} maximum initial velocity or rate of an enzyme-catalysed reaction.

water potential (Ψ) measure of the quantity of water compared to solutes, measured as the pressure created by the water molecules in kilopascals (kPa).

wild type the allele that codes for the most common phenotype in a natural population.

xerophytes plants with adaptations that enable them to survive in dry habitats or habitats where water is in short supply in the environment.

xylem plant transport tissue that carries water and minerals from the roots to the other parts of the plant as a result of physical forces.

zygote the initial diploid cell formed when two gametes are joined by means of sexual reproduction. Earliest stage of embryonic development.

Answers

Download all answers to end-of-topic questions
for free from our website at:

www.oxfordsecondary.co.uk/ocrbiologyanswers.

Alternately you can scan our unique QR code
which will take you straight to the answers page.

Index

Appendix (Statistics data tables)

▼ *Table of values of* t

Degree of freedom (df)	p values			
	0.10	0.05	0.01	0.001
1	6.31	12.71	63.66	636.60
2	2.92	4.30	9.92	31.60
3	2.35	3.18	5.84	12.92
4	2.13	2.78	4.60	8.61
5	2.02	2.57	4.03	6.87
6	1.94	2.45	3.71	5.96
7	1.89	2.36	3.50	5.41
8	1.86	2.31	3.36	5.04
9	1.83	2.26	3.25	4.78
10	1.81	2.23	3.17	4.59
12	1.78	2.18	3.05	4.32
14	1.76	2.15	2.98	4.14
16	1.75	2.12	2.92	4.02
18	1.73	2.10	2.88	3.92
20	1.72	2.09	2.85	3.85
α	1.64	1.96	2.58	3.29

▼ *Critical values for Spearman's rank correlation coefficient,* r_s

	$p = 0.1$	$p = 0.05$	$p = 0.02$	$p = 0.01$			$p = 0.1$	$p = 0.05$	$p = 0.02$	$p = 0.01$
	5%	$2\frac{1}{2}$%	1%	$\frac{1}{2}$%	1-Tail Test		5%	$2\frac{1}{2}$%	1%	$\frac{1}{2}$%
	10%	5%	2%	1%	2-Tail Test		10%	5%	2%	1%
n						n				
1	–	–	–	–		31	0.3012	0.3560	0.4185	0.4593
2	–	–	–	–		32	0.2962	0.3504	0.4117	0.4523
3	–	–	–	–		33	0.2914	0.3449	0.4054	0.4455
4	1.0000	–	–	–		34	0.2871	0.3396	0.3995	0.4390
5	0.9000	1.0000	1.0000	–		35	0.2829	0.3347	0.3936	0.4328
6	0.8286	0.8857	0.9429	1.0000		36	0.2788	0.3300	0.3882	0.4268
7	0.7143	0.7857	0.8929	0.9286		37	0.2748	0.3253	0.3829	0.4211
8	0.6429	0.7381	0.8333	0.8810		38	0.2710	0.3209	0.3778	0.4155
9	0.6000	0.7000	0.7833	0.8333		39	0.2674	0.3168	0.3729	0.4103
10	0.5636	0.6485	0.7455	0.7939		40	0.2640	0.3128	0.3681	0.4051
11	0.5364	0.6182	0.7091	0.7545		41	0.2606	0.3087	0.3636	0.4002
12	0.5035	0.5874	0.6783	0.7273		42	0.2574	0.3051	0.3594	0.3955
13	0.4835	0.5604	0.6484	0.7033		43	0.2543	0.3014	0.3550	0.3908
14	0.4637	0.5385	0.6264	0.6791		44	0.2513	0.2978	0.3511	0.3865
15	0.4464	0.5214	0.6036	0.6536		45	0.2484	0.2945	0.3470	0.3822
16	0.4294	0.5029	0.5824	0.6353		46	0.2456	0.2913	0.3433	0.3781
17	0.4142	0.4877	0.5662	0.6176		47	0.2429	0.2880	0.3396	0.3741
18	0.4014	0.4716	0.5501	0.5996		48	0.2403	0.2850	0.3361	0.3702
19	0.3912	0.4596	0.5351	0.5842		49	0.2378	0.2820	0.3326	0.3664
20	0.3805	0.4466	0.5218	0.5699		50	0.2353	0.2791	0.3293	0.3628

▼ *continued*

	p = 0.1	p = 0.05	p = 0.02	p = 0.01
21	0.3701	0.4364	0.5091	0.5558
22	0.3608	0.4252	0.4975	0.5438
23	0.3528	0.4160	0.4862	0.5316
24	0.3443	0.4070	0.4757	0.5209
25	0.3369	0.3977	0.4662	0.5108
26	0.3306	0.3901	0.4571	0.5009
27	0.3242	0.3828	0.4487	0.4915
28	0.3180	0.3755	0.4401	0.4828
29	0.3118	0.3685	0.4325	0.4749
30	0.3063	0.3624	0.4251	0.4670

	p = 0.1	p = 0.05	p = 0.02	p = 0.01
51	0.2329	0.2764	0.3260	0.3592
52	0.2307	0.2736	0.3228	0.3558
53	0.2284	0.2710	0.3198	0.3524
54	0.2262	0.2685	0.3168	0.3492
55	0.2242	0.2659	0.3139	0.3460
56	0.2221	0.2636	0.3111	0.3429
57	0.2201	0.2612	0.3083	0.3400
58	0.2181	0.2589	0.3057	0.3370
59	0.2162	0.2567	0.3030	0.3342
60	0.2144	0.2545	0.3005	0.3314

▼ *Table of values of chi-squared*

df	p values								df
	0.99	0.95	0.90	0.50	0.10	0.05	0.01	0.001	
1	0.0001	0.0039	0.016	0.46	2.71	3.84	6.63	10.83	1
2	0.02	0.10	0.21	1.39	4.60	5.99	9.21	13.82	2
3	0.12	0.35	0.58	2.37	6.25	7.81	11.34	16.27	3
4	0.30	0.71	1.06	3.36	7.78	9.49	13.28	18.46	4
5	0.55	1.14	1.61	4.35	9.24	11.07	15.09	20.52	5
6	0.87	1.64	2.20	5.35	10.64	12.59	16.81	22.46	6
7	1.24	2.17	2.83	6.35	12.02	14.07	18.48	24.32	7
8	1.65	2.73	3.49	7.34	13.36	15.51	20.09	26.12	8
9	2.09	3.32	4.17	8.34	14.68	16.92	21.67	27.88	9
10	2.56	3.94	4.86	9.34	15.99	18.31	23.21	29.59	10
11	3.05	4.58	5.58	10.34	17.28	19.68	24.72	31.26	11
12	3.57	5.23	6.30	11.34	18.55	21.03	26.22	32.91	12
13	4.11	5.89	7.04	12.34	19.81	22.36	27.69	34.53	13
14	4.66	6.57	7.79	13.34	21.06	23.68	29.14	36.12	14
15	5.23	7.26	8.55	14.34	22.31	25.00	30.58	37.70	15
16	5.81	7.96	9.31	15.34	23.54	26.30	32.00	39.29	16
17	6.41	8.67	10.08	16.34	24.77	27.59	33.41	40.75	17
18	7.02	9.39	10.86	17.34	25.99	28.87	34.80	42.31	18
19	7.63	10.12	11.65	18.34	27.20	30.14	36.19	43.82	19
20	8.26	10.85	12.44	19.34	28.41	31.41	37.57	45.32	20
21	8.90	11.59	13.24	20.34	29.62	32.67	38.93	46.80	21
22	9.54	12.34	14.04	21.34	30.81	33.92	40.29	48.27	22
23	10.20	13.09	14.85	22.34	32.01	35.17	41.64	49.73	23
24	10.86	13.85	15.66	23.34	33.20	36.42	42.98	51.18	24
25	11.52	14.61	16.47	24.34	34.38	37.65	44.31	52.62	25
26	12.20	15.38	17.29	25.34	35.56	38.88	45.64	54.05	26
27	12.88	16.15	18.11	26.34	36.74	40.11	46.96	55.48	27
28	13.56	16.93	18.94	27.34	37.92	41.34	48.28	56.89	28
29	14.26	17.71	19.77	28.34	39.09	42.56	49.59	58.30	29
30	14.95	18.49	20.60	29.34	40.26	43.77	50.89	59.70	30
40	22.16	26.51	29.05	39.34	51.81	55.76	63.69	73.40	40
60	37.48	43.19	46.46	59.33	74.40	79.08	88.38	99.61	60
80	53.54	60.39	64.28	79.33	96.58	101.88	112.33	124.84	80
100	70.06	77.93	82.36	99.33	118.50	124.34	135.81	149.45	100

Acknowledgements

COVER: Valentina Razumova / Shutterstock **p2-3** & **4-5**: Science Photo/Shutterstock; **p6-7**: Sashkin/Shutterstock; **p10**(L): Dr. Jeremy Burgess/Science Photo Library; **p10**(R): Dr. Jeremy Burgess/Science Photo Library; **p12**(T): Biophoto Associates/Science Photo Library; **p12**(B): CNRI/Science Photo Library; **p18**: Biophoto Associates/Science Photo Library; **p12**(C): Dr. Frederick Skvara, Visuals Unlimited/Science Photo Library; **p14**: John Burbidge/Science Photo Library; **p16**: Victor Shahin, Prof. Dr. H. Oberleithner, University Hospital of Muenster/Science Photo Library; **p19**(T): Eye of Science/Science Photo Library; **p19**(B): David Scharf/Science Photo Library; **p20**: Marilyn Schaller/Science Photo Library; **p22**: Heiti Paves/Science Photo Library; **p23**: Victor Shahin, Prof. Dr. H.Oberleithner, University Hospital Of Muenster/Science Photo Library; **p24**(L): IBM and Nature Chemistry; **p24**(R): Peter Eaton-Requimte/University of Porto, Visuals Unlimited /Science Photo Library; **p25**(L): Alfred Pasieka/Science Photo Library; **p25**(R): National Cancer Institute/Science Photo Library; **p27**: Biophoto Associates/Science Photo Library; **p28**(T): Alfred Pasieka/Science Photo Library; **p28**(B): Dr. Gopal Murti/Science Photo Library; **p30**(B): Biomedical Imaging Unit, Southampton General Hospital/Science Photo Library; **p33**: Marilyn Schaller/Science Photo Library; **p34**(L): Dr. Kari Lounatmaa/Science Photo Library; **p30**(T): Dr. Gopal Murti/Science Photo Library; **p34**(R): Dr. David Furness, Keele University/Science Photo Library; **p31**(T): Steve Gschmeissner/Science Photo Library; **p31**(B): Biophoto Associates/Science Photo Library; **p37**: Biology Media/Science Photo Library; **p43**: Science Photo Library; **p45**: Hermann Eisenbess/Science Photo Library; **p51**: Martyn F. Chillmaid/Science Photo Library; **p52**: Martyn F. Chillmaid/Science Photo Library; **p64**(T): lculig/Shutterstock; **p64**(B): Indigo Molecular Images/Science Photo Library; **p65**: Steve Gschmeissner/Science Photo Library; **p71**: Philippe Psaila/Science Photo Library; **p89**: Vasiliy Koval/Shutterstock; **p97**: Laguna Design/Science Photo Library; **p115**: Dr. Stanley Flegler/Science Photo Library; **p124**: Power and Syred/Science Photo Library; **p125**(T): Steve Gschmeissner/Science Photo Library; **p125**(B): Steve Gschmeissner/Science Photo Library; **p126**(T): Steve Gschmeissner/Science Photo Library; **p126**(B): Steve Gschmeissner/Science Photo Library; **p127**(T): Dr. Gopal Murti/Science Photo Library; **p127**(B): Steve Gschmeissner/Science Photo Library; **p129**: Ed Reschke/Photolibrary/Getty Images; **p130**(T): Ed Reschke/Photolibrary/Getty Images; **p130**(M): Ed Reschke/Photolibrary/Getty Images; **p130**(B): Ed Reschke/Photolibrary/Getty Images; **p131**(T): Ed Reschke/Photolibrary/Getty Images; **p131**(CT): Ed Reschke/Photolibrary/Getty Images; **p131**(CB): Ed Reschke/Photolibrary/Getty Images; **p131**(B): Ed Reschke/Photolibrary/Getty Images; **p135**(T): Herve Conge, ISM/Science Photo Library; **p135**(B): Eric Grave/Science Photo Library; **p136**(T): Marek Mis/Science Photo Library; **p136**(C): Dr. Keith Wheeler/Science Photo Library; **p136**(B): Dr. Keith Wheeler/Science Photo Library; **p137**: Eye of Science/Science Photo Library; **p138**: Science Photo Library; **p139**(T): Biophoto Associates/Science Photo Library; **p139**(B): Steve Gschmeissner/Science Photo Library; **p140**: Dr. John Runions/Science Photo Library; **p158**: Four Oaks/Shutterstock; **p160**: Dr. Fred Hossler, Visuals Unlimited/Science Photo Library; **p158**: Dr. Fred Hossler, Visuals Unlimited/Corbis; **p159**: Biophoto Associates/Science Photo Library; **p158**: Vivid Pixels/Shutterstock; **p159**: Image Point Fr/Shutterstock; **p161**: Martin Dohrn/Science Photo Library; **p174**: Anthony Short; **p174**: Steve Gschmeissner/Science Photo Library; **p179**: Ed Reschke/Getty Images; **p176**: Biophoto Associates/Science Photo Library; **p182**: Biophoto Associates/Science Photo Library; **p195**: Photographee.eu/Shutterstock; **p198**: Anthony Short; **p199**(T): Dr. Keith Wheeler/Science Photo Library; **p199**(C): Stan Elems/Visuals Unlimited/Corbis; **p199**(B): Jubal Harshaw/Shutterstock; **p201**: Anthony Short; **p204**: David Cavagnaro/Visuals Unlimited/Science Photo Library; **p209**(L): D. Fischer; **p209**(R): D. Fischer; **p209**(L): D. Fischer; **p214**: Anthony Short; **p211**(T): Power and Syred/Science Photo Library; **p211**(C): Anthony Short; **p211**(B): Anthony Short; **p216**(T): Anthony Short; **p216**(B): Anthony Short; **p217**(TL): Martin Fowler/Shutterstock; **p217**(TR): Dr. Keith Wheeler/Science Photo Library; **p213**(B): Digital Vision; **p218**: Dr. Keith Wheeler/Science Photo Library; **p232**(L): Chantelle Bosch/Shutterstock; **p232**(C): Lenkadan/Shutterstock; **p224**(R): DragoNika/Shutterstock; **p233**(L): Howard Marsh/Shutterstock; **p225**(C): Stephen Rees/Shutterstock; **p233**(R): Olha Insight/Shutterstock; **p234**(T): Beata Aldridge/Shutterstock; **p234**(B): Kamonrat/Shutterstock; **p235**: Scimat/Science Photo Library; **p236**(T): Lebendkulturen.de/Shutterstock; **p236**(C): Power and Syred/Science Photo Library; **p236**(B): Liza1979/Shutterstock; **p238**: Jim David/Shutterstock; **p241**: Gary Hincks/Science Photo Library; **p242**(T): Nicku/Shutterstock; **p242**(B): Dr. Jeremy Burgess/Science Photo Library; **p243**: Vladimir Sazonov/Shutterstock; **p244**: Sinclair Stammers/Science Photo Library; **p245**: Paul D. Stewart/Science Photo Library; **p246**: Eric Isselee/Shutterstock; **p247**(CL): Bernadette Heath/Shutterstock; **p247**(C): Grigoriy Pil/Shutterstock; **p247**(CR): Bildagentur Zoonar GmbH/Shutterstock; **p247**(TR): Jeanne White/

Science Photo Library; **p247**(BR): Dr. Keith Wheeler/Science Photo Library; **p248**: Jeff Foott/Getty Images; **p249** (L): Debu55y/Shutterstock; **p249**(R): Diana Taliun/Shutterstock; **p252**: TAGSTOCK1/Shutterstock; **p253**: Madlen/Shutterstock; **p255**: Dr. Linda Stannard, UCT/Science Photo Library; **p259**(T): Krzysztof Wiktor/Shutterstock; **p259**(C): Matt Jeppson/Shutterstock; **p259**(B): Matt Jeppson/Shutterstock; **p260**(T): Dr. Keith Wheeler/Science Photo Library; **p260**(B): Frans Lanting/Mint Images/Science Photo Library; **p253**(T): Diccon Alexander/Science Photo Library; **p261**(C): Eric Isselee/Shutterstock; **p261**(B): BMJ/Shutterstock; **p262**(T): Michael & Patricia Fogden/Corbis; **p262**(BL): Grimplet/Shutterstock; **p262**(BR): Adrian Thomas/Science Photo Library; **p263**(T): Nelik/Shutterstock; **p263**(C): Karel Gallas/Shutterstock; **p263**(B): Adam Van Spronsen/Shutterstock; **p265**(T): Michael W. Tweedie/Science Photo Library; **p265**(B): Michael W. Tweedie/Science Photo Library; **p266**: Louise Murray/Science Photo Library; **p267**: GJones Creative/Shutterstock; **p270**(T): Vlad61/Shutterstock; **p270**(B): Pchais/Shutterstock; **p271**(T): Anneka/Shutterstock; **p271**(B): Marietjie/Shutterstock; **p272**: Public Health England/Science Photo Library; **p273**: Martyn F. Chillmaid/Science Photo Library; **p274**(T): Martyn F. Chillmaid/Science Photo Library; **p274**(B): Nigel Cattlin/Science Photo Library; **p276**: Chris Johnson/Alamy; **p280**: Anthony Short; **p281**(T): Sinclair Stammers/Science Photo Library; **p281**(B): Mary Terriberry/Shutterstock; **p286**: Frans Lanting/Mint Images/Science Photo Library; **p287**(T): Aseph/Shutterstock; **p287**(B): Alan Bryant/Shutterstock; **p288**: Art Wolfe/Mint Images/Science Photo Library; **p289**(T): Agap/Shutterstock; **p289**(B): PHOTO FUN/Shutterstock; **p290**(T): Helen Hotson/Shutterstock; **p290**(B): Piotr Krzeslak/Shutterstock; **p291**(T): Calin Tatu/Shutterstock; **p291**(B): Nickolay Khoroshkov/Shutterstock; **p293**: David Wrobel/Visuals Unlimited/Science Photo Library; **p296**(T): Nsemprevivo/Shutterstock; **p296**(B): Henk Bentlage/Shutterstock; **p295**: Nelik/Shutterstock; **p296**(T): Helen Hotson/Shutterstock; **p296**(C): Louise Murray/Science Photo Library; **p296**(B): A40757/Shutterstock; **p297**: Frans Lanting/Mint Images/Science Photo Library; **p298**: Ronald van der Beek/Shutterstock; **p299**: 360b/Shutterstock; **p305**(B): Nigel Cattlin/Visuals Unlimited/Getty Images; **p306**(TL): Niko Grigorieff/Visuals Unlimited, Inc./Science Photo Library; **p306**(TR): Norm Thomas/Science Photo Library; **p306**(CT): Vadym Zaitsev/Shutterstock; **p310**(T): Amir Ridhwan/Shutterstock; **p310**(B): Eye of Science/Science Photo Library; **p311**(T): Denis Kuvaev/Shutterstock; **p311**(CT): Bork/Shutterstock; **p311**(C): Slawomir Fajer/Shutterstock; **p311**(CB): Reddogs/Shutterstock; **p311**(B): Rio Patuca/Shutterstock; **p312**: Amir Ridhwan/Shutterstock; **p313**: Schubbel/Shutterstock; **p315**: James King-Holmes/Science Photo Library; **p317**: Steve Gschmeissner/Science Photo Library; **p318**(T): Don W.Fawcett/Science Photo Library; **p318**(B): Steve Gschmeissner/Science Photo Library; **p326**: Monkey Business Images/Shutterstock; **p329**(T): Brian Maudsley/Shutterstock; **p329**(C): Jassada Watt/Shutterstock; **p329**(B): AN NGUYEN/Shutterstock; **p171**(L): Astrid & Hanns-Frieder Michler/Science Photo Library; **p171**(B): Dr. Keith Wheeler/Science Photo Library; **p171**(R): Herve Conge, ISM/Science Photo Library; **p190**(T): Neil Carveth, Allison Daley and Nathalie Feiner from Department of Zoology, University of Oxford; **p190**(B): Neil Carveth, Allison Daley and Nathalie Feiner from Department of Zoology, University of Oxford; **p228-229**: Pablo Hidalgo - Fotos593/Shutterstock; **p283**: Steve McWilliam/Shutterstock; **p305**(T): 1000 Words/Shutterstock; **p306**(B): Chantal de Bruijne/Shutterstock; **p306**(CB): Nigel Cattlin/Visuals Unlimited/Corbis; **p307**: Eduard Kyslynskyy/Shutterstock; **p309**: Raj Creationzs/Shutterstock; **p166**: Neil Carveth, Allison Daley and Nathalie Feiner from Department of Zoology, University of Oxford; **p169**(T): Neil Carveth, Allison Daley and Nathalie Feiner from Department of Zoology, University of Oxford; **p169**(C): Neil Carveth, Allison Daley and Nathalie Feiner from Department of Zoology, University of Oxford; **p165**(B): Neil Carveth, Allison Daley and Nathalie Feiner from Department of Zoology, University of Oxford; **p176**: Neil Carveth, Allison Daley and Nathalie Feiner from Department of Zoology, University of Oxford; **p152-153**: Anthony Short; **p231**(L): MidoSemsem/Shutterstock; **p231**(R): Nicku/Shutterstock; **p223**: PhotographybyMK/Shutterstock; **p327**: Nagui Antoun; **p38**: Claire Ting/Science Photo Library; **p269**: Becky Stares/Shutterstock; **p268**: Eye of Science/Science Photo Library; **p340-341**: Bruce Rolff/Shutterstock; **p342**(B): Vishnevskiy Vasily/Shutterstock; **p342**(T): Ilia Torlin/Shutterstock; **p343**: Simon Fraser/Science Photo Library; **p344**: Steve Gschmeissner/Science Photo Library; **p345**: Thomas Deerinck, NCMIR/Science Photo Library; **p346**: Blueringmedia/Shutterstock; **p347**: Anatomical Travelogue/Science Photo Library; **p359**: John Greim/Science Photo Library; **p363**(C): Cginspiration/Shutterstock; **p363**(L): Science Photo Library; **p363**(R): Oliversved/Shutterstock; **p364**: Natural History Museum, London/Science Photo Library; **p370**(C): Innerspace Imaging/Science Photo Library; **p370**(L): Eric Grave/Science Photo Library; **p370**(R): Science Photo Library; **p371**: Biology Media/Science Photo Library; **p372**: Microscape/Science